Serotonin and Melatonin

Their Functional Role in Plants, Food, Phytomedicine, and Human Health

Serotonin and Melatonin
Their Functional Role in Plants, Food, Phytomedicine, and Human Health

Edited by
Gokare A. Ravishankar
Akula Ramakrishna

CRC Press
Taylor & Francis Group
Boca Raton London New York

CRC Press is an imprint of the
Taylor & Francis Group, an **informa** business

CRC Press
Taylor & Francis Group
6000 Broken Sound Parkway NW, Suite 300
Boca Raton, FL 33487-2742

First issued in paperback 2021
First issued in hardback 2019

ISBN-13: 978-1-03-209744-2 (pbk)
ISBN 13: 978-1-4987-3905-4 (hbk)

Visit the Taylor & Francis Web site at
http://www.taylorandfrancis.com

and the CRC Press Web site at
http://www.crcpress.com

Contents

Section I Phytoserotonin and Phytomelatonin: Occurrence, Plant Growth and Development, Environmental Adaptations

Section II Horticultural and Agricultural Aspects

Section III Medicinal Plants: Occurrence and Efficacy in Humans

Section IV Food: Occurrence and Dietary Implications

Section V Human Health: Brain, Behavior, Neurological Disorders, Sleep Disorders, Depression

Section VI Receptors, Transporters, and Signaling

Preface

Neurotransmitter molecules such as serotonin and melatonin have been well studied for their defined role in animal systems. These molecules have also been reported in various genera of plants. In recent years, the precise function of these compounds in plant systems has been a subject of intense study. The understanding of the biosynthetic pathway of serotonin and melatonin in higher plants has been unfolding. The derivatives of serotonin and melatonin also play a significant role in plant adaptations. In recent years, the role of melatonin and serotonin in plant morphogenesis has been gathering attention. There is speculation of the possible role of serotonin and melatonin as candidate hormones owing to their structural similarity with the plant hormone, indole acetic acid. Their role in adaptive functions to environmental conditions and stresses in plants is of continued research interest.

This book makes a bold attempt to bring together current understanding of and developments in serotonin and melatonin biosynthesis; and their role in ontogeny and environmental adaptation in plants; levels in food and medicinal plants; involvement and implication of endogenous serotonin and melatonin, and from dietary sources to human health, based on the world literature and authors' data from their laboratories.

The above aspects have been described in detail in this book, and the following points provide the focus at a glance.

- The involvement of serotonin and melatonin in complex processes, such as plant growth regulation, flowering, xylem sap exudation, ion permeability, antioxidant, and detoxification, delay of senescence, circadian rhythm, and adaptation to environmental changes.
- The localization of serotonin and melatonin in various tissues and also intracellularly.
- The analytical trends in the determination of serotonin and melatonin levels, to help researchers to adopt the most appropriate method(s) in their investigations.
- The role of melatonin in regulating the ability of plants to optimize levels of photosynthesis.
- The detoxification mechanism in plants mediated by serotonin accumulation including the influence of serotonin in the prevention or inhibition of the shikimate pathway during seed development through a reduction in the accumulation of ammonia. The protection of the seed during seed desiccation by the foregoing mechanism coupled to hyperaccumulation of melatonin.
- The calcium-mediated signaling in plants under the influence of serotonin and melatonin brought about using *in vitro* tissue culture experiments in *Coffea canephora* and *Mimosa pudica*. Also, the role of serotonin in the enhancement of vegetative growth and seed production elucidated through a model system, namely, *Arabidopsis* sps.
- Melatonin is reported to upregulate certain genes of secondary metabolite pathways in plants resulting in tolerance to salt, drought, and so on. Several examples of adaptation to environmental stresses and the enhancement of productivity in horticultural and agricultural crops using exogenous administration of melatonin have been provided.
- The possibility of developing transgenic plants with enhanced serotonin and melatonin levels for conferring environmental tolerance and improved productivity is discussed along with the possibility to obtain cold-resistant plants.
- With the gaining popularity through evidenced-based use of traditional Ayurveda and Chinese medicines, herbal drugs are receiving worldwide attention in health care and therapeutics. This book also deals with the presence of melatonin in traditional herbal medicines from India,

Thailand, and Brazil, which are used in sleep disorders, inflammation, antimicrobial formulations, and immunomodulatory effects.

- The presence of melatonin and serotonin in various food-value plants has kindled lot of interest in their dietary role in humans. Dietary sources of serotonin and melatonin and tryptophan-rich foods and their bioavailability have been presented. The possible effects of such foods on mood, sleep, and behavior have been described. Specifically, the occurrence of melatonin in wine and in yeast, as well as the bioavailability in humans is of interest to brewers and consumers.
- The metabolism of melatonin to 6-hydroxymelatonin in the liver by P450 $1A_2$, subsequently O-demethylation by CYP2C19 or CYPI A2 extra-hepatically has been evident. The drugs that are substrates of CYPI A2 such as paracetamol, theophylline, and fluvoxamine are predicted to increase melatonin circulation on consumption of melatonin-containing foods. Coffee has been shown to be an inhibitor of CYPI A2, which increases melatonin circulation when taken orally. Also, the levels of serotonin and melatonin in coffee brew are well known as reported by us.
- The frequency of mutation in CYP1 A2 in Asians was found to be high. Hence, dietary melatonin could result in varied circulatory levels in different ethnic groups.
- Studies on the functional relationship of melatonin in humans relating to the regulation of sleep, appetite, immune system, and defense against oxidative stress, learning, and memory will continue to be a subject of intensive research. Melatonin is a powerful molecule in the regulation of circadian rhythms, hence it is of relevance in treating the effects of jet lag, night shift, and neuropsychiatric disorders. Evidently, the US FDA has categorized melatonin as a dietary supplement.
- Therapies related to 5-hydroxy tryptamine have been successful in the treatment of neurological conditions such as akinesia, tardive dyskinesia, and spinal trauma.
- The relevance of Ayurvedic plants in the management of sleep disorders and depression for possible linkages with serotonin and melatonin is being investigated.
- A detailed account of Alzheimer's disease and neurodegenerative and psychiatric conditions with reference to the regulation of melatoninergic pathways has been included, which presents a comprehensive treatise on their relationships.
- The beneficial effects of serotonin and melatonin in cancer treatment are beginning to unfold.
- An overview of biochemical and genetic evidence for the association of melatonin in autism has been provided in a dedicated chapter. Also, melatonin in oral health has been discussed. Finally, the aspects related to current understanding of melatonin receptors in the reproductive behavior of animals, and serotonin transporters in neuropsychiatric disorders have been described.

The chapters written by experts in their respective fields from 19 countries provide a vast coverage of areas on the topics including the latest developments in the field. Each chapter has been presented in a stand-alone manner for comprehensive understanding with minimal interdependence. They are arranged in the following sections to benefit readers who are interested in specific areas.

- Phytoserotonin and Phytomelatonin: Occurrence, Plant Growth and Development, Environmental Adaptations (Chapters 1 through 10)
- Horticultural and Agricultural Aspects (Chapters 11 through 13)
- Medicinal Plants: Occurrence and Efficacy in Humans (Chapters 14 through 16)
- Food: Occurrence and Dietary Implications (Chapters 17 through 22)
- Human Health: Brain, Behavior, Neurological Disorders, Sleep Disorders, Depression (Chapters 23 through 33)
- Receptors, Transporters, and Signaling (Chapters 34 through 36)

This volume is expected to attract scientists, academicians, and professionals from all over the world who are engaged in plant/animal/microbial physiology and biochemistry, ecology, life sciences, agricultural sciences, and nutrition, and health experts, as well as medical doctors and practitioners of integrative medicines.

Gokare A. Ravishankar
Akula Ramakrishna

Acknowledgment

The editors would like to thank all the contributors to this volume, which is a comprehensive treatise on the topic. We understand that the contributors are active scientists and have volunteered to share their findings and perspectives with the readers, who will benefit from learning about the implications and applications of the subject matter of the book as emerging areas of science.

We are grateful to John Sulzycki, Jill J. Jurgensen, and the team from Taylor & Francis, as well as Michelle van Kampen and the Deanta team, for their help in bringing out the volume in a beautiful and appealing manner.

We are both thankful to our families, who have encouraged us to undertake this task and permitted us to take time off from the family. G.A.R. thanks his wife Shyla, son Prashanth, daughter-in-law Vasudha, and daughter Apoorva. A.R. thanks his wife Venkatalaxmi, and children Hasini and Jashvitha, and brother Mr. Pakkirappa.

G.A.R. is also thankful to the management of Dayananda Sagar Institutions for granting permission to take time out for this work. A.R. is thankful to Monsanto Crop Breeding center, Bengaluru, for their support.

We thank the Department of Science and Technology, Government of India, for providing the grant for our research carried out on this topic by us at the Central Food Technological Research Institute (CFTRI) at Mysore. We also thank the staff and students of the Plant Cell Biotechnology Department, CFTRI; and other colleagues, especially Dr. P. Giridhar and Dr. K. Udaysankar; and Dr. Vinod Kumar, Kuwait Institute for Scientific Research, for their active collaboration.

Last, but not least, G.A.R. would like to thank the *TRendys in Biochemistry* forum of Professor T. Ramasarma at the Indian Institute of Science, Bangalore, for providing a platform to enter into the research of this exciting field of phytoserotonin and melatonin.

Gokare A. Ravishankar
Akula Ramakrishna

Editors

Gokare A. Ravishankar is presently the vice president of R&D in life sciences at Dayanada Sagar Institutions, Bengaluru, India; professor of biotechnology, and chairman of the board of studies in biotechnology of Visvesvaraya Technological University, Karnataka, India. He had a distinguished research career of over 28 years working at the Central Food Technological Research Institute (CFTRI), Mysuru. As an international expert in the areas of plant secondary metabolites, plant biotechnology, algal biotechnology, and food biotechnology and postharvest technologies, and he served as the visiting professor to universities in Japan, Korea, Taiwan, and Russia.

Dr Ravishankar holds a master's and a PhD degree from M. S. University of Baroda; guided 25 students to PhD degree, 48 masters students, 6 postdocs, and 8 international guest scientists; authored over 250 peer-reviewed research papers in international and national journals, 45 reviews, 55 patents in India and abroad, with a h-index 48. He has presented over 200 lectures in various scientific meetings in India and abroad, including visits to over 25 countries.

Dr Ravishankar received several coveted awards and recognitions as follows: Young Scientist award (Botany) by the then prime minister of India in 1992; National Technology Day Award of Government of India in 2003; Laljee Goodhoo Smarak Nidhi award for food biotechnology R&D of industrial relevance; and the prestigious Professor V. Subramanyan Food-Industrial Achievement Award; Professor S. S. Katiyar Endowment Lecture Award in New Biology by Indian Science Congress; Professor Vyas Memorial Award of Association of Microbiologists of India; Professor V. N. Raja Rao Endowment Lecture Award in Applied Botany, University of Madras; and Life Time Achievement Award by the Society for Applied Biotechnologists.

He is honored as a fellow of several national organizations viz. National Academy of Sciences, India; National Academy of Agricultural Sciences, India; Association of Microbiologists of India; Society of Agricultural Biochemists; Society for Applied Biotechnology; Indian Botanical Society, and the Association of Food Scientists and Technologists of India. Several premier international bodies have honored him with a fellowship viz. the International Academy of Food Science and Technology (Canada), the Institute of Food Technologists (USA), the Institute of Food Science and Technology (UK), and Certified Food Scientist of the United States. Dr Ravishankar has been actively involved as a member of task forces in organizations of the government of India, and has served as an expert in the selection committees for the appointment of professors, scientists, and research students in universities, as well as R&D institutions.

Akula Ramakrishna is currently a scientist at Monsanto Crop Breeding Centre, Bangalore, India. Dr Ramakrishna holds a master's degree from Sri Krishna Devaraya University, Anantapur, India. He started his research career in 2005 at the Department of Plant Cell Biotechnology, CFTRI, Mysuru, in the research group of Dr G. A. Ravishankar. He is the recipient of a senior research fellow (CSIR, New Delhi). He obtained his PhD (biochemistry from the University of Mysore, Mysuru) in the area of development of high-frequency somatic embryogenesis and the regulation of secondary metabolites in *Coffea canephora*. He worked extensively on the role of serotonin, melatonin vis-à-vis calcium-mediated signaling in plants. His research includes contributions to the metabolic engineering of secondary metabolites from plants and abiotic stress adaptations; tissue culture; *in vitro* production and regulation of plant secondary metabolites from food-value plants that include natural pigments caffeine, steviosides, anthocyanins, and carotenoids. He also worked on the transformation of coffee to regulate caffeine biosynthesis, somatic embryogenesis in coffee, transformation methods, and the analysis of caffeine alkaloids. He attended the Fifth International Symposium on Plant Neurobiology, held during May 25–29, 2009 at Florence, Italy. He is currently the author of 12 peer-reviewed publications, 2 reviews, 4 book chapters and 18 articles published at Monsanto Scientific Literature. He is a member of the Society for Biotechnologists (India). He is a recipient of the fellow of Society for Applied Biotechnology, India (2012), three global technology awards, Rapid Recognition Award, Test Master, and Asia Veg R&D quarterly recognition from the Monsanto Company. He also attended the Technical Community of Monsanto (TCM) Conference, held during June 7–8, 2016, at St. Louis, Missouri.

Contributors

Aruna Agrawal
Institute of Medical Sciences
Banaras Hindu University
Varanasi, India

Ann Mary Alex
Division of Neurobiology and Genetics
Human Molecular Genetics Laboratory
Rajiv Gandhi Center for Biotechnology
Thiruvananthapuram, India

George Anderson
CRC Scotland & London
London, England

Dhara Arora
Laboratory of Plant Physiology and Biochemistry
Department of Botany
University of Delhi
Delhi, India

Moinak Banerjee
Division of Neurobiology and Genetics
Human Molecular Genetics Laboratory
Rajiv Gandhi Center for Biotechnology
Thiruvananthapuram, India

Gemma Beltran
Departament de Bioquímica i Biotecnologia
Universitat Rovira i Virgili
Tarragona, Spain

Satish Chander Bhatla
Laboratory of Plant Physiology and Biochemistry
Department of Botany
University of Delhi
Delhi, India

Atanu Bhattacharjee
Department of Pharmacognosy
Nitte Gulabi Shetty Memorial Institute of
    Pharmaceutical Sciences
Karnataka, India

Paramita Bhattacharjee
Department of Food Technology and Biochemical
    Engineering
Jadavpur University
Kolkata, India

Amnon Brzezinski
Department of Obstetrics and Gynecology
Hadassah-Hebrew University Medical Center
Jerusalem, Israel

Ana B. Cerezo
Facultad de Farmacia
Universidad de Sevilla
Seville, Spain

Soledad Cerutti
Instituto de Química San Luis
Universidad Nacional de San Luis
San Luis, Argentina

Vijay Kumar Chava
Department of Periodontics
Narayana Dental College and Hospital
Nellore, India

Hui-Yen Chuang
Department of Biomedical Imaging and
    Radiological Sciences
National Yang-Ming University
Taipei, Taiwan

Mona Gergis Dawood
Botany Department
National Research Centre
Cairo, Egypt

Govind Prasad Dubey
Institute of Medical Sciences
Banaras Hindu University
Varanasi, India

Mohamed El-Awadi
Botany Department
National Research Centre
Cairo, Egypt

Chun-Kai Fang
Department of Psychiatry
Mackay Memorial Hospital
Taipei, Taiwan

Gaia Favero
Division of Anatomy and Physiopathology
Department of Clinical and Experimental
    Sciences
University of Brescia
Brescia, Italy

Eduardo Luzía França
Institute of Health and Biological Science
Federal University of Mato Grosso
Mato Grosso, Brazil

Yumi Fukuda
Department of Environmental Science
Fukuoka Women's University
Fukuoka, Japan

Cruz García
Department of Biochemistry and Molecular
    Biology III
School of Medicine
Complutense University of Madrid
Madrid, Spain

M. Carmen Garcia-Parrilla
Facultad de Farmacia
Universidad de Sevilla
Seville, Spain

Probir Kumar Ghosh
Department of Food Technology and
    Biochemical Engineering
Jadavpur University
Kolkata, India

Sarvajeet S. Gill
Stress Physiology and Molecular Biology
    Laboratory
Centre for Biotechnology
Maharshi Dayanand University
Haryana, India

Federico J. V. Gomez
Instituto de Biología Agrícola de Mendoza
Universidad Nacional de Cuyo
Mendoza, Argentina

Biao Gong
State Key Laboratory of Crop Biology
College of Horticulture Science and Engineering
Shandong Agricultural University
Tai'an, China

Michelangelo Bauwelz Gonzatti
Institute of Health and Biological Science
Federal University of Mato Grosso
Mato Grosso, Brazil

Jose Manuel Guillamon
Department Biotecnologia de Alimentos
Instituto de Agroquímica y Tecnología de los
    Alimentos
Valencia, Spain

Yang-Dong Guo
College of Horticulture
China Agricultural University
Beijing, China

Ismael Gatica Hernández
Instituto de Biología Agrícola de Mendoza
Universidad Nacional de Cuyo
Mendoza, Argentina

Adenilda Cristina Honorio-França
Institute of Health and Biological Science
Federal University of Mato Grosso
Mato Grosso, Brazil

Jeng-Jong Hwang
Department of Biomedical Imaging and
    Radiological Sciences
National Yang-Ming University
Taipei, Taiwan

Marcello Iriti
Department of Agricultural and Environmental
    Sciences
Milan State University
Milan, Italy

Rama Jayasundar
Department of NMR
All India Institute of Medical Sciences
New Delhi, India

Jeffrey Johns
Melatonin Research Group
Faculty of Pharmaceutical Sciences
Khon Kaen University
Khon Kaen, Thailand

Nutjaree Pratheepawanit Johns
Melatonin Research Group
Faculty of Pharmaceutical Sciences
Khon Kaen University
Khon Kaen, Thailand

Cristina Filomena Justo
Institute of Health and Biological Science
Federal University of Mato Grosso
Mato Grosso, Brazil

Rinki Kumari
Institute of Medical Sciences
Banaras Hindu University
Varanasi, India

Bindu M. Kutty
Human Sleep Research Laboratory
Department of Neurophysiology
National Institute of Mental Health and
    Neurosciences
Karnataka, India

Robert Lalonde
Department of Psychology
University of Rouen
Rouen, France

José López-Bucio
Instituto de Investigaciones Químico-Biológicas
Universidad Michoacana de San Nicolás de Hidalgo
Michoacán, Mexico

Michael Maes
Department of Psychiatry
Deakin University
Geelong, Australia

Małgorzata Maria Posmyk
Department of Ecophysiology and Plant
    Development
University of Lodz
Lodz, Poland

Albert Mas
Departament de Bioquímica i Biotecnologia
Universitat Rovira i Virgili
Tarragona, Spain

Takeshi Morita
Department of Environmental Science
Fukuoka Women's University
Fukuoka, Japan

Soumya Mukherjee
Laboratory of Plant Physiology and
    Biochemistry
Department of Botany
University of Delhi
Delhi, India

Ravindra P. Nagendra
Gadag Institute of Medical Sciences
Karnataka, India

Lorenzo Nardo
Department of Radiology and Biomedical Imaging
University of California
San Francisco, CA

Vitor Antunes Oliveira
Department of Biochemistry and Molecular
    Biology
Federal University of Santa Maria
Santa Maria, Brazil

Adejoke Onaolapo
Department of Human Anatomy
Ladoke Akintola University of Technology
Ogbomosho, Nigeria

Olakunle Onaolapo
Department of Pharmacology and Therapeutics
Ladoke Akintola University of Technology
Oshogbo, Nigeria

Tanit Padumanonda
Faculty of Pharmaceutical Sciences
Khon Kaen University
Khon Kaen, Thailand

Mitradas M. Panicker
National Centre for Biological Sciences
GKVK Campus
Bengaluru, India

Sergio D. Paredes
Department of Physiology
School of Medicine
Complutense University of Madrid
Madrid, Spain

Ramón Pelagio-Flores
Instituto de Investigaciones
    Químico-Biológicas
Universidad Michoacana de San Nicolás de
    Hidalgo
Michoacán, Mexico

Akula Ramakrishna
Monsanto Crop Breeding Centre
Bangalore, India

Lisa Rancan
Department of Biochemistry and Molecular
    Biology III
School of Medicine
Complutense University of Madrid
Madrid, Spain

Gokare A. Ravishankar
Dr. C. D. Sagar Center for Life Sciences
Dayananda Sagar Institutions
Bengaluru, India

Rita Rezzani
Division of Anatomy and Physiopathology
Department of Clinical and Experimental Sciences
University of Brescia
Brescia, Italy

Luigi Fabrizio Rodella
Division of Anatomy and Physiopathology
Department of Clinical and Experimental
    Sciences
University of Brescia
Brescia, Italy

Victoria V. Roshchina
Laboratory of Microspectral Analysis of Cells
    and Cellular Systems
Russian Academy of Sciences Institute of Cell
    Biophysics
Moscow Region, Russia

Santanu Saha
Department of Pharmacognosy
Nitte Gulabi Shetty Memorial Institute of
    Pharmaceutical Sciences
Karnataka, India

Virginia N. Sarropoulou
Department of Horticulture
Aristotle University
Thessaloniki, Greece

Eirini Sarrou
Department of Aromatic and Medicinal Plants
Hellenic Agricultural Organization DEMETER
Institute of Plant Breeding and Genetic
    Resources
Thessaloniki, Greece

Vijay Kumar Saxena
Molecular Physiology Laboratory
Division of Physiology and Biochemistry
ICAR-Central Sheep and Wool Research Institute
Rajasthan, India

Krishna K. Sharma
Department of Microbiology
Maharshi Dayanand University
Haryana, India

Shastry Chakrakodi Shashidhara
Department of Pharmacology
Nitte Gulabi Shetty Memorial Institute of
    Pharmaceutical Sciences
Karnataka, India

Qinghua Shi
State Key Laboratory of Crop Biology
College of Horticulture Science and
    Engineering
Shandong Agricultural University
Tai'an, China

María Fernanda Silva
Instituto de Biología Agrícola de Mendoza
Facultad de Ciencias Agrarias
Universidad Nacional de Cuyo
Mendoza, Argentina

Gur Prit Inder Singh
Centre for Interdisciplinary Biomedical
    Research
Adesh University
Bathinda, India

Praveen K. Singh
Advanced Centre for Research in AYUSH
    Systems of Medicine
Adesh University
Bathinda, India

Shishu Pal Singh
National Centre for Biological Sciences
GKVK Campus
Bengaluru, India

Rekha S. Singhal
Research Consultancy and Research Mobilization
Institute of Chemical Technology
Mumbai, India

Shuchita Soman
National Centre for Biological Sciences
GKVK Campus
Bengaluru, India

Catherine Strazielle
Stress, Immunity & Pathogens Laboratory
University of Lorraine
Vandoeuvre-les-Nancy, France

B. Swathy
Division of Neurobiology and Genetics
Human Molecular Genetics Laboratory
Rajiv Gandhi Centre for Biotechnology
Thiruvananthapuram, India

Katarzyna Szafrańska
Department of Ecophysiology and Plant
    Development
University of Lodz
Lodz, Poland

Maria Jesus Torija
Departament de Bioquímica i
    Biotecnologia
Universitat Rovira i Virgili
Tarragona, Spain

Jesús A. F. Tresguerres
Department of Physiology
School of Medicine
Complutense University of Madrid
Madrid, Spain

Ana M. Troncoso
Facultad de Farmacia
Universidad de Sevilla
Seville, Spain

Narendra Tuteja
Amity Institute of Microbial Technology
Amity University
Uttar Pradesh, India

Elena Vara
Department of Biochemistry and Molecular
    Biology III
School of Medicine
Complutense University of Madrid
Madrid, Spain

Elena Maria Varoni
Department of Biomedical, Surgical, and Dental
    Sciences
Milan State University
Milan, Italy

Manvi Vernekar
Department of Food Engineering and Technology
Institute of Chemical Technology
Mumbai, India

Hai-Jun Zhang
College of Horticulture
China Agricultural University
Beijing, China

Na Zhang
College of Horticulture
China Agricultural University
Beijing, China

Section I

Phytoserotonin and Phytomelatonin: Occurrence, Plant Growth and Development, Environmental Adaptations

1

Melatonin in Plants

Katarzyna Szafrańska and Małgorzata Maria Posmyk
University of Lodz
Lodz, Poland

CONTENTS

ABSTRACT   For many decades, melatonin (MEL) has been considered as a typically animal hormone, but its wide occurrence in the plant kingdom has now been well recognized. Its content in plants varies not only between species but also between organs of the same plant. Although the mechanism of MEL action is not precisely elucidated in plants, its role seems to be very important. In this chapter we (1) describe the history of MEL discovery, (2) indicate its presence especially in phototrophic organisms, and (3) gather the available knowledge on its biosynthetic pathways and localization in plants. Although plants and animals share a common scheme of MEL biosynthetic pathways, recently collected data revealed some plant-specific steps. Additionally, the site of MEL biosynthesis in plant cells (mitochondria and chloroplasts), as well as the availability of a MEL precursor—tryptophan––may account for the much higher level of this indoleamine in plants than in animals.

KEY WORDS: *phytomelatonin, localization, biosynthesis.*

Abbreviations

5-MT: 5-methoxytryptamine
AADC: 5-hydroxytryptophan decarboxylase (aromatic-L-amino-acid decarboxylase)
AANAT: arylalkylamine-N-acetyltransferase
ABA: abscisic acid
ASMT: acetylserotonin-N-methyltransferase
COMT: 3-O-methyl caffeic acid transferase
ER: endoplasmic reticulum
FR: free radicals
GNAT: superfamily of GCN5-related N-acetyltransferases
HIOMT: hydroxyindole-O-methyltransferase
HPLC: high-performance liquid chromatography
IAA: indole-3-acetic acid
MeJA: methyl jasmonate
MEL: melatonin, N-acetyl-5-methoxytryptamine

MT15:	transgenic *E. coli* strain number 15 (among 18 analyzed), which harbors a cDNA coding for the ASMT enzyme
NAS:	*N*-acetylserotonin
OMT:	*O*-methyltransferase genes family
RIA:	radioimmunoassay
ROS:	reactive oxygen species
SA:	salicylic acid
SNAT:	serotonin *N*-acetyltransferase
T5H:	tryptamine 5-hydroxylase
TDC:	tryptophan decarboxylase
TP5H:	tryptophan 5-hydroxylase

Melatonin in Plant Kingdom

Melatonin (MEL) was isolated for the first time from an ox pineal gland by Lerner et al. in 1958. Then it was identified as *N*-acetyl-5-methoxy-tryptamine. Its common name comes from melanin, a dye present in melanophores, cells of amphibians determining their skin color. *N*-acetyl-5-methoxy-tryptamine proved to be a factor decoloring the skin of amphibians (Lerner et al. 1958).

For nearly four decades, MEL was considered mainly as an animal neurohormone (Reiter 1991). Currently, it is known that its functions in animals are associated with modulation of sleep and mood, retinal physiology, sexual behavior, reproductive seasonality, circadian rhythms, and stimulation of the immune system (Arnao and Hernandez-Ruiz 2006; Janas et al. 2005; Posmyk and Janas 2009). In the meantime, presence of this inoleamine in other organisms has been also reported. In the mid-1970s MEL was detected as a by-product of decaffeination of coffee beans, but it was thought to be the result of a chemical reaction accompanying this process. Nobody suspected that MEL could be synthesized in plants (Tan et al. 2012). In the 1980s, MEL was discovered in invertebrates (insects, crustaceans, e.g., planarians). The circadian rhythm of MEL occurrence with night maximum was described in insects, although arrhythmicity or high daily MEL levels were observed in many other species (Hardeland and Poeggeler 2003; Ramakrishna et al. 2011b). It is supposed that due to the conservative nature of this molecule the rhythmic changes in MEL content resulted from evolutionary pathways differentiating species as well as their habitats, and their circadian and seasonal behaviors are the result of these factors (Bentkowski and Markowska 2007). In 1991, MEL was detected in *Lingulodinium polyedrum* Stein (formerly *Gonyaulax polyedra*), autotrophic unicellular dinoflagellates (*Dinoflagellata*) currently classified as *Protista*, flagellates and green algae (Balzer and Hardeland 1991; Hardeland and Fuhrberg 1996). The next major discovery was made in 1992, when it turned out that MEL, known as a neurohormone, was also a highly effective antioxidant, scavenging free radicals (FR) and other reactive oxygen species (ROS) (Tan et al. 1993). Antioxidant protection of cells and tissues is a conservative feature of all aerobic organisms including plants, therefore it was not surprising that over time research concerning the presence of MEL in phototrophic plants became intensive. In 1995 there appeared two independent papers confirming the existence of this indoleamine in plants (Dubbels 1995; Hattori et al. 1995). Although the scientists used different methods of extraction and detection both groups confirmed the presence of MEL in crops such as tobacco, rice, corn, wheat, and banana. Since then MEL presence in higher plants became the focus of interest. Reddish quinoa (*Chenopodium rubrum* L.) was the next species in which this indoleamine was discovered (Kolař et al. 1995). Interestingly, in medicinal herbs, including pyrethrum maruna (*Tanacetum parthenium* L.) and St. John's wort (*Hypericum perforatum* L.), MEL levels were several times higher than those in the blood of animals (Murch et al. 1997). When Kolář et al. (1997) reported a circadian rhythm of MEL occurrence in reddish quinoa, with high content at night and low during the day, a group of Murch et al. (2000) questioned these results, showing that the MEL content in St. John's wort during a day was positively correlated with the intensity of light to which the plants were exposed. Generally, there is a theory that the physiological circadian rhythm of MEL content with the night peak and day decline is associated with the photooxidation processes activated by daylight—as shown in reddish quinoa by Kolář et al. (1997) under optimum photoperiodic conditions. Similarly the light/

dark cycle has a profound influence on the indoleamine profile in *Dunaliella bardawil* (Ramakrishna et al. 2011b). Probably, the group of Murch et al. (2000), using high-intensity light, had provoked stress, which might have led to MEL overproduction in St John's wort during exposure phases. It is known that plants synthetizing high amounts of endogenous MEL in response to intensive light, particularly UV, also exhibit increased content of methoxyindole (a derivative of tryptophan—probably a metabolite of MEL oxidation) (Conti et al. 2002; Afreen et al. 2006). This fact may support the theory that implicates the photoprotective role of MEL capable of eliminating FR and singlet oxygen (Behrmann et al. 1997; Hardeland 2008). It is known that plants rich in MEL and those able to biosynthesize it under harmful environmental conditions tolerate stress better (Zhang et al. 2015). Therefore discrepancies between the results describing MEL content in plants presented by different groups of researchers may have arisen not only from various techniques of extraction and detection applied but also from the fact that: (1) species and even varieties of the same plant species (Tan et al. 2012) contain different amounts of MEL (from pico- to micrograms per gram of tissue), (2) the biosynthesis and metabolism of this indoleamine are affected and modified by environmental conditions including stresses responses, and (3) MEL levels change during plant ontogenesis (Okazaki and Ezura 2009). It was also found that different organs of the same plant contained various amounts of MEL with the biggest content in seeds (see Table 1.1).

Studies on 108 species of herbs commonly used in Chinese medicine revealed the presence of MEL ranging from a few to several thousand nanograms per gram of tissue (Chen et al. 2003) which means they are among the best natural sources of health-related molecules. Its presence has been confirmed in many edible plants (Huang and Mazza 2011; Manchester et al. 2000; Ramakrishna et al. 2012a,b; Ramakrishna 2015). Radioimmunoassays (RIA) and high-performance liquid chromatography (HPLC) with electrochemical and fluorescent detection are the main techniques used to determine endogenous MEL levels. However, a combination of HPLC with mass spectrometry is the key and the most powerful tool allowing precise determination of MEL and its metabolite contents in plant samples (Mercolini et al. 2012; Boccalandro et al. 2011; Ramakrishna et al. 2012; Ramakrishna et al. 2009).

Phytomelatonin: Localization and Biosynthetic Pathway

The biosynthetic pathway of MEL in animals is well known. Its precursor—tryptophan—is converted to 5-hydroxytryptophan by tryptophan 5-hydroxylase (TP5H). Next it is decarboxylated to 5-hydroxytryptamine (serotonin) by 5-hydroxytryptophan decarboxylase (aromatic-L-amino-acid decarboxylase [AADC]) (see Figure 1.1). Due to *N*-acetylation of serotonin, catalyzed by arylalkylamine-*N*-acetyltransferase (AANAT), *N*-acetylserotonin (NAS) is produced and then converted to *N*-acetyl-5-methoxytryptamine (MEL) in a reaction catalyzed by acetylserotonin-*N*-methyltransferase (ASMT), known also as hydroxyindole-*O*-methyltransferase (HIOMT) (Park 2011). The biosynthetic pathway of MEL in plants appears to be similar to that defined in animals. In particular, the first two enzymes were also identified in plant tissues (Facchini et al. 2000; Fujiwara et al. 2010), but studies on the leaves of rice revealed that the first enzymatic step of MEL synthesis in plants consisted rather in decarboxylation of tryptophan (involving tryptophan decarboxylase [TDC]) than in its hydroxylation, as is the case in the pineal gland of vertebrates, resulting in tryptamine not in 5 hydroxytryptophan formation (Figure 1.1). Subsequently, tryptamine is hydroxylated to serotonin by tryptamine 5-hydroxylase (T5H) (Park et al. 2012; Ramakrishna et al. 2011a).

Although there are reports on cloning the AANAT gene from *Chlamydomonas reinhardtii* and from humans and its overexpression in tomato and in rice, respectively (Kang et al. 2010; Okazaki et al. 2009), still little is known about homologs of the enzyme naturally occurring in plants. Therefore, in the work of Kang et al. (2013) it was assumed that the gene of serotonin *N*-acetyltransferase (SNAT) belonging to the family of genes (GNAT) responsible for the transfer of an acetyl group from acetyl-CoA to numerous molecules such as histones, aminoglycosides, arylalkylamines (Dyda et al. 2000) can be a substitute for the animal AANAT gene. For this purpose, 31 GNAT family genes were cloned (cDNA GNAT5, GNAT6, GNAT7, GNAT9) and expressed in *E. coli*, to check the activity of recombinant SNAT proteins catalyzing the transfer of an acetyl group onto tryptamine. In the presence of tryptamine (a substrate for MEL biosynthesis), only 5 out of the 31 transgenic strains of *E. coli* produced increased amounts of *N*-acetyltryptamine, suggesting that the GNAT5 cDNA of rice may encode a SNAT protein. However,

TABLE 1.1

Melatonin (MEL) Contents in Some Plant Organs

Common Name	Scientific Name	Organ	MEL(ng g⁻¹ FW [DW] tissue)	Literature
Coffee Robusta	*Coffea canephora Pierr.*	Bean	5800 (DW)	Ramakrishna et al 2012a
Coffee Arabica	*Coffea arabica* (L.)	Bean	6800 (DW)	Ramakrishna et al. 2012a
Onion	*Allium cepa* (L.)	Bulb	0.032	Hattori et al. 1995
White radish	*Raphanus sativus* (L.)	Bulb	485 (DW)	Chen et al. 2003
St. John's wort	*Hypericum perforatum* (L.)	Flower	4	Murch et al. 1997
Banana	*Musa acuminata* Colla	Fruit	0.00046	Dubbels et al. 1995
Banana	*Musa paradisiaca* (L.)	Fruit	0.002	Dubbels et al. 1995
Grapevine	*Vitis vinifera* (L.)	Fruit	0.005–0.965	Iriti et al. 2006
Sweet cherry	*Prunus avium* (L.)	Fruit	0.006–0.224	Gonzalez-Gomez et al. 2009
Banana	*Musa acuminata* Colla	Fruit	0.009	Sae-Teaw et al. 2012
Strawberry	*Fragaria x ananassa* (Duch.)	Fruit	0.012	Hattori et al. 1995
Wild strawberry	*Fragaria ananassa* (Duch.)	Fruit	0.012	Hattori et al. 1995
Tomato	*Solanum lycopersicum* (L.)	Fruit	0.02–0.016	Van Tassel et al. 2001
Kiwi	*Actinidia deliciosa* (Liang-Ferg.)	Fruit	0.024	Hattori et al. 1995
Cucumber	*Cucumis sativus* (L.)	Fruit	0.025	Hattori et al. 1995
Tomato	*Lycopersicon esculentum* (Mill.)	Fruit	0.032	Hattori et al. 1995
Pineapple	*Ananas comosus* (Stickm.) Merill.	Fruit	0.036	Hattori et al. 1995
Apple	*Malus domestica* (Borkh)	Fruit	0.048	Hattori et al. 1995
Pineapple	*Ananas comosus* (L.)	Fruit	0.302	Sae-Teaw et al. 2012
Tomato	*Lycopersicon esculentum* (Mill.)	Fruit	0.5	Dubbels et al. 1995
Pomegranate	*Punica granatum* (L.)	Fruit	0.54–5.5	Mena et al. 2012
Grapevine	*Vitis vinifera* (L.)	Fruit	0.6–1.2	Stege et al. 2010
Tomato	*Solanum lycopersicum* (L.)	Fruit	0.616–1.068	Pape and Lüning 2006
Strawberry	*Fragaria x ananassa* (Duch.)	Fruit	1.4–11.26	Stürtz et al. 2011
Tart cherry	*Prunus cerasus* (L.)	Fruit	1–19.5	Burkhardt et al. 2001
Tomato	*Solanum lycopersicum* (L.)	Fruit	2.5	Okazaki and Ezura 2009
Grapevine	*Vitis vinifera* (L.)	Fruit	3–18	Vitalini et al. 2011
Cherry	*Prunus cerasus* (L.)	Fruit	18	Burkhardt et al. 2001
Tomato	*Solanum lycopersicum* (L.)	Fruit	4.1–114.5	Stürtz et al. 2011
Sweet cherry	*Prunus avium* (L.)	Fruit	8–120	Zhao et al. 2013
Wolf berry (goji berry)	*Lycium barbarum* (L.)	Fruit	530 (DW)	Chen et al. 2003
Morning glory	*Pharbitis nil* Choisy	Leaf	0.0005	Van Tassel et al. 2001
Indian spinach	*Basella alba* (L.)	Leaf	0.039	Hattori et al. 1995
Cabbage	*Brassica oleracea* (L.)	Leaf	0.107	Hattori et al. 1995
Chine cabbage	*Brassica chinensis* (Juslen)	Leaf	0.112	Hattori et al. 1995
Feverfew	*Tanacetum parthenium* (L.)	Leaf	2	Murch et al. 1997
St. John's wort	*Hypericum perforatum* (L.)	Leaf	2	Murch et al 1997
Sesban	*Sesbania sesban* (L.) Merr.	Leaf	8.7 (DW)	Padumanonda et al. 2014
Java bean	*Senna tora* (L.) Roxb.	Leaf	10.5 (DW)	Padumanonda et al. 2014
Bitter melon	*Momordica charantia* (L.)	Leaf	21.4 (DW)	Padumanonda et al. 2014
Agati	*Sesbania glandiflora* (L.) Desv.	Leaf	26.3 (DW)	Padumanonda et al. 2014
Burmese grape	*Baccaurea ramiflora* Lour.	Leaf	43.2 (DW)	Padumanonda et al. 2014
Black pepper	*Piper nigrum* (L.)	Leaf	1093 (DW)	Padumanonda et al. 2014
Beet	*Beta vulgaris* (L.)	Root	0.002	Dubbels et al. 1995
Carrot	*Daucus carota*	Root	0.055	Hattori et al. 1995
Mung bean	*Vigna radiata* (L.)	Root	0.24 (DW)	Szafrańska et al. 2014

TABLE 1.1 (CONTINUED)

Melatonin (MEL) Contents in Some Plant Organs

Common Name	Scientific Name	Organ	MEL(ng g⁻¹ FW [DW] tissue)	Literature
Ginger	*Zingiber officinale* (Roscoe)	Root	0.583	Hattori et al. 1995
Red radish	*Raphanus sativus* (L.)	Root	0.6	Hattori et al. 1995
Curcuma	*Curcuma aeruginosa* Roxb.	Root	120 (DW)	Chen et al. 2003
Red pepper	*Capsicum annuum* (L.)	Seed	0.180 (DW)	Huang and Mazza 2011
Barley	*Hordeum vulgare* (L.)	Seed	0.378	Hattori et al. 1995
Green pepper	*Capsicum annuum* (L.)	Seed	0.521 (DW)	Huang and Mazza 2011
Barley	*Hordeum vulgare* (L.)	Seed	0.58	Hernandez-Ruiz and Arnao 2008
Orange pepper	*Capsicum annuum* (L.)	Seed	0.581 (DW)	Huang and Mazza 2011
Rice	*Oryza sativa japonica* (L.)	Seed	1	Hattori et al. 1995
Corn	*Zea mays* (L.)	Seed	1.366	Hattori et al. 1995
Oat	*Avena sativa* (L.)	Seed	1.796	Hattori et al. 1995
Milk thistle	*Silybum marianum* (L.)	Seed	2 (DW)	Manchester et al. 2000
Walnut	*Juglans regia* (L.)	Seed	3.5 (DW)	Reiter et al. 2005
Lupin	*Lupinus albus* (L.)	Seed	3.830	Hernandez-Ruiz and Arnao 2008
Tall fescue	*Festuca arundinacea*	Seed	5.288	Hattori et al. 1995
Cucumber	*Cucumis sativus (L)*	Seed	11–80	Kołodziejczyk et al. 2015
Corn	*Zea mays* (L)	Seed	14–53	Kołodziejczyk et al. 2015
Poppy	*Papaver somniferum* (L.)	Seed	6 (DW)	Manchester et al. 2000
Anise	*Pimpinela anisum* (L.)	Seed	7 (DW)	Manchester et al. 2000
Celery	*Apium graveolens* (L.)	Seed	7 (DW)	Manchester et al. 2000
Coriander	*Coriandrum sativum* (L.)	Seed	7 (DW)	Manchester et al. 2000
Flax	*Linum usitatissimum* (L.)	Seed	12 (DW)	Manchester et al. 2000
Linseed (flax)	*Linum usitatissimum* (L.)	Seed	12 (DW)	Manchester et al. 2000
Green cardamom	*Elettaria cardamomum* (White et Maton)	Seed	15 (DW)	Manchester et al. 2000
Alfalfa	*Medicago sativum* (L.)	Seed	16 (DW)	Manchester et al. 2000
Fennel	*Foeniculum vulgare* (Gilib.)	Seed	28 (DW)	Manchester et al. 2000
Sunflower	*Helianthus annuus* (L.)	Seed	29 (DW)	Manchester et al. 2000
Almond	*Prunus amygdalus* (Batsch)	Seed	39 (DW)	Manchester et al. 2000
Fenugreek	*Trigonella foenum-graecum* (L.)	Seed	43 (DW)	Manchester et al. 2000
Wolf berry (goji berry)	*Lycium barbarum*	Seed	103 (DW)	Manchester et al. 2000
Black mustard	*Brassica nigra* (L.)	Seed	129 (DW)	Manchester et al. 2000
White mustard	*Sinapis alba* (L.)	Seed	189 (DW)	Manchester et al. 2000
Morning glory	*Pharbitis nil* Choisy	Shoot	0.004	Van Tassel et al. 2001
Taro	*Colocasia esculenta* (L.)	Shoot	0.055	Hattori et al. 1995
Asparagus	*Asparagus officinalis* (L.)	Shoot	0.01	Hattori et al. 1995
Red pigweed	*Chenopodium rubrum* (L.)	Shoot	0.2	Kolář et al. 1997
Japanese butterbur	*Petasites japonicus*	Shoot	0.495	Hattori et al. 1995

Source: Modified from Posmyk, M. M., and Janas, K. M. *Acta Physiol Plant.* 2009. 31: 1–11.

SNAT turned out to be a new *N*-acetyltransferase, not related to the family of the AANAT gene, and such homologous genes are present in many organisms from cyanobacteria to higher plants (Kang et al. 2013). Lee et al. (2014) cloned SNAT from *Arabidopsis thaliana* (*AtSNAT*) and functionally characterized this enzyme for the first time in dicotyledonous plants. Other SNAT genes were cloned from various plant species including pine (Park et al. 2014) and laver (Byeon et al. 2013). High conservation of SNAT

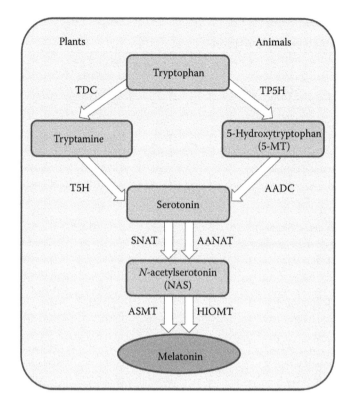

FIGURE 1.1 Biosynthetic pathway of MEL in plants and animals. TDC, tryptophan decarboxylase; T5H, tryptamine 5-hydroxylase; TP5H, tryptophan 5-hydroxylase; AADC, aromatic-L-amino-acid decarboxylase; AANAT, arylalkyl-amine *N*-acetyltransferase; SNAT, serotonin-*N*-acetyltransferase; HIOMT, hydroxyindole-*O*-methyltransferase; ASMT, *N*-acetylserotonin methyltransferase. (Modified From Park, W. J. *J Plant Biol.* 2011. 54: 143–149.)

genes among a wide range of organisms suggests that, as the penultimate gene in the MEL biosynthesis pathway, SNAT plays a pivotal role in production of MEL, and it may act as an essential molecule in plant growth and development.

A phenotype of transgenic tomato plants overexpressing oAANAT and oHIOMT, homologous genes of sheep, responsible for the last two steps of MEL synthesis was also examined. In the plants overexpressing the oAANAT gene the content of MEL was higher but that of IAA was lower than in the control plants, probably due to the competition between pathways for tryptophan, which is a common precursor of both indoleamines. In addition, these lines lost "apical dominance" (the result of decreased auxin level), which may indicate that despite the structural analogy to IAA MEL did not compensate for auxin activity. In the lines overexpressing the oHIOMT gene the content of MEL was much higher than in the lines overexpressing the oAANAT gene, which can confirm an important role of ASMT in the biosynthesis of MEL in plants (Wang et al. 2014).

Attempts to clone homologues of the ASMT gene from rice and their expression in *Escherichia coli* were made for the first time by Kang et al. (2011; 2013). Databases were searched for such fragments of rice cDNA that showed high similarity to wheat 3-*O*-methyl caffeic acid transferase (COMT) belonging to the family of *O*-methyltransferase (OMT) genes encoding the enzymatic proteins carrying a methyl group. A total of 18 cDNA fragments was selected, cloned and expressed in *E. coli*. Among all analyzed strains of *E. coli* only MT15 synthesized MEL in the presence of *N*-acetylserotonin, which suggests that it carries the ASMT gene encoding this enzyme. Subsequent studies showed that the rice ASMT gene was present in at least three copies (ASMT1–ASMT3), of which only ASMT1 gene was functionally identified (Park et al. 2013a). Due to the absence of functional expression of the purified ASMT2

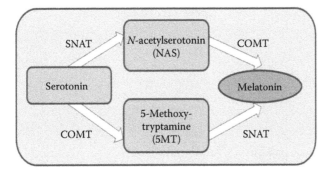

FIGURE 1.2 Schematic representation of MEL biosynthesis by SNAT (serotonin-*N*-acetyltransferase) and COMT (caffeic acid *O*-methyltransferase). (Modified from Lee, H. Y., et al. *J Pineal Res*. 2014. 57: 262–268.)

and ASMT3 recombinants in *E. coli* they were over-expressed in callus of transgenic rice lines where independent expression of all three genes was observed resulting in increased activity of ASMT enzyme compared to the wild type (Park et al. 2013b). However, in the seedlings of these transgenic lines MEL level was not higher, suggesting that the increased activity of ASMT is not directly associated with the synthesis of MEL. When the seedlings were treated with NAS, a direct substrate for ASMT, MEL level rapidly increased, which may indicate that MEL synthesis in rice is regulated by both ASMT activity and the concentration of the substrate (Park et al. 2013b). Moreover, in mature rice tissues the levels of ASMT1 and ASMT2 gene transcripts were higher than in seedlings and their largest amounts were found in stems and flowers, while ASMT3 gene transcripts were hardly detected in any organ. The increased expression of all ASMT genes was induced by such hormones as abscisic acid (ABA) and methyl jasmonate (MeJA), while the levels of ASMT2 and ASMT3 gene transcripts were reduced by 2-chloroethylphosphonic acid (ethephon—plant growth retardant), and salicylic acid (SA). Therefore, the action of the ASMT gene family appears to be tissue- and hormone-dependent, suggesting its possible role in the growth and development of rice (Park et al. 2013b).

Because many plants lack ASMT homologous genes, it is believed that COMT is responsible for the last step of MEL biosynthesis in plants. Lee et al. (2014) found that MEL was produced *in vitro* from serotonin in the presence of both AtSNAT and AtCOMT, thus confirming COMT involvement in MEL synthesis in plants. Their data suggest that serotonin can be converted either into NAS by SNAT or into 5-methoxytryptamine (5-MT) by COMT, and next metabolized into MEL by COMT or SNAT, respectively (Figure 1.2). These results provide biochemical evidence for the presence of a serotonin *O*-methylation pathway in plant MEL biosynthesis (Lee et al. 2014).

Possible Site of Melatonin Biosynthesis in Plant Cell

In all organisms the primary function of MEL consists in detoxification of FR and ROS generated mainly in mitochondria and chloroplasts during aerobic metabolism, thus these organelles are hypothesized as the sites of MEL biosynthesis in plants (Tan et al. 2013). Evolutionary mitochondria probably originated from purple nonsulfur bacteria, particularly *Rhodospirillum rubrum*, and chloroplasts likely originated from photosynthetic cyanobacteria. It is believed that they transformed into eukaryotic cell organelles by endosymbiosis. According to this hypothesis of Tan et al. (2013), in most organisms, if not all, the capacity of MEL biosynthesis was transferred from the mitochondria and chloroplasts to other cell compartments such as cytosol. However, at the early stage of biosymbiosis mitochondria and chloroplasts were the only sites of MEL synthesis in cells. Genes involved in MEL biosynthesis found in *R. rubrum* or other nonsulfur purple bacteria and cyanobacteria were gradually integrated into the nuclear genome during evolution. It is assumed that other cellular components adopted the capacity of MEL biosynthesis in this way. During evolution the genes of MEL synthesis might have been transferred from the mitochondrial genome into a nucleus but they also could have been lost (Adams and Palmer 2003).

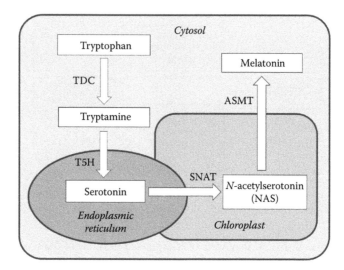

FIGURE 1.3 Site of MEL biosynthesis. TDC, tryptophan decarboxylase; T5H, tryptamine 5-hydroxylase; SNAT, serotonin-*N*-acetyltransferase, ASMT, *N*-acetylserotonin methyltransferase. (Modified From Tan, D. X., et al. *J Pineal Res.* 2013. 54: 127–138.)

It seems that the hypothesis about the original location of MEL biosynthesis in mitochondria and chloroplasts can help to answer the question why MEL content in green plants is much higher than in animals. The presence of two organelles (mitochondria and chloroplasts) able to synthesize MEL in plants in comparison to animals that have only mitochondria may be a contributory factor to explain this phenomenon. However, the proposed theory of MEL synthesis in the above organelles does not exclude its synthesis in other cell compartments (Tan et al. 2013). The availability of the MEL precursor, tryptophan, seems to further explain its higher level in plants. Plants are able to synthesize this amino acid; therefore, it is theoretically permanently available for further transformations, while in animals it can be only supplied with food.

Experiments with transgenic rice plants seem to confirm the hypothesis concerning correlation of MEL biosynthesis with endosymbiont theory showing that SNAT was located in chloroplasts while ASMT was located in cytoplasm (Byeon et al. 2014). Similar to rice SNAT, AtSNAT was also localized in chloroplasts, while AtCOMT was localized in cytoplasm (Lee et al. 2014). Although there is no direct evidence for TDC involvement in MEL biosynthesis in plants, indirect evidence suggests such correlation (Zhao et al. 2013; Kang et al. 2007). TDC is localized in cytoplasm (Stevens et al. 1993; De Luca and Cutler 1987), while the second enzyme on the MEL biosynthetic pathway (tryptamine 5-hydroxylase [T5H]) in endoplasmic reticulum (ER) (Fujiwara et al. 2010). Therefore, it is suggested that MEL biosynthesis in plants takes place successively in three cellular compartments: cytoplasm, ER and chloroplasts (Figure 1.3). Overexpression of ASMT or T5H genes in transgenic rice plants only slightly increased MEL level (Park et al. 2013b,c), so it can be assumed that their role in regulation of the biosynthesis of this indoleamine in plants is not crucial. Therefore, to be able to tell which of these four enzymes plays a key role in MEL biosynthesis in plants, additional tests on overexpression or suppression of genes encoding them are necessary.

Conclusion

The use of molecular biology techniques allows for precise study of genes potentially involved in MEL biosynthesis. The mode of action at the gene level needs detailed study. The role of MEL in plant signaling remains to be understood. In general, this topic of research would provide enormous new understanding of the hitherto unexplored action of the novel metabolites of melatonin.

REFERENCES

Adams, K. L., and Palmer, J. D. Evolution of mitochondrial gene content: Gene loss and transfer to the nucleus. *Mol Phylogenet Evol.* 2003. 29: 380–395.

Afreen, F., Zobayed, S. M., and Kozai, T. Melatonin in *Glycyrrhiza uralensis*: Response of plant roots to spectral quality of light and UV-B radiation. *J Pineal Res.* 2006. 41: 108–115.

Arnao, M. B., and Hernandez-Ruiz, J. The physiological function of melatonin in plants. *Plant Signal Behav.* 2006. 1: 89–95.

Balzer, I., and Hardeland, R. Photoperiodism and effects of indoleamines in a unicellular alga, *Gonyaulax polyedra*. *Science.* 1991. 253: 795–797.

Behrmann, G., Fuhrberg, B., Hardeland, R., Uria, H., and Poeggeler, B. Photooxidation of melatonin, 5-methoxytryptamine and 5-methoxytryptophol: Aspects of photoprotection by periodically fluctuating molecules? *Biometeorology.* 1997. 14: 258–263.

Bentkowski, P., and Markowska, M. Ewolucja fizjologicznych funkcji melatoniny u bezkręgowców. *Kosmos.* 2007. 56: 383–391.

Boccalandro, H. E., González, C.V., Wuderlin, D. A., and Silva, M. F. Melatonin levels, determined by LC-ESI-MS/MS, fluctuate during the day/night cycle in *Vitis vinifera* cv Malbec: Evidence of its antioxidant role in fruits. *J Pineal Res.* 2011. 51: 226–232.

Burkhardt, S., Tan, D. X., Manchester, L. C., Hardeland, R., and Reiter, R. J. Detection and quantification of the antioxidant melatonin in Montmorency and Balaton tart cherries (*Prunus cerasus*). *J Agr Food Chem.* 2001. 49: 4898–4902.

Byeon, Y., Lee, H. Y., Lee, K., Park, S., and Back, K. Cellular localization and kinetics of the rice melatonin biosynthetic enzymes SNAT and ASMT. *J Pineal Res.* 2014. 56: 107–114.

Byeon, Y., Lee, K., Park, Y. I., Park, S., and Back, K. Molecular cloning and functional analysis of serotonin *N*-acetyltransferase from the cyanobacterium *Synechocystis sp.* PCC 6803. *J Pineal Res.* 2013. 55: 371–376.

Chen, G., Huo, Y., Tan, D. X., Liang, Z., Zhang, W., and Zhang, Y. Melatonin in Chinese medicinal herbs. *Life Sci.* 2003. 73: 19–26.

Conti, A., Tettamanti, C., Singaravel, M., Haldar, C., Pandi-Perumal, S. R., and Maestroni, G. J. M. Melatonin: An ubiquitous and evolutionary hormone. In Haldar, C., Singaravel, M., and Maitra, S. K., editors, *Treatise on Pineal Gland and Melatonin.* 2002. 105–143. Enfield, New Hampshire: Science Publishers.

De Luca, V., and Cutler, A. J. Subcellular localization of enzymes involved in indole alkaloid biosynthesis in *Catharanthus roseus*. *Plant Physiol.* 1987. 85: 1099–1102.

Dubbels, R., Reiter, R. J., Klenke, E., Goebel, A., Schnakenberg, E., Ehlers, C., Schiwara, H. W., and Schloot, W. Melatonin in edible plants identified by radioimmunoassay and by high performance liquid chromatography–mass spectrometry. *J Pineal Res.* 1995. 18: 28–31.

Dyda, F., Klein, D. C., and Hickman, A. B. GCN5-related *N*-acetyltransferases: A structural overview. *Ann Rev Biophys Biomol Struct.* 2000. 29: 81–103.

Facchini, P. J., Huber-Allanach, K. L., and Tari, L. W. Plant aromatic L-Amino acid decarboxylases: Evolution, biochemistry, regulation, and metabolic engineering applications. *Phytochemistry.* 2000. 54: 21–138.

Fujiwara, T., Maisonneuve, S., Isshiki, M., Mizutani, M., Chen, L., Wong, H. I., Kawasaki, T., and Shimamoto, K. Sekiguchi lesion gene encodes a cytochrome P450 monooxygenase that catalyzes conversion of tryptamine to serotonin in rice. *J Biol Chem.* 2010. 285: 11308–11313.

Gonzalez-Gomez, D., Lozano, M., Fernandez-Leon, M. F., Ayuso, M. C., Bernalte, M. J., and Rodrıguez, A. B. Detection and quantification of melatonin and serotonin in eight sweet cherry cultivars (*Prunus avium* L.). *Eur Food Res Technol.* 2009. 229 (2): 223–229.

Hardeland, R. Melatonin, hormone of darkness and more: Occurrence, control mechanisms, actions and bioactive metabolites. *Cell Mol Life Sci.* 2008. 65: 2001–2018.

Hardeland, R., and Fuhrberg, B. Ubiquitous melatonin: Presence and effects in unicells, plants and animals. *Trends Comp Biochem Physiol.* 1996. 2: 25–45.

Hardeland, R., and Poeggeler, B. Non-vertebrate melatonin. *J Pineal Res.* 2003. 34: 233–241.

Hattori, A., Migitaka, H., Iigo, M., Itoh, M., Yamamoto, K., Ohtani-Kaneko, R., Hara, M., Suzuki, T., and Reiter, R. J. Identification of melatonin in plants and its effects on plasma melatonin levels and binding to melatonin receptors in vertebrates. *Biochem Mol Biol Int.* 1995. 35: 627–634.

Hernandez-Ruiz, J., and Arnao, M. B. Distribution of melatonin in different zones of lupin and barley plants at different ages in the presence and absence of light. *J Agr Food Chem*. 2008. 56 (22): 10567–10573.

Huang, X., and Mazza, G. Simultaneous analysis of serotonin, melatonin, piceid and resveratrol in fruits using liquid chromatography tandem mass spectrometry. *J Chromatogr A*. 2011. 1218: 3890–3899.

Iriti, M., Rossoni, M., and Faoro, F. Melatonin content in grape: Myth or panacea? *J Sci Food Agr*. 2006. 86 (10): 1432–1438.

Janas, K. M., Szafrańska, K., and Posmyk, M. M. Melatonina w roślinach. *Kosmos*. 2005. 54: 251–258.

Kang, K., Kong, K., Park, S., Natsagdorj, U., Kim, Y. S., and Back, K. Molecular cloning of a plant *N*-acetylserotonin methyltransferase and its expression characteristics in rice. *J Pineal Res*. 2011. 50: 304–309.

Kang, K., Lee, K., Park, S., Byeon, Z., and Back, K. Molecular cloning of rice serotonin *N*-acetyltransferase, the penultimate gene in plant melatonin biosynthesis. *J Pineal Res*. 2013. 55: 7–13.

Kang, K., Lee, K., Park, S., Kim, Y. S., and Back, K. Enhanced production of melatonin by ectopic overexpression of human serotonin *N*-acetylotransferase plays a role in cold resistance in transgenic rice seedlings. *J Pineal Res*. 2010. 49: 176–182.

Kang, S., Kang, K., Lee, K., Back, K. Characterization of rice tryptophan decarboxylases and their direct involvement in serotonin biosynthesis in transgenic rice. *Planta*. 2007. 227: 263–272.

Kolář, J., Macháčková, I., Eder, J., Prinsen, E., Van Dongen, W., Van Onckelen, H.A., and Illnerová, H. Melatonin: Occurrence and daily rhythm in *Chenopodium rubrum*. *Phytochem*. 1997. 44: 1407–1413.

Kolař, J., Macháčková, I., Illnerova, H., Prinsen, E., Van Dongen, W., and Van Onckelen, H. A. Melatonin in higher plants determined by radioimmunoassay and liquid chromatography–mass spectrometry. *Bio. Rhythm Res*. 1995. 26: 406–409.

Kołodziejczyk, I., Bałabusta, M., Szewczyk, R., and Posmyk, M. M. The levels of melatonin and its metabolites in conditioned corn (*Zea mays* L.) and cucumber (*Cucumis sativus* L.) seeds during storage. *Acta Physiol Plant*. 2015. 37: 105.

Lee, H. Y., Byeon, Y., and Back, K. Melatonin as a signal molecule triggering defence responses against pathogen attack in *Arabidopsis* and tobacco. *J Pineal Res*. 2014. 57: 262–268.

Lerner, A. B., Case, J. D., Takahashi, Y., Lee, T. H., and Mori, W. Isolation of melatonin, a pineal factor that lightens melanocytes. *J Am Chem Soc*. 1958. 80: 2587–2587.

Manchester, L. C., Tan, D. X., Reiter, R. J., Park, W., Monis, K., and Qi, W. B. High levels of melatonin in the seeds of edible plants: Possible function in germ tissue protection. *Life Sci*. 2000. 67: 3023–3029.

Mena, P., Gil-Izquierdo, A., Moreno, D. A., Mart, N., and Garcıa-Viguera, C. Assessment of the melatonin production in pomegranate wines. *LWT: Food Science and Technology*. 2012. 47: 13–18.

Mercolini, L., Mandrioli, R., and Raggi, M. A. Content of melatonin and other antioxidants in grape-related foodstuffs: Measurement using a MEPS-HPLC-F method. *J Pineal Res*. 2012. 53: 21–28.

Murch, S. J., Krishna, R. S., and Saxena, P. K. Tryptophan is a precursor for melatonin and serotonin biosynthesis in *in vitro* regenerated St. John's wort (*Hypericum perforatum* L. cv. Anthos) plants. *Plant Cell Reps*. 2000. 19: 698–704.

Murch, S. J., Simmons, C. B., and Saxena, P. K. Melatonin in feverfew and other medicinal plants. *Lancet*. 1997. 350: 1598–1599.

Okazaki, M., and Ezura, H. Profiling of melatonin in the model tomato (*Solanum lycopersicum* L.) cultivar Micro-Tom. *J Pineal Res*. 2009. 46: 338–343.

Okazaki, M., Higuchi, K., Hanawa, Y., Shiraiwa, Y., and Ezura, H. Cloning and characterization of a *Chlamydomonas reinhardtii* cDNA arylalkylamine *N*-acetyltransferase and its use in the genetic engineering of melatonin content in the Micro-Tom tomato. *J Pineal Res*. 2009. 46: 373–382.

Padumononda, T., Johns, J., Sangkasat, A., and Tiyaworanant, S. Determination of melatonin content in traditional Thai herbal remedies used as sleeping aids. *DARU J Pharm Sci*. 2014. 22: 6.

Pape, C., and Lüning, K. Quantification of melatonin in phototrophic organisms. *J Pineal Res*. 2006. 41 (2): 157–165.

Park, S., Byeon, Y., and Back, K. Functional analyses of three ASMT gene family members in rice plants. *J Pineal Res*. 2013b. 55: 409–415.

Park, S., Byeon, Y., and Back, K. Transcriptional suppression of tryptamine 5-hydroxylase, a terminal serotonin biosynthetic gene, induces melatonin biosynthesis in rice (*Oryza sativa* L.). *J Pineal Res*. 2013c. 55: 131–137.

Park, S., Byeon, Y., Kim, Y. S., and Back, K. Kinetic analysis of purified recombinant rice *N*-acetylserotonin methyltransferase and peak melatonin production in etiolated rice shoots. *J Pineal Res.* 2013a. 54: 139–144.

Park, S., Byeon, Y., Lee, H. Y., Kim, Y. S., Ahn, T., and Back, K. Cloning and characterization of a serotonin *N*-acetyltransferase from a gymnosperm, loblolly pine (*Pinus taeda*). *J Pineal Res.* 2014. 57: 348–355.

Park, S., Lee, K., Kim, Y. S., and Back, K. Tryptamine 5-hydroxylase-deficient Sekiguchi rice induces synthesis of 5-hydroxytryptophan and *N*-acetyltryptamine but decreases melatonin biosynthesis during senescence process of detached leaves. *J Pineal Res.* 2012. 52: 211–216.

Park, W. J. Melatonin as an endogenous plant regulatory signal: Debates and perspectives. *J Plant Biol.* 2011. 54: 143–149.

Posmyk, M. M., and Janas, K. M. Melatonin in plants. *Acta Physiol Plant.* 2009. 31: 1–11.

Ramakrishna, A. Indoleamines in edible plants: Role in human health effects. In Catalá, A., Biochemistry Research Trends Series, editor, *Indoleamines: Sources, Role in Biological Processes and Health Effects.* 2015. 279. New York: Nova Publishers.

Ramakrishna, A., Dayananda, C., Giridhar, P., Rajasekaran, T., and Ravishankar, G. A. Photoperiod influences endogenous indoleamines in cultured green alga *Dunaliella bardawil. Indian J Exp Biol.* 2011b. 49: 234–240.

Ramakrishna, A., Giridhar, P., Jobin, M., Paulose, C. S., and Ravishankar, G. A. Indoleamines and calcium enhance somatic embryogenesis in cultured tissues of *Coffea canephora* P ex Fr. *Plant Cell Tiss Org Cult.* 2012. 108: 267–278.

Ramakrishna, A., Giridhar, P., and Ravishankar, G. A. Indoleamines and calcium channels influence morphogenesis in *in vitro* cultures of *Mimosa pudica* L. *Plant Signal Behav.* 2009. 12: 1136–1141.

Ramakrishna, A., Giridhar, P., and Ravishankar, G. A. Phytoserotonin: A review. *Plant Signal Behav.* 2011a. 6: 800–809.

Ramakrishna, A., Giridhar, P., Sankar, K. U, and Ravishankar, G. A. Melatonin and serotonin profiles in beans of *Coffea* species. *J Pineal Res.* 2012a. 52: 470–476.

Ramakrishna, A., Giridhar, P., Sankar, K. U., and Ravishankar, G. A. Endogenous profiles of indoleamines: Serotonin and melatonin in different tissues of *Coffea canephora* P ex Fr. as analyzed by HPLC and LC-MS-ESI. *Acta Physiol Plant.* 2012b. 34: 393–396.

Reiter, R. J. Pineal melatonin: Cell biology of its synthesis and of its physiological interactions. *Endocr Rev.* 1991. 12: 151–180.

Reiter, R. J., Manchester, L. C., and Tan, D. X. Melatonin in walnuts: Influence on levels of melatonin and total antioxidant capacity of blood. *Nutrition.* 2005. 21 (9): 920–924.

Sae-Teaw, M., Johns, J., Johns, N. P., and Subongkot, S. Serum melatonin levels and antioxidant capacities after consumption of pineapple, orange, or banana by healthy male volunteers. *J Pineal Res.* 2012. 55 (1): 58–64.

Stege, P. W., Sombra, L. L., Messina, G., Martinez, L. D., and Silva, M. F. Determination of melatonin in wine and plant extracts by capillary electro chromatography with immobilized carboxylic multi-walled carbon nanotubes as stationary phase. *Electrophoresis.* 2010. 31 (13): 2242–2248.

Stevens, L. H., Blom, T. J. M., and Verpoorte, R. Subcellular localization of tryptophan decarboxylase, strictosidine synthase and strictosidine glucosidase in suspension cultured cells of *Catharanthus roseus* and *Tabernaemontana divaricata. Plant Cell Rep.* 1993. 12: 573–576.

Stürtz, M., Cerezo, A. B., Cantos-Villar, E., and Garcia-Parrilla, M. C. Determination of the melatonin content of different varieties of tomatoes (*Lycopersicon esculentum*) and strawberries (*Fragaria ananassa*). *Food Chem.* 2011. 127 (3): 1329–1334.

Szafrańska, K., Szewczyk, R., and Janas, K. M. Involvement of melatonin applied to *Vigna radiata* L. seeds in plant response to chilling stress. *Cent Eur J Biol.* 2014. 9 (11): 1117–1126.

Tan, D. X., Chen, L. D., Poeggeler, B., Manchester, L., and Reiter, R. J. Melatonin: A potent, endogenous hydroxyl radical scavenger. *Endocr J.* 1993. 1: 57–60.

Tan, D. X., Hardeland, R., Manchester, L. C., Korkmaz, A., Ma, S., Rosales-Corral, S., and Reiter, R. J. Functional roles of melatonin in plants, and perspectives in nutritional and agricultural science. *J Exp Bot.* 2012. 63: 577–597.

Tan, D. X., Manchester, L. C., Liu, X., Rosales-Corral, S. A., Acuna-Castroviejo, D., and Reiter, R. J. Mitochondria and chloroplasts as the original sites of melatonin synthesis: A hypothesis related to melatonin's primary function and evolution in eukaryotes. *J Pineal Res.* 2013. 54: 127–138.

Van Tassel, D. L., Roberts, N., Lewy, A., and O'Neil, S. D. Melatonin in plant organs. *J Pineal Res*. 2001. 31: 8–15.

Vitalini, S., Gardana, C., Zanzotto, A., Simonetti, P., Faoro, F., Fico, G., and Iriti, M. The presence of melatonin in grapevine (*Vitis vinifera* L.) berry tissues. *J Pineal Res*. 2011. 51(3): 331–337.

Wang, L., Zhao, Y., Reiter, R. J., He, C. H., Liu, G. Lei, Q., Zuo, B., Zheng, X. D., Li, Q., and Kong, J. Changes in melatonin levels in transgenic Micro-Tom tomato overexpressing ovine AANAT and ovine HIOMT genes. *J Pineal Res*. 2014. 56: 134–142.

Zhang, N., Sun, Q., Zhang, H., Cao, Y., Weeda, S., Ren, S., and Guo, Y. D. Roles of melatonin in abiotic stress resistance in plants. *J Exp Bot*. 2015. 66 (3): 647–656.

Zhao, Y., Tan, D. X., Lei, Q., Chen, H., Wang, L., Li, Q., Gao, Y., and Kong, J. Melatonin and its potential biological functions in the fruits of sweet cherry. *J Pineal Res*. 2013. 55: 79–88.

2

Occurrence of Serotonin, Melatonin, and Their Derivatives in Plants

Sergio D. Paredes, Lisa Rancan, Cruz García, Elena Vara, and Jesús A. F. Tresguerres
Complutense University of Madrid
Madrid, Spain

CONTENTS

ABSTRACT   Serotonin (5-hydroxytryptamine) and melatonin (*N*-acetyl-5-methoxytryptamine) have been detected in a number of plant species. The occurrence of its precursor, tryptophan, has also been reported in plants. Studies indicate that serotonin and melatonin are unequally distributed in plant parts, their concentration varies extremely widely in different plants, and both amines act promoting and/or modulating a variety of plant physiological mechanisms. Serotonin is involved in growth regulation, flowering, xylem sap exudation, ion permeability, and plant morphogenesis, as well as exhibiting antioxidant properties. The functions of melatonin in plants are varied and diverse, including seed germination or improvement of plant maturation and development. A plethora of studies indicate, however, that the primary function of melatonin in plants may be to serve as the first line of defense against oxidative stresses that are a result of internal and environmental insults. Similar to serotonin and melatonin, phenylpropanoid amides of serotonin, for example, *N*-p-coumaroylserotonin, and N^1-acetyl-N^2-formyl-5-methoxykynuramine (AFMK), a derivative from melatonin metabolism, have been shown to exert important protective roles in plants. This chapter reviews the investigations that have been performed to date on the occurrence of serotonin, melatonin, and their derivatives in plants; this would certainly assist in encouraging plant scientists to investigate the aspects that remain still unknown or to be clarified in this research topic.

KEY WORDS: *melatonin, serotonin, phytomelatonin, phytoserotonin, plant.*

Abbreviations

AFMK: N^1-acetyl-N^2-formyl-5-methoxykynuramine
HPLC: high-performance liquid chromatography
mRNA: messenger ribonucleic acid

Introduction

Serotonin is a physiologically active amine and well-known neurotransmitter that regulates mood, sleep, and anxiety in mammals (Veenstra-VanderWeele et al. 2000). It was initially identified as a vasoconstrictor substance in blood serum and later chemically identified as 5-hydroxytryptamine by Rapport et al. (1948). Serotonin regulates numerous biological processes including cardiovascular function, bowel motility, ejaculatory latency, and bladder control. Additionally, it has been suggested that serotonin may regulate some processes, including platelet aggregation, by receptor-independent, transglutaminase-dependent covalent linkage to cellular proteins (Berger et al. 2009).

The identification of serotonin in plants was reported in the 1950s, when Bowden et al. (1954) described its occurrence in cowhage (*Mucuna pruriens* [L.] DC.), a widely naturalized and cultivated tropical legume native to Africa and tropical Asia. Despite the diverse functions of serotonin in animal organisms, its roles in plants are not known precisely. Studies indicate that serotonin seems to be involved in a variety of plant physiological mechanisms, including growth regulation (Murch et al. 2001), flowering, xylem sap exudation (Kang et al. 2007), ion permeability (Odjakova and Hadjiivanova 1997), and plant morphogenesis (Csaba and Pal 1982; Ramakrishna et al. 2009, 2012c). In addition, serotonin plays a role as an antioxidant in plant cells by scavenging reactive oxygen species, and shows strong *in vitro* antioxidant activity. For example, it has been reported that serotonin relieves the accumulation of the toxic metabolite tryptamine, and maintains the reducing potential of cells through its powerful antioxidant activity in senesced rice leaves (Kang et al. 2009).

Although the identification of the tryptophan derivative, *N*-acetyl-5-methoxytryptamine (melatonin), in bovine pineal tissue was a major discovery in the late 1950s (Lerner et al. 1958, 1959), it was portrayed as being exclusively an animal hormone for nearly four decades. The chemical characterization of melatonin along with discoveries related to its synthesis (Axelrod and Weissbach 1960, 1961; Quay and Halevy 1962; Wurtman et al. 1963; Reiter et al. 2015), and evidence of its actions in mammals (Reiter and Hester 1966), established the pineal gland, an organ previously thought to be vestigial by most scientists, as a legitimate member of the endocrine hierarchy in vertebrates, particularly mammals. Since the pineal complex only exists in the nervous system of vertebrates (Quay and Renzoni 1963; Gundy and Wurst 1976; Oksche and Hartwig 1979; McNulty 1984), and melatonin's described actions were endocrine in nature (Reiter and Fraschini 1969; Roche et al. 1970), melatonin was referred to initially as a neurohormone, and it was assumed to be only produced in the pineal gland. Subsequently, melatonin has been uncovered in numerous organs (Reiter et al. 2013) and, theoretically at least, it may be synthesized in every vertebrate cell (Tan et al. 2013). However, evidence shows that melatonin is a constitutive molecule present in plants as well. Melatonin aids seeds in germinating, improves plant development and maturation of both the root system and above-ground tissues (Katul et al. 2003; Afreen et al. 2006; Meng et al. 2014; Weeda et al. 2014), protects plants from abiotic (Melchiorri et al. 1996; Abrahamsson et al. 2003; Posmyk et al. 2008; Li et al. 2012) and biotic stresses (Yin et al. 2013) which, because plants are sessile, cannot be avoided. In doing so plants, because they contain melatonin, exhibit an increased tolerance to environmental insults and can, in fact, acutely improve their defensive posture by upregulating endogenous melatonin production (Posmyk and Janas 2009; Arnao and Hernandez-Ruiz 2014).

As for the precursor of both serotonin and melatonin, the amino acid tryptophan, Mahan and Escott-Stump (2007) have reported a wide variety of plant foodstuffs that contain this essential amino acid, including cereals (rice, soybean, wheat, barley), vegetables (beans, lentils, peas, lettuce, tomato), nuts (almonds, walnuts, peanuts, dates), and fruits (bananas and cherries). In some cases, the occurrence of tryptophan in plants coincides with that of serotonin and melatonin, for example, in Jerte Valley cherries (Cubero et al. 2010). Currently, the list of plant foodstuffs rich in tryptophan is expanding greatly due to the development of more precise and effective techniques for the detection and quantification of this amino acid. Some of these plants are listed in Table 2.1. Moreover, occurrence of serotonin and melatonin in edible plants and processed foods has been reported (Tan et al. 2012; Janas and Posmyk 2013; Ramakrishna et al. 2015). This chapter is constructed to provide the reader with a state-of-the-art view of the current knowledge of the occurrence of serotonin, melatonin, and their derivatives in plants, in the hope of stimulating studies that will clarify the exact physiological roles of these molecules in plants.

TABLE 2.1

Levels of Tryptophan in Different Plants

Common Name	Scientific Name	[Tryptophan]	Reference
Sweet cherry	*Prunus avium* (L.) L.	3.6–8.2 mg/100 g (fresh weight)	Cubero et al. 2010
Banana	*Musa x paradisiaca* L.	9 mg/100 g (fresh weight)	USDA
Wild strawberry	*Fragaria vesca* L.	5 mg/100 g (fresh weight)	USDA
Mandarin orange	*Citrus reticulata* Blanco	2 mg/100 g (fresh weight)	USDA
Pear	*Pyrus communis* L.	2 mg/100 g (fresh weight)	USDA
Apple	*Pyrus malus* L.	1 mg/100 g (fresh weight)	USDA
Corn	*Zea mays* L.	67 mg/100 g (fresh weight)	USDA
Rice	*Oryza sativa* L.	75 mg/100 g (fresh weight)	USDA
Chickpea	*Cicer arietinum* L.	185 mg/100 g (fresh weight)	USDA
Lentil	*Lens culinaris* Medik.	232 mg/100 g (fresh weight)	USDA
White asparagus	*Asparagus albus* L.	27 mg/100 g (fresh weight)	USDA

Occurrence of Serotonin in Plants

As mentioned previously in this chapter, serotonin was first identified in the legume *Mucuna pruriens* (L.) DC. (Bowden et al. 1954). To date, it has been reported in over 40 plant species from 20 families (Roshchina 2001; Ramakrishna et al. 2011a). The highest values of serotonin have been found in walnuts (*Juglans regia* L.) and hickory (*Carya* sps. Nutt.), with levels ranging from 24 mg/kg to 400 mg/kg. Moderate levels of this molecule have been reported in plantain, pineapple, plums, banana, kiwifruit, and tomatoes, with values in the range of 3–30 mg/kg (Feldman and Lee 1985). Table 2.2 shows some plants where the presence of serotonin has been described.

It is well established that the levels of serotonin vary in different plant parts and, interestingly, they also vary within different tissues of a given plant. For example, in the West African leguminous plant *Griffonia simplicifolia* (M. Vahl ex DC.) Baill., serotonin was reported to be 0.007 µg/g fresh weight in leaves, while its amount augmented to 2000 µg/g in seeds (Fellows and Bell 1971). Moreover, high quantities of serotonin have been shown to increase during the reproductive period of this plant species (i.e., March), up to 1.2–1.3 µmole g/fresh weight tissue, whereas in the vegetative period (i.e., November and December) they accumulate only up to 0.3 µmole g/fresh weight of tissue. Seeds, and also fruits, are the major tissues where serotonin can be detected in significantly higher concentrations (Ramakrishna et al. 2011a). Other plant organs where serotonin has been identified include the sting of the nettle *Urtica dioica* L. (Collier and Chesher 1956), and the pods of cowhage (Bowden et al. 1954), in which the mono-amine has been suggested to play a protective role against predators. Another fruit in which serotonin is not uniformly distributed is the pineapple (*Ananas comosus* [L.] Merr.), which has greater levels in its outer skin and considerably lower values in the inner skin and pulp (Udenfriend et al. 1955). Ripening appears as a factor that decreases serotonin levels in this fruit, although the inverse is true for other species, including the tomato (Foy and Parratt 1961). Regarding the levels in fruits of the temperate species *Juglans regia* L., serotonin levels were found to be lowest in spring. In fall, however, a sharp increase occurred, with levels of the monoamine exceeding those measured in animal tissues (Erspamer 1961). This phenomenon coincided with the seed dispersal period (Bergmann et al. 1970). Thus, serotonin mainly accumulated during the process of fruit abscission. This abscission period was accompanied by proteolysis and deamination of amino acids giving rise to ammonium accumulation in walnut seeds. It is known that, to circumvent the toxic accumulation of ammonia, glutamine synthase assimilates ammonia together with glutamic acid via the synthesis of glutamine, which directly serves as a substrate for tryptophan synthesis. Serotonin synthesis is closely associated either with ripening or maturation of plant organs or with the accumulation of ammonia, which occurs predominantly during the process of plant senescence (Peeters and Van Laere 1992). Serotonin has also been quantified in walnut embryos, with levels around 0.4–0.6 µg/g fresh mass (Bergmann et al. 1970).

TABLE 2.2

Levels of Serotonin in Different Plants

Common Name	Scientific Name	[Serotonin]	Reference
Pineapple	*Ananas comosus* (L.) Merr.	1.5 µg/g (fresh weight)	West 1960
Runner bean	*Phaseolus multiflorus* Willd.	0.6 µg/g (fresh weight; in leaf)	Applewhite 1973
Pea	*Pisum sativum* L.	0.9–1 µg/g (fresh weight; in leaf and stem)	Applewhite 1973
Black walnut	*Juglans nigra* L.	180 µg/g (fresh weight; in seed)	Grobe 1982
Coffee	*Coffea arabica* L.	2.5 mg/kg (dry weight)	Stranc 1993
Welsh onion	*Allium fistulosum* L.	8 µg/g (fresh weight)	Ly et al. 2008
Chinese cabbage	*Brassica rapa* L.	90–130 µg/g (fresh weight)	Ly et al. 2008
Lettuce	*Lactuca serriola* L.	2.7–3.9 µg/g (fresh weight)	Ly et al. 2008
Tomato	*Lycopersicon esculentum* Mill.	221.9 µg/g (fresh weight)	Ly et al. 2008
Banana	*Musa × sapientum* L.	40–150 µg/g (fresh weight; in peel)	Ly et al. 2008
Strawberry	*Fragaria × ananassa* (Weston) Duchesne ex Rozier	3–4.5 µg/g (fresh weight)	Ly et al. 2008
Sweet cherry	*Prunus avium* (L.) L.	2.8–37.6 ng/100 g (fresh weight)	González-Gómez et al. 2009

In *Sedum morganianum* E. Walther and *Sedum pachyphyllum* Rose, serotonin seems to peak in the light time, and a lower concentration is found during darkness. In rice, the induced synthesis of serotonin has been closely associated with symptoms of senescence (Kang et al. 2007). Finally, serotonin occurrence has been described in *Mimosa pudica* L. and *Mimosa verrucosa* Benth., with concentrations varying from 0.06 to 0.86% dry weight (Nicasio et al. 2005). In the particular case of *Mimosa pudica* L., Ramakrishna et al. (2009) reported high levels of serotonin in seeds (80.4 µg/g fresh weight) compared to *in vitro* or *ex vitro* leaves (8.3 µg/g and 17.3 µg/g fresh weight, respectively).

Murch et al. (2009) have reported the occurrence of serotonin in *Datura metel* L. The flowers and smallest flower buds of this shrub-like perennial herb contained the highest concentrations of the molecule. The exposure of the flower buds to a cold stress significantly increased the concentrations of serotonin in the youngest buds at the most sensitive stage of reproductive development. It is worth mentioning that the same results were also found for melatonin (Murch et al. 2009). Using high-performance liquid chromatography (HPLC) with mass spectrometry detection, serotonin has also been detected and quantified in a variety of sweet cherry (*Prunus avium* L.) cultivars. The highest serotonin levels were found in the cultivar Ambrunés (37.6 ± 1.4 ng/100 g of fresh fruit) (González-Gómez et al. 2009). Furthermore, changes in the levels of serotonin during ripening of wine grapes have been detected (Murch et al. 2010).

It has been reported that pulp of underripe and ripe yellow banana contains serotonin at concentrations of 31.4 and 18.5 ng/g, respectively (Badria 2002; Vetorazzi 1974). However, it seems that serotonin levels in the Cavendish variety decrease during ripening (Vetorazzi 1974). Compared to the pulp, Udenfriend et al. (1959) reported that higher concentrations of serotonin were quantified in the peel (24,000 ng/g and 150,000 ng/g, respectively). During ripening, concentrations of serotonin increase significantly in both the inner and outer banana peel. For example, in Prata bananas (*Musa × sapientum* L.), serotonin levels were between 15,000 and 20,000 ng/g until the 14th day of storage, where serotonin levels decreased to about 7500 ng/g (Waalkes et al. 1958).

Localization

Considering the tissues where serotonin has been measured, immunolocalization analysis has revealed that serotonin was abundant in the vascular parenchyma cells, including companion and xylem cells of rice, suggesting its involvement in maintaining the cellular integrity for facilitating efficient nutrient recycling from senescing leaves to sink tissues during senescence (Kang et al. 2007; Ramakrishna et al.

2011a). Synthesis of serotonin also occurs in rice seedlings, as well as roots (6.5 mg/g fresh weight) and stem (1 mg/g fresh weight), and requires tryptamine as a substrate for the tryptamine-5-hydroxylase enzymatic activity (Kang et al. 2007), which is required to form serotonin from tryptamine (Schroder et al. 1999). Kimura (1968) demonstrated the presence of serotonin in banana, plantain, onion and lemon by means of fluorescence histochemical techniques. The banana fruit tissues showed yellow specific fluorescence of serotonin, which was especially abundant in the peel parenchymal cells. It was found that specific fluorescence was also distributed in the banana leaf and the stem tissues. In the plantain, specific fluorescence was seen in vascular bundle cells rather than in other cells. In the onion tissues, the fluorescence of serotonin was seen in epidermal cells. However, no specific fluorescence of serotonin was found in the lemon fruit peel (Kimura 1968).

Determination

HPLC has been widely used for serotonin determination in different samples, usually on reversed-phase or cation exchange analytical columns, including electrochemical or fluorimetric quantitation (Battini et al. 1989; Lagana et al. 1989; Kele and Ochmacht 1996; Cao et al. 2006). Accurate analysis of serotonin in plant samples using liquid chromatography–tandem mass spectrometry has been proposed. Serotonin was detected and quantified in the electrospray ionization, atmospheric pressure chemical ionization and atmospheric pressure photoionization positive mode (Cao et al. 2006). The studies on the influence of serotonin on physiological processes are governed by the ability to detect minute endogenous levels. Hence, the detection method plays a crucial role. Nevertheless, the amount of serotonin in these natural products has not been perfectly established due to the great variability of this amine in different fruits, and also to the degree of ripeness during the harvest period (Garcia and Marine 1983).

Finally, the endogenous pool of serotonin seems to vary under changes in photoperiod. In particular, in unicellular green alga *Dunaliella bardawii* Ben-Amotz & Avron, maximum biomass and carotenoid contents were found when cultures were grown in light (intensity of 2.0 Klux) at a photoperiod of 16:8 h light:dark cycle (Ramakrishna et al. 2011b). There was a profound influence on serotonin, and also melatonin, production of tested light:dark photoperiod conditions, including 8:16, 10:14, and 12:12 h, and continuous light, as estimated by HPLC and confirmed by mass spectral data obtained from liquid chromatography–mass spectrometry-electrospray ionization studies. Serotonin level increased from 908 to 1765 pg/g fresh weight with an increase proportional to light duration (Ramakrishna et al. 2011b).

Occurrence of Melatonin in Plants

More than two decades ago, Poeggeler et al. reported the presence of melatonin in the dinoflagellate *Lingulodinium polyedrum* (F. Stein) J. D. Dodge, 1989 (1989; 1991). This revealed that melatonin also occurred in phototropic organisms (Poeggeler and Hardeland 1994) and prompted research of melatonin in the plant kingdom. From that discovery on, the number of articles published in relation to melatonin in plants has largely increased (Paredes et al. 2009; Iriti et al. 2010; Reiter et al. 2015).

Although the first complete publications showing that melatonin indeed existed in plants were independently provided by Dubbels et al. (1995), and Hattori et al. (1995), there were some preliminary indications. Thus, early observations were made on cytoskeletal effects of melatonin in plants, namely, in endosperm cells of the amaryllidacean *Scadoxus multiflorus* (Martyn) Raf. (Jackson 1969), and in epidermal cells of *Allium cepa* L. (Banerjee and Margulis 1973). Changes were observed in mitotic spindles, which could lead to mitotic arrest. In addition, in the mid- to late 1970s, melatonin was removed as a by-product of the decaffeination process of coffee beans. This was presumably the first indication that plants or plant products contain melatonin. At that time, it was not imagined that melatonin existed in plants. As a consequence of the decaffeination process, it was presumed that melatonin was a result of an unexpected chemical reaction (Tan et al. 2012). This episode, however, had little impact on the subsequent discovery of melatonin in plants (Tan et al. 2012). In the early 1990s, Van Tassel et al. claimed that melatonin was detected in *Pharbitis nil* (L.) Choisy (Japanese morning glory) using radioimmunoassay as the method of measurement (Van Tassel et al. 1993, 1995). Melatonin occurrence was also observed in a higher plant, *Chenopodium rubrum* L. (Kolar and Machackova 1994; Kolar et al. 1995).

Currently, melatonin has been detected in over 100 plant species from numerous families. Among these, species with astonishingly high concentrations have been identified, often in the range of micrograms per gram, including *Hypericum* (Murch et al. 1997), numerous Chinese herbs (Chen et al. 2003), and alpine as well as Mediterranean plants (Conti et al. 2002). Particularly, in the flowering plants or angiosperms, the presence of melatonin has been described in a number of families belonging to both the Magnoliopsida and Liliopsida classes (Paredes et al. 2009). These include the Apiaceae (or Umbelliferae), Asteraceae (or Compositae), Brassicaceae (or Cruciferae), Fabaceae (or Leguminosae), Lamiaceae, Rosaceae, and Solanaceae in terms of dicotyledons; and Alliaceae, Poaceae (or Gramineae), and Zingiberaceae for the monocotyledons. Thus, melatonin may be viewed as a molecule with a widespread occurrence in the Magnoliphyta division (Paredes et al. 2009). A selection of edible vegetables, fruits, and plant-derived beverages where melatonin has been measured can be seen in Tables 2.3 and 2.4.

Excluding flowering plants, the available information on the presence of melatonin is rather scarce. The other divisions comprising the superdivision Spermatophyta (Pinophyta or conifers, Cicadophyta, Ginkgophyta, and Gnetophyta) have not been examined in such a systematic way for their melatonin content. This is also the case for the Marchantiophyta (liverworts), Anthocerotophyta (hornworts),

TABLE 2.3

Levels of Melatonin in Different Plants

Common Name	Scientific Name	[Melatonin]	Reference
Apple	*Pyrus malus* L.	48 pg/g (fresh weight)	Hattori et al. 1995
Rice	*Oryza sativa* L.	1 ng/g (in seed)	Hattori et al. 1995
Corn	*Zea mays* L.	1.3 ng/g (in seed)	Hattori et al. 1995
Oat	*Avena sativa* L.	1.8 ng/g (in seed)	Hattori et al. 1995
Almond	*Prunus amygdalus* Batsch	39.9 ng/g (dry weight; in seed)	Manchester et al. 2000
Sunflower	*Helianthus annuus* L.	29 ng/g (dry weight; in seed)	Manchester et al. 2000
White mustard	*Brassica hirta* Moench.	189 ng/g (dry weight; in seed)	Manchester et al. 2000
Tart cherry	*Prunus cerasus* L.	2–15 ng/g	Burkhardt et al. 2001
Banana	*Musa x paradisiaca* L.	0.5 ng/g	Hardeland and Pandi-Perumal 2005
Walnut	*Juglans regia* L.	3.5 ng/g (in seed)	Reiter et al. 2005
Grape	*Vitis vinifera* L.	5–965 pg/g (in skin)	Iriti et al. 2006
Tomato	*Lycopersicon esculentum* Mill.	1068.5 pg/g (fresh weight)	Pape and Lüning 2006
Lupin	*Lupinus albus* L.	16.2–18.4 ng (fresh weight; in root)	Arnao and Hernández-Ruiz 2007
Sweet cherry	*Prunus avium* (L.) L.	0–20 ng/100 g (fresh weight)	González-Gómez et al. 2009
Coffee	*Coffea canephora* Pierre ex A.Froehner	115 µg/g (fresh weight; in seed)	Ramakrishna et al. 2012b

TABLE 2.4

Levels of Melatonin in Different Plant-Derived Beverages

Beverages	[Melatonin]	Reference
Olive oil	53–119 pg/ml	de la Puerta et al. 2007
Beer	51–196 pg/ml	Maldonado et al. 2009
Must	1.1 ng/ml	Mercolini et al. 2012
Italian pomace brandy (Grappa)	0.3 ng/ml	Mercolini et al. 2012
Grape juice	0.5 ng/ml	Mercolini et al. 2012
Coffee	3–3.9 µg/50 ml	Ramakrishna et al. 2012a

Bryophyta (mosses), Lycopodiophyta (clubmosses), and Pteridophyta (ferns and horsetails). However, recent reports indicate that this indoleamine is also present in conifers. Thus, Park et al. (2014) observed that serotonin *N*-acetyltransferase mRNA, the penultimate enzyme in melatonin biosynthesis in both animals and plants, was constitutively expressed in all tissues, including leaf, bud, flower, and pinecone, whereas the corresponding protein was detected only in leaf. In accordance with the exclusive serotonin *N*-acetyltransferase protein expression in leaf, melatonin was detected only in leaf at 0.45 ng per gram fresh weight.

Algae, however, are an exception, as melatonin has been found in members of the Chlorophyta or green algae (including genera of the unicellular organisms *Chlamydomonas* Ehrenberg, *Dunaliella* Teodoresco, or *Acetabularia* J. V. F. Lamouroux, and in the macroalga *Ulva lactuca* L.), as well as in the Rhodophyta (red algae; *Palmaria palmata* [L.] Kuntze, *Porphyra umbilicalis* [L.] Kützing, *Chondrus crispus* Stackhouse) and Phaeophyta (brown algae; *Pterygophora californica* Ruprecht, and *Petalonia fascia* [O. F. Müller] Kuntze) (Balzer and Hardeland 1996; Fuhrberg et al. 1996; Balzer et al. 1998; Lorenz and Lüning 1998; Hardeland 1999; Pape and Lüning 2006).

Unicellular photosynthesizing organisms from the phylum Dinoflagellata, including *Pyrocystis acuta* Kofoid or the afore-mentioned *Lingulodinium polyedrum* (F. Stein) J. D. Dodge, 1989 (Poeggeler et al. 1991; Poeggeler and Hardeland 1994; Hardeland and Fuhrberg 1996; Fuhrberg et al. 1997; Hardeland and Poeggeler 2003) have also been reported to contain melatonin. Besides dinoflagellates, the presence of melatonin has been demonstrated in numerous evolutionarily different groups containing phototrophic species, including Alphaproteobacteria, Cyanobacteria (Tan et al. 2012), Euglenoidea, Rhodophyta, Phaeophyta (Hardeland and Fuhrberg 1996; Hardeland 1999), and Viridiplantae (Hardeland et al. 2007).

Localization

Melatonin has been identified in the roots, stems, leaves, flowers, and seeds of plants (Manchester et al. 2000; Paredes et al. 2009; Iriti et al. 2010) but, similarly to serotonin, its content varies considerably among species, from a few picograms to several micrograms per gram, an indication for different actions of this indoleamine (Tettamanti et al. 2000; Paredes et al. 2009; Posmyk and Janas 2009). At elevated levels, the common and presumably ancient property as an antioxidant may prevail (Hardeland 2015). In fact, evidence indicates that the initial and primary function of melatonin was to serve as the first line of defense against oxidative stress, and all other functions were acquired during evolution either by the process of adoption or by the extension of its antioxidative capacity (Tan et al. 2014). For example, seeds have been reported to possess high levels of melatonin, typically in the ng/g tissue range, although with considerable interspecific disparity, from 189 ng/g dry seed found in *Brassica hirta* Moench (white mustard) to only 2 ng/g dry seed described for *Silybum marianum* (L.) Gaertn. (milk thistle) (Manchester et al. 2000; Reiter et al. 2005). It is speculated that the high melatonin concentrations, because of the antioxidant activity of the indoleamine, aids in the germination of the seed. Seeds are rich in easily oxidized fats, and high concentrations of a potent antioxidant such as melatonin would be highly beneficial in preventing molecular damage and maintaining the ability of a seed to successfully germinate (Reiter et al. 2011).

Melatonin is an amphiphilic molecule (Shida et al. 1994; Ceraulo et al. 1999), which can freely cross cell membranes and distribute to any aqueous compartment including the cytosol, nucleus, and mitochondria (Menendez-Pelaez et al. 1993; Acuna-Castroviejo et al. 2001, 2003). In plants, kelps, and other algae, the distribution of melatonin between cytoplast/symplast, vacuoles, and the apoplast has been discussed (Hardeland 1999; Hardeland and Poeggeler 2003; Hardeland et al. 2007), but specific details have not yet been elaborated. Chloroplasts, however, are believed to be a primary site of melatonin production. The serotonin *N*-acetyltransferase has been localized in chloroplasts of the rice plant (Byeon and Back 2014). This is not surprising, since cyanobacteria are considered to be the precursors of chloroplasts, and these organisms already have the capacity to synthesize melatonin (Tan et al. 2013). In fact, Byeon et al. (2015) have recently reported that chloroplasts of red algae encode the melatonin synthetic enzyme gene, serotonin *N*-acetyltransferase. This observation provides direct evidence that melatonin synthesis occurs in the chloroplast.

Determination

It has been consistently reported that melatonin levels in plants are higher than those measured in the blood of animals. Initially, differences were interpreted to be due to the different methodologies used by researchers in terms of plant melatonin extraction and measurement (Reiter et al. 2007; Arnao and Hernandez-Ruiz 2009a). This led to the design of techniques able to obtain rapid and accurate results. Cao et al. (2006) designed a quick method for the reproducible concurrent detection and quantification of melatonin by means of chromatography–tandem mass spectrometry with electrospray ionization, atmospheric pressure chemical ionization, and atmospheric pressure photoionization. Electrospray ionization proved more reliable in routine tests to resist the effect of the sample matrix, especially for low melatonin contents. The limit of detection of melatonin in the plant extract was 5 pg ml^{-1} and the limit of quantification was 0.02 ng ml^{-1}. Other techniques were developed to measure the melatonin in a given plant sample, adjusting the methodological steps to its characteristics. This is the case of the work by Mercolini et al. (2008), who optimized an analytical method based on HPLC coupled to fluorescence detection for the determination of melatonin in red and white wine. Furthermore, Stege et al. (2010) created a new method for determination of melatonin in complex food matrices by capillary electrochromatography with immobilized carboxylic multilayer carbon nanotubes as stationary phase. The results yielded high electrochromatographic resolution, good capillary efficiencies, and improved sensitivity with respect to those obtained with conventional capillaries. In addition, it demonstrated highly reproducible results between runs, days, and columns.

The much higher melatonin levels in plants than those found in animals may reflect the absence of limits in tryptophan availability, which is much better in organisms producing aromatic amino acids via the shikimic acid pathway than in vertebrates, which are devoid of this metabolic route (Tan et al. 2012). These arguments are also valid for other high-melatonin organisms, such as dinoflagellates (Fuhrberg et al. 1997), euglenoids (Hardeland 1999), and yeast (Sprenger et al. 1999). However, the divergence in melatonin content between plants and animals may also be interpreted in terms of greater needs for resisting damage from harsh environments, which animals can actively avoid by behavioral changes. Considering the elevated free radical production that occurs during photosynthesis, it is easy to understand the advantage of increased melatonin production when plants are exposed to light, that is, high levels of melatonin are required to scavenge the large number of free radicals and preserve the photosynthetic function of plants. Many studies in fact have reported that exogenously applied or endogenously produced melatonin protects against chlorophyll degradation and preserves the photosynthetic function of plants due to a variety of stressors. These stressors include extremely cold or hot environments (Lei et al. 2004; Posmyk et al. 2009), UV irradiation (Afreen et al. 2006), copper (Tan et al. 2007a), cadmium, salted soil and hydrogen peroxide toxicity (Arnao and Hernandez-Ruiz 2009b). In relation to this, the highest levels of melatonin measured in any plant organ to date are in the pistachio nut (*Pistacia vera* L.), where the reported values are in the range of μg/mL methanolic extract (Oladi et al. 2014). Given that the pistachio is a desert plant, survives long periods without water (drought), is highly tolerant of salty soils, and can survive environmental temperatures ranging from $-10°C$ to $48°C$, it is possible that the pistachio tree is able to survive and thrive under desert conditions because of its high melatonin levels. Similarly, other plant species where melatonin levels are several orders of magnitude higher than those found in the serum of animals include *Glycyrrhiza uralensis* Fisher (Afreen et al. 2006), and several medicinal herbs (Chen et al. 2003). It can therefore be predicted that extremophiles (organisms which thrive under physically or geochemically extreme conditions) are possibly melatonin-enriched relative to the levels of the indole in mesophiles or neutrophils (organisms that grow best in moderate temperatures or thrive in neutral pH environments, respectively).

Another characteristic responsible for the seemingly divergent findings may be the presence in the plant of a circadian rhythm of melatonin synthesis. In the water hyacinth, *Eichhornia crassipes* (Mart.) Solms, melatonin levels vary throughout the light:dark cycle (Tan et al. 2007b). The peak melatonin levels, however, are not linked to darkness as in most vertebrates, but rather occur near the end of the light phase. Tan et al. (2007b) surmised that melatonin increases during the day to scavenge free radicals produced as a consequence of the process of photosynthesis. In the short-flowering plant, *Chenopodium rubrum* Fischer, a rhythm of melatonin has also been reported, but contrary to the case of the common hyacinth, high values are reached during darkness, while low levels were attained during the day (Kolar et al. 1997; Wolf et al. 2001). Finally, in at least two varieties of cherry fruits melatonin levels vary over a 24-hour period and also with fruit development (Zhao et al. 2013). In "Hongdeng" (*Prunus avium* L. cv. Hongdeng) and in "Rainier" (*Prunus avium* L. cv. Rainier) cherries, the 24-hour pattern of melatonin exhibited two peaks. The nighttime peak occurred at roughly 05:00 h, and in the day the peak was in the late afternoon, corresponding to the highest temperature and light intensity. The highest melatonin levels in the "Hongdeng" cherry were about 7.7 µg/g fresh weight while in "Rainier" the values approached 20 µg/g fresh weight. A similar relationship in "Red Fuji" apple (*Malus domestica* Borkh. cv. Red) has been reported by Lei et al. (2013). As in the cherry, highest melatonin levels coexisted with the periods of the most rapid expansion and increased respiration in the apple, points at which free radical generation is maximal.

Occurrence of Serotonin and Melatonin Derivatives with Potential Actions in Plants

As previously mentioned, the initial precursor of serotonin and melatonin biosynthesis is an amino acid, tryptophan. In animals, melatonin is formed via several enzymatic steps including tryptophan 5-hydroxylation, decarboxylation to form serotonin, *N*-acetylation and *O*-methylation. In that sequence, *N*-acetyl-5-methoxytryptamine (melatonin) is synthesized. Alternatively, but at lower flux rates, melatonin can be formed via *O*-methylation of serotonin and subsequent *N*-acetylation of 5-methoxytryptamine, or by *O*-methylation of tryptophan followed by decarboxylation and *N*-acetylation (Tan et al. 2007c). In plants, a major difference exists in the formation of serotonin from tryptophan. Thus, in the plants studied so far, the amino acid is first decarboxylated by tryptophan decarboxylase and the resulting amine is subsequently 5-hydroxylated by a cytochrome P450 monooxygenase, also known as tryptamine 5-hydroxylase (Hardeland 2015). The steps from serotonin to melatonin are identical to those of other organisms. Subsequently, both serotonin and melatonin can be converted into derivatives that may be potentially involved in plant physiological processeces or for ecophysiological adaptations. For example, phenylpropanoid amides of serotonin including *N*-p-coumaroylserotonin and *N*-feruloylserotonin have been shown to accumulate in witches' broom diseased bamboo, an infection caused by the ascomycetes *Aciculosporium take*. It has been reported that at least *N*-p-coumaroylserotonin possesses antifungal activity against this plant pathogen (Tanaka et al. 2003). Safflower (*Carthamus tinctorius* L.), is a highly branched, herbaceous, thistle-like annual plant where the afore-mentioned derivatives of serotonin have been detected, particularly in its seeds. Anti-inflammatory, anti-tumor, and anti-bacterial activities have been observed in animals administered with extracts and oils of this plant (Ramakrishna et al. 2011a). Moreover, *N*-p-coumaroyl serotonin and *N*-feruloyl serotonin isolated from safflower are also active as α-glucosidase inhibitors, and may aid in controlling diabetes (Takahashi and Miyazawa 2012).

Oxidative pyrrole ring cleavage represents the first step of a ubiquitous pathway of melatonin metabolism. The product formed by this reaction, N^1-acetyl-N^2-formyl-5-methoxykynuramine, is generated by numerous enzymatic, pseudoenzymatic, and nonenzymatic reactions, by free radicals, and by other oxidants such as singlet oxygen (Hardeland et al. 2009). AFMK can be also formed nonenzymatically from cyclic 3-hydroxymelatonin. Among the enzymes capable of producing AFMK from melatonin, an indoleamine 2,3-dioxygenase has been identified in rice and used for transgenic modulation of melatonin levels in tomatoes (Okazaki and Ezura 2009). The only study in which AFMK formation in plants has been more profoundly analyzed was conducted in the water hyacinth, where a 24-hour rhythm of this substituted kynuramine roughly paralleling that of melatonin has been described (Tan et al. 2007b). Thus, the temporal pattern was characterized by a gradual increase in both compounds over the photophase, with maxima around the light/dark transition. This may indicate a dependence of AFMK formation on melatonin concentrations, a conclusion supported by experiments using tryptophan supplementation, which caused substantial rises in both melatonin and AFMK (Hardeland 2015). AFMK, as melatonin, has resulted to act as a potent free radical scavenger. It has been shown to scavenge reactive oxygen and nitrogen species and to protect tissues from damage due to reactive intermediates in various models (Hardeland et al. 2009). AFMK reacts with ·OH at diffusion-limited rates, regardless of the polarity of the environment, which supports its excellent ·OH radical scavenging activity. Radical ·OOCCl3, whose reactivity is between those of ·OH and ·OOH and represents halogenated peroxyl radicals, has been reported to be also scavenged by AFMK very efficiently (Galano et al. 2013). In plants, biosynthesis of this melatonin derivative may assist them in coping with harsh environmental insults, including soil and water pollutants (Tan et al. 2007a). High levels of AFMK, and melatonin, in water hyacinth may explain why this plant more easily tolerates environmental pollutants, including toxic chemicals and heavy metals, and is successfully used in phytoremediation. Secondary products derived from AFMK have not yet been studied in plants.

In other phototrophs, such as dinoflagellates, the phaeophyte *Pterygophora*, or the chlorophycean *Chlorogonium*, AFMK formation has been observed only under rather nonphysiological conditions, such as in cell extracts exposed to light or after exposure to high exogenous melatonin (Hardeland 2015).

Conclusion

Occurrence of serotonin and melatonin in plants has been reported in a variety of species. The results of these investigations have uncovered several facts: Both amines are present in a wide number of plant products; the concentration of serotonin and melatonin varies extremely widely in different plants; and serotonin and melatonin are unequally distributed in plant parts (Paredes et al. 2009). However, researchers agree that there are still questions to be explored. In the case of plant serotonin, although interesting results have been obtained in the past years, data are incomplete, and more detailed studies are needed to elucidate the role of this compound in plants. For example, more studies should be performed to determine clearly the parts of plants where serotonin synthesis occur, and whether the levels of the molecule change on a seasonal basis when plants are grown in their natural habitat. Moreover, variations in the levels of plant serotonin in species grown at different latitudes should be studied (Ramakrishna et al. 2011a). As for melatonin, although several species of different taxonomic positions exhibit 24-hour rhythms, a truly circadian nature has been only demonstrated exceptionally. Moreover, the phase positions of maxima within the daily cycle turned out to differ considerably among organisms. Therefore, melatonin cannot be regarded as a general signal of darkness in plants, although this may not be ruled out in some species (Hardeland 2015). Studies indicate that the primary function of melatonin in plants is to serve as the first line of defense against oxidative stresses that are a result of internal and environmental insults. In fact, some plants have been shown to produce remarkable quantities of melatonin, and thus direct oxidant detoxification may occur and be relevant for these organisms. Nonetheless, the molecular signals that initiate the upregulation of melatonin biosynthesis are still unknown. Furthermore, the conditions under which melatonin concentrations are influenced by phytohormones should be addressed (Hardeland 2015). Finally, serotonin derivatives N-p-coumaroylserotonin and N-feruloylserotonin and melatonin derivative AFMK seem to play a role in plants. Other molecules derived from these amines may also be involved in plant physiology, but information of their nature and mechanisms of action is still unclear.

Clarification of the occurrence and function of serotonin, melatonin, and their derivatives in plants is, therefore, a major area of research. Clearly, the mechanisms whereby serotonin, melatonin and their derivatives occur and carry out their multiple functions in plants still require extensive investigative effort.

Acknowledgment

The authors gratefully acknowledge the support of Red Cooperativa de Envejecimiento y Fragilidad 2013–2016. Instituto de Salud Carlos III, Madrid (Spain).

REFERENCES

Abrahamsson, K., Choo, K. S., Pedersen, M., Johansson, G., and Snoeijs, P. Effects of temperature on the production of hydrogen peroxide and volatile halocarbons by brackish-water algae. *Phytochemistry.* 2003. 64: 725–734.

Acuna-Castroviejo, D., Escames, G., Leon, J., Carazo, A., and Khaldy, H. Mitochondrial regulation by melatonin and its metabolites. *Adv Exp Med Biol.* 2003. 527: 549–557.

Acuna-Castroviejo, D., Martin, M., Macias, M., Escames, G., Leon, J., Khaldy, H., and Reiter, R. J. Melatonin, mitochondria, and cellular bioenergetics. *J Pineal Res.* 2001. 30: 65–74.

Afreen, F., Zobayed, S. M. A., and Kozai, T. Melatonin in *Glycyrrhiza uralensis*: Response of plant roots to spectral quality of light and UV-B radiation. *J Pineal Res.* 2006: 108–115.

Applewhite, P. B. Serotonin and norepinephrine in plant tissues. *Phytochemistry.* 1973. 12: 191–192.

Arnao, M. B, and Hernández-Ruiz J. Melatonin promotes adventitious- and lateral root regeneration in etiolated hypocotyls of *Lupinus albus* L. *J Pineal Res.* 2007. 42: 147–152.

Arnao, M. B., and Hernandez-Ruiz, J. Assessment of different sample processing procedures applied to the determination of melatonin in plants. *Phytochem Anal.* 2009a. 20: 14–18.

Arnao, M. B., and Hernandez-Ruiz, J. Chemical stress by different agents affects the melatonin content of barley roots. *J Pineal Res.* 2009b. 46: 295–299.

Arnao, M. B., and Hernandez-Ruiz, J. Melatonin: Plant growth regulator and/or biostimulator during stress? *Trends Plant Sci.* 2014. 19: 789–797.

Axelrod, J., and Weissbach, H. Purification and properties of hydroxyindole-*O*-methyl transferase. *J Biol Chem.* 1961. 236: 211–213.

Axelrod, J., and Weissbach, H. Enzymatic *O*-methylation of *N*-acetylserotonin to melatonin. *Science.* 1960. 131: 1312.

Badria, F. A. Melatonin, serotonin and tryptamine in some Egyptian food and medicinal plants. *J Med Food.* 2002. 5: 153–157.

Balzer, I., Bartolomaeus, B., and Höcker, B. Circadian rhythm of melatonin content in *Chlorophyceae. Proceedings of the Conference News from the Plant Chronobiology Research.* Markgrafenheide, Germany. 1998. 55–56.

Balzer, I., and Hardeland, R. Melatonin in algae and higher plants: Possible new roles as a phytohormone and antioxidant. *Botanica Acta.* 1996. 109: 180–183.

Banerjee, S., and Margulis, L. Mitotic arrest by melatonin. *Exp Cell Res.* 1973. 78: 314–318.

Battini, M. L., Careri, M., Casoli, A., Mangia, A., and Lugari, M. T. Determination of *N*-alkanoyl-5-hydroxytryptamides (C5HT) in coffee beans by means of HPLC and TLC. *Ann Chim.* 1989. 79: 369–377.

Berger, M., Gray, J. A., and Roth, B. L. The expanded biology of serotonin. *Annu Rev Med.* 2009. 60: 355–366.

Bergmann, L., Grosse, W., and Ruppel, H. G. Serotonin in *Juglans regia* L. *Planta.* 1970. 94: 47–59.

Bowden, K., Brown, B. G., and Batty, J. E. 5-hydroxytriptamine: Its occurrence in cowhage (*Mucuna pruriens*). *Nature.* 1954. 174: 925–926.

Burkhardt, S., Tan, D. X., Manchester, L. C., Hardeland, R., and Reiter, R. J. Detection and quantification of the antioxidant melatonin in Montmorency and Balaton tart cherries (*Prunus cerasus*). *J Agric Food Chem.* 2001. 49: 4898–4902.

Byeon, Y., and Back, K. Melatonin synthesis in rice seedlings *in vivo* is enhanced at high temperatures and under dark conditions due to increased serotonin *N*-acetyltransferase and *N*-acetylserotonin methyltransferase activities. *J Pineal Res.* 2014. 56: 189–195.

Byeon, Y., Lee, H. Y., Choi, D. W., and Back, K. Chloroplast-encoded serotonin *N*-acetyltransferase in the red alga (*Pyropia yezoensis*): Gene transition to the nucleus from chloroplasts. *J Exp Bot.* 2015. 66: 709–171.

Cao, J., Murch, S. J., O'Brien, R., and Saxena, P. K. Rapid method for accurate analysis of melatonin, serotonin and auxin in plant samples using liquid chromatography tandem mass spectrometry. *J Chromatogra A.* 2006. 1134: 333–337.

Ceraulo, L., Ferrugia, M., Tesoriere, L., Segreto, S., Livrea, M. A., and Turco Liveri, V. Interactions of melatonin with membrane models: Portioning of melatonin in AOT and lecithin-reversed micelles. *J Pineal Res.* 1999. 26: 108–112.

Chen, G., Huo, Y., Tan, D. X., Liang, Z., Zhang, W., and Zhang, Y. Melatonin in Chinese medicinal herbs. *Life Sci.* 2003. 73: 19–26.

Collier, H. O. J., and Chesher, G. B. Identification of 5-hydroxytryptamine in the sting of the nettle (*Urtica dioica*). *Br J Pharmacol Chemother.* 1956. 11: 186–189.

Conti, A., Tettamanti, C., Singaravel, M., Haldar, C., Pandi-Perumal, S. R., and Maestroni, G. J. M. Melatonin: An ubiquitous and evolutionary hormone. In Haldar, C., Singaravel, M., and Maitra, S. K. (editors), *Treatise on Pineal Gland and Melatonin.* Enfield, New Hampshire: Science Publishers. 2002. 105–143.

Csaba, G., and Pal, K. Effect of insulin triiodothyronine and serotonin on plant seed development. *Protoplasma.* 1982. 110: 20–22.

Cubero, J., Toribio, F., Garrido, M., Hernández, M., Maynar, J., Barriga, C., and Rodríguez, A. Assays of the amino acid tryptophan in cherries by HPLC-fluorescence. *Food Anal Method.* 2010. 3: 36–39.

de la Puerta, C., Carrascosa-Salmoral, M. P., García-Luna, P. P., Lardone, P. J., Herrera, J. L., Fernández-Montesinos, R., Guerrero, J. M., and Pozo, D. Melatonin is a phytochemical in olive oil. *Food Chem.* 2007. 104: 609–612.

Dubbels, R., Reiter, R. J., Klenke, E., Goebel, A., Schnakenberg, E., Ehlers, C., Schiwara, H. W., and Schloot, W. Melatonin in edible plants identified by radioimmunoassay and by high performance liquid chromatography–mass spectrometry. *J Pineal Res.* 1995. 18: 28–31.

Erspamer, V. Recent research in the field of 5-hydroxytryptamine and related indolealkylamines. In Jucker, E. (editor), *Progress in Drug Research.* Basel, Switzerland: Birkhaüser Verlag. 1961. 151–367.

Feldman, J. M., and Lee, E. M. Serotonin content of foods: Effect on urinary excretion of 5-hydroxy indoleacetic acid. *Am J Clin Nutr.* 1985. 42: 639–643.

Fellows, L. E., and Bell, E. A. Indole metabolism in *Piptadenia peregrina. Phytochemistry.* 1971. 10: 2083–2091.

Foy, J. M., and Parratt, J. R. 5-hydroxytryptamine in pineapples. *J Pharm Pharmacol.* 1961. 13: 382–383.

Fuhrberg, B., Balzer, I., Hardeland, R., Wemer, A., and Lüning, K. The vertebrate pineal hormone melatonin is produced by the brown alga *Pterygophora californica* and mimics dark effects on growth rate in the light. *Planta.* 1996. 200: 125–131.

Fuhrberg, B., Hardeland, R., Poeggeler, B., and Behrmann, G. Dramatic rises of melatonin and 5-methoxytryptamine in *Gonyaulax* exposed to decreased temperature. *Biol Rhythm Res.* 1997. 28: 144–150.

Galano, A., Tan, D. X., and Reiter, R. J. On the free radical scavenging activities of melatonin's metabolites, AFMK and AMK. *J Pineal Res.* 2013. 54: 245–257.

Garcia C., and Marine, A. Serotonin contents in fresh and processed foods. *Rev Agroquim Tecnol Aliment.* 1983. 23: 60–70.

González-Gómez, D., Lozano, M., Fernández-León, M. F., Ayuso, M. C., Bernalte, M. J., and Rodríguez, A. B. Detection and quantification of melatonin and serotonin in eight sweet cherry cultivars (*Prunus avium* L.). *Eur Food Res Tech.* 2009. 229: 223–229.

Grobe, W. Function of serotonin in seeds of walnuts. *Phytochemistry.* 1982. 21: 819–822.

Gundy, G. C., and Wurst, G. Z. Parietal eye-pineal morphology in lizards and its physiological implications. *Anat Rec.* 1976. 185: 419–431.

Hardeland, R. Melatonin and 5-methoxytryptamine in non-metazoans. *Reprod Nutr Dev.* 1999. 39: 399–408.

Hardeland, R. Melatonin in plants and other phototrophs: Advances and gaps concerning the diversity of functions. *J Exp Bot.* 2015. 66: 627–646.

Hardeland, R., and Fuhrberg, B. Ubiquitous melatonin: Presence and effects in unicells, plants and animals. *Trends Comp Biochem Physiol.* 1996. 2: 25–45.

Hardeland, R., and Pandi-Perumal, S. R. Melatonin, a potent agent in antioxidative defense: Actions as a natural food constituent, gastrointestinal factor, drug and prodrug. *Nutr Metab.* 2005. 10: 2–22.

Hardeland, R., Pandi-Perumal, S. R., and Poeggeler, B. Melatonin in plants: Focus on a vertebrate night hormone with cytoprotective properties. *Funct Plant Sci Biotechnol.* 2007. 1: 32–45.

Hardeland, R., and Poeggeler, B. Non-vertebrate melatonin. *J Pineal Res.* 2003. 34: 233–241.

Hardeland, R., Tan, D. X., and Reiter, R. J. Kynuramines, metabolites of melatonin and other indoles: The resurrection of an almost forgotten class of biogenic amines. *J Pineal Res.* 2009. 47: 109–126.

Hattori, A., Migitaka, H., Iigo, M., Itoh, M., Yamamoto, K., Ohtani-Kaneko, R., Hara, M., Suzuki, T., and Reiter, R. J. Identification of melatonin in plants and its effects on plasma melatonin levels and binding to melatonin receptors in vertebrates. *Biochem Mol Biol Int.* 1995. 35: 627–634.

Iriti, M., Rossoni, M., and Faoro, F. Melatonin content in grape: Myth or panacea? *J Sci Food Agric.* 2006. 86: 1432–1438.

Iriti, M., Varoni, E. M., and Vitalini, S. Melatonin in traditional Mediterranean diets. *J Pineal Res.* 2010. 49: 101–105.

Jackson, W. T. Regulation of mitosis. II. Interaction of isopropyl *N*-phenylcarbamate and melatonin. *J Cell Sci.* 1969. 5: 745–755.

Janas, K. M., and Posmyk, M. M. Melatonin, an underestimated natural substance with great potential for agricultural application. *Acta Physiol Plant.* 2013. 35: 3285–3292.

Kang, S., Kang, K., Lee, K., and Back, K. Characterization of tryptamine 5-hydroxylase and serotonin synthesis in rice plants. *Plant Cell Rep.* 2007. 26: 2009–2015.

Kang, K., Kim, Y. S., Park, S., and Back, K. Senescence-induced serotonin biosynthesis and its role in delaying senescence in rice leaves. *Plant Physiol.* 2009. 150: 1380–1393.

Katul, G., Leuning, R., and Oren, R. Relationship between plant hydraulic and biochemical properties derived from a steady-state coupled water and carbon transport model. *Plant Cell Environ.* 2003. 26: 339–350.

Kele, M., and Ochmacht, R. Determination of serotonin released from coffee wax by liquid chromatography. *J Chromatogra A.* 1996. 730: 59–62.

Kimura, M. Fluorescence histochemical study on serotonin and catecholamine in some plants. *Jap J Pharmacol.* 1968. 18: 162–168.

Kolar, J., and Machackova, I. Melatonin: Does it regulate rhythmicity and photoperiodism also in higher plants? *Flower Newsletter.* 1994. 17: 53–54.

Kolar, J., Machackova, I., Eder, J., Prinsen, E., Van Dogen, W., Van Onckelen, H., and Illnerova, H. Melatonin: Occurrence and daily rhythm in *Chenopodium rubrum.* *Phytochemistry.* 1997. 44: 1407–1413.

Kolar, J., Machackova, I., Illnerova, H., Prinsen, E., Van Dongen, W., and Van Onckelen, H. A. Melatonin in higher plants determined by radioimmunoassay and liquid chromatography–mass spectometry. *Biol Rhythm Res.* 1995. 26: 406.

Lagana, A., Curini, L., de Angelis Curtis, S., and Marino, A. Rapid liquid chromatographic analysis of carboxylic acid 5-hydroxytryptamides in coffee. *Chromatographia.* 1989. 28: 593–596.

Lei, Q., Wang, L., Tan, D. X., Zhao, Y., Zheng, X. D., Chen, H., Li, Q. T., Zuo, B. X., and Kong J. Identification of genes for melatonin synthetic enzymes in "Red Fuji" apple (*Malus domesticus* Borkh. cv. Red) and their expression and melatonin production during fruit development. *J Pineal Res.* 2013. 55: 443–451.

Lei, X. Y., Zhu, R. Y., Zhang, G. Y., and Dai, Y. R. Attenuation of cold-induced apoptosis by exogenous melatonin in carrot suspension cells: The possible involvement of polyamines. *J Pineal Res.* 2004. 36: 126–131.

Lerner, A. B., Case, J. D., and Heinzelman, R. V. Structure of melatonin. *J Am Chem Soc.* 1959. 81: 6084–6085.

Lerner, A. B., Case, J. D., Takahashi, Y., Lee, Y., and Mori, W. Isolation of melatonin, the pineal gland factor that lightens melanocytes. *J Am Chem Soc.* 1958. 81: 2587.

Li, C., Wang, P., Wei, Z., Liang, D., Liu, C., Yin, L., Jia, D., Fu, M., and Ma, F. The mitigation effects of exogenous melatonin on salinity-induced stress in *Malus hupehensis.* *J Pineal Res.* 2012. 53: 298–306.

Lorenz, M., and Lüning, K. Detection of endogenous melatonin in the marine red macroalgae *Porphyra umbilicalis* and *Palmaria palmate* by enzyme-linked immunoassay (ELISA) and effects of melatonin administration on algal growth. *Proceedings of the Conference News from the Plant Chronobiology Research.* Markgrafenheide, Germany. 1998. 42–43.

Ly, D., Kang, K., Choi, J. Y., Ishihara, A., Back, K., and Lee, S. G. HPLC analysis of serotonin, tryptamine, tyramine and the hydroxycinnamic acid amides of serotonin and tyramine in food vegetables. *J Med Food.* 2008. 11: 385–389.

Mahan, L. K., and Escott-Stump, S. *Krause's Food and Nutrition Therapy.* Philadelphia, Pennsylvania: Saunders. 2007.

Maldonado, M. D., Moreno, H., and Calvo, J. R. Melatonin present in beer contributes to increase the levels of melatonin and antioxidant capacity of the human serum. *Clin Nutr.* 2009. 28: 188–191.

Manchester, L. C., Tan, D. X., Reiter, R. J., Park, W., Monis, K., and Qi, W. High levels of melatonin in the seeds of edible plants: Possible function in germ tissue protection. *Life Sci.* 2000. 67: 3023–3029.

McNulty, J. A. Functional morphology of the pineal complex in cyclostomes, elasmobranches and bony fishes. *Pineal Res Rev.* 1984. 2: 1–40.

Melchiorri, D., Reiter, R. J., Sewerynek, E., Hara, M., Chen, L., and Nistico, G. Paraquat toxicity and oxidative damage: Reduction by melatonin. *Biochem Pharmacol.* 1996. 51: 1095–1099.

Menendez-Pelaez, A., Poeggeler, B., Reiter, R. J., Barlow-Walden, L., Pablos, M. I., and Tan, D. X. Nuclear localization of melatonin in different mammalian tissues: Immunocytochemical and radioimmunoassay evidence. *J Cell Biochem.* 1993. 53, 373–382.

Meng, J. F., Xu, T. F., Wang, Z. Z., Fang, Y. L., Xi, Z. M., and Zhang, Z. W. The ameliorative effects of exogenous melatonin on grape cuttings under water-deficient stress: Antioxidant metabolites, leaf anatomy, and chloroplast morphology. *J Pineal Res.* 2014. 57: 200–212.

Mercolini, L., Addolorata Saracino, M., Bugamelli, F., Ferranti, A., Malaguti, M., Hrelia, S., and Raggi, M. A. HPLC-F analysis of melatonin and resveratrol isomers in wine using an SPE procedure. *J Sep Sci.* 2008. 31: 1007–1014.

Mercolini, L., Mandrioli, R., and Raggi, M. A. Content of melatonin and other antioxidants in grape-related foodstuffs: Measurement using a MEPS-HPLC-F method. *J Pineal Res.* 2012. 53: 21–28.

Murch, S. J., Alan, A. R., Cao, J., and Saxena, P. K. Melatonin and serotonin in flowers and fruits of *Datura metel* L. *J Pineal Res.* 2009. 47: 277–283.

Murch, S. J., Campbell, S. S. B., and Saxena, P. K. The role of serotonin and melatonin in plant morphogenesis: Regulation of auxin-induced root organogenesis in *in vitro*-cultured explants of St. John's wort (*Hypericum perforatum* L.). *In Vitro Cell Dev Biol Plant.* 2001. 37: 786–793.

Murch, S. J., Hall, B. A., Le, C. H., and Saxena, P. K. Changes in the levels of indoleamine phytocemicals during veraison and ripening of wine grapes. *J Pineal Res.* 2010. 49: 95–100.

Murch, S. J., Simmons, C. B., and Saxena, P. K. Melatonin in feverfew and other medicinal plants. *Lancet.* 1997. 350: 1598–1599.

Nicasio, M. D. P., Villrreal, M. L., Gillet, F., Bensaddek, L., and Fliniaux, M. A. Variation in the accumulation levels of N,*N*-Dimethyltryptamine in micropropagated trees and in *in vitro* cultures of *Mimosa Tenuiflore.* *Nat Prod Res.* 2005. 19: 61–67.

Odjakova, M., and Hadjiivanova, C. Animal neurotransmitter substances in plants. *Bulg J Plant Physiol.* 1997. 23: 94–102.

Okazaki, M., and Ezura, H. Profiling of melatonin in the model tomato (*Solanum lycopersicum* L.) cultivar Micro-Tom. *J Pineal Res.* 2009. 46: 338–343.

Oksche, A., Hartwig, H. G. Pineal sense organs: Components of photoneuroendocrine systems. *Prog Brain Res.* 1979. 52: 113–130.

Oladi, E., Mohamadi, M., Shamspur, T., and Mostafavi, A. Spectrofluorimetric determination of melatonin in kernels of four different *Pistacia* varieties after ultrasound-assisted solid-liquid extraction. *Spectrochim Acta A Mol Biomol Spectrosc.* 2014. 132: 326–329.

Pape, C., and Lüning, K. Quantification of melatonin in phototrophic organisms. *J Pineal Res.* 2006. 41: 157–165.

Paredes, S. D., Korkmaz, A., Manchester, L. C., Tan, D. X., and Reiter, R. J. Phytomelatonin: A review. *J Exp Bot.* 2009. 60: 57–69.

Park, S., Byeon, Y., Lee, H. Y., Kim, Y. S., Ahn, T., and Back, K. Cloning and characterization of a serotonin *N*-acetyltransferase from a gymnosperm, loblolly pine (*Pinus taeda*). *J Pineal Res.* 2014. 57: 348–355.

Peeters, K. M. U., and Van Laere, A. J. Ammonium and amino acid metabolism in excised leaves of wheat (*Triticum aestivum*) senescing in the dark. *Physiol Plant.* 1992. 84: 243–249.

Poeggeler, B., Balzer, I. Fischer, J., Behrmann, G., and Hardeland, R. A role of melatonin in dinoflagellates? *Acta Endocrinol. (Copenh).* 1989. 120 suppl. 1: 97.

Poeggeler, B., Balzer, I., Hardeland, R., and Lerchl, A. Pineal hormone melatonin oscillates also in the dinoflagellate *Gonyaulax polyedra. Naturwissenschaften.* 1991. 78: 268–269.

Poeggeler, B., and Hardeland, R. Detection and quantification of melatonin in a dinoflagellate, *Gonyaulax polyedra.* Solutions to the problem of methoxyindole destruction in non-vertebrate material. *J Pineal Res.* 1994. 17: 1–10.

Posmyk, M. M., Balabusta, M., Wieczorek, M., Sliwinska, E., and Jana, K. M. Melatonin applied to cucumber (*Cucumis sativus* L.) seeds improves germination during chilling stress. *J Pineal Res.* 2009. 46: 214–223.

Posmyk, M. M., and Janas, K. M. Melatonin in plants. *Acta Physiol Plant.* 2009. 31: 1–11.

Posmyk, M. M., Kuran, H., Marciniak, K., and Janas, K. M. Presowing seed treatment with melatonin protects red cabbage seedlings against toxic copper concentrations. *J Pineal Res.* 2008. 45: 24–31.

Quay, W. B., and Halevy, A. Experimental modification of the rat pineal's content of serotonin and related indoleamines. *Physiol Zool.* 1962. 35: 1–7.

Quay, W. B., and Renzoni A. Comparative and experimental study of the structure and cytology of the pineal body in the Passeriformes. *Riv Biol.* 1963. 56: 363–407.

Ramakrishna, A. Indoleamines in edible plants: Role in human health effects. In Angel Catalá, Biochemistry Research Trends Series, editor, *Indoleamines: Sources, Role in Biological Processes and Health Effects.* New York: Nova Publishers. 2015. 279.

Ramakrishna, A., Dayananda, C., Giridhar, P., Rajasekaran, T., and Ravishankar, G. A. Photoperiod influences endogenous indoleamines in cultured green alga *Dunaliella bardawil. Indian J Exp Biol.* 2011b. 49: 234–240.

Ramakrishna, A., Giridhar, P., Jobin, M., Paulose, C. S., and Ravishankar, G. A. Indoleamines and calcium enhance somatic embryogenesis in cultured tissues of *Coffea canephora* P ex Fr. *Plant Cell Tiss Organ Cult.* 2012c. 108: 267–278.

Ramakrishna, A., Giridhar, P., and Ravishankar, G. A. Indoleamines and calcium channels influence morphogenesis in *in vitro* cultures of *Mimosa pudica* L. *Plant Signal Behav.* 2009. 4: 1136–1141.

Ramakrishna, A., Giridhar, P., and Ravishankar, G. A. Phytoserotonin: A review. *Plant Signal Behav.* 2011a. 6: 800–809.

Ramakrishna, A., Giridhar, P., Sankar, K. U, and Ravishankar, G. A. Melatonin and serotonin profiles in beans of *Coffea* species. *J Pineal Res.* 2012a. 52: 470–476.

Ramakrishna, A., Giridhar, P., Sankar, K. U, and Ravishankar, G. A. Endogenous profiles of indoleamines: Serotonin and melatonin in different tissues of *Coffea canephora* P ex Fr. as analyzed by HPLC and LC-MS-ESI. *Acta Physiol Plant.* 2012b. 34: 393–396.

Rapport, M. M., Green, A. A., and Page, I. H. Crystalline serotonin. *Science.* 1948. 108: 329–330.

Reiter, R. J., Coto-Montes, A., Boga, J. A., Fuentes-Broto, L., Rosales-Corral, S., and Tan, D. X. Melatonin: New applications in clinical and veterinary medicine, plant physiology and industry. *Neuro Endocrinol Lett.* 2011. 32: 575–587.

Reiter, R. J., and Fraschini, F. Endocrine aspects of the mammalian pineal gland: A review. *Neuroendocrinology.* 1969. 5: 219–255.

Reiter, R. J., and Hester, R. J. Interrelationships of the pineal gland, the superior cervical ganglia and the photoperiod in the regulation of the endocrine systems of hamsters. *Endocrinology.* 1966. 79: 1168–1170.

Reiter, R. J, Manchester, L. C., Tan, D. X. Melatonin in walnuts: Influence on levels of melatonin and total antioxidant capacity of blood. *Nutrition.* 2005. 21: 920–924.

Reiter, R. J., Tan, D. X., Manchester, L. C., Simopoulos, A. P., Maldonado, M. D., Flores, L. J., and Terron, M. P. Melatonin in edible plants (phytomelatonin): Identification, concentrations, bioavailability and proposed functions. *World Rev Nutr Diet.* 2007. 97: 211–230.

Reiter, R. J., Tan, D. X., Rosales-Corral, S. A., and Manchester, L. C. The universal nature, unequal distribution and antioxidant functions of melatonin and its derivatives. *Mini Rev Med Chem.* 2013. 13: 373–384.

Reiter, R. J., Tan, D. X., Zhou, Z., Cruz, M. H., Fuentes-Broto, L., and Galano, A. Phytomelatonin: Assisting plants to survive and thrive. *Molecules.* 2015. 20: 7396–7437.

Roche, J. F., Karsch, F. J., Foster, D. L., Takagi, S., and Dziuk, P. J. Effect of pinealectomy on estrus, ovulation and luteinizing hormone in ewes. *Biol Reprod.* 1970. 2: 251–254.

Roshchina, V. V. *Neurotransmitters in Plant Life.* Enfield, New Hampshire: Science Publishers. 2001. 4–81.

Schroder, P., Abele, C., Gohr, P., Stuhlfauth-Roisch, U., and Grosse, W. Latest on the enzymology of serotonin biosynthesis in walnut seeds. *Adv Exp Med Biol.* 1999. 467: 637–644.

Shida, C. S., Castrucci, A. M., and Lamy-Freund, M. T. High melatonin solubility in aqueous medium. *J Pineal Res.* 1994. 16: 198–201.

Sprenger, J., Hardeland, R., Fuhrberg, B., and Han, S. Z. Melatonin and other 5-methoxylated indoles in yeast: Presence in high concentrations and dependence on tryptophan availability. *Cytologia.* 1999. 64: 209–213.

Stege, P. W., Sombra, L. L., Messina, G., Martinez, L. D., and Silva, M. F. Determination of melatonin in wine and plant extracts by capillary electrochromatography with immobilized carboxylic multi-walled carbon nanotubes as stationary phase. *Electrophoresis.* 2010. 31: 2242–2248.

Stranc, A. Astra: A natural coffee with a reduced irritant content. *Przemysl Spozywezy.* 1993. 47: 76–77.

Takahashi, T., and Miyazawa, M. Potent α-glucosidase inhibitors from safflower (*Carthamus tinctorius* L.) seed. *Phytother Res.* 2012. 26: 722–726.

Tan, D. X., Hardeland, R., Manchester, L. C., Korkmaz, A., Ma, S., Rosales-Corral, S., and Reiter, R. J. Functional roles of melatonin in plants, and perspectives in nutritional and agricultural science. *J Exp Bot.* 2012. 63: 577–597.

Tan, D. X., Manchester, L. C., Di Mascio, P., Martinez, G. R., Prado, F. M., and Reiter, R. J. Novel rhythms of N^1-acetyl-N^2-formyl-5-methoxykynuramine and its precursor melatonin in water hyacinth: Importance for phytoremediation. *FASEB J.* 2007b. 21: 1724–1729.

Tan, D. X., Manchester, L. C., Helton, P., and Reiter, R. J. Phytoremediative capacity of plants enriched with melatonin. *Plant Signal Behav.* 2007a. 2: 514–516.

Tan, D. X., Manchester, L. C., Rosales-Corral, S. A., Liu, X. Y., Acuna-Castroviejo, D., and Reiter, R. J. Mitochondria and chloroplasts as the original sites of melatonin synthesis: A hypothesis related to melatonin's primary function and evolution in eukaryotes. *J Pineal Res.* 2013. 54: 127–138.

Tan, D. X., Manchester, L. C., Terron, M. P., Flores, L. J., and Reiter, R. J. One molecule, many derivatives: A never-ending interaction of melatonin with reactive oxygen and nitrogen species? *J Pineal Res.* 2007c. 42: 28–42.

Tan, D. X., Zheng, X., Kong, J., Manchester, L. C., Hardeland, R., Kim, S. J., Xu, X., and Reiter, R. J. Fundamental issues related to the origin of melatonin and melatonin isomers during evolution: Relation to their biological functions. *Int J Mol Sci.* 2014. 15: 15858–15890.

Tanaka, E., Tanaka, C., Mori, N., Kuwahara, Y., and Tsuda, M. Phenylpropanoid amides of serotonin accumulate in witches' broom diseased bamboo. *Phytochemistry.* 2003. 64: 965–969.

Tettamanti, C., Cerabolini, B., Gerola, P., and Conti, A. Melatonin identification in medicinal plants. *Acta Phytotherapeutica.* 2000. 3: 137–144.

Udenfriend, S., Lovenberg, W., and Sjoerdsma, A. Physiologically active amines in common fruits and vegetables. *Arch Biochem Biophys.* 1959. 85: 487–490.

Udenfriend, S., Weissbach, H., and Clark, C. T. The estimation of 5-hydroxytryptamine (serotonin) in biological tissues. *J Biol Chem.* 1955. 215: 337–344.

USDA National Nutrient Database for Standard Reference. 2016. Available online at http://ndb.nal.usda.gov/.

Van Tassel, D., Li, J., and O'Neill, S. Melatonin: Identification of a potential dark signal in plants. *Plant Physiol.* 1993. 102: 659.

Van Tassel, D., Roberts, N., and O'Neill, S. Melatonin from higher plants: Isolation and identification of *N*-acetyl-5-methoxytryptamine. *Plant Physiol.* 1995. 108: 101.

Veenstra-VanderWeele, J., Anderson, G. M., and Cook Jr, E. H., Pharmacogenetics and the serotonin system: Initial studies and future directions. *Eur J Pharmacol.* 2000. 410: 165–181.

Vetorazzi, G. 5-hydroxytryptamine content of bananas and banana products. *Food Cosm Toxicol.* 1974. 12: 107–113.

Waalkes, T. P., Sjoerdsma, A., Creveling, C. R., Weissbach, H., and Udenfriend, S. Serotonin, norepinephrine and related compounds in bananas. *Science.* 1958. 127: 648–650.

Weeda, S., Zhang, N., Zhao, X., Ndip, G., Guo, Y., Buck, G. A., Fu, C., and Ren, S. *Arabidopsis* transcriptome analysis reveals key roles of melatonin in plant defense systems. *PLoS ONE.* 2014. 9: e93462.

West, G. B. Carcinoid tumors and pineapples. *J Pharm Pharmacol.* 1960. 12: 768–769.

Wolf, K., Kolar, J., Witters, E., Van Dogen, W., Van Onckelen, H., and Machackova, I. Daily profile of melatonin levels in *Chenopodium rubrum* L. depends on photoperiod. *J Plant Physiol.* 2001. 158: 1491–1493.

Wurtman, R. J., Axelrod, J., and Phillips, L. W. Melatonin synthesis in the pineal gland: Control by light. *Science.* 1963. 142: 1071–1073.

Yin, L., Wang, P., Li, M., Ke, X., Li, C., Liang, D., Wu, S., Ma, X., Li, C., Zou, Y., and Ma, F. Exogenous melatonin improves *Malus* resistance to Marssonina apple blotch. *J Pineal Res.* 2013. 54: 426–434.

Zhao, Y., Tan, D. X., Lei, Q., Chen, H., Wang, L., Li, Q. T., Gao, Y., and Kong, J. Melatonin and its potential biological functions in the fruits of sweet cherry. *J Pineal Res.* 2013. 55: 79–88.

3

Analytical Trends for the Determination of Melatonin and Precursors in Plants

Federico J. V. Gomez, Ismael Gatica Hernández, and María Fernanda Silva
Universidad Nacional de Cuyo, CONICET
Mendoza, Argentina

Soledad Cerutti
Universidad Nacional de San Luis, CONICET
San Luis, Argentina

CONTENTS

ABSTRACT Nowadays, several groups are devoted to the development of reliable analytical methodologies such as microchip electrophoresis (ME), capillary electrophoresis (CE), liquid chromatography (LC), coupled to different detection strategies, and, particularly, the application of nanotechnologies and nanomaterials in the development of sensors for the determination of melatonin (MEL), its isomers, and precursors in different plant tissues and food samples. These approaches provide useful tools and knowledge to demonstrate the great potential of this natural substance for agricultural applications, improving crop production and quality of functional foods. The analysis of MEL in plants and foods represents a highly challenging task due to its low concentration range, the difficulty in the selection of the extraction solvents because of its amphipathic nature, and the fact that it reacts quickly with other matrix components. Chemical complexity of plant extract can also interfere with MEL determinations, giving false positive results if methods from vertebrate MEL research are directly adopted. Indeed, the additional challenges related to the vast variety of matrices as well as the extremely wide concentration range (ppm–ppq levels) should be considered. Therefore, the main objective of this chapter is to take a closer look at the recent trends in analytical tools employed to contribute to the elucidation of the biological role of MEL in plants, and its implications in human health.

KEY WORDS: *analytical methodologies, sample treatment, melatonin, foods, plants, MEL isomers.*

Abbreviations

5-HT: serotonin
Ab: antibody
Ag: antigen
APCI: atmospheric pressure chemical ionization
BGE: background electrolyte
CD: cyclodextrin
CE: capillary electrophoresis
CEC: capillary electrochromatography
c-MWNT: carboxylic multiwalled carbon nanotube
CNT: carbon nanotube
CSPEs: carbon screen–printed electrodes
DSE: direct-sample extraction
ECD: electrochemical detection
EDTA: ethylenediaminetetraacetic acid
EIA: enzyme immunoassay
ELISA: enzyme-linked immunoabsorbent assay
ESI: electrospray ionization
FD: fluorescence detector
FI: fluorescence
GC: gas chromatography
GON: graphene oxide nanoribbons
GRN: graphene-reduced nanoribbons
HILIC: hydrophilic interaction liquid chromatography
HSE: homogenized-sample extraction
LC: liquid chromatography
LLE: liquid–liquid extraction
LOD: limit of detection
LOQ: limit of quantification
MAE: microwave-assisted extraction
ME: microchip electrophoresis
MEKC: micellar electrokinetic chromatography
MEL: melatonin
MRM: multiple selected reaction-monitoring mode
MS: mass spectrometry
MWCNT: multi-walled carbon nanotube
PDA: photodiode array
RIA: radioimmunoassay
SDS: sodium dodecyl sulfate
SPE: solid-phase extraction
SWCNT: single-walled carbon nanotube
SW-PTEs: single-walled press-transferred carbon nanotube electrodes
Trp: tryptophan
UPLC: ultra-performance liquid chromatography

Introduction

Melatonin (*N*-acetyl-5-methoxytryptamine, MEL) is a neurohormone synthesized from L-tryptophan via serotonin (5-HT). Originally described in the pineal gland, its role in the regulation of circadian rhythm is well accepted. Indeed, melatonin supplements sold over the counter are used to treat sleeping disorders or jet lag. In addition to the high content of melatonin already present in the gastrointestinal tract, food intake and certain nutrients such as tryptophan can serve as stimuli for melatonin release (Hardeland and Pandi-Perumal 2005). Melatonin is classified as a potent, naturally occurring antioxidant and a signaling molecule based on its primary and secondarily evolved functions in organisms (Tan et al. 2010). In animals, this molecule is formed exclusively from the essential amino acid tryptophan. Meanwhile, in other organisms including bacteria, several protists, fungi, macroalgae, and plants, which possess the shikimic acid pathway, tryptophan can be synthesized starting with D-erythrose-4-phosphate, and phosphoenolpyruvate, in phototrophs ultimately with carbon dioxide (Bochkov et al. 2012). Therefore, melatonin production is not limited by tryptophan uptake in these species and can, thus, lead to remarkably high levels often exceeding those found in vertebrates by orders of magnitude (Chen et al. 2003; Fuhrberg et al. 1997).

Ever since the presence of melatonin in higher plants has been reported (Murch et al. 1997; Chen et al. 2003; Paredes et al. 2009; Ramakrishna et al. 2009; Ramakrishna et al. 2012b,c) new perspectives have broadened out for exogenous melatonin as a natural component of food intake (Tan et al. 2012; Janas and Posmyk 2013; Ramakrishna et al. 2012a; Ramakrishna 2015). In order to evaluate the contribution of this compound as a bioactive agent, it is necessary to set up and validate suitable analytical methods for plants and plant derivates analysis (Garcia-Parrilla et al. 2009). The analysis of melatonin in plants and its derivates presents some difficulties. First, the content of melatonin in some plants is in the picogram to microgram per gram range. Thus, any analytical method must be sensitive to these variations. Second, the amphipathic characteristic of the molecule makes it difficult to choose a solvent yielding a complete recovery and accurate results. Third, melatonin is a potent antioxidant and reacts quickly with other matrix constituents; careful handling of the sample is thus a prerequisite (Tan et al. 1993). The analytical method must therefore take these constraints into account in a vegetal tissue by using an adequate matrix and by showing sufficient sensitivity and specificity.

Further evaluation of MEL occurrence and function in higher plants proceeds slowly because of problems related to extraction and reliable quantification of this hormone (Boccalandro et al. 2011; Poeggeler and Hardeland 1994; Van Tassel and O'Neill 2001a; Van Tassel et al. 2001b). Chemical complexity of plant extract, which often contains large amount of carbohydrates, lipids, and pigments, can interfere with MEL determinations, giving false positive results if methods from vertebrate MEL research are directly adopted, for example, because of coelution in LC or cross-reactivity with antibodies of immunological methods such as radioimmunoassay (RIA) or enzyme-linked immunoabsorbent assay (ELISA) (Cassone and Natesan 1997; Pape and Lüning 2006; Van Tassel and O'Neill 2001a). Van Tassel and O'Neill (2001a) observed that validation according to standard criteria, linear serial dilution and parallel inhibition for immunoassays, and coelution with authentic standard for high-performance liquid chromatography (HPLC), are insufficient to ensure the identity of MEL. Gas chromatography (GC) and LC combined with mass spectrometry (MS) detection have proven to be trustworthy, but these could be costly and labor-intensive methods (Kolář and Macháčková 2005; Van Tassel and O'Neill 2001a; Van Tassel et al. 2001b). This might explain several contradictory results for plant material from the same species quantified with different methods. For example, ginger tubers were reported to contain between 584 and 1423 pg/g fresh weight (Badria 2002; Hattori et al. 1995), whereas Van Tassel and O'Neill (2001a) found only trace amounts of MEL in this plant organ. Differences in MEL degradation and extraction efficiency could also result in variation in content (Gomez et al. 2013).

Considering the above points, various methods for melatonin extraction, purification, and determination have been developed in recent years. In the case of GC-MS, derivatization of MEL prior to analysis using silanizing agents or pentafluoropropionic anhydride to form trimethylsilyl or pentafluoropropionyl melatonin, respectively, was found necessary (Hevia et al. 2010). HPLC is a widely used technique for the analysis of melatonin; isocratic elution with a reversed phase is the most commonly used. In this case, different detectors such as fluorescence (FI) (Hamase et al. 2004; Kulczykowska and Iuvone 1998; Mills et al. 1991; Minami et al. 2009), electrochemical detection (ECD) (Raynaud and Pevet 1991; Vieira et al. 1992), and UV detection (Ayano et al. 2007; Brömme et al. 2008; Hevia et al. 2008) have been used. With the development of MS interface technology, high-performance liquid chromatography coupled to mass spectrometry (HPLC–MS) has increased in popularity, and has proved to be a powerful tool for complex sample analysis (Gomez et al. 2012, 2013; Rodriguez-Naranjo et al. 2011a,b; Vitalini et al. 2013). The figure below shows the current most popular techniques used for melatonin analysis (Figure 3.1).

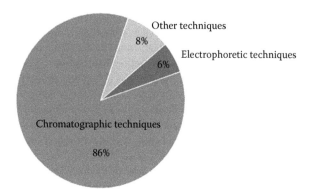

FIGURE 3.1   Analytical techniques applied to MEL analysis in plant and derivate products in the last 10 years.

Sample Pretreatment

The presence and role of melatonin in plants are still under debate owing to difficulties of identification and quantification. The analysis of melatonin is a challenging task. Solutions of melatonin are light-sensitive and subject to oxidation. In addition, its amphipathic nature and its solubility (0.1 mg/mL in water and 8 mg/mL in ethanol) make it difficult to choose a solvent reaching complete recovery. Indeed, it reacts quickly with other matrix components, considering its antioxidant capabilities.

Thus, sample cleanup is still one of the most important bottlenecks of the analytical process. Effective sample preparation is essential for achieving good analytical results because matrix-related compounds may also coextract and interfere in the analysis (Garcia-Parrilla et al. 2009).

On the other hand, the matrix effects should be carefully assessed when vegetal tissues are involved; the cleanup, chromatographic, and detection conditions have to be evaluated for each sample. Cleanup and preconcentration are usually necessary when low detection limits are required.

Several approaches have been proposed for the extraction, however, all the processes in the procedures should be carried out under dim artificial light in order to prevent melatonin degradation.

Direct Solvent Extraction

Boccalandro et al. (2011) analyzed grapes from *Vitis vinifera* cv. Malbec by LC-ESI-MS/MS after ultrasound-mediated extraction with methanol. Results suggest that melatonin fluctuation could be the result of the action of a plant biological clock, combined with a light-induced depletion of melatonin levels. Melatonin levels in berry skins collected at harvest were low, but detectable, during the night (10.9 and 8.9 ng/g at 02:30 and 06:00 h, respectively) and reached a strong peak at dawn (158.9 ng/g at 09:30 h). After this peak, melatonin levels deeply decayed to undetectable values both at noon and during the afternoon.

Immunoprecipitation and ELISA were combined for melatonin determination (de la Puerta et al. 2007). A combined method was used for melatonin extraction, based on sequential methanol and chloroform extraction. Melatonin was found in olive oil at higher levels in extra-virgin olive oil than in refined olive or sunflower oil samples.

Arnao and Hernández-Ruiz (2009) compared two different sample extraction procedures: (1) a direct-sample extraction (DSE), and (2) a homogenized-sample extraction (HSE). The approaches involved extraction with chloroform and ethyl acetate before HPLC with fluorescence detection. For the DSE technique, the system was left overnight in darkness with shaking. In the HSE procedure, the whole procedure took only 12 minutes. In some cases, an ultrasonic treatment was included, which resulted in different levels of efficiency, depending on the sample. They reached detection limits up to 0.012 and 0.045 ng/g for DSE and HSE, respectively. Differential melatonin concentrations were found for the studied vegetal tissues from *Lupinus albus* and *Hordeum vulgare* evaluated, within the range 0.58 ng/g (leaves) to 20.20 ng/g (roots).

Kocadağli et al. (2014) developed a method for the determination of melatonin and its isomers in various food products. The method involved an ethanol extraction from solid samples (or dilution of liquid samples) prior to liquid chromatography coupled to triple quadruple mass spectrometry (LC–MS/MS). Black and green tea, green coffee, walnut, cacao powder, black olive, tomato, frozen sour cherry, probiotic yogurt, kefir, bread, red wine, and beer were analyzed. Limits of quantification (LOQs) were within the range 32.10 to 176.50 pg/g.

Stege et al. (2010) presented a direct extraction of medicinal herbs and wines prior to capillary electrochromatography (CEC) with immobilized carboxylic multi-walled carbon nanotubes (c-MWNT) as stationary phase. Samples were dried under nitrogen gas, grounded, and extracted with methanol. Ultrasonication was employed to assist and accelerate the extraction of MEL from vegetal tissues. The limit of detection (LOD) for melatonin was 0.01 ng/mL.

Korkmaz et al. (2014) determined melatonin content in pepper plant samples. Extraction was performed with ethylacetate overnight at 4°C in darkness with shaking. The supernatant was transferred to another tube, and the remaining plant residue was washed with ethyl acetate. The extract and washing from each were evaporated to dryness, redissolved in methanol, and analyzed using HPLC with fluorescence detection. The authors found a higher concentration of melatonin in seedlings at the cotyledon stage with a progressive decline as the plants matured.

Setyaningsih et al. (2012) presented a microwave-assisted extraction (MAE) for melatonin from rice grains. MAE shortened extraction times for different kinds of compounds in vegetables and related foods. In MAE, the rapid generation of heat and pressure forces compounds from within the matrix and produces good quality extracts with better target compound recovery. The efficiency of the MAE process depends on time, temperature ratio, and the type and composition of solvent used. These authors applied a chemometric approach based on the use of an experimental design to evaluate the variables that affect MAE. They reached an LOD of 1.70 μg/kg. Several kinds of rice grains were extracted using the developed method, and the extracts were analyzed by HPLC with fluorescence detection. The highest concentration of melatonin was obtained in exotic rice followed by integral brown whole short-grain rice, and then other rice samples. This trend of melatonin concentration in the rice products corresponds to their protein contents; the exotic rice has the highest amount of proteins. The same group developed a pressurized liquid approach for the extraction of melatonin from rice, also prior to HPLC with fluorescence detection (Setyaningsih et al. 2015). The LOD was slightly smaller (1.15 μg/kg). The highest amounts of melatonin were found for pigmented rice samples.

Tal et al. (2011) presented a dispersed solid phase extraction with acid-treated glass beads prior to preparative thin-layer chromatography for the determination of melatonin in algal material was adopted. Bands corresponding to the melatonin standard retention factor were scraped off and suspended in ethanol, and quantified by fluorescence detection. The method was applied to quantify melatonin in intertidal zone green macroalgae (*Ulva sp.*) under different environmental conditions.

Oladi et al. (2014) proposed an ultrasound-assisted solid–liquid strategy for the extraction of melatonin in kernels of four different *Pistacia* varieties. A detection limit of 0.0036 μg/mL was reached. Samples of pistachio kernels were crushed and accurately weighed, and extracted with methanol facilitated by sonication. The method was validated by comparison with GC/MS.

Solid-Phase Extraction

A solid-phase (SPE) approach prior to HPLC with fluorescence detection was developed for the determination of trans- and cis-resveratrol, and of melatonin in red and white wine (Mercolini et al. 2008). Sample pretreatment was developed, using SPE with C_{18} cartridges. The LOD for melatonin was 0.01 ng/mL.

Sample preparation using ultrasonic technique in combination with solid-phase extraction was presented in 2011 for the determination of melatonin in mulberry leaves (Pothinuch and Tongchitpakdee 2011). Results were compared with homogenization in combination with liquid–liquid extraction (LLE) approach. Determination was performed by HPLC combined with fluorescence detector. The recovery rate of ultrasonic/SPE and homogenization/LLE procedures were 76% and 12%, respectively. The melatonin contents of all cultivars tested were highest in the tip of the leaves, followed by that in the young leaves, whereas the lowest was found in the old leaves. Heat treatment during tea processing decreased the melatonin content in mulberry leaves when compared to that of the fresh leaves. However, there were no significant differences between the melatonin contents of the mulberry leaf tea produced with blanching (mulberry green tea) and those produced without blanching (mulberry black tea).

Tomatoes and strawberries were analyzed after ultrasound-mediated extraction and SPE (C_{18} cartridges) prior to liquid chromatography–mass detection. Melatonin concentration ranged from 4.11 ng/g to 114.52 ng/g (fresh weight) in tomato samples and from 1.38 to 11.26 ng/g (fresh weight) in strawberry samples (Stürtz et al. 2011). Interestingly, the varieties of strawberries collected during two harvests showed greater similarity in melatonin contents.

Gomez et al. (2013) studied different approaches for the determination of melatonin and its isomer in different vegetal/food samples. LODs in the low picograms range were reached. A sonication-mediated extraction step was necessary for grape skin (100% [v/v] methanol) and plant tissues (50% [v/v] methanol), while wine and must required a solid-phase extraction preconcentration step. Proposed methods were validated. The optimized approaches were applied to the determination of melatonin and its isomer in different vegetal/food samples; levels found were within the range: 4.9–440 ng g^{-1}. Results clearly demonstrate that optimal conditions were quite different for the vegetal/food matrices under study; matrix effects up to 88.4% were found.

Liquid Chromatographic Separation Strategies for Melatonin and Associated Derivatives

In the last few years, a considerable variety of analytical methods have been made available (Garcia-Parrilla et al. 2009). The choice of the analytical method depends upon various factors, such as sample complexity, target analyte(s) nature, levels, and lability, goal of the analysis, and the availability of resources (human and instrumental). In this section, current liquid separation methodologies used in melatonin-based analysis in plant samples, and the advantages and drawbacks of each one, will be discussed.

In recent years, chromatographic techniques have proved to be a great alternative, and even better, than many others, such as radioimmunoassay and enzyme-linked immunoabsorbent assay techniques, for melatonin determination (Huang and Mazza 2011; Vitale et al. 1996). For RIA and ELISA methods, a variety of molecules in plants may cross-react with the related antibodies and enzymes, and this can lead to potential over-estimation of the actual melatonin values.

In the case of vegetal matrixes, high-performance liquid chromatography with electrochemical detector (HPLC-ECD), fluorescence detector (HPLC-FD), UV/VIS photodiode array (PDA), or HPLC-MS have been proposed (Arnao and Hernández-Ruiz 2007; Iriti et al. 2006; Lu et al. 2002; Muszyńska et al. 2014; Ramakrishna et al. 2012a,b). These are powerful and precise methods that do not require derivatization.

In general, an HPLC system consisting of a C_{18} or a C_8 reverse-phase column (2–4.6 mm × 100–250 mm), combined with a mobile phase made of either in 0.1 mM potassium phosphate buffer (pH 4.5) with acetonitrile (20%) (Reiter et al. 2005) or a mixture (40:60, v/v) of methanol:50 mM Na_2HPO_4/H_3PO_4 pH 4.5 (Chen et al. 2003), has been mostly used for the analysis of melatonin in plant tissues with two different detectors, ECD and FD (Feng et al. 2014). Indeed, the first finding of melatonin in the brown alga *Pterygophora californica* Rupr. was identified by HPLC with ECD, and melatonin was well separated from the other constituents of the extract (Fuhrberg et al. 1996). Although not specific (plants may contain a variety of compounds with comparable oxidation potentials and retention times similar to melatonin), binary elution systems and isocratic mobile phases are widely used for the ECD detector. The aqueous is usually acidified with acetate or phosphate, and the organic phase consisted of methanol or acetonitrile. The concentration of the organic solvent is usually prepared with no more than 20% (v/v), since the high concentration of organic solvent in the mobile phase may shorten the lifetime of the electrode in the detector. The applied potential is very important for the determination. Usually, this potential is set at greater than 800 mV, with 850 mV being used predominantly since it provided a higher sensitivity for melatonin detection. Among the several and relevant examples of the HPLC–ECD coupling, it is possible to cite the detection of melatonin in rice (Byeon et al. 2012), in the seeds of 15 edible plants (Manchester et al. 2000), walnuts (*Juglans regia* L.) (Reiter et al. 2005), and sour cherries (*Prunus cerasus* L.) (Burkhardt et al. 2001). Separation was achieved using isocratic elution, and the potential was increased in a gradient. In the same way, Hernandez-Ruiz et al. determined melatonin and indole-3-acetic acid in white lupin (*Lupinus albus* L.) hypocotyls, canary grass (*Phalaris canariensis* L.), wheat (*Triticum aestivum* L.), barley (*Hordeum vulgare* L.), and oat (*Avena sativa* L.) by HPLC-ECD (Hernández-Ruiz et al. 2004, 2005). A polar mobile phase in isocratic elution was selected. These authors also used LC-MS to confirm the presence of melatonin and indole-3-acetic acid.

On the other hand, HPLC-FD has been found to be sensitive and versatile to quantify melatonin (and related compounds) in plants, and also gave low and suitable limits of detection and quantitation (Garcia-Parrilla et al. 2009; Venegas et al. 2011; Zhao et al. 2013). For this detection approach, the excitation and emission wavelengths are two important parameters to be optimized; in general, melatonin excitation is accomplished between 280 and 298 nm and emission between 345 and 386 nm. Many methods have been reported for the HPLC-FD detection of melatonin and related indolamines in plants (Adão and Glória 2005; Engstrom et al. 1992; Hattori et al. 1995; Pape and Lüning 2006). For these methods, isocratic elution and acidic aqueous solvents were commonly used. In Pape and Luning's study (Pape and Lüning 2006), melatonin was quantified in tomatoes (*Lycopersicon esculentum* Mill.), ginger (*Zingiber officinale* Roscoe), and the marine green macroalga *Ulva lactuca* L. The mobile phase used was relatively complex and included citric acid, NaH_2PO_4/Na_2HPO_4 buffer, octane sulfonic acid, EDTA, and methanol. The same HPLC system used for the purification of extracts was used for parallel quantification after derivatization of melatonin under alkaline conditions in the presence of hydrogen peroxide. The method gave results with a high correlation and the detection limit was 2.5 pg for melatonin. Melatonin was determined in white lupin (*Lupinus albus*) and barley (*Hordeum vulgare*) using HPLC with FD (Arnao and Hernández-Ruiz 2013; Hernández-Ruiz and Arnao 2008a,b). The isocratic mobile phase consisted of water and acetonitrile with the proportion of 60:40 or 75:25 at a flow rate between 0.2 and 0.5 mL/min.

The mobile phase composition is exceedingly varied in different published methods. It usually consisted of acetonitrile and aqueous phase with the different proportions. Unlike ECD, there is no limit for the concentration of organic solvents in HPLC-FD; consequently, it is better to increase the concentration of organic solvents in the mobile phase to shorten the retention time (Huang and Mazza 2011). Therefore, HPLC-FD offers a better opportunity for detecting melatonin in comparison with HPLC-ECD. The aqueous phase is generally adjusted to pH 2.8–5.5 with acetate or phosphate. Isocratic-, linear-, or step-gradient elution have been used. In addition, a variety of columns have been reported for different applications. Thus Mercolini et al. (2012) reported an original analytical method for the determination of resveratrol and melatonin in wine, this research proved that a MV-Rainin column instead of Zorbax was more selective toward matrix compounds present in must and grappa.

UV/VIS detector is the most common HPLC detector used in plant analysis. In the HPLC with UV detection methods for melatonin and derivatives, the wavelength generally used is 280 nm. The mobile phases containing acidic water adjusted with weak acid (trifluoroacetic acid or acetic acid) and methanol or acetonitrile are most commonly selected. Isocratic and gradient elutions are employed. In this way, determination of melatonin in the seed, leaf, stem, and root tissues of Chinese licorice (*Glycyrrhiza uralensis* Fisch.) plants was performed using an HPLC-UV/VIS photodiode array method (Afreen et al. 2006). The detection limit was 10 pg/μL. The melatonin content in the different plant organs was confirmed by HPLC-FD analysis. Additionally, using the HPLC-UV method, melatonin and serotonin in purple wheat were analyzed by Hosseinian et al. (2008). Among the different strategies, the liquid chromatography–mass spectrometric (LC-MS) methods offer high sensitivity and excellent detection specificity. Different types of mass analyzers have been used for melatonin and indolamines analysis in plants. Thus LC-MS determination includes single quadrupole, triple quadrupole, ion trap, and TOF analyzer, with the triple quadrupole being the analyzer of choice. For the ionization, ambient techniques including ESI, APCI, and APPI have been selected in recent research, with the ESI ion source being the one most commonly employed. When tandem mass spectrometry (MS/MS) is applied, the multiple selected reaction-monitoring mode (MRM) provides more specificity and reliability, allowing the analysis of multiple specific-reaction paths at the same time, with the production of several fragment ions from their respective defined precursor ions (Boccalandro et al. 2011; Gomez et al. 2012; Riga et al. 2014; Vigentini et al. 2015).

As mentioned, triple quadrupole is the most popular for quantitative and structural melatonin analysis in plants, while liquid chromatography with time-of-flight mass spectrometry is commonly used for melatonin identification, the accurate mass being the criterion for compound assignment.

For LC-MS (MS/MS) determination, analytical conditions are usually optimized using melatonin and related compounds standards. Deuterated standards are commercially available. In order to enhance the ionization process and the LC separation, acidic aqueous mobile phase and gradient elution are mainly used. In LC-MS determination, reported limits of detection for melatonin are in the ppb to ppt range, LC-MS being more sensitive than HPLC-ECD, FD, or UV. Because of the wide range of LC systems, MS techniques, column types, and solvent mixtures, no single available method is suitable for all applications. Both GC-MS and LC-MS protocols have been applied to melatonin analysis in plants (Cao et al. 2006; Van Tassel and O'Neill 2001a; Van Tassel et al. 2001b). Unlike gas chromatography, liquid chromatography avoids the derivatization step, which is an important advantage and makes it suitable for a wider range of applications.

The mobile-phase flow rate used in LC-MS detection is lower than that used in conventional LC and is dependent on the ionization source selected. Sometimes splitting of the flow is needed and this allows for the optimum flow rate into the MS system. In addition, the pH value of the mobile phase should be adjusted using organic acids or alkalis or salts, which should be compatible with the MS detection (Huang and Mazza 2011). As commented above, there are many alternatives regarding ionization approaches for the analysis of melatonin. In addition, the sources can be configured in positive or negative polarity modes. Results have demonstrated that the positive ion mode is more sensitive than the negative one for the determination of melatonin and indoleamines.

In order to enhance chromatographic performances in terms of efficiency and rapidity, LC has recently evolved in the development of short columns packed with small particles (sub-2 μm) working at high pressures (Nguyen et al. 2006). Advantages of these can be discussed in terms of sensitivity, efficiency, resolution, and analysis time. Several applications have highlighted the advantages related to the use of UPLC systems for melatonin analysis in plants (Gomez et al. 2013; Hosseinian et al. 2008). Recently, Gomez et al. (2013) proposed a compendium of analytical methods based on the development of different UPLC chromatographic strategies, such as hydrophilic interaction liquid chromatography (HILIC) and reverse phase (C_8 and C_{18}), associated with electrospray ionization (ESI) and atmospheric pressure chemical ionization (APCI) in both negative and positive modes coupled to tandem mass spectrometry for the quantitative analysis of melatonin and its isomers in plants. The optimized voltages were determined to monitor the precursor ions, m/z 233, for melatonin and its fragments, m/z 216 and 174. Quite different optimal conditions were achieved for each of the matrices under study (grape skin, vegetal tissues, and derivatives, among others). Thus, sonication and solid-phase extraction approaches were proposed for the analyte's removal and preconcentration prior analysis. In the same context, HILIC-(+) APCI ionization demonstrated to be better for melatonin standards, while $C_{(8)}$-(+) APCI was the best choice for grape skin and $C_{(18)}$-(+ESI) was suitable for wine. On the other hand, $C_{(8)}$-(+)ESI was the most appropriate for vegetal tissues of *Arabidopsis thaliana*. The above-mentioned proposed methods were validated, and the LODs were in the low picograms range. The so-called *matrix effect* was also observed during the MS/MS detection. As known, this effect reduces or enhances the ionization efficiency, modifying the analytes' response. Thus the Gomez et al. (2015) study demonstrated that matrix effects should be carefully assessed when plant tissues are involved, and cleanup procedure, and chromatographic and detection conditions, have to be evaluated for each sample. This conclusion is extremely important since in the literature many chromatographic and methodological conditions for the MEL analysis of animal tissues or fluids have been directly applied to plants, without considering the nature of the sample itself.

Capillary and Microchip Electrophoresis

MEL is an indoleamine, whose two functional groups are not only decisive for specificity of receptor binding, but also for its amphiphilicity, allowing the molecule to enter any cell, compartment or body fluid. Definitely, a careful handling of the sample is essential because MEL is rather unstable in plant extracts and reacts quickly with other sample components. As established, solutions of MEL are light-sensitive and subject to oxidation. Consequently, the whole procedure should be as fast and simple as possible.

Gomez et al. (2015) have developed a new method based on solid-phase extraction followed by micellar electrokinetic chromatography (MEKC) for the extraction, preconcentration, and simultaneous determination of melatonin, tryptophan, serotonin, and indole-3-acetic acid in plant material. The method was robust, reliable, low-cost, quick, and simple. The authors proposed the use of a dual pseudostationary phase system, involving a mixture of sodium dodecyl sulfate (SDS) anionic micelles and β–cyclodextrin (CD), enabled to reach adequate selectivity. A BGE of 10 mM sodium tetraborate (pH 9.2), containing 20 mM β–CD, 20 mM SDS, and 10% (v/v) of acetonitrile, allowed baseline separation in less than 10 minutes for the four target analytes. The proposed methodology provided limits of detection down to low ppb levels. Under the optimal conditions, a successful application on *Arabidopsis* tissue, green, and linden tea leaves confirmed the validity of the method for food and plant analysis.

Saracino et al. (2008) developed three analytical methods and compared them for the quality control of a new formulation (Soymen GN® capsules) containing soy extract and melatonin for the treatment of menopausal symptoms. The first method is based on MEKC with diode-array detection, using a mixture of basic carbonate buffer (95%) and methanol (5%), containing 55 mM SDS, as the BGE. The second method is an HPLC method with UV detection at 260 nm. The third method is an HPLC method coupled to amperometric detection. The results obtained with the three methods are in good agreement with each other and highly satisfactory for the analysis of the sample.

Another alternative is the use of capillary electrochromatography that consists of a hybrid technique between CE and HPLC; this technique has evolved as a powerful tool in the analysis of complex matrices. Four different types of CEC are generally distinguished, based on the capillary column used: packed-column CEC, open tubular CEC, monolithic-column CEC, and pseudostationary phase CEC. Applications of nanoparticles are of rising interest in separation science, due to their favorable surface-to-volume ratio as well as their applicability in miniaturization. Taking advantage of these benefits, Stege et al. (2010) implemented a new method for the determination of MEL in red and white wines, grape skin, and plant extracts by CEC with immobilized c-MWNT as the stationary phase. The obtained LOD for melatonin was 0.01 ng/mL. The feasibility of using c-MWNT immobilized fused-silica capillary through covalent modification of bare capillaries for the determination of MEL in complex food samples was demonstrated. The straightforwardness and reproducibility of the extraction procedure is accompanied by the high electrochromatography resolution, robustness, and high sensitivity obtained, providing a useful tool for further research into MEL in foods of vegetal origin.

Gomez et al. (2015) developed a microchip electrophoresis technology, which offers fast analysis times with low sample and reagents consumption. They proposed a novel single-walled press-transferred carbon nanotube electrodes (SW-PTEs) on PMMA substrates for electrochemical microfluidic sensing (Vilela et al. 2012, 2014) and explore the analytical possibilities of the press-transfer technology coupled to ME for the fast and reliable electrochemical sensing of MEL and their precursor 5-HT and Trp. In this approach, the electronic transfer is carried out directly on the target nanomaterial where single-wall carbon nanotubes (SWCNT) is the only one transducer in the electrochemical sensing, and consequently the analytical advantages of these nanomaterials are directly exploited. SW-PTEs were simply fabricated by press-transferring a filtered dispersion of single-wall carbon nanotubes on a nonconductive PMMA substrate, where single-wall carbon nanotubes act as exclusive transducers. The coupling of ME–SW-PTEs allowed the fast detection of MEL, Trp, and 5-HT in less than 150 s with excellent analytical features. This work reveals the analytical merits of ME coupled to the SW-PTEs, where the fast and reliable determination of MEL and its precursors were performed directly on the nanoscale CNT detectors without any other electrochemical transducer in under 150 s with excellent analytical performance.

Other Techniques Applied to Determination of
Melatonin and Serotonin in Vegetal Tissues

Formerly, studies of analysis of melatonin were carried out to assess its presence in the pineal gland, saliva, or other biological fluids (Kennaway et al. 1977). Immunological techniques were considered appropriate, and were successfully applied to animal fluids–derived matrices. Therefore, a first approach consisted in applying those available methods of common use in these biological samples to plant matrices. Main techniques involved both radioimmunoassay (RIA) and enzyme immunoassay (EIA). RIA is successfully applied to determine quantitatively melatonin in biological samples such as human serum, plasma, and saliva (McIntyre et al. 1987). Indeed, there are commercial kits available for these purposes. They are based on the competition principle. A limited amount of specific antibody (Ab) reacts with the corresponding antigen (*Ag) labeled with ^{125}I. Upon addition of an increasing amount of the Ag (sample) a correspondingly decreasing fraction of *Ag is bound to the antibody. Bound and free *Ag can be separated by centrifugation. The bound radioactivity is subsequently determined in a Gamma counter, and results are obtained directly using a standard curve. EIA has also been applied to melatonin determination. Indeed, the commercial kit based on a competitive enzyme immunoassay (Melatonin ELISA, IBL-Hamburg) was applied to determine melatonin in olive oil for the first time (de la Puerta et al. 2007) and it was also used to determine melatonin in grape skin extracts from different cultivars (Iriti et al. 2006).

However, Van Tassel and O'Neill (2001a) reported their concerns when attempting to validate RIA in tomato (*Lycopersicum esculentum*) and *Pharbitis nil* extract samples. In their experiments, a prior fractionation step was performed by HPLC. More than one single fraction gave response to RIA, and only one fraction increased its response when spiking with melatonin standard. In addition, the amount of melatonin determined by RIA was higher than that determined by GC/MS. Therefore, one conclusion of their work is the evidence of overestimation of RIA analysis, and the possibility of false positive findings. This result could be explained by the presence of a compound or compounds that cross-react(s) with melatonin RIA antiserum. Nevertheless, Johns et al. (2013) reported that using the ELISA technique should not affect the melatonin levels in fruits due to the low cross-reactivity. They also suggested the possibility of the coelution of another compound with the same retention time as melatonin when using HPLC was ruled out by the ELISA test, as reported by Iriti et al. (2006).

A chemiluminescence method was proposed for the determination of melatonin (Lu et al. 2002). The principle of the assay is that ultra-weak chemiluminescence is emitted after the reaction of melatonin, H_2O_2, and acetonitrile, under alkaline conditions. The formation of singlet oxygen in the reaction implies that other compounds apart from melatonin, phenolics, and terpenes among them, can interfere with the analysis. Therefore, it is not advisable to use this method with plant tissue where these substances are widespread.

Chromatographic methods have been the most widely used separation technique for melatonin in recent years. GC/MS method had been applied to determine melatonin in different fruit sources (Badria 2002; Dubbels et al. 1995). Although GC/MS has high sensitivity and specificity, derivatization of MEL using silanizing agents or pentafluoropropionic anhydride to form trimethylsilyl or pentafluoropropionyl melatonin respectively, prior to analysis, is necessary (Hevia et al. 2010); the need for derivatization presents disadvantages (Feng et al. 2014). In these cases, pentafluoropropionic anhydride was used for the derivatization process.

Recently carbon nanomaterial sensors had been proposed as a novel tool for rapid and cheap analysis, becoming a valuable screening technique. Gomez et al. (2015) explored a set of carbon nanomaterial–based electrodes for the detection of melatonin and serotonin in herb capsules. In this work they demonstrated that carbon nanomaterial–based screen-printed electrodes provide an easy, fast, and reliable analytical tool for the simultaneous detection and determination of the coexistent serotonin and melatonin in biological systems and samples. Two groups of carbon screen-printed electrodes (CSPEs) with identical geometry (4 mm in diameter) were systematically explored: commercial single-walled carbon nanotube, multi-walled carbon nanotube, and graphene electrodes, as well as in-lab drop casting modified electrodes with SWCNT, MWCNT, and graphene oxide nanoribbons (GON) and graphene-reduced (GRN) nanoribbons. Carbon nanomaterials enhanced the electroactive area following the order

CSPE<MWCNTs<SWCNTs<graphene, allowing the simultaneous detection of serotonin and melatonin with enhanced sensitivity in comparison with the CSPEs. This method is clearly competitive in comparison with the other electrochemical approaches such as cyclic voltammetry and square-wave voltammetry, since it offers a very fast response, low sample consumption (50 μL) and single use (disposability). In comparison with other methodologies such as HPLC or CE, which require skilled personnel manipulating sophisticated and expensive instrumentation, electrochemical techniques possess analytical and economic advantages. These features include simple instrumentation and low-cost, high sensitivity, fast response, and easy operation.

Conclusion

Taken together, the development of reliable analytical methodologies for the extraction, purification, and determination of MEL in vegetal tissues is of upmost importance. Indeed, it should be considered the additional challenges related to the vast variety of matrices as well as the extremely wide concentration range (ppm to ppq levels).

It can be concluded that the analysis of melatonin in plants and foods represents a highly challenging task due to its wide concentration range, the difficulty in the selection of the extraction solvents because of its amphipathic nature, and the fact that it reacts quickly with other matrix components. Thus, sample processing factors, preparation/cleanup procedures, and detection parameters should be evaluated.

It has been shown that melatonin is widespread in different varieties of plants, and that its exact amount is influenced by many factors. Thus melatonin contents reported in various plant parts as determined by LC, GC, CE, and other techniques with different kinds of detectors show discordance and inaccuracy. Therefore, validated analytical methods with adequate sample treatment are required, which remains the most challenging in the study of melatonin in plants.

On the other hand, melatonin and melatonin isomers exist and/or coexist in living organisms, including plants (Diamantini et al. 1998; Rodriguez-Naranjo et al. 2011b; Tan et al. 2014). The first publication referring to naturally occurring melatonin isomers appeared in 2011, and the unknown compound was identified by LC-ESI-MS/MS (Rodriguez-Naranjo et al. 2011b). Subsequently, several reports have confirmed that melatonin isomers are indeed present in wine (Gomez et al. 2012, 2013; Vitalini et al. 2013), and also in fermented orange juice (Fernández-Pachón et al. 2014). All melatonin isomers are structural isomers and their functional groups are joined at different sites on the indole nucleus. Theoretically, based on the possible positions of the two side chains on the indole nucleus of melatonin, it has been calculated that as many as 42 melatonin isomers could exist (Tan et al. 2014). Therefore, separation and identification of melatonin isomers are new frontiers to be explored.

REFERENCES

Adão, R. C., and Glória, M. B. A. Bioactive amines and carbohydrate changes during ripening of "Prata" banana (*Musa acuminata* x *M. balbisiana*). *Food Chem.* 2005. 90: 705–711.

Afreen, F., Zobayed, S. M. A., and Kozai, T. Melatonin in *Glycyrrhiza uralensis*: Response of plant roots to spectral quality of light and UV-B radiation. *J Pineal Res.* 2006. 41: 108–115.

Arnao, M. B., and Hernández-Ruiz, J. Melatonin promotes adventitious- and lateral root regeneration in etiolated hypocotyls of *Lupinus albus* L. *J Pineal Res.* 2007. 42: 147–152.

Arnao, M. B., and Hernández-Ruiz, J. Assessment of different sample processing procedures applied to the determination of melatonin in plants. *Phytochem Anal.* 2009. 20: 14–18.

Arnao, M. B., and Hernández-Ruiz, J. Growth conditions determine different melatonin levels in *Lupinus albus* L. *J Pineal Res.* 2013. 55: 149–155.

Ayano, E., Suzuki, Y., Kanezawa, M., Sakamoto, C., Morita-Murase, Y., Nagata, Y., Kanazawa, H., Kikuchi, A., and Okano, T. Analysis of melatonin using a pH- and temperature-responsive aqueous chromatography system. *J Chromatogr A.* 2007. 1156: 213–219.

Badria, F. A. Melatonin, serotonin, and tryptamine in some Egyptian food and medicinal plants. *J Med Food.* 2002. 5: 153–157.

Boccalandro, H. E., Gonzalez, C. V., Wunderlin, D. A., and Silva, M. F. Melatonin levels, determined by LC-ESI-MS/MS, fluctuate during the day/night cycle in *Vitis vinifera* cv Malbec: Evidence of its antioxidant role in fruits. *J Pineal Res.* 2011. 51: 226–232.

Bochkov, D. V., Sysolyatin, S. V., Kalashnikov, A. I., and Surmacheva, I. A. Shikimic acid: Review of its analytical, isolation, and purification techniques from plant and microbial sources. *J Chem Biol.* 2012. 5: 5–17.

Brömme, H. J., Peschke, E., and Israel, G. Photo-degradation of melatonin: Influence of argon, hydrogen peroxide, and ethanol. *J Pineal Res.* 2008. 44: 366–372.

Burkhardt, S., Tan, D. X., Manchester, L. C., Hardeland, R., and Reiter, R. J. Detection and quantification of the antioxidant melatonin in Montmorency and Balaton tart cherries (*Prunus cerasus*). *J Agric Food Chem.* 2001. 49: 4898–4902.

Byeon, Y., Park, S., Kim, Y. S., Park, D. H., Lee, S., and Back, K. Light-regulated melatonin biosynthesis in rice during the senescence process in detached leaves. *J Pineal Res.* 2012. 53: 107–111.

Cao, J., Murch, S. J., O'Brien, R., and Saxena, P. K. Rapid method for accurate analysis of melatonin, serotonin and auxin in plant samples using liquid chromatography–tandem mass spectrometry. *J Chromatogr A.* 2006. 1134: 333–337.

Cassone, V. M., and Natesan, A. K. Time and time again: The phylogeny of melatonin as a transducer of biological time. *J Biol Rhythms.* 1997. 12: 489–497.

Chen, G., Huo, Y., Tan, D. X., Liang, Z., Zhang, W., and Zhang, Y. Melatonin in Chinese medicinal herbs. *Life Sci.* 2003. 73: 19–26.

de la Puerta, C., Carrascosa-Salmoral, M P., García-Luna, P. P., Lardone, P. J., Herrera, J. L., Fernández-Montesinos, R., Guerrero, J. M., and Pozo, D. Melatonin is a phytochemical in olive oil. *Food Chem.* 2007. 104: 609–612.

Diamantini, G., Tarzia, G., Spadoni, G., D'Alpaos, M., and Traldi, P. Metastable ion studies in the characterization of melatonin isomers. *Rapid Commun Mass Sp.* 1998. 12: 1538–1542.

Dubbels, R., Reiter, R. J., Klenke, E., Goebel, A., Schnakenberg, E., Ehlers, C., Schiwara, H. W., and Schloot, W. Melatonin in edible plants identified by radioimmunoassay and by high performance liquid chromatography–mass spectrometry. *J Pineal Res.* 1995. 18: 28–31.

Engstrom, K., Lundgren, L., and Samuelsson, G. Bioassay-guided isolation of serotonin from fruits of *Solanum tuberosum* L. *Acta Pharm Nord.* 1992. 4:91–92.

Feng, X., Wang, M., Zhao, Y., Han, P., and Dai, Y. Melatonin from different fruit sources, functional roles, and analytical methods. *Trends Food Sci Tech.* 2014. 37: 21–31.

Fernández-Pachõn, M. S., Medina, S., Herrero-Martín, G., Cerrillo, I., Berná, G., Escudero-Lõpez, B., Ferreres, F., Martín, F., García-Parrilla, M. C., and Gil-Izquierdo, A. Alcoholic fermentation induces melatonin synthesis in orange juice. *J Pineal Res.* 2014. 56: 31–38.

Fuhrberg, B., Balzer, I., Hardeland, R., Werner, A., and Lüning, K. The vertebrate pineal hormone melatonin is produced by the brown alga *Pterygophora californica* and mimics dark effects on growth rate in the light. *Planta.* 1996. 200: 125–131.

Fuhrberg, B., Hardeland, R., Poeggeler, B., and Behrmann, G. Dramatic rises of melatonin and 5-methoxytryptamine in *Gonyaulax* exposed to decreased temperature. *Biol Rhythm Res.* 1997. 28: 144–150.

Garcia-Parrilla, M. C., Cantos, E., and Troncoso, A. M. Analysis of melatonin in foods. *J Food Compos Anal.* 2009. 22: 177–183.

Gomez, F., Martín, A., Silva, M., and Escarpa, A. Screen-printed electrodes modified with carbon nanotubes or graphene for simultaneous determination of melatonin and serotonin. *Microchim Acta.* 2015. 1–7.

Gomez, F. J. V., Hernández, I. G., Martinez, L. D., Silva, M. F., and Cerutti, S. Analytical tools for elucidating the biological role of melatonin in plants by LC-MS/MS. *Electrophoresis.* 2013. 34: 1749–1756.

Gomez, F. J. V., Raba, J., Cerutti, S., and Silva, M. F. Monitoring melatonin and its isomer in *Vitis vinifera* cv. Malbec by UHPLC-MS/MS from grape to bottle. *J Pineal Res.* 2012. 52: 349–355.

Hamase, K., Hirano, J., Kosai, Y., Tomita, T., and Zaitsu, K. A sensitive internal standard method for the determination of melatonin in mammals using precolumn oxidation reversed-phase high-performance liquid chromatography. *J Chromatogr B.* 2004. 811: 237–241.

Hardeland, R., and Pandi-Perumal, S. R. Melatonin, a potent agent in antioxidative defense: Actions as a natural food constituent, gastrointestinal factor, drug and prodrug. *Nutr Metab.* 2005. 2: 22.

Hattori, A., Migitaka, H., Iigo, M., Itoh, M., Yamamoto, K., Ohtani-Kaneko, R., Hara, M., Suzuki, T., and Reiter, R. J. Identification of melatonin in plants and its effects on plasma melatonin levels and binding to melatonin receptors in vertebrates. *Biochem Mol Biol Int.* 1995. 35: 627–634.

Hernández-Ruiz, J., and Arnao, M. B. Distribution of melatonin in different zones of lupin and barley plants at different ages in the presence and absence of light. *J Agric Food Chem.* 2008a. 56: 10567–10573.

Hernández-Ruiz, J., and Arnao, M. B. Melatonin stimulates the expansion of etiolated lupin cotyledons. *Plant Growth Regul.* 2008b. 55: 29–34.

Hernández-Ruiz, J., Cano, A., and Arnao, M. B. Melatonin: A growth-stimulating compound present in lupin tissues. *Planta.* 2004. 220: 140–144.

Hernández-Ruiz, J., Cano, A., and Arnao, M. B. Melatonin acts as a growth-stimulating compound in some monocot species. *J Pineal Res.* 2005. 39: 137–142.

Hevia, D., Botas, C., Sainz, R. M., Quiros, I., Blanco, D., Tan, D. X., Gomez-Cordoves, C., and Mayo, J. C. Development and validation of new methods for the determination of melatonin and its oxidative metabolites by high performance liquid chromatography and capillary electrophoresis, using multivariate optimization. *J Chrom A.* 2010. 1217: 1368–1374.

Hevia, D., Sainz, R. M., Blanco, D., Quirós, I., Tan, D. X., Rodríguez, C., and Mayo, J. C. Melatonin uptake in prostate cancer cells: Intracellular transport versus simple passive diffusion. *J Pineal Res.* 2008. 45: 247–257.

Hosseinian, F. S., Li, W., and Beta, T. Measurement of anthocyanins and other phytochemicals in purple wheat. *Food Chem.* 2008. 109: 916–924.

Huang, X., and Mazza, G. Application of LC and LC-MS to the analysis of melatonin and serotonin in edible plants. *Crit Rev Food Sci.* 2011. 51: 269–284.

Iriti, M., Rossoni, M., and Faoro, F. Melatonin content in grape: Myth or panacea? *J Sci Food Agric.* 2006. 86: 1432–1438.

Janas, K. M., and Posmyk, M. M. Melatonin, an underestimated natural substance with great potential for agricultural application. *Acta Physiol Plant.* 2013. 35: 3285–3292.

Johns, N. P., Johns, J., Porasuphatana, S., Plaimee, P., and Sae-Teaw, M. Dietary intake of melatonin from tropical fruit altered urinary excretion of 6-sulfatoxymelatonin in healthy volunteers. *J Agric Food Chem.* 2013. 61: 913–919.

Kennaway, D. J., Frith, R. G., Phillipou, G., Matthews, C. D., and Seamark, R. F. A specific radioimmunoassay for melatonin in biological tissue and fluids and its validation by gas chromatography mass spectrometry. *Endocrinology.* 1977. 101: 119–127.

Kocadağli, T., Yilmaz, C., and Gökmen, V. Determination of melatonin and its isomer in foods by liquid chromatography tandem mass spectrometry. *Food Chem.* 2014. 153: 151–156.

Kolář, J., and Macháčková, I. Melatonin in higher plants: Occurrence and possible functions. *J Pineal Res.* 2005. 39: 333–341.

Korkmaz, A., Değer, Ö., and Cuci, Y. Profiling the melatonin content in organs of the pepper plant during different growth stages. *Sci Hortic.* 2014. 172: 242–247.

Kulczykowska, E., and Iuvone, P. M. Highly sensitive and specific assay of plasma melatonin using high-performance liquid chromatography with fluorescence detection preceded by solid-phase extraction. *J Chromatogr Sci.* 1998. 36: 175–178.

Lu, J., Lau, C., Lee, M. K., and Kai, M. Simple and convenient chemiluminescence method for the determination of melatonin. *Anal Chim Acta.* 2002. 455: 193–198.

Manchester, L. C., Tan, D. X., Reiter, R. J., Park, W., Monis, K., and Qi, W. High levels of melatonin in the seeds of edible plants: Possible function in germ tissue protection. *Life Sci.* 2000. 67: 3023–3029.

McIntyre, I. M., Norman, T. R., Burrows, G. D., and Armstrong, S. M. Melatonin rhythm in human plasma and saliva. *J Pineal Res.* 1987. 4:177–183.

Mercolini, L., Mandrioli, R., and Raggi, M. A. Content of melatonin and other antioxidants in grape-related foodstuffs: Measurement using a MEPS-HPLC-F method. *J Pineal Res.* 2012. 53: 21–28.

Mercolini, L., Saracino, M. A., Bugamelli, F., Ferranti, A., Malaguti, M., Hrelia, S., and Raggi, M. A. HPLC-F analysis of melatonin and resveratrol isomers in wine using an SPE procedure. *J Sep Sci.* 2008. 31: 1007–1014.

Mills, M. H., Finlay, D. C., and Haddad, P. R. Determination of melatonin and monoamines in rat pineal using reversed-phase ion-interaction chromatography with fluorescence detection. *J Chrom B.* 1991. 564: 93–102.

Minami, M., Takahashi, H., Inagaki, H., Yamano, Y., Onoue, S., Matsumoto, S., Sasaki, T., and Sakai, K. Novel tryptamine-related substances, 5-sulphatoxydiacetyltryptamine, 5-hydroxydiacetyltryptamine, and reduced melatonin in human urine and the determination of those compounds, 6-sulphatoxymelatonin, and melatonin with fluorometric HPLC. *J Chrom B.* 2009. 877: 814–822.

Murch, S. J., Simmons, C. B., and Saxena, P. K. Melatonin in feverfew and other medicinal plants. *Lancet.* 1997. 350: 1598–1599.

Muszyńska, B., Ekiert, H., Kwiecień, I., Maślanka, A., Zodi, R., and Beerhues, L. Comparative analysis of therapeutically important indole compounds in *in vitro* cultures of *Hypericum perforatum* cultivars by HPLC and TLC analysis coupled with densitometric detection. *Nat Prod Commun.* 2014. 9: 1437–1440.

Nguyen, D. T. T., Guillarme, D., Rudaz, S., and Veuthey, J. L. Fast analysis in liquid chromatography using small particle size and high pressure. *J Sep Sci.* 2006. 29: 1836–1848.

Oladi, E., Mohamadia, M., Shamspura, T., and Mostafavia, A. Spectrofluorimetric determination of melatonin in kernels of four different *Pistacia* varieties after ultrasound-assisted solid–liquid extraction. *Spectrochim Acta A.* 2014. 132: 326–329.

Pape, C., and Lüning, K. Quantification of melatonin in phototrophic organisms. *J Pineal Res.* 2006. 41: 157–165.

Paredes, S. D., Korkmaz, A., Manchester, L. C., Tan, D. X., and Reiter, R. J. Phytomelatonin: A review. *J Exp Bot.* 2009. 60: 57–69.

Poeggeler, B., and Hardeland, R. Detection and quantification of melatonin in a dinoflagellate, *Gonyaulax polyedra*: Solutions to the problem of methoxyindole destruction in non-vertebrate material. *J Pineal Res.* 1994. 17: 1–10.

Pothinuch, P., and Tongchitpakdee, S. Melatonin contents in mulberry (Morus spp.) leaves: Effects of sample preparation, cultivar, leaf age and tea processing. *Food Chem.* 2011. 128: 415–419.

Ramakrishna, A. Indoleamines in edible plants: Role in human health effects. In Angel Catalá, Biochemistry Research Trends Series (editor), *Indoleamines: Sources, Role in Biological Processes and Health Effects.* New York: Nova Publishers. 2015. 279.

Ramakrishna, A., Giridhar, P., Jobin, M., Paulose, C. S., and Ravishankar, G. A. Indoleamines and calcium enhance somatic embryogenesis in cultured tissues of *Coffea canephora* P ex Fr. *Plant Cell Tiss Organ Cult.* 2012a. 108: 267–278.

Ramakrishna, A., Giridhar, P., and Ravishankar, G. A. Indoleamines and calcium channels influence morphogenesis in *in vitro* cultures of *Mimosa pudica* L. *Plant Signal Behav.* 2009. 4: 1136–1141.

Ramakrishna, A., Giridhar, P., Sankar, K. U, and Ravishankar, G. A. Endogenous profiles of indoleamines: Serotonin and melatonin in different tissues of *Coffea canephora* P ex Fr. as analyzed by HPLC and LC-MS-ESI. *Acta Physiol Plant.* 2012b. 34: 393–396.

Ramakrishna, A., Giridhar, P., Sankar, K. U, and Ravishankar, G. A. Melatonin and serotonin profiles in beans of Coffea species. *J Pineal Res.* 2012c. 52: 470–476.

Raynaud, F. and Pevet, P. Determination of 5-methoxyindoles in pineal gland and plasma samples by high-performance liquid chromatography with electrochemical detection. *J Chrom B* 1991;564:103–113.

Reiter, R. J., Manchester, L. C., and Tan, D. X. Melatonin in walnuts: Influence on levels of melatonin and total antioxidant capacity of blood. *Nutrition.* 2005. 21: 920–924.

Riga, P., Medina, S., García-Flores, L. A., and Gil-Izquierdo, Á. Melatonin content of pepper and tomato fruits: Effects of cultivar and solar radiation. *Food Chem.* 2014. 156: 347–352.

Rodriguez-Naranjo, M. I., Gil-Izquierdo, A., Troncoso, A. M., Cantos-Villar, E., and Garcia-Parrilla, M. C. Melatonin is synthesised by yeast during alcoholic fermentation in wines. *Food Chem.* 2011a. 126: 1608–1613.

Rodriguez-Naranjo, M. I., Gil-Izquierdo, A., Troncoso, A. M., Cantos, E., and Garcia-Parrilla, M. C. Melatonin: A new bioactive compound in wine. *J Food Compos Anal.* 2011b. 24: 603–608.

Saracino, M. A., Mercolini, L., Musenga, A., Bugamelli, F., and Raggi, M. A. Comparison of analytical methods for the quality control of a new formulation containing soy extract and melatonin. *J Sep Sci.* 2008. 31: 1851–1859.

Setyaningsih, W., Palma, M., and Barroso, C. G. A new microwave-assisted extraction method for melatonin determination in rice grains. *J Cereal Sci.* 2012. 56: 340–346.

Setyaningsih, W., Saputro, I. E., Barbero, G. F., Palma, M., and García Barroso, C. Determination of melatonin in rice (*Oryza sativa*) grains by pressurized liquid extraction. *J Agric Food Chem.* 2015. 63: 1107–1115.

Stege, P. W., Sombra, L. L., Messina, G., Martinez, L. D., and Silva, M. F. Determination of melatonin in wine and plant extracts by capillary electrochromatography with immobilized carboxylic multi-walled carbon nanotubes as stationary phase. *Electrophoresis.* 2010. 31: 2242–2248.

Stürtz, M., Cerezo, A. B., Cantos-Villar, E., and Garcia-Parrilla, M. C. Determination of the melatonin content of different varieties of tomatoes (*Lycopersicon esculentum*) and strawberries (*Fragaria ananassa*). *Food Chem.* 2011. 127: 1329–1334.

Tal, O., Haim, A., Harel, O., and Gerchman, Y. Melatonin as an antioxidant and its semi-lunar rhythm in green macroalga *Ulva* sp. *J Exp Bot.* 2011. 62: 1903–1910.

Tan, D. X., Chen, L. D., Poeggeler, B., Manchester, L. C., and Reiter, R. J. Melatonin: A potent, endogenous hydroxyl radical scavenger. *Endocr J.* 1993. 1: 57–60.

Tan, D. X., Hardeland, R., Manchester, L. C., Paredes, S. D., Korkmaz, A., Sainz, R. M., Mayo, J. C., Fuentes-Broto, L., and Reiter, R. J. The changing biological roles of melatonin during evolution: From an antioxidant to signals of darkness, sexual selection and fitness. *Biol Rev Camb Philos Soc.* 2010. 85: 607–623.

Tan, D. X., Hardeland, R., Manchester, L. C., Korkmaz, A., Ma, S., Rosales-Corral, S., and Reiter, R. J. Functional roles of melatonin in plants, and perspectives in nutritional and agricultural science. *J Exp Bot.* 2012. 63: 577–597.

Tan, D. X., Zheng, X., Kong, J., Manchester, L. C., Hardel, R., Kim, S. J., Xu, X., and Reiter, R. J. Fundamental issues related to the origin of melatonin and melatonin isomers during evolution: Relation to their biological functions. *Int J Mol Sci.* 2014. 15: 15858–15890.

Van Tassel, D. L., and O'Neill, S. D. Putative regulatory molecules in plants: Evaluating melatonin. *J Pineal Res.* 2001a. 31: 1–7.

Van Tassel, D. L., Roberts, N., Lewy, A., and O'Neill, S. D. Melatonin in plant organs. *J Pineal Res.* 2001b. 31: 8–15.

Venegas, C., Cabrera-Vique, C., García-Corzo, L., Escames, G., Acuña-Castroviejo, D., and López, L. C. Determination of coenzyme Q10, coenzyme Q9, and melatonin contents in virgin argan oils: Comparison with other edible vegetable oils. *J Agric Food Chem.* 2011. 59: 12102–12108.

Vieira, R., Miguez, J., Lema, M., and Aldegunde, M. Pineal and plasma melatonin as determined by high-performance liquid chromatography with electrochemical detection. *Anal Biochem.* 1992. 205: 300–305.

Vigentini, I., Gardana, C., Fracassetti, D., Gabrielli, M., Foschino, R., Simonetti, P., Tirelli, A., and Iriti, M. Yeast contribution to melatonin, melatonin isomers and tryptophan ethyl ester during alcoholic fermentation of grape musts. *J Pineal Res.* 2015. 58: 388–396.

Vilela, D., Garoz, J., Colina, A., González, M. C., and Escarpa, A. Carbon nanotubes press-transferred on PMMA substrates as exclusive transducers for electrochemical microfluidic sensing. *Anal Chem.* 2012. 84: 10838–10844.

Vilela, D., Martín, A., González, M. C., and Escarpa, A. Fast and reliable class-selective isoflavone index determination on carbon nanotube press-transferred electrodes using microfluidic chips. *Analyst.* 2014. 139: 2342–2347.

Vitale, A. A., Ferrari, C. C., Aldana, H., and Affanni, J. M. Highly sensitive method for the determination of melatonin by normal-phase high-performance liquid chromatography with fluorometric detection. *J Chrom B.* 1996. 681: 381–384.

Vitalini, S., Gardana, C., Simonetti, P., Fico, G., and Iriti, M. Melatonin, melatonin isomers and stilbenes in Italian traditional grape products and their antiradical capacity. *J Pineal Res.* 2013. 54: 322–333.

Zhao, Y., Tan, D. X., Lei, Q., Chen, H., Wang, L., Li, Q., Gao, Y., and Kong, J. Melatonin and its potential biological functions in the fruits of sweet cherry. *J Pineal Res.* 2013. 55: 79–88.

4

Serotonin in Living Cells: Fluorescent Tools and Models to Study

Victoria V. Roshchina
Russian Academy of Sciences Institute of Cell Biophysics
Pushchino, Russia

CONTENTS

ABSTRACT Studies of serotonin function and localization in unicellular organisms, animal embryos, and plants are the subject of scientists' attention today. Fluorescence as a parameter for observation and measurement is believed to be suitable for such analyses. Visible emission of forming products, fluorescent-labeling or fluorescent model cellular systems are used. In this short review some examples illustrating these possibilities have been considered, based on the work of the author.

KEY WORDS: *agonists, antagonists, fluorescent drugs, histochemical reactions, laser-scanning confocal microscopy, luminescence microscopy, microspectrofluorimetry, model systems.*

Abbreviations

BODIPY:	boron-dipyrromethene
FRET:	fluorescence resonance energy transfer
HEK293:	human embryonic kidney 293 cells
NBD:	7-nitrobenz-2-oxa-1,3-diazol-4-yl
7TM:	seven-transmembrane

Introduction

Serotonin (5-hydroxytryptamine) was first discovered in animals by Erspamer in 1940, and by Rappoport with coworkers in 1948, but, as described in some monographs and reviews, today is also found in living animal cells, including embryos of mammalians (Buznikov 1967, 1987, 1990; Buznikov et al. 1996; Schmukler and Nikishin 2012), in plants (Roshchina 1991, 2001, 2010; Ramakrishna et al. 2011b), and in microorganisms (Oleskin et al. 2010; Oleskin 2012). This is a very important bioactive compound known as a neurotransmitter that has a profound effect on many aspects of human and mammalian behavior and health modulating neural activity and a wide range of neuropsychological processes (Berger et al. 2009). There are special receptors of serotonin—M, D, and T-serotoninic receptors—found in animals. Each group of receptors has agonists (mainly tryptophan derivatives including the product of the serotonin methylation) and antagonists (chlorpromazine, its derivatives and fluoridated peridoles, liserginic acid known as a hallucinaceous poison). Drugs that target 15 serotonin receptors (Kroeze et al. 2002) are used widely in psychiatry and neurology as antidepressants (see Roth 2006).

Especially great interest in the compound was connected with finding serotonin in plants because it may be applied in food and pharmacology as natural drug-containing preparations. Wide-ranging investigations of the compound's distribution in plants were undertaken (Regula 1970, 1986; Regula and Devide 1980; Regula et al. 1989; Grosse 1982, 1984; Grosse et al. 1983; Lembeck and Skofitsch 1984; Roshchina 1991, 2001; Murch et al. 2000, 2009, 2010; Ramakrishna et al. 2009; Ramakrishna et al. 2011a; Ramakrishna et al. 2012; Ramakrishna et al. 2012a,b; Ramakrishna 2015). The amount of serotonin varies dependent on the taxonomic position of the plant, and is especially great in representatives of families Juglandaceae, Fabaceae, and Lygophyllaceae. In comparison with animals, its quantities in plants are the same (average) or lesser than in mammalians, whereas in some species of mollusk (*Octopus*) it correlates with values observed in fruits and seeds (Roshchina 1991, 2001). Fruit food enriched in serotonin may be a beneficial treatment for patients because it can show them how daily exercise, along with a good diet, can help to raise serotonin levels without relying on any type of prescription drugs. This may be useful for a wide range of health problems derived from a lack of brain serotonin such as insomnia, depression, obesity, eating disorders, panic attacks, alcoholism, anxiety disorders, and bulimia. Serotonin precursor, 5-hydroxytryptophan, has been also used as a natural replacement for prescription antidepressants.

Today neurotransmitters, in particular serotonin, may also play an important role in plant physiology (Turi et al. 2014). Serotonin accumulation and distribution in plants and its role in stressful conditions (Kang et al. 2007, 2009; Mukherjee et al. 2014) is of relevance for agriculture. Serotonin is connected with the plant–microbe interactions that relate us to evolutionary aspects of the serotonin functions (Roshchina 2010).

At the cellular level there are various methods to study serotonin that have been applied, mainly to animal neuronal cells, but the modes may be applied to living cells, including plant matter. The aim of this review is to consider some fluorescent tools to study serotonin in living cellular systems sensitive to the neurotransmitter including those used in our laboratory and those that may be useful for scientists in future.

Approaches to Study of Serotonin in Cells Using Fluorescence

The problem is, how to study cellular effects of serotonin and its agonists or antagonists in any cells and what approaches may be applied to the research. Besides radioactive-labeled reagents, fluorescence serves to study. Acidic solutions of serotonin fluoresced in the green region with peak emission near 550 nm that is widely used for the fluorometric assay of serotonin (Chen 1968). After absorbing a photon, serotonin becomes protonated, and it is the protonated form that is responsible for the visible fluorescence in acid solutions. Today fluorescence of forming products, fluorescent-labeling or fluorescent model cellular systems are used for cellular analysis. Listed here are examples to the approaches in serotonin studies of living cells with use of fluorescence parameters:

1. Reactions with serotonin to see its location in cells and tissues
2. Reactions with agonist or antagonist of serotonin binding with the serotonin receptor
3. Reactions for the serotonin transporters
4. Autofluorescence of cells sensitive to serotonin
5. Fluorescing drugs, their effects and binding in single cells as model systems

These approaches also may be applied to study other neurotransmitters at their interactions with living cells.

Earlier reactions with special histochemical reagents on serotonin, mainly aldehydes, leads to formation of fluorescent indoleamines products with maximum emission at 520–527 nm in animal neuronal cells (Falck 1962) and blastocysts (Leonov and Budantsev 1971). Today, too, fluorescent-labeled probes with emission in the green region due to boron-dipyrromethene (BODIPY) fluorophores (Bezuglov et al. 2004) or fluorescent proteins are used for various investigations (Herrick-Davis et al. 2013). Due to the analysis, the receptors' diffusion in plasmatic membrane and transport of serotonin observation were established. A new aspect of the problem is in genetic manipulations (Tanaka et al. 2012).

Human serotonin receptors have a lot of subtypes that may be studied in cells with the fluorescent probes. A recent review of some fluorescent molecule ligands (Vernall et al. 2013) adds further strength to the argument that fluorescent ligand design and synthesis requires examples illustrating the selection of the correct fluorescent dye, selectivity, and potency of the final conjugate when compared with its unlabeled precursor. Moreover, the fluorescent ligands permit the use of flow cytometry, fluorescence microscopy, fluorescence resonance energy transfer (FRET), and scanning confocal microscopy.

The 5-HT1A receptor is the most thoroughly studied among the neurotransmitter receptor subtypes. A pathophysiological role of serotonin receptors (a seven-transmembrane [7TM] domain receptor negatively coupled with adenylyl cyclase) may be related to neuroprotection, cognitive impairment, and pain, as well as anxiety and depression. The fluorescent ligands offer a possibility for exploring multiple areas of research such as the mechanism of ligand binding, the movement and internalization of receptors in living cells, as seen in a special review devoted to new fluorescent ligands (Hovius 2013). It is possible to visualize ligand–receptor interactions in real time in live cells. First, a fluorescent probe for serotonin has been performed on the base of the serotonin antagonist (Berque-Bestel et al. 2003). This has been done using fluorescent antagonists for human 5-HT(4) receptors synthesized based on ML10302 1, a potent 5-HT(4) receptor agonist, and on piperazine analog 2. These molecules were derived with three fluorescent moieties, dansyl, naphthalimide, and 7-nitrobenz-2-oxa-1,3-diazol-4-yl (NBD), through alkyl chains. For the human 5-HT1A receptor subtype two new fluorescent ligands (based on 1-arylpiperazine) emitting in red have been chosen (Lacivita et al. 2010). This receptor also has been cloned. The fluorescence correlation spectroscopy showed the diffusion of receptor species within defined membrane microdomains can be monitored in real time, although nonspecific binding was seen too. High nonspecific binding has been related to high lipophilicity of the probe. Specific information, from 2000 and later years, about commercial fluorescent ligands dealing with serotonin (relating mainly to animals with neuronal system) is in Haugland (2010).

Besides the probes, C-terminally labeling of the serotonin receptors (G protein-coupled receptors from the 5-HT$_{2A}$ receptor families) with green or yellow fluorescent proteins was also used (Herrick-Davis et al. 2013). The target of the proteins was plasma membrane that was confirmed by confocal microscopy. The heterodimerization of the brain serotonin 5-HT$_{2A}$ receptor and dopamine D$_2$R receptor was also analyzed by fluorescence spectroscopy and fluorescence lifetime microscopy to determine the degree of fluorescence resonance energy transfer between cyan and yellow fluorescent protein labeled receptor variants coexpressed in human embryonic kidney 293 cells [HEK293] (Łukasiewicz et al. 2010). The data demonstrated the existence of energy transfer between the wild-type forms of the receptors, possibilities of their interactions, and having a common motif at colocalization.

As serotonin transporters, the fluorescent labeling drug citalopram (Zhang et al. 2013; Banala et al. 2013; Kumar et al. 2014) or 4-(4-[Dimethylamino] phenyl)-1-methylpyridinium (APP$^+$) and APP$^+$ analogs (Wilson et al. 2014) were also used. In the first case bright red emission prevailed due to rhodamine fluorophores (Zhang et al. 2013), while in the second, blue-green emissions prevailed, decreasing due to phenyl-1-methylpyridinium. In all cases, fluorescence was concentrated near plasmatic membranes on the cell periphery. Using fluorescently labeled antibodies for living systems has no perspective due to a specificity of the receptor proteins (occuring only in own species) isolated from the Vertebrata tissues.

In most cases the experiments with fluorescent ligands related to serotonin were performed on animals with neuronal systems, and only a few data were represented for nonneuronal living cells that will be considered in a specific section of this chapter.

Chemoreception of neurotransmitters as chemosignals studied with their agonists and antagonists and first response to neurotransmitters shows alterations in cellular fluorescence excited by UV (360–380 nm) light. The fluorescence of the cells was estimated by spectrofluorimetry and microspectrofluorimetry after the treatment with substances analyzed. In this case, suitable samples in which emission changed in response to the antitransmitter compound may be considered as sensitive models (Roshchina 2014).

Autofluorescence of chlorophyll-containing cells with maximum at 675–680 nm are shown to respond to the serotonin addition. Vegetative microspores of horsetail *Equisetum arvense*, the algae thallus of *Chara carolina* and leaf of *Lemna minor* were tested as possible models (Roshchina 2004; Roshchina et al. 2009, 2013). In the first two, serotonin increased the fluorescence intensity at 680 nm to almost double in comparison with the control, and this reaction is suitable for observation. Moreover, the fluorescence could be decreased if in the medium with vegetative microspores of *Equisetum arvense* cytochalasin B (inhibitor of the actin polymerization) was added, that showed the possible transfer of external serotonin chemosignal within cells with the participation of contractile protein actin (Roshchina 2005). The above-mentioned were used as biosensors or models (Roshchina 2003, 2006, 2014).

Histochemical Dyes on Serotonin That Form Fluorescent Products

The first experiments in serotonin histochemistry were performed on animal cells (Falck 1962) and on plant cells (Kimura 1968). The fluorescence related to the neurotransmitters, mainly yellow or green-yellow, was seen in mammalian neuronal cells (Falck 1962), and embryonal living cells (Leonov and Budantsev 1971). Fruits of banana *Musa paradisiaca* L., and *Musa basjoo,* as well as onion *Allium cepa* and *Citrus lemon* have been analyzed (Kimura 1968). Specific green or yellow fluorescence of serotonin was scarcely found in the cytoplasm of the epidermal cells of the fruit peel of *Musa* species and the bulb of the onion, while not in *Citrus*. The fruit pulp contained a considerably large amount of serotonin, but the amount was less than that of the fruit peel. Most of the fluorescent materials in the parenchymal cells were starch grain-like, but some were diffuse. The maturation of the fruit gave rise to an increase in fluorescence in the cells of the fruit pulp and peel. The leaf blade of the banana did not exhibit specific fluorescence, while the leaf venation exhibited a considerably abundant density of specific fluorescence in the vascular bundle of its venation. The intercellular distribution of yellow fluorescence, especially in the vascular bundle, was observed in the inner layer of the fruit wall in *Musa basjoo*.

Agonists as Fluorescent Probes for Analysis of Serotonin Functioning outside and within Cells

Lipophilic analogs of neurotransmitters mainly served as the algorithm for fluorescent analysis. Lipophilic derivatives of serotonin (BODIPY-5-HT), dopamine (BODIPY-DA), and acetylcholine (BODIPY-ACH) containing the 4,4-difluoro-5,7-dimethyl-4-bora-3a,4a-diaza-2 s-indacene-3-dodecanoic acid residue were synthesized from the corresponding amines and fluorescent dodecanoic acid by Bezuglov with coworkers (2004) with the method of mixed anhydrides, and used as fluorescent probes.

The excitation wavelengths were 360–380, 410–440, 460, and 480 nm. Upon excitation by actinic light, these compounds fluoresce in the green part of the spectrum (maximum at 512–518 nm). An example of the probe on serotonin is shown in Figure 4.1.

Binding of BODIPY-serotonin (10^{-7}–10^{-6} M) with embryos from mice, pollen from plantain *Plantago lanceolatum* and wild chamomile *Matricaria chamomilla,* or vegetative microspores of horsetail *Equisetum arvense*, as well as from sea urchins *Paracentrotus lividus* and *Lytechinus variegatus* during the first stage of development, has been studied (Roshchina et al. 2003, 2005). The visible fluorescence at 512–518 nm observed (see Figure 4.1) in the cover surface of cells studied was especially well seen on optical slices under a confocal microscope. Pollen and the early embryos of sea urchins *Lytechinus variegatus* and *Paracentrotus lividus* (just after fertilization) demonstrated the most pronounced green emission in the primembrane layer and, apparently, on the surface of the membrane (a marked asymmetry of staining of the surface membrane of the embryos).

After cell incubation with the fluorescent ligand the emission spectra may be recorded by microspectrofluorimetry (Roshchina et al. 2003, 2005) as shown in Figure 4.2. The cells were treated both with the

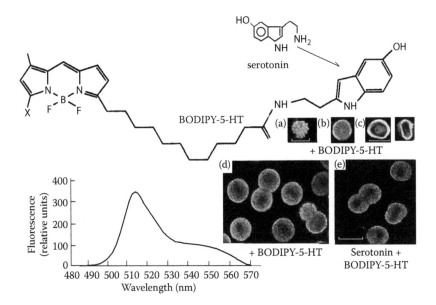

FIGURE 4.1 Fluorescent agonist of serotonin—BODIPY-5-HT—and its interaction with living cells. (Adapted from Roshchina, V. V. et al. In V.P. Zinchenko (editor), *Materials of International Symposium: "Reception and Intracellular Signaling".* 2005, Suppl. color. Pushchino: ONTI. 399–402.) Formulas (top), fluorescence spectrum after excitation (480 nm) (bottom left), and images of cells treated with the agonist under laser-scanning confocal microscope with laser excitation (488 nm) (bottom right). Binding of fluorescent-labeled serotonin BODIPY (10^{-6} M) with (a) pollen from wild chamomile *Matricaria chamomilla,* (b) its optical slice, and (c) nuclei from while petals of wild chamomile *Matricaria chamomilla* flowers (optical slices where green lightening surface of cover is seen); (d) embryos from sea urchins *Paracentrotus lividus* (in a period of first devision) treated with 10^{-6} M BODIPY-5-HT (the green emission, which is absent in the variant without any treatment, is seen), and (e) pretreated with arachinodoyl serotonin 10^{-6} M, then treated with 10^{-6} M BODIPY-5-HT (the decreased green emission in comparison with [d]-variant is seen).

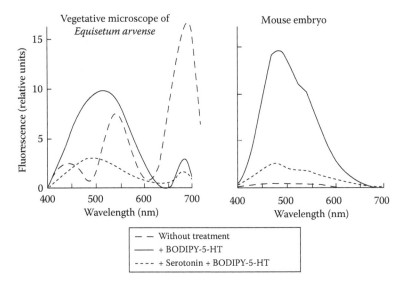

FIGURE 4.2 Fluorescence spectra of cells treated with BODIPY-5-HT measured by microspectrofluorimeter, (1) without treatment, (2) after the 20-minute incubation with BODIPY-5-HT 10^{-6} M, and (3) after preincubation with serotonin chloride 10^{-6} M, then addition of BODIPY-5-HT 10^{-6} M.

fluorescent probes and nonfluorescent serotonin at concentrations of 10^{-7}–10^{-5} M. The emission in green at 518–520 nm increased many times in comparison with a control (without treatment), as seen for both vegetative microspores of *Equisetum arvense* and the first embryonal cells of mice *Mus*. In vegetative microspores containing chloroplasts fluorescing with maxium 675–680 nm related to chlorophyll, the height of the peak decreased. If the cells had been preliminarily treated with serotonin and then BODIPY-5-HT was added, the emission at 518 nm was much smaller than without preincubation with neurotransmitter. This shows that the binding of fluorescent ligand and serotonin occurred at one and the same site—the serotonin receptor. However, it cannot be ruled out that these compounds may be bound inside the cell, because fluorescent probes contain a lipophilic moiety and, therefore, should penetrate into the cell.

After the treatment, fluorescence of vegetative microspores of horsetail *Equisetum arvense* and mouse embryo cells was analyzed at the development of the cells. Dry plant microspores had weak self-fluorescence; in the case of mouse embryos, any self-fluorescence in the visible part of the spectrum was absent at all developmental stages. Histochemical staining of both (animal and plant) types of cells with BODIPY-5-HT resulted in the appearance or enhancement of fluorescence intensity in the green part of the spectrum at 500–530 nm. But the fluorescence intensity at 500–530 nm increased slightly if the cells were preliminarily treated with nonfluorescent serotonin. This assumes that neurotransmitters and their fluorescent analogs bind mostly to the same sites in the cell. Green fluorescence was especially strong in peripheral cell areas, where the plasma membrane (and the cellulose wall in plant cells) is located. Laser-scanning confocal microscopy allows us to see the concentration of the green fluorescence at the pollen surface of *Plantago lanceolata* and sea urchin eggs *Paracentrotus lividus* (Roshchina et al. 2005). The location of fluorescent transmitters predominantly outside the cell and a significant decrease in fluorescence intensity after the pretreatment of cells with serotonin (also shown for BODIPY-dopamine) suggests that fluorescent derivatives are bound on the cell surface by the structures that bind natural neurotransmitters (e.g., receptors or transporters). This binding (possibly fairly specific) prevents a large part of lipophilic BODIPY derivatives from penetrating into the cell. Due to the presence of a lipophilic fluorophore, these fluorescent probes should penetrate across the plasma membrane; however, they concentrate outside the cell. Thus, similar to nonfluorescent serotonin and dopamine, BODIPY-neurotransmitters are bound on the cell surface (possibly by the corresponding plasma-membrane receptors).

The binding took place not only plasmalemma where the receptor is located, but also within the cell, which has been shown in the experiments with isolated nuclei (Roshchina et al. 2005) as shown in Figure 4.1. In this case, the intracellular interaction of serotonin is possible. Earlier, the stimulation of RNA synthesis in isolated nuclei from rat brain by serotonin has been demonstrated (Arkhipova et al. 1988). The experiments were in accordance with data (Roshchina et al. 2003) received on the model systems and analyzed the fluorescence of individual cell components (albumin, DNA, RNA, and β-carotene, which often binds to chromatin). Modeling with the individual cellular compounds (Figure 4.3), albumin, DNA, and RNA showed that BODIPY-5-HT has the most affinity to protein in comparison with nucleic acids. There is a possiblility to consider: the existence of serotonin receptor on the surface membrane of the nucleus. The fluorescence of water-wetted preparations (except for β-carotene) was weak or absent. The addition to these preparations of 10^{-6}–10^{-7} M aqueous solution of a fluorescent probe always enhanced fluorescence in the green part of the spectrum. The effects of BODIPY-5-HT and BODIPY-dopamine differed in these reactions with individual compounds. At the interaction with albumin, the first compound has a pronounced maximum emission at 465 nm and a not very pronounced maximum at 500–600 nm, while the second compound has a narrow maximum at 517–530 nm and a not very pronounced maximum at 465 nm. Two maxima at 517–530 and 465 nm were detected in the fluorescence spectra of DNA and RNA in the experiments with both neurotransmitters. The value of the maximum at 518 nm sharply increased in the presence of albumin. Taking into account that all serotonin and dopamine receptors are of protein nature, the experiments with albumin assume that the compounds studied possibly bind with cell receptors. The determination of the fluorescence intensities of individual cell sites after histochemical treatment showed that as much as 70%–90% of fluorescent neurotransmitters bind on the cell surface, mainly with receptors or within cells, for example with nuclei (see Figure 4.1). Currently, the use of such fluorescent probes may be recommended for histochemical studies on the regulation of the cell system development in plants and animals by biogenic amines.

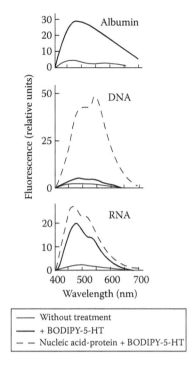

FIGURE 4.3 Modeling with the cellular compounds as the fluorescence spectra of individual cell components (albumin, DNA, and RNA) after the treatment with 10^{-7} M fluorescent serotonin BODIPY-5-HT (excitation wavelength at 360–380 nm). (Adapted from Roshchina, V. V. et al. In V. P. Zinchenko (editor), *Materials of International Symposium: "Reception and Intracellular Signaling".* 2005. Pushchino: ONTI. 399–402.) (1) Fluorescence of untreated component (moistening with water, 2 mg/0.02 mL; the control), (2) after treatment with BODIPY-neurotransmitter, and (3) fluorescence of the nucleic acid–protein mixture (ratio 1:1) after the treatment with BODIPY-neurotransmitter.

Fluorescent agonist of serotonin may be also used for analysis of cell development. It is demonstrated on the mouse embryos (Figure 4.4), after incubation of mice the embryos (from the stage of two blastomeres to blastocyst), with the fluorescent ligand of observed luminescence of surface membrane blastomeres, as well as the intercellular contacts. At the stage of four blastomeres, light intensity increased 4–5 times in comparison with that observed in zygotes and at the stage of two blastomeres. At late blastocyst stage, fluorescence is observed in the cells of the inner cell mass. Pretreatment of embryos with lipophilic analogs of serotonin caused reduction of BODIPY-5-HT emission. Using fluorescence microscopy, microspectrofluorimetry and laser-scanning confocal microscopy in this work shows that BODIPY-serotonin emitted in green on the surface of cells, and the fluorescence significantly reduced following pretreatment of these cells nonfluorescing serotonin. On using fluorescence microscopy, confocal laser microscopy, and microspectrofluorimetry BODIPY-5-HT emission in green has been shown on the surface of vegetative microspores from *Equisetum arvense* and pollen from *Plantago lanceolatum* (Roshchina et al. 2003, 2005); this fluorescence is greatly diminished after the pretreatment of samples with serotonin (10^{-7}–10^{-6} m) or arachidonoyl-serotonin. Decrease in fluorescence in all investigated sites while making the ligand shows the specificity of the binding observable and, presumably, the existence of specific receptors.

A similar intracellular binding of the serotonin fluorescent analog was shown after histochemical staining of the nuclei isolated from white flower petals of *Philadelphus grandiflorus* and *Matricaria chamomilla* with BODIPY-5-HT. The nuclei fluoresced in green, and on an optical slice obtained by laser-scanning confocal microscope one can see concentration of the dye emission on the surface of the organelles (see Figure 4.1). Intracellular binding of neurotransmitters located in secretory vesicles within cells is possible too, when the compounds are released internally. If the compounds are released from secretory vesicles within the cell they also can be bonded with the organelles (Buznikov 1987; Buznikov

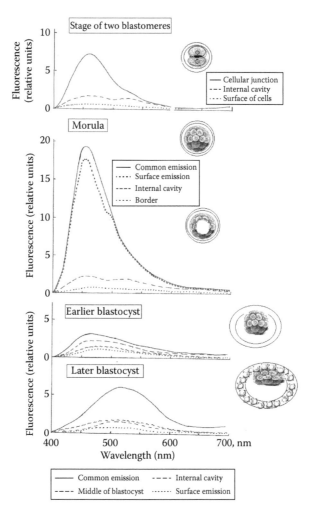

FIGURE 4.4 Fluorescence spectra of developed mouse embryo from two blastomeres—cells produced by cell division of the zygote after fertilization—then through the stage of multicellular morula to a blastocyst, having inner cell mass. Green emission was induced by the treatment with BODIPY-5-HT. This shows the changes on the surface of the embryonic mouse blastomeres and their fluorescence spectra after treatment with BODIPY-serotonin.

et al. 1996; Roshchina 1991, 2001). Nonfluorescent serotonin does not penetrate into the cells, but stimulates the synthesis of RNA in isolated nuclei by binding to nuclear proteins (Arkhipova et al. 1988; Tretyak and Arkhipova 1992). Experiments with isolated nuclei from the white petals of *Philadelphus grandiflorus* showed that both artificial and natural agonists and antagonists of some neurotransmitters may fluoresce on the surface of the nuclei, as was seen for plasmalemma of various living cells (Roshchina et al. 2005).

Fluorescence is used in the studies of serotonin reception on embryos with fluorescing lipophiling analogs (serotoninamides of polyenoic fatty acids), as in the work of Willows et al. (2012) on early embryos of *Tritonia diomedea*, a nudibranch mollusk, as a model system. In late embryos (veliger stage), serotonin and to a lesser extent its lipophilic analogs strongly increase embryonic motility.

Fluorescent Drugs: Antagonists of Serotonin for the Study of Reception

For studies of serotonin reception, fluorescent drugs that act as its antagonists could be used for experiments to analyze cellular mechanisms of the pharmaceuticals. In our laboratory three of them

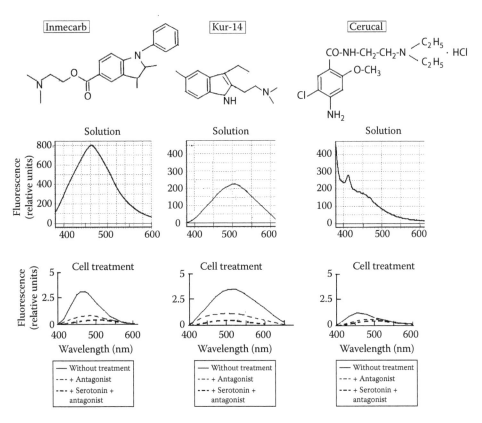

FIGURE 4.5 Fluorescence spectra of serotonin antagonists. (Partly adapted from Roshchina, V. V. *Fluorescing World of Plant Secreting Cells.* 2008. Enfield, New Hampshire: Science Publishers, and from unpublished data.) Solutions of antagonists—10^{-4} M in 1 cm^{-3} cuvette. Cell treatment of pollenfrom *Hippeastrum hybridum* with inmecarb or Kur-14, and treatment of pollen from *Haemanthus katharinae* with Cerucal. (1) control, without treatment, (2) after addition of antagonist, and (3) after addition of serotonin 10^{-5} M, then antagonist 10^{-4} M.

were studied (Figure 4.5). Among the drugs acting on the human serotonin receptors are inmecarb (β-dimethylaminoethyl l-benzyl-2,3-dimethylindole-5-carbonate hydrochloride), having a putative anti-alcoholic effect, its analog Kur-14 (5-methyl-2-α-dimethyl-aminoethyl-3-β-ethylindole), and Cerucal (4-amino-5-chloro-N-[2-diethylamino] ethyl-2-methoxybenzamide), which is used for the treatment of nausea, retching, and vomiting, motility disturbances of the upper gastro-intestinal tract, diabetic gastroparesis, and so on.

Maxima of the emission in solutions after the excitation (369 nm) were 460 nm for inmecarb, 520 nm for Kur-14, and 450 for Cerucal. Self-fluorescence of studied pure compounds added on the subject glass was small, but the emission increased 5 to 10 times after their interactions with the microspores. The green or blue rings of the studied agonists or antagonists were seen in the luminescent microscope on the cellular surface where they were bound with proposed receptors. Maximal emission was observed after the treatment with antagonist of serotonin Kur-14 (Roshchina 2008), and to a lesser degree in the case of inmecarb. Weaker fluorescence was characteristic for Cerucal addition to the model cells.

These fluorescent drugs were also tested on the main reaction to serotonin for most living cells—growth reaction. The seed germination stimulated by serotonin was sensitive to their antagonists (Roshchina 1992). If seeds of *Raphanus sativus* were preliminarily treated with 10^{-5} M antagonists of serotonin Kur-14 or inmecarb, the stimulation changed to the inhibition over 24 h of the development after moistening with the tested solution (Figure 4.6). In the case of Cerucal the same reaction occurred after 48 h. Root formation was stimulated by serotonin (and depressed by Kur-14). The root system is

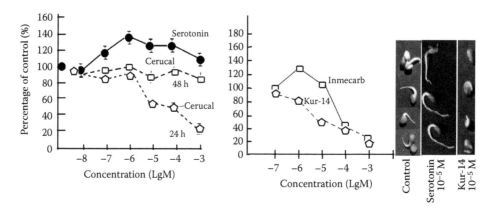

FIGURE 4.6   Effects of serotonin and its antagonists on the germination of seeds from *Raphanus sativus* over 24 h. For Cerucal, inhibition occurred after 48 h.

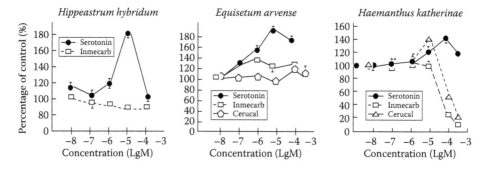

FIGURE 4.7   Effects of serotonin and its antagonists on the germination of pollen from different species.

known to be most sensitive to serotonin (Pelagio-Flores et al. 2011) and may be a model for the study of antiserotonin drugs testing.

Serotonin at concentrations 10^{-6}–10^{-4} M also stimulated the pollen germination of *Hippeastrum hybridum* (Roshchina et al. 1998) and Kur-14 inhibited the process (Roshchina 2004). Figure 4.7 shows how inmecarb and Cerucal diminished the stimulation-induced serotonin in pollen from *Hippeastrum hybridum* and *Haemanthus katherinae* as well as in vegetative microspores of horsetail *Equisetum arvense*. For *H. katherinae*, the inhibition was seen at high concentrations.

The mechanism of action of a substance on growth consists either in activating the proteins involved in growth, including the enzymes of phytohormone synthesis, or in affecting directly the genetic apparatus, that is, DNA and RNA synthesis, as well as the processes of transcription and translation, repression or activation of separate genes (Pelagio-Flores et al. 2011).

First of all this is a role of light-activated forms of serotonins in living systems. For example, in *Xenopus laevis* embryos, light (365 or 740 nm) activated serotonin and increased the occurrence of left–right asymmetry in embryonic development patterning defects (Rea et al. 2013). Serotonin shares structural similarity with indolyl-acetic acid, a known phytohormone, it but it differs in its effects on growth and development. As regards the photoperiod, effects on the indoleamines are also important (Ramakrishna et al. 2011a). This is of interest not only to plant investigators. Forms of indoleamines similar to serotonin, as well as the serotonin antagonists inmecarb, ritanserin, and hydrochloride, are studied with fluorescing lipophiling analogs (serotoninamides of polyenoic fatty acids), such as in the work of Willows et al. (2012) on early embryos of *Tritonia diomedea*, a nudibranch mollusk. In late embryos (veliger stage), serotonin, and to a lesser extent its lipophilic analogs, strongly increases embryonic motility.

Conclusion

Use of fluorescent tools appears to have perspectives for study in living cells if suitable sensitive and well-visible model material can be found. Most of the new fluorescent compounds synthesized on the base of green- or red-emitted artificial fluorophore confirm the binding of the serotonin receptor in the plasmatic membrane. Blue-fluorescing ligands are rarely used thus far, although many drugs acting on the receptor potentially may emit in both the blue or blue-green range that make their binding visible. Some of the examples were represented by inmecarb or Cerucal. It is also necessary to look for the strengthening of their emission in order to analyze their binding at the surface and within a cell. The choice of the fluorescence-labeled compounds such as BODIPY derivatives (agonists of serotonin) permitted their application to analysis of the serotonin reception in plasmatic membrane and isolated nuclei. This approach should lead to an understanding of the mechanisms of the intracellular effects of serotonin, and finding an exact binding of the neurotransmitter with some of the cellular organelles.

REFERENCES

Arkhipova, L. V., Tretyak, T. M., and Ozolin. O. N. 1988. The influence of catecholamines and serotonin on RNA-synthesizing capacity of isolated nuclei and chromatin of brain and rat liver. *Biochemistry (USSR)*. 53. 1078–1081.

Banala, A. K, Zhang, P., Plenge, P., Cyriac, G., Kopajtic, T., Katz, J. L., Loland, C. J., and Newman, A. H. 2013. Design and synthesis of 1-(3-[dimethylamino]propyl)-1-(4-fluorophenyl)-1,3-dihydroisobenzofuran-5-carbonitrile (Citalopram) analogues as novel probes for the serotonin transporter S1 and S2 binding sites. *J Med Chem*. 56 (23): 9709–9724.

Berger, M., Gray, J. A., and Roth, B. L. 2009. The expanded biology of serotonin. *Ann Rev Med*. 60: 355–366.

Berque-Bestel, I., Soulier, J. L., Giner, M., Rivail, L., Langlois, M., and Sicsic, S. 2003. Synthesis and characterization of the first fluorescent antagonists for human 5-HT4 receptors. *J Med Chem*. 46 (13): 2606–2620.

Bezuglov, V. V., Gretskaya, N. M., Esipov, S. S., Polyakov, N. B., Nikitina, L. A., Buznikov, G. A., and Lauder, J. (2004). Fluorescent lipophilic analogs of serotonin, dopamine and acetylcholine: Synthesis, mass-spectrometry and biological activity. *Bioorg Chem. (Russia)*. 30 (5): 512–519.

Buznikov, G. A. 1967. *Low Molecular Regulators of Embryonal Development*. Moscow: Nauka.

Buznikov, G. A. 1987. *Neurotransmitters in Embryogenesis*. Moscow: Nauka.

Buznikov, G. A. 1990. *Neurotransmitters in Embryogenesis*. Chur: Harwood Academic Press. 526.

Buznikov, G. A., Shmukler, Y. B., and Lauder, J. M. 1996. From oocyte to neuron: Do neurotransmitters function in the same way throughout development? *Cell Mol Neurobiol*. 16 (5): 532–559.

Chen, R. F. 1968. Fluorescence of protonated excited-state forms of 5-hydroxytryptamine (serotonin) and related indoles. *PNAS*. 60 (2): 598–605.

Falck, B. 1962. Observations on the possibilities of the cellular localization of monoamines by a fluorescene method. *Acta Physiol Scand*. 56. Suppl. 197: 5–25.

Grosse, W. 1982. Function of serotonin in seeds of walnuts. *Phytochemistry*. 21: 819–822.

Grosse, W. 1984. Biosynthesis of serotonin and auxin in seeds of *Juglans regia*. In Schlossberger, H., W. Kochen, B. Linzen, and H. Steinhart (editors), *Progress in Tryptophan and Serotonin Research*. Berlin, New York: de Gruyter. 803–806. (Proceedings of the Fourth Meeting of the International Study Group for Tryptophan Research. Martinsried, Federal Republic of Germany. April 19–22, 1983).

Grosse, W., Karisch, M., and Schroder, P. 1983. Serotonin biosynthesis and its regulation in seeds of *Juglans regia* L. *Zeischrift der Pflanzenphysiologie*. 110: 221–229.

Haugland, R. P. 2010. *Handbook of Fluorescent Probes and Research Chemicals*. Leiden, Netherlands: Molecular Probes.

Herrick-Davis, K., Grinde, E., Cowan, A., and Mazurkiewicz, J. E. 2013. Fluorescence correlation spectroscopy analysis of serotonin, adrenergic, muscarinic, and dopamine receptor dimerization: The oligomer number puzzle. *Mol Pharmacol*. 84 (4): 630–642.

Hovius, R. 2013. Characterization and validation of fluorescent receptor ligands: A case study of the ionotropic serotonin receptor. In M. R. Banghart (editor), *Chemical Neurobiology: Methods and Protocols*, vol. 995, 161–178, New York: Springer Science+Business Media.

Kang, S., Kang, K., Lee, K., and Back, K. 2007. Characterization of tryptamine 5-hydroxylase and serotonin synthesis in rice plants. *Plant Cell Rep.* 26: 2009–2015.

Kang, K., Kim, Y. S., Park, S., and Back, K. 2009. Senescence-induced serotonin biosynthesis and its role in delaying senescence in rice leaves. *Plant Physiol.* 150 (3): 1380–1393.

Kimura, M. 1968. Fluorescence histochemical study on serotonin and catecholamine in some plants. *Jap J Pharmacol.* 18. 162–168.

Kroeze, W. K., Kristiansen, K., and Roth, B. L. 2002. Molecular biology of serotonin receptors: Structure and function at the molecular level. *Curr Top Med Chem.* 2: 507–528.

Kumar, V., Rahbek-Clemmensen, T., Billesbølle, C. B., Jorgensen, T. N., Gether, U., and Newman, A. M. 2014. Novel and high affinity of fluorescent ligands for the serotonin transporter based on (S)-citalopram. *ACS Med Chem Lett.* 5 (6): 696–699.

Lacivita, E., Masotti, A. C., Jafurulla, M., Saxena, R., Rangaraj, N., Chattopadhyay, A., Colabufo, N. A., Berardi, F., Perrone, R., and Leopoldo, M. 2010. Identification of a red-emitting fluorescent ligand for *in vitro* visualization of human serotonin 5-HT1A receptors. *Bioorg Med Chem Lett.* 20: 6628–6632.

Lembeck, F., and Skofitsch, G. 1984. Distribution of serotonin in *Juglans regia* seeds during ontogenetic development and germination. *Zeischrift der Pflanzenphysiologie.* 114: 349–353.

Leonov, B. V., and Budantsev, A. Y. 1971. Detection of serotonin in blastocytes of rats by the method of microspectrofluorimetry. *Dokl USSR Acad Sci.* 198 (3): 734–736.

Łukasiewicz, S., Polit, A., Kędracka-Krok, S., Wędzony, K., Maćkowiak, M., and Dziedzicka-Wasylewska, M. 2010. Hetero-dimerization of serotonin 5-HT$_{2A}$ and dopamine D$_2$ receptors. *Biochim Biophys Acta.* 1803 (12): 1347–1358.

Mukherjee, S., David, A., Yadav, S., Baluška, F., and Bhatla, S.C. 2014. Salt stress-induced seedling growth inhibition coincides with differential distribution of serotonin and melatonin in sunflower seedling roots and cotyledons. *Physiol Plant.* 152 (4): 714–728.

Murch, S. J., Alan, A. R., Ca, J., and Saxena, P. K. 2009. Melatonin and serotonin in flowers and fruits of *Datura metel* L. *J Pineal Res.* 47: 277–283.

Murch, S. J., Hall, B. A., Le, C. H., and Saxena, P. K 2010. Changes in the levels of indoleamine phytocemicals during veraison and ripening of wine grapes. *J Pineal Res.* 49: 95–100.

Murch, S. J., Krishna, R. S., and Saxena, P. K. 2000. Tryptophan is a precursor for melatonin and serotonin biosynthesis in *in vitro* regenerated St. John's wort (*Hypericum perforatum* L. cv. Anthos) plants. *Plant Cell Rep.* 19: 698–704.

Oleskin, A.V. 2012. *Biopolytics: The Political Potential of the Life Sciences.* New York: Nova Science Publishers.

Oleskin, A.V., Shishov, V.I., and Malikina, K. D. (2010). *Symbiotic Biofilms and Brain Neurochemistry.* New York: Nova Science Publishers.

Pelagio-Flores, R., Ortız-Castro, R., Mendez-Bravo, A., Macıas-Rodrıguez, L., and Lopez-Bucio, J. 2011. Serotonin, a tryptophan-derived signal conserved in plants and animals, regulates root system architecture probably acting as a natural auxin inhibitor in *Arabidopsis thaliana*. *Plant Cell Physiol.* 52 (3): 490–508.

Ramakrishna, A. 2015. Indoleamines in edible plants: Role in human health effects. In Angel Catalá, Biochemistry Research Trends Series (editor), *Indoleamines: Sources, Role in Biological Processes and Health Effects*. New York: Nova Publishers. 279.

Ramakrishna, A., Dayananda, C., Giridhar, P., Rajasekaran, T., and Ravishankar, G. A. 2011a. Photoperiod influences endogenous indoleamines in cultured green alga *Dunaliella bardawil*. *Indian J Exp Biol.* 49: 234–240.

Ramakrishna, A., Giridhar, P., Jobin, M., Paulose, C. S., and Ravishankar, G. A. 2012. Indoleamines and calcium enhance somatic embryogenesis in cultured tissues of *Coffea canephora* P ex Fr. *Plant Cell Tiss Organ Cult.* 108: 267–278.

Ramakrishna, A., Giridhar, P., and Ravishankar, G. A. 2009. Indoleamines and calcium channels influence morphogenesis in *in vitro* cultures of *Mimosa pudica* L. *Plant Signal Behav.* 12: 1136–1141.

Ramakrishna, A., Giridhar, P., and Ravishankar, G. A. 2011b. Phytoserotonin. *Plant Signal Behav.* 6 (6): 800–809.

Ramakrishna, A., Giridhar, P., Sankar, K. U., and Ravishankar, G. A. 2012a. Endogenous profiles of indoleamines: Serotonin and melatonin in different tissues of *Coffea canephora* P ex Fr. as analyzed by HPLC and LC-MS-ESI. *Acta Physiol Plant.* 34: 393–396.

Ramakrishna, A., Giridhar, P., Sankar, K. U., and Ravishankar, G. A. 2012b. Melatonin and serotonin profiles in beans of *Coffea* sps. *J Pineal Res.* 52: 470–476.

Rea, A. C., Vandenberg, L. N., Ball, R. E., Snouffer, A. A., Hudson, A. G., Zhu, Y., McLain, D., et al. 2013. Light-activated serotonin for exploring its action in biological systems. *Chem Biol.* 20 (12): 1536–1546.

Regula, I. 1970. 5-Hidroksitriptamin u ljutoj Koprivi (*Urtica pilulifera* L.). *Acta Bot Croat.* 29: 69–74.

Regula, I. 1981. Serotonin in the tissues of *Loasa vulcanica* ed. André. *Acta Bot Croat.* 40: 91–94.

Regula, I. (1986). The presence of serotonin in embryo of black walnut (*Juglans nigra*). *Acta Bot Croat*, 45, 91–95.

Regula, I., and Devide, Z. 1980. The presence of serotonin in some species of genus *Urtica*. *Acta Bot Croat.* 39. 47–50.

Regula, I., Kolevska-Pleticapic, B., and Krsnik-Rasol, M. 1989. Presence of serotonin in *Juglans ailanthifolia* var ailanthifolia Carr and its physiological effects on plants. *Acta Bot Croat.* 48: 57–62.

Roshchina, V. V. 1991. *Biomediators in Plants: Acetylcholine and Biogenic Amines.* Pushchino: Biological Center of the Russion Academy of Sciences.

Roshchina, V. V. 1992. The action of neurotransmitters on the seed germination. *Biol Nauk.* 9: 124–129.

Roshchina, V. V. 2001. *Neurotransmitters in Plant Life.* Enfield, New Hampshire: Science Publishers.

Roshchina, V. V. 2003. Autofluorescence of plant secreting cells as a biosensor and bioindicator reaction. *J Fluoresc.* 13 (5): 403–420.

Roshchina, V. V. 2004. Cellular models to study the allelopathic mechanisms. *Allelopathy J.* 13 (1): 3–16.

Roshchina, V. V. 2005. Contractile proteins in chemical signal transduction in plant microspores. *Biol Bull Ser Biol.* 3: 281–286.

Roshchina, V. V. 2006. Plant microspores as biosensors. *Trends Mod Biol.* 126 (3): 262–274.

Roshchina, V. V. 2008. *Fluorescing World of Plant Secreting Cells.* Enfield, New Hampshire: Science Publishers.

Roshchina, V. V. 2010. Evolutionary considerations of neurotransmitters in microbial, plant and animal cells. In M. Lyte, and P. P. E. Freestone (editors), *Microbial Endocrinology: Interkingdom Signaling in Infectious Disease and Health.* New York: Springer. 17–52.

Roshchina, V. V. 2014. *Model Systems to Study Excretory Function of Higher Plants.* Dordrecht, Netherlands: Springer.

Roshchina, V. V., Bezuglov, V. V., Markova, L. N., Sakharova, N. Y., Buznikov, G. A., Karnaukhov, V. N., and Chailakhyan, L. M. 2003. Interaction of living cells with fluorescent derivatives of biogenic amines. *Dokl Russian Acad Sci.* 393 (6): 832–835.

Roshchina, V. V., Markova, L. N., Bezuglov, V. V., Buznikov, G. A., Shmukler, Y. B., Yashin, V. A., and Sakharova, N. Y. 2005. Linkage of fluorescent derivatives of neurotransmitters with plant generative cells and animal embryos. In V. P. Zinchenko (editor), *Materials of International Symposium: "Reception and Intracellular Signaling".* Pushchino, Russia: ONTI. 399–402; 418–420. Suppl. poster color figures, 318–402.

Roshchina, V. V., Popov, V. I., Novoselov, V. I., Melnikova, E. V, Gordon, R. Y., Peshenko, I. V., and Fesenko, E. E. 1998. Transduction of chemosignal in pollen. *Cytologia* (Russia). 40: 964–971.

Roshchina, V. V., Yashin, V. A., Yashina, A. V., and Gol'tyaev, M. V. 2013. Microscopy for modelling of cell–cell allelopathic interactions. In Z. A. Cheema, M. Farooq, and A. Wahid (editors), *Allelopathy. Current Trends and Future Applications.* Berlin: Springer. 407–427.

Roshchina, V. V., Yashin, V. A., Yashina, A. V., Gol'tyaev, M. V., and Manokhina, I. A. 2009. Microscopic objects for the study of chemosignaling. In V. P. Zinchenko, S. S. Kolesnikov, and A. V. Berezhnov (editors), *Reception and Intracellular Signalling.* Pushchino: Biological Center of the Russion Acadamy of Science. 699–703.

Roth, B. L. (editor). 2006. *The Serotonin Receptors: From Molecular Pharmacology to Human Therapeutics.* Totowa, New Jersey: Humana Press.

Schmukler, Y. B., Nikishin, D. A. 2012. Transmitters in blastomere interaction. In S. Gowder (editor), *Cell Interaction.* Rijeka, Croatia: InTech. 31–65.

Tanaka, K. F., Samuels, B. A., and Hen, R. 2012. Serotonin receptor expression along the dorsal–ventral axis of mouse hippocampus. *Philos Trans B.* 370 (1661).

Tretyak, T. M., and Arkhipova, L. V. 1992. Intracellular activity of mediators. *Uspekhi Sovremennoi Biologii (Trends Mod Biol.).* 11: 265–272.

Turi, C. E., Axwik, K. E., and Murch, S. J. 2014. *In vitro* conservation, phytochemistry, and medicinal activity of *Artemisia tridentata* Nutt: Metabolomics as a hypothesis-generating tool for plant tissue culture. *Plant Growth Regul.* 74: 239–250.

Vernall, A. J, Hill, S. J., and Kellam, B. 2013. The evolving small-molecule fluorescent-conjugate toolbox for Class A GPCRs. *Br J Pharmacol.* 171 (5): 1073–1084.

Willows, A. O. D., Nikitina, L. A., Bezuglov, V. V., Gretskaya, N. M., and Buznikov, G. A. 2012. Interaction between serotonin and neuropeptides in the control processes of embryogenesis. *Russ J Dev Biol.* 31 (2): 106–112.

Wilson, J. N., Ladefoged, L. K., Babinchak, W. M., and Schiøtt, B. 2014. Binding-induced fluorescence of serotonin transporter ligands: A spectroscopic and structural study of 4-(4-[Dimethylamino] phenyl)-1-methylpyridinium (APP⁺) and APP⁺ analogues. *ACS Chem Neurosci.* 5 (4): 296–304.

Zhang, P., Jørgensen, T. N., Loland, C. J., and Newman, A. H. 2013. A rhodamine-labeled citalopram analogue as a high-affinity fluorescent probe for the serotonin transporter. *Bioorg Med Chem Lett.* 23: 323–326.

5

Phytomelatonin: Physiological Functions

Katarzyna Szafrańska and Małgorzata Maria Posmyk
University of Lodz
Lodz, Poland

CONTENTS

ABSTRACT   The antioxidant properties of melatonin (MEL) and its role in the maintenance of proper reduction-oxidation (redox) state are considered to be one of the most important functions in living cells. The effects of its action are not only caused by the direct elimination of reactive oxygen species (ROS) but also by the stimulation of the antioxidant enzyme activities and possible activation of other metabolic pathways, for example, polyphenol synthesis. Melatonin also affects the efficiency of the photosynthesis process by (1) delaying chlorophyll degradation, (2) increasing the uptake of CO_2, and (3) accelerating the electron transport. This indoleamine, similar to indole-3-acetic acid (IAA), could be involved in the regulation of plant growth and development. There are also reports that indicate the role of melatonin on plant circadian rhythms and photoperiodic responses; however, to explain exactly this phenomenon, further investigations are necessary.

KEY WORDS: *phytomelatonin, antioxidant, bioregulator, circadian rhythms, seasonal rhythms, chlorophyll, photosynthesis, senescence.*

Abbreviations

ABA: abscisic acid
AFMK: N1-acetyl-N2-formyl-5-methoxykynuramine
APX: ascorbate peroxidase
ATG: autophagy gene
AXR3:   auxin-resistant 3 gene
bZIP: transcription factor binding to promoter regions of genes to control their expression
CAB: gene encoding proteins binding chlorophyll a/b
CAT: catalase
CLH1:   chlorophyllase gene
ETR: electron transport rate
FR: free radicals

Fv/Fm:	photochemical quantum yield of photosystem II
HXK1:	hexokinase-1 gene
IAA:	indole-3-acetic acid
IAA17:	indole-3-acetic acid–inducible 17 genes
IBA:	indole-3-butyric acid
MDC:	monodansylcadaverine
MEL:	melatonin, N-acetyl-5-methoxytryptamine
MPTP:	mitochondrial permeability transition pore
MYB:	transcription factor involved in the control of various processes, such as responses to biotic and abiotic stresses, development, differentiation, metabolism, defense, and so on
NAC:	transcription factor implicated in various aspects of plant development
NCED:	9-cis-epoxycarotenoid dioxygenase
NYC1:	non-yellow coloring 1gene
NYC3:	non-yellow coloring 3 gene
PAO:	pheide a oxygenase
PetF:	ferredoxin gene
PHL-C:	*Prunus avium* × *Prunus cerasus*
PS I:	photosystem I
PS II:	photosystem II
RBCS:	gene encoding the small subunit of Rubisco
Rubisco:	ribulose bisphosphate carboxylase/oxygenase
RCBs:	rubisco-containing bodies
RCCR1:	red chlorophyll catabolite reductase 1 gene
RNS:	reactive nitrogen species
ROS:	reactive oxygen species
SAG 12:	senescence-associated gene 12
SAGs:	senescence-associated genes
SER:	serotonin
SGR:	stay-green gene
SOD:	superoxide dismutase
TFs:	transcription factors
VTC4:	gene encoding L-galactose 1-P-phosphatase

Introduction

This chapter describes available knowledge on the most important functions that melatonin plays in plants, including its antioxidative properties, contribution to the process of photosynthesis, and involvement in the regulation of plant growth and development, as well as in the regulation of circadian rhythm and photoperiodic response.

MEL as a Universal Antioxidant

Protection against internal and external oxidizing agents is one of the most important functions of MEL in plants (Tan et al. 2012). Moreover, occurrence melatonin in edible plants and processed foods has been reported (Tan et al. 2012; Janas and Posmyk 2013; Ramakrishna et al. 2012; Ramakrishna et al. 2012 a,b; Ramakrishna 2015). Due to its small size, as well as solubility in both water and lipids, MEL can be a universal amphiphilic antioxidant freely penetrating all compartments of a cell. It seems probable that this unrestricted migration between cellular compartments makes MEL a much more potent antioxidant than vitamins C, E, and K, capable only of selective migration (Bonnefont-Rousselot and Collin 2010). However, it should be taken into account that endogenous MEL level (<1 μM) is negligible compared to the concentrations of other antioxidants such as ascorbate and glutathione. So the direct reactions of MEL with ROS are of marginal importance *in vivo* compared to the reactions of other antioxidants.

Thus, the beneficial effect of MEL seems to be rather the result of their synergistic interaction under oxidative stress (Gitto et al. 2001). In chilled cell suspensions of carrot, the positive effect of MEL on cell apoptosis correlated with the synthesis of polyamines was found (Lei et al. 2004). It seems that MEL acting as a signal molecule regulates various metabolic pathways, including the phenylpropanoid pathway, both under stress conditions and during recovery (Szafrańska et al. 2012, 2013; Turk et al. 2014).

Antioxidant properties of MEL consist in (1) direct scavenging of free radicals (FR), (2) induction of the activity of antioxidant enzymes, (3) stimulation of the activity of other antioxidants, (4) protection of antioxidant enzymes against oxidative damage, and (5) improving the efficiency of electron transport in the mitochondrial chain, due to the reduction of electron leakage, thereby limiting FR production (Wang et al. 2012). MEL as a natural electron donor is especially efficient at detoxification of reactive oxygen and nitrogen species (ROS and RNS) by (1) high affinity and selectivity for certain radicals, and (2) the ability to generate a radical scavenger cascade in which products of radical reactions contribute to the further elimination of ROS (Rosen et al. 2006). Unlike other antioxidants, MEL oxidation products such as β-hydroxymelatonin, cyclic β-hydroxymelatonin or cyclic melatonin (Kołodziejczyk et al. 2015), and also N1-acetyl-N2-formyl-5-methoxykynuramine (AFMK) (Tan et al. 2007), also show antioxidant activities. The cascade protection provided by MEL and its metabolites makes MEL, even at low concentrations, very effective in protecting organisms against oxidative stress (Galano et al. 2013) (Figure 5.1). MEL involvement in plant defense against various environmental stresses generating ROS was testified to by its elevated levels in the water hyacinth tolerant to contamination (*Eichornia crassipes* [Mart.] Solms) (Tan et al. 2007), as well as in the tomato cultivar (*Lycopersicon esculentum* Mill.) more resistant to ozone, in which MEL content was 5 times higher than in the nonresistant wild type (*Lycopersicon pimpinellifolium* L. Mill.) (Dubbels et al. 1995). Similar differences in MEL content were found in two varieties of tobacco (*Nicotiana tabacum* L.) with different sensitivity to ozone. A higher content of MEL in ripe tomato fruits was also reported, which may be associated with their protection against increased ROS production during maturation (Dubbels et al. 1995).

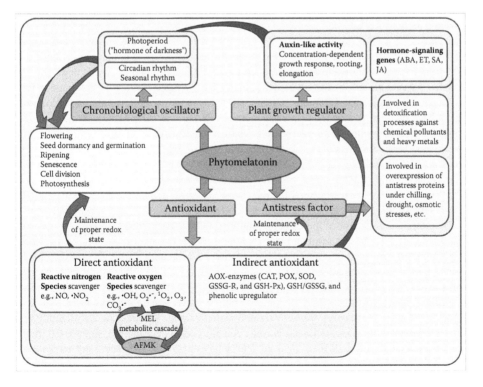

FIGURE 5.1 The phytomelatonin role and mechanism of action.

In addition to direct elimination of FR, MEL also affects gene expression and activity of antioxidant enzymes such as superoxide dismutase (SOD), catalase (CAT), and ascorbate peroxidase (APX), which was shown in transgenic rice with increased synthesis of this indoleamine (Park et al. 2013), and in leaves of apple trees and wheat seedlings treated with MEL (Wang et al. 2012, 2013a; Turk et al. 2014).

Mitochondria and chloroplasts are particularly vulnerable to oxidative stress, so they require special protection. Excessive amounts of toxic products of respiration and photosynthesis can lead to their morphological and functional damage. Thus the main task of MEL at the mitochondrial level is not only to directly interact with ROS and FR but also to increase the efficiency of ATP production by accelerating electron flow in an electron transport chain by a mechanism that promotes activities of complexes I, III, and V. MEL also contributes to the maintenance of an optimal membrane potential across the inner mitochondrial membrane by regulating the mitochondrial permeability transition pore (MPTP). MEL actually exerts dual effects on the MPTP. Under optimal conditions it activates MPTP and mildly decreases the mitochondrial membrane potential, while under oxidative stress which disrupts MPTP function, MEL significantly inhibits MPTP and thus preserves the membrane potential to avoid mitochondrial collapse. Because of these MEL properties it can be concluded that its fundamental role in mitochondria is to preserve physiological functions of this organelle both under optimal and stressful conditions (Tan et al. 2013).

In chloroplasts, which are also particularly susceptible to oxidative stress, the role of MEL has not been thoroughly studied, but it has been suggested that it effectively prevents stress-induced chlorophyll degradation. Preserving the integrity of chloroplast membranes, MEL supports the course of photosynthesis—a fundamental process for plant life (Tan et al. 2013).

MEL-Triggered Improvement of Photosynthetic Apparatus

Photosynthesis is a basic and most important process of organic biomass production indispensable for the functioning of the Earth's biosphere. The yield of photosynthesis is associated with the vitality of leaves, and degradation and loss of chlorophyll is one of the biochemical markers of aging. Chloroplast dysfunction and their conversion into gerontoplasts leads to progressive loss of proteins, such as Rubisco (ribulose bisphosphate carboxylase/oxygenase, an enzyme crucial for CO_2 assimilation) and others necessary for chlorophyll functioning (Sarropoulou et al. 2012a). Although degradation of these proteins started in the early stages of aging, long-term application of MEL significantly slowed down this process in apple leaves (Wang et al. 2013b). Simultaneously, in the MEL-treated plants the expression of RBCS, a gene encoding the small subunit of Rubisco, and of CAB gene-encoding proteins binding chlorophyll a/b which form a light-harvesting antenna of photosystem II (PS II), was inhibited much more slowly than in the control nontreated group. Additionally, in the plants treated with this indoleamine photosynthesis was more efficient, concentrations of starch, sorbitol and sucrose were higher, and chlorophyll degradation during the aging process was slower, while the expression of pheide a oxygenase (PAO) gene (a key gene for the chlorophyll degradation process), and senescence-associated gene 12 (SAG12), was inhibited. Moreover, exogenous MEL inhibited the expression of sugar-sensing and senescence-associated hexokinase-1 gene (HXK1) (Wang et al. 2013b), and effectively prevented the accumulation of H_2O_2, which is a characteristic indicator of the cell aging process (Wang et al. 2012, 2013a).

In *A. thaliana,* MEL treatment inhibited chlorophyllase (CLH1) gene expression, an enzyme regulated by light and also involved in chlorophyll degradation (Weeda et al. 2014). Decreased chlorophyll degradation in barley leaves treated with MEL was also observed and its effects were more pronounced than those invoked by cytokines, widely accepted as anti-aging hormones (Arnao and Hernandez-Ruiz 2009). A similar impact of exogenous MEL on chlorophyll protection against the effects of oxidative stress was observed in the leaves of *A. thaliana* treated with paraquat (Weeda et al. 2014). Studies on rice revealed that treatment with MEL significantly reduced chlorophyll degradation, suppressed transcripts of senescence-associated genes (SAGs), delayed leaf senescence, and enhanced salt stress tolerance (Liang et al. 2015). The authors investigated four senescence-associated genes involved in chlorophyll degradation: stay-green (SGR), non-yellow coloring 1 (NYC1) and 3 (NYC3) genes, and red chlorophyll catabolite reductase 1 (RCCR1), as well as four senescence-induced genes—OsNAP, Osh36, Osh69, and OsI57—which are widely used as age-dependent or dark-induced leaf senescence markers in rice. In comparison with the control, relative expression of all these genes under continuous dark and salinity stresses was suppressed by exogenous MEL, which was accompanied by increased chlorophyll and reduced H_2O_2 contents. These results suggest that MEL prevents leaf senescence, thus promoting longevity. According to Liang et al. (2015), MEL regulates the expression of transcription factors (TFs), including bZIP (binding to promoter regions of genes to control their expression), NAC (implicated in various aspects of plant development), and MYB (involved in the control of various processes such as responses to biotic and abiotic stresses, development, differentiation, metabolism, defense, etc.), which, in turn, directly mediate expression of H_2O_2-scavenging enzyme-encoding genes or via TFs-DREBs-HSFs transcriptional cascade to regulate their transcription (Figure 5.2). Sarropoulou et al. (2012a) showed that in shoot tip explants of cherry rootstock *Prunus avium* × *Prunus cerasus* (PHL-C), low doses of exogenous MEL slightly enhanced the content of photosynthetic pigments, and total biomass, as well as total carbohydrates, and reduced the proline content in roots, indicating MEL involvement in the plant stress metabolism.

More recently, the expression level of auxin-resistant 3 (AXR3)/indole-3-acetic acid–inducible 17 (IAA17) gene was significantly downregulated by exogenous MEL treatment (Shi et al. 2015). The authors discovered that *AtIAA17*-overexpressing plants showed early leaf senescence with lower chlorophyll content in rosette leaves compared with the wild-type plants, while *AtIAA17*-knockout mutants

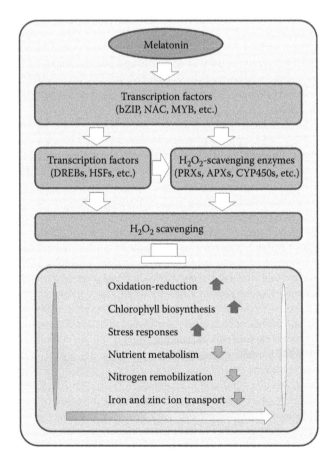

FIGURE 5.2 A proposed model of MEL-mediated delay of leaf senescence and cell death in rice. MEL directly or indirectly increases the expression of some TFs, such as bZIP, MAC, and MYB TFs, which activate the genes encoding the H_2O_2-scavenging enzymes or regulate their transcription via TFs-DREBs-HSFs transcriptional cascade. Consequently, the expression of genes involved in multiple biological processes is altered, leading to delayed leaf senescence and cell death. (Modified from Liang et al. *J Pineal Res.* 2015. 59 (1): 91–101.)

displayed delayed leaf senescence with higher chlorophyll content. These results may indicate that *AtIAA17* is a positive modulator of natural leaf senescence, which provides a direct link between MEL and natural leaf senescence in *Arabidopsis*.

Transcriptome analysis of soybean plants revealed that seed-coating with MEL solution improved their tolerance to salt and drought stress, probably due to enhanced expression of genes related to photosynthesis, carbohydrate/fatty acid metabolism, and ascorbate biosynthesis (Wei et al. 2015). Among interesting genes upregulated by MEL, there are two subunits of photosystem I (PS I) (PsaK and PsaG), two elements (PsbO and PsbP) related to the oxygen-evolving complex of PS II (oxygen-evolving enhancer proteins), the ferredoxin gene PetF, and the VTC4 gene, encoding the L-galactose 1-P-phosphatase involved in ascorbate biosynthesis (Wei et al. 2015).

Autophagy is one of the intracellular processes of vacuolar degradation of cytoplasmic components that are required for nutrient recycling in cells, and for maintenance of normal cellular homeostasis. Chloroplast proteins can be degraded through autophagy via Rubisco-containing bodies (RCBs), a type of autophagic bodies containing chloroplast stroma. Expression of an autophagy gene (ATG) associated with rapid degradation of proteins at the later stage of leaf senescence was significantly lower in the apple leaves treated with MEL than in nontreated ones, suggesting a positive role of MEL in delaying the autophagy process in plants (Wang et al. 2013b). These observations were confirmed in subsequent studies on autophagy induced by oxidative stress in *Arabidopsis* seedlings (Wang et al. 2015). Pretreatment with 5 or 10 μM MEL induced relatively strong autophagy, as evidenced by the number of monodansylcadaverine (MDC)-stained autophagosomes in root samples. Strong upregulation of AtATG8-PE genes, which enhanced capacity for autophagy, was also noticed. Exogenous MEL upregulated the expression of AtAPX1 and AtCAT genes that are involved in H_2O_2-scavenging but in fact did not affect ROS content under oxidative stress. MEL-triggered activation of autophagy not accompanied by alteration in ROS production may be part of a survival strategy that is enhanced by MEL after cellular damage. Therefore, it represents a kind of secondary defense that removes damaged proteins when antioxidant activities are compromised (Wang et al. 2015).

Proteomic analysis performed during leaf senescence of *Malus hupehensis* also brought very interesting results (Wang et al. 2014). MEL applied to roots for 2 months significantly delayed senescence in apple. Proteomic data revealed that 622 proteins were altered by the natural senescence process, and 309 after MEL treatment. In general, MEL changed the expression of many proteins involved in senescence, and led to downregulation of proteins that are normally upregulated during this process. Therefore, it is not surprising that according to many authors MEL improves photosynthesis, since MEL is able to upregulate three chloroplastic ATP synthases, two Rubisco small subunits, one Rubisco-interacting protein, and one PS I reaction center subunit, which results in more efficient photosynthetic activity. MEL also downregulates proteins involved in starch granule breakdown. This phenomenon is associated with photosynthesis decrease and marked starch degradation during leaf senescence, which is alleviated by MEL. Additionally, several proteins involved in protein folding and posttranslational modifications, such as some MAP kinases, serine/threonine protein kinases/phosphatases, among others (many of which are engaged in senescence and stress responses), were downregulated by MEL (Wang et al. 2014).

The analysis of chlorophyll fluorescence is a very useful tool to determine the efficiency of photosynthetic apparatus. The maximum photochemical quantum yield of PS II (Fv/Fm), as well as the electron transport rate (ETR) in the MEL-treated apple leaves, were markedly higher under dark and drought conditions than in the control leaves (Wang et al. 2012, 2013a). A significant increase in Fv/Fm after exogenous MEL application was also observed in characeae (*Chara australis* R. Brown). However, this increased quantum efficiency of photosynthesis seems to be rather the result of greater number of open reaction centers of PS II than their higher productivity. The greater number of open reaction centers reflects greater efficiency of all components of the photosynthetic electron transport chain, suggesting that MEL protection against ROS includes not only chlorophyll but also other proteins involved in photosynthesis (Lazár et al. 2013).

MEL not only protects chlorophyll against degradation but also contributes to photosynthesis improvement by increasing CO_2 assimilation. Limited diffusion of CO_2 from the atmosphere is considered to be one of the main factors that reduce efficiency of photosynthesis. Stomata are sensitive to changing CO_2 concentrations and are regarded as pressure regulators, which reduce stomatal conductance even at slight soil dehydration. The process of opening and closing of stomata is regulated by abscisic acid (ABA), a growth inhibitor and stress hormone synthesized in leaves or transported there from dehydrated roots (drought stress signal), and may be controlled by a number of external and internal factors (Pinheiro and Chaves 2011). Wang et al. (2013a,b) found that MEL application to soil improved uptake and assimilation of CO_2 in leaves. It seems that MEL absorbed by roots delays aging, and can affect the ABA biosynthesis that enhances this process (Guo and Gan 2005). The antagonism between these two molecules was clearly observed during senescence of detached barley leaves (Arnao and Hernandez-Ruiz 2009). Interestingly, in leaves pretreated with MEL, ABA level under drought stress was approximately half that in MEL-untreated leaves (Li et al, 2015). MEL downregulated 9-cis-epoxycarotenoid dioxygenase (NCED), a key enzyme in ABA biosynthesis in plants, and upregulated two CYP707 monooxygenases, key catabolic enzymes involved in the ABA metabolism under drought conditions. According to the authors, the lower ABA and H_2O_2 levels in the MEL-treated plants improved stomatal performance, counteracting the stress conditions. However, the possibility of crosstalk or some other interaction between these two types of plant molecules requires further investigation.

The functioning of the stomata in two *Malus* species was improved by MEL treatment (Li et al. 2015). They were longer, wider and with larger apertures, in both control and drought-stressed plants. Several leaf parameters, such as the relative water content, electrolyte leakage, chlorophyll content, photosynthetic efficiency, stomatal conductance, hydrogen peroxide levels, and antioxidant enzyme activities, were optimized by exogenous MEL application through roots (Li et al. 2015).

A beneficial MEL impact on leaf morphology and anatomy is worth noting (see Figure 5.1). Under water-deficiency stress, MEL-treated grape plants maintained chlorophyll content and the efficiency of PS II at a similar level as nonstressed plants (Meng et al. 2014). Anatomically, water-deficiency stress triggered leaf thickening, with thicker cuticles and smaller stomata, especially in mature leaves. Significant deformations of palisade and spongy tissues were also observed. However, in the MEL-treated grape plants greater stomatal pore length and width were correlated with a wider stomata opening. Although the leaves, cuticles, palisade, and spongy tissues were not as thick as in the well-watered plants, they were much thicker than in the stressed plants. MEL also optimized the photosynthetic processes, especially CO_2 assimilation, by promoting osmotically active metabolite levels (as proline) and maintenance of high cell turgor, which causes maximal stomata opening and increase in CO_2 uptake. In the grape plants chloroplast morphology was strongly affected under water stress. The chloroplasts became round in shape by the increase in width and decrease in length, chloroplast membrane systems were damaged, starch grains disappeared, and thylakoids were dilated, loosened, and distorted. In contrast, after MEL treatment the chloroplasts showed a very well preserved internal lamellar system, and all the damage and destructions were seriously mitigated (Meng et al. 2014). All these reports suggest that MEL plays a very important role in photosynthesis optimization and has the significant potential to streamline this process under stress conditions.

MEL as a Plant Growth Regulator

MEL, similarly to auxins, originates from tryptophan, and its structure is very similar to indole-3-acetic acid (Figure 5.3). Both molecules have a common indole ring but different side chains at C3: IAA has an acidic, while MEL an N-acetyl group. Additionally, in position 5 of MEL the methoxy group is attached. However, the distance between the indole ring and the substituents of both these molecules is very similar (~ 0.50 nm), suggesting that under certain conditions MEL can simulate the auxin effects (Arnao 2014). Since one of the main functions of IAA is to stimulate root formation, it was interesting to check whether MEL acted alike.

Evidence for the involvement of MEL in root formation in *Hypericum perforatum* L. (Murch et al. 2001), and etiolated hypocotyls of *Lupinus albus* L. (Arnao and Hernandez-Ruiz 2007), was provided. The authors showed that exogenously applied MEL stimulated root formation in a dose-dependent manner. Although auxins stimulate root formation, they also inhibit growth of existing ones. Studies on sweet cherry rootstocks with MEL and auxins, IAA, and indole-3-butyric acid (IBA), showed that low concentrations of MEL stimulated root formation, while high concentrations inhibited this process

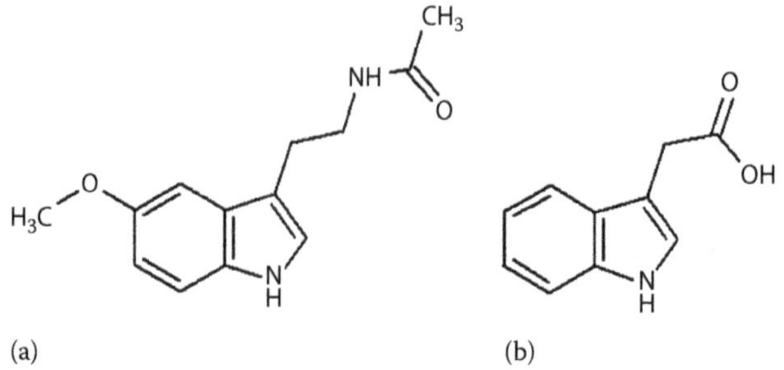

(a)   (b)

FIGURE 5.3 The structures of (a) MEL and (b) IAA.

(Sarropoulou et al. 2012b). A similar effect was observed in the roots of mustard (*Brassica juncea* [L.] Czern.) after exogenous MEL application (Chen et al. 2009), which was accompanied by an increase in IAA level. Still, it is unclear whether MEL induces auxin biosynthesis, or whether it may be metabolized and converted to IAA, or whether by structural analogy to IAA MEL exerts auxin-like effects. It was also shown that MEL, in a dose-dependent manner, stimulated elongation of hypocotyls (Hernández-Ruiz et al. 2004) and cotyledons of lupin (*Lupinus albus* L.) (Hernández-Ruiz and Arnao 2008), similarly to IAA. Moreover, in some monocotyledonous species MEL also stimulated the coleoptile growth and cell wall acidification, just as IAA did. (Hernández-Ruiz et al. 2005). So it seems that both these compounds may participate in plant growth regulation (see Figure 5.1), although recent studies of Pelagio-Flores et al. (2012) on *Arabidopsis thaliana* seedlings indicated that the effects of MEL and IAA might be regulated by different mechanisms. Detailed analysis of morphological parameters showed that MEL modulated root system architecture by stimulating lateral and adventitious root formation but minimally affected primary root growth or hairy root development. Studies on transgenic *A. thaliana* lines indicated that root developmental changes elicited by MEL were independent of auxin signaling. Interestingly, the exogenous MEL stimulated lateral root formation at concentrations of 100 μM or higher, which are much greater than those characteristic of IAA or other auxins (Woodward and Bartel 2005). So it seems that although IAA and MEL regulate lateral root formation, the mechanisms of their action may differ slightly. These results are in agreement with findings that auxin-inducible DR5:GUS reporter gene cannot be induced by MEL, and the function of MEL is independent of auxin (Pelagio-Flores et al. 2012). Although MEL stimulated growth of adventitious roots this process was not influenced by the increase in the expression of the gene mentioned above. Transcriptome analysis of *A. thaliana* did not show changes in the expression of genes associated with the biosynthesis of auxins under influence of MEL, which may confirm their independent actions (Weeda et al. 2014). The dose-dependent effects of MEL and IAA on mustard root growth were similar—low doses increased root growth, while high doses inhibited it (Chen et al. 2009). MEL at low concentration (0.1 μM) had a positive impact on root elongation of 2-day-old seedlings, which was accompanied by an increase in the endogenous IAA. Therefore, it is possible that the final effect was actually caused by IAA, whose biosynthesis was stimulated by MEL. However, at high MEL concentrations a significant increase in the level of IAA was not observed, while the root elongation was strongly inhibited. Thus, the inhibitory effect of MEL on root growth seems to involve mechanisms independent of IAA signaling, but the specific relationship between these two compounds remain unknown. The results concerning transgenic rice seedlings with increased biosynthesis of MEL (Park and Back 2012) confirmed that the participation of MEL in auxin biosynthesis or auxin signaling pathways was unlikely (Pelagio-Flores et al. 2012). Although it is known that auxins are involved in the formation and growth of adventitious roots (Hochholdinger and Zimmermann 2008), in these seedlings they were not stimulated, and only the main root development was observed. Therefore, further research is needed to confirm either synergistic or independent effects of MEL and auxins. Moreover, exogenously administered calcium and calcium ionophore A23187 enhanced *in vitro* shoot multiplication and root induction in cultures of *M. pudica* (Ramakrishna et al. 2009). In addition, exogenous supplementation with 100 μM SER/MEL, induced somatic embryos from each callus with concomitant enhancement of endogenous pools of serotonin (SER) and IAA (Ramakrishna et al. 2012).

MEL as a Circadian and Seasonal Oscillator in Plants

MEL occurring in animals is known as a dark hormone because of its highest content at night, and hardly detectable level during the day (in most species). A similar scheme of MEL synthesis was found not only in certain flagellates but also in some photosynthetic plants (see Figure 5.1) (Balzer and Hardeland 1991; Poeggeler et al. 1991). Poeggeler et al. (1991) suggested participation of MEL in circadian cycle regulation in unicellular photosynthesizing dinoflagellate *Lingulodinium polyedrum*. Bioluminescence observed in this organism is a classic example of diurnal changes because it emits strong light at night, while daylight inhibits this process. It is associated with a high MEL content at night and low during the day. A similar rhythm of MEL occurrence was reported in short-day plants (*Chenopodium rubrum*) (Wolf et al. 2001), while in tomato daily changes in MEL content were not observed (Van Tassel et al. 2001). Moreover, the endogenous pool of serotonin seems to vary under changes in photoperiod. In particular, in unicellular green alga *Dunaliella bardawil*, maximum biomass and carotenoid contents were found when cultures were grown in light (intensity of 2.0 Klux) at a photoperiod of 16:8 h light:dark cycle (Ramakrishna et al. 2011).

The research of Tan et al. (2007) on the water hyacinth revealed the highest content of MEL and AFMK just before sunset, which might confirm that MEL synthesis depends on the intensity of light. This differentiates plants from animals, where light suppresses melatonin production in pineal glands (Murch et al. 2000). It seems that MEL content in plants is regulated not only by its biosynthesis rate but also by the intensity of its conversion or degradation. There were reports on two maxima of MEL content (approx. hrs. 5:00 and 14:00), in mature cherry fruits (Zhao et al. 2013), or in reddish quinoa grown in artificially controlled light/dark cycle (Kolář et al. 1997). Probably, the first maximum of MEL content is induced by the rate of its synthesis or limitation of its consumption in the dark, and the second by additional stimulation of its biosynthesis by ROS accumulated as a result of high temperature, light intensity, and UV. However, in grape fruits (*Vitis vinifera* cv Malbec.) the maximum of MEL concentration observed in early morning could be related to the increased temperature at sunrise following the night cold stress (Boccalandro et al. 2011).

An interesting phenomenon associated with the biological clock was observed in green macroalgae *Ulva* sp., where increasing MEL content was correlated with expected spring tides. This phenomenon can be interpreted as the preparation of these organisms to the expected outflows during which they are vulnerable to rising temperatures, drought, and salinity stresses, inducing oxidative stress. Interestingly, these algae showed the characteristic rhythm of the moon even without being exposed to real modifications in day length or water level changes during experiments (Tal et al. 2011).

Thus, although the correlation of rhythmic MEL content changes with the circadian rhythm depending on the photoperiod was confirmed in some plants, this phenomenon seems to be more complex, and susceptible to the influence of additional factors. In general, rhythmic changes in MEL content in plants, as in other organisms, may be evolutionarily related to the photoperiod. MEL synthesized and stored in the dark is consumed in the light in order to reduce photo-oxidative damage, so its level during the day falls. It seems that repeated rhythmic fluctuations associated with the synthesis and utilization of MEL could be used by organisms to biochemically measure duration of the day, and consequently to determine changes of the seasons. However, it should be taken into account that during evolution plants developed a number of other photoreceptors (phytochrome, cryptochrome, light-dependent photosynthetic apparatus), which effectively regulate the processes of morphogenesis.

It is worth mentioning that research focused on changes in MEL content in dry cucumber and corn seeds stored for one year (Kołodziejczyk et al. 2015) showed seasonal changes in this indoleamine concentration. Despite constant storage conditions (tightly sealed plastic bags, darkness, room temperature) during winter months in both species a significant increase in MEL content was recorded. These seasonal changes may suggest the existence of an endogenous circannual rhythm regulated or set by MEL—independently of environmental conditions. According to the definition, endogenous rhythm is the term describing biological processes that alter periodically although external conditions remain constant.

Until now, the possible role of MEL as a regulator of light/dark cycles in plants has been studied, pointing to its possible role as a regulator of light/dark-dependent processes. The changes mentioned above suggest the existence of some sort of independent "chemical memory" in plants that is not suscep-tible to extrinsic factors. Seasonal fluctuations of MEL content in seeds seem to indicate the existence of an endogenous biological clock which functions as a biochemical calendar (probably evolutionarily developed in cold- and temperate-climate species) precisely regulated by this indoleamine. However, to confirm this theory further observations and investigations are needed.

The information contained in this chapter clearly demonstrate that endogenous MEL influences the growth and development of plants in many different ways as depicted in Figure 5.1.

Conclusion

Due to the antioxidant potential of MEL, it may be concluded that it could impact the photosynthetic apparatus, influence plant growth and development, and also act as a circadian and seasonal oscillator.

REFERENCES

Arnao, M. B. Phytomelatonin: Discovery, content, and role in plants. *Adv Bot.* 2014. Article ID 815769, doi:10.1155/2014/815769.

Arnao, M. B., and Hernandez-Ruiz, J. Melatonin promotes adventitious- and lateral root regeneration in etio-lated hypocotyls of *Lupinus albus* L. *J Pineal Res.* 2007. 42: 147–152.

Arnao, M. B., and Hernandez–Ruiz, J. Protective effect of melatonin against chlorophyll degradation during senescence of barley leaves. *J Pineal Res.* 2009. 46: 58–63.

Balzer, I., and Hardeland, R. Photoperiodism and effects of indoleamines in a unicellular alga. *Gonyaulax polyedra. Science.* 1991. 253: 795–797.

Boccalandro, H. E., González, C. V., Wuderlin, D. A, and Silva, M. F. Melatonin levels, determined by LC-ESI-MS/MS, fluctuate during the day/night cycle in *Vitis vinifera* cv Malbec: Evidence of its antioxi-dant role in fruits. *J Pineal Res.* 2011. 51: 226–232.

Bonnefont-Rousselot, D., and Collin, F. Melatonin: Action as antioxidant and potential applications in human disease and aging. *Toxicology.* 2010. 278: 55–67.

Chen, Q., Qi, W. B., Reiter, R. J., Wei, W., and Wang, B. M. Exogenously applied melatonin stimulates root growth and raises endogenous indoleacetic acid in roots of etiolated seedlings of *Brassica juncea. J Plant Physiol.* 2009. 166: 324–328.

Dubbels, R., Reiter, R. J., Klenke, E., Goebel, A., Schnakenberg, E., Ehlers, C., Schiwara, H. W., and Schloot, W. Melatonin in edible plants identified by radioimmunoassay and by high performance liquid chroma-tography–mass spectrometry. *J Pineal Res.* 1995. 18: 28–31.

Galano, A., Tan, D. X., and Reiter, R. J. On the free radical scavenging activities of melatonin's metabolites, AFMK and AMK. *J Pineal Res.* 2013. 54: 245–257.

Gitto, E., Tan, D. X., Reiter, R. J., Karbownik, M., Manchester, L. C., Cuzzocrea, S., Fulia, F., and Barberi, I. Individual and synergistic antioxidative actions of melatonin: Studies with vitamin E, vitamin C, gluta-thione and desferrioxamine (desferoxamine) in rat river homogenates. *J Pharm Pharmacol.* 2001. 53: 1393–1401.

Guo, Y. F., and Gan, S. S. Leaf senescence: Signals, execution, and regulation. *Curr Top Dev Biol.* 2005. 71: 83–112.

Hernández-Ruiz, J., and Arnao, M. B. Melatonin stimulates the expansion of etiolated lupin cotyledons. *Plant Growth Regul.* 2008. 55: 29–34.

Hernández-Ruiz, J., Cano, A., and Arnao, M. B. Melatonin: A growth-stimulating compound present in lupin tissues. *Planta.* 2004. 220 (1): 140–144.

Hernández-Ruiz, J., Cano, A., and Arnao, M. B. Melatonin acts as a growth-stimulating compound in some monocot species. *J Pineal Res.* 2005. 39: 137–142.

Hochholdinger, F., and Zimmermann, R. Conserved and diverse mechanisms in root development. *Curr Opin Plant Biol.* 2008. 11: 70–74.

Janas, K. M., and Posmyk, M. M. Melatonin, an underestimated natural substance with great potential for agricultural application. *Acta Physiol Plant.* 2013. 35: 3285–3292.

Kolář, J., Macháčková, I., Eder, J., Prinsen, E., Van Dongen, W., Van Onckelen, H., and Illnerová, H. Melatonin: Occurrence and daily rhythm in *Chenopodium rubrum. Phytochem.* 1997. 44: 1407–1413.

Kołodziejczyk, I., Bałabusta, M., Szewczyk, R., and Posmyk, M. M. The levels of melatonin and its metabolites in conditioned corn (*Zea mays* L.) and cucumber (*Cucumis sativus* L.) seeds during storage. *Acta Physiol Plant.* 2015. 37: 105.

Lazár, D., Murch, S. J., Beilby, M. J., and Khazaaly, S. Exogenous melatonin affects photosynthesis in *Characeae Chara australis. Plant Signal Behav.* 2013. 8 (3): E23279.

Lei, X. Y., Zhu, R. Y., Zhang, G. Y., and Dai, Y. R. Attenuation of cold-induced apoptosis by exogenous melatonin in carrot suspension cells: The possible involvement of polyamines. *J Pineal Res.* 2004. 36: 126–131.

Li, C., Tan, D. X., Liang, D., Chang, C., Jia, D., and Ma, F. Melatonin mediates the regulation of ABA metabolism, free radical scavenging, and stomatal behavior in two *Malus* species under drought stress. *J Exp Bot.* 2015. 66: 669–680.

Liang, C., Zheng, G., Li, W., Wang, Y., Wu, H., Qian, Y., Zhu, X. G., Tan, D. X., Chen, S. Y., and Chu, C. Melatonin delays leaf senescence and enhances salt stress tolerance in rice. *J Pineal Res.* 2015. 59 (1): 91–101.

Meng, J. F., Xu, T. F., Wang, Z. Z., Fang, Y. L., Xi, Z. M., and Zhang, Z. W. The ameliorative effects of exogenous melatonin on grape cuttings under water-deficient stress: Antioxidant metabolites, leaf anatomy, and chloroplast morphology. *J Pineal Res.* 2014. 57: 200–212.

Murch, S. J., Campbell, S. S. B., and Saxena, P. K. The role of serotonin and melatonin in plant morphogenesis: Regulation of auxin-induced root organogenesis in *in vitro*-cultured explants of St. John's wort (*Hypericum perforatum* L.). *In Vitro Cell Dev Biol.* 2001. 37: 786–793.

Murch, S. J., Krishna, R. S., and Saxena, P. K. Tryptophan is a precursor for melatonin and serotonin biosynthesis in *in vitro* regenerated St. John's wort (*Hypericum perforatum* L. cv. Anthos) plants. *Plant Cell Rep.* 2000. 19: 698–704.

Park, S., and Back, K. Melatonin promotes seminal root elongation and root growth in transgenic rice after germination. *J Pineal Res.* 2012. 53: 385–385.

Park, S., Lee, D. E., Jang, H., Byeon, Y., Kim, Y. S., and Back, K. Melatonin-rich transgenic rice plants exhibit resistance to herbicide-induced oxidative stress. *J Pineal Res.* 2013. 54: 258–263.

Pelagio-Flores, R., Munoz-Parra, E., Ortiz-Castro, R., and López-Bucio, J. Melatonin regulates *Arabidopsis* root system architecture likely acting independently of auxin signaling. *J Pineal Res.* 2012. 53: 279–288.

Pinheiro, C., and Chaves, M. M. Photosynthesis and drought: Can we make metabolic connections from available data? *J Exp Bot.* 2011. 62: 869–882.

Poeggeler, B., Balzer, I., Hardeland, R., and Lerchi, A. Pineal hormone melatonin oscillates also in the dinoflagellate *Gonyaulax polyedra. Naturwissenschaften.* 1991. 78: 268–269.

Ramakrishna, A. Indoleamines in edible plants: Role in human health effects. In Angel Catalá, Biochemistry Research Trends Series (editor), *Indoleamines: Sources, Role in Biological Processes and Health Effects.* 2015. 279. New York: Nova Publishers.

Ramakrishna, A., Dayananda, C., Giridhar, P., Rajasekaran, T., and Ravishankar, G. A. Photoperiod influences endogenous indoleamines in cultured green alga *Dunaliella bardawil. Indian J Exp Biol.* 2011. 49: 234–240.

Ramakrishna, A., Giridhar, P., Jobin, M., Paulose, C. S., and Ravishankar, G. A. Indoleamines and calcium enhance somatic embryogenesis in cultured tissues of *Coffea canephora* P ex Fr. *Plant Cell Tiss Org.* 2012. 108: 267–278.

Ramakrishna, A., Giridhar, P., and Ravishankar, G. A. Indoleamines and calcium channels influence morphogenesis in *in vitro* cultures of *Mimosa pudica* L. *Plant Signal Behav.* 2009. 4: 1136–1141.

Ramakrishna, A., Giridhar, P., Sankar, K. U, and Ravishankar, G. A. Melatonin and serotonin profiles in beans of *Coffea* species. *J. Pineal Res.* 2012a. 52: 470–476.

Ramakrishna, A., Giridhar, P., Sankar, K. U, and Ravishankar, G. A. Endogenous profiles of indoleamines: serotonin and melatonin in different tissues of *Coffea canephora* P ex Fr. as analyzed by HPLC and LC-MS-ESI. *Acta Physiol Plant.* 2012b. 34: 393–396.

Rosen, J., Than, N. N., Koch, D., Poeggeler, B., Laatsch, H., and Hardeland, R. Interactions of melatonin and its metabolites with the ABTS cation radical: Extension of the radical scavenger cascade and formation of a novel class of oxidation products, C2-substituted 3-indolinones. *J Pineal Res.* 2006. 41: 374–381.

Sarropoulou, V., Dimassi-Theriou, K., Therios, I., and Koukourikou-Petridou, M. Melatonin enhances root regeneration, photosynthetic pigments, biomass, total carbohydrates and proline content in the cherry rootstock PHL-C *(Prunus avium* × *Prunus cerasus). Plant Physiol Biochem.* 2012a. 61: 162–168.

Sarropoulou, V. N., Therios, I. N., and Dimassi-Theriou, K. N. Melatonin promotes adventitious root regeneration in *in vitro* shoot tip explants of the commercial sweet cherry rootstocks CAB-6P (*Prunus cerasus* L.), Gisela 6 (*P. cerasus* × *P. canescens*), and MxM 60 (*P. avium* × *P. mahaleb*). *J Pineal Res.* 2012b. 52: 38–46.

Shi, H., Reiter, R. J., Tan, D. X., and Chan, Z. Indole-3-acetic acid–inducible 17 positively modulates natural leaf senescence through melatonin-mediated pathway in *Arabidopsis. J Pineal Res.* 2015. 58: 26–33.

Szafrańska, K., Glińska, S., and Janas, K. M. Changes in the nature of phenolic deposits after re-warming as a result of melatonin pre-sowing treatment of *Vigna radiata* seeds. *J Plant Physiol.* 2012. 169: 34–40.

Szafrańska, K., Glińska, S., and Janas, K. M. Ameliorative effect of melatonin on meristematic cells of chilled and re-warmed *Vigna radiata* roots. *Biol Plant.* 2013. 57 (1): 91–96.

Tal, O., Haim, A., Harel, O., and Gerchman, Y. Melatonin as an antioxidant and its semi-lunar rhythm in green macroalga *Ulva* sp. *J Exp Bot.* 2011. 62: 1903–1911.

Tan, D. X., Hardeland, R., Manchester, L. C., Korkmaz, A., Ma, S., Rosales-Corral, S., and Reiter, R. J. Functional roles of melatonin in plants, and perspectives in nutritional and agricultural science. *J Exp Bot.* 2012. 63 (2): 577–597.

Tan, D. X., Manchester, L. C., Di Mascio, P., Martinez, G. R., Prado, F. M., and Reiter, R. J. Novel rhythms of N1-acetyl-N2-formyl-5-methoxykynuramine and its precursor melatonin in water hyacinth: Importance for phytoremediation. *FASEB J.* 2007. 21: 1724–1729.

Tan, D. X., Manchester, L. C., Liu, X., Rosales-Corral, S. A., Acuna-Castroviejo, D., and Reiter, R. J. Mitochondria and chloroplasts as the original sites of melatonin synthesis: A hypothesis related to melatonin's primary function and evolution in eukaryotes. *J Pineal Res.* 2013. 54: 127–138.

Turk, H., Erdal, S., Genisel, M., Atici, O., Demir, Y., and Yanmis, D. The regulatory effect of melatonin on physiological, biochemical and molecular parameters in cold-stressed wheat seedling. *Plant Growth Regul.* 2014. 74 (2): 139–152.

Van Tassel, D. L., Roberts, N., Lewy, A., and O'Neill, S. D. Melatonin in plant organs. *J Pineal Res* 2001. 31: 8–15.

Wang, P., Sun, X., Li, C. H., Wei, Z., Liang, D., and Ma, F. Long-term exogenous application of melatonin delays drought-induced leaf senescence in apple. *J Pineal Res.* 2013a. 54: 292–302.

Wang, P., Sun, X., Chang, C., Feng, F., Liang, D., Cheng, L., and Ma, F. Delay in leaf senescence of *Malus hupehensis* by long-term melatonin application is associated with its regulation of metabolic status and protein degradation. *J Pineal Res.* 2013b. 55: 424–434.

Wang, P., Sun, X., Xie, Y., Li, M., Chen, W., Zhang, S., Liang, D., and Ma, F. Melatonin regulates proteomic changes during leaf senescence in *Malus hupehensis. J Pineal Res.* 2014. 57: 291–307.

Wang, P., Sun, X., Wang, N., Tan, D. X., and Ma, F. Melatonin enhances the occurrence of autophagy induced by oxidative stress in *Arabidopsis* seedlings. *J Pineal Res.* 2015. 58: 479–489.

Wang, P., Yin, L., Liang, D., Li, C. H., Ma, F., and Yue, Z. Delayed senescence of apple leaves by exogenous melatonin treatment: Toward regulating the ascorbate–glutathione cycle. *J Pineal Res.* 2012. 53: 11–20.

Weeda, S., Zhang, N., Zhao, X., Ndip, G., Guo, Y., Buck, G. A., Fu, C., and Ren, S. *Arabidopsis* transcriptome analysis reveals key roles of melatonin in plant defense systems. *PLOS ONE.* 2014. 9 (3): E93462.

Wei, W., Li, Q. T., Chu, Y. N., Reiter, R. J., Yu, X. M., Zhu, D. H., Zhang, W. K., et al. Melatonin enhances plant growth and abiotic stress tolerance in soybean plants. *J Exp Bot.* 2015. 66 (3): 695–707.

Wolf, K., Kolář, J., Witters, E., Van Dongen, W., Van Onckelen, H., and Macháčková, I. Daily profile of melatonin levels in *Chenopodium rubrum L.* depends on photoperiod. *J Plant Physiol.* 2001. 158: 1491–1493.

Woodward, A. W., and Bartel, B. Auxin: Regulation, action, and interaction. *Ann Bot.* 2005. 95: 707–735.

Zhao, Y., Tan, D. X., Lei, Q., Chen, H., Wang, L., Li, Q., Gao, Y., and Kong, J. Melatonin and its potential biological functions in the fruits of sweet cherry. *J Pineal Res.* 2013. 55: 79–88.

6

Serotonin and Melatonin as Metabolic Signatures for the Modulation of Seed Development, Seedling Growth, and Stress Acclimatization

Soumya Mukherjee, Dhara Arora, and Satish Chander Bhatla
University of Delhi
Delhi, India

CONTENTS

ABSTRACT   Seed development involves highly coordinated signaling pathways associated with desiccation phase, accumulation and mechanisms of scavenging of reactive oxygen species (ROS), and differential activity of a variety of growth regulators. Seed germination and seedling establishment involves mobilization of reserves accompanied with long-distance signaling of biomolecules from roots to cotyledons. Serotonin and melatonin, two major indoleamines in plants, exhibit a crucial role in the plethora of events involved in these biological processes. The serotonin and auxin biosynthetic pathway is regulated by tryptophan metabolism in developing seeds, thus affecting serotonin accumulation in seedlings. Various pharmacological treatments have demonstrated a concentration-dependent regulation of seedling growth by exogenous application of serotonin and melatonin. Abiotic stress acclimatization in seedlings has been observed to be modulated by endogenous and exogenous levels of these indoleamines. Melatonin structurally consists of free radical scavenging sites and regulates metabolomic changes associated with hormones and antioxidant enzymes during seedling growth. Abiotic stress–induced modulation of melatonin accumulation by hydroxyindole-*O*-methyltrasferase (HIMOT) activity during seedling growth is likely to be effective in providing stress tolerance in various plant systems. Oilseed crops are unique model systems to decipher lipolytic events associated with seed germination and seedling growth. Such events, under the impact of various environmental cues, are positively modulated by these indoleamines, thus imparting abiotic stress tolerance and longevity to young seedlings. Genetic manipulation of indoleamine biosynthetic pathway is likely to provide new information on the mechanisms of stress management in oilseed crops. Biotechnological implications of these indoleamines have been suggested with reference to crop management in response to adverse effects of biotic and abiotic stress.

Identification and characterization of serotonin and melatonin receptors in plants will help in deciphering their mode of action and signaling cascade.

KEY WORDS: *serotonin, melatonin, seed development, seed germination, oilseed, oleosin.*

Abbreviations

ABA: abscisic acid
GA: gibberellic acid
HIMOT: hydroxyindole-O-methyltrasferase
NaCl: sodium chloride
OB: oil body
PAL: L-phenylalanine ammonia-lyase
ROS: reactive oxygen species
SNAT: serotonin N-acetyltransferase
WT: wild type

Introduction

Seed development and seedling growth involve a set of complex and dynamic events of hormonal crosstalk leading to tissue differentiation. Seed development begins with the process of cell proliferation followed by seed filling and seed desiccation. Polarized growth of embryo involves an interplay of hormonal gradients. The seed filling stage involves maximum accumulation of proteins, carbohydrates and lipid reserves. The desiccation stage involves a slowing-down of metabolism followed by the activation of antioxidant defense mechanisms that facilitate the survival of seed in a dry and physiologically inactive state. The nutritive value of seeds is mostly due to protein, lipids and carbohydrates. Oilseed crops contain an abundance of unsaturated and saturated fatty acids. Oil content and its quality are of prime importance for seed productivity. Seed germination and seedling growth are characterized with resumption of enzyme activity and enhanced metabolism for mobilization of food reserves.

Physiological aspects of seed development and seedling growth involve a conserved set of metabolic signatures associated with early and late signaling responses brought about by various biomolecules. Serotonin and melatonin are the two essential indoleamines in plants which bring about a plethora of effects on growth, morphogenesis and differentiation of plant organs (Murch et al. 2001; Park and Back 2012; Ramakrishna et al. 2009; Ramakrishna et al. 2012). Various investigations have reported the presence of these biomolecules in seeds and various regions of young seedlings. A crosstalk between auxins and these indoleamines has been established in *Arabidopsis* and *Brassica* (Chen et al. 2009; Pelagio-Flores et al. 2011). This chapter reviews various physiological and biomolecular aspects associated with serotonin- and melatonin-modulated regulation of seed development and seedling growth. An interplay among generation of ROS and potent antioxidant defense mechanisms provides a multidimensional route to signaling and crosstalk during seed development and seedling growth. Moreover, the occurrence of serotonin and melatonin in edible plants and processed foods has been reported (Tan et al. 2012; Janas and Posmyk 2013; Ramakrishna et al. 2012a,b; Ramakrishna 2015). In this chapter, these aspects have been analyzed with respect to serotonin and melatonin in relation to seed development, seedling growth, establishment, photomodulation, and abiotic stress tolerance. In this context, work on oilseed crops has been highlighted with special reference to their economic importance.

Serotonin and Melatonin Accumulation Accompanying Seed Development

Seed development involves coordination among various physiological and biomolecular signaling events associated with protein modifications, metabolism, and detoxification mechanisms. Seed filling and desiccation stages involve protein accumulation, their posttranslational modifications, and activation of antioxidant systems (Thakur and Bhatla 2014). Recalcitrant seeds are mostly deficient in antioxidant enzymes, namely peroxidase, catalase, and superoxide dismutase, which regulate the pool of hydrogen peroxide and other reactive species produced as a by-product of various primary metabolic pathways. Among several biomolecules, serotonin and melatonin are two indoleamines reported to regulate seed development. Presence of serotonin has earlier been listed among various secondary metabolites, such as tannins, alkaloids and phenols, which are present in the fruits and seeds of plants species, and are possibly associated with defense mechanisms (Grobe 1982). Reports from various plant species have suggested the presence of these amines and their modulation with ontogenic stages of seed development. Relative serotonin content is as high as 2 mg.gm^{-1} FW in the seeds of *Griffonia simplicifolia* in comparison with its leaves (Fellows and Bell 1971). High levels of serotonin in *Mimosa pudica* seeds suggest its possible role in maintaining seed viability and facilitating seed germination. Safflower seeds contain serotonin derivatives (feruloyl-serotonin and coumaryl-serotonin) possessing anti-inflammatory, anti-bacterial, and anti-stress properties. In the post-pollination phases of fruit development, which are followed by fruit abscission, the seeds cease to undertake any exchange of metabolites from the plant body. The development of cotyledons, which serve as the major storage organs for the future seedling, undergo accumulation of proteins and lipid bodies. Presence of serotonin in protein bodies has been reported to be associated with the loss of vacuolation in mature walnut (*Juglans regia*) seeds. It involves temporal differences in serotonin accumulation at various stages of seed development (Lembeck and Skofitsch 1984). The embryo accumulates a high amount of serotonin. The seed coat, however, contains the highest serotonin levels among all the seed components.

Regulation of tryptophan metabolism appears to be a major factor for modulating serotonin accumulation during seed development. Shikimate pathway involves the biosynthesis of tryptophan from anthranilate. Anthranilate synthase possesses a tryptophan-sensitive feedback inhibition domain (Taiz et al. 2015). Mature seeds lack vacuoles and exhibit reduced glutamine synthetase activity. This results in the accumulation of ammonia as a result of less glutamine formation. Increased serotonin accumulation in walnut seeds shows spatial distribution, which prevents the inhibition of shikimate pathway, thus resulting in less ammonia accumulation in the tissue. Serotonin evokes morphogenic responses similar to auxin, and shows higher accumulation in fruits and seeds in comparison to auxin. This suggests its protective role in the developing embryos and associated seed components. Serotonin derivatives, namely N-(trans-p-coumaroyl)-serotonin and N-(trans-feruloyl)- serotonin, extracted from the seeds of *Centaurea nigra,* possess antibacterial and free radical scavenging properties (Kumarasamy et al. 2003). Melatonin is biosynthesized from serotonin further downstream in the biosynthetic pathway. Its levels in the seeds of various plants suggest its protective role during seed development by protection of the germ tissue (Manchester et al. 2000; Murch et al. 2009; Garcia-Parrilla et al. 2009; Ramakrishma et al. 2012a,b). Developing seeds at desiccation stage have been reported to be better protected by melatonin application (Janas and Posmyk 2013).

Modulation of Seed Germination and Seedling Growth by Serotonin and Melatonin

Seed germination and seedling growth are affected by various environmental and physiological factors that involve hormonal regulation, molecular crosstalk, and a variety of biochemical signaling events. Radicle emergence from the seeds is followed by polarized growth and morphogenesis of seedlings, and subsequent mobilization of food reserves from the cotyledons. Polar growth of seedlings is regulated by acropetal and basipetal gradients of auxin distribution. Serotonin and melatonin, (biosynthesized from tryptophan) share various functional similarities with auxin. Modulation of the endogenous levels of these molecules affects auxin biosynthesis and its activity. Seedlings provide a unique system to investigate the early signaling events associated with these indoleamines.

Serotonin and melatonin act as growth regulators and activate seed germination and rhizogenesis. High serotonin levels are reported in the seeds of *Malus pudica* L. Serotonin stimulates seed germination in radish, and growth of roots and coleoptile hook in oats (Ramakrishna et al. 2011). At concentrations as low as 10^{-8} M, it significantly enhances the longitudinal growth of barley roots in seedlings but does not have much influence on the bulk growth of roots (Csaba and Pal 1982). Dual effects of exogenous serotonin application have been observed in *Arabidopsis thaliana* seedlings whereby lower concentrations (10 to 160 mM) stimulate lateral root growth while higher concentrations are inhibitory to lateral root growth, primary root growth, and root hair development. The repressing effect of serotonin on primary root growth has been attributed to its effect on cell division and elongation. Though serotonin enhances lateral root growth, *de novo* lateral root primordia initiation due to serotonin has not been observed. Probably, serotonin induces the maturation of preformed lateral root primordia from the pericycle cells, and also enhances root branching (Pelagio-Flores et al. 2011).

The growth-promoting effect of melatonin has been reported in various seedling tissues. Lupin cotyledons exhibit expansion on application of exogenous melatonin (Hernandez-Ruiz et al. 2004). Similarly, melatonin (10^{-7} and 10^{-6} M) stimulates coleoptile growth in *Phalaris canariensis* and *Hordeum vulgare* (Hernandez-Ruiz et al. 2005). Wei et al. (2014) have reported melatonin-induced seedling growth in soybean. Melatonin (50 μM)-treated seedlings are relatively healthy, and treatment also leads to faster development of trifoliates. Developmental stage–dependent modulation of seedling growth by melatonin has been observed in 2-day-old etiolated seedlings of wild leaf mustard, where melatonin (0.1 mM) stimulates root growth, while at 100 mM it is inhibitory. Four-day-old seedlings are less susceptible to the growth stimulatory and inhibitory effects of melatonin. These dosage-dependent differential responses to exogenous melatonin concentrations may be related to marked differences in endogenous melatonin concentrations (Chen et al. 2009). Promotory effect of exogenous melatonin on primary root growth and lateral root formation has been observed in various plants. Treatment of 10^{-8}–10^{-4} M melatonin to seedlings leads to increase in root growth in *Phalaris canariensis* and *Avena sativa*, while dosage-dependent response has been observed in *Triticum aestivum*, where higher concentrations (10^{-6}–10^{-4} M) of melatonin are inhibitory (Hernandez-Ruiz et al. 2005). In liquid cultures of 7-day-old seedlings of *Arabidopsis thaliana*, an increase in lateral root number is observed upon addition of 50–500 μM melatonin (Koyama et al. 2013). A twofold increase in the growth of seminal roots of transgenic rice seedlings is also evident following overexpression of serotonin *N*-acetyltransferase (SNAT, the penultimate enzyme of the melatonin biosynthetic pathway). Also, overall seedling growth is accelerated with enhanced biomass (twofold) with respect to the wild type (WT) (Arnao and Hernandez-Ruiz 2014; Byeon and Back 2014a,b). A possible association of melatonin with auxin and phytohormones with respect to root development has been proposed in rice seedlings (Park and Back 2012).

When sprayed at a concentration of 25–100 μM, melatonin enhances net photosynthetic rate in cucumber seedlings (Tan et al. 2012). Melatonin levels are not altered in *Pharbitis nil* seedlings by changing the light/dark photoperiod. (Tassel et al. 2001). The steady levels of melatonin are, however, inhibited by light in rice seedlings. This inhibitory effect of light may be due to oxidative stress or photochemical destruction during the day, leading to enhanced consumption of melatonin (Byeon and Back 2014a,b). Priming of negatively photoblastic and thermosensitive seeds of *Phacelia tanacetifolia* with melatonin leads to elevation of germination response. Application of 6 μM melatonin enhances germination response from 2.5% to 52% in the dark, and 1 μM melatonin results in highest response (21%) in the presence of light at 30°C (Tiryaki and Keles 2012). Melatonin-induced changes in membrane and protein peroxidation are suggested to influence the role of melatonin in seed germination (Zhang et al. 2013).

Exogenous applications of serotonin and melatonin have thus been found to bring about differential, development stage–dependent, and species-specific response with respect to germination response, seedling growth, and lateral root formation (Figure 6.1). Various mechanisms by which these molecules manifest the said responses are yet to be analyzed. Nevertheless, it is evident from the currently available information that these indoleamines and their photomodulatory effects play a crucial regulatory role in seedling growth and establishment.

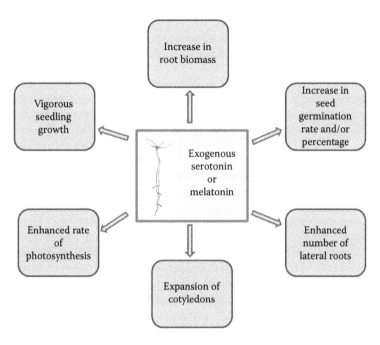

FIGURE 6.1 Effects of exogenous serotonin or melatonin application on seedling growth.

Physiological Implications of Serotonin and Melatonin in Oilseed Crops with Relation to Lipid Mobilization

Serotonin and melatonin contents are of prime importance in edible seeds, which form a part of our daily diet. However, the protective roles of these indoleamines with reference to oilseed crops have been an emerging field of research in the recent past. The oilseed crops of sunflower, brassica, soybean, and so on, are cultivated worldwide. Seedling growth and establishment of oilseed crops is important with regard to agricultural productivity measured in terms of seed oil content and oil quality (saturated and unsaturated fatty acids). Lipid metabolism and oil body (OB) mobilization lead to the liberation of primary storage lipids (di- and tri-acylglycerols––DAGs and TAGs), which are the major sources of energy for growing oil crop seedlings (David 2012). OB mobilization is brought about by the precise activity of lipases and several proteases involved in OB membrane lipid and protein hydrolysis, respectively. Serotonin and melatonin have been reported to be associated with oilseed sprouts (Hardeland et al. 2007; Cho et al. 2008). The potent antioxidative properties of these biomolecules are effective in the scavenging of free radicals generated as by-products of various primary metabolic pathways operative in seeds and cotyledons. *Brassica juncea* seedlings show alterations in the levels of endogenous auxins upon melatonin treatment at 0.1 μM (Chen et al. 2009). This concentration of exogenous melatonin is also stimulatory for root growth in the young seedlings of Brassica, two days after seed germination. The stimulatory effect of melatonin is, however, less at later stages of seedling growth. Other investigations have shown relative melatonin content in two *Brassica* species to be higher than in other plants, such as grasses and oats (Manchester et al. 2000; Hernandez-Ruiz et al. 2005). The physiological significance of an interrelationship between melatonin and auxin levels has earlier been reported by Hernandez-Ruiz et al. (2004, 2005). High concentrations of exogenous melatonin inhibit root growth in Brassica seedlings, and this has been attributed to the depletion of endogenously produced ROS (Chen et al. 2009). This explains the possibility of a regulatory role of melatonin in Brassica seedlings. Melatonin-induced improvement in abiotic stress tolerance in soybean plants is evident in seeds coated with melatonin (Wei et al. 2014). Molecular analysis of melatonin-induced gene induction has revealed enhanced expression of genes associated with fatty acid biosynthesis (Wei et al. 2014). This suggests the potential effect of melatonin in improving plant yield in soybean to be due to an increase in fatty acid content. Development of salt tolerance in soybean is, thus, of prime importance for its use as an oil- and protein-yielding crop in the tropics.

Endogenous serotonin and melatonin levels in sunflower seedlings have been observed to be regulated by sodium chloride (NaCl) stress (120 mM). Two-day-old sunflower seedling cotyledons exhibit significant enhancement in melatonin content as an effect of NaCl stress (Mukherjee et al. 2014). Serotonin and melatonin show differential distribution in seedling roots and cotyledons. Seedling cotyledons at 2-, 4-, and 6-day stages of germination show presence of these biomolecules associated with OB-containing cells (Mukherjee et al. 2014). NaCl stress induced slower OB mobilization manifested by greater retention of oleosin (OB membrane protein), as has also been reported in sunflower seedlings (David et al. 2010). Salinity-induced alterations in oleic and linoleic acid contents have also been reported (Di Caterina et al. 2007). Exogenous serotonin and melatonin promote seedling growth in sunflower. An interrelationship between stress-induced alterations in the rate of auxin biosynthesis and serotonin/melatonin accumulation stands out as major crosstalk events in sunflower seedlings (Mukherjee et al. 2014). Following the event of abiotic stress-induced melatonin accumulation in the cotyledons of oilseeds, it is worth mentioning that melatonin has been observed to modulate various biochemical events associated with OB mobilization, thus leading to better protection of oilseeds during germination (Figure 6.2). Melatonin accumulation in sunflower seedling cotyledons is thus anticipated to be beneficial in providing adaptive responses to young plants. Seed development involves profuse accumulation of OBs. Further research will reveal the effect of serotonin and melatonin as growth supplements and their physiological effects in improving oil yield and quality during production of oilseeds.

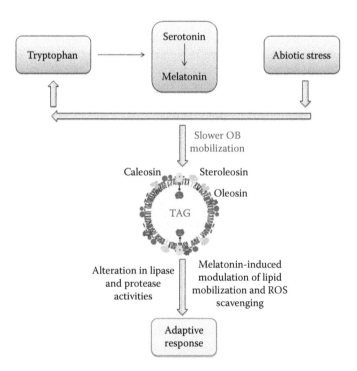

FIGURE 6.2 Modulation of OB mobilization in cotyledons of germinating seedlings is associated with abiotic stress–induced enhanced melatonin accumulation. (Molecular structure of OB adapted from Bhatla et al. *Plant Signal Behav.* 2009. 4: 176–182.)

Protective Roles of Serotonin and Melatonin during Stress Acclimatization

One of the major targets of research in molecular physiology has been to identify varied methods to improve and enhance seed germination under unfavorable conditions. Germinating seeds face numerous physiological and environmental stress conditions leading to enhanced ROS production, reduced growth, less yield, and detrimental alterations in metabolic pathways. Plant growth regulators play significant roles in stress-combating mechanisms adopted by plants, and indoleamines have emerged as a unique group of molecules in this regard. Research on different abiotic stress conditions supports the possible positive roles of serotonin and melatonin in the seedlings exposed to unfavorable conditions.

Hydroxyindole-*O*-methyltrasferase (HIMOT), a crucial enzyme involved in melatonin biosynthesis, is induced by abiotic stress. Heat stress studies in rice seedlings have shown elevation in the activities of SNAT and HIMOT (Byeon and Back 2014a,b). In etiolated, two-day-old sunflower seedling cotyledons, NaCl (120 mM) stress induces a 72% increase in HIMOT activity (Mukherjee et al. 2014). Higher serotonin accumulation has been observed in the vascular cells of primary roots and OB-containing cells of cotyledons in sunflower seedlings in NaCl-induced stress conditions (Mukherjee et al. 2014). High serotonin and melatonin accumulation in seedlings during abiotic stress are associated with reduced auxin gradient, thereby causing growth inhibition (Figure 6.3). During high temperature stress, exogenous application of melatonin exhibits a protective role on seed germination of photosensitive and thermosensitive seeds of *Phacelia tanacetifolia* Benth. (Tiryaki and Keles 2012). Melatonin has shown a positive involvement in seed germination in cucumber seedlings exposed to NaCl (150 mM) stress (Zhang et al. 2014). Genes involved in ROS and plant hormone metabolism are regulated by exogenous application of melatonin (1 μM), thereby reducing the inhibitory effect of salt stress on seed germination. Melatonin action is explained by two mechanisms: (1) property of melatonin to alleviate ROS damage, and (2) positive upregulation of gibberellic acid (GA) biosynthesis and abscisic acid (ABA) catabolism by melatonin (Zhang et al. 2014). Under NaCl (1% w/v) stress, melatonin-treated soybean seedlings show an increase in plant height, larger leaves, and reduction of biomass, relative to control seedlings. A higher level of drought tolerance is shown by soybean seedlings, with respect to plant height and biomass, when seeds are coated with melatonin (Wei et al. 2014).

Zhang et al. (2013) reported alleviation of water stress–induced inhibition of seed germination with melatonin treatment of seeds. Exogenous application of 100 μM melatonin to seeds reverses the negative effects of water by stimulation of root generation, increase in root:shoot ratio, and reduction in chlorophyll degradation. It also enhances the levels and activities of ROS scavenging enzymes, namely superoxide dismutase, peroxidase, and catalase. Alleviation of cold stress–induced symptoms by melatonin in wheat seedlings has also been reported (Turk et al. 2014). Cold stress leads to a reduction in leaf surface area, water content, and photosynthetic pigment content, and enhances the accumulation of ROS. Pretreatment of ten-day-old seedlings with 1 mM melatonin for 12 h before exposure to cold stress leads to relatively less damage, with respect to these parameters, in comparison with untreated seedlings.

Application of melatonin (50 μM) to *Vigna radiata* seeds has shown positive impact on plant acclimation to chilling stress. Relatively higher activity of L-phenylalanine ammonia-lyase (PAL) and a significant decline in electrolyte leakage has been observed in melatonin-treated seedlings under stress conditions (Szafranska et al. 2014). Melatonin leads to differential and dosage-dependent response under chilling stress in *Cucumis sativus* L. While it acts as a protective agent against membrane damage, at higher concentrations it provokes oxidative changes in proteins in the seedlings. Exogenous melatonin improves seed germination response under chilling stress (Posmyk et al. 2009). Melatonin plays a possible role in alleviating metal stress as well. Hydropriming of *Brassica oleracea* seeds with melatonin eliminates the inhibitory effect of toxic copper ions on seedling growth and germination response. (Posmyk et al. 2008).

Thus, potential protective and stress-alleviating roles of serotonin and melatonin have been observed in plants subjected to abiotic stress conditions. The roles of these indoleamines to deal with negative impact in stressed seedlings are dependent on their variable dosages. Morphological and physiological effects of stress in seedlings have been reversed by exogenous application of these indoleamines, and they clearly play a regulatory role at genetic and biochemical level as well.

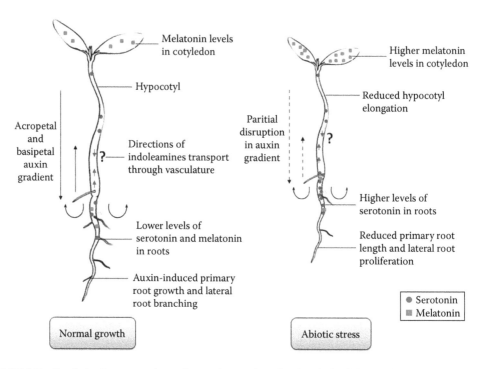

FIGURE 6.3 Correlation between auxin gradient and serotonin and melatonin levels in seedling roots and cotyledons subjected to abiotic stress.

Conclusion

Ever since their discovery in plants, serotonin and melatonin have been observed to be involved in modulating various physiological and biochemical pathways affecting plant growth, metabolism, and nutritive value of edible plant parts. These indoleamines, designated as plant growth regulators in plants, require further investigations to understand their mode of action, transport, and sequestration in plants. In this context, the isolation and characterization of specific plant transporters for serotonin and melatonin is necessary to unveil their mechanism of action and differential pattern of accumulation in seedlings. Food biotechnology and nutritive aspects of edible seeds and their nutritious contents are of prime concern in the current area of research. Exogenous application of serotonin and melatonin in various plant systems has lead to significant pharmacological observations that provide a base for future investigations on their effect on the growth and productivity of crops. Seedlings raised for cultivation are required to attain sufficient tolerance or management to pests, pathogens, and adverse soil conditions, such as salinity or water stress. In this context it is worth mentioning that serotonin and melatonin possess ample potential in modulating biotic or abiotic stress-mediated genetic pathways in plants. Physiological and molecular mechanisms of their actions in inducing various genes have been analyzed in details. Transgenic crops with increased biosynthesis and accumulation of serotonin and melatonin in some edible seeds are likely to be beneficial for dietary importance. Biotechnology of oilseed crops, and modulation of OB membrane proteins and associated enzymes by serotonin and melatonin, appears to be significant in restoring oil quality by providing antioxidative defense and longevity of seedlings subjected to adverse conditions. Seed sowing can be preceded by seed coat priming with serotonin and melatonin, which can potentially alter various physiological responses in young seedlings in terms of reduced levels of ROS. Future research in the fields of molecular physiology or pharmacognosy and food biotechnology must integrate the aspect of serotonin and melatonin-induced regulation of seed development and seedling growth in plants.

Acknowledgments

Thanks are due to the Council of Scientific and Industrial Research (CSIR), New Delhi, for providing research fellowship to Soumya Mukherjee and Dhara Arora. This work is also supported by the University of Delhi, through a Research and Development grant and Promotion of University Research and Scientific Excellence (PURSE) grant from the Department of Science and Technology, New Delhi.

REFERENCES

Arnao, M. B., and Hernandez-Ruiz, J. Melatonin: plant growth regulator and/or biostimulator during stress. *Trend Plant Sci.* 2014. 19 (12): 789–797.

Bhatla, S. C., Vandana, S., and Kasuhik, V. Recent developments in the localization of oil body–associated signaling molecules during lipolysis in oilseeds. *Plant Signal Behav.* 2009. 4: 176–182.

Byeon, Y., and Back, K. An increase in melatonin in transgenic rice causes pleiotropic phenotypes, including enhanced seedling growth, delayed flowering, and low grain yield. *J Pineal Res.* 2014a. 56: 408–414.

Byeon, Y., and Back, K. Melatonin synthesis in rice seedlings *in vivo* is enhanced at high temperatures and under dark conditions due to increased serotonin *N*-acetyltransferase and *N*-acetylserotonin methyltransferase activities. *J Pineal Res.* 2014b. 56: 189–195.

Chen, Q., Qi, W., Reiter, R. J., Wei, W., and Wang, B. Exogenously applied melatonin stimulates root growth and raises endogenous indoleacetic acid in roots of etiolated seedlings of *Brassica juncea*. *J Plant Physiol.* 2009. 166: 324–328.

Cho, M. H., No, H. K., and Prinyawiwatkul, W. Chitosan treatments affect growth and selected quality of sunflower sprouts. *J Food Sci.* 2008. 73: S70–S77.

Csaba, G., and Pal, K. Effects of insulin, triiodothyronine, and serotonin on plant seed development. *Protoplasma.* 1982. 110: 20–22.

David, A. Involvement of nitric oxide and associated biomolecules in sunflower seedling growth in response to salt stress. PhD thesis. 2012. Delhi: University of Delhi.

David, A., Yadav, S., and Bhatla, S. C. Sodium chloride stress induces nitric oxide accumulation in root tips and oil body surface accompanying slower oleosin degradation in sunflower seedlings. *Physiol Plant.* 2010. 140: 342–354.

Di Caterina, R., Giuliani, M. M., Rotunno, T., De Caro, A., and Flagella, Z. Influence of salt stress on seed yield and oil quality of two sunflower hybrids. *Annals Appl Biol.* 2007. 151: 145–154.

Fellows, L. E., and Bell, E. A. Indole metabolism in *Piptadenia peregrina*. *Phytochem.* 1971. 10: 2083–2091.

Garcia-Parrilla, M. C., Cantos, E., and Troncoso, A. M. Analysis of melatonin in foods. *J Food Comp Anal.* 2009. 22: 177–183.

Grobe, W. Function of serotonin in seeds of walnuts. *Phytochem.* 1982. 21 (4): 819–822.

Hardeland, R., Pandi-Perumal, S. R., and Poeggeler, B. Melatonin in plants: Focus on a vertebrate night hormone with cytoprotective properties. *Funct Plant Sci Biotechnol.* 2007. 1: 32–45.

Hernandez-Ruiz, J., Cano, A., and Arnao, M. B. Melatonin: A growth-stimulating compound present in lupin tissues. *Planta.* 2004. 220: 140–144.

Hernandez-Ruiz, J., Cano, A., and Arnao, M. B. Melatonin acts as a growth-stimulating compound in some monocot species. *J Pineal Res.* 2005. 39: 137–142.

Janas, K. M., and Posymk, M. M. Melatonin, an underestimated natural substance with great potential for agricultural application. *Acta Physiol Plant.* 2013. 35: 3285–3292.

Koyama, F. C., Carvalho, T. L. G., Alves, E., da Silva, H. B., de Azevedo, M. F., Hemerly, A. S., and Gracia, C. R. S. The structurally related auxin and melatonin tryptophan-derivatives and their roles in *Arabidopsis thaliana* and in the human malaria parasite *Plasmodium falciparum*. *J Euk Microbiol.* 2013. 60: 646–651.

Kumarasamy, Y., Middletona, M., Reida, R. G., Naharb, L., and Sarkera, S. D. Biological activity of serotonin conjugates from the seeds of *Centaurea nigra*. *Fitoterapia.* 2003. 74: 609–612.

Lembeck, F., and Skofitsch, G. Distribution of serotonin in *Juglans regia* seeds during ontogenetic development and germination. *Zeitschrift fur Pflanzenphysiologie.* 1984. 114 (4): 349–353.

Manchester, L. C., Tan, D. X., Reiter, R. J., Park, W., Monis, K., and Qi, W. High levels of melatonin in the seeds of edible plants: Possible function in germ tissue protection. *Life Sciences*. 2000. 67: 3023–3029.

Mukherjee, S., David, A., Yadav, S., Baluska, F., and Bhatla, S. C. Salt stress–induced seedling growth inhibition coincides with differential distribution of serotonin and melatonin in sunflower seedling roots and cotyledons. *Physiol Plant*. 2014. 152: 714–728.

Murch, S. J., Alan, A. R., Cao, J., and Saxena, P. K. Melatonin and serotonin in flowers and fruits of *Datura metel* L. *J Pineal Res*. 2009. 47: 277–283.

Murch, S. J., Campbell, S. S. B., and Saxena, P. K. The role of serotonin and melatonin in plant morphogenesis: Regulation of auxin-induced root organogenesis in *in vitro*-cultured explants of St. John's wort (Hypericum perforatum L.). *In Vitro Cell Dev Biol Plant*. 2001. 37: 786–793.

Park, S., and Back, K. Melatonin promotes seminal root elongation and root growth in transgenic rice after germination. *J Pineal Res*. 2012. 53: 385–389.

Pelagio-Flores, R. P., Castro, R. O., Bravo, A. M., Rodriguez, L. M., and Bucio, J. L. Serotonin, a tryptophan-derived signal conserved in plants and animals, regulates root system architecture probably acting as a natural auxin inhibitor in *Arabidopsis thaliana*. *Plant Cell Physiol*. 2011. 52 (3): 490–508.

Posmyk, M. M., Balabusta, M., Wleczorek, M., Silwinkska, E., and Janas, K. M. Melatonin applied to cucumber (*Cucumis sativus* L.) seeds improves germination during chilling stress. *J Pineal Res*. 2009. 46: 214–223.

Posmyk, M. M., Kuran, H., Marciniak, K., and Janas, K. M. Presowing seed treatment with melatonin protects red cabbage seedlings against toxic copper ion concentrations. *J Pineal Res*. 2008. 45: 24–31.

Ramakrishna, A. Indoleamines in edible plants: Role in human health effects. In Angel Catalá, Biochemistry Research Trends Series (editor), *Indoleamines: Sources, Role in Biological Processes and Health Effects*. 2015. 279. New York: Nova Publishers.

Ramakrishna, A., Giridhar, P., Jobin, M., Paulose, C. S., and Ravishankar, G. A. Indoleamines and calcium enhance somatic embryogenesis in cultured tissues of *Coffea canephora* P ex Fr. *Plant Cell Tiss Org*. 2012. 108: 267–278.

Ramakrishna, A., Giridhar, P., and Ravishankar, G. A. Indoleamines and calcium channels influence morphogenesis in *in vitro* cultures of *Mimosa pudica* L. *Plant Signal Behav*. 2009. 4: 1136–1141.

Ramakrishna, A., Giridhar, P., Sankar, K. U., and Ravishankar, G. A. Phytoserotonin. *Plant Signa Behav*. 2011. 6: 800–809.

Ramakrishma, A., Giridhar, P., Sankar, K. U., and Ravishankar, A. Endogenous profiles of indoleamines: Serotonin and melatonin in different tissues of *Coffea canephora* P ex Fr. as analyzed by HPLC and LC-MS-ESI. *Acta Physiol Plant*. 2012a. 34: 393–396.

Ramakrishna, A., Giridhar, P., Sankar, K. U., and Ravishankar, A. Melatonin and serotonin profiles in beans of Coffea species. *J Pineal Res*. 2012b. 52: 470–476.

Szafranska, K., Szewczyk, R., and Janas, K. M. Involvement of melatonin applied to *Vigna radiata* L. seeds in plant response to chilling stress. *Cent Eur J Biol*. 2014. 9 (11): 1117–1126.

Taiz, L., Zeiger, E., Moller, I. M., and Murphy, A. *Plant Physiology and Development*. Sixth edition. Massachusetts, U.S.A.: Sinaur Associates. 2015.

Tan, D. X, Hardeland, R., Manchester, L. C., Korkmaz, A., Ma, S., Corral, S. R., and Reiter, R. J. Functional roles of melatonin in plants, and perspectives in nutritional and agricultural science. *J Pineal Res*. 2012. 63 (2): 577–597.

Tassel, D. L. V., Roberts, N., Lewy, A., and O'Neill, S. D. Melatonin in plant organs. *J Pineal Res*. 2001. 31: 8–15.

Thakur, A., and Bhatla, S. C. A probable crosstalk between Ca^{+2}, reactive oxygen species accumulation and scavenging mechanisms and modulation of protein kinase C activity during seed development in sunflower. *Plant Signal Behav*. 2014. 9: e27900.

Tiryaki, I., and Keles, H. Reversal of the inhibitory effect of light and high temperature on germination of *Phacelia tanacetifolia* seeds by melatonin. *J Pineal Res*. 2012. 52: 332–339.

Turk, H., Erdal, S., Genisel, M., Atici, O., Demir, Y., and Yanmis, D. The regulatory effect of melatonin on physiological, biochemical and molecular parameters in cold-stressed wheat seedlings. *Plant Growth Regul*. 2014. 74: 139–152.

Wei, W., Li, Q. T., Chu, Y. N., Reiter, R. J., Yu, X. M., Zhu, D. H., Zhang, et al. Melatonin enhances plant growth and abiotic stress tolerance in soybean plants. *J Exp Bot*. 2014. 66(3): 695–707.

Zhang, H. J., Zhang, N., Yang, R. C., Wang, L., Sun, G. G., Li, O. B., Cao, Y. Y., et al. Melatonin promotes seed germination under high salinity by regulating systems, ABA and GA4 interaction in cucumber (*Cucumis sativus* L.). *J Pineal Res.* 2014. 57: 269–279.

Zhang, N., Zhao, B., Zhang, H. J., Weeda, S., Yang, C., Ren, S., and Guo, Y. D. Melatonin promotes water-stress tolerance, lateral root formation, and seed germination in cucumber (*Cucumis sativus* L.). *J Pineal Res.* 2013. 54: 15–23.

7

Indoleamines (Serotonin and Melatonin) and Calcium-Mediated Signaling in Plants

Akula Ramakrishna
Monsanto Crop Breeding Centre
Bangalore, India

Sarvajeet S. Gill
Centre for Biotechnology, MD University
Haryana, India

Krishna K. Sharma
Department of Microbiology, MD University
Haryana, India

Narendra Tuteja
Amity Institute of Microbial Technology, Amity University
Uttar Pradesh, India

Gokare A. Ravishankar
Dayananda Sagar Institutions
Bangalore, India

CONTENTS

ABSTRACT This chapter deals with the influence of indoleamines in plant morphogenesis both *in vivo* and *in vitro*, involvement of calcium under the influence of indoleamines in morphogenesis, and the present understanding of signaling mechanisms.

KEY WORDS: *indoleamines, melatonin, serotonin, calcium signaling, plant morphogenesis.*

Abbreviations

ABA: abscisic acid
Ca^{2+}: calcium

CaBP: calcium-binding protein
CaCl$_2$:   calcium choloride
CaM: calmodulin
CBLs: calcineurin B–like proteins
CDPKs: Ca^{2+}-dependent protein kinases (CDPKs)
CIPKs: Ca^{2+}-interacting kinases
cGMP: cyclic guanosine monophosphate
CMLs: CAM-like
EG: embryogenic medium
EGTA: ethylene glycol tetraacetic acid
IAA: indole-3-acetic acid
IP3:   inositol trisphosphate
MEL: melatonin
NAD(P)H: nicotinamide adenine dinucleotide phosphate
p-CPA: p-chloro phenyl alanine
(ROS): reactive oxygen species
SER: serotonin
TDZ: thidiazuron

Introduction

Neurotransmitters such as serotonin (SER), melatonin (MEL), dopamine, noradrenaline, and adrenaline have a defined role in animal systems. Their role has also been demonstrated in plants (Odjakova et al. 1997). SER and MEL, potent neurotransmitters in animal systems, are present in various plant species, structurally and functionally (Tretyn and Kendrick 1991; Ramakrishna et al. 2011c, 2012a,b). Moreover, SER, and MEL have been reported in various edible plants (Huang and Mazza 2011; Paredes et al. 2009; Manchester et al. 2000; Ramakrishna et al. 2012a,b; Ramakrishna 2015). The precise function of these compounds in plant systems is actively being researched and is now beginning to unfold. Partial characterization of the biosynthetic pathway of MEL and SER from tryptophan in cultured cells of *Hypericum perforatum* L. have also been reported (Murch et al. 2000). Both MEL and SER, structurally related to indole-3-acetic acid (IAA), probably have a role in plant morphogenesis (Murch and Saxena 2002; Ramakrishna et al. 2012a,b). Preliminary reports suggest the possible role of these compounds as candidate hormones. SER may act as a growth regulator, and stimulates the growth of roots (Csaba and Pal 1982; Ramakrishna et al. 2009) and the hook of oat coleoptiles (Niaussat et al. 1958). It also stimulates the germination of both radish seeds and the pollen of *Hippeastrum hybridum* (Roshchina 2001).

Calcium (Ca$^{2+)}$ is crucial for plant growth and development (White 2000), particularly in the initiation of a wide array of signal transduction processes in cells of higher plants (Sudha and Ravishankar 2002, 2003). Ca^{2+} is vital for plant growth and development, as it exerts a profound influence on various biological processes (Poovaiah and Reddy 1993; Bush 1995; White and Broadley 2003; White 2000). Ca^{2+} is also an important intracellular messenger in plants. In *Echinacea purpurea* L., enhanced levels of MEL, SER, and IAA have been observed upon activation of Ca^{2+} channels (Jones et al. 2007). Further, no direct link has been established between the Ca^{2+} and SER/MEL-mediated morphogenesis in plants. In this chapter, the role of indoleamines on morphogenesis, the relationship between the Ca^{2+} and indoleamines-mediated morphogenesis, and calcium-mediated signaling mechanisms have been discussed.

Influence of Indoleamines in Plant Morphogenesis

MEL was reported to promote the vegetative growth in lupin (*Lupinus albus* L.) hypocotyls, coleoptiles of wheat, barley, canary grass, and oat in a similar manner to IAA (Hernandez-Ruiz et al. 2004). The endogenous indoleamines were increased by exposure to growth regulator thidiazuron (TDZ) in tissues of *Echinacea purpurea* L. (Jones et al. 2007). Exogenously applied, MEL induces lateral leaflet regeneration and inhibition of flowering (Kolar et al. 2003), stimulates coleoptiles elongation in barley (Hernandez-Ruiz and Arnao 2008), induces shoot and root multiplication (Murch and Saxena. 2002; Ramakrishna et al. 2009), and improves seed germination in cucumber (Posmyk et al. 2009). Moreover, we have reported the morphogenetic potential of indoleamines with reference to *in vitro* shoot multiplication in *Mimosa pudica* L (Ramakrishna et al. 2009). Similarly, the light/dark cycle has a profound influence on indoleamine profile in *Dunaliella bardawil* (Ramakrishna et al. 2011a).

Chen et al. (2009) have reported that exogenous application of 0.1 mM MEL raised the endogenous levels of free IAA in the roots of etiolated seedlings of *Brassica juncea*. As an indoleamine, MEL may be able to interfere with typical auxin functions such as the regulation of morphogenesis *in vitro*, promotion of coleoptiles growth, and inhibition of root elongation. In St. John's wort (*Hypericum perforatum*), a specific ratio of endogenous SER and MEL regulates morphogenesis *in vitro* while MEL enhances *de novo* root formation (Murch and Saxena 2002). Similar to IAA, MEL may stimulate growth in etiolated lupines (*L. albus*) and coleoptiles of canary grass (*Phalaris canariensis*), wheat (*Triticum aestivum*), barley (*Hordeum vulgare*), and oats (*Avena sativa*), with a relative auxinic activity of 10% to 55% (Hernandez-Ruiz et al. 2005). Based on the functional relationship between MEL and IAA as reported in previous studies (Hernandez-Ruiz et al. 2004, 2005; Arnao and Hernandez-Ruiz 2006), it has been hypothesized that exogenous MEL might cause changes in the concentration of endogenous free IAA. Arnao and Hernandez-Ruiz (2006) have reported the effect of MEL on the regeneration of lateral and adventious roots in etiolated hypocotyls of *L. albus* L. compared with the effect of IAA. Auxin-induced root and cytokinin-induced shoot organogenesis are reportedly inhibited by alterations in the endogenous concentration of MEL and inhibitors of the transport of SER and MEL (Murch and Saxena 2002). Posmyk et al. (2008) have reported that MEL pretreatment increases the germination of seeds from *Brassica oleracea rubrum* L. by about 17% in water and by about 12% to 14% in the presence of copper. MEL has been shown to exert a suppressive effect on the flowering of both *Chenopodium rubrum* L. and in the long-day plant *Arabidopsis thaliana* (L.) (Kolar et al. 2003; Kolar and Machackova 2005). MEL reportedly stimulates the microtubule assembly in *Haemanthus katharinae* (Baker) whereas, in contrast, it caused microtubule depolymerization, disrupting the mitotic apparatus, in the onion root tip (Kolar and Machackova 2005). Chen et al. (2009) described similar effects in *B. juncea* L. Here, 0.1 μM MEL has exhibited a stimulatory effect on root growth, while 100 μM was inhibitory. Recently, Arnao and Hernandez-Ruiz (2009) have reported that barley leaves treated with MEL solutions clearly slowed the senescence process, as estimated from the chlorophyll lost in leaves; this protective effect against senescence appeared to be MEL concentration–dependent. Moreover, pretreatment with MEL for five days is found to significantly attenuate cold-induced apoptosis in *Daucus carota* L. (carrot) root cell suspensions. Another area in which MEL apparently promotes plant growth seems to emerge in seeds, seedlings, and in the root system. MEL promotes seed germination as reported in cucumber (Posmyk et al. 2009) and in red cabbage (Posmyk et al. 2008). The possible role of MEL in plant tissue and organ culture is presented in Table 7.1.

Calcium-Mediated Morphogenetic Responses in Plants

Earlier reports suggest that increase of calcium regulates calcium signaling and acts as a good stimulator for somatic embryogenesis (Montoro et al. 1995; Vanderluit et al. 1999; Arruda et al. 2000; Ramakrishna et al. 2011b). An increase in intracellular calcium levels was involved in a large number of physiological processes coupled to the calcium-binding protein (CaBP) calmodulin (CaM) (Poovaiah and Reddy 1993). A predominant influx of calcium leads to organogenesis and plays an important role in establishment of polarity in higher plants (Hush et al. 1991). Available reports on the possible involvement of calcium in somatic embryogenesis cover carrot (Overvoorde and Grimes 1994; Jansen et al. 1990; Takeda et al. 2003; Ramakrishna et al. 2012), sandalwood (Anil and Rao 2000), and wheat (Mahalakshmi et al. 2007). Moreover, exogenously administered calcium and calcium ionophore A23187 enhance *in vitro* shoot multiplication and root induction in cultures of *M. pudica* (Ramakrishna et al. 2009). Calcium ionophore A23187 has been used to alter intracellular and extracellular calcium levels to study changes

TABLE 7.1

Possible Role of Melatonin in Plant Tissue and Organ Culture

Plant Species	Function	Reference
Hypercom perforatum	Shoot multiplication and root induction	Murch et al. 2000
Chenopodium rubrum	Suppress flowering	Kolar et al. 2003
Arabidopsis thaliana	Suppress flowering	Kolar and Machackova 2005
Lupinus albus	Elongation of hypocotyls	Hernandez Ruiz et al. 2004
Lupinus albus	Promotes adventitious- and lateral root regeneration	Arnao and Hernandez-Ruiz 2006
Brassica oleracea	Improve seed germination	Posmyk et al. 2008
Lupinus albus	Stimulates the expansion of etiolated cotyledons	Hernandez-Ruiz and Arnao 2008
Mimosa pudica	Shoot multiplication and root induction	Ramakrishna et al. 2009
Brassica Juncea	Stimulate root growth	Chen et al. 2009
Cucumis sativus	Improve seed germination	Posmyk et al. 2009
Vitis vinifera	Seed growth and development	Murch et al. 2010
Rhodiola crenulata	Improve the survival of cryopreserved callus	Zhao et al. 2011
Coffea canephora	Somatic embryogenesis	Ramakrishna et al. 2012a,b

in calcium associated with cellular activities (Wu et al. 1992). Similarly, the calcium ionophore A23187 elicits anthocyanin in callus cultures of *D. carota* (Sudha and Ravishankar 2003). Calcium ionophore A23187 stimulates cytokinin production and is involved in mitosis leading to bud formation in *Funaria hygrometrica* by increasing the intracellular calcium levels (Saunders and Helper 1982). In tobacco cells, the calcium ionophore A23187 is known to enhance intracellular Ca^{2+} levels (Chandra and Low 1997). Earlier reports have suggested that increase in cytosolic calcium triggers numerous cellular processes by modulation of protein kinases, ion channels, and other cellular proteins (White 2000).

Role of Indoleamines and Calcium in Plant Morphogenesis

Understanding morphogenetic response of cultured cells is a special area of intense research, while studies have focused on plant growth regulators' influence on morphogenesis. The activation of calcium channels that change cell polarity has resulted in an increase in MEL in *Echinacea purpurea* L. explants, together with an inhibition in TDZ-induced callus induction (Jones et al. 2007). As of now, reports on the influence of calcium on somatic embryogenesis in *C. canephora* are not known. High concentration of calcium promotes embryo formation from the callus (Montoro et al. 1995). We have reported the influence of indoleamines on somatic embryogenesis in *in vitro* cultures of *C. canephora* (robusta). Exogenously supplemented with 100 μM SER, induced 84 embryos from each callus with concomitant enhancement of endogenous pools of SER and IAA (Ramakrishna et al. 2012a,b).

Incorporation of MEL produced 62 embryos per callus with enhanced endogenous MEL and a concomitant increase of IAA content. Incorporation of SER–MEL conversion inhibitor (40 μM p-choro phenyl alanine [p-CPA]) into embryogenic (EG) medium devoid of IAA led to a drastic reduction in somatic embryo production (Figure 7.1). Supplementing with 100 μM SER and 5 mM calcium chloride ($CaCl_2$) induced 105 ± 6 embryos from each callus by increasing endogenous MEL and SER. Addition of 100 μM calcium ionophore and 100 μM SER induced 155 ± 15 somatic embryos per culture, and augmented the endogenous MEL and IAA levels. Similarly, 100 μM MEL and 5 mM $CaCl_2$ treatment produced 78 ± 2 embryos from each callus (see Figure 7.1). *In vitro*-grown *C. canephora* plants, supplemented with 5 mM calcium and 100 μM calcium ionophore A23187, induced rooting (Figure 7.2), whereas supplementation

FIGURE 7.1 The influence of SER, MEL, and their inhibitors (p-CPA and Prozac) on somatic embryo induction in *C. canephora* under dark and light conditions: (a) formation of yellow friable callus, (b) induction of somatic embryos by 100 μM SER, (c) somatic embryos induction by 100 μM MEL, and (d) suppression of somatic embryo induction by 40 μM p-CPA. (Adapted from Ramakrishna et al. *Plant Cell Tiss Organ Cult.* 2012. 108: 267–278.)

FIGURE 7.2 The effect of indoleamines, calcium, and calcium ionophore (A23187) on somatic embryogenesis in *in vitro* cultures of *C. canephora* under dark conditions: (a) 100 μM MEL+5 mM calcium, (b) 100 μM calcium ionophore A23187+100 μM verapamil, (c) 100 μM SER+100 μM calcium ionophore A23187, and (d) *in vitro* rooting by 100 μM calcium ionophore A23187. (Adapted from Ramakrishna et al. *Plant Cell Tiss Organ Cult.* 2012. 108: 267–278.)

of calcium channel blockers effectively reduce root induction. Similarly, calcium channel blockers have been administered to study the involvement of calcium during the MEL- and SER-mediated somatic embryogenesis. This response may be due to auxin-like activity of indoleamines (Hernandez-Ruiz et al. 2004; Ramakrishna et al. 2009; Murch and Saxena 2002).

Furthermore, we have reported the morphogenetic potential of indoleamines with reference to *in vitro* shoot multiplication in *M. pudica* L (Ramakrishna et al. 2009). Our earlier report showed that indoleamines (SER and MEL), calcium, and calcium ionophore A23187 induce morphogenesis in *in vitro* cultures of *M. pudica*, whereas verapamil hydrochloride and EGTA could significantly suppress shoot multiplication and root induction (Ramakrishna et al. 2009) (Figures 7.3 and 7.4).

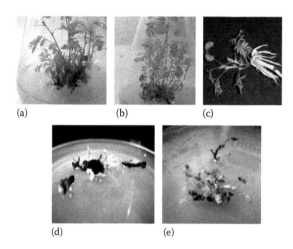

FIGURE 7.3 The effect of indoleamines and their inhibitors on shoot multiplication in *in vitro* cultures of *M. pudica*: (a) elongation of shoots in *M. pudica* by melatonin (100 μM), (b) shoot induction by serotonin (100 μM), (c) root induction by serotonin/melatonin (100 μM), (d) suppression of shoot elongation by p-CPA at 40 μM, and (e) suppression of shoot elongation by Prozac at 20 μM. (Adapted from Ramakrishna et al. *Plant Signal Behav.* 2009. 12: 1136–1141.)

FIGURE 7.4 The effect of calcium and calcium channel inhibitors on shoot multiplication in *in vitro* cultures of *M. pudica:* (a) multiple shoot induction in *M. pudica* under the influence of calcium (5 mM), (b) root induction by calcium ionophore A23187 at 100 μM, (c) suppression of shoot elongation by verapamil hydrochloride at 1 mM, and (d) suppression of shoot elongation by EGTA at 100 μM. (Adapted from Ramakrishna et al. *Plant Signal Behav.* 2009. 12: 1136–1141.)

Calcium-Mediated Signaling Mechanisms in Plants

Being sessile, plants face a plethora of environmental disturbances during different stages of their growth and development. To sense various developmental, environmental, and hormonal changes, plants have evolved robust metabolic and biochemical responses. In response to environmental changes, the plant cell reprograms the cellular metabolism by initiating the signaling events (perception of stress signal at membrane level and their transduction for cellular response). Transduction of signals in the cell requires the proper spatial and temporal coordination of all signaling molecules. Among various secondary signaling molecules (Ca^{2+}, IP3, or cGMP) of the signal transduction pathway, Ca^{2+} has been established as

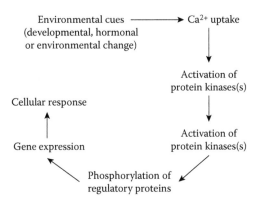

FIGURE 7.5 A generic pathway for Ca^{2+}-mediated signaling mechanisms in plants. In response to developmental, hormonal or environmental change, the Ca^{2+} level rises in the plant cell (Ca^{2+} signature), which is recognized by Ca^{2+} ion sensors or calcium-binding proteins (calcium-modulating protein or calmodulin [CaM]; CaM-like [CML] and other EF hand–containing Ca^{2+}-binding proteins; calcineurin B–like proteins [CBLs]), and activates the Ca^{2+}-dependent protein kinases (CDPKs). The kinases regulate the function of many genes, and results in cellular responses. (Modified from Tuteja and Mahajan. *Plant Signal Behav.* 2007. 2: 79–85.)

being implicated in diverse signaling pathways. Therefore, the status of Ca^{2+} in the plant cell has received considerable attention in mediating plant responses to environmental fluctuations (Tuteja and Mahajan 2007). Apart from being an essential plant macronutrient, the role of Ca^{2+} as secondary messenger has been extensively studied (Tuteja and Mahajan 2007; Virdi et al. 2015). It has been reported that extracellular signals and biotic and abiotic stresses elicit a change in the level of Ca^{2+} in plant cells. The Ca^{2+} ions are found to be important signaling molecules and regarded as a convergent point of many signaling pathways. A transient increase in Ca^{2+} in the cell is also considered significant for physiological responses in plants; however, elevation in its concentration in the cytoplasm is regarded as a universal response to stress. A generic pathway for Ca^{2+}-mediated stress signaling mechanisms in plants is shown in Figure 7.5.

In response to developmental, hormonal or environmental change, the Ca^{2+} level rises in the plant cell (the Ca^{2+} signature), which gets recognized by Ca^{2+} ion sensors or calcium-binding proteins (calcium-modulating protein or calmodulin [CaM]; CaM-like [CML] and other EF hand–containing Ca^{2+}-binding proteins; calcineurin B–like proteins [CBLs]) and activates the Ca^{2+}-dependent protein kinases (CDPKs). The kinases regulate the function of many genes and results in cellular responses. It has been reported that many genes get upregulated in response to the level of Ca^{2+} in the cell. Kaplan et al. (2006) performed a transcriptome study and noted a change in the gene expression pattern of ~230 genes in a Ca^{2+}-responsive manner, which in general includes early stress responsive genes. Recently, Liu et al. (2015) noted a nonlinear amplification of Ca^{2+} signals during the interaction of Ca^{2+}-CaM with target proteins that imparted greater versatility to cells in deciphering different Ca^{2+} signatures for changes in gene expression (Liu et al. 2015).

Calcium-Mediated Signaling Mechanisms during Plant Morphogenesis

The plant-decoding Ca^{2+} ions include several families of Ca^{2+} sensors such as CMLs, CaM, CBLs, CDPKs, and their interacting kinases (CIPKs) (Batistič and Kudla 2012; Virdi et al. 2015). These Ca^{2+} sensors are encoded by complex a gene family that forms robust signaling networks in plants. Also, a major group of Ca^{2+} sensor proteins possesses classical helix–loop–helix EF-hand motifs, which bring about Ca^{2+} binding of these proteins that result in Ca^{2+}-dependent conformational changes (Batistič and Kudla 2009), although several of these proteins can be categorized into protein families that acquire specific structural rearrangements or appear to be specific to plants and lower protists (Batistič and Kudla 2009). The large number of the Ca^{2+} binding proteins compared with other organisms, the unique structural composition of Ca^{2+} binding proteins, and the complexity of the target proteins regulated by the Ca^{2+} sensors allows the plant to tightly control the appropriate adaptation to its ever-changing environment (Batistič and Kudla 2009).

Regulation of the cellular Ca^{2+} is an essential cell function that is accomplished by a complex of processes collectively called the *Ca^{2+} homeostat*. Therefore, Ca^{2+} is the most versatile signaling molecule in the plant kingdom. The plant cells respond to numerous extracellular stimuli with changes in cytosolic calcium concentration, which eventually controls many integrated physiological processes (Bush 1995). Entry of Ca^{2+} into the cytoplasm, mediated by ion channels, is due either to release from vacuoles and the endoplasmic reticulum or to entry from the cell wall space, that is, apoplasm. Ca^{2+} is removed from the cytosol by ATP-hydrolyzing pumps and calcium exchangers (Bush 1995). Cooperation between channel and pump activities mediates important spatial and temporal changes in calcium concentration and may elicit localized bursts in Ca^{2+} or induce Ca^{2+} oscillations that are crucial to the specificity of Ca^{2+} signaling pathways. Furthermore, calcium plays a key role in plant growth and development because changes in cellular Ca^{2+}, acting through Ca^{2+}-modulated proteins and their target molecules, regulate a variety of cellular functions. The regulatory actions of Ca^{2+} range from control of ion transport to gene expression, and are possible because of a homeostatic system that regulates Ca^{2+} levels. The development of this homeostatic system is evolutionarily primitive, and probably reflects a biochemical necessity of maintaining low levels of Ca^{2+} in the phosphate-rich environment of the cytosol (Bush 1995).

The organization of the cytoskeleton is crucial in plant cell signaling and morphogenesis, and for the regulation of membrane protein activities and spatial distribution. It is driven by actin-dependent translocation of plasma membrane proteins to form a polar growth axis from which the rhizoid develops from the zygote cell. The presence of F-actin generates a locally specialized cortical domain that allows Ca^{2+} channel blocker–binding proteins to be translocated and concentrated at the tip of the rhizoid (Shaw and Quatrano 1996). Accumulation of Ca^{2+} channels and the subsequent increase of cytosolic Ca^{2+} concentration in a particular site of the zygote might control growth orientation of the future rhizoid. In pollen tubes, a steep cortical Ca^{2+} gradient, most concentrated at the apex, has been observed in several plant species (Hepler 1997). It has been suggested that these intracellular Ca^{2+} gradients are closely correlated to the presence of a higher amount of active Ca^{2+} channels in the pollen tube tip (Shaw and Quatrano 1996). Different reported examples suggest that the elements of cytoskeleton in plant cells might be able to mobilize or concentrate Ca^{2+} channels to particular plasma membrane domains.

Evidences are emerging that reactive oxygen species (ROS) can also function as cellular second messengers that are likely to modulate many different proteins leading to a variety of responses. Interestingly, the target of ROS signal transduction is the activation of Ca^{2+}-permeable channels in plant membranes (Mori and Schroeder 2004). ROS activation of Ca^{2+} channels may be a central step in many ROS-mediated processes, such as stress and hormone signaling, polar growth, and development (Mori and Schroeder 2004). A class of hyperpolarization-activated Ca^{2+}-permeable cation channels in several types of plant cells have been identified and characterized. In *Fucus* rhizoid cells, there is a local oxidative burst at the growing rhizoid tip. Moreover, ROS activation of rhizoid apex Ca^{2+} channels and a tip-focused Ca^{2+} gradient after hyperosmotic treatment has been reported (Coelho et al. 2002). Moreover, there are genetic evidences for the role of membrane-bound NAD(P)H oxidases in abscisic acid (ABA)-ROS signal transduction in guard cells and root hair growth. Hyperpolarization-activated Ca^{2+} channels are activated by the hydroxyl radical (OH·) in epidermal cells of the *Arabidopsis* root elongation zone (Foreman et al. 2003). Loss-of-function mutations in the NAD(P)H oxidase gene, *atrbohC*, caused a short root hair phenotype (Foreman et al. 2003). Polar growth is associated with tip-localized Ca^{2+} influx and cytosolic Ca^{2+} elevations (Malho and Trewavas 1996; Holdaway-Clarke et al. 1997; Messerli et al. 2000; Plieth and Trewavas 2002). Therefore, it can be concluded that signal transduction mechanisms may activate NAD(P)H oxidases, causing localized ROS production and eventually local activation of Ca^{2+}-permeable channels in plant membranes.

Conclusion

Recent advances in understanding the regulation of plant Ca^{2+} channels provide new insights into mechanisms of plant cell signaling. Further studies of the regulatory mechanisms of individual Ca^{2+} channel types by G proteins, cytoskeletal elements, second messengers, and other effectors will allow a cell biological understanding of important elements in early signal transduction cascades.

The role of indole amines, serotonin, and melatonin in the morphogenetic response in plants is amply demonstrated. Calcium signaling in plants is a fairly well-understood process and the new dimension of the influence of indole amines appears to enhance our understanding of their inter-relationships. Recent advances in understanding the regulation of Ca^{2+} channels provide new insights into mechanisms of plant cell signaling. Further studies of the regulatory mechanisms of individual Ca^{2+} channel types of G-proteins, cytoskeletal elements, second messengers, and other effectors will enhance our understanding of the important elements in early signal transduction cascades.

REFERENCES

Anil, V. S. and Rao, K. S. Calcium-mediated signaling during sandalwood somatic embryogenesis. Role for exogenous calcium as second messenger. *Plant Physiol.* 2000. 123: 1301–1312.

Arnao, M. B. and Hernandez Ruiz, J. The physiological function of melatonin in plants. *Plant Signal Behav.* 2006. 1: 89–95.

Arnao, M. B. and Hernandez-Ruiz, J. Protective effect of melatonin against chlorophyll degradation during the senescence of barley leaves. *J Pineal Res.* 2009. 46: 58–63.

Arruda, S. C., Souza, G. M., Almeida, M. and Cocal, S. A. Anatomical and biochemical characterization of the calcium effect on *Eucalyptus uopylla* callus morphogenesis in vitro. *Plant Cell Tissue Organ Cult.* 2000. 63: 143–154.

Batistič, O. and Kudla, J. Plant calcineurin B–like proteins and their interacting protein kinases. *Biochim Biophys Acta.* 2009. 1793: 985–992.

Batistič, O. and Kudla, J. Analysis of calcium signaling pathways in plants. *Biochim Biophys Acta.* 2012. 1820: 1283–1293.

Bush, D. S. Calcium regulation in plant cells and its role in signaling. *Ann Rev Plant Physiol Plant Mol Biol.* 1995. 46: 95–122.

Chandra, S. and Low, P. S. A23187: A divalent cation ionophore. *J Biol Chem.* 1997. 272: 28274–28280.

Chen, Q., Qi, W-B., Reiter, R. J., Wei, W., and Wang, B. M. Exogenously applied melatonin stimulates root growth and raises endogenous indoleacetic acid in roots of etiolated seedlings of *Brassica juncea. J Plant Physiol.* 2009. 166: 324–328.

Coelho, S. M., Taylor, A. R., Ryan, K. P., Sousa-Pinto, I., Brown, M. T., and Brownlee, C. Spatiotemporal patterning of reactive oxygen production and Ca^{2+} wave propagation in *Fucus* rhizoid cells. *Plant Cell.* 2002. 14: 2369–2381.

Csaba, G. and Pal, K. Effect of insulin triodothyronine and serotonin on plant seed development. *Protoplasma.* 1982. 110: 20–22.

Foreman, J., Demidchik, V., Bothwell, J. H. F., Mylona, P., Miedema, H., Torres, M. A., Linstead, P., Costa, S., Brownlee, C., and Jones, J. D. G. Reactive oxygen species produced by NADPH oxidase regulate plant cell growth. *Nature.* 2003. 422: 442–446.

Hepler, P. K. Tip growth in pollen tubes: Calcium leads the way. *Trends Plant Sci.* 1997. 2: 79–80.

Hernandez-Ruiz, A. and Arnao, M. B. Melatonin stimulates the expansion of etiolated lupin cotyledons. *Plant Growth Reg.* 2008. 55: 29–34.

Hernandez-Ruiz, J., Cano, A., and Arnao, M. B. Melatonin: A growth-stimulating compound present in lupin tissues. *Planta.* 2004. 220: 140–144.

Hernandez-Ruiz, A., Cano, A., and Arnao, M. B. Melatonin acts as a growth-stimulating compound in some monocot species. *J Pineal Res.* 2005. 39: 137–142.

Holdaway-Clarke, T. L., Feijo, J. A., Hackett, G. R., Kunkel, J. G., and Hepler, P. K. Pollen tube growth and the intracellular cytosolic calcium gradient oscillate in phase while extracellular calcium influx is delayed. *Plant Cell.* 1997. 9: 1999–2010.

Huang, X. and Mazza, G. Application of LC and LC-MS to the analysis of melatonin and serotonin in edible plants. *Crit Rev Food Sci Nutr.* 2011. 51: 269–284.

Hush, J. M., Overall, R. I., and Newman, I. A. Calcium influx precedes organogenesis in Graptopetalum. *Plant Cell Environ.* 1991. 14: 657–665.

Jansen, M. A. K., Booij, H., Schell, J. H. N., and De Vries, S. C. Calcium increases the yield of somatic embryos in carrot embryogenic suspension cultures. *Plant Cell Rep.* 1990. 9: 221–223.

Jones, M. P. A., Cao, J., O'Brien, R., Murch, S. J., and Saxena, P. K. The mode of action of thidiazuron: Auxins, indoleamines, and ion channels in the regeneration of *Echinacea purpurea* L. *Plant Cell Rep.* 2007. 26: 1481–1490.

Kaplan, B., Davydov, O., Knight, H., Galon, Y., Knight, M. R., Fluhr, R., and Fromm, H. Rapid transcriptome changes induced by cytosolic Ca^{2+} transients reveal ABRE-related sequences as Ca^{2+}-responsive cis elements in *Arabidopsis. Plant Cell.* 2006. 18: 2733–2748.

Kolar, J., Johnson, H., and Machackova, I. Exogenously applied melatonin (N-acetyl-5-methoxytryptamine) affects flowering of the short-day plant *Chenopodium rubrum. Physiol Plant.* 2003. 118: 605–612.

Kolar, J. and Machackova, I. Melatonin in higher plants: Occurrence and possible functions. *J Pineal Res.* 2005. 39: 333–341.

Liu, J., Whalley, H. J., and Knight, M. R. Combining modelling and experimental approaches to explain how calcium signatures are decoded by calmodulin-binding transcription activators (CAMTAs) to produce specific gene expression responses. *New Phytol.* 2015. 208: 174–187.

Mahalakshmi, A., Singla, B., Khurana, J. P., and Khurana, P. Role of calcium–calmodulin in auxin-induced somatic embryogenesis in leaf base cultures of wheat (*Triticum aestivum* var. HD 2329). *Plant Cell Tiss Organ Cult.* 2007. 88: 167–174.

Malho, R. and Trewavas, A. J. Localized apical increases of cytosolic free calcium control pollen tube orientation. *Plant Cell.* 1996. 8: 1935–1949.

Manchester, L. C., Tan, D. X., Reiter, R. J., Park, W., Monis, K., and Qi, W. B. High levels of melatonin in the seeds of edible plants: Possible function in germ tissue protection. *Life Sci.* 2000. 67: 3023–3029.

Messerli, M. A., Creton, R., Jaffe, L. F., and Robinson, K. R. Periodic increases in elongation rate precede increases in cytosolic Ca^{2+} during pollen tube growth. *Dev Biol.* 2000. 222: 84–98.

Montoro, P., Etienne, H., and Carron, M. P. Effect of calcium on callus friability and somatic embryogenesis in *Hevea brasiliensis* Müll. Arg.: Relations with callus mineral nutrition, nitrogen metabolism and water parameters. *J Exp Bot.* 1995. 46:255–261.

Mori, I. C. and Schroeder, J. I. Reactive oxygen species activation of plant Ca^{2+} channels. A signaling mechanism in polar growth, hormone transduction, stress signaling, and hypothetically mechanotransduction. *Plant Physiol.* 2004. 135: 702–708.

Murch, S. and Saxena, P. K. Melatonin: A potential regulator of plant growth and development? *In Vitro Cell Dev Biol Plant.* 2002. 38: 531–536.

Murch, S. J., Krishna, Raj, S., and Saxena, P. K. Tryptophan is a precursor for melatonin and serotonin biosynthesis in *in vitro*-regenerated St. John's wort (*Hypericum perforatum* L. cv. Anthos) plants. *Plant Cell Rep.* 2000. 19: 698–704.

Murch, S. J., Hall, B. A., Le, C. H., and Saxena, P. K. Changes in the levels of indoleamine phytocemicals during veraison and ripening of wine grapes. *J Pineal Res.* 2010. 49: 95–100.

Niaussat, P., Laborit, H., Dubois, C., and Hiaussat, M. Action de la serotonine sur la croissance des jeunes plantules d'Avoine. *Compt Rend Soc Biol.* 1958. 152: 945–947.

Odjakova, M. and Hadjiivanova, C. Animal neurotransmitter substances in plants. *Bulg J Plant Physiol.* 1997. 23: 94–102.

Overvoorde, P. J. and Grimes, H. D. The role of calcium and calmodulin in carrot somatic embryogenesis. *Plant Cell Physiol.* 1994. 34: 135–144.

Paredes, S. D., Korkmaz, A., Manchester, L. C., Tan, D. X., and Reiter, R. J. Phytomelatonin: A review. *J Exp Bot.* 2009. 60: 57–69.

Plieth, C. and Trewavas, A. J. Reorientation of seedlings in the earth's gravitational field induces cytosolic calcium transients. *Plant Physiol.* 2002. 129: 786–796.

Poovaiah, B. W. and Reddy, A. S. N. Calcium and signal transduction in plants. *CRC Crit Rev Plant Sci.* 1993. 12: 185–211.

Posmyk, M. M., Bałabusta, M., Wieczorek, M., Sliwinska, E., and Janas, K. M. Melatonin applied to cucumber (*Cucumis sativus* L.) seeds improves germination during chilling stress. *J Pineal Res.* 2009. 46: 214–223.

Posmyk, M. M., Kuran, H., Marciniak, K., and Janas, K. M. Presowing seed treatment with melatonin protects red cabbage seedlings against toxic copper ion concentrations. *J Pineal Res.* 2008. 45: 24–31.

Ramakrishna, A. Indoleamines in edible plants: Role in human health effects. In Angel Catalá, Biochemistry Research Trends Series (editor), *Indoleamines: Sources, Role in Biological Processes and Health Effects.* 2015. 279. New York: Nova Publishers.

Ramakrishna, A., Dayananda, C., Giridhar, P., Rajasekaran, T., and Ravishankar, G. A. Photoperiod influences endogenous indoleamines in cultured green alga *Dunaliella bardawil*. *Indian J Exp Biol*. 2011a. 49: 234–240.

Ramakrishna, A., Giridhar, P., Jobin, M., Paulose, C. S., and Ravishankar, G. A. Indoleamines and calcium enhance somatic embryogenesis in cultured tissues of *Coffea canephora* P ex Fr. *Plant Cell Tiss Organ Cult*. 2012. 108: 267–278.

Ramakrishna, A., Giridhar, P., and Ravishankar, G. A. Indoleamines and calcium channels influence morphogenesis in *in vitro* cultures of *Mimosa pudica* L. *Plant Signal Behav*. 2009. 12: 1136–1141.

Ramakrishna, A., Giridhar, P., and Ravishankar, G. A. Calcium and calcium ionophore A23187 induce high frequency somatic embryogenesis in cultured tissues of *Coffea canephora* P ex Fr. *In Vitro Cell Dev Biol Plant*. 2011b. 47: 667–673.

Ramakrishna, A., Giridhar, P., and Ravishankar, G. A. Phytoserotonin: A review. *Plant Signal Behav*. 2011c. 6: 800–809.

Ramakrishna, A., Giridhar, P., Sankar, K. U., and Ravishankar, G. A. Endogenous profiles of indoleamines: Serotonin and melatonin in different tissues of *Coffea canephora* P ex Fr. as analyzed by HPLC and LC-MS-ESI. *Acta Physiol Plant*. 2012a. 34: 393–396.

Ramakrishna, A., Giridhar, P., Sankar, K. U., and Ravishankar, G. A. Melatonin and serotonin profiles in beans of *Coffea* sps. *J Pineal Res*. 2012b. 52: 470–476.

Roshchina, V. V. *Neurotransmitters in Plant Life*. 2001. 4–81. Enfield, New Hampshire: Science Publishers.

Saunders, M. J. P. and Helper, K. Calcium ionophore A23187 stimulates cytokinin-like mitosis in *Funaria*. *Science*. 1982. 217: 943–945.

Shaw, S. L. and Quatrano, R. S. Polar localization of a dihydropyridine receptor on living *Fucus* zygote. *J Cell Sci*. 1996. 109: 335–342.

Sudha, G. and Ravishankar, G. A. Influence of calcium channel modulators in capsaicin production by cell suspension cultures of *Capsicum frutescens* Mill. *Curr Sci*. 2002. 83: 480–484.

Sudha, G. and Ravishankar, G. A. Elicitation of anthocyanin production in callus cultures of *Daucus carota* and involvement of calcium channel modulators. *Curr Sci*. 2003. 84: 775–779.

Takeda, T., Hideyuki, I., and Hiroshi, M. Stimulation of somatic embryogenesis in carrot cells by the addition of calcium. *Biochem Eng J*. 2003. 14: 143–148.

Tretyn, A. and Kendrick, R. E. Acetylcholine in plants: Presence, metabolism and mechanism of action. *Bot Rev*. 1991. 57: 33–73.

Tuteja, N. and Mahajan, S. Calcium signaling network in plants: An overview. *Plant Signal Behav*. 2007. 2: 79–85.

Van der Luit, A. H., Olivari, C., Haley, A., Knight, M. R., and Trewavas, A. J. Distinct calcium signalling pathways regulate calmodulin gene expression in tobacco. *Plant Physiol*. 1999. 121: 705–714.

Virdi, A. S., Singh, S., and Singh, P. Abiotic stress responses in plants: Roles of calmodulin-regulated proteins. *Front Plant Sci*. 2015. 6: 809.

White, P. J. Calcium channels in higher plants. *Biochim Biophys Acta*. 2000. 1465: 171–189.

White, J. W. and Broadley, M. R. Calcium in plants. *Ann Bot*. 2003. 92: 487–511.

Wu, F. S., Cheng, C. M., and Tsong-Teh, T. T. Effect of Ca++ chelator, calcium++ ionophore, and heat shock pretreatment on *in vitro* protein phosphorylation of rice suspension culture cells. *Bot Bull Acad Sin*. 1992. 33: 151–159.

Zhao, Y., Qi, L. W., Wang, W. M., Saxena, P. K. and Liu, C. Z. Melatonin improves the survival of cryopreserved callus of Rhodiola crenulata. *J Pineal Res*. 2011. 50: 83–88.

8

Serotonin and Melatonin in Plant Growth and Development

Ramón Pelagio-Flores and José López-Bucio
Universidad Michoacana de San Nicolás de Hidalgo, Instituto de Investigaciones Químico-Biológicas
Michoacán, México

CONTENTS

ABSTRACT   Serotonin and melatonin are two indoleamines widely distributed in plants, well known due to their function as neurotransmitters in mammals. Since they share structural similarity with indole-3-acetic acid (IAA), previous reports suggested a possible auxin activity for these compounds. However, detailed analyses of root architecture have shown that in contrast to serotonin, which represses primary root growth, lateral root formation, and root hair development in high concentrations, melatonin had the opposite effect, promoting lateral and adventitious root formation without affecting primary root growth, and minimally affecting root hair development. These data indicate that indoleamines affect root morphogenesis in a differential and contrasting manner. Accumulating information further suggests a critical role of serotonin and melatonin in abiotic stress tolerance, regulation of senescence, activation of defense responses, plant productivity, and developmental phase transitions, which may be related to antioxidant and growth-promoting properties. In this chapter, we discuss possible mechanisms of serotonin and melatonin sensing in model and crop plants.

KEY WORDS: *serotonin, melatonin, plant growth, stress, senescence, plant productivity.*

Abbreviations

ABA: abscisic acid
AADC: aromatic L-amino acid decarboxylase
AANAT: arylalkylamine *N*-acetyltransferase
AXR3:   auxin-resistant 3

IAA:	indole-3-acetic acid
IAA17:	indole-3-acetic acid–inducible 17
JA:	jasmonic acid
SNAT:	*N*-acetyltransferase
ASMT:	*N*-acetylserotonin methyltransferase
NAA:	naphthyl acetic acid
PR:	pathogenesis-related
ROS:	reactive oxygen species
SA:	salicylic acid
SAGs:	senescence associated genes
SHY2:	short hypocotyl 2
TDC:	tryptophan decarboxylase
T5H:	tryptamine 5-hydroxilase
TPH:	tryptophan hydroxylase

Introduction

Serotonin (5-hydroxytryptamine) and melatonin (*N*-acetyl-5-methoxytryptamine) are two bioactive indoleamines derived from tryptophan, which modulate important physiological processes in mammals such as circadian rhythms, mood, sleep, anxiety, body temperature, sexual behavior, and reproduction (Veenstra-VanderWeele et al. 2000; Reiter 1993; Galano et al. 2011). These compounds are present in organisms belonging to evolutionarily distant taxonomic groups including Bacteria, Cyanobacteria, Dinoflagellata, Euglenoidea, Rhodophyta, Phaeophyta, and Plantae (Odjakova and Hadjiivanova 1997; Roshchina 2001; Murch et al. 2001; Kang et al. 2009a; Paredes et al. 2009; Ramakrishna et al. 2011a,b). Indoleamine levels vary considerably among plant species, and their different biological activities impact growth, development, defense, and adaptation to abiotic stress. Serotonin and melatonin are structurally related to indole-3-acetic acid (IAA), the main and most abundant auxin present naturally in plants. IAA is the most widely investigated phytohormone, involved in virtually every aspect of growth and development throughout the plant life cycle, including tropic responses toward light, gravity, and touch stimuli, as well as in root and shoot system establishment (Woodward and Bartel 2005). IAA is a tryptophan-derived compound for which transporters, receptors, and transcription factors mediate its role in plant morphogenesis. Since the molecular structure of IAA is closely related to serotonin and melatonin, the possibility that these compounds could act in similar ways in plant signaling was proposed (Murch et al. 2001; Murch and Saxena 2002). Although research on serotonin and melatonin in plants is still in its infancy, in the last 10 years important advances have been made in this field that provide the basis not only for understanding the mechanisms of action of these compounds, but also regarding their possible applications as phytostimulants.

Serotonin and Melatonin Distribution in Plants

Serotonin was identified in plants for the first time by Bowden et al. (1954), in the legume *Mudica pruriens*, and almost forty years later Van Tassel et al. (1993) described the presence of melatonin in the ivy morning glory (*Pharbitis nil*) plant and in tomato (*Solanum lycopersicum*) (Dubbels et al. 1995; Van Tassel et al. 1995; Kolar et al. 1995). Since then, serotonin and melatonin have been identified in an increasing number of species from different plant families, including edible and medicinal plants used in traditional medicine to treat a variety of chronobiological disorders or degenerative diseases (Feldman and Lee 1985; Roshchina 2001; Murch et al. 1997; Badria 2002; Chen et al. 2003; Ramakrishna et al. 2011a, 2012a,b; Ramakrishna 2015). Accordingly, some research has been conducted to identify plants with high levels of beneficial metabolites, which may have applications on human health or agriculture. Up to 400 µg/g serotonin have been found in walnuts (*Juglans regia*), while melatonin content varies from 6 pg/g in the shoots of *Ipomoea nil* to 230 μg/g from *Pistachio* kernels, although lower levels can be found in different edible plant parts (Paredes et al. 2009; Ramakrishna et al. 2011a; Arnao 2014; Hardeland 2015). In spite of the apparent ubiquity of these molecules in the plant kingdom, the precise roles and/or the importance of serotonin and melatonin for plant functional processes still needs to be explored.

Biosynthesis of Serotonin and Melatonin

The serotonin and melatonin biosynthetic pathways were first characterized in humans. Serotonin is synthesized from tryptophan in a short biosynthetic route that involves only two enzymatic steps where tryptophan is converted into 5-hydroxytriptophan and then into serotonin by the enzymes tryptophan hydroxylase (TPH) and aromatic L-amino acid decarboxylase (AADC), respectively (Veenstra-VenderWeele et al. 2000). Melatonin synthesis continues from serotonin by two following reactions catalyzed by the arylalkylamine *N*-acetyltransferase (AANAT) and the *N*-acetylserotonin methyltransferase (ASMT) to form *N*-acetylserotonin and melatonin, respectively (Yu and Reiter 1993; Arnao and Hernández-Ruiz 2006). In plants, serotonin and melatonin biosynthesis also occurs from L-tryptophan through a reaction that first converts L-tryptophan to tryptamine by the tryptophan decarboxylase (TDC) and then the tryptamine is converted to serotonin by the tryptamine 5-hydroxilase (T5H), followed by serotonin conversion to *N*-acetylserotonin and then into melatonin by the enzymes serotonin *N*-acetyltransferase (SNAT) and *N*-acetylserotonin *O*-methyltransferase (ASMT), respectively (Figure 8.1) (Tan et al. 2014). Interestingly, auxin biosynthesis also has been proposed to occur from a tryptamine pathway (Quittenden et al. 2009). Most genes involved in serotonin and melatonin biosynthesis (TDC, T5H, SNAT, and ASMT) have been cloned and characterized mostly from rice plants (Kang et al. 2007a,b, 2011, 2013). Seven TDC genes are present in the genome of rice plants and three of them have been demonstrated to be functional. A T5H gene identified as cytochrome P450 monooxygenase subform CYP71P1 is involved in serotonin synthesis (Fujiwara et al. 2010). Plant SNAT seems to be unrelated to the SNAT from animals due to the lack of homology or cellular localization. Recent studies further showed that plant SNAT is located in chloroplasts, where it acts as a rate-limiting enzyme in melatonin biosynthesis. In contrast, plant and animal ASMT share homology and have a cytoplasmic localization; plant ASMT is highly expressed in the root system, suggesting organ-specific roles (Byeon et al. 2014). The evidence already available thus suggests that serotonin and melatonin biosynthesis in plants and animals occurs in a different manner, as chloroplasts are plant-specific organelles. However, some reports have shown some commonalities, since both plant and animals produce 5-hydroxytryptophan as intermediate compound in serotonin and melatonin biosynthesis (Murch et al. 2000; Park et al. 2012; Tan et al. 2012, 2013).

Serotonin and Melatonin as Plant Growth-Promoting Compounds

Due to the structural similarity with auxin, initial attempts to explain the biological activities of serotonin and melatonin, particularly in root and shoot developmental processes, were attributed to possible auxin-like activity (Csaba and Pal 1982; Murch and Saxena 2002; Arnao and Hernández-Ruiz 2006; Ramakrishna et al. 2009, 2011a). The function of serotonin and melatonin on plant growth regulation has been widely investigated. Csaba and Pal (1982) found that serotonin treatment increases root length in barley, showing a higher activity than the synthetic auxin naphthylacetic acid (NAA) and IAA. Years later, it was reported that both serotonin and melatonin have an important role on St. John's wort (*Hypericum performatum* L.) morphogenesis, since increased root and shoot formation were correlated with high levels of melatonin and serotonin, respectively, suggesting that a balance in the serotonin–melatonin ratio may affect plant morphogenesis (Murch et al. 2001). Melatonin promotes vegetative growth in dark-grown *Lupinus albus* L. hypocotyls, increases cotyledon expansion, and stimulates adventitious- and lateral root formation in this plant species in a similar manner to IAA, which led

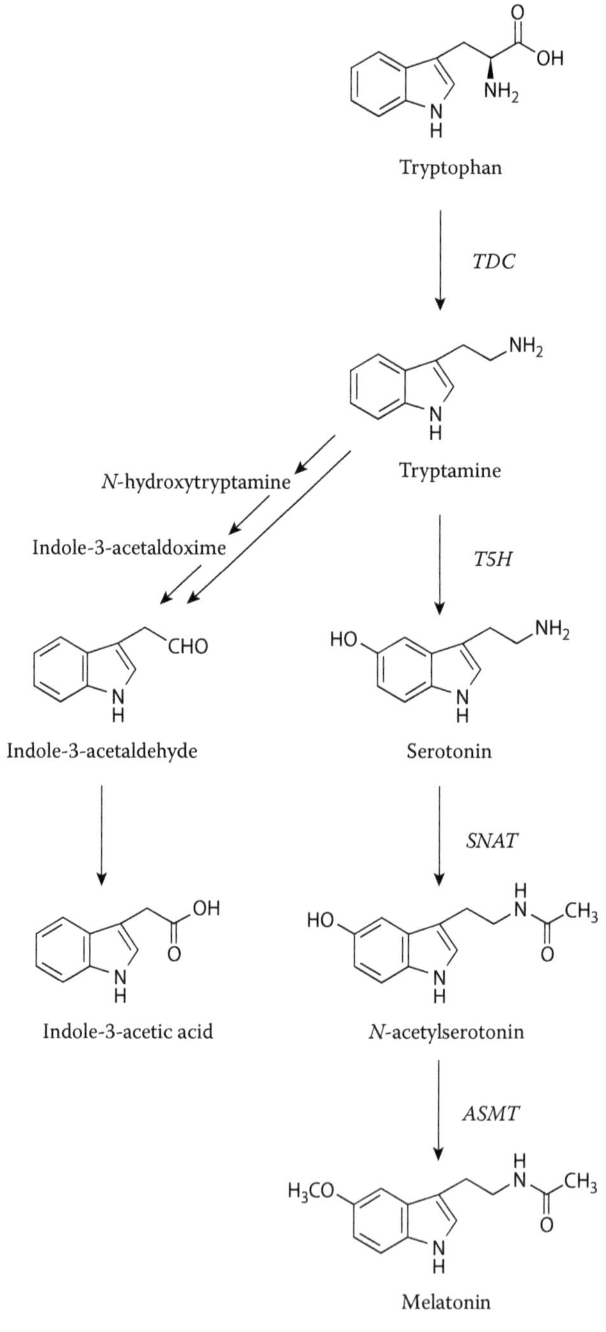

FIGURE 8.1 The proposed pathway for serotonin and melatonin biosynthesis in rice plants. The specific reactions catalyzed by tryptophan decarboxylase (TDC), tryptamine 5-hydroxylase (T5H), serotonin *N*-acetyltransferase (SNAT), and *N*-acetylserotonin-*O*-methyltransferase (ASMT) are shown.

the authors to suggest an auxinic activity for melatonin on plant growth (Hernández-Ruiz et al. 2004; Arnao and Hernández Ruiz 2007; Hernández-Ruiz and Arnao 2008). Similarly, melatonin had a growth-promoting effect in coleoptiles of wheat, barley, canary grass, and oat, but in this case the melatonin activity was lower than that of IAA. In contrast, both melatonin and IAA showed an inhibitory effect in root growth in these monocotyledonous plants (Hernández-Ruiz et al. 2005). In agreement with the above-described results, melatonin applied to etiolated seedlings of wild leaf mustard (*Brassica juncea*) stimulated root growth (Chen et al. 2009). Despite these seminal reports, until recently, clear evidence was lacking demonstrating the auxin activities of serotonin and melatonin at the molecular level. It was Pelagio-Flores and associates (2011, 2012) who characterized the effects of both indoleamines on the architecture of the root system in *Arabidopsis thaliana* seedlings, as well as their relationship with the auxin signaling pathway by analyzing auxin-inducible gene expression in transgenic seedling expressing the DR5:uidA and BA3:uidA gene markers. Both serotonin and melatonin modulated root system architecture in a highly specific and dose-responsive manner, affecting different processes of growth and development (Figure 8.2a). Serotonin was found to modulate primary root growth, lateral and adventitious root formation, and root hair development (Pelagio-Flores et al. 2011). Melatonin regulated root system architecture by stimulating lateral and adventitious root formation but minimally affected primary root growth or root hair development (Pelagio-Flores et al. 2012). Intriguingly, the melatonin response was independent of the auxin signaling pathway, because melatonin treatments that promote lateral root growth did not affect auxin-responsive gene expression. In contrast, serotonin inhibited the expression of DR5:uidA marker (Figure 8.2b), indicating that it likely acts as an endogenous auxin inhibitor. Other

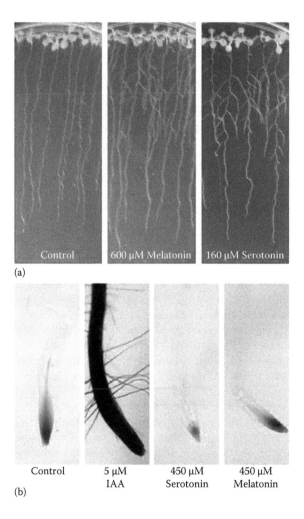

FIGURE 8.2 The effects of melatonin and serotonin on *Arabidopsis* root system architecture and auxin-regulated gene expression: (a) melatonin and serotonin effects on primary and lateral root formation and (b) comparative effect of IAA, melatonin, and serotonin on the auxin-responsive marker DR5:uidA. Notice that neither serotonin nor melatonin activated auxin-inducible gene expression; in contrast, serotonin decreases expression of the markers in both shoots and roots. Representative photographs of *Arabidopsis* seedlings treated with the indicated compounds are shown.

recent reports have evidenced the rhizogenic capacity of melatonin; a pair of these reports were published by Sarropoulou et al. (2012a,b), who observed an auxinic response to melatonin in explants of the sweet cherries, cherry rootstock PHL-C (*Prunus avium* × *Prunus cerasus*), cherry rootstocks CAB-6P (*Prunus cerasus* L.), Gisela 6 (*P. cerasus* × *P. canescens*), and MxM 60 (*P. avium* × *P. mahaleb*) on adventitious root formation and root growth. Moreover, the use of transgenic rice lines that overexpress SNAT from sheep revealed that enhanced melatonin levels in these transgenic lines correlated with increasing adventitious root lengths and improved growth (Park and Back 2012). In cucumber (*Cucumis sativus*), melatonin concentration of 300 μM promoted lateral root growth (Zhang et al. 2013, 2014). The beneficial effects of melatonin to plants via promotion of root branching were recently confirmed in *Punica granatum* cv. (Sarrou et al. 2014). An analysis of most reports about the effects of serotonin and melatonin on root growth and lateral and/or adventitious roots thus indicates that these indolamines may affect developmental processes either acting as auxin inhibitors or more likely in an auxin-independent pathway (Pelagio-Flores et al. 2011, 2012; Koyama et al. 2013). Thus, it is tempting to speculate that specific signal transduction pathways may exist, which control the cellular and developmental responses to internal levels of serotonin and melatonin. In agreement with this speculation, analysis of genes differentially expressed in response to melatonin in rice (Byeon et al. 2013), cucumber (Zhang et al. 2014) and *Arabidopsis* (Weeda et al. 2014) have shown that auxin-related genes are minimally represented or are mostly downregulated after melatonin treatment, thus providing support to an auxin-unrelated or antagonist signaling pathway mediating melatonin response.

Abiotic Stress Tolerance

Oxidative stress is an important inducer of damage and cellular death caused by an increase in the level of reactive oxygen species (ROS) and other free radicals. ROS play an important role in regulating plant growth and development, defense, stomatal closure, and cell signaling (Wrzaczek et al. 2013; Kangasjärvi and Kangasjärvi 2014). Plants as sessile organisms are exposed to diverse biotic and abiotic stresses that in most cases induce ROS accumulation, which together with the ROS generated as result of metabolism and the high photosynthetic activity of plants could lead to an oxidative stress that needs to be controlled to avoid damage of plant tissues (Foyer and Noctor 2013; Considine and Foyer 2013; Ramakrishna and Ravishankar 2011, 2013).

Serotonin and melatonin possess a high antioxidant capacity that overpasses that of the antioxidants ascorbic acid and tocopherol (Poeggeler et al. 2002; Kang et al. 2009a). In this regard, many beneficial effects of melatonin in response to the majority of abiotic stresses have been evidenced. Increased melatonin levels are found in tomato (*Lycopersicon esculentum* Mill.), tobacco (*Nicotiana tabacum* L.), and in alpine and Mediterranean plants naturally exposed to high UV levels when compared to the same species exposed to lower UV radiation (Dubbels et al. 1995; Tettamanti et al. 2000). Melatonin levels correlated with tolerance to ozone, a free radical generator, suggesting its antioxidant role (Hardeland and Pandi-Perumal 2005). Similar data were obtained in *Glycyrrhiza uralensis*, in which plants grown in high-intensity UV-B radiation showed an enhanced melatonin accumulation in roots when compared with plants grown in low-intensity radiation, suggesting that the high-intensity UV-B radiation increases the melatonin levels as a natural response to protect the plants from oxidative stress (Afreen et al. 2006). In addition, the detrimental effects induced by abiotic stresses, such as drought, cold, heat, salinity, alkalinity, chemical pollutants, and herbicide treatments on plant growth can be alleviated by melatonin via ROS scavenging, and regulating the expression and activity of antioxidant enzymes (Kang et al. 2010; Li et al. 2012; Park et al. 2013; Bajwa et al. 2014; Arnao and Hernández-Ruiz 2014; Zhang et al. 2015; Liu et al. 2015).

The molecular mechanisms by which melatonin enhance stress tolerance have been poorly studied. However, recent findings indicate that melatonin decreases the drought stress in two *Malus* species by regulating expression of genes involved in ABA metabolism and decreasing the ABA content, and that it also acts as a scavenger of H_2O_2, and improves the stomatal function maintaining open stomata (Li et al. 2015). Multiple genes involved in diverse plant processes such as nitrogen and carbohydrate metabolism, the tricarboxylic acid cycle, transport, hormone metabolism, metal homeostasis, redox reactions, and secondary metabolism are upregulated in bermudagrass (*Cynodon dactylon* [L.] Pers.) by melatonin (Shi et al. 2015). All the above-mentioned evidences indicate that environmental stress can increase the level of endogenous melatonin in plants. Concomitantly, overexpression of the melatonin biosynthetic genes in crops could represent a promising avenue toward adapting plants to changing and/ or challenging environments.

Regulation of Senescence

Serotonin and melatonin modulate plant senescence via their antioxidant properties. The phytohormones abscisic acid (ABA), salicylic acid (SA), and jasmonic acid (JA) promote senescence, while cytokinins delay it. Serotonin and melatonin also delay the senescence process in different plants and under diverse growth conditions. Melatonin showed a clear protective effect delaying the senescence process in detached barley (*Hordeum vulgare* L.) leaves, evidenced by the greater chlorophyll content, where it could act through specific control of genes associated to senescence (SAGs) or via its role as antioxidant, preventing oxidative stress (Arnao and Hernández-Ruiz 2009). Similarly, serotonin delayed senescence induced by nutrient deprivation or leaf detachment in rice. Transgenic rice plants with high serotonin levels showed delayed senescence, while plants with suppressed serotonin production senesced faster than the wild type, and these effects were associated with the high antioxidant capacity of serotonin suggesting that it may protect the cellular integrity via facilitating efficient nutrient recycling from senescing leaves to sink tissues (Kang et al. 2009a).

Melatonin regulates senescence of apple leaves by modulating the levels and activity of antioxidative enzymes and suppression of senescence genes (Wang et al. 2012). Similar outcomes were observed in apple leaf senescence under drought conditions (Wang et al. 2013a) and in leaf senescence of *Malus hupehensis* (Wang et al. 2013b). In conclusion, the above-described effects of melatonin on senescence regulation have been closely related with their properties as free radical scavengers, preventing the ROS generation or by affecting directly or indirectly the antioxidant enzymes and/or senescence genes (Tan et al. 2002, 2012; Kang et al. 2009a; Galano et al. 2011, 2013). Although the molecular mechanisms of serotonin and melatonin action on senescence regulation remain unclear, a recent transcriptome analysis from *Arabidopsis* in response to melatonin indicated that the transcription factors auxin-resistant 3 (AXR3)/indole-3-acetic acid–inducible 17 (IAA17) and short hypocotyl 2 (SHY2)/IAA3, related to auxin signaling, were downregulated by melatonin (Weeda et al. 2014). Shi et al. (2014) found that melatonin levels in *Arabidopsis* rosette leaves were induced in an age-dependent manner and that leaf senescence is delayed by exogenous melatonin, which correlated with downregulated expression of AXR3/IAA17. Besides, AtIAA17-overexpressing plants and AtIAA17 knockout mutants showed contrasting effects with early and delayed leaf senescence, respectively, indicating that AXR3/IAA17 plays an essential role in senescence regulation mediated by melatonin probably affecting the expression of senescence-related SEN4 and SAG12 genes (Shi et al. 2014).

Activation of Defense Responses

The plant defense includes physical and chemical barriers. The protective effects of serotonin and melatonin to biotic stress have also been documented in plant defense both against pathogens and herbivores. Serotonin may be part of defense responses as it accumulates in trichomes of the bullnettle plant (*Cnidoscolus texanus*), as part of protective chemicals (Lookadoo and Pollard 1991). Some hydroxycinnamic acid amides have been implicated in the inducible defense of plants and are considered as phytoalexins; in this case, the hydroxycinnamic acid amides of serotonin such as *N-p*-coumaroylserotonin and *N*-feruloylserotonin, which are present in low amounts on healthy bamboo plants and are accumulated in diseased plants in response to pathogenic fungi, play a role in defense via their antifungal activity (Tanaka et al. 2003).

Serotonin plays an important role in defense against the pathogen fungus *Bipolaris oryzae*, since serotonin and its hydroxycinnamic acid amides are accumulated and incorporated into the cell wall at damaged areas in infected rice leaves, possibly serotonin polymers form a physical barrier, which prevents the spread of the pathogen at the site of infection (Ishihara et al. 2008a). In another report, Ishihara et al. (2008b) evaluated the plant response against herbivore attack using the striped rice stem borer (*Chilo suppressalis*). The authors found increased levels of serotonin and *p*-coumaroylserotonin in the larvae-fed leaves, similarly to other reports of plants grown in natural conditions or upon pathogen infection (Ly et al. 2008; Kang et al. 2009b). Thus, the defense properties of serotonin are explained via reinforcement of cell walls, its strong antioxidant activity, and antifungal activity. As serotonin, melatonin can modulate defense responses in apple plants (*Malus prunifolia* [Willd.] Borkh. cv. Donghongguo) against the damage caused by the fungus *Diplocarpon mali*. Pretreatment of plants with melatonin improved the resistance to *Marssonina* apple blotch, by reducing the number of lesions, inhibiting pathogen expansion, maintaining a high potential efficiency of photosystem II, improving the total chlorophyll content, modulating the oxidation–reduction system and enhancing the activities of plant defense–related enzymes such as chitinase and β-1,3-glucanasa (Yin et al. 2013). Melatonin treatments limit the propagation of *Pseudomonas syringae* DC3000 in *Arabidopsis* leaves inhibiting bacterial growth tenfold compared to untreated leaves. These effects were related to induction of pathogenesis-related (PR) genes, as well as several genes induced in defense responses mediated by SA and ethylene. Further analysis of *Arabidopsis* mutants suggested that the mechanisms by which melatonin modulate the plant defense responses are mediated by the SA and ethylene signaling pathways (Lee et al. 2014).

Serotonin and Melatonin in Plant Productivity

Yields of the major cereal crops such as rice, wheat, and maize have increased steadily over the past years. However, to meet cereal production demand in the next decade, increase of yields must continue. Crop yield might be influenced by several factors including root system architecture, environmental conditions, and the senescence process, all of which are regulated by serotonin and melatonin (Janas and Posmyk 2013). Thus, it is not surprising that these indoleamines could have a positive effect on crop yield. Moreover, crop yields are limited by a combination of biotic stresses, abiotic stresses, and nutritional factors. Various analyses have suggested that drought, heat, cold, or salinity are the major factors that prevent crops from realizing their full yield. Recent interest in renewable fuels has led to a substantial increase in ethanol production from plant materials. Although the initial emphasis has been on using starch from corn, and to a lesser extent from other food grains, this is unlikely to be sustainable in the long term as maize is of primary importance for human nutrition; alternatives such as cellulose sources from crops with high biomass production such as sugarcane are likely to be a more important substrate for ethanol production. The challenge here is to increase total vegetative biomass rather than seed yield. It is expected that by increasing total leaf area and leaf number, while decreasing senescence, the photosynthetic capacity of plants as well their overall yield should increase.

The potential of serotonin and melatonin to increase biomass production of plants has been found to depend on the plant species, growth conditions, and time of application. Contrasting effects on yield have been reported for serotonin. A study conducted by Kanjanaphachoat et al. (2012) showed that in rice plants overexpression of two putative tryptophan decarboxylase genes, TDC-1 and TDC-3, involved in serotonin synthesis, increases serotonin levels, causing a stunted growth, low fertility, and dark brown color. Enzymatic assays of both TDC-1 and TDC-3 recombinant proteins showed tryptophan decarboxylase (TDC) activities that converted tryptophan to tryptamine, which could be converted to serotonin by a constitutively expressed tryptamine 5-hydroxylase (T5H). A mass-spectrometry assay demonstrated that the dark brown leaves in the TDC-overexpressing lines were caused by the accumulation of serotonin but not tryptamine. These results represent the first evidence that over-expression of TDC results in deleterious effects to plants. In another report, the roots and shoot of rice seedlings grown in the presence of tryptamine exhibited a dose-dependent increase in serotonin in parallel with enhanced T5H enzyme activity. However, no detrimental growth was evidenced and the seedlings retained a normal phenotype (Kang et al. 2007a). *In vitro* experiments using *Arabidopsis* seedlings germinated and grown in media supplied with increasing concentrations of serotonin revealed that serotonin indeed represses growth and causes dark brown color in roots and shoots (Pelagio-Flores et al. 2011). More recent studies

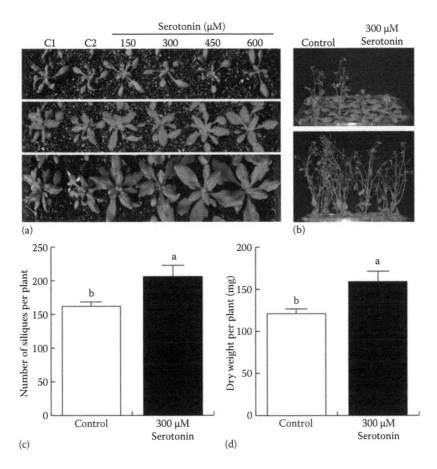

FIGURE 8.3 The effect of serotonin on *Arabidopsis* growth. *Arabidopsis* seedlings were grown 10 days *in vitro* at indicated serotonin concentrations and then transferred to soil to complete their life cycle. Note the positive impact of serotonin in rosette leaves (a and b), seed, and biomass production (c and d, respectively). Plants were grown with a photoperiod of 16 h light/8 h darkness at 22°C in a growth chamber.

in which *Arabidopsis* seedlings were pretreated for 10 days with serotonin and then transferred to soil indicated a delayed life cycle, and enhanced leaf biomass production and seed yield (Figure 8.3a and b). Interestingly, serotonin pretreatment delayed the transition from vegetative growth to flowering in plants transferred to soil, thereby extending the period of vegetative growth of plants, which leads to greater biomass, larger leaves, thicker stems, and more fruits and seeds (Figure 8.3c and d). These results are promising toward the use of serotonin, which may be now incorporated into different agricultural biostimulants to increase leaf biomass and photosynthesis in crops for more extensive evaluation under field conditions.

Melatonin supply to cucumber and corn seeds had a beneficial effect on the growth of seedlings and crop production of plants, especially when subjected to cold (Posmyk et al. 2009) and water stress (Zhang et al. 2013). In experiments conducted under field conditions, plants from seeds of corn (*Zea mays* L.), mung bean (*Vigna radiata* L.), and cucumber (*Cucumis sativus* L.) pretreated with melatonin had higher yields than untreated plants (Janas and Posmyk 2013). Treatment of seeds with melatonin also increased leaf size, plant height, seed production, and fatty acid content in soybean. Melatonin also improved soybean tolerance to salt and drought stresses through upregulation of genes related to secondary metabolic processes, cell division, photosynthesis, carbohydrate metabolism, fatty acid biosynthesis, and ascorbate metabolism, which are normally repressed by salt stress. These results show the potential of melatonin for improvement of growth and seed production in different crops (Wei et al. 2015).

Selection and breeding for high biomass–yielding varieties is a traditional strategy to increase total plant biomass. Alternatively, some effort is now going into the use of transgenic approaches to increase total plant biomass; in this regard, the genes for serotonin and melatonin biosynthesis are promising targets for potential biotechnological applications.

Conclusion

Serotonin and melatonin are emerging signals for the regulation of plant growth and development that play an important role in plant adaptation to stressing environmental conditions. Melatonin acts as a circadian regulator, cytoprotector, and growth-promoting substance. It also acts in rhizogenesis, cellular expansion, and stress-protection. Serotonin modulates root branching, affecting both *de novo* formation of organs and cell differentiation of root epidermal cells. Currently, the main aspects of serotonin and melatonin research are to: (1) elucidate the signal transduction pathways and mechanisms of action of these indoleamines, (2) explore the activity of these compounds in crops and vegetables, and (3) evaluate the impact of their application as biostimulators in agriculture. Regarding the first aspect, the use of *Arabidopsis* reporter and mutant lines indicates that serotonin and melatonin did not increase auxin response; instead, serotonin actually inhibits auxin-regulated gene expression. Still remaining to be investigated is the possible crosstalk of serotonin and melatonin response with ethylene, cytokinin, JA, or SA signaling, particularly considering their proposed roles in defense and senescence. Application of melatonin improves root development and adaptation to soil stress such as drought, salinity, and pollutants. This may be explained because roots are the critical sites of water- and nutrient uptake by plants. Melatonin thus may enhance the rate of germination and modulate developmental transitions during the plant life cycle, with positive impact on plant productivity. This may be due to melatonin acting as a retardant of stress-induced leaf senescence. From this information, it is tempting to speculate that increasing melatonin levels in crops or obtaining melatonin-overproducing plants may be a promising strategy to improve plant productivity under adverse environmental conditions that affect harvest index, or to better adapt plants to climate changes. Another proposal concerns the use of plants rich in melatonin as a tool in phytoremediation of heavy metal–contaminated soils, a strategy that certainly deserves further attention. An interesting point to underline is that, although serotonin and melatonin appear to modulate the same physiological processes such as root development, they seem to act through different mechanisms. Future studies should be conducted to determine the factors that increase serotonin and melatonin biosynthesis and the relationship with auxin biosynthesis, particularly considering that all three compounds are produced from tryptophan as precursor.

REFERENCES

Afreen, F., Zobayed, S. M., and Kozai, T. Melatonin in *Glycyrrhiza uralensis*: Response of plant roots to spectral quality of light and UV-B radiation. *J Pineal Res*. 2006. 41: 108–115.

Arnao, M. B. Phytomelatonin: Discovery, content, and role in plants. *Adv Bot*. 2014. 2014: 11.

Arnao, M. B., and Hernández-Ruiz, J. The physiological function of melatonin in plants. *Plant Signal Behav*. 2006. 1: 89–95.

Arnao, M. B., and Hernández-Ruiz, J. Melatonin promotes adventitious- and lateral root regeneration in etiolated hypocotyls of *Lupinus albus* L. *J Pineal Res*. 2007. 42: 147–152.

Arnao, M. B., and Hernández-Ruiz, J. Protective effect of melatonin against chlorophyll degradation during the senescence of barley leaves. *J Pineal Res*. 2009. 46: 58–63.

Arnao, M. B., and Hernández-Ruiz, J. Melatonin: Possible role as light-protector in plants. In Radosevich, J. A., Physics Research & Technology Series (editor), *UV Radiation: Properties, Effects, and Applications*. 2014. 79–92. New York: Nova Science.

Badria, F. A. Melatonin, serotonin and tryptamine in some Egyptian food and medicinal plants. *J Med Food*. 2002. 5: 153–157.

Bajwa, V. S., Shukla, M. R., Sherif, S. M., Murch, S. J., and Saxena, P. K. Role of melatonin in alleviating cold stress in *Arabidopsis thaliana*. *J Pineal Res*. 2014. 56: 238–245.

Bowden, K., Brown, B. G., and Batty, J. E. 5-hydroxytryptamine: Its occurrence in cowhage (*Mucuna pruriens*). *Nature*. 1954. 174: 925–926.

Byeon, Y., Lee, H. Y., Lee, K., Park, S., and Back, K. W. Cellular localization and kinetics of the rice melatonin biosynthetic enzymes SNAT and ASMT. *J Pineal Res*. 2014. 56: 107–114.

Byeon, Y., Park, S., Kim, Y. S., and Back, K. W. Microarray analysis of genes differentially expressed in melatonin-rich transgenic rice expressing a sheep serotonin *N*-acetyltransferase. *J Pineal Res*. 2013. 55: 357–363.

Chen, G. F., Huo, Y. S., Tan, D. X., Liang, Z., Zhang, W. B., and Zhang, Y. K. Melatonin in Chinese medicinal herbs. *Life Sci*. 2003. 73: 19–26.

Chen, Q., Qi, W. B., Reiter, R. J., Wei, W., and Wang, B. M. Exogenously applied melatonin stimulates root growth and raises endogenous indoleacetic acid in roots of etiolated seedlings of *Brassica juncea*. *J Plant Physiol*. 2009. 166: 324–328.

Considine, M. J., and Foyer, C. H. Redox regulation of plant development. *Antioxid Redox Signal*. 2013. 21 (9): 1305–1326.

Csaba, G., and Pal, K. Effect of insulin triiodothyronine and serotonin on plant seed development. *Protoplasma* 1982. 110: 20–22.

Dubbels, R. Melatonin in edible plants identified by radioimmunoassay and by HPLC-MS. *J Pineal Res*. 1995. 18: 28–31.

Feldman, J. M., and Lee, E. M. Serotonin content of foods: Effect on urinary excretion of 5-hydroxy indoleacetic acid. *Am J Clin Nutr*. 1985. 42: 639–643.

Foyer, C. H., and Noctor, G. D. Redox signaling in plants. *Antioxid Redox Signal*. 2013. 18: 2087–2090.

Fujiwara, T., Maisonneuve, S., Isshiki, M., Mizutani, M., Chen, L., Wong, H. L., Kawasaki, T., and Shimamoto, K. Sekiguchi lesion gene encodes a cytochrome P450 monooxygenase that catalyzes conversion of tryptamine to serotonin in rice. *J Biol Chem*. 2010. 285: 11308–11313.

Galano, A., Tan, D. X., and Reiter, R. J. Melatonin as a natural ally against oxidative stress: A physicochemical examination. *J Pineal Res*. 2011. 51: 1–16.

Galano, A., Tan, D. X., and Reiter, R. J. On the free radical scavenging activities of melatonin's metabolites, AFMK and AMK. *J Pineal Res*. 2013. 54: 245–257.

Hardeland, R. Melatonin in plants and other phototrophs: Advances and gaps concerning the diversity of functions. *J Exp Bot*. 2015. 66: 627–646.

Hardeland, R., and Pandi-Perumal, S. R. Melatonin, a potent agent in antioxidative defense: Actions as a natural food constituent, gastrointestinal factor, drug and prodrug. *Nutr Metab Lond*. 2005. 2: 22.

Hernández-Ruiz, J., and Arnao, M. B. Melatonin stimulates the expansion of etiolated lupin cotyledons. *Plant Growth Regul*. 2008. 55: 29–34.

Hernández-Ruiz, J., Cano, A., and Arnao, M. B. Melatonin: A growth-stimulating compound present in lupin tissues. *Planta*. 2004. 220: 140–144.

Hernández-Ruiz, J., Cano, A., and Arnao, M. B. Melatonin acts as a growth-stimulating compound in some monocot species. *J Pineal Res*. 2005. 39: 137–142.

Ishihara, A., Hashimoto, Y., Miyagawa, H., and Wakasa, K. Induction of serotonin accumulation by feeding of rice striped stem borer in rice leaves. *Plant Signal Behav*. 2008b. 3: 714–716.

Ishihara, A., Hashimoto, Y., Tanaka, C., Dubouzet, J. G., Nakao, T., Matsuda, F., Nishioka, T., Miyagawa, H., and Wakasa, K. The tryptophan pathway is involved in the defense responses of rice against pathogenic infection via serotonin production. *Plant J*. 2008a. 54: 481–495.

Janas, K., and Posmyk, M. Melatonin, an underestimated natural substance with great potential for agricultural application. *Acta Physiol Plant*. 2013. 35: 3285–3292.

Kanjanaphachoat, P., Wei, B. Y., Lo, S. F., Wang, I. W., Wang, C. S., Yu, S. M., Yen, M. L., Chiu, S. H., Lai, C. C., and Chen, L. J. Serotonin accumulation in transgenic rice by over-expressing tryptophan decarboxylase results in a dark brown phenotype and stunted growth. *Plant Mol Biol*. 2012. 78: 525–543.

Kang, S., Kang, K., Lee, K., and Back, K. Characterization of tryptamine-5-hydroxylase and serotonin synthesis in rice plants. *Plant Cell Rep*. 2007a. 26: 2009–2015.

Kang, S., Kang, K., Lee, K., and Back, K. Characterization of rice tryptophan decarboxylases and their direct involvement in serotonin biosynthesis in transgenic rice. *Planta*. 2007b. 227: 263–272.

Kang, K., Kim, Y. S., Park, S., and Back, K. Senescence-induced serotonin biosynthesis and its role in delaying senescence in rice leaves. *Plant Physiol*. 2009a. 150: 1380–1393.

Kang, K., Kong, K., Park, S., Natsagdorj, U., Kim, Y. S., and Back, K. Molecular cloning of a plant *N*-acetylserotonin methyltransferase and its expression characteristics in rice. *J Pineal Res*. 2011. 50: 304–309.

Kang, K., Lee, K., Park, S., Byeon, Y., and Back, K. Molecular cloning of rice serotonin *N*-acetyltransferase, the penultimate gene in plant melatonin biosynthesis. *J Pineal Res*. 2013. 55: 7–13.

Kang, K., Lee, K., Park, S., Kim, Y. S., and Back, K. Enhanced production of melatonin by ectopic overexpression of human serotonin *N*-acetyltransferase plays a role in cold resistance in transgenic rice seedlings. *J Pineal Res*. 2010. 49: 176–182.

Kang, K., Park, S., Kim, Y. S., Lee, S., and Back, K. Biosynthesis and biotechnological production of serotonin derivatives. *Appl Microbiol Biotechnol*. 2009b. 83: 27–34.

Kangasjärvi, S., and Kangasjärvi, J. Towards understanding extracellular ROS sensory and signaling systems in plants. *Adv Bot*. 2014. 2014: 10.

Kolar, J., Machackova, I., Illnerova, H., Prinsen, E., Van Dogen, W., and Van Onckelen, H. A. Melatonin in higher plants determined by radioimmunoassay and liquid chromatography mass spectrometry. *Biol Rhythm Res*. 1995. 26: 406–409.

Koyama, F. C., Carvalho, T. L., Alves, E., da Silva, H. B., de Acevedo, M. F., Hemerly, A. S., and Garcia, C. R. S. The structurally related auxin and melatonin tryptophan-derivatives and their roles in *Arabidopsis thaliana* and in the human malaria parasite *Plasmodium falciparum*. *J Eukaryot Microbiol*. 2013. 60: 646–651.

Lee, H. Y., Byeon, Y., and Back, K. Melatonin as a signal molecule triggering defense responses against pathogen attack in *Arabidopsis* and tobacco. *J Pineal Res*. 2014. 57: 262–268.

Li, C., Liang, D., Chang, C., Jia, D., and Ma, F. Melatonin mediates the regulation of ABA metabolism, free radical–scavenging, and stomatal behavior in two *Malus* species under drought stress. *J Exp Bot*. 2015. 66: 669–680.

Li, C., Wang, P., Wei, Z., Liang, D., Liu, C., Yin, L., Jia, D., Fu, M., and Ma, F. The mitigation effects of exogenous melatonin on salinity-induced stress *in Malus hupehensis*. *J Pineal Res*. 2012. 53: 298–306.

Liu, N., Jin, Z., Wang, S., Gong, B., Wen, D., Wang, X., Wei, M., and Shi, Q. Sodic alkaline stress mitigation with exogenous melatonin involves reactive oxygen metabolism and ion homeostasis in tomato. *Sci Hortic*. 2015. 181: 18–25.

Lookadoo, S. E., and Pollard, A. J. Chemical contents of stinging trichomes of *Cnidoscolus texanus*. *J Chem Ecol*. 1991. 17: 1909–1916.

Ly, D., Kang, K., Choi, J. Y., Ishihara, A., Back, K., and Lee, S. G. HPLC analysis of serotonin, tryptamine, tyramine and the hydroxycinnamic acid amides of serotonin and tyramine in food vegetables. *J Med Food*. 2008. 11: 385–389.

Murch, S. J., Campbell, S. S. B., and Saxena, P. K. The role of serotonin and melatonin in plant morphogenesis: Regulation of auxin-induced root organogenesis in *in vitro*-cultured explants of St. John's wort (*Hypericum perforatum* L.). *In Vitro Cell Dev Biol Plant*. 2001. 37: 786–793.

Murch, S. J., Krishna, R. S., and Saxena, P. K. Tryptophan is a precursor for melatonin and serotonin biosynthesis in *in vitro*-regenerated St. John's wort (*Hypericum perforatum* L. cv. Anthos) plants. *Plant Cell Rep*. 2000. 19: 698–704.

Murch, S. J., and Saxena, P. K. Melatonin: A potential regulator of plant growth and development? *In Vitro Cell Dev Biol Plant*. 2002. 38: 531–536.

Murch, S. J., Simmons, C. B., and Saxena, P. K. Melatonin in feverfew and other medicinal plants. *Lancet*. 1997. 350: 1598–1599.

Odjakova, M., and Hadjiivanova, C. Animal neurotransmitter substances in plants. *Bulg J Plant Physiol*. 1997. 23: 94–102.

Paredes, S. D., Korkmaz, A., Manchester, L. C., Tan, D. X., and Reiter, R. J. Phytomelatonin: A review. *J Exp Bot*. 2009. 60: 57–69.

Park, S., and Back, K. Melatonin promotes seminal root elongation and root growth in transgenic rice after germination. *J Pineal Res*. 2012. 53: 385–389.

Park, S., Lee, D. E., Jang, H., Byeon, Y., Kim, Y. S., and Back, K. Melatonin-rich transgenic rice plants exhibit resistance to herbicide-induced oxidative stress. *J Pineal Res*. 2013. 54: 258–263.

Park, S., Lee, K., and Kim, Y. S. Tryptamine 5-hydroxylase–deficient Sekiguchi rice induces synthesis of 5-hydroxytryptophan and *N*-acetyltryptamine but decreases melatonin biosynthesis during senescence process of detached leaves. *J Pineal Res*. 2012. 52: 211–216.

Pelagio-Flores, R., Muñoz-Parra, E., Ortiz-Castro, R., and López-Bucio, J. Melatonin regulates *Arabidopsis* root system architecture likely acting independently of auxin signaling. *J Pineal Res.* 2012. 53: 279–288.

Pelagio-Flores, R., Ortiz-Castro, R., Méndez-Bravo, A., Macías-Rodríguez, L., and López-Bucio, J. Serotonin, a tryptophan-derived signal conserved in plants and animals, regulates root system architecture probably acting as a natural auxin inhibitor in *Arabidopsis thaliana*. *Plant Cell Physiol.* 2011. 52: 490–508.

Poeggeler, B., Thuermann, S., Dose, A., Schoenke, M., Burkhardt, S., and Hardeland, R. Melatonin's unique radical scavenging properties: Roles of its functional substituents as revealed by a comparison with its structural analogs. *J Pineal Res.* 2002. 33: 20–30.

Posmyk, M. M., Bałabusta, M., Wieczorek, M., Sliwinska, E., and Janas, K. M. Melatonin applied to cucumber (*Cucumis sativus* L.) seeds improves germination during chilling. *J Pineal Res.* 2009. 46: 214–223.

Quittenden, L. J., Davies, N. W., Smith, J. A., Molesworth, P. P., Tivendale, N. D., and Ross, J. J. Auxin biosynthesis in pea: Characterization of the tryptamine pathway. *Plant Physiol.* 2009. 151: 1130–1138.

Ramakrishna, A. Indoleamines in edible plants: Role in human health effects. In Angel Catalá, Biochemistry Research Trends Series (editor), *Indoleamines: Sources, Role in Biological Processes and Health Effects*. 2015. 279. New York: Nova Science Publishers.

Ramakrishna, A., Dayananda, C., Giridhar, P., Rajasekaran, T., and Ravishankar, G. A. Photoperiod influences endogenous indoleamines in cultured green alga *Dunaliella bardawil*. *Indian J Exp Biol.* 2011b. 49: 234–240.

Ramakrishna, A., Giridhar, P., and Ravishankar, G. A. Indoleamines and calcium channels influence morphogenesis in *in vitro* cultures of *Mimosa pudica* L. *Plant Signal Behav.* 2009. 12: 1–6.

Ramakrishna, A., Giridhar, P., and Ravishankar, G. A. Phytoserotonin: A review. *Plant Signal Behav.* 2011a. 6: 800–809.

Ramakrishna, A., Giridhar, P., Sankar, K. U., and Ravishankar, G. A. Endogenous profiles of indoleamines: Serotonin and melatonin in different tissues of *Coffea canephora* P ex Fr. as analyzed by HPLC and LC-MS-ESI. *Acta Physiol Plant.* 2012a. 34: 393–396.

Ramakrishna, A., Giridhar, P., Sankar, K. U., and Ravishankar, G. A. Melatonin and serotonin profiles in beans of *Coffea* species. *J Pineal Res.* 2012b. 52: 470–476.

Ramakrishna, A., and Ravishankar, G. A. Influence of abiotic stress signals on secondary metabolites in plants. *Plant Signal Behav.* 2011. 6: 1720–1731.

Ramakrishna, A., and Ravishankar, G. A. Role of plant metabolites in abiotic stress tolerance under changing climatic conditions with special reference to secondary compounds. In Narendra Tuteja and Sarvajeet S. Gill (editors), *Climate Change and Abiotic Stress Tolerance*. 2013. Weinheim, Germany: Wiley.

Reiter, R. J. The melatonin rhythm: Both a clock and a calendar. *Experientia.* 1993. 49: 654–664.

Roshchina, V. V. *Neurotransmitters in Plant Life*. 2001. 4–81. Enfield, New Hampshire: Science Publishers.

Sarropoulou, V., Dimassi-Theriou, K., Therios, I., and Koukourikou-Petridou, M. Melatonin enhances root regeneration, photosynthetic pigments, biomass, total carbohydrates and proline content in the cherry rootstock PHL-C (*Prunus avium × Prunus cerasus*). *Plant Physiol Biochem.* 2012a. 61: 162–168.

Sarropoulou, V. N., Therios, I. N., and Dimassi-Theriou, K. N. Melatonin promotes adventitious root regeneration in *in vitro* shoot tip explants of the commercial sweet cherry rootstocks CAB-6P (*Prunus cerasus* L.), Gisela 6 (*P. cerasus × P. canescens*), and MxM 60 (*P. avium × P. mahaleb*). *J Pineal Res.* 2012b. 52: 38–46.

Sarrou, E., Therios, I., and Dimassi-Theriou, K. Melatonin and other factors that promote rooting and sprouting of shoot cuttings in *Punica granatum cv.* Wonderful. *Turk J Bot.* 2014. 38: 293–301.

Shi, H., Jiang, C., Ye, T., Tan, D. X., Reiter, R. J., Zhang, H., Liu, R., and Chan, Z. Comparative physiological, metabolomic and transcriptomic analyses reveal mechanisms of improved abiotic stress resistance in Bermuda grass (*Cynodon dactylon* [L.] Pers.) by exogenous melatonin. *J Exp Bot.* 2015. 66: 681–694.

Shi, H., Reiter, R. J., Tan, D. X., and Chan, Z. Indole-3-acetic acid–inducible 17 positively modulates natural leaf senescence through melatonin-mediated pathway in *Arabidopsis*. *J Pineal Res.* 2014. 58: 26–33.

Tan, D. X., Hardeland, R., Manchester, L. C., Korkmaz, A., Ma, S., Rosales-Corral, S., and Reiter, R. J. Functional roles of melatonin in plants, and perspectives in nutritional and agricultural science. *J Exp Bot.* 2012. 63: 577–597.

Tan, D. X., Manchester, L. C., Liu, X. Y., Rosales-Corral, S. A., Acuna-Castroviejo, D., and Reiter, R. J. Mitochondria and chloroplasts as the original sites of melatonin synthesis: A hypothesis related to melatonin's primary function and evolution in eukaryotes. *J Pineal Res.* 2013. 54: 127–138.

Tan, D. X., Reiter, R. J., Manchester, L. C., Yan, M. T., El-Sawi, M., Sainz, R. M., Mayo, J. C., Kohen, R., Allegra, M. C., and Hardeland, R. Chemical and physical properties and potential mechanisms: Melatonin as a broad spectrum antioxidant and free radical scavenger. *Curr Top Med Chem.* 2002. 2:1 81–197.

Tan, D. X., Zheng, X., Kong, J., Manchester, L. C., Hardeland, R., Kim, S. J., Xu, X., and Reiter, R. J. Fundamental issues related to the origin of melatonin and melatonin isomers during evolution: Relation to their biological functions. *Int J Mol Sci.* 2014. 15: 15858–15890.

Tanaka, E., Tanaka, C., Mori, N., Kuwahara, Y., and Tsuda, M. Phenylpropanoid amides of serotonin accumulate in witches' broom diseased bamboo. *Phytochemistry.* 2003. 64: 965–969.

Tettamanti, C., Cerabolini, B., Gerola, P., and Conti, A. Melatonin identification in medicinal plants. *Acta Phytotherapeutica.* 2000. 3: 137–144.

Van Tassel, D. L., Li, J., and O'Neill, S. D. Melatonin: Identification of a potential dark signal in plants. *Plant Physiol.* 1993. 102 (1 suppl.): 659.

Van Tassel, D. L., Roberts, N. J., and O'Neill, S. D. Melatonin from higher plants: Isolation and identification of *N*-acetyl-5-methoxytryptamine. *Plant Physiol.* 1995. 108 (2 suppl.): 101.

Veenstra-VanderWeele, J., Anderson, G. M., and Cook, E. H. Pharmacogenetics and the serotonin system: Initial studies and future directions. *Eur J Pharm.* 2000. 410: 165–181.

Wang, P., Sun, X., Chang, C., Feng, F., Liang, D., Cheng, L., and Ma, F. W. Delay in leaf senescence of *Malus hupehensis* by long-term melatonin application is associated with its regulation of metabolic status and protein degradation. *J Pineal Res.* 2013b. 55: 424–434.

Wang, P., Sun, X., Li, C., Wei, Z., Liang, D., and Ma, F. W. Long-term exogenous application of melatonin delays drought-induced leaf senescence in apple. *J Pineal Res.* 2013a. 54: 292–302.

Wang, P., Yin, L., Liang, D., Li, C., Ma, F., and Yue, Z. Delayed senescence of apple leaves by exogenous melatonin treatment: Toward regulating the ascorbate–glutathione cycle. *J Pineal Res.* 2012. 53: 11–20.

Weeda, S., Zhang, N., Zhao, X., Ndip, G., Guo, Y., Buck, G. A., Fu, C., and Ren, S. X. *Arabidopsis* transcriptome analysis reveals key roles of melatonin in plant defense systems. *PLoS One.* 2014. 9: e93462–e93462.

Wei, W., Li, Q. T., Chu, Y. N., Reiter, R. J., Yu, X. M., Zhu, D. H., Zhang, W. K., et al. Melatonin enhances plant growth and abiotic stress tolerance in soybean plants. *J Exp Bot.* 2015. 66: 695–707.

Woodward, A. W., and Bartel, B. Auxin: Regulation, action, and interaction. *Ann Bot.* 2005. 95: 707–735.

Wrzaczek, M., Brosché, M., and Kangasjärvi, J. ROS signaling loops: Production, perception, regulation. *Curr Opin Plant Biol.* 2013. 16:575–582.

Yin, L., Wang, P., Li, M., Ke, X., Li, C., Liang, D., Wu, S., Ma, X., Zou, Y., and Ma, F. Exogenous melatonin improves *Malus resistance* to *Marssonina* apple blotch. *J Pineal Res.* 2013. 54: 426–434.

Yu, H. S., and Reiter, R. J. *Melatonin: Biosynthesis, Physiological Effects and Clinical Applications.* 1993. Boca Raton, FL: CRC Press.

Zhang, N., Sun, Q. Q., Zhang, H. J., Cao, Y., Weeda, S., Ren, S., and Guo, Y. D. Roles of melatonin in abiotic stress resistance in plants. *J Exp Bot.* 2015. 66: 647–656.

Zhang, N., Zhang, H. J., Zhao, B., Sun, Q. Q., Cao, Y. Y., Li, R., Wu, X. X., et al. The RNA-seq approach to discriminate gene expression profiles in response to melatonin on cucumber lateral root formation. *J Pineal Res.* 2014. 56: 39–50.

Zhang, N., Zhao, B., Zhang, H. J., Weeda, S., Yang, C., Yang, Z. C., Ren, S. X., and Guo, Y. D. Melatonin promotes water-stress tolerance, lateral root formation, and seed germination in cucumber (*Cucumis sativus* L.). *J Pineal Res.* 2013. 54: 15–23.

9

The Roles of Melatonin in Osmotic Stress in Plants

Hai-Jun Zhang, Na Zhang, and Yang-Dong Guo
China Agricultural University
Beijing, China

CONTENTS

ABSTRACT: In recent years, the roles of melatonin in plants have become a research highlight. Evidences indicate that exogenously applied melatonin can improve the tolerance of plants to abiotic stresses. This chapter mainly focuses on the regulatory effects of melatonin against osmotic stress, such as drought, salinity, and freezing. Previous efforts have demonstrated that exogenous melatonin can promote seed germination and root rhizogenesis, and regulate plant growth and photosynthesis under osmotic stress. Overexpression of the melatonin biosynthetic genes increases melatonin levels in transgenic plants. The transgenic plants show enhanced tolerance to osmotic stresses. Melatonin also regulates many genes involved in plant hormone biosynthesis and metabolism, ion homeostasis, and ROS scavenging. Possible mechanisms of melatonin against osmotic stress are also discussed herein.

KEY WORDS: *melatonin, mechanism, osmotic stress, plant growth, tolerance.*

Abbreviations

AANAT: arylalkylamine *N*-acetyltransferase
ABA: abscisic acid

APX:	ascorbate peroxidase
ASC:	ascorbate
ASMT:	*N*-aceylserotonin methyltransferase
CAT:	catalase
EL:	electrolyte leakage
ETH:	ethylene
Fv/Fm:	maximum efficiency of PSII
GA:	gibberellic acid
GPX:	glutathione peroxidase
GSH:	glutathione
HIOMT:	hydroxyindole-*O*-methyltransferase
JA:	jasmonic acid
PAO:	pheide a oxygenase
PEG:	polyethylene glycol
POD:	peroxidase
PSI:	photosystem I
PSII:	photosystem II
ROS:	reactive oxygen species
SA:	salicylic acid
SNAT:	Serotonin *N*-acetyltransferase
sl:	Sekiguchi lesion
SOD:	superoxide dismutase
T5H:	tryptamine 5-hydroxylase
TDC:	tryptophan decarboxylase

Introduction

Melatonin (*N*-acetyl-5-methoxytryptamine) is a well-known regulator of circadian rhythm and sleep in animals (Hardeland et al. 2012). The initial discovery of the melatonin was in the bovine pineal gland in 1958 (Lerner et al. 1958). Melatonin could lighten skin in certain fish, reptiles, and amphibians (Lerner et al. 1958). In animals, melatonin has many physiological roles and acts as a regulator of mood, sleep, locomotor activity, sexual behavior, seasonal reproduction, circadian rhythms, body temperature, and immunology (Arnao and Hernandez-Ruiz 2006; Jan et al. 2009; Reiter et al. 2010; Hardeland et al. 2012; Carrillo-Vico et al. 2013; Cipolla-Neto et al. 2014). In recent years, more and more efforts have focused on the roles of melatonin in plants.

The determination of melatonin in Japanese morning glory and some edible plants solidly demonstrated the existence of melatonin in higher plants in 1993 and 1995, respectively (Dubbels et al. 1995; Hattori et al. 1995; Van Tassel et al. 1995). Since then, melatonin has been detected in many plants, and it is now accepted that melatonin is synthesized throughout the plant kingdom (Hernandez-Ruiz et al. 2004; Zhang et al. 2013; Arnao and Hernandez-Ruiz 2007; Ramakrishna et al. 2012a,b; Ramakrishna 2015). In the past 5 years, a number of publications have elucidated the roles of melatonin in plants, underlining a hot interest in this topic. Unequivocal evidence indicates that melatonin acts as a growth promoter and rooting activator (Hernandez-Ruiz et al. 2004; Arnao and Hernandez-Ruiz 2007; Sarrou et al. 2014; Zhang et al. 2014b; Ramakrishna et al. 2009; Ramakrishna et al. 2012). Various studies have suggested the important roles of melatonin in plant stress defense. Plants may frequently encounter abiotic stresses such as drought, salinity, chemical pollutants, extreme temperature, and radiation, all of which lead to reactive oxygen species (ROS) generation (Ramakrishna and Ravishankar, 2011a, 2013). Some plant species rich in endogenous melatonin have shown a higher tolerance to stress. Exogenous melatonin treatment or ectopic overexpression of melatonin biosynthetic genes can also increase the resistance against abiotic stresses in plants. The mechanism of this phenomenon is mainly due to the natural antioxidant capacity of melatonin. To understand these phenomena and utilize melatonin in agriculture, we must

TABLE 9.1

Functions of Melatonin in Osmotic Stress in Plants

Physiological Action of Melatonin	References
Regulates plant growth	Li et al. 2012; Zhang et al. 2013
Promotes seed germination	Posmyk et al. 2009; Zhang et al. 2013; Zhang et al. 2014a
Enhances photosynthesis	Kostopoulou et al. 2015; Li et al. 2012; Li et al. 2015; Liu et al. 2015; Meng et al. 2014; Wang et al. 2013; Wei et al. 2015; Zhang et al. 2013
Promotes root rhizogenesis	Mukherjee et al. 2014; Zhang et al. 2013; Zhang et al. 2014b; Szafranska et al. 2013
Scavenges ROS	Bajwa et al. 2014; Kostopoulou et al. 2015; Li et al. 2012; Li et al. 2015; Liu et al. 2015; Meng et al. 2014; Wang et al. 2013; Zhang et al. 2013; Zhang et al. 2014a
Maintains ion homeostasis of seedling leaves	Li et al. 2012; Kostopoulou et al. 2015; Liu et al. 2015
Regulates the biosynthesis and metabolism of plant hormones	Li et al. 2015; Zhang et al. 2014

figure out the reason for melatonin existence and make clear the biochemical mechanisms in response to abiotic stress.

Drought, salinity, and freeze-induced dehydration cause osmotic stresses (Oktem et al. 2006). Increasing plant resistance to osmotic stress is an important and long-held goal of agriculture. Some different functions of melatonin have been investigated in higher plants under osmotic stress, but the data are still scarce (Table 9.1). We begin this review by combining existing data with our work and emphasizing the contribution of melatonin to alleviate osmotic stress in plants. Possible mechanisms of melatonin against osmotic stress are also illustrated in Figure 9.1.

Functions of Melatonin in Plants under Osmotic Stress

Melatonin Regulates the Plant Growth

Roots and shoots together constitute the entire structure of higher plants. The root/shoot ratio (fresh weight ratio) is an indicator of nutrient allocation in plants. Melatonin treatment increased the root/shoot ratio sevenfold in cucumber under drought stress (Zhang et al. 2013). This suggested that melatonin treatment promoted the flow of nutrients toward the roots and improved the cucumber seedlings' growth. High salinity obviously inhibited the growth of *Malus hupehensis* seedlings. However, application of 0.1 μM melatonin to the roots noticeably increased shoot heights, leaf numbers, and fresh and dry weights compared with control (Li et al. 2012).

Melatonin Promotes Seed Germination under Osmotic Stress

Seed germination is a crucial process in the life cycle of higher plants and determines seedling establishment. However, it is vulnerable by osmotic stress such as drought or salinity. Exogenous melatonin treatment can alleviate the inhibitory effect of osmotic stress on seed germination. Polyethylene glycol (PEG) stress significantly decreased the seed germination rate from 97% to 87.5% in cucumber (*Cucumis sativus* L.) (Zhang et al. 2013). After seeds were imbibed with different concentrations of melatonin (50, 100, 300, and 500 μM), the inhibitory effects were significantly alleviated. The addition of 100 μM melatonin showed the greatest reversal of PEG-induced inhibition of germination

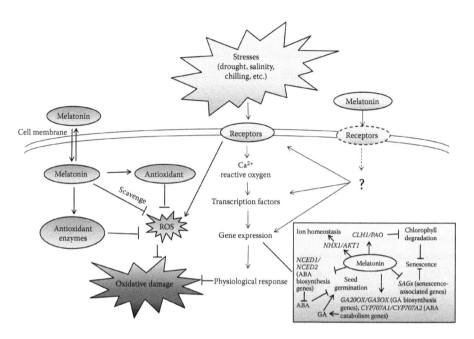

FIGURE 9.1   An overview of cellular responses to osmotic stresses following treatment with exogenous melatonin. (Modified from Zhang et al. 2015. *J Exp Bot.* 66: 647–656, with permission.)

(94%) (Zhang et al. 2013). This stimulus of germination was similarly reflected under salt stress in cucumber seeds. NaCl treatment (150 mM) significantly delayed seed germination. Whereas lower concentrations of melatonin (0.1–100 μM) reduced the inhibitory effects of high salinity, higher concentration of melatonin (500 μM) had no discernible effects, and even inhibited germination (Zhang et al. 2014a). It suggested that melatonin affected seed germination in a dosage-dependent manner under high salinity (Zhang et al. 2014a). Chilling can indirectly constitute osmotic stress via effects on water uptake and loss (Oktem et al. 2006). Nearly 99% of seeds germinated at 25°C, whereas the percentage of germination was only 30% at 15°C, and almost no germination (4%) occurred at 10°C. Addition of melatonin increased seed germination up to 98% at 15°C and 83% at 10°C, respectively (Posmyk et al. 2009). Endogenous melatonin content was significantly increased after addition of melatonin treatment (Posmyk et al. 2009; Zhang et al. 2014a), indicating uptake and utilization of melatonin by germinating seeds.

Melatonin Promotes Root Rhizogenesis under Osmotic Stress

Roots are very important for plants due to their control of the water and nutrient uptake from the soil. Lateral roots determine the final shape of the root system and constitute the ability to uptake nutrients and water (Vilches-Barro and Maizel 2015). Drought stress significantly decreased lateral root number in cucumber seedlings (Zhang et al. 2013). However, this phenomenon was mitigated by melatonin treatment along with the increasing root viability (Zhang et al. 2013). It is proposed that melatonin improved the root diameter rather than the root length (Zhang et al. 2013). Further experiment revealed that melatonin improved the initiation of lateral root primordia in the main root (Zhang et al. 2014b). Melatonin also facilitated the primary root growth and lateral root branching in sunflower seedlings under salt stress (Mukherjee et al. 2014). Primary root length showed a 13% increase in salt-stressed seedlings after addition of melatonin (Mukherjee et al. 2014). Similar results were observed in *Vigna radiata* seedlings under chilling stress. When the *Vigna radiata* seedlings were rewarmed after chilling stress, melatonin-priming seedlings exhibited about 20% increase in root length compared to the control (Szafranska et al., 2013).

Melatonin Affects Photosynthesis under Osmotic Stress

Under osmotic stress, the net photosynthetic rates were markedly inhibited and chlorophyll content significantly declined in plants. However, pretreatment with melatonin significantly reverses this inhibition. Melatonin improved the rate of net photosynthesis under salt stress, a crucial index of biomass performance (Li et al. 2012). Meanwhile, to improve photosynthetic capacity, melatonin treatment significantly reduced chlorophyll metabolism (Li et al. 2012; Zhang et al. 2013; Kostopoulou et al. 2015; Liu et al. 2015) and improved the maximum photochemistry efficiency of photosystem II (PSII) (Fv/Fm) under osmotic stress (Wang et al. 2013; Zhang et al. 2013; Liu et al. 2015). Relative electrolyte leakage (EL) of leaves, an indicator of the extent of membrane damage, was significantly decreased by exogenous melatonin application under drought and salt stress, respectively (Li et al. 2012; Zhang et al. 2013). Osmotic stress generally affected chloroplast morphology and then inhibited photosynthesis. Water-deficient stress also significantly changed the internal structure of the chloroplasts. Under osmotic stress, the chloroplasts' lengths decreased and their widths increased, showing a round shape (Meng et al. 2014). Internal structure of the chloroplasts was also markedly changed, such as damaged membrane systems and dilated, loosened, distorted thylakoids. After melatonin treatment, these phenomena were significantly mitigated in chloroplasts (Meng et al. 2014). Under salt stress, melatonin upregulated gene expressions such as PsaA, PsaF, PsaG, PsaH, PsaK, and PsaO in photosystem I (PSI), and PsbE, PsbO, PsbP, PsbQ, PsbY, PsbZ, and Psb28 in PSII (Wei et al. 2015). Transcript levels of pheide a oxygenase (PAO), a key enzyme involved in chlorophyll degradation, were inhibited by exogenous melatonin under drought stress (Wang et al. 2013). Long-term exogenous melatonin treatment suppressed the upregulation of senescence-associated gene 12 (SAG12) and delayed drought-induced leaf senescence in apple (Wang et al. 2013). Electron transporter genes in photosynthesis, PetF family, and an F-type ATPase gene, ATPF1A, were also upregulated by melatonin (Wei et al. 2015). In the Calvin cycle, rbcS, GAPC1, and GAPCP-2, which encoded glyceraldehyde-3-phosphate dehydrogenase, were upregulated by melatonin under salt stress (Wei et al. 2015). Current study also indicates an idea that melatonin increased stomatal length and width of plant leaves to maintain the change of air and water under drought stress (Meng et al. 2014; Li et al. 2015).

Site of Melatonin Biosynthesis

The primary location of melatonin biosynthesis may be the mitochondria and chloroplasts. (Tan et al. 2013). The root is most frequently considered as the potential site of melatonin biosynthesis. However, limited evidence supported this theory and the unequivocal site of melatonin biosynthesis remains unknown. Currently, there is no information on melatonin receptor(s) or binding site(s) in plants, thus it will be an interesting subject and hot point of research in the future. Since melatonin has a similar structure to auxin and performs auxin-like effects in plants, it could possibly interact with the auxin receptor. Another emerging area to explore is the study of melatonin isomers, which probably share many of the biological functions of melatonin (Tan et al. 2012).

Genetic Modification Research of Melatonin in Osmotic Stress

Melatonin Biosynthesis Pathway in Plants

Melatonin is synthesized via similar biosynthetic pathways in plants and animals (Zhang et al. 2015; Ramakrishna et al. 2011). Additionally, the enzymes that are involved in melatonin biosynthesis have been cloned. There are four consecutive steps for melatonin biosynthesis in plants (Kang et al. 2011). It has been suggested that the first step in melatonin biosynthesis is tryptophan decarboxylase (TDC) catalyzes the conversion of tryptophan into tryptamine. Then, the cytochrome P450 enzyme tryptamine 5-hydroxylase (T5H) hydroxylates the C-5 position of tryptamine to produce serotonin (Fujiwara et al. 2010). The final two enzymes on the melatonin biosynthetic pathway are arylalkylamine *N*-acetyltransferase (AANAT) and *N*-aceylserotonin methyltransferase (ASMT). Currently, TDC has been cloned in various plant species (De Luca et al. 1989; Di Fiore et al. 2002; Kang et al. 2007a; Park et al. 2009). T5H has been isolated in Sekiguchi lesion (sl) rice mutants (Fujiwara et al. 2010) and is constitutively expressed in healthy rice (*Oryza sativa*) (Kang et al. 2007b). There is not enough information about AANAT in higher plants. Serotonin *N*-acetyltransferase (SNAT) genes have been cloned in rice. ASMT, which was known as HIOMT (hydroxyindole-*O*-methyltransferase), has also been purified from rice (Kang et al. 2013). Subcellular localization experiment shows that SNAT and ASMT are localized in chloroplasts and cytoplasm, respectively (Byeon and Back 2014).

Transgenic Research of Melatonin Biosynthesis Genes in Plants under Osmotic Stress

Many efforts have been made to investigate the functions of melatonin biosynthesis genes by transgenic technology in plants. However, the transgenic research into melatonin biosynthesis genes against osmotic stress remains poorly understood. In 2014, transgenic Micro-Tom tomato plants overexpressing the homologous sheep oAANAT gene had higher melatonin levels and lost the "apical dominance." The oHIOMT lines showed enhanced drought tolerance (Wang et al. 2014). More work was needed to figure out the melatonin functions in plants.

Mechanisms of Melatonin-Mediated Stress Tolerance

As plants are sessile, they can only adjust their own physical conditions to resist adverse environments. So when faced with a harsh environment, a rapid and tremendous change must occur inside the plant cells in order to survive. A cell is separated from its surrounding environment by the plasma membrane. This membrane is selectively permeable to small molecules and ions. Melatonin is an amphipathic molecule that can easily diffuse through cell membranes into the cytoplasm and enter subcellular compartments.

Melatonin Scavenges Osmotic Stress-Induced ROS

Drought limits plant growth due to photosynthetic decline. Salinity interferes with plant growth as it leads to physiological drought and ion toxicity. All the abiotic stresses cause the accumulation of ROS (Prasad et al. 1994; Foyer et al. 1997) such as hydrogen peroxide (H_2O_2), superoxide anion ($O_2^{\cdot-}$), and hydroxyl radical ($\cdot$OH) (Mittler et al. 2004). ROS are clearly required for growth and function as secondary messengers in signal transduction (Baxter et al. 2014), which interact directly or indirectly with many signaling pathways, such as nitric oxide (NO) and the stress hormones salicylic acid (SA), jasmonic acid (JA), and ethylene (ETH) (Mittler et al. 2004). But high concentrations of ROS are detrimental to plants and can trigger genetically programmed cell suicide events (Foyer and Noctor 2005). Furthermore, excess levels of ROS can cause many detrimental effects such as lipid peroxidation in cellular membranes, DNA damage, protein denaturation, carbohydrate oxidation, pigment breakdown, and impaired enzyme activity (Bose et al. 2014). The ROS generation and scavenging are generally maintained at a delicate balance in plants. In this sense, antioxidants (e.g., ascorbate [ASC], glutathione [GSH], tocopherol) and ROS-scavenging enzymes (e.g., ascorbate peroxidase [APX], catalase [CAT], peroxidase [POD], superoxide dismutase [SOD], glutathione peroxidase [GPX]) are necessary for plants to detoxify ROS. Melatonin alleviated oxidative damage by enhancing antioxidants and antioxidant enzyme activities. This theory has been demonstrated in many experiments. Under osmotic stress, the activities of antioxidant enzymes were increased by melatonin in cucumber (Li et al. 2012; Zhang et al. 2013, 2014a), *Citrus aurantium* (Kostopoulou et al. 2015), tomato (Liu et al. 2015), apple (Wang et al. 2013; Li et al. 2015), and grape (Meng et al. 2014). Under salinity and drought stress, the exogenous application of melatonin also directly suppressed the production of H_2O_2 and hydroxyl radicals (Zhang et al. 2013; Zhang et al. 2014a). The transcript levels of the genes encoding SOD, APX, CAT, and peroxidase were also upregulated by melatonin (Zhang et al. 2014a). Melatonin upregulates the expression of the ROS-related antioxidant genes ZAT10 and ZAT12 under cold stress (Bajwa et al. 2014). The upregulation of cold-signaling genes by melatonin may contribute to the protection of the membrane structure against peroxidation during chilling.

Melatonin Regulates Biosynthesis and Metabolism of Plant Hormones

Plant hormones are important signaling molecules that relay information on changes in the environment to seed germination. Generally, abscisic acid (ABA) and gibberellic acid (GA) are viewed as the major hormones in germinating seeds (Footitt et al. 2011). The dynamic balance of synthesis and catabolism of ABA and GA is crucial for seeds to completely germinate (Footitt et al. 2011). NaCl stress significantly inhibited the ABA catabolism and GA biosynthesis and thus delayed the seed germination (Zhang et al. 2014a). However, melatonin acted as a signaling molecule to regulate seed germination by stimulating both ABA catabolism and GA biosynthesis under high salinity. Following application of exogenous melatonin, ABA catabolism genes (e.g., CsCYP707A1 and CsCYP707A2) were significantly upregulated and ABA biosynthesis genes (e.g., CsNECD2) were significantly downregulated, resulting in a rapid decrease of ABA content during the early stage of germination in cucumber (Zhang et al. 2014a). A similar phenomenon was confirmed in *Malus* species under drought stress (Li et al. 2015). Similarly, GA biosynthesis genes (e.g., GA20ox and GA3ox) were positively upregulated by exogenous melatonin application, contributing to a significant increase of GA (especially GA_4) content (Zhang et al. 2014a). Based on these observations, the authors proposed a regulatory role of melatonin on plant hormones under osmotic stress.

Melatonin Regulates Ion Homeostasis under Osmotic Stress

Ion homeostasis is considered to be an important mechanism underlying salt stress acclimation responses in plants. The maintenance of K^+ and Na^+ homeostasis is crucial under salt stress (Zhu 2003). At tissue level, the K^+/Na^+ ratio is considered as an excellent parameter to salinity tolerance with the model that higher ratio equals higher tolerance (Cakmak 2005; Chen et al. 2007; Genc et al. 2007). Exogenous application of melatonin enabled tomato seedlings to keep lower Na^+ content and higher K^+ content under sodic alkaline stress. The K^+/Na^+ ratio was relatively higher in the leaves (Liu et al. 2015). A similar ameliorative effect had been demonstrated in *Malus hupehensis* (Li et al. 2012). Combined ascorbic acid + melatonin treatment showed lower Cl^- accumulation in leaves of *Citrus aurantium* (Kostopoulou et al. 2015). SLAC1, an anion channel–associated gene, was downregulated by ascorbic acid + melatonin treatment under salinity which might lead to lower leaf Cl− level. The expression of Na^+ and K^+ transporters' genes (MdNHX1 and MdAKT1) were upregulated, which would all help to maintain ion homeostasis, thereby alleviating the saline-induced damage (Li et al. 2012). Based on these researches, melatonin was a positive regulator in maintaining ion homeostasis and played an important role against salt stress.

Conclusion

The protective effects of melatonin have been demonstrated in many plants. Almost all the abiotic stresses that cause oxidative stress can be alleviated by melatonin treatment at appropriate endogenous levels, in a concentration-dependent manner.

Melatonin is an amphiphilic molecule which can freely cross cell membranes. Melatonin can scavenge ROS directly. It also increases the levels of antioxidants and the activities of related enzymes to scavenge ROS. Stresses are first recognized by receptors, followed by a signal transduction cascade. Melatonin alters almost all of the transduction steps along the way by receptor-dependent or receptor-independent processes. Melatonin alters expression of genes related to stresses. However, further studies will be necessary to determine the functions of melatonin on osmotic stress. Although some studies have focused on the signaling pathway of melatonin and plant hormones, it still needs further investigation in the near future.

Acknowledgments

This work was partly supported by a grant to Y. D. Guo (2012AA101801, 2011BAD17B01, BLVT-03 and 2012CB113900). We appreciate all the help from the National Energy R & D Center for Non-food Biomass, China Agricultural University, Beijing.

REFERENCES

Arnao, M. B., and Hernandez-Ruiz, J. 2006. The physiological function of melatonin in plants. *Plant Signal Behav.* 1: 89–95.

Arnao, M. B., and Hernandez-Ruiz, J. 2007. Melatonin promotes adventitious- and lateral root regeneration in etiolated hypocotyls of *Lupinus albus* L. *J Pineal Res.* 42: 147–152.

Bajwa, V. S., Shukla, M. R., Sherif, S. M., Murch, S. J., and Saxena, P. K. 2014. Role of melatonin in alleviating cold stress in *Arabidopsis thaliana*. *J Pineal Res.* 56: 238–245.

Baxter, A., Mittler, R., and Suzuki, N. 2014. ROS as key players in plant stress signalling. *J Exp Bot.* 65: 1229–1240.

Bose, J., Rodrigo-Moreno, A., and Shabala, S. 2014. ROS homeostasis in halophytes in the context of salinity stress tolerance. *J Exp Bot.* 65: 1241–1257.

Byeon, Y., and Back, K. 2014. An increase in melatonin in transgenic rice causes pleiotropic phenotypes, including enhanced seedling growth, delayed flowering, and low grain yield. *J Pineal Res.* 56: 408–414.

Cakmak, I. 2005. The role of potassium in alleviating detrimental effects of abiotic stresses in plants. *J Plant Nutr Soil Sci.* 168: 521–530.

Carrillo-Vico, A., Lardone, P. J., Alvarez-Sanchez, N., Rodriguez-Rodriguez, A., and Guerrero, J. M. 2013. Melatonin: Buffering the immune system. *Int J Mol Sci.* 14: 8638–8683.

Chen, Z., Zhou, M., Newman, I. A., Mendham, N. J., Zhang, G., and Shabala, S. 2007. Potassium and sodium relations in salinized barley tissues as a basis of differential salt tolerance. *Funct Plant Biol.* 34: 150–162.

Cipolla-Neto, J., Amaral, F. G., Afeche, S. C., Tan, D. X., and Reiter, R. J. 2014. Melatonin, energy metabolism, and obesity: A review. *J Pineal Res.* 56: 371–381.

De Luca, V., Marineau, C., and Brisson, N. 1989. Molecular cloning and analysis of cDNA encoding a plant tryptophan decarboxylase: Comparison with animal dopa decarboxylases. *Proc Natl Acad Sci USA.* 86: 2582–2586.

Di Fiore, S., Li, Q. R., Leech, M. J., Schuster, F., Emans, N., Fischer, R., and Schillberg, S. 2002. Targeting tryptophan decarboxylase to selected subcellular compartments of tobacco plants affects enzyme stability and *in vivo* function and leads to a lesion-mimic phenotype. *Plant Physiol.* 129: 1160–1169.

Dubbels, R., Reiter, R. J., Klenke, E., Goebel, A., Schnakenberg, E., Ehlers, C., Schiwara, H. W., and Schloot, W. 1995. Melatonin in edible plants identified by radioimmunoassay and by high-performance liquid chromatography–mass spectrometry. *J Pineal Res.* 18: 28–31.

Footitt, S., Douterelo-Soler, I., Clay, H., and Finch-Savage, W. E. 2011. Dormancy cycling in *Arabidopsis* seeds is controlled by seasonally distinct hormone-signaling pathways. *Proc Natl Acad Sci USA.* 108: 20236–20241.

Foyer, C. H., Lopez-Delgado, H., Dat, J. F., and Scott, I. M. 1997. Hydrogen peroxide- and glutathione-associated mechanisms of acclimatory stress tolerance and signalling. *Physiol Plantarum.* 100: 241–254.

Foyer, C. H., and Noctor, G. 2005. Oxidant and antioxidant signalling in plants: A re-evaluation of the concept of oxidative stress in a physiological context. *Plant Cell Environ.* 28: 1056–1071.

Fujiwara, T., Maisonneuve, S., Isshiki, M., Mizutani, M., Chen, L., Wong, H. L., Kawasaki, T., and Shimamoto, K. 2010. Sekiguchi lesion gene encodes a cytochrome P450 monooxygenase that catalyzes conversion of tryptamine to serotonin in rice. *J Biol Chem.* 285: 11308–11313.

Genc, Y., McDonald, G. K., and Tester, M. 2007. Reassessment of tissue Na+ concentration as a criterion for salinity tolerance in bread wheat. *Plant Cell Environ.* 30: 1486–1498.

Hardeland, R., Madrid, J. A., Tan, D. X., and Reiter, R. J. 2012. Melatonin, the circadian multioscillator system and health: The need for detailed analyses of peripheral melatonin signaling. *J Pineal Res.* 52: 139–166.

Hattori, A., Migitaka, H., Iigo, M., Itoh, M., Yamamoto, K., Ohtanikaneko, R., Hara, M., Suzuki, T., and Reiter, R. J. 1995. Identification of melatonin in plants and its effects on plasma melatonin levels and binding to melatonin receptors in vertebrates. *Biochem Mol Biol Int.* 35: 627–634.

Hernandez-Ruiz, J., Cano, A., and Arnao, M. B. 2004. Melatonin: A growth-stimulating compound present in lupin tissues. *Planta.* 220: 140–144.

Jan, J. E., Reiter, R. J., Wasdell, M. B., and Bax, M. 2009. The role of the thalamus in sleep, pineal melatonin production, and circadian rhythm sleep disorders. *J Pineal Res.* 46: 1–7.

Kang, S., Kang, K., Lee, K., and Back, K. 2007a. Characterization of rice tryptophan decarboxylases and their direct involvement in serotonin biosynthesis in transgenic rice. *Planta.* 227: 263–272.

Kang, S., Kang, K., Lee, K., and Back, K. 2007b. Characterization of tryptamine 5-hydroxylase and serotonin synthesis in rice plants. *Plant Cell Rep.* 26: 2009–2015.

Kang, K., Kong, K., Park, S., Natsagdorj, U., Kim, Y. S., and Back, K. 2011. Molecular cloning of a plant *N*-acetylserotonin methyltransferase and its expression characteristics in rice. *J Pineal Res.* 50: 304–309.

Kang, K., Lee, K., Park, S., Byeon, Y., and Back, K. 2013. Molecular cloning of rice serotonin *N*-acetyltransferase, the penultimate gene in plant melatonin biosynthesis. *J Pineal Res.* 55: 7–13.

Kostopoulou, Z., Therios, I., Roumeliotis, E., Kanellis, A. K., and Molassiotis, A. 2015. Melatonin combined with ascorbic acid provides salt adaptation in *Citrus aurantium* L. seedlings. *Plant Physiol Bioch.* 86: 155–165.

Lerner, A. B., Case, J. D., Takahashi, Y., Lee, T. H., and Mori, W. 1958. Isolation of melatonin, the pineal gland factor that lightens melanocytes. *J Am Chem Soc.* 80: 2587–2587.

Li, C., Tan, D. X., Liang, D., Chang, C., Jia, D., and Ma, F. 2015. Melatonin mediates the regulation of ABA metabolism, free radical scavenging, and stomatal behaviour in two malus species under drought stress. *J Exp Bot.* 66: 669–680.

Li, C., Wang, P., Wei, Z., Liang, D., Liu, C., Yin, L., Jia, D., Fu, M., and Ma, F. 2012. The mitigation effects of exogenous melatonin on salinity-induced stress in *Malus hupehensis*. *J Pineal Res.* 53: 298–306.

Liu, N., Jin, Z., Wang, S., Gong, B., Wen, D., Wang, X., Wei, M., and Shi, Q. 2015. Sodic alkaline stress mitigation with exogenous melatonin involves reactive oxygen metabolism and ion homeostasis in tomato. *Sci Hortic.* 181: 18–25.

Meng, J. F., Xu, T. F., Wang, Z. Z., Fang, Y. L., Xi, Z. M., and Zhang, Z. W. 2014. The ameliorative effects of exogenous melatonin on grape cuttings under water-deficient stress: Antioxidant metabolites, leaf anatomy, and chloroplast morphology. *J Pineal Res.* 57: 200–212.

Mittler, R., Vanderauwera, S., Gollery, M., and Van Breusegem, F. 2004. Reactive oxygen gene network of plants. *Trends Plant Sci.* 9: 490–498.

Mukherjee, S., David, A., Yadav, S., Baluska, F., and Bhatla, S. C. 2014. Salt stress–induced seedling growth inhibition coincides with differential distribution of serotonin and melatonin in sunflower seedling roots and cotyledons. *Physiol Plantarum.* 152: 714–728.

Oktem, H. A., Eyidogan, F., Selcuk, F., Silva, J. A. T. D., Yucel, M., and Da Silva, J. A. T. 2006. Osmotic stress tolerance in plants: Transgenic strategies. In *Floriculture, Ornamental and Plant Biotechnology.* 194–208. Isleworth, England: Global Science Books.

Park, S., Kang, K., Lee, K., Choi, D., Kim, Y. S., and Back, K. 2009. Induction of serotonin biosynthesis is uncoupled from the coordinated induction of tryptophan biosynthesis in pepper fruits (*Capsicum annuum*) upon pathogen infection. *Planta.* 230: 1197–1206.

Posmyk, M. M., Balabusta, M., Wieczorek, M., Sliwinska, E., and Janas, K. M. 2009. Melatonin applied to cucumber (*Cucumis sativus* L.) seeds improves germination during chilling stress. *J Pineal Res.* 46: 214–223.

Prasad, T. K., Anderson, M. D., Martin, B. A., and Stewart, C. R. 1994. Evidence for chilling-induced oxidative stress in maize seedlings and a regulatory role for hydrogen peroxide. *Plant Cell.* 6: 65–74.

Ramakrishna, A. 2015. Indoleamines in edible plants: Role in human health effects. In Angel Catalá, Biochemistry Research Trends Series (editor), *Indoleamines: Sources, Role in Biological Processes and Health Effects.* 279. New York: Nova Publishers.

Ramakrishna, A., Giridhar, P., Jobin, M., Paulose, C. S., and Ravishankar, G. A. 2012. Indoleamines and calcium enhance somatic embryogenesis in cultured tissues of *Coffea canephora* P ex Fr. *Plant Cell Tiss Org.* 108: 267–278.

Ramakrishna, A., Giridhar, P., and Ravishankar, G. A. 2009. Indoleamines and calcium channels influence morphogenesis in *in vitro* cultures of *Mimosa pudica* L. *Plant Sig Behav.* 12: 1136–1141.

Ramakrishna, A., Giridhar, P., and Ravishankar, G. A. 2011. Phytoserotonin: A review. *Plant Sig Behav.* 6: 800–809.

Ramakrishna, A., Giridhar, P., Sankar, K. U., and Ravishankar, G. A. 2012a. Endogenous profiles of indoleamines: Serotonin and melatonin in different tissues of *Coffea canephora* P ex Fr. as analyzed by HPLC and LC-MS-ESI. *Acta Physiologia Plantarum.* 34: 393–396.

Ramakrishna, A., Giridhar, P., Sankar, K. U., and Ravishankar, G. A. 2012b. Melatonin and serotonin profiles in beans of *Coffea* sps. *J Pineal Res.* 52: 470–476.

Ramakrishna, A., and Ravishankar, G. A. 2011a. Influence of abiotic stress signals on secondary metabolites in plants. *Plant Sig Behav.* 6: 1720–1731.

Ramakrishna, A., and Ravishankar, G. A. 2013. Role of plant metabolites in abiotic stress tolerance under changing climatic conditions with special reference to secondary compounds. In Tuteja, N., and Gill, S. S. (editors), *Climate Change and Abiotic Stress Tolerance.* 705–726. Weinheim, Germany: Wiley.

Reiter, R. J., Tan, D. X., and Fuentes-Broto, L. 2010. Melatonin: A multitasking molecule. In Luciano, M., (editor) *Neuroendocrinology: The Normal Neuroendocrine System*. 181: 127–151. Amsterdam: Elsevier.

Sarrou, E., Therios, I., and Dimassi-Theriou, K. 2014. Melatonin and other factors that promote rooting and sprouting of shoot cuttings in *Punica granatum* cv. Wonderful. *Turk J Bot.* 38: 293–301.

Szafranska, K., Glinska, S., and Janas, K. M. 2013. Ameliorative effect of melatonin on meristematic cells of chilled and re-warmed *Vigna radiata* roots. *Biol Plantarum.* 57: 91–96.

Tan, D. X., Hardeland, R., Manchester, L. C., Rosales-Corral, S., Coto-Montes, A., Boga, J. A., and Reiter, R. J. 2012. Emergence of naturally occurring melatonin isomers and their proposed nomenclature. *J Pineal Res.* 53: 113–121.

Tan, D. X., Manchester, L. C., Liu, X. Y., Rosales-Corral, S. A., Acuna-Castroviejo, D., and Reiter, R. J. 2013. Mitochondria and chloroplasts as the original sites of melatonin synthesis: A hypothesis related to melatonin's primary function and evolution in eukaryotes. *J Pineal Res.* 54: 127–138.

Van Tassel, D. L., Roberts, N. J., and O'Neill, S. D. 1995. Melatonin from higher plants: Isolation and identification of *N*-acetyl-5-methoxytryptamine. *Plant Physiol.* 108: 101–101.

Vilches-Barro, A., and Maizel, A. 2015. Talking through walls: Mechanisms of lateral root emergence in *Arabidopsis thaliana*. *Curr Opin Plant Biol.* 23: 31–38.

Wang, L., Zhao, Y., Reiter, R. J., He, C., Liu, G., Lei, Q., Zuo, B., Zheng, X. D., Li, Q., and Kong, J. 2014. Changes in melatonin levels in transgenic "Micro-Tom" tomato overexpressing ovine AANAT and ovine HIOMT genes. *J Pineal Res.* 56: 134–142.

Wang, P., Sun, X., Li, C., Wei, Z., Liang, D., and Ma, F. 2013. Long-term exogenous application of melatonin delays drought-induced leaf senescence in apple. *J Pineal Res.* 54: 292–302.

Wei, W., Li, Q. T., Chu, Y. N., Reiter, R. J., Yu, X. M., Zhu, D. H., Zhang, W. K., et al. 2015. Melatonin enhances plant growth and abiotic stress tolerance in soybean plants. *J Exp Bot.* 66: 695–707.

Zhang, H. J., Zhang, N., Yang, R. C., Wang, L., Sun, Q. Q., Li, D. B., Cao, Y. Y., et al. 2014a. Melatonin promotes seed germination under high salinity by regulating antioxidant systems, ABA and GA4 interaction in cucumber (*Cucumis sativus* L.). *J Pineal Res.* 57: 269–279.

Zhang, N., Sun, Q., Zhang, H., Cao, Y., Weeda, S., Ren, S., and Guo, Y. D. 2015. Roles of melatonin in abiotic stress resistance in plants. *J Exp Bot.* 66: 647–656.

Zhang, N., Zhang, H. J., Zhao, B., Sun, Q. Q., Cao, Y. Y., Li, R., Wu, X. X., et al. 2014b. The RNA-seq approach to discriminate gene expression profiles in response to melatonin on cucumber lateral root formation. *J Pineal Res.* 56: 39–50.

Zhang, N., Zhao, B., Zhang, H. J., Weeda, S., Yang, C., Yang, Z. C., Ren, S., and Guo, Y. D. 2013. Melatonin promotes water-stress tolerance, lateral root formation, and seed germination in cucumber (*Cucumis sativus* L.). *J Pineal Res.* 54: 15–23.

Zhu, J. K. 2003. Regulation of ion homeostasis under salt stress. *Curr Opin Plant Biol.* 6: 441–445.

10

Physiological Role of Melatonin in Plants

Mona Gergis Dawood and Mohamed El-Awadi
National Research Centre
Cairo, Egypt

CONTENTS

ABSTRACT   Melatonin can act as a universal hydrophilic and hydrophobic antioxidant. Moreover, melatonin probably acts as a night signal, shows coordinating responses to diurnal and photoperiodic environmental cues, and acts as a plant growth regulator. A large portion of the melatonin in a plant is synthesized by the plant itself. All required enzymes for melatonin biosynthesis have been identified in plants. Moreover, soil microorganisms are rich in melatonin, so their decomposition releases melatonin into the soil and plants can absorb melatonin from the soil via roots. The main function of melatonin in higher plants is to act as an antistress agent against abiotic stressors, such as drought, salinity, low and high ambient temperatures, UV radiation, and toxic chemicals. Several investigators have reported that melatonin may possess a variety of functions in vascular plants, including regulation of mitosis, delaying flower induction, stimulation of hypocotyl, coleoptile, and root growth, stimulation of adventitious- and lateral root regeneration, protection against cold-induced apoptosis, lowering chlorophyll degradation, and enhanced plant tolerance to environmental stress. Additionally, melatonin is considered to be involved in the regulation of circadian rhythm. Compared to the classic antioxidants such as vitamin C, vitamin E, and glutathione, melatonin exhibits a more potent antioxidant capacity to reduce oxidative injury. This chapter includes the most updated results about the role of melatonin in the enhancement of plant tolerance to environmental stress.

KEY WORDS: *indoleamines, melatonin, serotonin, indole acetic acid, tryptophan, plant tolerance, environmental stress.*

Abbreviations

$1O_2$:   singlet oxygen
5HT: 5-hydroxytryptamine
AFMK: *N*1-acetyl- *N*2-formyl-5-methoxykynuramine

AMK:	*N*1-acetyl-5-methoxykynuramine
ASMT:	*N*-acetylserotonin *O*-methyltransferese
C-3HOM:	cyclic 3-hydroxymelatonin
CAT:	catalase
DW:	dry weight
GA$_3$:	gibberellic acid
GSSG-R:	glutathione reductase
H$_2$O$_2$:	hydrogen peroxide
HO˙:	hydroxyl radical
IAA:	indole-3-acetic acid
MPTP:	mitochondrial permeability transition pore
NADH:	nicotinamide adenine dinucleotide
NO:	nitric oxide
O^{-2}:	superoxide anion
ONOO⁻:	peroxynitrite anion
PAL:	*L*-phenylalanine ammonialyase
PAO:	pheophorbide a oxygenase
PEG:	polyethylene glycol
POX:	peroxidases
PSII:	photosystem II
RNS:	reactive nitrogen species
ROS:	reactive oxygen species
SAG:	senescence-associated gene
SDS-PAGE:	sodium dodecyl sulfate polyacrylamide gel electrophoresis
SOD:	superoxide dismutase
T5H:	tryptamine 5-hydroxylase
TDC:	tryptophan decarboxylase
Trp:	tryptophan

Introduction

Physiological Role of Melatonin

Several investigators have reported that melatonin may possess a variety of functions in vascular plants (Kolar and Machackova 2005; Uchendu et al. 2013), including regulation of mitosis (Banerjee and Margulis 1973), delaying flower induction (Kolar et al. 2003), stimulation of hypocotyl, coleoptile, and root growth (Hernandez-Ruiz et al. 2004, 2005; Chen et al. 2009), stimulation of adventitious- and lateral root regeneration (Arnao and Hernandez-Ruiz 2007), protection against cold-induced apoptosis (Lei et al. 2004), lowering chlorophyll degradation (Arnao and Hernandez-Ruiz 2009b), and enhanced plant tolerance to environmental stress. Additionally, melatonin is considered to be involved in the regulation of circadian rhythm as well as in many physiological processes, for example, root and shoot development (Murch et al. 2001; Park 2011), flowering, flower and fruit development, or delaying leaf senescence (Park 2011; Tal et al. 2011; Wang et al. 2012; Park et al. 2013).

Exogenously applied melatonin affects developmental processes during vegetative and reproductive growth and has the ability to regulate plant growth and enhance crop production. Hernandez-Ruiz et al. (2005) stated that melatonin promotes coleoptile growth in four monocot species including canary grass, wheat, barley, and oat. Melatonin-treated corn plants had greater production than nontreated plants (Tan et al. 2012).

Exogenous melatonin may cause increases in indole-3-acetic acid (IAA) (Chen et al. 2009) or polyamines (Lei et al. 2004). As plant hormones, melatonin displayed weak effects at higher concentrations or even had inhibitory actions (Hernandez-Ruiz et al. 2004).

The biogenic monoamines such as serotonin, Prozac, tryptamine, tyramine, and melatonin play a hormonal role in seedling growth and are involved in flowering, morphogenesis, and protection of plant

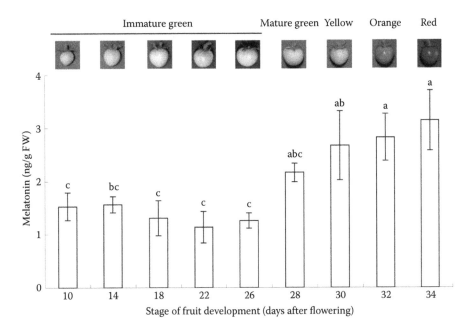

FIGURE 10.1 Melatonin contents in tomato seedlings at 0, 1, and 2 h after the addition of putative precursors. The values shown represent the means of duplicate parallel ± SD. Different letters represent statistically significant differences at $p < .05$ by Fisher's PLSD test. (From Okazaki, Dissection of melatonin accumulation and biosynthesis in tomato. PhD Thesis. Tsukuba, Japan: University of Tsukuba. 2009.)

from environmental changes (Hernandez-Ruiz et al. 2005; Arnao and Hernandez-Ruiz 2006; Paredes et al. 2009; Park 2011).

Okazaki (2009) observed that concentration of melatonin was much higher in mature tomato fruits than green ones, and showed that melatonin accumulated in fruits during development as shown in Figure 10.1. Melatonin may be related to the protection of fruit against intensive production of ROS during ripening, through its role as an antioxidant.

Melatonin effectively lowered chlorophyll degradation in aging leaves of barley (*Hordeum vulgare* L.) (Arnao and Hernandez-Ruiz 2009b) and detached leaves of apple (*Malus domestica* Borkh. cv. Golden Delicious), and protected the photosystems from damage (Wang et al. 2012) and increased photosynthetic efficiency of chlorophyll in plants (Tan et al. 2012). Exogenous melatonin at 10 mM delayed the process of dark-induced senescence in detached apple leaves, possibly through regulation of the ascorbate–glutathione cycle (Wang et al. 2012).

Melatonin and chlorophyll accumulations were found to be inversely related and the levels of melatonin would accumulate as chlorophyll development declined, as reported by Okazaki (2009) and shown in Figure 10.2.

Generally, melatonin exerts multiple functions on plant development that can be categorized into three categories: (1) Growth promoter as an auxin (acts as an independent plant growth regulator and may mediate the activities of other plant growth regulators) (Park 2011), (2) antioxidant for free radicals, and (3) other functions, such as signal molecule for circadian maintenance, regulation of flower development, or maintenance of developmental stage in fruits tissue (Paredes et al. 2009).

Kang et al. (2009a) reported that serotonin has been involved in several physiological roles in plants, including flowering, morphogenesis, and adaptation to environmental changes. They added that serotonin is greatly accumulated in rice (*Oryza sativa*) leaves undergoing senescence induced by either nutrient deprivation or detachment, and its synthesis is closely coupled with transcriptional and enzymatic induction of the tryptophan biosynthetic genes as well as tryptophan decarboxylase (TDC). Transgenic rice plants that overexpressed TDC accumulated higher levels of serotonin than the wild type and showed delayed senescence of rice leaves.

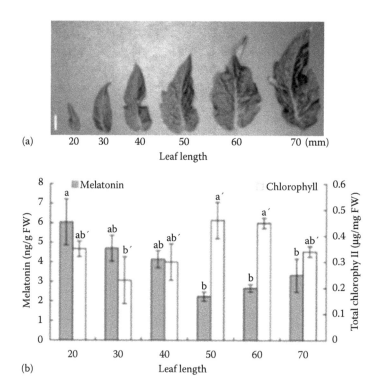

(a)   20   30   40   50   60   70 (mm)
Leaf length

(b)   Leaf length

FIGURE 10.2 Melatonin and chlorophyll contents in developing tomato leaves at different stages: (a) Micro-Tom leaf development. The white bar indicates 10 mm; and (b) melatonin (filled bar) and chlorophyll (open bar) concentrations in leaves at different stages. The values represent the mean of triplicate extractions ± SEM. Different letters represent statistically significant differences at $p < .05$ by Fisher's PLSD test. (From Okazaki, Dissection of melatonin accumulation and biosynthesis in tomato. PhD Thesis. Tsukuba, Japan: University of Tsukuba. 2009.)

Role of Melatonin in Plant Growth Regulation

The chemical structure of melatonin (indole derivative) and its biosynthetic pathway (from tryptophan [Trp]) indicated that the influence of melatonin and auxin on plant growth was similar (Arnao and Hernandez-Ruiz 2006). Melatonin at the low level (1 µM) caused auxinic response concerning the number and length of roots, but at the higher level (10 µM) it inhibited the root growth, as in sweet cherry rootstocks (Sarropoulou et al. 2012). Moreover, Hernandez-Ruiz et al. (2004) reported that melatonin at high concentrations acts as an inhibitor (probably reaching the toxic level in tissues), while at a lower concentration it induces the growth of *Lupinus albus* hypocotyl segments.

Similar to IAA, melatonin acts as a growth promoter, and stimulates growth in etiolated lupins (*Lupinus albus*) and coleoptiles of canary grass (*Phalaris canariensis*), wheat (*Triticum aestivum*), barley (*Hordeum vulgare*), and oats (*Avena sativa*); however, its activity in comparison with IAA ranged between 10% and 55%. Moreover, melatonin inhibited root elongation in some monocots even at very low concentrations. For canary grass and oats, 0.01 mM melatonin inhibited root growth (Hernandez-Ruiz et al. 2005), whereas, the maximum inhibitory effect of melatonin on wild leaf mustard roots occurred at 100 mM, which is considerably higher than that for canary grass and oat as reported by Manchester et al. (2000) and Hernandez- Ruiz et al. (2005).

Melatonin affects also the regeneration of lateral and adventitious roots in etiolated seedlings of *Lupinus albus*, which was observed by comparing the effect of different concentrations of melatonin and IAA on root promotion (Arnao and Hernandez-Ruiz 2007).

Several studies reported that melatonin enhanced root growth in different plants (Arnao and Hernandez-Ruiz 2007; Chen et al. 2009; Sarropoulou et al. 2012) and improved soybean growth at both the vegetative stage and the reproductive stage (Wei et al. 2015).

Arnao and Hernandez-Ruiz (2006) studied the distribution of melatonin and IAA in different zones of the hypocotyls (apical, central, and basal), demonstrating that melatonin and IAA coexist in the tissues, and that a similar concentration gradient existed for both indoles. They studied the stimulatory effect of melatonin on the growth at concentration ranged between 10 nM and 0.1 mM), and reported that the optimum growth-promoting effect was determined at 10 µM melatonin.

Chen et al. (2009) showed that 0.1 mM melatonin has a stimulatory effect on root growth of wild leaf mustard (*Brassica juncea*), while 100 mM has an inhibitory effect. Furthermore, the stimulatory effect was only detectable in young seedlings (2 days old). Older seedlings (4 days old) appear to be less susceptible to both the stimulatory and the inhibitory effect of melatonin. Exogenous application of 0.1 mM melatonin had the maximal positive effect on root elongation of 2-day-old mustard seedlings and raised the endogenous levels of free IAA in roots.

Based on the functional relationship between melatonin and IAA, it was obvious that exogenous melatonin might cause changes in the concentration of endogenous free IAA as reported by Chen et al. (2009) and shown in Figure 10.3. They mentioned that endogenous free IAA levels increased at low melatonin concentrations (0.1 and 0.2 mM). The stimulation of root growth at low concentrations of melatonin is actually triggered by melatonin-stimulated IAA synthesis, whereas, at high melatonin concentrations, IAA was not significantly increased, and root elongation was strongly inhibited. Thus, melatonin's suppressive effect on root growth seems to involve mechanisms not related to IAA, at least in the wild leaf mustard.

The regulatory role of melatonin on plant growth and development was shown by Murch et al. (2001) who observed that application of metabolism-inhibitors of auxins, serotonin, and melatonin caused increases in the endogenous melatonin concentration and induced root growth in *Hypericum perforatum* L. cultures *in vitro*, while accumulation of serotonin—the direct melatonin precursor—promotes shoot formation. They reported that auxin-promoted root formation was blocked by the inhibitors of serotonin and melatonin transport. On the other hand, the formation of the cytokinin-induced shoot is stimulated by the substances inhibiting serotonin conversion into melatonin (Murch and Saxena 2002). Thus, it seems that morphogenetic abilities of *Hypericum perforatum* L. cultures may depend on the proper serotonin/melatonin ratio similarly as in the case of plant growth modification caused by the changes in the ratio of auxins to other phytohormones, for example, cytokinins (Murch and Saxena 2002; Jones et al. 2007).

Serotonin, the precursor of melatonin, at low concentrations stimulated lateral root growth in *Arabidopsis thaliana* L., whereas, at higher concentrations, it inhibited primary and lateral root growth, and promoted formation of adventitious roots (Pelagio-Flores et al. 2011). In St. John's wort (*Hypericum perforatum*), a specific ratio of endogenous serotonin and melatonin regulates morphogenesis *in vitro* and increases *de novo* root formation (Murch et al. 2001). Morphogenetic potential of indoleamines

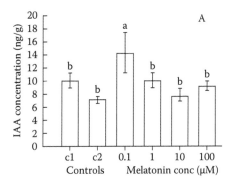

FIGURE 10.3 Free indoleacetic acid (IAA) in roots of 2-day-old seedlings of wild leaf mustard following treatment with different concentrations of melatonin: (A) IAA contents at different melatonin concentrations (0.1–100 mM); (B) Logit-transformed IAA contents at low melatonin concentrations (0.01–0.5 mM). The bars represent standard errors of the mean. The different superscripts represent statistically significant differences at Po0.05 for each treatment. The controls contained distilled water (C1) and 0.02% dimethyl sulfoxide (C2). (From Chen et al., *J Plant Physiol*, 166, 324–328, 2009.)

with reference to *in vitro* shoot multiplication in *Mimosa pudica* L. was reported (Ramakrishna et al. 2009). Moreover, indoleamines and calcium enhance somatic embryogenesis in *Coffea canephora* (Ramakrishna et al. 2012).

The effects of melatonin may be indirect by influencing the auxin levels, but during reproductive development it may act as a transition signal (Park 2011). The research on melatonin and serotonin in plants is still insufficient, and our understanding of their roles in plant growth and development is not clear enough.

Role of Melatonin in Circadian Cycle Regulation and Cell Protection

In animals, melatonin is involved in the regulation of the circadian rhythm. It was called *the hormone of darkness*, since its highest level is observed at night, while during the day it decreased to a hardly detectable level. Several attempts have been made to seek a role for this indolic compound as a photoperiodic and circadian regulator (Kolar et al. 1997; Van Tassel et al. 2001; Wolf et al. 2001).

The circadian rhythm is photoperiod-dependent and the elevated level of melatonin is related to the length of the night (Wolf et al. 2001). In *Chenopodium rubrum*, the level of melatonin shows diurnal fluctuation—an increase during the night and a decrease during the day (Kolar et al. 1997; Wolf et al. 2001). Similarly, the light/dark cycle has a profound influence on the indoleamine profile in *Dunaliella bardawil* (Ramakrishna et al. 2011).

Kolar et al. (1997) studied the possible changes in the melatonin levels in *Chenopodium rubrum* L. during light/dark cycles of 12 h/12 h. An oscillating behavior was observed, showing low or undetectable melatonin levels during the light period and a considerable increase in darkness. Nevertheless, in a subsequent work, Kolar et al. (2003) exposed plants to different photoperiodic profiles (6, 12, and 18 h of darkness) and observed no changes in the level of the melatonin with the photoperiod applied. Although the maximum level of melatonin always occurred after lights off, it is being concluded that melatonin synthesis is not directly light-regulated, but shows a circadian rhythm, as in animals.

Van Tassel and O'Neill (2001) investigated the possible variations in melatonin concentrations in *Pharbitis nil* Choisy seedlings under light/dark photoperiods, and showed no differences in the melatonin content with respect to the light/dark cycle and no conclusive evidence relating melatonin levels and photoperiod.

Murch and Saxena (2002) studied the possible protective role of melatonin during flower development of *Hypericum perforatum* L. A profile of indole levels (IAA, serotonin, and melatonin) throughout flower development indicated that concentrations of serotonin and melatonin were higher at given stages, and the higher melatonin level coincided with the maximum regeneration potential of isolated anthers. The authors suggested that melatonin could act as a stress-protecting agent, providing an adaptive mechanism to ensure reproduction.

An interesting effect of melatonin as a protector against cold-induced apoptosis in carrot suspension cells has been suggested by Lei et al. (2004), since the pretreatment of carrot suspension cells by melatonin significantly increased the level of the polyamines, putrescine, and spermidine, stimulated plant development (auxins, cytokinins, and gibberellins), and decreased the inhibitory hormones (abscisic acid and ethylene).

Indoleamines are important for detoxification of reproductive organ tissues. It was observed that serotonin present in walnut seeds takes part in detoxification of excess ammonium that accumulates during dehydration and seed desiccation (Murch and Saxena 2002).

Melatonin may protect lipids stored in seeds against peroxidation because of its lipophilic character and antioxidant properties, thus increasing seeds' viability and vigor (Manchester et al. 2000; Van Tassel and O'Neil 2001). The concentrations of melatonin in the seeds were very high, as mentioned previously, and germ tissue is highly vulnerable to oxidative stress and damage, so the high concentrations of melatonin detected in seeds presumably provides anti-oxidative defense. Moreover, exogenous melatonin application to the red cabbage seeds could be a good tool for seed vigor improvement (Posmyk et al. 2008). There are some reports that showed the effect of melatonin on seed dormancy (Balzer and Hardeland 1996), but the mechanisms of this regulation are unknown. ROS was defined as one of the factors regulating dormancy, for example, by protein oxidation and changes in protein carbonylation patterns (Oracz et al. 2007); therefore, the role of melatonin as antioxidant showed a regulating effect on ROS level and indirectly influenced seed dormancy.

Role of Melatonin as Antioxidant

One of the important roles of melatonin is its acting as an antioxidant and protecting plants against biotic or abiotic stress (Tan et al. 2012). This antioxidative effect has been detected in several plant species (Wang et al. 2012; Park et al. 2013; Vitalini et al. 2013; Yin et al. 2013) against biological free radicals, such as reactive oxygen (ROS) and nitrogen species (RNS), including the hydroxyl radical (HO·), singlet oxygen $1O_2$, peroxyl radical, hydrogen peroxide (H_2O_2), peroxynitrite anion (ONOO$^-$), and nitric oxide (NO) (Tan et al. 1993; Galano et al. 2011). The intermediate products of its metabolism also have antioxidant properties (Galano et al. 2013) and have synergistic action with other antioxidants, such as ascorbic acid, glutathione, polyamines, and so on (Gitto et al. 2001). The free radical scavenging capacity of melatonin extends to its secondary, tertiary, and quaternary metabolites (Hardeland et al. 2009). Bonnefont-Rousselot and Collin (2010) suggested that the efficiency of free radical scavenging by melatonin was highly dependent on their production site and thereby protecting lipids and/or protein against oxidation.

The free radicals are continuously generated during cellular respiration in mitochondria and during photosynthesis in chloroplasts. To protect these organelles against oxidative injuries and preserve their functions, the noxious free radicals must be detoxified via antioxidant processes. Melatonin has a pivotal role in defending against radicals at both the mitochondrial and chloroplast levels (Tan et al. 2013).

Hardeland et al. (2003) stated that melatonin may be a critical molecule that preserves mitochondrial integrity and physiology through (1) Protection against mitochondrial oxidative stress and apoptosis via scavenging ROS and RNS, (2) increasing the efficiency of ATP production by accelerating electron flow through a mechanism that promotes the activities of complexes I, III, and V (Martin et al. 2000; Acuna-Castroviejo et al. 2001; Martin et al. 2002), and (3) maintaining an optimal membrane potential across the inner mitochondrial membrane by regulating the mitochondrial permeability transition pore (MPTP) (Andrabi et al. 2004; Petrosillo et al. 2009).

Due to its antioxidant properties, melatonin may also stabilize cell redox status and protect tissues against stressful environments. Pretreatment of melatonin attenuates cold-induced apoptosis in carrot suspension cells (Lei et al. 2004), seed hydropriming with melatonin decreases lipid peroxidation caused by toxic copper ion in red cabbage seedlings (Posmyk et al. 2008), melatonin prevents chlorophyll degradation during the senescence of barley leaves (Arnao and Hernandez-Ruiz 2009b), and protects membrane structures against peroxidation during chilling stress and recovery in cucumber seeds (Posmyk et al. 2009). Numerous studies have demonstrated that melatonin can scavenge radicals, either directly or by regulating the activity of antioxidant enzymes such as peroxidases (POX), glutathione reductase (GSSG-R), superoxide dismutase (SOD), and catalase (CAT) (Allegra et al. 2003; Teixeira et al. 2003; Rodriguez et al. 2004; Reiter et al. 2007b).

Since melatonin at low concentrations is soluble in both water and lipids, it may be a hydrophilic and hydrophobic antioxidant. So, melatonin distributes in the aqueous cytosol and in lipid-rich membranes (Catala 2007; Venegas et al. 2012).

Melatonin may recycle several oxidized antioxidants including vitamin C, vitamin E, glutathione, and nicotinamide adenine dinucleotide (NADH), and these antioxidants also recycle the melatonin neutral radical (Tan et al. 2005). It seems that this bidirectional recycling by melatonin may depend on conditions, and it exhibits concentration equilibrium. This suggests that when antioxidants in the cytosol are exhausted by excessive cytosolic oxidative stress, the lipophilic antioxidants resident in other cellular compartments could provide antioxidative protection in the cytosol using melatonin as the bridge to recycle water-soluble antioxidants and vice versa. This would be classified as an antioxidant network as illustrated by Tan et al. (2013) and shown in Figure 10.4. The synergistic effects of melatonin with other antioxidants have been reported under several experimental conditions (Gitto et al. 2001; Lopez-Burillo et al. 2003).

Compared to the classic antioxidants such as vitamin C, vitamin E, and glutathione, melatonin exhibits a more potent antioxidant capacity to reduce oxidative injury, especially in *in vivo* conditions (Tunez

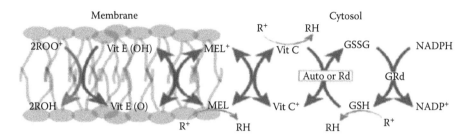

FIGURE 10.4 The proposed cellular antioxidant network. Melatonin is an amphiphilic molecule. The cellular hydrophilic antioxidants may be functionally connected with cellular lipophilic antioxidants within cells via melatonin as a bridge. MEL, melatonin; MEL·, melatonin neutral radical; Vit. E (O), oxidized vitamin E; Vit. E (OH), reduced vitamin E; Vit. C (O), oxidized vitamin C; Vit. C (OH), oxidized vitamin C; ROO·, proxy radical; R·, radical; RA, reduced agent; Auto or Rd, automatically or via reductase; GSH, reduced glutathione; GSSG, oxidized glutathione; GRd, glutathione reductase. (From Tan et al., *J Pineal Res*, 54, 127–138, 2013.)

et al. 2007). The superior antioxidant capacity of melatonin is, at least partially, attributed to what is referred to as the cascade reaction when scavenging free radicals (Tan et al. 2000). This specific cascade of scavenging reactions of melatonin is a result of its unique metabolism. Melatonin interacts with a variety of ROSs and RNSs including HO·, H_2O_2, $1O_2$, superoxide anion (O^{-2}), ONOO-, and NO (Hibaoui et al. 2009). The resulting products of these reactions are cyclic 3-hydroxymelatonin (C-3HOM) and other hydroxylated melatonin metabolites, N1-acetyl-N2-formyl-5-methoxyknuramine (AFMK), and N-acetyl-5-methoxyknuramine (AMK). These metabolites, as their precursor melatonin, also function as radical scavengers (Galano et al. 2013). Some of them, including C-3HOM and AMK, are more aggressive than melatonin regarding their capacity to scavenge oxidants (Ressmeyer et al. 2003; Lopez-Burillo et al. 2003). It is estimated that via the cascade reaction, one melatonin molecule may scavenge up to 10 free radicals (Tan et al. 2007b), which contrasts with the classic antioxidants because they typically detoxify one radical per molecule.

In addition, Poeggeler et al. (2002) showed that melatonin is a more powerful antioxidant than vitamins C, E, and K, and is able to protect biological tissue from the harmful effects of free radicals. This is probably because melatonin is a broad-spectrum antioxidant easily penetrating all cells, whereas the vitamins mentioned above are capable of only selective migration. Sliwinski et al. (2007) stated that melatonin reduced the oxidative damage of important molecules such as nucleic acids, proteins, and lipids.

The role of melatonin as an antioxidant can be classified into six categories, as reported by Posmyk and Janas (2009) and shown in Figure 10.5: (1) direct scavenging of free radicals, (2) stimulating activities of antioxidant enzymes, (3) stimulating the synthesis of glutathione, (4) augmenting the activities of other antioxidants, (5) protection of antioxidant enzymes from oxidative damage, and (6) increasing the efficiency of mitochondrial electron transport chain thereby lowering electron leakage and thus reducing free radical generation (Tan et al. 2002; Kładna et al. 2003; Rodriguez et al. 2004; Leon et al. 2005; Tan et al. 2010).

Role of Melatonin in Enhancement of Plant Tolerance to Environmental Stress

It can be speculated that melatonin is inexpensive and safe for animals and humans, and that its application as a biostimulator could be a good, feasible, and effective method used in agriculture to decrease environmental stress (Bonnefont-Rousselot and Collin 2010), as well as to increase food quality. There are several reports demonstrating the role of melatonin in alleviating the adverse effects of abiotic stresses such as extreme temperature (Lei et al. 2004; Xu 2010), cold (Munne-Bosch and Alegre 2002; Wang et al. 2012), heavy metals (Posmyk et al. 2008), and salinity (Li et al. 2012).

Exogenous melatonin could be a good biostimulator, improving not only seed germination and seedling/plant growth but also crop production, especially under stress conditions (Janas and Posmyk 2013). Melatonin not only enhanced the size of soybean seedlings, but also improved their growth

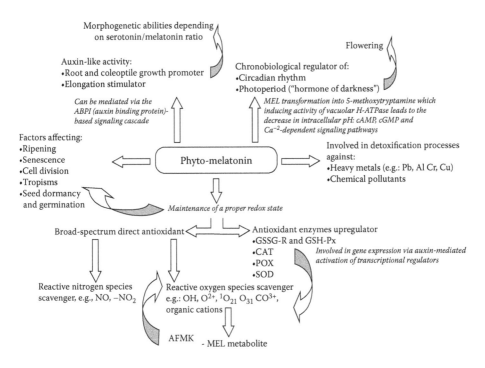

Morphogenetic abilities depending on serotonin/melatonin ratio

Auxin-like activity:
•Root and coleoptile growth promoter
•Elongation stimulator

Can be mediated via the ABPI (auxin binding protein)-based signaling cascade

Flowering

Chronobiological regulator of:
•Circadian rhythm
•Photoperiod ("hormone of darkness")

MEL transformation into 5-methoxytryptamine which inducing activity of vacuolar H-ATPase leads to the decrease in intracellular pH: cAMP, cGMP and Ca^{-2}-dependent signaling pathways

Factors affecting:
•Ripening
•Senescence
•Cell division
•Tropisms
•Seed dormancy and germination

Phyto-melatonin

Involved in detoxification processes against:
•Heavy metals (e.g.: Pb, Al Cr, Cu)
•Chemical pollutants

Maintenance of a proper redox state

Broad-spectrum direct antioxidant

Antioxidant enzymes upregulator
•GSSG-R and GSH-Px
•CAT
•POX
•SOD

Involved in gene expression via auxin-mediated activation of transcriptional regulators

Reactive nitrogen species scavenger, e.g., NO, $-NO_2$

Reactive oxygen species scavenger e.g.: OH, O^{2+}, $^1O_{21}$ O_{31} CO^{3+}, organic cations

AFMK - MEL metabolite

FIGURE 10.5 The phyto-melatonin role and mechanism of action. (From Posmyk et al., *Acta Physiol Plant*, 31, 1–11, 2009.)

rate and soybean yields (Wei et al. 2015). Hernandez-Ruiz et al. (2004) mentioned that a higher concentration of melatonin (200 μM) had no significant effect or even inhibitory effect on seed germination. However, lower concentrations of melatonin (50 or 100 μM) promoted seed germination as mentioned by Wei et al. (2015).

This indoleamine was found in ripe fruit of wild *Lycopersicon pimpinellifolium* L. Mill. and in common *Lycopersicon esculentum* Mill. However, in the latter—more resistant to ozone—the melatonin content was five times higher than in the sensitive wild type (Dubbels et al. 1995). Moreover, the level of melatonin in ripe tomato fruit is much higher than in green ones. This may be connected with protection of fruit against high ROS generation during ripening (Dubbels et al. 1995), also according to Van Tassel et al. (2001), who mentioned that melatonin may be involved in the maintenance of proper organ redox state during fruit ripening. Melatonin regulates transcript levels of many defense-related factors, including stress receptors, kinases, and transcription factors (Weeda et al. 2014).

Both abiotic and biotic stresses cause a significant increase in endogenous melatonin levels in plants. Melatonin regulates the growth of roots and shoots, activates seed germination, and delays leaf senescence of plants subjected to abiotic stress through its role as regulator of antistress genes and as a plant growth regulator (Arnao and Hernández-Ruiz 2014). The main function of melatonin in higher plants is acting as an antistress agent against abiotic stressors such as drought, salinity, low and high ambient temperatures, UV radiation, and toxic chemicals (Arnao and Hernández-Ruiz 2015; Zhang et al. 2015).

Plants possess a high content of melatonin to cope better under adverse environmental conditions compared to those having lower levels of this compound (Zhang et al. 2013). Melatonin is the first line of defense against the oxidative stress, since it protects different plant tissues and organs, particularly reproductive tissues, fruit, and germ tissues of the seed (Manchester et al. 2000; Van Tassel et al. 2001; Afreen et al. 2006; Tan et al. 2010).

Arnao and Hernandez-Ruiz (2013) showed that melatonin was accumulated in the leaves, stems, and roots of tomato plants cultivated in open field conditions, in comparison with those that were cultivated in chamber and *in vitro* cultures. Moreover, the water hyacinth—an aquatic plant—is highly tolerant of environmental pollutants because of its high levels of melatonin (Tan et al. 2007a). Transgenic rice plants

TABLE 10.1

Existence of Melatonin in Higher Plants

Plant Species	Organs	Concentration	Method of Melatonin Determination	References
5 edible species	Edible parts	2–510 pg/g FW	RIA; GC–MS	Dubbels et al. 1995
Tobacco	Leaves	40–100 pg/g FW		
24 edible plants	Edible parts	10–5,300 pg/g FW	RIA; HPLC–FD	Hattori et al.1995
Chenopodium rubrum up	Shoots	to 250 pg/g FW	LC–MS/MS; RIA	Kolár et al.1997
Feverfew	Leaves and flowers	1.4–2.5 μg/gDW		
St. John's wort	Leaves and flowers	1.8–4.4 μg/gDW	not reported	Murch et al.1997
Scutellaria species	Leaves	0.1–7.1 μg/g DW		
15 species of edible plants	Seeds	2–190 ng/g dry seed	RIA; HPLC–ECD	Manchester et al. 2000
Tomato	Fruits	up to 1.4 ng/g FW	HPLC, ELISA	Pape and Luning 2006
Morning glory	Shoots	up to 12 pg/g FW	GC–MS; RIA	Van Tassel et al. 2001
Chenopodium rubrum	Shoots	100 pg/g	FW LC–MS/MS	Wolf et al. 2001
Tart cherries	Fruits	2–13 ng/g FW	HPLC–ECD	Burkhardt et al. 2001
St. John's wort developing	Flowers	up to 4,000 nmol/g FW	HPLC–ECD; LC–MS/MS; RIA	Murch and Saxena 2002
Chinese medicinal herbs	Various organs	up to 3,800 ng/g DW	HPLC–FD; LC–MS/MS	Chen et al. 2003
Huang-qin	Shoots	9–44,000 nmol/g DW	HPLC–ECD; LC–MS/MS; RIA	Murch et al. 2004
Lupin	Hypocotyls	9–28 ng/g FW	HPLC–ECD; LC–MS/MS	Hernandez-Ruiz et al. 2004
Purslane	Leaves	19 ng/g FW	GC–MS	Simopoulos et al.2005
Grapes	Berry skins	up to 965 pg/g FW	HPLC, ELISA	Iriti et al. 2006
Glycyrrhizauralensis	Root, leaf	seeds 0.3–34 μg/g FW	HPLC	Afreen et al. 2006
Water hyacinth	Leaves, flowers	4–306 ng/g FW	LC/ESI/MS–MS	Tan et al. 2007
Lupin roots	Leaves, cotyledons, hypocotyls, seeds	0.5–38 ng/g FW	LC–TOF/MS	Hernandez-Ruiz and Arnao 2008
Barley	Roots, leaves, cotyledons, hypocotyls, seeds	0.1–9.6 ng/g FW		

Source: Okazaki, M. Dissection of melatonin accumulation and biosynthesis in tomato. PhD Thesis. Tsukuba, Japan: University of Tsukuba. 2009.

Note: FW: Fresh weight, DW: Dry weight, HPLC: High-performance liquid chromatography, ECD: Electrochemical detection, FD: Fluorescence detection, LC–MS/MS: Liquid chromatography—tandem mass spectrometry, GC–MS: Gas chromatography—mass spectrometry, RIA: Radioimmunoassay, LC–TOF/MS: Liquid chromatography with time-of-flight/mass spectrometry.

rich in melatonin were more resistant to butafenacil, a herbicide which induces oxidative stress (Park et al. 2013). Melatonin accumulated within different organs of higher plants such as leaves, stems, roots, flowers, fruits (pericarp and locular tissues), seedlings, and seeds at concentrations usually ranging from picograms to micrograms per gram of tissue as reported by Okazaki (2009) and shown in Table 10.1.

Beneficial effects of melatonin may result from its signaling function, through the induction of different metabolic pathways, and stimulate the production of various substances, preferably operating under stress (Tan et al. 2012). The osmotic adjustment in plants subjected to salt stress occurs by the accumulation of high concentrations of osmotically active compounds known as compatible solutes, such as proline, glycinebetaine, soluble sugars, free amino acids, and polyamines (Jagesh et al. 2010). Such substances play an important role in the adaptation of cells to various adverse environmental conditions through raising osmotic pressure in the cytoplasm, stabilizing proteins and membranes, and maintaining the relatively high water content obligatory for plant growth and cellular functions. The accumulation of ROS under stress was inhibited by melatonin applications due to direct scavenging and/or enhanced activities of antioxidant enzymes (Tan et al. 2000; Tan et al. 2007b).

The present collected knowledge justifies the belief that melatonin could be used as an effective biostimulator and protectant in agriculture that improves crop production under adverse environmental conditions.

Due to its antioxidant properties, melatonin protected the roots of barley from the damaging effects of NaCl, $ZnSO_4$, and H_2O_2 (Arnao and Hernández-Ruiz 2009a). Melatonin alleviates the effect of drought on seed germination and seedling growth in PEG-stressed cucumber (*Cucumis sativus* L.) (Zhang et al. 2013). Melatonin had a positive impact on cold-induced apoptosis in a suspension culture of carrot (*Daucus carota* L.), and positively correlated with polyamine synthesis (Lei et al. 2004). Posmyk et al. (2008) reported that the presowing seed treatment with melatonin protected red cabbage seedlings against toxic Cu ion concentrations as well as melatonin application to cucumber seeds had a beneficial effect on seed germination, the growth of seedlings, and crop production of plants, especially those subjected to cold- (Posmyk et al. 2009) and water stress (Zhang et al. 2013).

Arnao and Hernández-Ruiz (2009b) cited that melatonin treatments lowered chlorophyll degradation and slowed down the senescence process in barley plants, where 1 mM melatonin was optimal. Xu et al. (2010) mentioned that treatment with melatonin played an important role in the preservation of chlorophyll and promotion of photosynthesis due to raising the antioxidant enzyme activities and antioxidant contents and thus inhibiting the production of ROS. In addition, Zhang et al. (2013) stated that treatment with 100 μM melatonin significantly reduced chlorophyll degradation in cucumber seedlings and alleviated the effect of water stress. Furthermore, the ultrastructure of chloroplasts in water-stressed cucumber leaves was maintained after melatonin treatment. Melatonin molecule significantly reduced chlorophyll degradation and suppressed the upregulation of senescence-associated genes, and increased the photosynthetic efficiency of many plants (Tan et al. 2012; Wang et al. 2013).

Arnao and Hernández-Ruiz (2009a) showed that the melatonin content in barley roots increased due to chemical stress agents (sodium chloride, zinc sulfate, or H_2O_2), reaching up to six times the melatonin content of control roots. H_2O_2 (10 mM) and zinc sulfate (1 mM) were the most effective inducers. Such increases in melatonin probably play an important antioxidative role in the defense against chemical stress and other abiotic/biotic stresses.

Abdel-Monem et al. (2010) carried out a field experiment to study the effect of different concentrations of Trp and Prozac (5-hydroxytryptamine [5HT]) (0.00, 2.50, and 5.00 mg/l) on improving tolerance of two sunflower cultivars (Hysun 336 and Euroflor) grown under different saline conditions (1.56, 4.68, and 7.83 dS/m). Data indicated that application of either Trp or Prozac resulted in significant increases in plant growth, seed yield, and oil yield per feddan concomitantly with an increase in the level of IAA, photosynthetic pigments, total carbohydrates, and protein contents. The maximum increase was recorded in plants treated with 5.00 mg/l of either Trp or Prozac. The exogenous application of Trp or Prozac (serotonin) (5HT) mitigated partially or completely the adverse effects of salt stress on growth of the two sunflower cultivars through increasing photosynthetic pigments and endogenous promoters, especially IAA (see Table 10.2).

The stimulatory effect of Prozac or Trp on plants may be attributed to the role of Prozac or Trp as growth promoters that in turn alleviate the harmful effect of salinity. In this connection, Barazan and Friedman (2000) concluded that Trp induced an effect on chloroplast biosynthesis through its role in IAA biosynthesis, which was found to lessen the salt-induced decrease in chlorophyll content. Prozac

TABLE 10.2

Effect of Tryptophan And Prozac on Total Pigments, Indoles, Seed Yield, Seed Protein, and Oil Contents of Sunflower Cultivars (Hysun 336 and Euroflor) under Different Levels of Soil Salinity

Treatment		Total Pigment (mg/g fresh weight)		Total Indole (μg/g fresh weight)		Seeds Wt/Feddan (Ton)		Seed Protein (%)		Oil Yield/Feddan (kg)	
Salinity EC (dSm⁻¹)	Material (mg/l)	Hysun 336	Euroflor	Hysun 336	Euroflor	Hysun 336	Euroflor	Hysun 336	Euroflor	Hysun 336	Euroflor
1.56	0.0	1.052	0.993	43.900	41.900	0.88	0.83	14.17	14.00	188.00	175.33
	Tryptophan 2.5	1.203	1.109	48.770	46.500	1.00	0.91	15.17	14.87	201.00	191.33
	5.0	1.246	1.153	49.630	47.50	1.08	1.00	16.00	15.60	217.67	211.00
	Prozac 2.5	1.159	1.070	48.400	46.200	1.04	0.91	15.60	14.82	203.67	192.56
	5.0	1.219	1.123	49.470	46.800	1.11	1.05	16.37	15.93	217.33	212.67
4.68	0.0	0.926	0.820	38.690	36.030	0.75	0.62	13.93	13.57	161.00	148.33
	Tryptophan 2.5	1.032	0.889	43.130	40.270	0.85	0.70	14.67	14.33	173.67	160.33
	5.0	1.089	0.922	43.730	40.870	0.93	0.78	15.37	14.97	193.67	185.00
	Prozac 2.5	1.002	0.872	42.500	39.900	0.89	0.75	15.23	14.77	178.67	168.00
	5.0	1.056	0.919	43.330	40.630	0.96	0.83	15.73	15.23	197.33	189.33
7.83	0.0	0.739	0.666	32.870	30.830	0.26	0.10	13.17	12.67	135.33	126.67
	Tryptophan 2.5	0.829	0.740	36.970	34.740	0.49	0.37	13.83	13.33	148.33	137.67
	5.0	0.876	0.776	37.970	35.930	0.59	0.48	14.37	13.90	172.67	154.00
	Prozac 2.5	0.783	0.709	36.500	34.300	0.54	0.42	14.17	13.60	153.33	146.00
	5.0	0.853	0.766	37.200	34.970	0.61	0.49	14.53	14.10	184.67	163.33
LSD at 5%		0.004		2.122		0.0907		0.0907		7.833	

Source: Abdel-Monem et al., *Inter J Academic Res*, 2: 254–262, 2010.

led to an alternate metabolic pathway originated with Trp, and regulated the relative ratio of auxin (root inducer) to cytokinin (chloroplast inducer) (Murch et al. 2001). Moreover, Kang et al. (2009a,b) and Park et al. (2009) postulated that serotonin plays a hormonal role in plant seedling growth and plays a protective role as a senescence-retarding compound. Murch et al. (2001) proved that the endogenous concentration of IAA was higher in explants exposed to the medium supplemented with Prozac as compared with negative control. After conversion of Prozac to melatonin (*N*-acetyl-5-methoxytryptamine) (Murch et al. 2000), melatonin may be linked to secondary messenger systems and Ca^{2+}-signal transduction pathways (Murch and Saxena 2002). Biogenic monoamines such as serotonin, tryptamine, Prozac, and tyramine are considered as mitogenic factors that are involved in flowering, morphogenesis, and protection from and adaptation to environmental changes in plants (Dalin et al. 2008).

In addition, Abdel-Monem et al. (2010) mentioned that treatments with Prozac or Trp under different salinity levels resulted in highly significant increases in the contents of total carbohydrate and total protein contents of sunflower shoots. In this respect, they concluded that the marked increase in total carbohydrate and protein contents by Prozac or Trp treatment not only play a hormonal role by alleviating the inhibitory effect of salinity stress, via osmotic adjustment, or act in a protective role by conferring some desiccation resistance to plant cells, but also stimulated the accumulation of carbohydrates and nitrogenous contents (Kang et al. 2009a,b; Park et al. 2009). Abdel-Monem et al. (2010) added that the increase in yield components of sunflower in response to Prozac or Trp treatments relative to untreated plants might be resulting from increased head diameter, seed weight/head, and 100-seeds weight, under all salinity levels. These results are in harmony with those obtained by El-Bassiouny (2005) who concluded that increased levels of endogenous IAA, gibberellic acid (GA_3), and cytokinin in wheat plants treated with Trp contributed enhanced growth and yield. Plant growth regulators appear either to form a sink mobilizing the different nutrients, which are involved in building new tissues in wheat plants and/or enhance photosynthesis. Moreover, Zahir et al. (2005) and Ahmad et al. (2008) revealed that application of L-tryptophan significantly affected the maize crop growth, yield and total nitrogen uptake compared with an untreated control. Melatonin exogenously applied may exhibit some auxin-like effects in plants. It could hypothetically bind to auxin receptors and act directly as an agonist auxin (Kolar and Machackova 2005).

Spraying sunflower cultivars with 2.5 or 5 mg/l of either Trp or prozac induced significant increases in yield quality (protein% and oil yield) as shown in Table 10.2. The maximum increases were obtained in response to plants sprayed with 5 mg/l prozac under all salinity levels in both cultivars. The oil yield increased by 5 mg/l prozac at EC1.56 dSm^{-1} salt by 15.60% and 21.30%, 4.68 dSm^{-1} salt by 22.56% and 27.64% and at EC 7.83 dSm^{-1} salt by 36.46% and 28.94% at Hysun 336 and Euroflor cultivars, respectively.

Li et al. (2012) studied the melatonin role in regulating growth, ion homeostasis, and the response to oxidative stress in *Malus hupehensis* Rehd. under high-salinity conditions. Salinity stress reduced plant growth, net photosynthetic rates, and chlorophyll content. However, pretreatment with 0.1 µM melatonin significantly alleviated this growth inhibition and enabled plants to maintain an improvement of photosynthetic capacity. Moreover, addition of melatonin lowered the amount of oxidative damage due to salinity, perhaps by directly scavenging H_2O_2 or enhancing the activities of antioxidative enzymes such as ascorbate peroxidase, CAT, and POX. Melatonin possibly contributed to the maintenance of ion homeostasis, thus improved salinity-resistance of plants. They mentioned that melatonin might control the expression of ion channel genes under salinity, which may possibly contribute to the maintenance of ion homeostasis and thus improve the salinity-resistance of plants. The ability to limit Na^+ transport into the shoots and to reduce Na^+ accumulation in the rapidly growing shoot tissues is critically important for maintenance of high growth rates and protection of metabolic processes in elongating cells from the toxic effects of Na^+ (Roodbari et al. 2013).

Wang et al. (2013) reported that 100 µM melatonin decreased the oxidative stress and delayed leaf senescence of HanFu apple (Malus domestica Borkh.) when added to soils under drought conditions. This molecule significantly reduced chlorophyll degradation and enhanced the photosynthetic process as shown in Figure 10.6 through suppressed upregulation of the senescence-associated gene 12 (SAG12) and pheophorbide a oxygenase (PAO), and helped to maintain better function of photosystem II (PSII) under drought. The addition of melatonin also controlled the burst of H_2O_2, possibly through direct

FIGURE 10.6   The effects of melatonin on leaf yellowing after 90 days of treatment: (a) well-watered control, (b) drought control, (c) melatonin (MEL)-applied/well watered treatment, and (d) melatonin (MEL)-applied/drought treatment. (From Wang et al., *J Pineal Res*, 54, 292–302, 2013.)

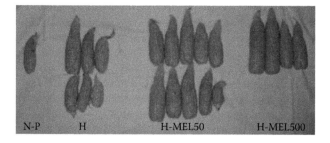

FIGURE 10.7   The effect of melatonin presowing treatment on corn (*Zea mays* L.) yield. Corn seeds were hydroprimed with melatonin at different concentrations (HMEL50 and HMEL500). Seeds hydroprimed with water (H) and nonprimed (N-P) were established as the control. Plants were grown in the field, under natural conditions, without further treatment. The presented yield originated from the same number of plants in each experimental variant. The experiments were conducted in 2009. (From Janas et al., *Acta Physiol Plant*, 35, 3285–3292, 2013.)

scavenging and by enhancing the activities of antioxidative enzymes and the capacity of the ascorbate–glutathione cycle.

The data collected by Janas and Posmyk (2013) showed that hydroprimed or osmoprimed seeds of corn (*Zea mays* L.), mung bean (*Vigna radiata* L.), and cucumber (*Cucumis sativus* L.) with melatonin had higher crop yield than the controls under field conditions (Figures 10.7 through 10.9). 50 and 500 lM melatonin–treated corn plants had more and larger cobs than those hydroprimed without melatonin and the nonprimed plants (see Figure 10.7). Similar results were observed with mung bean whose seeds were hydroprimed with melatonin at 20, 50 and 500 lM concentrations. The number of pods was more due to the seeds hydroprimed with 50 lM melatonin than those hydroprimed without melatonin and nonprimed ones. It seems that 500 lM melatonin concentration used in seed priming was too high, and the number of pods was lower in comparison to the seeds hydroprimed without melatonin and nontreated ones (see Figure 10.8). At harvesting, 50 lM melatonin osmoprimed cucumber plants had more fruits than those osmoprimed with melatonin 500 lM, osmoprimed without melatonin, or nontreated plants, and some fruits of melatonin-treated plants were larger than those osmoprimed without melatonin and non-osmoprimed ones (see Figure 10.9). It is surprising that the one-time melatonin application to the seeds gave a significantly positive effect on the yield of plants that grew in the field. Moreover, production of corn, cucumber, and mung bean primed with melatonin were about 10%–25% greater in comparison to those primed without melatonin, and it also depended on plant species.

Zhang et al. (2013) reported that 100 μM melatonin alleviated polyethylene glycol–induced inhibition of cucumber seed germination, showing the greatest germination rate and photosynthetic rate, and at the

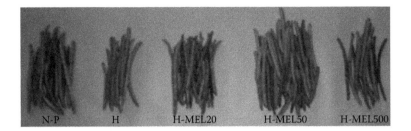

FIGURE 10.8 The effect of melatonin presowing treatment on mung bean (*Vigna radiata* L.) yield. Mung bean seeds were hydroprimed with melatonin at different concentrations (HMEL20, HMEL50, and HMEL200). Seeds hydroprimed with water (H) and nonprimed (N-P) were established as the control. Plants were grown in the field, in natural conditions, without further treatment. Presented yield originated from the same number of plants in each experimental variant. The experiments were conducted in 2011. (From Janas et al., *Acta Physiol Plant*, 35, 3285–3292, 2013.)

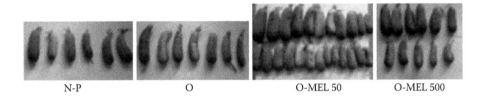

FIGURE 10.9 The effect of melatonin presowing treatment on cucumber (*Cucumis sativus* L.) yield. Cucumber seeds were osmoprimed with melatonin at different concentrations (OMEL50 and OMEL500). Seeds osmoprimed with water (O) and nonprimed (N-P) were established as the control. Plants were grown in the field, in natural conditions, without further treatment. Presented yield originated from the same number of plants in each experimental variant. The experiments were conducted in 2011. (From Janas et al., *Acta Physiol Plant*, 35, 3285–3292, 2013.)

same time significantly reduced chlorophyll degradation. Furthermore, the ultrastructure of chloroplasts in water-stressed cucumber leaves was improved after melatonin treatment, thus reversing the effect of water stress. Melatonin stimulated root generation and increased the root:shoot ratio. The antioxidant levels and activities of the ROS scavenging enzymes, i.e., SOD, POX, and CAT, were also increased by melatonin treatment. These results suggest that the adverse effects of water stress can be minimized by the application of melatonin.

Mukherjee et al. (2014) reported that salt stress for 48 h caused increases in endogenous serotonin and melatonin content in the roots and cotyledons of sunflower (*Helianthus annuus*) seedlings. Accumulation of serotonin and melatonin under salt stress exhibits differential distribution in the vascular bundles and cortex in the differentiating zones of the primary roots, suggesting their compartmentalization in the growing region of roots. Moreover, serotonin and melatonin accumulation in oil body–rich cells of salt-treated seedling cotyledons correlate with longer retention of oil bodies in the cotyledons. Exogenous serotonin and melatonin treatments (15 µM) regulate hypocotyl elongation and root growth of sunflower (*Helianthus annuus*) seedlings under NaCl stress. Salt stress–induced root growth inhibition thus pertains to partial impairment of auxin functions caused by increased serotonin biosynthesis. In seedling cotyledons, NaCl stress modulates the activity of *N*-acetylserotonin *O*-methyltransferase (ASMT [or HIOMT] EC 2.1.1.4), the enzyme responsible for melatonin biosynthesis from *N*-acetylserotonin.

Szafrańska et al. (2014) studied the impact of hydropriming *Vigna radiata* seeds with melatonin (50 µM L^{-1}) on phenolic content, *L*-phenylalanine ammonialyase (PAL) activity, melatonin level, and antioxidant properties, as well as electrolyte leakage from roots of chilled and rewarmed seedlings. The level of melatonin in roots derived from hydroprimed seeds with melatonin was seven times higher than in roots derived from nonprimed seeds. The level of melatonin in the roots of seedlings derived from nonprimed seeds (C) at 25°C was 0.237 ng g^{-1} dry weight (DW), and a slight decrease appeared after chilling (0.180 ng g^{-1} DW), reaching the lowest level after rewarming (0.090 ng g^{-1} DW) as shown in Figure 10.10a. Melatonin level in H roots at 25°C was 0.130 ng g^{-1} DW. PAL activity after rewarming

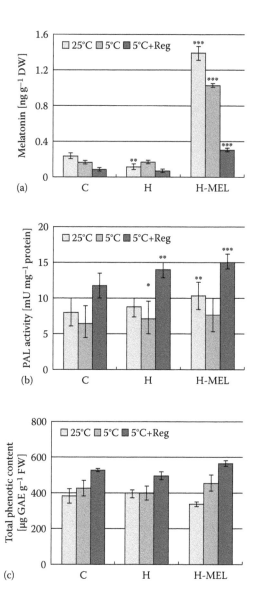

FIGURE 10.10   (a) Melatonin concentration in roots of *Vigna radiata* seedlings derived from C, H, and H-melatonin seeds, which were grown for 2 days at 25°C or at 5°C, and then rewarmed for 2 days at 25°C after chilling (5°C+Reg). (From Szafrańska et al., *Cent Eur J Biol*, 9, 1117–1126, 2014.) (b) Specific PAL activity in roots of *Vigna radiata* seedlings derived from C, H, and H-melatonin seeds, which were grown for 2 days at 25°C or at 5°C, and then rewarmed for 2 days at 25°C after chilling (5°C+Reg). (From Szafrańska et al., *Cent Eur J Biol*, 9, 1117–1126, 2014.). (c) Total phenolic contents in roots of *V. radiata* seedlings derived from C, H, and H-melatonin seeds, which were grown for 2 days at 25°C or at 5°C, and then rewarmed for 2 days at 25°C after chilling (5°C+Reg). The results are expressed as a GAE levels. (From Szafrańska et al., *Cent Eur J Biol*, 9, 1117–1126, 2014.)

has increased in all variants investigated but particularly in H-melatonin roots (Figure 10.10b), which can suggest that PAL was not only activated by the damage caused by chilling but also by the exogenous application of melatonin. Similar profiles of total phenolic content after rewarming (Figure 10.10c) may indicate that this significant phenolic accumulation in roots of *V. radiata* was not induced by chilling, but rather by the disappearance of the stressor. Moreover, the antioxidant capacity from chilled and rewarmed roots was correlated with phenolic content, while the reducing ability was correlated with

PAL activity. However, both parameters were higher in rewarmed roots with applied melatonin and accompanied by a significant decline in electrolyte leakage. Melatonin applied to the seeds of *V. radiata* can improve the defense mechanisms of plants during cold acclimation by increasing the activity of PAL after rewarming. This process, in turn, is accompanied by the induced accumulation of total phenolics by increased antioxidant capacity. Melatonin can play a positive role in plant acclimation to stressful conditions via activation of synthetic pathway of phenolic.

Wei et al. (2015) concluded that melatonin promoted soybean plant growth, increased yield, and improved salinity stress tolerance, since coating soybean seeds with melatonin significantly promoted plant growth, and reflected on soybean production and their fatty acid content. Transcriptome analysis revealed that salt stress inhibited expressions of genes related to oxido-reductase activity/processes, and secondary metabolic processes. On the other hand, melatonin regulated the expression of the genes inhibited by salt stress, and hence alleviated the inhibitory effects of salt stress on gene expression. Melatonin leads to enhancement of genes involved in cell division, photosynthesis, carbohydrate metabolism, fatty acid biosynthesis, ascorbate metabolism, and DNA replication. Moreover, melatonin also increased yield of soybeans both in the greenhouse and in the field, suggesting its potential application in agriculture.

They added that melatonin treatments alleviated the harmful effect of 1% (w/v) NaCl and drought (20%) on soybean plants, and increased plant height, leaf area, and plants' biomass chlorophyll content (Figures 10.11 and 10.12). Moreover, the control seedlings had higher H_2O_2 levels than the melatonin-treated seedlings, since the leaves of the control seedlings had a deeper brown color. The relative electrolyte leakage was lower in melatonin-treated seedlings compared with the control seedlings under salt stress. All these results imply that exogenous melatonin treatment increases salt tolerance of soybean plants. These effects are probably a result of the increased antioxidative ability and more stable membrane systems, as electrolyte leakage.

In addition, melatonin treatment caused increases in the accumulation of fatty acids in soybean as shown in Figure 10.13 and reported by Wei et al. (2015).

Dawood and El-Awadi (2015) concluded that melatonin treatments (100 mM and 500 mM) improved growth parameters, relative water content, photosynthetic pigments, total carbohydrate, total phenolic content, IAA, K^+, and Ca^{+2}, and reduced the levels of compatible solutes, Na^+, and Cl^- contents in leaf tissues of faba bean plants irrigated with diluted seawater (3.85 dS/m and 7.69 dS/m). Melatonin at 500 mM had a more pronounced effect in alleviating the adverse effects of the two salinity levels on the performance of faba bean plants than 100 mM melatonin. The beneficial effects of melatonin treatments in alleviating the harmful effect of salinity stress on the growth parameters was more pronounced in the plants grown under the higher salinity level (S2 = 7.69 dS/m) than those grown under lower salinity level (S1 = 3.85 dS/m) relative to corresponding controls. The increases in total photosynthetic pigments were 22.31%, 12.87%, and 15.85% in the plants treated with 500 mM melatonin and irrigated with tap water (S0), and diluted seawater at lower (S1) and higher (S2) concentrations, respectively, as compared with corresponding controls.

Melatonin concentrations (ME1 and ME2) in the absence of salinity caused gradual increases in total carbohydrate, phenolic content, and IAA accompanied by gradual decreases in soluble carbohydrate, free amino acid, and proline content relative to control plant (S0ME0) as shown in Table 10.3. Under salinity stress, the effect of melatonin on some biochemical constituents and compatible solutes of faba bean plants was more or less similar to its effect on the plants irrigated with tap water. It is worth mentioning that 500 mM melatonin was more effective than 100 mM melatonin either in enhancement of some parameters (total carbohydrate, phenolic content, and IAA) or inhibition of the others (soluble carbohydrate, free amino acid, and proline). Table 10.3 shows that salinity stress and/or melatonin treatments enhanced the phenolic content.

Moreover, both melatonin concentrations caused significant decreases in Na^+ and Cl^- percentages accompanied by significant increases in K^+ and K^+/Na^+ ratio, as well as nonsignificant increases in Ca^{+2}, relative to corresponding controls.

The sodium dodecyl sulfate polyacrylamide gel electrophoresis (SDS-PAGE) electrophoretic protein patterns of faba bean plants grown under the effect of two different concentrations of melatonin (100 mM and 500 mM), and irrigation with either tap water (S0) or diluted seawater at 3.85 dS/m (S1) or 7.69 dS/m

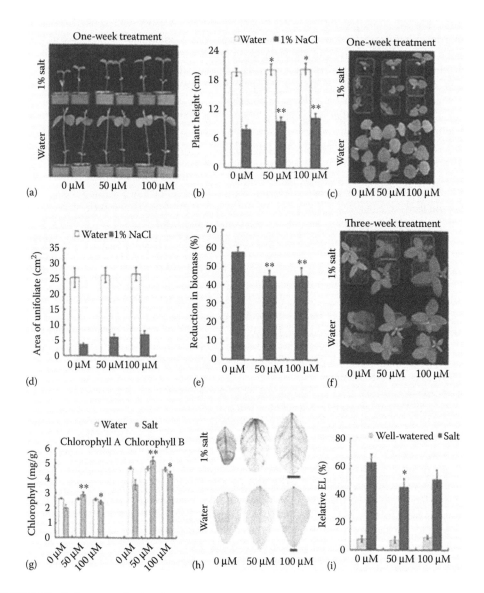

FIGURE 10.11 Performance of melatonin-treated seedlings in response to salt stress: (a) melatonin effects on seedlings treated with 1% salt for 1 week; (b) melatonin action on plant height after salt stress; (c) phenotypes of 1-week-old treated seedlings after melatonin and salt treatments; (d) comparison of leaf area after treatments; (e) reduction in biomass after treatments. Reduced proportion of biomass (dry weight) = [(biomass of well-watered plants)–(biomass of salt-treated plants)]/(biomass of well watered plants); (f) phenotypes of 3-week-old treated seedlings; (g) chlorophyll contents in soybean leaves after salt stress. The left-hand part represents the content of chlorophyll A, and right-hand part represents the content of chlorophyll B; (h) DAB staining. The brown color indicates accumulation of H_2O_2. Bars = 1 cm; (i) relative electrolyte leakage in treated plants. * and ** indicate significant differences ($p < .05$ and $p < 0.01$, respectively) compared with mock coating (0 μM). The bars indicate standard deviation. For leaf area and plant height, n = 36; for biomass analysis, n = 10; for chlorophyll test, n = 3; for relative electrolyte leakage, n = 5. (From Wei et al., *J Exp Botany*, 66, 695–707, 2015.)

(S2), are shown in (Figure 10.14). The control plants (S0ME0) exhibited seven protein bands (Mr: 125, 103, 72, 60, 44, 35, and 31 kDa, respectively). This number (seven bands) represented the least number of bands that were common among the different treatments. The high concentration of salinity (S2) alone or in combination with melatonin at the relatively low (ME1) or high (ME2) concentration induced a new protein band having a molecular rate 96 kDa. A unique protein band (Mr: 92 kDa) was induced by 500 mM melatonin (S0ME2). Furthermore, a protein band having a Mr 82 kDa molecular rate seemed to

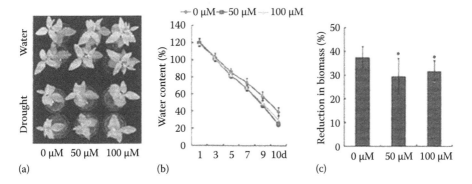

FIGURE 10.12 The growth of melatonin-treated seedlings in response to drought stress: (a) Performance of soybean seedlings grown in well watered soil or in soil supplied with 20% water for 1 week, (b) water content of the pot-grown plants after drought stress, and (c) reduction in biomass after drought stress. * indicates significant difference ($p < .05$) compared with mock coating (0 μM). The bars indicate standard deviation (n = 10). (From Wei et al., *J Exp Botany*, 66, 695–707, 2015.)

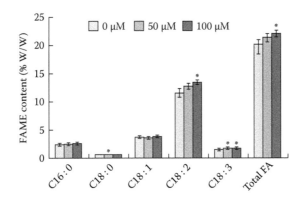

FIGURE 10.13 Fatty acid analysis in melatonin-treated plants. Seeds from field-grown plants were analyzed for their FA content (% w/w). The bars indicate standard deviation (n = 3). * indicates significant difference ($p < .05$) compared with mock coating (0 μM). (From Wei et al., *J Exp Botany*, 66, 695–707, 2015.)

be a marker for all melatonin treatments either alone (S0ME1 or S0ME2) or in combination with salinity at lower and higher levels (S1ME1, S1ME2, S2ME1, S2ME2). The low concentration of salinity (S1) alone or in combination with melatonin at the relatively low (ME1) or high (ME2) concentration induced two new protein bands having a molecular rate of 69 and 56 kDa, respectively.

The results showed the appearance of new protein bands in faba bean plants under the effect of salinity stress and/or melatonin treatments. The most important mechanism involved in cell protection against salt stress is the induction of *de novo* synthetic protein groups (Kermode 1997). These stress proteins may provide a storage form of nitrogen that may be reutilized when the stress is over (Badr et al. 1998).

It is suggested that these new proteins may play an important role in triggering a special system that helps the whole plant against salinity stress. These proteins may have an osmoprotection function (Dure 1993) or protect cellular structures (Close and Lammers 1993). Moreover, the melatonin-induced new protein bands, shown in the present work, might be attributed to the synthesis of new polypeptides or might represent degradative product(s) of proteins due to the effect of hydrolytic enzymes on high molecular weight proteins. This assumption might be reinforced by that of Posmyk et al. (2009) who reported that although melatonin protected membrane structure against peroxidation during chilling, excessive melatonin levels (~4 μg/g fresh weight) in cucumber seeds provoked oxidative changes in proteins.

TABLE 10.3

Effect of Melatonin on Some Biochemical Constituents and Compatible Solutes of Faba Bean Plants

Treatments		Total Carbohydrate (% Dry Weight)	Soluble Carbohydrate (% Dry Weight)	Total Phenolic Content (mg/g Dry Weight)	Free Amino Acid (mg/g Dry Weight)	Proline (mg/g Dry Weight)	Indole-3-Acetic Acid (µg/100 g Fresh Weight)
Salinity (dS/m)	Melatonin (µM)						
S_0	ME_0	16.27 ± 0.62^b	3.41 ± 0.26^d	26.89 ± 0.15^e	9.80 ± 0.17^d	0.47 ± 0.008^e	53.04 ± 0.03^c
	ME_1	18.54 ± 0.37^a	3.18 ± 0.18^{de}	27.86 ± 0.88^{de}	7.87 ± 0.41^{de}	0.40 ± 0.002^{ef}	61.73 ± 1.00^b
	ME_2	19.22 ± 0.40^a	2.61 ± 0.02^e	29.83 ± 0.24^{cd}	6.36 ± 0.11^e	0.34 ± 0.023^f	78.09 ± 0.34^a
S_1	ME_0	14.14 ± 0.32^c	5.12 ± 0.03^c	27.52 ± 0.29^{de}	16.90 ± 0.75^b	0.69 ± 0.020^{bc}	41.43 ± 0.40^d
	ME_1	16.84 ± 0.35^b	5.07 ± 0.04^c	28.72 ± 0.56^{de}	13.77 ± 0.19^c	0.61 ± 0.014^{cd}	50.58 ± 0.74^c
	ME_2	17.15 ± 0.34^b	4.82 ± 0.1^c	31.92 ± 1.06^{bc}	9.10 ± 0.06^d	0.57 ± 0.14^d	51.25 ± 0.43^c
S_2	ME_0	12.02 ± 0.47^d	6.60 ± 0.01^a	29.65 ± 0.74^d	20.07 ± 0.53^a	0.95 ± 0.023^a	25.07 ± 0.47^e
	ME_1	13.82 ± 0.59^c	6.11 ± 0.06^{ab}	32.21 ± 0.59^b	17.46 ± 0.43^b	0.89 ± 0.005^a	41.97 ± 0.44^d
	ME_2	14.92 ± 0.41^c	5.83 ± 0.03^b	34.73 ± 1.09^a	16.90 ± 0.23^c	0.73 ± 0.020^b	49.04 ± 0.52^c

Source: Dawood et al., *Acta Biol Colomb*, 20, 223–235, 2015.

Note: Means followed by the same letter for each tested parameter are not significantly different by Duncan's test ($p \leq .05$) and presented by $\pm$SE. Effect of melatonin at 100 Mm (ME1) and 500 Mm (ME2) on some biochemical constituents and compatible solutes of faba bean plants irrigated with tap water (S0) or diluted seawater at 3.85 Ds/M (S1) and 7.69 Ds/M (S2). The presented results are means of the measurements taken at two successive seasons (2011/2012 and 2012/2013) for six replicates at each season

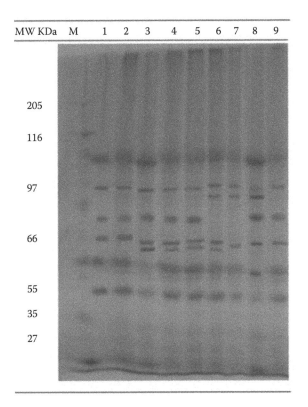

FIGURE 10.14 The effect of melatonin at 100 mM (ME1) and 500 mM (ME2) on SDS-PAGE electrophoretic protein patterns of faba bean plants irrigated with tap water (S0) or diluted seawater at 3.85 dS/m (S1) and 7.69 dS/m (S2). 1 = (S0ME0); 2 = (S0ME1); 3 = (S0ME2); 4 = (S1ME0); 5 = (S1ME1), 6 = (S1ME2); 7 = (S2ME0); 8 = (S2ME1); 9 = (S2ME2). (From Dawood et al., *Acta Biol Colomb*, 20, 223–235, 2015.)

Conclusion

Thus there are sufficient data on this role of melatonin on plant growth and development, especially on basic life cycle–related events such as cell division, root growth, senescence, and tolerance to environmental stress. The regulation of circadian rhythm has been investigated. The detoxification and antioxidant mechanisms influenced by melatonin add to the cellular functions and environmental adaptation of plants. Their role in phytoremediation needs to be explored in detail. The functional role of melatonin will be much more revealed by adopting the genomics and proteomic approaches to elucidate its influence on the regulatory pathways in plants.

REFERENCES

Abdel-Monem, A. A., H. M. S. El-Bassiouny, M. M. Rady, and M. S. Gaballah. The role of tryptophan and Prozac (5- hydroxy tryptophan) on the growth, some biochemical aspects and yield of two sunflower cultivars grown in saline soil. *Int J Academic Res*. 2010. 2: 254–262.

Acuna-Castroviejo, D., M. Martin, and M. Macias, G. Escames, J. Leon, H. Khaldy, and R. J. Reiter. Melatonin, mitochondria, and cellular bioenergetics. *J Pineal Res*. 2001. 30: 65–74.

Afreen, F., S. M. Zobayed, and T. Kozai. Melatonin in *Glycyrrhiza uralensis*: Response of plant roots to spectral quality of light and UV-B radiation. *J Pineal Res*. 2006. 41: 108–115.

Ahmad, R., A. A. Khalid, M. M. Arshad, A. Z. Zahir, and T. Mahmood. Effect of compost enriched with N and L-tryptophan on soil and maize. *Agron Sustain Dev*. 2008. 28: 299–305.

Allegra, M., R. J. Reiter, D. X. Tan, C. Gentile, L. Tesoriere, and M. A. Livrea. The chemistry of melatonin's interaction with reactive species. *J Pineal Res*. 2003. 34: 1–10.

Andrabi, S. A., I. Sayeed, D. Siemen, G. Wolf, and T. F. Horn. Direct inhibition of the mitochondrial permeability transition pore: A possible mechanism responsible for anti-apoptotic effects of melatonin. *FASEB J*. 2004. 18: 869–871.

Arnao, M. B., and J. Hernandez-Ruiz. The physiological function of melatonin in plants. *Plant Signal Behav*. 2006. 1: 89–95.

Arnao, M. B., and J. Hernandez-Ruiz. Melatonin promotes adventitious- and lateral root regeneration in etiolated hypocotyls of *Lupinus albus* L. *J Pineal Res*. 2007. 42: 147–152.

Arnao, M. B., and J. Hernández-Ruiz. Chemical stress by different agents affects the melatonin content of barley roots. *J Pineal Res*. 2009a. 46: 295–299.

Arnao, M. B., and J. Hernandez-Ruiz. Protective effect of melatonin against chlorophyll degradation during the senescence of barley leaves. *J Pineal Res*. 2009b. 46: 58–63.

Arnao, M. B., and J. Hernández-Ruiz. Growth conditions influence the melatonin content of tomato plants. *Food Chem*. 2013. 138: 1212–1214.

Arnao, M. B., and J. Hernández-Ruiz. Melatonin: Plant growth regulator and/or biostimulator during stress? *Trends Plant Sci*. 2014. 19: 789–797.

Arnao, M. B., and J. Hernández-Ruiz. Functions of melatonin in plants: A review. *J Pineal Res*. 2015. 59: 133–150.

Badr, A., A. Haider, S. Badr, and S. Radwan. Genotypic variation within Egyptian wheat in response to salt stress and heat shock. *Proc Intern Cong Mol Genetics*. 1998. 1: 11–18.

Balzer, I. R., and R. Hardeland. Melatonin in algae and higher plants: Possible new roles as a phytohormone and antioxidant. *Bot Acta*. 1996. 109: 180–183.

Banerjee, S., and L. Margulis. Mitotic arrest by melatonin. *Exp Cell Res*. 1973. 78: 314–318.

Barazan, O. Z., and J. Friedman. Effect of exogenously applied L-tryptophan on allelochemical activity of plant growth–promoting rhizobacteria (PGPR). *J Chem Ecol*. 2000. 26: 343–349.

Bonnefont-Rousselot, D., and F. Collin. Melatonin: Action as antioxidant and potential applications in human. *Toxicology*. 2010. 278: 55–67.

Catala, A. The ability of melatonin to counteract lipid peroxidation in biological membranes. *Curr Mol Med*. 2007. 7: 638–649.

Chen, Q., W. B. Qi, R. J. Reiter, W. Wei, and B. M. Wang. Exogenously applied melatonin stimulates root growth and raises endogenous indoleacetic acid in roots of etiolated seedlings of *Brassica juncea*. *J Plant Physiol*. 2009. 166: 324–328.

Close, T. J., and P. J. Lammers. An osmotic stress protein of cyanobacteria is immunologically related to plant dehydrins. *Plant Physiol*. 1993. 101: 773–779.

Dalin, L., K. Kang, J. Y. Choi, A. Ishihara, K. Back, and S. G. Lee. HPLC analysis of serotonin, tryptamine, tyramine, and the hydroxycinnamic acid amides of serotonin and tyramine in food vegetables. *J Med Food*. 2008. 11: 385–389.

Dawood, M. G., and M. E. El-Awadi. Alleviation of salinity stress on *Vicia faba* L. plants *via* seed priming with melatonin. *Acta biol Colomb*. 2015. 20: 223–235.

Dubbels, R., R. J. Reiter, E. Klenke, A. Goebel, E. Schnakenberg, C. Ehleers, H. W. Schiwara, and W. Schloot. Melatonin in edible plants identified by radioimmunoassay and by high performance liquid chromatography–mass spectrometry. *J Pineal Res*. 1995. 18: 28–31.

Dure, L. Structural motifs in LEA proteins. In Close, T. J., and E. A. Bray (editors), Plant responses to cellular dehydration during environmental stress. *Curr Topics Plant Physiol*. 1993. 10: 91–103.

El-Bassiouny, H. M. S. Physiological responses of wheat to salinity alleviation by nicotinamide and tryptophan. *Inter J Agric Biol*. 2005. 7: 653–659.

Galano, A., D. X. Tan, and R. J. Reiter. Melatonin as a natural ally against oxidative stress: A physicochemical examination. *J Pineal Res*. 2011. 51: 1–16.

Galano, A., D. X. Tan, and R. J. Reiter. On the free radical scavenging activities of melatonin's metabolites, AFMK and AMK. *J Pineal Res*. 2013. 54: 245–257.

Gitto, E., D. X. Tan, R. J. Reiter, M. Karbownik, L. C. Manchester, S. Cuzzocrea, F. Fulia, and I. Barberi. Individual and synergistic antioxidative actions of melatonin: Studies with vitamin E, vitamin C, glutathione and desferrioxamine (desferoxamine) in rat river homogenates. *J Pharm Pharmacol*. 2001. 53: 1393–1401.

Hardeland, R., A. Coto-Montes, and B. Poeggeler. Circadian rhythms, oxidative stress, and antioxidative defense mechanisms. *Chronobiol Int.* 2003. 20: 921–962.

Hardeland, R., D. X. Tan, and R. J. Reiter. Kynuramines, metabolites of melatonin and other indoles: The resurrection of an almost forgotten class of biogenic amines. *J Pineal Res.* 2009. 47: 109–126.

Hernandez-Ruiz, J., A. Cano, and M. B. Arnao. Melatonin: A growth-stimulating compound present in lupin tissues. *Planta.* 2004. 220: 140–144.

Hernandez-Ruiz, J., A. Cano, and M. B. Arnao. Melatonin acts as a growth-stimulating compound in some monocot species. *J Pineal Res.* 2005. 39: 137–142.

Hibaoui, Y., E. Roulet, and U. T. Ruegg. Melatonin prevents oxidative stress–mediated mitochondrial permeability transition and death in skeletal muscle cells. *J Pineal Res.* 2009. 47: 238–252.

Jagesh, K., A. D. Tiwari, R. K. Munshi, N. Raghu, A. J. S. Pandey, and A. K. S. Bhat. Effect of salt stress on cucumber: Na^+/K^+ ratio, osmolyte concentration, phenols and chlorophyll content. *Acta Physiol Plant.* 2010. 32: 103–114.

Janas, K. M., and M. M. Posmyk. Melatonin, an underestimated natural substance with great potential for agricultural application. *Acta Physiol Plant.* 2013. 35: 3285–3292.

Jones, M. P., J. Cao, R. O'Brien, S. J. Murch, and P. K. Saxena. The mode of action of thidiazuron: Auxins, indoleamines, and ion channels in the regeneration of *Echinacea purpurea* L. *Plant Cell Rep.* 2007. 26: 1481–1490.

Kang, K., S. Park, Y. S. Kim, S. Lee, and K. Back. Biosynthesis and biotechnological production of serotonin derivatives. *Appl Microbiol Biotechn.* 2009b. 83: 27–34.

Kang, K., Y. S. Kim, S. Park, and K. Back. Senescence-induced serotonin biosynthesis and its role in delaying senescence in rice leaves. *Plant Physiol.* 2009a. 150: 1380–1393.

Kermode, A. R. Approaches to elucidate the basis of desiccation-tolerance in seeds. *Seed Sci Res.* 1997. 7: 75–93.

Kładna, A., H. Y. Aboul-Enien, and I. Kruk. Enhancing effect of melatonin on chemiluminescence accompanying decomposition of hydrogen peroxide in the presence of copper. *Free Radic Biol Med.* 2003. 12: 1544–1554.

Kolar, J., C. H. Johnson, and I. Macháčková. Exogenously applied melatonin (*N*-acetyl-5-methoxytryptamine) affects flowering of the short-day plant *Chenopodium rubrum*. *Physiol Plant.* 2003. 118: 605–612.

Kolar, J., and I. Machackova. Melatonin in higher plants: Occurrence and possible functions. *J Pineal Res.* 2005. 39: 333–341.

Kolar, J., I. Machackova, J. Eder, E. Prinsen, W. Van Dongen, H Van Onckelen, and H. Illnerova. Melatonin: Occurrence and daily rhythm in *Chenopodium rubrum*. *Phytochem* 1997. 44: 1407–1413.

Lei, X. Y., R. Y. Zhu, G. Y. Zhang, and Y. R. Dai. Attenuation of cold-induced apoptosis by exogenous melatonin in carrot suspension cells: The possible involvement of polyamines. *J Pineal Res.* 2004. 36: 126–131.

Leon, J., D. Acuna-Castroviejo, G. Escames, D. X. Tan, and R. J. Reiter. Melatonin mitigates mitochondrial malfunction. *J Pineal Res.* 2005. 38: 1–9.

Li, C., P. Wang, Z. Wei, D. Liang, C. Liu, L. Yin, D. Jia, M. Fu, and F. Ma. The mitigation effects of exogenous melatonin on salinity-induced stress in *Malus hupehensis*. *J Pineal Res.* 2012. 53: 298–306.

Lopez-Burillo, S., D. X. Tan, J. C. Mayo, R. M. Sainz, L. C. Manchester, and R. J. Reiter. Melatonin, xanthurenic acid, resveratrol, EGCG, vitamin C and alphalipoic acid differentially reduce oxidative DNA damage induced by Fenton reagents: A study of their individual and synergistic actions. *J Pineal Res.* 2003. 34: 269–277.

Manchester, L. C., D. X. Tan, R. J. Reiter, W. Park, K. Monis, and W. Qi. High levels of melatonin in the seeds of edible plants: Possible function in germ tissue protection. *Life Sci.* 2000. 67: 3023–3029.

Martin, M., M. Macias, G. Escames, R. J. Reiter, M. T. Agapito, G. G. Oriz, and D. Acuna-Castroviejo. Melatonin-induced increased activity of the respiratory chain complexes I and IV can prevent mitochondrial damage induced by Ruthenium Red *in vivo*. *J Pineal Res* 2000. 28: 242–248.

Martin, M., M. Macias, J. Leon, G. Escames, H. Khaldy, and D. Acuna-Castroviejo. Melatonin increases the activity of the oxidative phosphorylation enzymes and the production of ATP in rat brain and liver mitochondria. *Int J Biochem Cell Biol.* 2002. 34: 348–357.

Mukherjee, S., A. David, S. Yadav, F. Baluska, and S. Bhatla. Salt stress–induced seedling growth inhibition coincides with differential distribution of serotonin and melatonin in sunflower seedling roots and cotyledons. *Physiologia Plantarum.* 2014. 152: 714–728.

Munne-Bosch, S., and L. Alegre. Plant aging increases oxidative stress in chloroplasts. *Planta*. 2002. 214: 608–615.

Murch, S., and P. K. Saxena. Melatonin: A potential regulator of plant growth and development? *In Vitro Cell Dev Biol Plant*. 2002. 38: 531–536.

Murch, S. J., S. Krishna Raj, and P. K. Saxena. Tryptophan is a precursor for melatonin and serotonin biosynthesis in *in vitro*-regenerated St. John's wort (*Hypericum perforatum* L. cv. Anthos) plants. *Plant Cell Rep*. 2000. 19: 698–704.

Murch, S. J., S. S. Campbell, and P. K. Saxena. The role of serotonin and melatonin in plant morphogenesis: Regulation of auxin-induced root organogenesis in *in vitro*-cultured explants of St. John's wort (*Hypericum perforatum* L.). *In Vitro Cell Dev Biol Plant*. 2001. 37: 786–793.

Okazaki, M. Dissection of melatonin accumulation and biosynthesis in tomato. PhD Thesis. Tsukuba, Japan: University of Tsukuba. 2009.

Oracz, K., H. El-Maarouf Bouteau, J. M. Farrant, K. Cooper, M. Belghazi, C. Job, D. Job, F. Corbineau, and C. H. Bailly. ROS production and protein oxidation as a novel mechanism for seed dormancy alleviation. *Plant J*. 2007. 50: 452–465.

Paredes, S. D., A. Korkmaz, L. C. Manchester, D. X. Tan, and R. J. Reiter. Phytomelatonin: A review. *J Exp Bot*. 2009. 60: 57–69.

Park, S., K. Kang, K. Lee, D. Choi, Y. S. Kim, and K. Back. Induction of serotonin biosynthesis is uncoupled from the coordinated induction of tryptophan biosynthesis in pepper fruits (*Capsicum annuum*) upon pathogen infection. *Planta*. 2009. 230: 1197–1206.

Park, S., T. N. N. Le, Y. Byeon, Y. S. Kim, and K. Back. Transient induction of melatonin biosynthesis in rice (*Oryza sativa* L.) during the reproductive stage. *J Pineal Res*. 2013. 55: 40–45.

Park, W. J. Melatonin as an endogenous plant regulatory signal: Debates and perspectives. *J Plant Biol*. 2011. 54: 143–149.

Pelagio-Flores, R., R. Ortiz-Castro, A. Mendez-Bravo, L. Macias-Rodriquez, and J. Lopez-Bucio. Serotonin, a tryptophan-derived signal conserved in plant and animals, regulates root system architecture probably acting as a natural auxin inhibitor in *Arabidopsis thaliana*. *Plant Cell Physiol*. 2011. 52: 490–508.

Petrosillo, G., N. Moro, F. M. Ruggiero, and G. Paradies. Melatonin inhibits cardiolipin peroxidation in mitochondria and prevents the mitochondrial permeability transition and cytochrome *c* release. *Free Radic Biol Med*. 2009. 47: 969–974.

Poeggeler, B., S. Thuermann, A. Dose, M. Schoenke, S. Burkhardt, and R. Hardeland. Melatonin's unique radical scavenging properties: Roles of its functional substituents as revealed by a comparison with its structural analogs. *J Pineal Res*. 2002. 33: 20–30.

Posmyk, M. M., M. Bałabusta, M. Wieczorek, E. Sliwinska, and K. M. Janas. Melatonin applied to cucumber (*Cucumis sativus* L.) seeds improves germination during chilling stress. *J Pineal Res*. 2009. 46: 214–223.

Posmyk, M. M., and K. M. Janas. Melatonin in plants. *Acta Physiol Plant*. 2009. 31:1–11.

Posmyk, M. M., H. Kuran, K. Marciniak, and K. M. Janas. Presowing seed treatment with melatonin protects red cabbage seedlings against toxic copper ion concentrations. *J Pineal Res*. 2008. 45: 24–31.

Ramakrishna, A., C. Dayananda, P. Giridhar, T. Rajasekaran, and G. A. Ravishankar. Photoperiod influences endogenous indoleamines in cultured green alga *Dunaliella bardawil*. *Indian J Exp Biol*. 2011. 49: 234–240.

Ramakrishna, A., P. Giridhar, and G. A. Ravishankar. Indoleamines and calcium channels influence morphogenesis in *in vitro* cultures of *Mimosa pudica* L. *Plant Sig Behav*. 2009. 12: 1136–1141.

Ramakrishna, A., P. Giridhar, M. Jobin, C. S. Paulose, and G. A. Ravishankar. Indoleamines and calcium enhance somatic embryogenesis in cultured tissues of *Coffea canephora* P ex Fr. *Plant Cell Tiss Org*. 2012. 108: 267–278.

Reiter, R. J., D. X. Tan, M. P. Terron, L. J. Flores, and Z. Czarnocki. Melatonin and its metabolites: New findings regarding their production and their radical scavenging actions. *Acta Biochimica Polonica*. 2007b. 54: 1–9.

Ressmeyer, A. R., J. C. Mayo, V. Zelosko, R. M. Sainz, D. X. Tan, B. Poeggeler, I. Antonin, B. K. Zsizsik, R. J. Reiter, and R. Hardeland. Antioxidant properties of the melatonin metabolite *N*1-acetyl-5-methoxykynuramine (AMK): Scavenging of free radicals and prevention of protein destruction. *Redox Rep*. 2003. 8: 205–213.

Rodriguez, C., J. C. Mayo, R. M. Sainz, I. Antolin, F. Herrera, V. Martin, and R. J. Reiter. Regulation of antioxidant enzymes: A significant role for melatonin. *J Pineal Res.* 2004. 36: 1–9.

Roodbari, N., S. Roodbari, A. Ganjali, F. S. Nejad, and M. Ansarifar. The effect of salinity stress on growth parameters and essential oil percentage of peppermint (*Mentha piperita* L.). *Int J Adv Biol Biom Res.* 2013. 1: 1009–1015.

Sarropoulou, V. N., I. N. Therios, and K. N. Dimassi-Theriou. Melatonin promotes adventitious root regeneration in *in vitro* shoot tip explants of the commercial sweet cherry rootstocks CAB-6P (*Prunus cerasus* L.), Gisela 6 (*P. cerasus* x *P. canescens*), and MxM 60 (*P. avium* x *P. mahaleb*). *J Pineal Res.* 2012. 52: 38–46.

Sliwinski, T., W. Rozej, A. Morawiec-Bajda, Z. Morawiec, R. J. Reiter, and J. Blasiak. Protective action of melatonin against oxidative DNA damage: Chemical inactivation *versus* base-excision repair. *Mutat Res.* 2007. 634: 220–227.

Szafrańska, K., R. Szewczyk, and K. M. Janas. Involvement of melatonin applied to *Vigna radiata* L. seeds in plant response to chilling stress. *Cent Eur J Biol.* 2014. 9: 1117–1126.

Tal, O., A. Haim, O. Harel, and Y. Gerchman. Melatonin as an antioxidant and its semi-lunar rhythm in green macroalga *Ulva* sp. *J Exp Bot.* 2011. 62: 1903–1910.

Tan, D. X., L. D. Chen, B. Poeggeler, L. C. Manchester, and R. J. Reiter. Melatonin: A potent, endogenous hydroxyl radical scavenger. *Endocrine.* 1993. 1: 57–60.

Tan, D. X., L. C. Manchester, P. Di Mascio, G. R. Martinez, F. M. Prado, and R. J. Reiter. Novel rhythms of N1-acetyl-N2-formyl-5-methoxykynuramine and its precursor melatonin in water hyacinth: Importance for phytoremediation. *FASEB J.* 2007a. 21: 1724–1729.

Tan, D. X., R. Hardeland, L. C. Manchester, A. Korkmaz, S. Ma, S. Rosales-Corral, and R. J. Reiter. Functional roles of melatonin in plants, and perspectives in nutritional and agricultural science. *J Exp Bot.* 2012. 63: 577–597.

Tan, D. X., R. Hardeland, L. C. Manchester, S. D. Paredes, A. Korkmaz, R. M. Sainz, J. C. Mayo, L. Fuentes-Broto, and R. J. Reiter. The changing biological roles of melatonin during evolution: From an antioxidant to signals of darkness, sexual selection and fitness. *Biol Rev.* 2010. 85: 607–623.

Tan, D. X., L. C. Manchester, X. Liu, D. Rosales-Corral, D. Acuna-Castroviejo, and R. J. Reiter. Mitochondria and chloroplasts as the original sites of melatonin synthesis: A hypothesis related to melatonin's primary function and evolution in eukaryotes. *J Pineal Res.* 2013. 54: 127–138.

Tan, D. X., L. C. Manchester, R. J. Reiter, W. Qi, M. Karbownik, and J. R. Calvo. Significance of melatonin in antioxidative defence system: Reactions and products. *Biol Signals Recept.* 2000. 9: 137–159.

Tan, D. X., L. C. Manchester, R. M. Sainz, J. C. Mayo, J. Leon, R. Hardeland, B. Poeggeler, and R. J. Reiter. Interactions between melatonin and nicotinamide nucleotide: NADH preservation in cells and in cell-free systems by melatonin. *J Pineal Res.* 2005. 39: 185–194.

Tan, D. X., L. C. Manchester, M. P. Terron, L. J. Flores, and R. J. Reiter. One molecule, many derivatives: A never-ending interaction of melatonin with reactive oxygen and nitrogen species? *J Pineal Res.* 2007b. 42: 28–42.

Tan, D. X., R. J. Reiter, L. C. Manchester, M. T. Yan, M. El-Sawi, R. M. Sainz, J. C. Mayo, R. Kohen, M. Allegra, and R. Hardeland. Chemical and physical properties and potential mechanisms: Melatonin as a broad-spectrum antioxidant and free radical scavenger. *Curr Top Med Chem.* 2002. 2: 181–197.

Teixeira, A., M. P. Morfim, C. A. S. De Cordova, C. C. T. Charaþo, V. R. de Lima, and T. B. Creczynski-Pasa. Melatonin protects against prooxidant enzymes and reduces lipid peroxidation in distinct membranes induced by the hydroxyl and ascorbyl radicals and by peroxynitrite. *J Pineal Res.* 2003. 35: 262–268.

Tunez, I., M. C. Munoz, F. J. Medina, M. Salcedo, M. Feijoo, and P. Montilla. Comparison of melatonin, vitamin E and L-carnitine in the treatment of neuro- and hepatotoxicity induced by thioacetamide. *Cell Biochem Funct.* 2007. 25: 119–127.

Uchendu, E. E., M. R. Shukla, B. M. Reed, and P. K. Saxena. Melatonin enhances the recovery of cryopreserved shoot tips of American elm (*Ulmus americana* L.). *J Pineal Res.* 2013. 55: 435–442.

Van Tassel, D. L., and S. D. O'Neill. Putative regulatory molecules in plants: Evaluating melatonin. *J Pineal Res.* 2001. 31: 1–7.

Van Tassel, D. L., N. J. Roberts, A. Lewy, and S. D. O'Neill. Melatonin in plant organs. *J Pineal Res.* 2001. 31: 8–15.

Venegas, C., J. A. Garcia, G. Escames, F. Ortiz, A. Lopez, C. Doerrier, L. Garcia-Corzo, L. C. Lopez, R. J. Reiter, and D. Acuna-Castroviejo. Extrapineal melatonin: Analysis of its subcellular distribution and daily fluctuations. *J Pineal Res.* 2012. 52: 217–227.

Vitalini, S., C. Gardana, P. Simonetti, G. Fico, and M. Iriti. Melatonin, melatonin isomers and stilbenes in Italian traditional grape products and their antiradical capacity. *J Pineal Res.* 2013. 54: 322–333.

Wang, P., L. Yin, D. Liang, C. Li, F. Ma, and Z. Yue. Delayed senescence of apple leaves by exogenous melatonin treatment: Toward regulating the ascorbate–glutathione cycle. *J Pineal Res.* 2012. 53: 11–20.

Wang, P., X. Sun, C. Li, Z. Wei, D. Liang, and F. Ma. Long-term exogenous application of melatonin delays drought-induced leaf senescence in apple. *J Pineal Res.* 2013. 54: 292–302.

Weeda, S., N. Zhang, X. L. Zhao, G. Ndip, Y. D. Guo, G. A. Fu, C. G. Buck, and S. X. Ren. *Arabidopsis* transcriptome analysis reveals key roles of melatonin in plant defense systems. *PLoS One.* 2014. 9: e93462.

Wei, W., L. Qing-Tian, C. Ya-Nan, R. J. Reiter, Y. Xiao-Min, Z. Dan-Hua, and W. Zhang. Melatonin enhances plant growth and abiotic stress tolerance in soybean plants. *J Exp Botany.* 2015. 66: 695–707.

Wolf, K., J. Kolar, E. Witters, W. Van Dongen, H. Van Onckelen, and I. Machackova. Daily profile of melatonin levels in *Chenopodium rubrum* L. depends on photoperiod. *J Plant Physiol.* 2001. 158: 1491–1493.

Xu, X. D. Effects of exogenous melatonin on physiological response of cucumber seedlings under high temperature stress. MA Thesis 2010. Yangling, China: Northwest Agriculture and Forestry University.

Yin, L., P. Wang, M. Li, X. Ke, C. Li, D. Liang, S. Wu, X. Ma, Y. Zou, and F. Ma. Exogenous melatonin improves *Malus* resistance to Marssonina apple blotch. *J Pineal Res.* 2013. 54: 426–434.

Zahir, Z. A., H. N. Asghar, M. J. Akhtar, and M. Arshad. Precursor (L-tryptophan)-inoculum (Azotobacter) interaction for improving yields and nitrogen uptake of maize. *J Plant Nutr.* 2005. 28: 805–817.

Zhang, N., B. Zhao, H. Zhang, S. Weeda, C. Yang, Z. Yang, S. Ren, and Y. Guo. Melatonin promotes water-stress tolerance, lateral root formation, and seed germination in cucumber (*Cucumis sativus* L.). *J Pineal Res.* 2013. 54: 15–23.

Zhang, N., Q. Sun, H. Zhang, Y. Cao, S. Weeda, S. Ren, and Y. D. Guo. Roles of melatonin in abiotic stress resistance in plants. *J Exp Bot.* 2015. 66: 647–656.

Section II

Horticultural and Agricultural Aspects

11

Exogenous Melatonin Influences Productivity of Horticultural and Agricultural Crops

Małgorzata Maria Posmyk and Katarzyna Szafrańska
University of Lodz
Lodz, Poland

CONTENTS

ABSTRACT   In recent years, plant melatonin (MEL) has been the focus of much interest. It appears that the direct and indirect effects of MEL are multidirectional; it is pleiotropic in character. As an amphiphilic, wide-spectral, and extremely effective antioxidant, phytomelatonin must have an impact on plant stress physiology. This chapter describes the role of MEL in the response of plants to stress, as well as its ability to increase their tolerance to changing environmental conditions. MEL has been hailed as a new plant biostimulator and its uses in horticulture and agriculture have proved to be extremely beneficial and desirable.

KEY WORDS: *phytomelatonin, biostimulators, environmental stresses.*

Abbreviations

AANAT: arylalkylamine N-acetyltransferase
AANAT: AANAT-encoding gene
AFMK: N1-acetyl-N2-formyl-5-methoxykynuramine
ANAT: N-acetyltransferase
ASMT: acetylserotonin methyltransferase
ASMT: ASMT-encoding gene
CLH1:   chlorophyllase
H: hydropriming
HIOMT: hydroxyindole-O-methyltransferase
HIOMT: HIOMT-encoding gene
IAA: indole-3-acetic acid
MEL: melatonin
O: osmopriming

PAO:	pheide *a* oxygenase
PSII:	photosystem II
RNS:	reactive nitrogen species
ROS:	reactive oxygen species
SNA/AANAT:	human serotonin N-acetyltransferase gene
T5H:	tryptamine 5-hydroxylase
T5H:	T5H-encoding gene
TDC:	tryptophan decarboxylase
TDC:	TDC-encoding gene

Introduction

Melatonin in Plant Stress Physiology

All living organisms, including plants, may be frequently exposed to suboptimal or harmful environmental conditions. Different intensities and durations of negative stimuli generate stress, which triggers a specific or nonspecific reaction. Plants have evolved several defense mechanisms to combat environmental stresses, viz. (1) the ability to avoid stress (e.g., morphological and biochemical barriers preventing or delaying stressor activity inside a cell; the adaptation of the life cycle to seasons) and (2) stress tolerance (e.g., alternative pathways allowing a cell to function under stressful conditions; the prevention of stress-generated changes; the tolerance of changes or mechanisms of fast damage repair) (Ramakrishna et al., 2011; 2013).

Evolutionary melatonin (MEL) is a highly conserved molecule whose primary role is cell protection. When we think about the exposure of archaic unicellular organisms to extremely high UV radiation, before the oxygen atmosphere and the ozonosphere were built by photoautotrophs, we can clearly see the urgent need for protection. This need resulted *inter alia* in high MEL production and concentration during the night and its decrease or absence during the day, when it was metabolized as a main anti-free-radical factor and antioxidant. It seems possible that recording the presence of MEL and the transduction of this signal enabled organisms to measure night length (Bentkowski and Markowska, 2007), which further led to seasonal recognition, and thus circadian and seasonal changes in behavior. Light, especially strong and very intensive light, can still generate photooxidation processes.

Free radicals can also be formed naturally as products of the biochemical reactions involved in normal metabolic functions, such as cellular respiration and photosynthesis, as well as cell wall biosynthesis and detoxification processes. Reactive oxygen species (ROS) are involved in oxygen metabolism, generating free radicals: hydroxy radical (OH*), phenoxy radicals (RO*), peroxy radicals (ROO*), superoxide radical anion (O_2-*), and nonradical compounds such as singlet oxygen ($^1O_2$) and hydrogen peroxide (H_2O_2). Nitrogen metabolites—reactive nitrogen species (RNS) (e.g., nitric oxide [NO*])—also have oxidative properties. Overproduction of these extremely reactive compounds induces oxidative stress when not followed by their neutralization.

Various stresses inhibit plant growth via different mechanisms, but all cause rises in ROS production and disturb redox homeostasis. It is well known that oxidative stress is a secondary effect of all biotic, and some abiotic, stresses. Thus, antioxidant behavior based on the capability of electron donation—resulting in the structural reconformation of antioxidants to prevent the secondary appearance of free radicals—is very desirable in plant cells.

Since melatonin is soluble in both water and lipids, it may be a hydrophilic and hydrophobic antioxidant. This fact, together with MEL's small size, makes it particularly able to migrate between cell compartments in order to protect them against excessive ROS and RNS. Thus, MEL is a broad-spectrum antioxidant. Moreover, recent evidence indicates that the primary MEL metabolites, especially N1-acetyl-N2-formyl-5-methoxykynuramine (AFMK), also have high antioxidant abilities. AFMK can be formed by numerous free radical reactions and thus, via AFMK, a single MEL molecule is reported to scavenge up to 10 ROS. It has been documented that the free radical–scavenging capacity of MEL extends to its secondary, tertiary, and quaternary metabolites (Tan et al., 2000; 2002; 2007c). This process is referred to as the free radical–scavenging cascade, and makes MEL even at low concentrations highly effective at protecting organisms against oxidative stress. This cascade reaction is characteristic of MEL, making it more efficient than other conventional antioxidants. It seems that, evolutionarily speaking, the strong antioxidant properties of MEL (Terrón et al., 2001; Maldonado et al., 2007) gave it its primary role in the defense against unfavorable conditions and in the stress tolerance of plants.

It has been mentioned in previous chapters that MEL concentrations differ not only from species to species but also among varieties of the same species. It has been detected and quantified in roots, shoots, leaves, flowers, and fruits, but its highest level is found in reproductive organs, particularly in seeds (Posmyk and Janas, 2009; Ramakrishna et al., 2012a,b; 2009; Janas and Posmyk, 2013; Ramakrishna, 2015). It has been suggested that variations in MEL content might result from different environmental impacts during plant growth and development, and also during consecutive stages of the morphological and physiological development of seeds.

An increase in melatonin content has been detected in sunflower seeds during sprouting (Cho et al. 2008). Since germ tissue is highly vulnerable to oxidative damage, melatonin—as a free radical scavenger in a dormant and more or less dry system, where enzymes are poorly effective and cannot be upregulated—might be an important component of an antioxidant defense system. Thus, melatonin in seeds may be essential for protecting plant germs and reproductive tissues from oxidative injuries (Manchester et al., 2000; Van Tassel and O'Neil, 2001). Generally, indoleamines are important for the detoxication of reproductive organ tissues. It has been observed that serotonin present in walnut seeds takes part in the detoxification of excess ammonium, which accumulates during dehydration and seed desiccation (Murch and Saxena, 2002).

In many studies, MEL has been observed to reduce the oxidative damage of important molecules such as nucleic acids, proteins, and lipids (Sliwinski et al., 2007). Its antioxidant activity seems to function in a number of ways: (1) as a direct free radical scavenger, (2) stimulating antioxidant enzymes, (3) stimulating the synthesis of glutathione due to its ability to (4) augment the activities of other antioxidants, (5) protecting antioxidant enzymes from oxidative damage, and (6) increasing the efficiency of the mitochondrial electron transport chain, thereby lowering electron leakage and reducing free radical generation (Kładna et al., 2003; Rodriguez et al., 2004; Leon et al., 2005; Tan et al., 2007c).

Generally, melatonin plays an important role in plant stress defense. Various plant species rich in MEL have shown a higher capacity for stress tolerance (Park et al., 2013b; Bajwa et al., 2014; Zhang et al., 2015). Melatonin is also involved in stress-affected developmental transitions, including flowering, fruiting, ripening, and senescence (Kolar et al., 2003; Arnao and Hernández-Ruiz, 2009a; Zhao et al., 2013; Byeon and Back, 2014). There are some reports that MEL is involved in seed dormancy (Balzer et al., 1993), but the mechanisms of this regulation are unknown. However, ROS has been defined as one of the factors regulating dormancy—for example, by protein oxidation and changes in protein carbonylation patterns (Oracz et al., 2007). Therefore, it is possible that MEL, as the antioxidant regulating ROS levels, can indirectly influence seed dormancy, which can be treated as an evolutionary adaptation to seasonal stresses (winter) occurring in a moderate climate zone.

It has been observed that plants grown in sunlight have 3 times more MEL in their roots and 2.5 times more in their leaves than those grown under moderate artificial light (Tan et al., 2007a). Similarly, in pepper fruit, solar irradiation causes an increase in MEL levels in comparison to shaded fruit (Riga et al., 2014). In ripe tomato fruit, the level of MEL is much higher than that in green ones (Okazaki and Ezura, 2009). This may be connected to the protection of the fruit against photooxidation during ripening (Dubbels et al., 1995). Also, according to Van Tassel and O'Neil (2001), MEL may be involved in the maintenance of a proper organ redox state during fruit ripening.

Similarly, MEL has been observed to be elevated in Alpine and Mediterranean plants exposed to strong UV irradiation in their natural habitat, a finding amenable to the interpretation that melatonin's antioxidant properties can antagonize damage caused by light-induced oxidants (Hardeland and Pandi-Perumal, 2005). The same species cultivated under lower UV exposure had lower MEL content (Simopoulos et al., 2005). Also, enhanced MEL levels have been observed in *Glycyrryza urealensis* roots exposed to UV-B irradiation, (Afreen et al., 2006), and plant species more resistant to ozone oxidation demonstrate higher MEL content than more sensitive ones (Dubbels et al., 1995).

Arnao and Hernández-Ruiz observed changes in MEL content in barley (2009b) and lupine roots (2013) under natural or artificially induced adverse conditions. Low temperatures and drought caused pronounced changes in the endogenous level of this indoleamine following $ZnSO_4$ or NaCl chemical stress. Restricted oxygen supply to roots slightly increased root MEL levels (hypoxia markedly decreased pH in cells). Similarly, direct acidification of the medium also resulted in a slight increase in root MEL content, whereas alkalization had the opposite result. Additionally, the responses in both barley and lupine were clearly dose and time dependent.

In green microalgae *Ulva* sp., temperature and heavy metals such as cadmium, lead, and zinc induce a rise in MEL levels (Tal et al., 2011). However, exposure to cadmium induces a significant rise in MEL content in this algae, while lead and zinc exposure also induce an increase in MEL levels but to a lesser extent. Rice seedlings at high temperatures and under dark conditions show enhanced melatonin synthesis due to increased serotonin N-acetyltransferase (ANAT) and N-acetylserotonin methyltransferase (ASMT) activities (Byeon and Back, 2013). Rice leaves subjected to accelerated aging also exhibit enhanced endogenous melatonin levels (Byeon et al., 2012). Intensive research on MEL biosynthesis pathways in plants, as well as on the enzymes and genes involved in them, allowed for the first genetic modifications aimed at intensified synthesis of this indoleamine in plants. The first melatonin biosynthesis gene, *TDC*, encoding tryptophan decarboxylase (TDC), was overexpressed in rice (Byeon et al., 2014). The obtained transgenic seeds exhibited melatonin concentrations about thirty times higher than those in wild-type seeds. Transgenic rice plants that overexpressed *TDC* exhibited delayed senescence of leaves (Kang et al., 2009), while the suppression of *TDC* by RNAi resulted in lower serotonin content (a direct MEL precursor) and promoted the senescence process.

Regardless of the fact that an AANAT gene homolog (encoding arylalkylamine N-acetyltransferase; AANAT) is absent from the higher plant genome, Okazaki et al. (2009) isolated *AANAT* from the unicellular green alga *Chlamydomonas reinhardtii*, then it was introduced into the Micro-Tom tomato genome. Its ectopic overexpression in tomato successfully resulted in higher MEL content. Wang et al. (2014) observed a loss of apical dominance and enhanced drought tolerance in transgenic Micro-Tom tomato plants overexpressing the homologs of ovine *AANAT* and *HIOMT* genes (encoding arylalkylamine N-acetyltransferase [AANAT] and hydroxyindole-O-methyltransferase [HIOMT], respectively).

In transgenic rice expressing the human serotonin N-acetyltransferase gene (*SNA/AANAT*), high levels of MEL and elevated chlorophyll synthesis during cold stress were noted, suggesting that MEL plays a role in cold stress tolerance (Kang et al., 2010). Transgenic rice seedlings expressing ovine *AANAT* display enhanced seminal root elongation (Park and Back, 2012), seedling growth (Byeon and Back, 2014), and resistance to herbicide-induced oxidative stress (Park et al., 2013b).

The biosynthesis of MEL was enhanced in transgenic *Nicotiana sylvestris* expressing the *AANAT* gene and the *HIOMT* gene, which protected them against UV-B-induced DNA damage following their exposure to ultraviolet irradiation (Zhang et al., 2012).

Moreover, it has been observed that *ASMT* mRNA in rice can be induced in treatments with abscisic acid (ABA) and methyl jasmonic acid (MeJA)—plant hormones known to be stress inducible/responsive. This partially explains higher levels of melatonin in response to various stresses (Park et al., 2013a).

Fungal infection of pepper fruit led to an increase in *TDC* gene expression (Park et al., 2009), and infection caused by *Magnaporthe grisea*, a causal pathogen of rice blast disease, resulted in high tryptamine 5-hydroxylase (T5H) activity and *T5H* gene expression (Fujiwara et al., 2010), indicating that MEL is responsive to pathogenic attacks. Moreover, *T5H* was constitutively expressed in healthy rice (*Oryza sativa*) plants (Kang et al., 2007). Further experiments showed that exogenously applied MEL improved resistance to Marssonina apple blotch (*Diplocarpon mali*), a serious disease leading to defoliation during the growth season (Yin et al., 2013). This also implies that MEL plays an important role in the innate immunity of plants.

Elevated levels of MEL probably help plants to protect themselves against environmental stress caused by water and soil pollutants. Different studies have shown that water hyacinth, rich in MEL and its metabolite AFMK, is able to tolerate contaminants in wastewater generated from industrial and agricultural sources (Trivedy and Pattanshetty, 2002; Singhal and Rai, 2003; Tan et al., 2007a; Munavalli and Saler, 2009), such as nitrogen and phosphorus (Jayaweera and Kasturiarachchi, 2004), ethion pesticide (Xia and Ma, 2006), mercury (Riddle et al., 2002), and arsenic (Misbahuddin and Fariduddin, 2002). Its supplementation with exogenous MEL can make this plant even more useful for phytoremediation. Tan et al. (2007b) investigated the potential relationship between MEL supplementation and the environmental tolerance of plants. Their results showed that MEL, when added to the soil, significantly enhanced the tolerance of pea plants to copper contamination and increased their survival under previously lethal copper concentrations. This is in agreement with the data obtained by Posmyk et al. (2008), where presowing seed treatment with MEL also protected red cabbage seedlings against toxic copper ion concentrations.

Other results showed that the application of MEL to the roots of *Malus hupehensis* Rehd. seedlings prior to salinity treatment partially alleviated the salt-induced inhibition of plant growth, and slowed down the decrease in photosynthesis rates and chlorophyll content (Li et al., 2012).

It should be pointed out that the effects of exogenously applied MEL range from significant amelioration to being ineffective or even toxic. This means that MEL impact may differ with low and high concentrations. Extreme concentrations may severely reduce ROS in cells, thereby affecting ROS-dependent signal transduction and inhibiting some important processes and cell growth (Afreen et al., 2006).

In the case of wild mustard leaves (*Brassica juncea*), a low level of melatonin (0.1 mM) stimulated root growth, while a high level (100 mM) inhibited it (Chen et al., 2009). Additionally, melatonin promoted rooting at low concentrations but inhibited growth at high concentrations in cherry tissue culture (Sarropoulou et al., 2012a).

On a molecular level, it has been shown that gene expression modulated by MEL is dose dependent. Not all genes regulated by low MEL content are regulated by high doses (Weeda et al., 2014). This further confirms that MEL influence vary depending on concentrations.

Plant species vary in their sensitivity to MEL. For example, lupine roots have been shown to be more sensitive to melatonin than barley; to show a similar response, the former needed 24 hours, while the latter needed 72 (Arnao and Hernández-Ruiz, 2007). Melatonin also modulates Arabidopsis root system architecture, stimulating lateral and adventitious root formation (Pelagio-Flores et al., 2012; Koyama et al., 2013). Moreover, in tissue culture systems, MEL was involved in the root regeneration of woody plant explants (Murch et al., 2001; Sarropoulou et al., 2012b; Sarrou et al., 2014).

These data suggest that an increase in endogenous MEL may be the adaptive reaction of plants to adverse conditions; thus, pretreatment with exogenous MEL could effectively increase plant stress tolerance.

Plant pretreatment with MEL significantly alleviates growth inhibition and slows the decline in net photosynthetic rates and chlorophyll content caused by high salinity and drought conditions (Li et al., 2012; Zhang et al., 2013, 2014).

Exposure to low temperatures triggers biochemical and physiological changes in plants and causes a loss of vigor and reduced growth rates, mostly due to chilling injuries to the cell membrane. The protective effect of exogenous MEL during chilling stress has been widely described. Mung bean (*Vigna radiata* L.), a plant originating from warm climate zones, is highly vulnerable to chilling (Hung et al., 2007). Three-day-old seedlings were exposed to 5°C for two days then regenerated at optimal temperature (25°C). The seedlings from seeds hydroprimed with MEL showed a 20% increase in root length and had less disorganized cell ultrastructure (Szafranska et al., 2013). In MEL-treated mung beans, the accumulation of total phenolic compounds and proline was also higher (Szafranska et al., 2012). Cucumber seeds treated with melatonin exhibited an improved germination rate during chilling stress (Posmyk et al., 2009a,b). Similar effects were also found in cucumber under osmotic stress (Zhang et al., 2013). Pretreatment with melatonin attenuated apoptosis induced by cold temperatures in cultured carrot suspension cells (Lei et al., 2004). The survival rate of cryopreserved *Rhodiola crenulata* callus was ~60%, but when the callus was pretreated with 0.1 μM MEL prior to freezing in liquid nitrogen, the survival rate significantly increased. Melatonin also significantly enhanced the recovery of cryopreserved shoot tips of American elm (*Ulmus americana* L.; Uchendu et al., 2013). Shoot explants grown in MEL-supplemented media showed increased regrowth. Melatonin significantly reversed the inhibitory effects of light and high temperature on the germination of photosensitive and thermosensitive *Phacelia tanacetifolia* Benth. seeds (Tiryaki and Keles, 2012). This indoleamine also delayed drought- and dark-induced leaf senescence in apple by maintaining photosystem II (PSII) function under stress and delaying the typical reduction in chlorophyll content (Wang et al., 2012; Wang et al., 2013a,b). In barley leaves treated with MEL solutions, chlorophyll loss, which is one of the main processes to occur during leaf senescence, was significantly slower (Arnao and Hernández-Ruiz, 2009a). Melatonin can also alleviate paraquat-induced photobleaching. Leaves treated with paraquat in the absence of MEL became completely photobleached, while those treated with 1 mM MEL remained green, similar to leaves nontreated with paraquat (Weeda et al., 2014). MEL significantly downregulated the expression of chlorophyllase (CLH1), a light-regulated enzyme involved in chlorophyll degradation (Weeda et al., 2014), and inhibited transcript levels of pheide *a* oxygenase (PAO), another key enzyme involved in chlorophyll degradation (Wang Li et al., 2013b). In MEL-rich transgenic rice lines, senescence-associated proteins were significantly downregulated (Byeon et al., 2013). These findings prove that MEL preserves chlorophyll content in leaves, enhances photosynthetic rates, and delays senescence. Furthermore, cell death–associated genes were mostly downregulated by melatonin treatment (Weeda et al., 2014), which further explains MEL-related antiapoptotic processes.

Our latest results (unpublished data) indicate that exogenous MEL expediently modifies the proteome of a maize (*Zea mays* L.) embryo during seed germination. Generally, hydroconditioning treatment generated 14 new proteins in embryonic axes of the seeds germinated at optimal temperature compared to the controls. When this presowing treatment was supplemented with MEL application, an additional 14 proteins appeared. Moreover, in the seed variants H-MEL50 (optimal MEL concentration of 50 µM) and H-MEL500 (overdose of 500 µM), 3 and 23 characteristic spots were noticed, respectively. The majority of additional proteins were energy metabolism enzymes; proteins involved in proteome plasticity by improving protein synthesis, folding, destination, and storage; and—most importantly—defense, antistress, and detoxifying proteins. This explains why the hydroprimed seeds, especially those hydroprimed with MEL, and the seedlings grown from them were stronger in comparison to the nontreated ones and how they so quickly and efficiently adapted to changing environmental conditions (Posmyk et al., 2009a; Janas et al., 2009). They could be *a priori* prepared to cope with potentially harmful conditions; in embryonic axes during the initial stage of growth, even under optimal conditions, a number of antioxidative, detoxifying, and chaperon proteins were synthesized. Moreover, the supply of energy from seed storage substances intensified.

Although protein biosynthesis was significantly inhibited under prolonged cold stress conditions (14 days/5°C), positive changes in corn embryo proteome were already observed in hydroconditioned seeds but especially in those hydroconditioned with MEL. In the seeds germinated at chilling temperature, hydroconditioning generated five new proteins in corn embryonic axes compared to the controls. When this presowing treatment was supplemented with MEL application, an additional six proteins appeared. Moreover, in the seed variants H-MEL50 (optimal MEL concentration of 50 µM) and H-MEL500 (overdose of 500 µM), two and five characteristic spots were noticed, respectively. Modifications related to the improvement of the respiratory/energetic metabolism of conditioned seeds were crucial and typical of stress amelioration (Grover et al., 2001). Moreover, in the seeds pretreated with MEL, many more stress-related, defense, and detoxifying proteins were synthesized. A plant's ability to survive abiotic stresses such as salinity, radiation, heavy metals, nutrient deprivation, cold, and drought heavily depends on proteomic plasticity, including protein synthesis *de novo* and their posttranslational modifications and folding, as well as transport to their activity sites and their specific degradation and turnover. Protein ubiquitination plays a crucial role, allowing plants to alter their proteome in order to effectively and efficiently perceive and respond to environmental stresses (Lyzenga and Stone, 2012). In seeds hydroconditioned with MEL, some characteristic proteins that could play the role of E3 ligase were noticed, as well as extended protein ubiquitination. Plant response to adverse environmental conditions is a complex and coordinated process involving the activation of signaling networks and changes in the expression of hundreds of genes. By modulating the amount of transcription factors, protein ubiquitination may affect the changes in gene expression required to mitigate the potentially negative effects of environmental stress.

These results partially explain how melatonin acts in plant stress defense and why various plant species rich in melatonin have shown a higher capacity for stress tolerance (Zhang et al., 2015).

To sum up: (1) MEL can be synthesized and taken up by plants; (2) genetic modification can enhance melatonin synthesis in transgenic plants and improve their resistance to adverse conditions; (3) plants accumulate high levels of melatonin when faced with harsh environments; and (4) exogenously applied melatonin helps improve tolerance to stresses.

Further research in this area could provide valuable information to improve organic farming, environmental phytoremediation, and plant-derived dietary supplements. However, the presented knowledge already indicates that MEL could be used as an effective biostimulator.

Melatonin: Horticultural and Agricultural Applications

Biostimulators and Their Application Methods

Plant crop production methods based only on improving agricultural technology (e.g., tillage, recultivation, fertilization, irrigation, etc.) are limited due to the inability to effectively use the biological potential of the cultivated variety. Plant production and protection face the difficult task of preventing damage in field crops caused by harmful organisms or abiotic stresses, thus stimulating plant growth and development and reducing the hazards presented to humans and the environment, as well as securing the production of safe, high-quality agricultural products at the same time. The best solution to the urgent need for alternative organic methods based on new safe, biologically active, environmentally friendly substances seems to be the application of biostimulators.

Plant biostimulators (*phytostimulators*) are various kinds of nontoxic substances of natural origin that improve and stimulate plant life processes differently to fertilizers or phytohormones. Their influence on plants is not the consequence of direct metabolic regulatory properties; their actions can be multidirectional and they influence metabolism more generally. The crucial point is that biostimulators, in contrast to bioregulators, improve plant metabolic processes without changing their natural pathway (Figure 11.1).

The quality of seed material is a primary and basic condition determining a good harvest. Thus, finding effective methods to improve sowing material by applying biostimulators to seeds is a crucial problem.

The known techniques of seed priming—hydro- and osmoconditioning—were tested by Posmyk et al. (Posmyk and Janas 2007; Posmyk et al., 2008, 2009a,b; Janas et al., 2009). Different presowing seed treatments effectively counteract diseases and pests as well as improve seed viability and seedling vigor per se (Taylor et al., 1998). They are based on an understanding of the physiology, biochemistry, and anatomy of plants that explains the mechanisms of bioprocesses and gives them practical application.

Our previous experiments (Posmyk and Janas 2007; Posmyk et al., 2008; Posmyk et al., 2009a,b; Janas et al., 2009) proved that exogenous melatonin applied to seeds by presowing treatment improved their vigor and germination as well as seedling growth. Since melatonin is safe for animals and humans as well as being inexpensive, a conditioning technique using this indoleamine may be a reliable, feasible, and cost-effective tool for positive seed quality modification and may be economically beneficial for organic farming (Posmyk and Janas 2009; Janas and Posmyk 2013).

Seed Priming with Melatonin: Advantages in Abiotic Stress Conditions

Recent trends in plant neurobiology based on auxin-signaling investigations (Brenner et al., 2006) focused our attention on melatonin, which demonstrates parallelism with plant auxin (indole-3-acetic acid; IAA). Resumption of the research on classic plant models used in plant physiology is still necessary to clarify the role and mechanism of action of melatonin (1) as an independent plant growth regulator, (2) as a factor mediating the activity of other substances influencing plant growth, and (3) as a substance involved in growth regulation but whose activity generally is ascribed to other compounds. Recent knowledge qualifies MEL as a biostimulator. Its advantages are as follows: (1) it is of natural origin but it can be easily synthesized in laboratories; (2) it is inexpensive; (3) it dissolves in different solvents (water and alcohols, but also lipids), which facilitates the use of various application methods; (4) it could be actively taken up by plants from the environment; (5) it is a small molecule and easily penetrates cell compartments; and (6) it has strong antioxidant properties.

Generally, physiological concentrations of MEL in seeds are very high; for example, in white and black mustard seeds they were shown to be 129 and 189 ng g^{-1}, respectively (Hattori et al., 1995), much

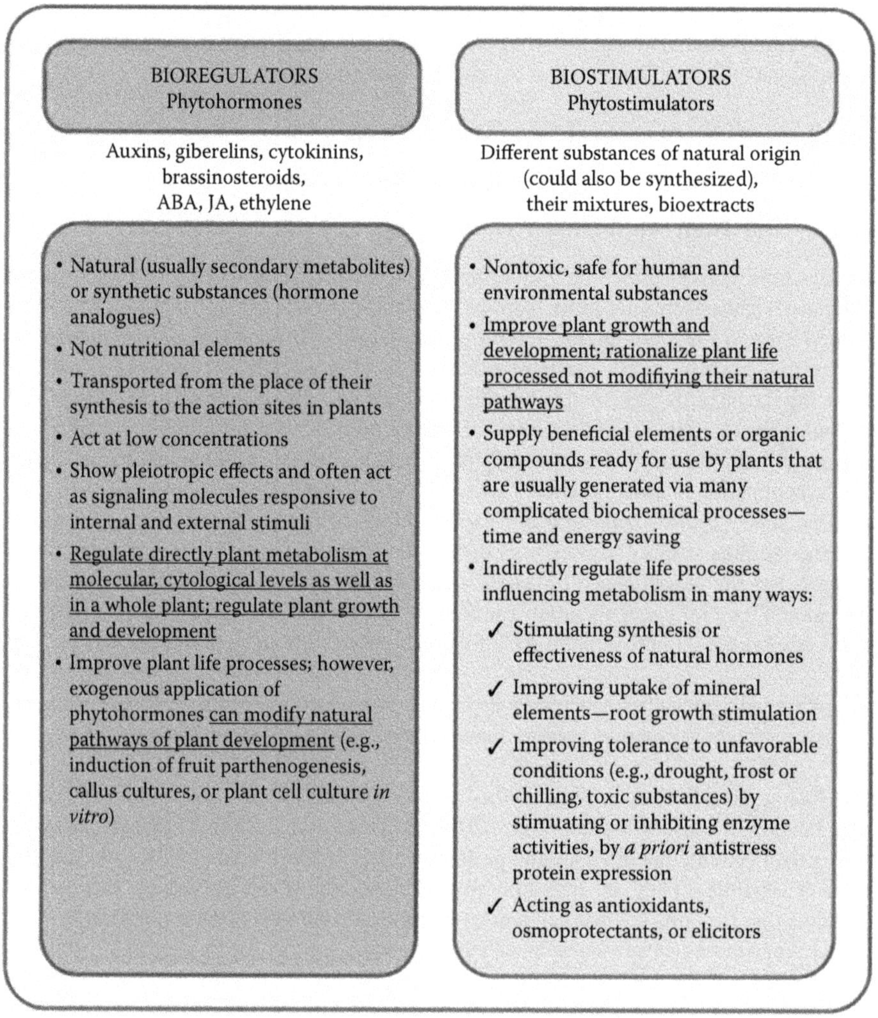

BIOREGULATORS Phytohormones	BIOSTIMULATORS Phytostimulators
Auxins, giberelins, cytokinins, brassinosteroids, ABA, JA, ethylene	Different substances of natural origin (could also be synthesized), their mixtures, bioextracts

• Natural (usually secondary metabolites) or synthetic substances (hormone analogues)
• Not nutritional elements
• Transported from the place of their synthesis to the action sites in plants
• Act at low concentrations
• Show pleiotropic effects and often act as signaling molecules responsive to internal and external stimuli
• <u>Regulate directly plant metabolism at molecular, cytological levels as well as in a whole plant; regulate plant growth and development</u>
• Improve plant life processes; however, exogenous application of phytohormones <u>can modify natural pathways of plant development</u> (e.g., induction of fruit parthenogenesis, callus cultures, or plant cell culture *in vitro*)
	• Nontoxic, safe for human and environmental substances
	• <u>Improve plant growth and development; rationalize plant life processed not modifiying their natural pathways</u>
	• Supply beneficial elements or organic compounds ready for use by plants that are usually generated via many complicated biochemical processes—time and energy saving
	• Indirectly regulate life processes influencing metabolism in many ways:
	  ✓ Stimulating synthesis or effectiveness of natural hormones
	  ✓ Improving uptake of mineral elements—root growth stimulation
	  ✓ Improving tolerance to unfavorable conditions (e.g., drought, frost or chilling, toxic substances) by stimulating or inhibiting enzyme activities, by *a priori* antistress protein expression
	  ✓ Acting as antioxidants, osmoprotectants, or elicitors

FIGURE 11.1 Differences between bioregulators and biostimulators.

higher than the known physiological blood concentrations of many vertebrates. Since a seed, particularly its germ tissue, is highly vulnerable to oxidative stress and damage, we suppose that the high concentrations of MEL detected in seeds may provide an antioxidative defense.

Indeed, hydropriming red cabbage seeds with MEL proved to be a good tool to improve their vigor (Posmyk et al., 2008). The positive effects of this treatment were clearly visible, especially under copper stress conditions. Similarly, experiments with cucumber seeds osmoprimed with MEL (Posmyk et al., 2009a,b) and with corn seeds hydroprimed with MEL (Janas et al., 2009) proved the thesis.

The beneficial effects of priming were not visible in cucumber and corn seed germination tests performed under optimal temperature conditions. However, subsequent experiments showed that seedlings grown from seeds conditioned with melatonin tolerated stresses of suboptimal temperature (10°C) and heavy metal contamination (2.5 mM Cu^{2+}) extremely well and regenerated much better after the relief of stress. This was manifested by better growth (greater weight of seedlings) and higher chlorophyll content and phenolic synthesis in the seedlings grown from the seeds hydroprimed with melatonin (Janas et al., 2009).

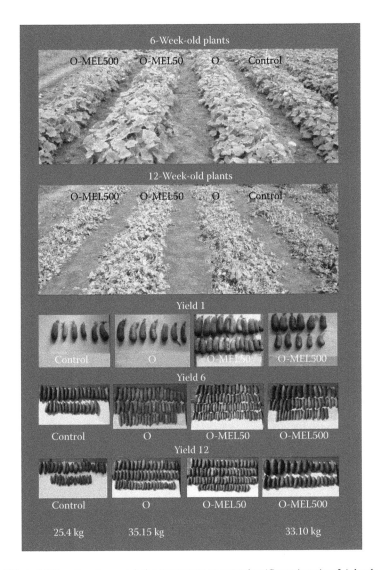

FIGURE 11.2 Effect of different seed osmopriming treatments on cucumber (*Cucumis sativus* L.) development and yield. Control: non-primed seeds; O: seeds primed with water; O-MEL50 and O-MEL500: seeds primed with melatonin water solutions 50 μM and 500 μM, respectively. Yields from 25 m^2 experimental fields.

The cucumber seeds osmoprimed with MEL started to germinate at 10°C, which was impossible for the untreated controls (Posmyk et al., 2009a,b). We also obtained interesting results when the corn seeds were germinated at 10°C (Janas et al., 2009).

Field tests unequivocally proved that MEL applied in a specific way to seeds is a perfect biostimulator, enhancing plant growth and development as well as increasing yield. Field experiments were performed with the following seeds: *sativus* L. (Figure 11.2; cucumber seed variants: nontreated [Control], osmoprimed with water [O], osmoprimed with MEL water solution of 50 and 500 μM [O-MEL50 and O-MEL500]), *Z. mays* L. (Figure 11.3; corn seed variants: nontreated [Control], hydroprimed with water [H], hydroprimed with MEL water solution of 50 and 500 μM [H-MEL50 and H-MEL500]), and *V. radiata* L. (Figure 11.4; mung bean seed variants: nontreated [Control], hydroprimed with water [H], hydroprimed with MEL water solution of 20, 50, and 200 μM [H-MEL20, H-MEL50, and H-MEL200]). All seeds were primed and redryed under laboratory conditions up to one month before sewing in fields.

FIGURE 11.3 Effect of different seed hydropriming treatments on corn (*Zea mays* L.) development and yield. Control: non-primed seeds; H: seeds primed with water; H-MEL50 and H-MEL500: seeds primed with melatonin water solutions 50 μM and 500 μM, respectively. Yields from 4 m^2 experimental fields.

During vegetation, plants were not supplemented by fertilizers or pesticides; experiments were performed as organic farming.

The experiments, which were conducted in the years 2009 and 2011 under open field conditions, showed that plants from the MEL-treated seeds of corn, mung bean, and cucumber were better developed and had a higher crop yield than the controls (Figures 11.2 through 11.4). At harvesting, cucumber plants osmoprimed with MEL 50 μM had more fruits than those osmoprimed with MEL 500 μM, osmoprimed without MEL or nontreated plants. We observed that the fruits of some MEL-treated plants were larger than those osmoprimed without MEL and those that were non-primed (Figure 11.2). MEL-treated (50–500 μM) corn plants had more and larger cobs than those hydroprimed without MEL and those that were non-primed (Figure 11.3). Similar results we observed with mung beans whose seeds were hydroprimed with MEL at concentrations of 20, 50, and 500 μM. The number of pods was greater in plants grown from the seed hydroprimed with 50 μM MEL than those hydroprimed without MEL those that were non-primed (Figure 11.4). MEL at 500 μM was not favorable.

FIGURE 11.4 Effect of different seed hydropriming treatments on mung bean (*Vigna radiata* L.) development and yield. Control: non-primed seeds; H: seeds primed with water; H-MEL20, H-MEL50, and H-MEL200: seeds primed with melatonin water solutions 20 μM, 50 μM, and 200 μM, respectively. Yields from 5 m² experimental fields.

Conclusion

Presently available information suggests that melatonin exhibits positive influences on the tolerance of plants to harmful environmental conditions. Plant varieties rich in MEL exhibit higher resistance to stress. It has been noted that especially high melatonin levels in tissues and organs are associated with reproduction (flowers, seeds, young seedlings). Melatonin has a positive effect on plant development and yield because it protects the plants against various types of stresses and also acts as a biostimulator.

REFERENCES

Afreen F, Zobayed SM, Kozai T. Melatonin in *Glycyrrhiza uralensis*: Response of plant roots to spectral quality of light and UV-B radiation. *J Pineal Res* 2006; 41:108–115.

Arnao MB, Hernández-Ruiz J. Melatonin promotes adventitious and lateral root regeneration in etiolated hypocotyls of *Lupinus albus* L. *J Pineal Res* 2007; 42:147–152.

Arnao MB, Hernández-Ruiz J. Protective effect of melatonin against chlorophyll degradation during the senescence of barley leaves. *J Pineal Res* 2009a; 46:58–63.

Arnao MB, Hernández-Ruiz J. Chemical stress by different agents affects the melatonin content of barley roots. *J Pineal Res* 2009b; 46:295–299.

Arnao MB, Hernández-Ruiz J. Growth conditions determine different melatonin levels in *Lupinus albus* L. *J Pineal Res* 2013; 55:149–155.

Bajwa VS, Shukla MR, Sherif SM, Murch SJ, Saxena PK. Role of melatonin in alleviating cold stress in *Arabidopsis thaliana*. *J Pineal Res* 2014; 56:238–245.

Balzer I, Poeggeler B, Hardeland R. Circadian rhythms of indoleamines in a dinoflagellate *Gonyaulax polyedra*: Persistence of melatonin rhythm in constant darkness and relationship to 5-methoxytryptamine. In: Touitou Y, Arendt J, Pevet P, eds. *Melatonin and the Pineal Gland: From Basic Science to Clinical Application*. Excerpta Medica, Amsterdam, 1993; 183–186.

Bentkowski P, Markowska M. Ewolucja fizjologicznych funkcji melatoniny u bezkręgowców. *Kosmos* 2007; 56(3–4):383–391.

Brenner ED, Stahlberg R, Mancuso S, Vivanco J, Baluška F, Van Volkenburgh E. Plant neurobiology: An integrated view of plant signaling. *Trends Plant Sci* 2006; 11:413–419.

Byeon Y, Back KW. Melatonin synthesis in rice seedlings *in vivo* is enhanced at high temperatures and under dark conditions due to increased serotonin N-acetyltransferase and N-acetylserotonin methyltransferase activities. *J Pineal Res* 2013; 56:189–195.

Byeon Y, Back KW. An increase in melatonin in transgenic rice causes pleiotropic phenotypes, including enhanced seedling growth, delayed flowering, and low grain yield. *J Pineal Res* 2014; 56:380–414.

Byeon Y, Park S, Kim YS, Back KW. Microarray analysis of genes differentially expressed in melatonin-rich transgenic rice expressing a sheep serotonin N-acetyltransferase. *J Pineal Res* 2013; 55:357–363.

Byeon Y, Park S, Kim YS, Park DH, Lee S, Back KW. Light-regulated melatonin biosynthesis in rice during the senescence process in detached leaves. *J Pineal Res* 2012; 53:107–111.

Byeon Y, Park S, Lee HY, Kim YS, Back KW. Elevated production of melatonin in transgenic rice seeds expressing rice tryptophan decarboxylase. *J Pineal Res* 2014; 56:275–282.

Chen Q, Qi WB, Reiter RJ, Wei W, Wang BM. Exogenously applied melatonin stimulates root growth and raises endogenous indoleacetic acid in roots of etiolated seedlings of *Brassica juncea*. *J Plant Physiol* 2009; 166:324–328.

Cho MH, No HK, Prinyawiwatkul W. Chitosan treatments affect growth and selected quality of sunflower sprouts. *J Food Sci* 2008; 73:70–77.

Dubbels R, Reiter RJ, Klenke E, Goebel A, Schnakenberg E, Ehlers C, Schiwara HW, Chloot W. Melatonin in edible plants identified by radioimmunoassay and by high performance liquid chromatography-mass spectrometry. *J Pineal Res* 1995; 18:28–31.

Fujiwara T, Maisonneuve S, Isshiki M, Mizutani M, Chen L, Wong HL, Kawasaki T, Shimamoto K. Sekiguchi lesion gene encodes a cytochrome P450 monooxygenase that catalyzes conversion of tryptamine to serotonin in rice. *J Biol Chem* 2010; 285:11308–11313.

Grover A, Kapoor A, Lakshmi OS, Agarwal S, Sahi C, Katiyar-Agarwal S, Agarwal M, Dubey H. Understanding molecular alphabets of the plant abiotic stress responses. *Current Sci* 2001; 80:206–216.

Hardeland R, Pandi-Perumal SR. Melatonin, a potent agent in antioxidative defense: Actions as a natural food constituent, gastrointestinal factor, drug and prodrug. *Nutr Metab (Lond)* 2005; 2:22.

Hattori A, Migitaka H, Masayaki I, Itoh M, Yamamoto K, Ohtani-Kaneko R, Hara M, Suzuki T, Reiter RJ. Identification of melatonin in plants and its effects on plasma melatonin levels and binding to melatonin receptors in vertebrates. *Biochem Mol Biol Int* 1995; 35:627–634.

Hung S, Wang C, Ivanov S, Alexieva V, Yu C. Repetition of hydrogen peroxide treatment induces a chilling tolerance comparable to cold acclimation in mung bean. *J Am Soc Hort Sci* 2007; 132:770–776.

Janas KM, Ciupińska E, Posmyk MM. Melatonin applied by hydropriming as a biostimulator improving sweet corn (*Zea mays* L.) seedling growth at abiotic stress conditions. In: Li S, Wang Y, Cao F, Huang P, Zhang Y, eds. *Progress in Environmental Science and Technology* Vol. II A. Science Press USA, Princeton Junction, NJ, 2009; 383–388.

Janas KM, Posmyk MM. Melatonin, an underestimated natural substance with great potential for agricultural application. *Acta Physiol Plant* 2013; 35:3285–3292.

Jayaweera M, Kasturiarachchi J. Removal of nitrogen and phosphorus from industrial wastewaters by phytoremediation using water hyacinth (*Eichhornia crassipes* (Mart.) Solms). *Water Sci Technol* 2004; 50:217–225.

Kang K, Kim YS, Park S, Back KW. Senescence-induced serotonin biosynthesis and its role in delaying senescence in rice leaves. *Plant Physiol* 2009; 150:380–1393.

Kang K, Lee K, Park S, Kim YS, Back KW. Enhanced production of melatonin by ectopic overexpression of human serotonin N-acetyltransferase plays a role in cold resistance in transgenic rice seedlings. *J Pineal Res* 2010; 49:176–182.

Kang S, Kang K, Lee K, Back KW. Characterization of tryptamine 5-hydroxylase and serotonin synthesis in rice plants. *Plant Cell Rep* 2007; 26:2009–2015.

Kładna A, Aboul-Enien Hy, Kruk I. Enhancing effect of melatonin on chemiluminescence accompanying decomposition of hydrogen peroxide in the presence of copper. *Free Rad Biol Med* 2003; 12:1544–1554.

Kolar J, Johnson C, Machackova I. Exogenously applied melatonin (N-acetyl-5-methoxytryptamine) affects flowering of the short day plant *Chenopodium rubrum. Physiol Plant* 2003; 118:605–612.

Koyama FC, Carvalho TLG, Alves E, da Silva HB, de Azevedo MF, Hemerly AS, Garcia CRS. The structurally related auxin and melatonin tryptophan-derivatives and their roles in *Arabidopsis thaliana* and in the human malaria parasite *Plasmodium falciparum. J Eukaryotic Microbiol* 2013; 60:646–651.

Lei X, Zhu R, Zhang G, Dai Y. Attenuation of cold-induced apoptosis by exogenous melatonin in carrot suspension cells: The possible involvement of polyamines. *J Pineal Res* 2004; 36:126–131.

Leon J, Acuna-Castroviejo D, Escames G, Tan DX, Reiter RJ. Melatonin mitigates mitochondrial malfunction. *J Pineal Res* 2005; 38:1–9.

Li C, Wang P, Wei Z, Liang D, Liu C, Yin L, Jia D, Fu M, Ma F. The mitigation effects of exogenous melatonin on salinity-induced stress in *Malus hupehensis. J Pineal Res* 2012; 53:298–306.

Lyzenga WJ, Stone SL. Abiotic stress tolerance mediated by protein ubiquitination. *J Exp Bot* 2012; 63:599–616.

Maldonado MD, Murillo-Cabezas F, Terron MP, Tan DX, Manchester LC, Reiter RJ. The potential of melatonin in reducing morbidity–mortality after craniocerebral trauma. *J Pineal Res* 2007; 42:1–11.

Manchester LC, Tan D-X, Reiter RJ, Park W, Monis K, Qi W. High levels of melatonin in the seeds of edible plants: Possible function in germ tissue protection. *Life Sci* 2000; 67:3023–3029.

Misbahuddin M, Fariduddin A. Water hyacinth removes arsenic from arsenic-contaminated drinking water. *Arch Environ Health* 2002; 57:516–518.

Munavalli GR, Saler PS. Treatment of dairy wastewater by water hyacinth. *Water Sci Technol* 2009; 59:713–722.

Murch SJ, Campbell SSB, Saxena PK. The role of serotonin and melatonin in plant morphogenesis: Regulation of auxin-induced root organogenesis in *in vitro*–cultured explants of St. John's wort (*Hypericum perforatum* L.). *In Vitro Cell Dev-Pl* 2001; 37:786–793.

Murch SJ, Saxena PK. Melatonin: A potential regulator of plant growth and development? *In vitro* Cell Dev Biol Plant 2002; 38:531–536.

Okazaki M, Ezura H. Profiling of melatonin in the model tomato (*Solanum lycopersicum* L.) cultivar Micro-Tom. *J Pineal Res* 2009; 46:338–343.

Okazaki M, Higuchi K, Hanawa Y, Shiraiwa Y, Ezura H. Cloning and characterization of a Chlamydomonas reinhardtii cDNA arylalkylamine N-acetyltransferase and its use in the genetic engineering of melatonin content in the Micro-Tom tomato. *J Pineal Res* 2009; 46:373–382.

Oracz K, El-Maarouf Bouteau H, Farrant JM, Cooper K, Belghazi M, Job C, Job D, Corbineau F, Bailly C. ROS production and protein oxidation as a novel mechanism for seed dormancy alleviation. *Plant J* 2007; 50:452–465.

Park S, Back KW. Melatonin promotes seminal root elongation and root growth in transgenic rice after germination. *J Pineal Res* 2012; 53:385–389.

Park S, Byeon Y, Back KW. Functional analyses of three ASMT gene family members in rice plants. *J Pineal Res* 2013a; 55:409–415.

Park S, Kang K, Lee K, Choi D, Kim Y, Back K. Induction of serotonin biosynthesis is uncoupled from the coordinated induction of tryptophan biosynthesis in pepper fruits (*Capsicum annuum*) upon pathogen infection. *Planta* 2009; 230:1197–1206.

Park S, Lee DE, Jang H, Byeon Y, Kim YS, Back K. Melatonin-rich transgenic rice plants exhibit resistance to herbicide-induced oxidative stress. *J Pineal Res* 2013b; 54:258–263.

Pelagio-Flores R, Munoz-Parra E, Ortiz-Castro R, Lopez-Bucio J. Melatonin regulates Arabidopsis root system architecture likely acting independently of auxin signaling. *J Pineal Res* 2012; 53:279–288.

Posmyk MM, Bałabusta M, Janas KM. Melatonin applied by osmopriming as a biostimulator improving cucumber (*Cucumis sativus* L.) seedling growth at abiotic stress conditions. In: Li S, Wang Y, Cao F, Huang P, Zhang Y, eds. *Progress in Environmental Science and Technology* Vol. II A. Science Press USA, Princeton Junction, NJ, 2009a; 362–369.

Posmyk MM, Bałabusta M, Wieczorek M, Śliwinska E, Janas KM. Melatonin applied to cucumber (*Cucumis sativus* L.) seeds improves germination during chilling stress. *J Pineal Res* 2009b; 46:214–223.

Posmyk MM, Janas KM. Effect of seed hydropriming in presence of exogenous proline on chilling injury limitation in *Vigna radiata* L. seedlings. *Acta Physiol Plant* 2007; 29:509–517.

Posmyk MM, Janas KM. Melatonin in plants. *Acta Physiol Plant* 2009; 31:1–11.

Posmyk MM, Kuran H, Marciniak K, Janas KM. Pre-sowing seed treatment with melatonin protects red cabbage seedlings against toxic copper ion concentrations. *J Pineal Res* 2008; 45:24–31.

Ramakrishna A. Indoleamines in edible plants: Role in human health effects. In Angel Catalá, ed. *Indoleamines: Sources, Role in Biological Processes and Health Effects*, Biochemistry Research Trends, Nova Publishers, Hauppauge, NY, 2015; 279.

Ramakrishna A, Giridhar P, Ravishankar GA. Indoleamines and calcium channels influence morphogenesis in *in vitro* cultures of *Mimosa pudica* L. *Plant Sig Behav* 2009; 12:1136–1141.

Ramakrishna A, Giridhar P, Udaya Sankar K and Ravishankar GA. Melatonin and serotonin profiles in beans of *Coffea* sps. *J Pineal Res* 2012a; 52:470–476.

Ramakrishna A, Giridhar P, Udaya Sankar K and Ravishankar GA. Endogenous profiles of indoleamines: Serotonin and melatonin in different tissues of *Coffea canephora* P ex Fr. as analyzed by HPLC and LC-MS-ESI. *Acta Physiol Plant* 2012b; 34:393–396.

Ramakrishna A, Ravishankar GA. Influence of abiotic stress signals on secondary metabolites in plants. *Plant Sig Behav* 2011; 6:1720–1731.

Ramakrishna A, Ravishankar GA. Role of plant metabolites in abiotic stress tolerance under changing climatic conditions with special reference to secondary compounds. In Tuteja N, Gill SS, eds. *Climate Change and Abiotic Stress Tolerance*, Wiley-VCH, Weinheim, Germany, 2013; 705–726.

Riddle S, Tran H, Dewitt J, Andrews J. Field, laboratory, and X-ray absorption spectroscopic studies of mercury accumulation by water hyacinths. *Environ Sci Technol* 2002; 36:1965–1970.

Riga P, Medina S, Garcia-Flores LA, Gil-Izquierdo A. Melatonin content of pepper and tomato fruits: Effects of cultivar and solar radiation. *Food Chem* 2014; 156:347–352.

Rodriguez C, Mayo JC, Sainz RM, Antolin I, Herrera F, Martin V, Reiter RJ. Regulation of antioxidant enzymes: A significant role for melatonin. *J Pineal Res* 2004; 36:1–9.

Sarropoulou VN, Dimassi-Theriou K, Therios I, Koukourikou-Petridou M. Melatonin enhances root regeneration, photosynthetic pigments, biomass, total carbohydrates and proline content in the cherry rootstock PHL-C (*Prunus avium* × *Prunus cerasus*). *Plant Physil Biochem* 2012a; 61:162–168.

Sarropoulou VN, Therios IN, Dimassi-Theriou K. Melatonin promotes adventitious root regeneration in *in vitro* shoot tip explants of the commercial sweet cherry rootstocks CAB-6P (*Prunus cerasus* L.), Gisela 6 (*P. cerasus P. canescens*), and MxM 60 (*P. avium P. mahaleb*). *J Pineal Res* 2012b; 52:38–46.

Sarrou E, Therios I, Dimassi-Theriou K. Melatonin and other factors that promote rooting and sprouting of shoot cuttings in *Punica granatum* cv. Wonderful. *Turk J Bot* 2014; 38:293–301.

Simopoulos A, Tan DX, Manchester LC, Reiter RJ. Purslane: A plant source of omega-3 fatty acids and melatonin. *J Pineal Res* 2005; 39:331–332.

Singhal V, Rai J. Biogas production from water hyacinth and channel grass used for phytoremediation of industrial effluents. *Bioresour Technol* 2003; 86:221–225.

Sliwinski T, Rozej W, Morawiec-Bajda A, Morawiec Z, Reiter RJ, Blasiak J. Protective action of melatonin against oxidative DNA damage: Chemical inactivation versus base-excision repair. *Mutat Res* 2007; 634:220–227.

Szafranska K, Glinska S, Janas KM. Changes in the nature of phenolic deposits after re-warming as a result of melatonin pre-sowing treatment of *Vigna radiata* seeds. *J Plant Physiol* 2012; 169:34–40.

Szafranska K, Glinska S, Janas KM. Ameliorative effect of melatonin on meristematic cells of chilled and re-warmed *Vigna radiata* roots. *Biol Plant* 2013; 57:91–96.

Tal O, Haim A, Harel O, Gerchman Y. Melatonin as an antioxidant and its semi-lunar rhythm in green mac-roalga *Ulva* sp. *J Exp Bot* 2011; 62:1903–1910.

Tan DX, Manchester LC, Di Mascio P, Martinez GR, Prado FM, Reiter RJ. Novel rhythms of N^1-acetyl-N^2-formyl-5-methoxykynuramine and its precursor melatonin in water hyacinth: Importance for phytore-mediation. *FASEB J* 2007a; 21:1724–1729.

Tan DX, Manchester LC, Helton P, Reiter RJ. Phytoremediative capacity of plants enriched with melatonin. *Plant Signal Behav* 2007b; 2:514–516.

Tan DX, Manchester LC, Reiter RJ, Qi W-B, Karbownik M, Calvo JR. Significance of melatonin in antioxida-tive defense system: Reactions and products. *Biol Signals Recept* 2000; 9:137–159.

Tan DX, Manchester LC, Terron MP, Flores LJ, Reiter RJ. One molecule, many derivatives: A never-ending interaction of melatonin with reactive oxygen and nitrogen species? *J Pineal Res* 2007c; 42:28–42.

Tan DX, Reiter RJ, Manchester LC, Yan M-T, El-Sawi M, Sainz RM, Mayo JC, Kohen R, Allegra M, Hardeland R. Chemical and physical properties and potential mechanisms: Melatonin as a broad spectrum antioxi-dant and free radical scavenger. *Curr Top Med Chem* 2002; 2:181–197.

Taylor AG, Allen PS, Bennett MA, Bradford KJ, Burris JS, Misra MK. Seed enhancement. *Seed Sci Res* 1998; 8:245–256.

Terrón MP, Marcgena J, Shadi F, Harvey S, Lea RW, Rodriguez AB. Melatonin: An antioxidant at physiologi-cal concentrations. *J Pineal Res* 2001; 31:95–96.

Tiryaki I, Keles H. Reversal of the inhibitory effect of light and high temperature on germination of *Phacelia tanacetifolia* seeds by melatonin. *J Pineal Res* 2012; 52:332–339.

Trivedy R, Pattanshetty S. Treatment of dairy waste by using water hyacinth. *Water Sci Technol* 2002; 45:329–334.

Uchendu EE, Shukla MR, Reed BM, Saxena PK. Melatonin enhances the recovery of cryopreserved shoot tips of American elm (*Ulmus americana* L.). *J Pineal Res* 2013; 55:435–442.

Van Tassel DL, O'Neil SD. Putative regulatory molecules in plants: Evaluating melatonin. *J Pineal Res* 2001; 31:1–7.

Wang L, Zhao Y, Reiter RJ, He C, Liu G, Lei Q, Zuo B, Zheng XD, Li Q, Kong J. Changes in melatonin levels in transgenic "Micro-Tom" tomato overexpressing ovine AANAT and ovine HIOMT genes. *J Pineal Res* 2014; 56:134–142.

Wang P, Sun X, Chang C, Feng F, Liang D, Cheng L, Ma FW. Delay in leaf senescence of Malus hupehensis by long-term melatonin application is associated with its regulation of metabolic status and protein degradation. *J Pineal Res* 2013a; 55:424–434.

Wang P, Sun X, Li C, Wei Z, Liang D, Ma FW. Long-term exogenous application of melatonin delays drought-induced leaf senescence in apple. *J Pineal Res* 2013b; 54:292–302.

Wang P, Yin L, Liang D, Li C, Ma F, Yue Z. Delayed senescence of apple leaves by exogenous melatonin treat-ment: Toward regulating the ascorbate–glutathione cycle. *J Pineal Res* 2012; 53:11–20.

Weeda S, Zhang N, Zhao X, Ndip G, Guo Y, Buck GA, Fu C, Ren SX. *Arabidopsis* transcriptome analysis reveals key roles of melatonin in plant defense systems. *PLoS One* 2014; 9:e93462–e93462.

Xia H, Ma X. Phytoremediation of ethion by water hyacinth (*Eichhornia crassipes*) from water. *Bioresour Technol* 2006; 97:1050–1054.

Yin L, Wang P, Li M, Ke X, Li C, Liang D, Wu S, Ma X, Li C, Zou Y, Ma F. Exogenous melatonin improves *Malus* resistance to Marssonina apple blotch. *J Pineal Res* 2013; 54:426–434.

Zhang L, Jia J, Xu Y, Wang Y, Hao J, Li T. Production of transgenic *Nicotiana sylvestris* plants expressing melatonin synthetase genes and their effect on UV-B-induced DNA damage. *In Vitro Cell Dev Biol Plant* 2012; 48:275–282.

Zhang N, Sun Q, Zhang H, Cao Y, Weeda S, Ren S, Guo YD. Roles of melatonin in abiotic stress resistance in plants. *J Exp Bot* 2015; 66(3):647–656.

Zhang N, Zhang HJ, Zhao B, Sun QQ, Cao YY, Li R, Wu XX, Weeda S, Li L, Ren S, Reiter RJ, Guo Y-D. The RNA-seq approach to discriminate gene expression profiles in response to melatonin on cucumber lateral root formation. *J Pineal Res* 2014; 56:39–50.

Zhang N, Zhao B, Zhang HJ, Weeda S, Yang C, Yang ZC, Ren SX, Guo YD. Melatonin promotes water-stress tolerance, lateral root formation, and seed germination in cucumber (*Cucumis sativus* L.). *J Pineal Res* 2013; 54:15–23.

Zhao Y, Tan DX, Lei Q, Chen H, Wang L, Li QT, Gao Y, Kong J. Melatonin and its potential biological functions in the fruits of sweet cherry. *J Pineal Res* 2013; 55:79–88.

12

Functional Role of Exogenously Applied Melatonin on Plant Adaption under Stress Conditions

Eirini Sarrou
Hellenic Agricultural Organization DEMETER
Thessaloniki, Greece

CONTENTS

ABSTRACT   Abiotic stress adversely limits plant growth and affects the world's agriculture, while adaption to environmental stress factors is crucial for plant survival. Under both biotic and abiotic stress conditions, plants are forced to develop many enzymatic and nonenzymatic mechanisms in order to resist the production of toxic free radicals and enhance their defense systems. This chapter summarizes the melatonin-induced mechanisms of plant adaption under stress conditions such as heat, chilling, drought, salinity, irradiation, herbicides, heavy metals, pathogens, and senescence. Thus, exogenously applied melatonin may influence plant adaptation to stress by (a) upregulating physiological parameters like water, monitoring concentrations of inorganic elements, enhancing ion homeostasis and chlorophyll content, improving photosynthetic efficiency, maintaining cell turgor and membrane selectivity, accumulating net solutes, and acting as an osmoprotectant; (b) scavenging reactive oxygen species by reducing membrane damage, maintaining intracellular H_2O_2 concentrations, and keeping the oxidation–reduction system at steady, safe levels through an enzymatic or nonenzymatic mechanism; (c) regulating the phenylpropanoid pathway, stimulating the production of antioxidant compounds; (d) improving the activity of some stress-related enzymes; and (e) altering the expression of stress-response genes involved in all steps along the way from receptors through transcription factors.

KEY WORDS: *melatonin, plant adaption, stress.*

Abbreviations

ABA: abscisic acid
APX: ascorbate peroxidase
CAT: catalase
DHAR: dehydroascorbate reductase

GA4:   gibberellic acid
GPX: glutathione peroxidase
IAA: indole-3-acetic acid
MEL: melatonin
POD: peroxidase
RNS: reactive nitrogen species
ROS: reactive oxygen species
SOD: superoxide dismutase

Introduction

Abiotic stress adversely limits plant growth and affects the world's agriculture. To date, there are at least three efficient approaches to improving plant resistance to stress conditions. The first is screening tolerant varieties from multiple plant species for their naturally occurring genetic variations, second are agronomical practices like the exogenous application of chemicals that contribute to the protection of plants from stress, and the third is genetic engineering via modulating certain core genes that have a more attractive and rapid approach to improving stress tolerance (Ramakrishna and Ravishankar, 2011; 2013). Present engineering strategies rely on the transfer of one or several genes that encode either biochemical pathways or endpoints of signaling pathways that are controlled by a constitutively active promoter (Nelson et al., 1998; Holmberg and Bülow, 1998; Smirnoff, 1998; Bohnert and Sheveleva, 1998).

MEL has been reported in various plant species (Huang and Mazza, 2011; Paredes et al., 2009; Manchester et al., 2000; Ramakrishna et al., 2009; 2011; 2012b,c; 2015), though the precise function of MEL in plant systems has not yet been brought out. Recent reports indicate that either exogenously applied or endogenously elevated melatonin enhances plant adaption under various stress factors such as drought, salinity, diseases, and photobleaching (Yin et al., 2013; Wei et al., 2014; Wang et al., 2014; Liang et al., 2015). The effect of melatonin on plant adaption is due to its ability to (1) maintain the photosynthetic efficiency of photosystem II (PSII) through the higher production of chlorophyll, the prevention of carotenoid degradation, and the enhancement of plant ion homeostasis; (2) scavenge reactive oxygen species (ROS) by reducing membrane damage and maintaining intracellular H_2O_2 concentrations and the oxidation–reduction system at steady, safe levels through an enzymatic or nonenzymatic mechanism; (3) maintain cell turgor and accumulate net solutes while acting as an osmoprotectant of osmoregulators such as proline; (4) regulate the phenylpropanoid pathway; (5) improve the activity of some stress-related enzymes; and (6) alter the expression of stress-response genes involved in all steps along the way from receptors through transcription factors (Meng et al., 2014; Shi and Chan, 2014; Wei et al., 2014; Weeda et al., 2014; Kostopoulou et al., 2015; Sarrou et al., 2015).

Chilling and Heat Stress

Chilling stress causes physiological and biochemical changes in plants, promoting loss of vigor and reducing growth rate, while cell membrane systems (cell structures, cell membranes, and cell walls) are the main sites of chilling injuries (Kratsch and Wise, 2000). The range of temperature and the time of exposure affect the damage caused to the plant materials. In recent years, many researchers have studied the mode of action of MEL. Bajwa et al. (2014) found that, under the influence of cold stress, the exogenous application of MEL significantly increased plant growth characteristics (fresh weight, shoot height, and primary root length) in *Arabidopsis* plants. To aid in the understanding of the role of MEL in alleviating cold stress, Bajwa et al. (2014) further investigated the expression of cold-related genes under the effects of MEL treatment. In this study, MEL upregulated *COR15a*, a cold-responsive gene, *CAMTA1*, a transcription factor involved in freezing and drought stress, and *ZAT10* and *ZAT12*, transcription activators of ROS-related antioxidant genes. The authors suggested that these changes in cold-signaling genes may stimulate the biosynthesis of cold-protecting compounds and contribute to the increased growth

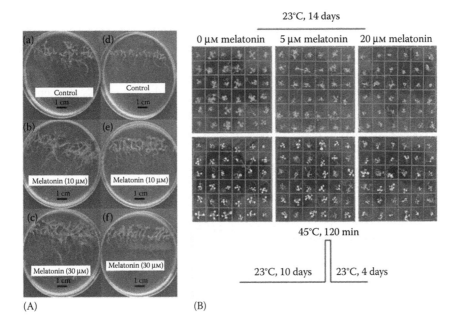

FIGURE 12.1 (A) *Arabidopsis* seedlings grown for 5 days on medium supplemented with melatonin. (a,d) Control 0 lM, (b,e) 10 lM, and (c,f) 30 lM, then exposed to 4°C for 72 (a,b,c) or 120 (d,e,f) hours, followed by two days of recovery under ambient conditions. (From Bajwa et al., 2014.) (B) Exogenous melatonin pretreatment conferred thermotolerance in Arabidopsis. Thermotolerance of WT seedlings without and with melatonin pretreatment. Before heat stress treatment, eight-day-old MS agar plate-grown WT (Col-0) seedlings at 23°C were transferred to new MS agar plates with different concentrations of melatonin for another two days. (From Shi et al., *J. Pineal Res.* 58, 335–342, 2015.)

characteristics of MEL-treated plants under cold stress (Figure 12.1A). In common terms, freezing stress resistance conferred by exogenous MEL was modulated by *AtZAT6* in *Arabidopsis* (Shi and Chan, 2014). Moreover, Szafrańska et al. (2014) observed that 50 µM L^{-1} MEL can play a positive role in the acclimation of *Vigna radiata* seedlings to cold stress conditions and the activation of the phenolic pathway. In this experiment, when MEL was incorporated with hydroprimed seeds, after two days of chilling and two days of rewarming at 25°C, the level of MEL in the roots was seven times higher than in those derived from nonprimed seeds. The results indicate that MEL can enhance the defense mechanisms of plants during cold acclimation by promoting the activity of phenylalanine ammonia-lyase (PAL) after rewarming. Additionally, after rewarming, the level of MEL rapidly decreased, which indicates that the enhanced chilling tolerance was rather caused by the activation of the phenylpropanoid pathway than by the direct action of MEL. This theory is also supported by the induced accumulation of total phenolics and increased antioxidant capacity. A year before, a similar experiment on *V. radiata* roots under chilling stress showed that MEL positively affected root elongation and protected plastids after rewarming (Szafrańska et al., 2013). Melatonin (1 mM) application significantly ameliorated the cold tolerance of maize by upregulating physiological parameters like water and chlorophyll content and the concentrations of inorganic elements: potassium, phosphorus, sulfur, magnesium, iron, copper, manganese, zinc, calcium, and boron. Moreover, MEL stimulated the activity of antioxidant enzymes such as SOD, guaiacol peroxidase, CAT, APX, and glutathione reductase, while concentrations of superoxide, hydrogen peroxide, and malondialdehyde decreased. It is at least possible that MEL facilitated cold tolerance in maize plants by maintaining membrane selectivity and protecting it from oxidative damage, directly and/or indirectly (Turk and Erdal, 2015). This study is supported by the earlier experiments of Turk et al. (2014), who examined the effect of MEL on biochemical, physiological, and molecular parameters in cold-stressed wheat seedlings. They suggested this substance could improve plant resistance by (1) directly scavenging ROS, (2) reactivating metabolic reactions (through the management of water intake and affecting root cell membranes), (3) stimulating the biosynthesis of osmotic substances that reduce the freezing point, and (4) modulating redox balance and other defense mechanisms. Exogenous application of MEL in cucumber seeds promoted germination at 15°C and 10°C to 98% and 83%, respectively (Posmyk et al., 2009). Lei et al. (2004) observed an antiapoptotic effect in carrots cells pretreated with MEL, as a result of its antioxidant actions. However, MEL seemed to stimulate the accumulation of putrescine and spermidine, which may be responsible for the alleviation of cold-induced apoptosis. Zhao et al. (2011) evaluated the role of MEL in cold stress protection during the process of cryopreservation using callus of an endangered plant species, *Rhodiola crenulata*. According to their results, pretreatment for five days with 0.1 μM MEL before freezing in liquid nitrogen significantly increased the survival rate of the cryopreserved callus, while biochemical analysis of the tissue showed a significant reduction in malondialdehyde production and the enhancement of peroxidase and catalase activity. Similar effects have also been found in cryopreserved shoot tips of American elm (*Ulmus americana* L.), whereby MEL enhanced their recovery (Uchendu et al., 2013).

Although the *in vivo* role of melatonin in alleviating cold stress in plants is evident, its involvement in heat stress is largely unknown. Shi et al. (2015a) observed that heat stress significantly induced endogenous melatonin concentration in *Arabidopsis* leaves, and that exogenous MEL treatment conferred improved thermotolerance in *Arabidopsis* (Figure 12.1B). The transcript levels of class-A1 heat-shock factors (*HSFA1s*), which serve as the master regulators of heat stress responses, were significantly upregulated by heat stress and exogenous MEL treatment in *Arabidopsis*. In particular, exogenous melatonin–stimulated thermotolerance was largely elevated in *HSFA1* quadruple knockout (QK) mutants, and *HSFA1*-activated transcripts of heat-responsive genes ([*HSFA2*], heat stress–associated protein 32 [*HSA32*], and heat-shock proteins 90 [*HSP90*] and 101 [*HSP101*]) might be attributed to MEL-mediated thermotolerance.

Salinity, Alkalinity, and Drought Stress

Arguably, the main environmental factor limiting plant production worldwide, especially in semiarid areas, is low water availability. Under drought, decreased water availability and salinity stress cause what has been called the *osmotic* or *water deficit effect* (Munns, 1993; Lawlor, 1995). Biological processes like seed germination, seedling growth and vigor, vegetative growth, flowering, and fruit set are adversely affected by high salt concentration, diminishing economic yields and the quality of produce (Sairam and Tyagi, 2004).

In addition to genetic improvements to create stress-tolerant plants, researchers have focused on the influence of certain molecules—namely, proline (Nounjan et al., 2012), ascorbic acid (Gallie, 2013), salicylic acid (Barba-Espín et al., 2011), and methyl jasmonate (Abdelgawad et al., 2014). However, it seems that the effect of MEL on salinity and drought stress response is of current interest. Mukherjee et al. (2014) have indicated the possible role of serotonin and melatonin in regulating root growth during salt stress in sunflowers. In their study, the application of 120 mM NaCl for 48 h increased endogenous serotonin and melatonin content in sunflower roots and cotyledons, and modulated the activity of the enzyme responsible for melatonin biosynthesis from N-acetylserotonin, indicating the involvement of serotonin and melotonin in salt-induced long-distance signaling from roots to cotyledons. Moreover, 15 μM exogenously applied MEL led to variable effects on hypocotyl elongation and root growth under NaCl stress. Additionally, it has been demonstrated that MEL has significant potential for improvement of soybean growth and seed production, since the exogenous application significantly improved leaf size, plant height, fatty acid content, and salt and drought tolerance (Wei et al., 2014). At the molecular level, MEL upregulated expressions of the genes inhibited by salinity, and hence alleviated the inhibitory effects of salt stress. According to Wei et al. (2014), the promotional role of MEL in soybean is probably due to the enhancement of genes involved in cell division, photosynthesis, fatty acid biosynthesis, and carbohydrate and ascorbate metabolism. Mineral analysis of faba beans under salt stress revealed that MEL treatment reduced the levels of compatible solutes as well as Na^+ and $Cl-$, and improved the levels of K^+ and Ca^{+2} as well as the ratios of K^+/Na^+ and Ca^{+2}/Na^+ (Dawood and El-Awadi, 2015). It is well known that Ca^{2+} levels and transport under salinity conditions and K^+/Na^+ ratios are important indicators of salt tolerance, given that Na^+ competes with K^+ for binding sites essential to cellular function but cannot substitute for K^+ in the activation of functional enzymes (Tester and Davenport, 2003). Hence, MEL may control the expression of ion channel genes under salinity, which may possibly contribute to the maintenance of ion homeostasis, thus improving salinity resistance (Li et al., 2012).

These data stand in agreement with those of Liu et al. (2015), who also observed MEL-mediated ion homeostasis with decreased Na^+ and increased K^+ content in tomato leaves under alkaline stress. According to Liu et al., the positive effects of exogenously applied MEL depend on concentration and are a result of the induced activities of antioxidant enzymes (Figure 12.2A) and the accumulation of

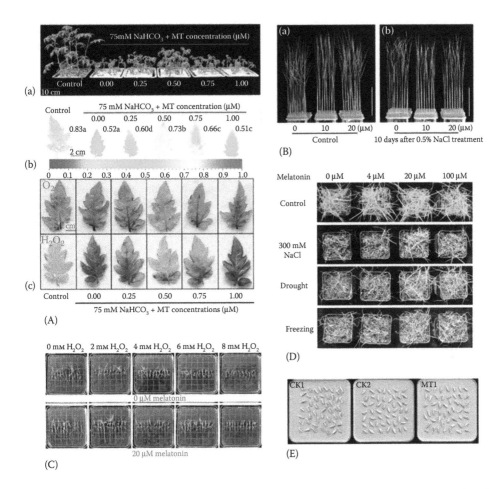

FIGURE 12.2 (A) Effects of different concentrations of melatonin (MT) on phenotype (a). Images and value of maximum photochemistry efficiency of PSII (Fv/Fm) of tomato seedlings under $NaHCO_3$ stress (b). The different color bar representing the value for maximum photochemistry efficiency of PSII is shown at the bottom of the images. The effects of different concentrations of melatonin on accumulation of superoxide radical ion and hydrogen peroxide in leaves (c) were detected by nitrobluetetrazolium and 3,3-diaminobenzidine tetrahydrochloride staining, respectively. Scale bars for superoxide radical ion and hydrogen peroxide represent 1 cm. (From Liu et al., *Scientia Horticulturae* 181, 18–25, 2015.) (B) Exogenous application of melatonin enhances salinity stress tolerance and delays salt-induced leaf senescence in rice. (a) Untreated control and (b) phenotypes of rice seedlings after 10-day 0.5% NaCl treatment. Scale bar = 5 cm (From Liang et al., *J. Pineal Res.* 59, 91–101, 2015). (C) Application of exogenous melatonin conferred oxidative stress resistance in Bermuda grass. The representative picture of 24-day-old plants without (top) and with (bottom) melatonin pretreatments under oxidative stress conditions. For the assays, 10-day-old Bermuda grass seedlings grown on MS agar plate were transferred to new MS agar plates with different concentrations of melatonin and H_2O_2 for another 14 days. (From Shi et al., *J. Pineal Res.* 58, 335–342, 2015.) (D) Application of exogenous melatonin improved abiotic stress resistance in bermudagrass. Growth of 28-day-old plants with different melatonin treatments and under control, salt, drought, or freezing stress conditions. (From Shi et al., *J. Pineal Res.* 59, 120–131, 2015.) (E) Melatonin (1 μM) promotes seed germination under NaCl stress. (From Zhang et al., *J. Pineal Res.* 57, 269–279, 2014.)

ascorbic acid and glutathione, which could be partly responsible for protecting chlorophyll degradation and maintaining higher PSII efficiency (Fv/Fm). Similarly, a combined treatment of MEL and ascorbic acid was capable of inducing salinity adaptation in *Citrus aurantium* L. seedlings. This synergistic mechanism of MEL and ascorbic acid was able to regulate *CaMIPS*, *CaSLAH1*, and *CaMYB73* expression, indicating that sugar metabolism, carbohydrate and proline accumulation, nonenzymatic and enzymatic antioxidant activity, myo-inositol biosynthesis, Cl⁻ ion homeostasis, and transcription regulation were triggered by the combined treatment (Kostopoulou et al., 2015). Furthermore, Liang et al. (2015) indicated that exogenous application of melatonin enhances salinity stress tolerance (Figure 12.2B) by acting as a potent agent that delays salt-induced leaf senescence and cell death in rice. This is exemplified by molecular analysis, according to which the expression levels of genes involved in multiple biological processes, such as oxidation–reduction, chlorophyll biosynthesis, pigment biosynthesis, stress responses, nutrient metabolism, and remobilization, are altered by MEL treatment. In general, the protective role of MEL on rice is due to its scavenging activity and ability to directly or indirectly counteract the cellular accumulation of H_2O_2 by inducing the expression of transcription factors such as bZIP, MAC, and MYB. These transcription factors activate the genes encoding the H_2O_2 scavenging enzymes, possibly regulating the expression of DREB and HSF transcription factors, thereby establishing a transcriptional cascade to reduce intracellular H_2O_2 levels. Shi et al. 2015c found that exogenously applied MEL significantly alleviated hydrogen peroxide; modulated plant growth, cell damage, and ROS accumulation; enhanced several metabolic pathways (including major carbohydrate metabolism, photosynthesis, redox, amino acid polyamine metabolism, and the ribosome pathway); and at the molecular level influenced 76 proteins in bermudagrass plants (Figure 12.2C). Additionally, abiotic stress induced the endogenous level of melatonin and exogenously applied melatonin improved abiotic stress resistance (Figure 12.2D) by promoting the accumulation of ROS, organic acids, sugars, and alcohols in bermudagrass (Shi et al., 2015b). Thus, the protective role of MEL in bermudagrass under abiotic stress conditions, via the modulation of metabolic homeostasis (nitrogen, hormone, and carbohydrate transformation, metal handling, and secondary metabolism; Shi et al., 2015c), has been well demonstrated. A correlation between MEL synthesis and seed germination in cucumber has been shown, while MEL-pretreated seeds showed enhanced germination rates under the influence of 150 mM NaCl stress (Figure 12.2E; Zhang et al., 2014). In this study, Zhang et al. indicated two apparent mechanisms by which MEL alleviated the salinity-induced inhibition of seed germination by (1) regulating the biosynthesis and catabolism of abscisic acid (ABA) and gibberellic acid (GA_4) and (2) enhancing the gene expression of antioxidant compounds and affecting the activities of antioxidant enzymes including SOD, CAT, and peroxidase (POD). Thus, MEL decreased ABA content during the early stages of germination by upregulating catabolism genes *CsCYP707A1* and *CsCYP707A2* (3.5- and 105-fold increases, respectively, at 16 h, compared to NaCl treatment) and downregulating biosynthesis gene *CsNECD2* (a 0.29-fold decrease at 16 h). In contrast, MEL contributed to an increase of GA_4 content by upregulating biosynthesis genes *GA20ox* and *GA3ox* (2.3- and 3.9-fold increases at 0 and 12 h, respectively) under salinity stress.

Apart from salinity resistance, MEL also alleviated the effect of water stress during seed germination, seedling growth, and metabolism in cucumber, having comprehensive physiological actions in plant tissues, such as improved seed germination, root generation, enhanced photosynthesis, and protection against environmental stress. The protective role of exogenously applied MEL on cucumber seedlings was due to reduced chlorophyll degradation, improved chlorophyll fluorescence, and protection of the ultrastructure of chloroplasts, which resulted in a higher photosynthetic rate (Zhang et al., 2013). Pretreatment with MEL also improved the tolerance of both drought-tolerant *Malus prunifolia* and drought-sensitive *Malus hupehensis* plants via two mechanisms: (1) reducing ABA content by downregulating *MdNCED3*, an ABA synthesis gene, and upregulating its catabolic genes, and (2) scavenging H_2O_2 and enhancing the activities of antioxidant enzymes to detoxify H_2O_2 under drought stress. These two mechanisms work synergistically, improving stomata functions, and inducing water conservation in leaves, less electrolyte leakage, steady chlorophyll content, and greater photosynthetic performance (Figure 12.3). Transgenic *Arabidopsis* plants overexpressed *MzASMT1*, producing higher levels of MEL and, furthermore, having greater tolerance to drought in contrast to the wild type. This elevated drought tolerance in transgenic *Arabidopsis* was attributed to the higher MEL content resulting from the overexpression of *MzASMT1*, which significantly decreased ROS levels. At the very least,

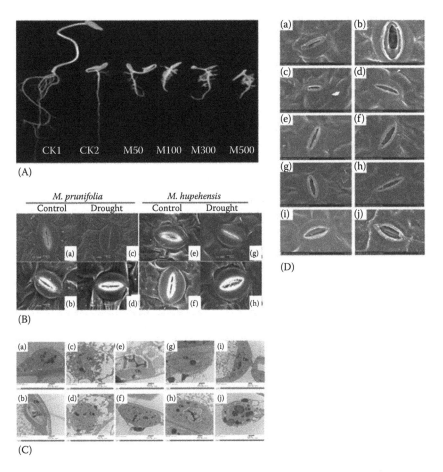

FIGURE 12.3 (A) The effect of melatonin on root development in polyethylene glycol (PEG)-treated cucumber seedlings (five days). (CK1) germinated with water; (CK2) germinated with 18% PEG; (M50, M100, M300, M500) germinated with PEG plus melatonin (50, 100, 300, 500 μM). (From Zhang et al., *J. Pineal Res.* 54, 15–23, 2013.) (B) SEM images of stomata from leaves of *M. prunifolia* and *M. hupehensis*: (a and e) control leaves in which stomata are open; (b and f) closed stomata in leaves exposed to drought stress for five days; (c and g) leaves from control plants that were pretreated with 100 μM melatonin; (d and h) leaves from drought-stressed, pretreated plants for which all stomata are open. Magnification × 3000; scale bars = 5 μm. (From Li et al., *J. Pineal Res.* 53, 298–306, 2012.) (C) Effects of melatonin treatment on ultrastructure of chloroplasts in grape leaves under drought stress induced by PEG. Top row (a, c, e, g, and i) bottom row (b, d, f, h, and j) represent young leaves and mature leaves, respectively, from subgroups, including those grown with half-strength nutrient solution only, half-strength nutrient solution plus 10% PEG treatment, half-strength nutrient solution plus 50 nmol/L melatonin and 10% PEG treatment, half-strength nutrient solution plus 100 nmol/L melatonin and 10% PEG treatment, and half-strength nutrient solution plus 200 nmol/L melatonin and 10% PEG treatment. Ch: chloroplast; SG: starch grain; scale bars in a–j = 500 nm. D) Effects of melatonin treatment on stomatal characteristics of grape leaves under drought stress induced by PEG. Left column (a, c, e, g, and i) and right column (b, d, f, h, and j) represent young leaves and mature leaves, respectively, from subgroups grown with half-strength nutrient solution only, half-strength nutrient solution plus 10% PEG treatment, half-strength nutrient solution plus 50 nmol/L melatonin and 10% PEG treatment, half-strength nutrient solution plus 100 nmol/L melatonin and 10% PEG treatment, and half-strength nutrient solution plus 200 nmol/L melatonin and 10% PEG treatment. Magnification × 93,000; scale bars = 10.0 lm. (From Meng et al., *J. Pineal Res.* 57, 200–212, 2014.)

these results elucidate the important role that membrane-located MEL synthase plays in drought tolerance (Zuo et al., 2014). Meng et al. (2014) have further described the ameliorative effects of exogenously applied MEL on grape cuttings under water deficit stress. Exogenously applied MEL could improve the resistance of wine grape seedlings grown from cuttings to polyethylene glycol–induced water deficit stress. After an application of 10% polyethylene glycol, Meng et al. observed that MEL partially alleviated oxidative injury to cuttings, slowed the decline in the potential efficiency of PSII, and limited the effects of stress on leaf thickness, spongy tissue, and stoma size. Additionally, MEL treatment helped preserve the internal lamellar system of chloroplasts and alleviated the ultrastructural damage induced by drought stress.

Plant Pathogens

Numerous crop diseases and yield losses are caused by various plant pathogens, which are associated with the production of mycotoxins. Plants utilize highly complex and effective strategies to counteract the entry of pathogens. It is well known that MEL can act as a regulator in the innate immunity of animals (Calvo et al., 2013); this fact prompted plant scientists to investigate the biological role of MEL in the innate immunity of plants (Yin et al., 2013; Lee et al., 2014; Zhao et al., 2015; Shi et al., 2015d). Innate plant immunity plays an essential role in plant–pathogen interaction. Pattern recognition receptors localized in plant plasma membrane can recognize pathogen-associated molecular patterns and trigger immunity. In addition, pathogen recognition mediated by resistance genes results in effector-triggered immunity, and as a result, a burst of ROS and reactive nitrogen species (RNS) generation, the hypersensitive response, and the transcriptional activation of defense genes (Shi et al., 2015d). The main problem limiting the production of transgenic plants in the agricultural biotechnology industry is the browning and death of cells transformed with *Agrobacterium tumefaciens*. Melatonin is suggested to be one of the most efficient compounds for increasing the stable transformation frequency in the tomato cv. Micro-Tom transformation system, while T-DNA integration is not affected by its application (Dan et al., 2015).

The first report on the protective role of MEL against pathogens concerned the disease Marssonina apple blotch (*Diplocarpon mali*), which leads to premature defoliation in the main regions of apple production (Figure 12.4A; Yin et al., 2013). The exogenous application of MEL to apple trees prior to inoculation with *D. mali* alleviates disease damage by reducing the number of lesions and inhibiting their expansion. The main reasons for the improved resistance of apples to the pathogen are that the potentially high efficiency of PSII is maintained, their normal chlorophyll contents are retained, intracellular H_2O_2 concentrations are kept at steady-state levels, and the activity of plant defense-related enzymes is increased. According to Lee et al. (2014), MEL may also be a novel defense-signaling molecule in plant–pathogen interactions. Exogenous application of a 10 lM concentration on *Arabidopsis* and tobacco leaves induced various pathogenesis-related (PR) genes, which caused an increase in resistance against the virulent bacterial pathogen *Pseudomonas syringae* DC3000 by suppressing its multiplication (Figure 12.4B). Zhao et al. (2015) suggested that MEL, in optimized doses, promoted the growth and development of *Arabidopsis* seedlings due to its positive effects on sucrose metabolism, invertase activity, and the transcription of invertase-related genes (Figure 12.4C). Moreover, MEL-induced resistance to *P. syringae* pv. tomato DC3000 in *Arabidopsis* was due to the activation of cell wall invertase–dependent and salicylic acid–dependent pathogen defense pathways. These data stand in agreement with those of Shi et al. (2015d), who reported that MEL-induced nitric oxide was responsible for the innate immunity response of *Arabidopsis* against *P. syringe* pv. tomato DC3000 infection and furthermore conferred improved disease resistance in *Arabidopsis* (Figure 12.4D). According to the molecular pathway, regarding the protective role of melatonin in response to bacterial pathogen infection in *Arabidopsis*, MEL stimulates nitric oxide production in pathogen-infected plants and the elevated nitric oxide production upregulates the transcription of salicylic acid–related genes to mediate innate immunity against bacterial infection.

Senescence

Leaf senescence is an integral process to plant development. This process has great practical value because, during senescence, nutrients are recycled to other parts of the plant. On the other hand, from an agricultural point of view, leaf senescence may limit yields in certain crops and may also contribute to the postharvest loss of vegetable crops. Like many other developmental processes in plants, it is a genetically controlled program regulated by a variety of environmental and endogenous factors (and their interactions), including stresses such as extremes of temperature, drought, nutrient deficiency, pathogen

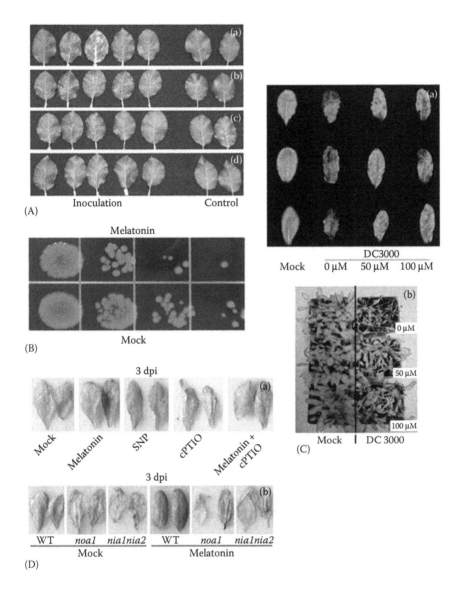

FIGURE 12.4 (A) Effect of melatonin concentration on apple leaf phenotype at 20 dpi. Fully mature leaves were collected from plants pretreated with (a) water only and no exogenous melatonin, (b) 0.05 mM melatonin, (c) 0.1 mM melatonin, or (d) 0.5 mM melatonin (d). (From Yin et al., *J. Pineal Res.* 54, 426–434, 2013.) (B) Growth of Pst DC3000. Five-week-old *Arabidopsis* Col-O leaves pretreated with 10 μM melatonin or mock (1 mM MES pH 5.6) were inoculated with Pst DC3000. (From Lee et al., *J. Pineal Res.* 57, 262–268, 2014.) (C) Effect of melatonin on susceptibility of 5-week-old *Arabidopsis* seedlings to Pst DC3000 after vacuum-infiltration (105 CFU/mL). (a) Leaf phenotypes and (b) growth performance were recorded at 72 hpi. (From Zhao et al., 2015.) (D) (a) Symptoms in plant leaves with different treatments at 3 dpi of bacterial infection. For the assays, 27-day-old soil-grown WT (Col-0) plants were watered with nutrient solution containing 0 lM melatonin, 20 lM melatonin, 500 lM SNP, 500 lM cPTIO, and 20 lM melatonin plus 500 lM cPTIO from below in pots with plants for one day, respectively; then, the 28-day-old plants were used for disease resistance assay. (b) Symptoms in plant leaves with different treatments at 3 dpi of bacterial infection. For the assays, 27-day-old soil-grown WT (Col-0) plants and noa1 and nia1nia2 mutants were watered with nutrient solution containing 0 lM and 20 lM melatonin for one day, respectively; then, the 28-day-old plants were used for disease resistance assay against Pst DC3000. (From Shi et al., *J. Pineal Res.* 59, 102–8, 2015.)

infection, high or low solar radiation, wounding, and ozone, whereas autonomous factors include age, reproductive development, and phytohormone levels (Munné-Bosch and Alegre, 2004).

Therefore, investigating the effect of MEL on leaf senescence could lead to new ways of manipulating senescence for agricultural applications. For example, barley, *Arabidopsis*, tomato, and apple leaves treated with melatonin clearly delayed the senescence process, as estimated from several parameters, while this effect was dose dependent most of the time (Figure 12.5A–C; Arnao and Hernández-Ruiz, 2009; Okazaki and Ezura, 2009; Wang et al., 2012; 2014; Shi et al., 2015e). Byeon et al. (2012) also noticed that control rice leaves contained an undetectable level of melatonin, which increased greatly during the senescence process, suggesting that melatonin acts as a stress-adaptive metabolite that slows down senescence through its antioxidant protection (Galano et al., 2011; Park et al., 2012; Wang et al.,

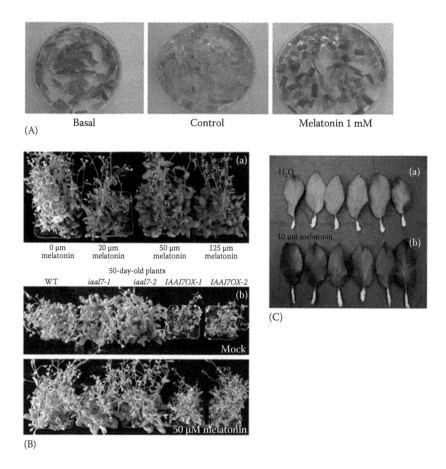

FIGURE 12.5 (A) Representative images of the effect of 1 mM melatonin treatment with respect to control (buffer solution) on senescence of barley leaf sections after 48 h in the dark at 24°C. Basal corresponds to leaves not treated and not subjected to the 48-h dark period. (From Arnao and Hernández-Ruiz, *J. Pineal Res.* 46, 58–63, 2009.) (B) (a) Exogenous melatonin pretreatment delayed leaf senescence in *Arabidopsis*. The picture shows 50-day-old soil-grown WT (Col-0) plants that were pretreated with different concentrations of melatonin. For the assay, 21-day-old soil-grown plants were watered with a nutrient solution containing 0, 20, 50, and 125 lM melatonin from below in pots with plants for seven days. (b) The involvement of AtIAA17 in melatonin-mediated leaf senescence in *Arabidopsis*. The picture shows 50-day-old soil grown plants that were pretreated with different concentrations of melatonin. For the assay, 21-day-old soil-grown plants were watered with a nutrient solution containing 0 and 50 lM melatonin from below in pots with plants for seven days. (From Shi et al., *J. Pineal Res.* 58, 26–33, 2015.) (C) Effects of melatonin on phenotype of detached, fully mature apple leaves at day 12 after dark treatment. (a) Leaves treated with double-distilled water as control. (b) Leaves treated with melatonin. Before being transferred into 10 mL centrifuge tubes containing 10 mL of either double-distilled water or 10 mm melatonin, petioles were cut short under water to prevent gas bolt. All leaves were incubated in the dark in a chamber at 25°C and 80% relative humidity. (From Wang et al., *J. Pineal Res.* 53, 11–20, 2012.)

2012). The role of MEL as an antioxidant—preventing the accumulation of free radicals (ROS, RNS, and lipid radicals) and providing extra time for the relocation of hydrolyzed proteins and lipids—is probably the main reason for the delayed senescence (Arnao and Hernández-Ruiz, 2009). Nevertheless, the activities of chlorophyllase (chlorophyll hydrolase), pheophorbide a oxygenase (ferredoxin-dependent monooxygenase), and red chlorophyll catabolite reductase (also ferredoxin dependent) were altered under the influence of MEL, whereas the breakdown of chlorophylls and the activity of these enzymes are important in programmed cell death, regulating plant defense pathways through the detoxification of free chlorophylls (and their phototoxic intermediates), as previously suggested (Hőrtensteiner, 2006). Furthermore, the expression level of *AUXIN-RESISTANT 3 (AXR3)/INDOLE-3-ACETIC ACID INDUCIBLE 17 (IAA17)* indicated that *AtIAA17* is a positive modulator of natural leaf senescence and provides a direct link between melatonin and *AtIAA17* in *Arabidopsis* (Shi et al., 2015e). In addition, biochemical and molecular parameters such as the delay of normal reduction in chlorophyll content and maximum potential Fv/Fm indicated the involvement of MEL in delaying dark-induced senescence in detached apple leaves (Wang et al., 2012). This effect was probably due to the ability of MEL to scavenge ROS (reducing the accumulation of H_2O_2), controlling the expression of senescence genes, elevating antioxidative enzymes in the AsA–GSH cycle (enhancing APX, which acts on mRNA and protein activity levels), and maintaining high levels of AsA and GSH. In addition, proteomic analysis in apple leaves showed that 622 and 309 proteins were altered by senescence and melatonin, respectively (Wang et al., 2014). Specifically, MEL treatment downregulated proteins that are involved in protein folding and posttranslational modifications, including three histone proteins and one histone chaperone (NAP1-related protein) involved in nucleosome assembly, as well as one histone acetyltransferase HAC1-like protein that acetylates histones to give a specific tag for transcriptional activation.

Irradiation, Herbicides, and Heavy Metals

Decreases in the ozone layer due to greenhouse gases could lead to a significant increase in UV-B radiation reaching the surface of the earth. It seems that plants are widely affected by irradiation, mostly in their biomass production, growth characteristics, and biochemical changes (Kramer et al., 1991). The damages UV-B irradiation can cause in plants are summarized as follows: (1) damage influencing PSII (thus affecting photosynthesis), the reduction of chlorophyll content (possibly by reducing the expression of genes encoding chlorophyll-binding proteins), the disturbance of membrane permeability, and damage to RuBP carboxylase; (2) changes in phytohormones (e.g., reduced levels or inactivation of IAA), leading to inhibition of cell expansion; (3) membrane damage, photoabsorption in membranes, and peroxidation (unsaturated fatty acids and changes in the membrane lipid composition); and (4) DNA damage. Higher plants have developed defense mechanisms against irradiation stress such as scattering and reflecting UV radiation with epidermal and cuticular structures; other optical leaf properties such as waxy layers, leaf hairs, and leaf bladders; the absorption of UV-B radiation by pigments (flavonoids, anthocyanins); excision repair of DNA damage; scavenging radicals formed by the absorption of UV-B photons by SOD and CAT (flavonoids are also involved in neutralizing radicals); and ameliorating UV damage to membrane with the use of polyamines (Rozema et al., 1997).

There are few reports on the influence of MEL on plant adaption to irradiation. It seems that the accumulation of MEL in plants depends both on the duration and intensity of UV radiation. The melatonin content of *Glycyrrhiza uralensis* roots varied depending on the wavelength of the light spectrum (Afreen et al., 2006). The increment in MEL content was proportional to the intensity of UV exposure. Dubbels et al. (1995) found that a strain of tobacco with high MEL content had more resistance to UV irradiation than a strain with low MEL content, indicating that MEL could play a role in antioxidant defense. Moreover, it has been suggested that MEL alleviates herbicide-induced oxidative stress in transgenic rice plants (Park et al., 2013). Wang et al. (2015) observed the beneficial effects of MEL against mitochondrial dysfunction and a strong induction of autophagy in *Arabidopsis* seedlings; these changes are attributed to the upregulation of genes involved in the *AtATG8-PE* conjugation pathway, which enhances the capacity for autophagy and the expression of two genes for H_2O_2-scavenging enzymes, *AtAPX1* and *AtCATs*. According to Park et al. (2013), melatonin-rich transgenic (MRT) seedlings treated with 0.1 µM butafenacil were resistant to the herbicide, and contained high levels of chlorophyll and low levels of malondialdehyde and hydrogen peroxide compared with control plants; both the MRT and control seedlings produced more MEL (by inducing MEL biosynthetic genes, including tryptophan decarboxylase, tryptamine 5-hydroxylase, and N-acetylserotonin methyltransferase). These are the only reports indicating resistance to peroxidizing herbicides, which act by generating ROS that kill plants. The possible mechanism of plant adaptation involves stimulating melatonin, which acts as an ROS scavenger and reduces lipid peroxidation, or antioxidative enzymes such as SOD and CAT.

Posmyk et al. (2008) have reported on the protective role of MEL in plants under the influence of heavy metals. Specifically, MEL applied at low concentrations (1–10 μM) protected cells against injury from Cu^{2+} in red cabbage seedlings, while hydropriming seeds with melatonin seems to be a good method for improving seed quality and seedling vigor in optimal and stress conditions, such as heavy metal toxicity. The results of this article confirm the suggestion of Manchester et al. (2000) that the tolerance of plants to toxic elements may be related to their intrinsic levels of MEL.

Conclusion

The objective of this review was to update the reader on the biological role of MEL in plant stress tolerance. This review has summarized the possible MEL-induced mechanisms of plants under the influence of environmental stress factors such as drought, salinity, chilling, heat, pathogens, and senescence, as well as irradiation, metal toxicity, and herbicides. We hope this review will stimulate further research in the promising areas of plant physiology and molecular biotechnology.

REFERENCES

Abdelgawad, Z. A., Khalafaallah, A. A., and Abdallah, M. M. 2014. Impact of methyl jasmonate on antioxidant activity and some biochemical aspects of maize plant grown under water stress condition. *Agric. Sci.* 5: 1077–1088.

Afreen, F., Zobayed, S. M. A., and Kozai, T. 2006. Melatonin in Glycyrrhiza uralensis: Response of plant roots to spectral quality of light and UV-B radiation. *J. Pineal Res.* 41: 108–115.

Arnao, M. B., and Hernández-Ruiz, J. 2009. Protective effect of melatonin against chlorophyll degradation during the senescence of barley leaves. *J. Pineal Res.* 46: 58–63.

Bajwa, V. S., Shukla, M. R., Sherif, S. M., Murch, S. J., and Saxena, P. K. 2014. Role of melatonin in alleviating cold stress in *Arabidopsis thaliana*. *J. Pineal Res.* 56: 238–245.

Barba-Espín, G., Clemente-Moreno, M. J., Alvarez, S., García-Legaz, M. F., Hernandez, J. A., and Díaz-Vivancos, P. 2011. Salicylic acid negatively affects the response to salt stress in pea plants. *Plant Biol.* 13: 909–917.

Bohnert, H. J., and Sheveleva, E. 1998. Plant stress adaptations: Making metabolism move. *Curr. Opin. Plant Biol.* 1: 267–274.

Byeon, Y., Park, S., Kim, Y. S., Park, D. H., Lee, S., and Back, K. 2012. Light-regulated melatonin biosynthesis in rice during the senescence process in detached leaves. *J. Pineal Res.* 53: 107–111.

Calvo, J. R., Gonzales-Yanes, C., and Maldonado, M. D. 2013. The role of melatonin in the cells of the innate immunity: A review. *J. Pineal Res.* 55: 103–120.

Dan, Y., Zhang, S., Zhong, H., Yi, H., and Sainz, M. B. 2015. Novel compounds that enhance *Agrobacterium*-mediated plant transformation by mitigating oxidative stress. *Plant Cell Rep.* 34: 291–309.

Dawood, M. G., and El-Awadi, M. E. 2015. Alleviation of salinity stress on *Vicia faba* L. plants via seed priming with melatonin. *Acta Biolo. Colomb.* 20: 223–235.

Dubbels, R., Reiter, R. J., Klenke, E., Goebel, A., Schnakenberg, E., Ehlers, C., Shiwara, H. W., and Schloot, W. 1995. Melatonin in edible plants identified by radioimmunoassay and by high performance liquid chromatography–mass spectrometry. *J. Pineal Res.* 18: 28–31.

Galano, A., Tan, D. X., and Reiter, R. J. 2011. Melatonin as a natural ally against oxidative stress: A physiochemical examination. *J. Pineal Res.* 51: 1–16.

Gallie, D. R. 2013. The role of L-ascorbic acid recycling in responding to environmental stress and in promoting plant growth. *J. Exp. Bot.* 64: 433–443.

Holmberg, N., and Bülow, L. 1998. Improving stress tolerance in plants by gene transfer. *Trends Plant Sci.* 3: 61–66.

Hőrtensteiner, S. 2006. Chlorophyll degradation during senescence. *Annu. Rev. Plant Biol.* 57: 55–77.

Huang X., and Mazza, G. 2011. Application of LC and LC-MS to the analysis of melatonin and serotonin in edible plants. *Crit. Rev. Food Sci. Nutr.* 51: 269–284.

Kostopoulou, Z., Therios, I., Roumeliotis, E., Kanellis, A. K., and Molassiotis, A. 2015. Melatonin combined with ascorbic acid provides salt adaptation in *Citrus aurantium* L. seedlings. *Plant Physiol. Biochem.* 86: 155–165.

Kramer, G. F., Norman, H. A., Krizek, D. T., and Mirecki, R.M. 1991. Influence of UV-B radiation on polyamines, lipid peroxidation and membrane lipids in cucumber. *Phytochemistry* 30: 2101–2108.

Kratsch, H. A., and Wise, R. R. 2000. The ultrastructure of chilling stress. *Plant Cell Environ.* 23: 337–350.

Lawlor, D. W. 1995. The effects of water deficit on photosynthesis. In: N. Smirnoff (Ed.): *Environment and Plant Metabolism, Flexibility and Acclimation*, 129–160. BIOS Scientific Publishers, Oxford, UK.

Lee, H. Y., Byeon, Y., and Kyoungwhan, B. 2014. Melatonin as a signal molecule triggering defense responses against pathogen attack in Arabidopsis and tobacco. *J. Pineal Res.* 57: 262–268.

Lei, X. Y., Zhu, R. Y., Zhang, G. Y., and Dai, Y. R. 2004. Attenuation of cold-induced apoptosis by exogenous melatonin in carrot suspension cells: The possible involvement of polyamines. *J. Pineal Res.* 36: 126–131.

Li, C., Wang, P., Wei, Z., Liang, D., Liu, C., Yin, L., Jia, D., Fu, M., and Ma, F. 2012. The mitigation effects of exogenous melatonin on salinity-induced stress in *Malus hupehensis. J. Pineal Res.* 53: 298–306.

Liang, C., Zheng, G., Li, W., Wang, Y., Hu, B., Wang, H., Wu, H., Qian, Y., Zhu, X. G., Tan, D. X., Chen, S. Y., and Chu, C. 2015. Melatonin delays leaf senescence and enhances salt stress tolerance in rice. *J. Pineal Res.* 59: 91–101.

Liu, N., Jin, Z., Wang, S., Gong, B., Wen, D., Wang, X., Wei, M., and Shi, Q. 2015. Sodic alkaline stress mitigation with exogenous melatonin involves reactive oxygen metabolism and ion homeostasis in tomato. *Sci. Hortic.* 181: 18–25.

Manchester, L. C., Tan, D. X., Reiter, R. J., Park, W., Monis, K., and Qi, W. B. 2000. High levels of melatonin in the seeds of edible plants: Possible function in germ tissue protection. *Life Sci.* 67: 3023–3029.

Meng, J. F., Xu, T. F., Wang, Z. Z., Fang, Y. L., Xi, Z. M., and Zhang, Z. W. 2014. The ameliorative effects of exogenous melatonin on grape cuttings under water-deficient stress: Antioxidant metabolites, leaf anatomy, and chloroplast morphology. *J. Pineal Res.* 57: 200–212.

Mukherjee, S., David, A., Yadav, S., Baluška, F., and Bhatla, S. C. 2014. Salt stress-induced seedling growth inhibition coincides with differential distribution of serotonin and melatonin in sunflower seedling roots and cotyledons. *Physiol. Plant.* 152: 714–728.

Munné-Bosch, S., and Alegre, L. 2004. Die and let live: Leaf senescence contributes to plant survival under drought stress. *Funct. Plant Biol.* 31: 203–216.

Munns, R. 1993. Physiological processes limiting plant growth in saline soils: Some dogmas and hypotheses. *Plant. Cell Env.* 16: 15–24.

Nelson, D. E., Shen, B., and Bohnert, H. J. 1998. Salinity tolerance: Mechanistic models, and the metabolic engineering of complex traits. *Genet. Eng.* 20: 153–176.

Nounjan, N., Nghia, P. T., and Theerakulpisut, P. 2012. Exogenous proline and trehalose promote recovery of rice seedlings from salt-stress and differentially modulate antioxidant enzymes and expression of related genes. *J. Plant Physiol.* 169: 596–604.

Okazaki, M., and Ezura, H. 2009. Profiling of melatonin in the model tomato (*Solanum lycopersicum* L.) cultivar Micro-Tom. *J. Pineal Res.* 46: 338–343.

Paredes, S. D., Korkmaz, A., Manchester, L. C., Tan, D. X., and Reiter, R. J. 2009. Phytomelatonin: A review. *J. Exp. Bot.* 60: 57–69.

Park, S., Lee, K., Kim, Y. S., and Back, K. 2012. Tryptamine 5-hydroxylase-deficient Sekiguchi rice induces synthesis of 5-hydroxytryptophan and N-acetyltryptamine but decreases melatonin biosynthesis during senescence process of detached leaves. *J. Pineal Res.* 52: 211–216.

Park, S., Lee, D. E., Jang, H., Byeon, Y., Kim, Y. S., and Back, K. 2013. Melatonin-rich transgenic rice plants exhibit resistance to herbicide-induced oxidative stress. *J. Pineal Res.* 54: 258–263.

Posmyk, M. M., Bałabusta, M., Wieczorek, M., Sliwinska, E., and Janas, K. M. 2009. Melatonin applied to cucumber (*Cucumis sativus* L.) seeds improves germination during chilling stress. *J. Pineal Res.* 46: 214–223.

Posmyk, M. M., Kuran, H., Marciniak, K., and Janas, K. M. 2008. Presowing seed treatment with melatonin protects red cabbage seedlings against toxic copper ion concentrations. *J. Pineal Res.* 45: 24–31.

Ramakrishna, A. 2015. Indoleamines in edible plants: Role in human health effects. In Angel Catalá, ed. *Indoleamines: Sources, Role in Biological Processes and Health Effects*, Biochemistry Research Trends, Nova Publishers, Hauppauge, NY, 279.

Ramakrishna, A., Giridhar, P., Jobin, M., Paulose, C. S., and Ravishankar, G. A. 2012a. Indoleamines and calcium enhance somatic embryogenesis in cultured tissues of *Coffea canephora* P ex Fr. *Plant Cell Tiss. Org.* 108: 267–278.

Ramakrishna, A., Giridhar, P., and Ravishankar, G. A. 2009. Indoleamines and calcium channels influence morphogenesis in *in vitro* cultures of *Mimosa pudica* L. *Plant. Sig. Behav.* 12: 1136–1141.

Ramakrishna, A., Giridhar, P., and Ravishankar, G. A. 2011. Calcium and calcium ionophore A23187 induce high frequency somatic embryogenesis in cultured tissues of *Coffea canephora* P ex Fr. *In Vitro Cell. Dev. Biol. Plant* 47: 667–673.

Ramakrishna, A., Giridhar, P., Udaya Sankar, K., and Ravishankar, G. A. 2012b. Endogenous profiles of indoleamines: Serotonin and melatonin in different tissues of *Coffea canephora* P ex Fr. as analyzed by HPLC and LC-MS-ESI. *Acta Physiologia Plantarum.* 34: 393–396.

Ramakrishna, A., Giridhar, P., Udaya Sankar, K., and Ravishankar, G. A. 2012c. Melatonin and serotonin profiles in beans of *Coffea* sps. *J. Pineal Res.* 52: 470–476.

Ramakrishna, A., and Ravishankar, G. A. 2011. Influence of abiotic stress signals on secondary metabolites in plants. *Plant Sig. Behav.* 6: 1720–1731.

Ramakrishna, A., and Ravishankar, G. A. 2013. Role of plant metabolites in abiotic stress tolerance under changing climatic conditions with special reference to secondary compounds. In N. Tuteja and S. S. Gill, eds. *Climate Change and Abiotic Stress Tolerance*, Wiley, Weinheim, Germany.

Rozema, J., van de Staaij, J., Björn, L. O., and Caldwell, M. 1997. UV-B as an environmental factor in plant life: Stress and regulation. *Trends Ecol. Evol.* 12: 22–28.

Sairam, R. K., and Tyagi, A. 2004. Physiology and molecular biology of salinity stress tolerance in plants. *Curr. Sci., Bangalore.* 86: 407–421.

Sarrou, E., Chatzopoulou, P., Therios, I., Dimassi-Theriou, K., and Koularmani, A. 2015. Effect of melatonin, salicylic acid and gibberellic acid on leaf essential oil and other secondary metabolites of bitter orange young seedlings. *JEOR.* 27: 487–496.

Shi, H., and Chan, Z. 2014. The cysteine2/histidine2-type transcription factor zinc finger of Arabidopsis thaliana 6-activated C-repeat-binding factor pathway is essential for melatonin-mediated freezing stress resistance in Arabidopsis. *J. Pineal Res.* 57: 185–191.

Shi, H., Chen, Y., Tan, D. X., Reiter, R. J., Chan, Z., and He, C. 2015d. Melatonin induces nitric oxide and the potential mechanisms relate to innate immunity against bacterial pathogen infection in Arabidopsis. *J. Pineal Res.* 59: 102–8.

Shi, H., Jiang, C., Ye, T., Tan, D. X., Reiter, R. J., Zhang, H., Liu, R., and Chan, Z. 2015c. Comparative physiological, metabolomic, and transcriptomic analyses reveal mechanisms of improved abiotic stress resistance in bermudagrass [*Cynodon dactylon* (L). Pers.] by exogenous melatonin. *J. Exp. Bot.* 66: 681–694.

Shi, H., Reiter, R. J., Tan, D. X., and Chan, Z. 2015e. Indole-3-Acetic Acid Inducible 17 positively modulates natural leaf senescence through melatonin-mediated pathway in Arabidopsis. *J. Pineal Res.* 58: 26–33.

Shi, H., Tan, D. X., Reiter, R. J., Ye, T., Yang, F., and Chan, Z. 2015a. Melatonin induces class A1 heat-shock factors (HSFA1s) and their possible involvement of thermotolerance in Arabidopsis. *J. Pineal Res.* 58: 335–342.

Shi, H., Wang, X., Tan, D. X., Reiter, R. J., and Chan, Z. 2015b. Comparative physiological and proteomic analyses reveal the actions of melatonin in the reduction of oxidative stress in Bermuda grass (Cynodon dactylon (L). Pers.). *J. Pineal Res.* 59: 120–131.

Smirnoff, N. 1998. Plant resistance to environmental stress. *Curr. Opin. Biotechnol.* 9: 214–219.

Szafrańska, K., Glińska, S., and Janas, K. M. 2013. Ameliorative effect of melatonin on meristematic cells of chilled and re-warmed *Vigna radiata* roots. *Biol. Plant.* 57: 91–96.

Szafrańska, K., Szewczyk, R., and Janas, K. 2014. Involvement of melatonin applied to *Vigna radiata* L. seeds in plant response to chilling stress. *Open Life Sci.* 9: 1117–1126.

Tester, M., and Davenport, R. 2003. Na$^+$ tolerance and Na$^+$ transport in higher plants. *Annals Bot. 91*: 503–527.

Turk, H., and Erdal, S. 2015. Melatonin alleviates cold-induced oxidative damage in maize seedlings by up-regulating mineral elements and enhancing antioxidant activity. *J. Plant Nutr. Soil Sci.* 178: 433–439.

Turk, H., Erdal, S., Genisel, M., Atici, O., Demir, Y., and Yanmis, D. 2014. The regulatory effect of melatonin on physiological, biochemical and molecular parameters in cold-stressed wheat seedlings. *Plant Gr. Reg.* 74: 139–152.

Uchendu, E. E., Shukla, M. R., Reed, B. M., and Saxena, P. K. 2013. Melatonin enhances the recovery of cryopreserved shoot tips of American elm (*Ulmus americana* L.). *J. Pineal Res.* 55: 435–442.

Wang, P., Sun, X., Wang, N., Tan, D. X., and Ma, F. 2015. Melatonin enhances the occurrence of autophagy induced by oxidative stress in Arabidopsis seedlings. *J. Pineal Res.* 58: 479–489.

Wang, P., Sun, X., Xie, Y., Li, M., Chen, W., Zhang, S., Liang, D., and Ma, F. 2014. Melatonin regulates proteomic changes during leaf senescence in *Malus hupehensis*. *J. Pineal Res.* 57: 291–307.

Wang, P., Yin, L., Liang, D., Li, C., Ma, F., and Yue, Z. 2012. Delayed senescence of apple leaves by exogenous melatonin treatment: Toward regulating the ascorbate–glutathione cycle. *J. Pineal Res.* 53: 11–20.

Weeda, S., Zhang, N., Zhao, X., Ndip, G., Guo, Y., Buck, G. A., Fu, C., and Ren, S. 2014. Arabidopsis transcriptome analysis reveals key roles of melatonin in plant defense systems. *PloS One.* 2014: 9: e93462.

Wei, W., Li, Q. T., Chu, Y. N., Reiter, R. J., Yu, X. M., Zhu, D. H., Zhang, W. K., Ma, B., Lin, Q., Zhang, J. S., and Chen, S. Y. 2014. Melatonin enhances plant growth and abiotic stress tolerance in soybean plants. *J. Exp. Bot.* 2014: doi: 10.1093/jxb/eru392.

Yin, L., Wang, P., Li, M., Ke, X., Li, C., Liang, D., Wu, S., Ma, X., Li, C., Zou, Y., and Ma, F. 2013. Exogenous melatonin improves Malus resistance to Marssonina apple blotch. *J. Pineal Res.* 54: 426–434.

Zhang, H. J., Zhang, N., Yang, R. C., Wang, L., Sun, Q. Q., Li, D. B., Cao, Y. Y., Weeda, S., Zhao, B., Ren, S., and Guo, Y. D. 2014. Melatonin promotes seed germination under high salinity by regulating antioxidant systems, ABA and GA4 interaction in cucumber (*Cucumis sativus* L.). *J. Pineal Res.* 57: 269–279.

Zhang, N., Zhao, B., Zhang, H. J., Weeda, S., Yang, C., Yang, Z. C., Ren, S., and Guo, Y. D. 2013. Melatonin promotes water-stress tolerance, lateral root formation, and seed germination in cucumber (*Cucumis sativus* L.). *J. Pineal Res.* 54: 15–23.

Zhao, H., Xu, L., Su, T., Jiang, Y., Hu, L., and Ma, F. 2015. Melatonin regulates carbohydrate metabolism and defenses against *Pseudomonas syringae* pv. tomato DC3000 infection in *Arabidopsis thaliana*. *J. Pineal Res.* 59: 109–119.

Zhao, Y., Qi, L. W., Wang, W. M., Saxena, P. K., and Liu, C. Z. 2011. Melatonin improves the survival of cryopreserved callus of *Rhodiola crenulata*. *J. Pineal Res.* 50: 83–88.

Zuo, B., Zheng, X., He, P., Wang, L., Lei, Q., Feng, C., Zhou, J., Li, Q., Han, Z., and Kong, J. 2014. Overexpression of *MzASMT* improves melatonin production and enhances drought tolerance in transgenic *Arabidopsis thaliana* plants. *J. Pineal Res.* 57: 408–417.

13

Role of Serotonin and Melatonin in Agricultural Science and Developments in Transgenics

Biao Gong and Qinghua Shi
Shandong Agricultural University
Tai'an, China

CONTENTS

ABSTRACT This review will attempt to provide an overview of serotonin and melatonin, and to compile a practically complete account of the roles of these two compounds in agricultural science and developments in transgenics. The common biosynthetic pathways shared by auxin and melatonin possibly suggest a coordinated regulation in plants. More specifically, our knowledge to date of the role of melatonin in the stress tolerance and developmental physiology of plants is presented in detail. The most interesting and promising aspects for future physiological studies are also presented.

KEY WORDS: *agricultural science, melatonin, serotonin, transgenics.*

Abbreviations

AcSNMT:	acetylserotonin N-methyltransferase
CRTISO:	carotenoid isomerase
HIOMT:	hydroxyindole-O-methyltransferase
IAA:	indole-3-acetic acid
IAAld:	indole-3-acetaldehyde
IAM:	indole-3-acetamide
IAN:	indole-3-acetonitrile
IBA:	indole-3-butric

IPA: indole-3-pyruvic acid
PSY1:   phytoene synthase1
RNA-seq: RNA deep-sequencing technology
SAM: S-adenosyl-L-methionine
SNAT: serotonin N-acetyltransferase
T5H: tryptamine-5-hydroxylase
TDA: tryptamine deaminase
TrpDC: L-tryptophan decarboxylase

Introduction

We begin this review by emphasizing the roles of serotonin and melatonin, and their contributions, especially melatonin, to agricultural science and genetic engineering in the future. With this aim in mind, several questions regarding their developmental tendencies remain unanswered and should be reviewed. Thus, the origins and metabolism of serotonin and melatonin in plants, their potential roles in agricultural production, their perspectives in nutritional and agricultural science, and the transgenic modification of their metabolism in plants will be discussed and reviewed in this chapter.

Serotonin and Melatonin in Plants: Origins and Metabolism

Phytomelatonin was first identified in higher plants in 1993 by radioimmunoassay and gas chromatography with mass spectrometry in *Convolvulaceae* ivy morning glory (*Pharbitis nil* L., syn. *Ipomoea nil* L.) and in tomato fruits (*Solanum lycopersicum* L.) (Vantassel et al., 1993). Since then, successive studies have quantified the presence of melatonin in many plants (Table 13.1) and it is now accepted that melatonin is present in animals, plants, and all the other kingdoms (Arnao, 2014; Vantassel et al., 1993; Ramakrishna et al., 2012b,c; Ramakrishna, 2015).

As the conjoint precursor, L-tryptophan is converted to serotonin and melatonin via tryptamine in plants (but not 5-hydroxytryptophan, as in animals) in the following process: L-tryptophan decarboxylase (*TrpDC*) converts L-tryptophan to tryptamine, which is catalyzed to serotonin by tryptamine-5-hydroxylase (*T5H*); next, serotonin N-acetyltransferase (*SNAcT*) catalyzes the conversion of serotonin to N-acetylserotonin, and this is changed to melatonin by acetyloserotonin N-methyltransferase (*AcSNMT*) or hydroxyindole-O-methyltransferase (*HIOMT*) (Figure 13.1). Genes of *TrpDC*, *T5H*, and *AcSNMT* have been cloned and expressed, while *SNAcT* has not yet been characterized in plants (Kang et al., 2009; Park, 2011; Park et al., 2013a; Ramakrishna et al., 2011). It has been suggested that in plants the pathway of this indoleamine could be more complex than was earlier thought (Park et al., 2012). With regard to the respective amounts of products, it has been concluded that the dominant pathway is that of tryptophan via 5-hydroxytryptophan to serotonin, and then N-acetyloserotonin formed by *SNAcT* and converted to melatonin by HIOMT, with the use of S-adenosyl-L-methionine (SAM) as a methyl group donor (Murch and Saxena, 2002). However, at a certain rate, an alternative sequence, serotonin via 5-methoxytryptamine to melatonin, is also possible (Hardeland and Poeggeler, 2003).

Importantly, L-tryptophan is not only the precursor of serotonin and melatonin but also of IAA and other auxins, including indole-3-butyric (IBA) and p-hydroxyphenylacetic acids. IAA may be synthesized via several L-tryptophan-dependent pathways (Figure 13.1) with (1) indole-3-pyruvic acid (IPA) and indole-3-acetaldehyde (IAAld), (2) indole-3-acetamide (IAM), or (3) indole-3-acetonitrile (IAN) as intermediate products. In addition, IAA can also be synthesized from tryptamine catalyzed by tryptamine deaminase (TDA), and this intermediate is directly connected with serotonin and melatonin biosynthesis, too.

TABLE 13.1

Data on the Presence and Content of Melatonin in Higher Plants

Family	Common Name	Species Name	Melatonin Content (pg g^{-1} FW)
Adoxaceae	Laurestine	*Viburnum tinus* L.	633 L
Amaranthaceae	Red goosefoot	*Chenopodium rubrum* L.	240 S
Anacardiaceae	Pistachio	*Pistacia lentiscus* L.	581 L; 536 F
	Cyprus turpentine	*Pistacia palaestina* Boiss.	498 L
Arecaceae	Date palm	*Phoenix dactylifera* L.	469 L
Asparagaceae	Wild asparagus	*Asparagus horridus* L.	142 L
	Jew's myrtle	*Ruscus aculeatus* L.	954 L
Asteraceae	Feverfew	*Tanacetum parthenium* L.	1,3–1,7 (µg g^{-1}) L
	Yarrow	*Achillea millefolium* L.	340,000 L + S
Brassicaceae	Thale cress	*Arabidopsis thaliana* L.	480,000 L; 4,400 Se
Caprifoliaceae	Honeysuckle	*Lonicera etrusca* Santi	521 L
Convolvulaceae	Ivy morning glory	*Ipomoea nil* L.	6 S
Ephedraceae	Joint fir	*Ephedra campylopoda* L.	178 L
Fabaceae	Chinese licorice	*Glycyrrhiza uralensis* Fisch.	250 L; 34 (µg g^{-1}) R
	White lupin	*Lupinus albus* L.	8,000–55,000 R; 10–75,000 L; 3,830 Se
Hypericaceae	St John's wort	*Hypericum perforatum* L.	1,8–23 (µg g^{-1}) L; 1,7 (µg g^{-1}) Fl
Lamiaceae	Sage	*Salvia officinalis* L.	280–400 L
Lauraceae	Laurel	*Laurus nobilis* L.	8,331 L; 3,710 F
Meliaceae	White cedar	*Melia azedarach* L.	1,579 L; 585 F
Moraceae	Common fig	*Ficus carica* L.	12,915 L; 3,959 F
	Mulberries	*Morus* spp. L.	990 L; 1,000–33,000 L
Myrtaceae	Pineapple guava	*Feijoa sellowiana* O. Berg	1,529 L
	True myrtle	*Myrtus communis* L.	291 L
		Myrtus spp.	490 L
Oleaceae	Olive	*Olea europaea* L.	4,306 L; 532 F
	Mock privet	*Phillyrea latifolia*	6,337 L; 589 F
Poaceae	Rice plant	*Oryza sativa* L.	100 L; 500 S; 200 R; 400 Fl
	Barley plant	*Hordeum vulgare* L.	500–12,000 R; 82,300 S
	Canary grass	*Phalaris canariensis* L.	26,700 S
	Wheat plant	*Triticum aestivum* L.	124,700 S
	Oat plant	*Avena sativa* L.	90,600 S
Pontederiaceae	Water hyacinth	*Eicchornia crassipes* Marth.	2,900–48,000 L
Portulacaceae	Purslane	*Portulaca oleracea* L.	19,000 L
Resedaceae	Arabic desert shrub	*Ochradenus baccatus* De L.	474 L
Rhamnaceae	Mediterranean	*Rhamnus alaternus* L.	584 L; 306 F
	Buckthorn	*Rhamnus palaestina* Boiss.	1,167 L; 907 F
	Christ's thorn jujube	*Ziziphus spina-christi* L	1,324 L
Rosaceae	Mediterranean medlar	*Crataegus azarolus* L.	435 L
	Mediterranean medlar	*Crataegus aronia* L.	341 L
	Shrubby blackberry	*Rubus fruticosus* Hegetschw.	805 L
Rubiaceae	Narrow-leaved madder	*Rubia tenuifolia* D'Urv.	905 L
Santalaceae	Osyris	*Osyris alba* L.	844 L
Smilacaceae	Sarsaparilla	*Smilax aspera* L.	443 L
Solanaceae	Silverleaf nightshade	*Solanum elaeagnifolium*	7,895 F
	Black nightshade	*Solanum nigrum* L.	323 F
	Tomato plant	*Solanum lycopersicum* L.	15,000–142,000 L; 1,400–33,100 S;
	Tobacco	*Nicotiana tabacum* L.	7,100–10,200 R
	Devil's trumpet	*Datura metel* L.	50 L
			1,500 F; 15,000 Fl
Styracaceae	Drug snowbell	*Styrax officinalis* L.	4,069 L
Verbenaceae	Common lantana	*Lantana camara* L.	389 L

Source: Arnao, *Advances in Botany*, 2014, e815769(2).

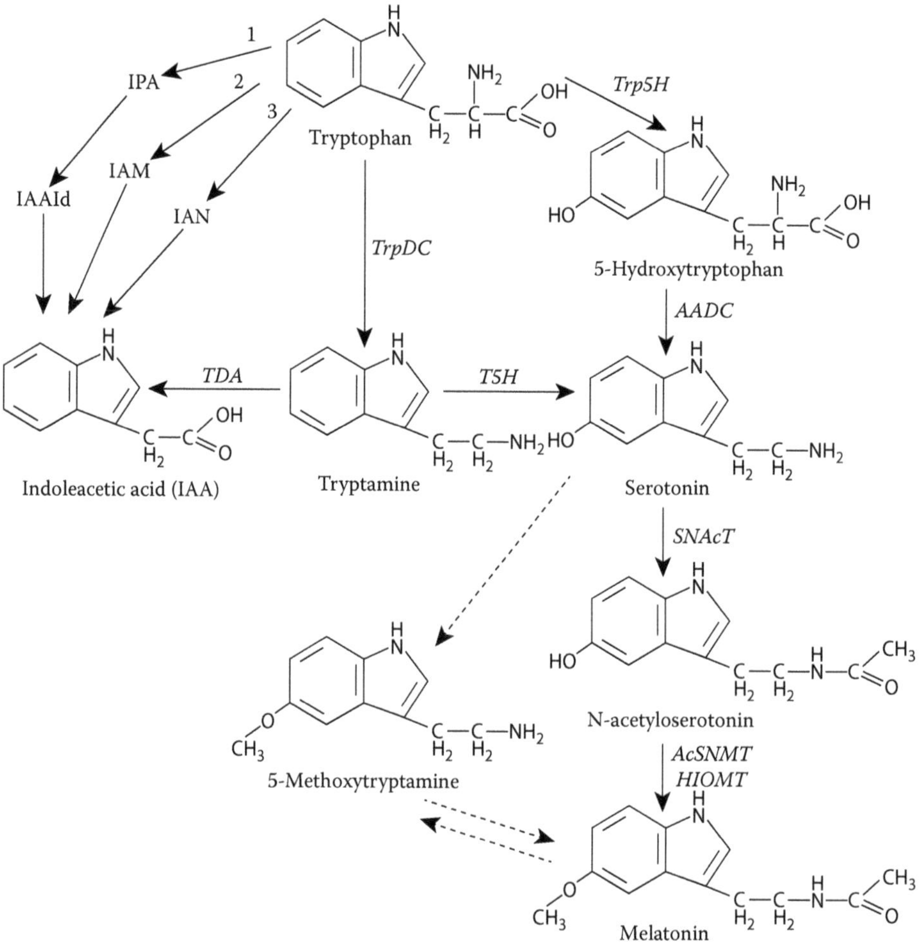

FIGURE 13.1 Biosynthetic pathway of serotonin and melatonin in plants. Enzymes involved: tryptophan 5-hydroxylase (*Trp5H*), aromatic L-amino acid decarboxylase (*AADC*), L-tryptophan decarboxylase (*TrpDC*), tryptamine 5-hydroxylase (*T5H*), tryptamine deaminase (*TDA*), serotonin N-acetyltransferase (*SNAcT*), acetyloserotonin N-methyltransferase (*AcSNMT*), hydroxyindole-O-methyltransferase (*HIOMT*). The basic tryptophan-dependent pathways of IAA biosynthesis with intermediates IPA, IAAld, IAM, and IAN are numbered. (From Posmyk et al., *Acta Physiologiae Plantarum*, 31, 1–11, 2009.)

Potential Roles of Serotonin and Melatonin in Improving Agricultural Production

Recent studies have indicated that serotonin and melatonin are ubiquitous in living organisms and exhibit pleiotropic biological activities in plant species. However, research on melatonin is in its infancy in agricultural science, and the functional significance of this indoleamine remains unestablished in diverse plant tissues. As a result, the roles of serotonin and melatonin in agricultural science are still ambiguous and they are difficult to apply to agricultural production. So, we would like to arrange the relative information about serotonin and melatonin in agricultural production systems. The limited number of papers published on the subject cover the various approaches taken toward the discovery of a possible role for serotonin and melatonin in plants. These approaches look at their roles in (1) growth and development, (2) abiotic stress tolerance, and (3) controlling plant diseases and insect pests. In this section, we present the most relevant data for each.

Regulation of Plants Growth and Development

The chemical structure (indole derivative) and biosynthetic pathway (from tryptophan) of melatonin suggest that it might be similar to IAA in several functions. IAA is an auxin involved in many physiological actions and is also a growth promoter (Arnao and Hernández-Ruiz, 2006; Ramakrishna et al., 2009, 2012a).

As regards the numerous hypotheses concerning melatonin's possible role as a hormonal agent, we recently demonstrated that melatonin does in fact act as a growth promoter in plants. It acts as a growth promoter in etiolated lupin (*Lupinus albus*) in a similar way to auxin (IAA), and induces the active growth of hypocotyls at micromolar concentrations while having an inhibitory effect at high concentrations. The growth-promoting effect of melatonin is 63% that of IAA, which is a considerable auxinic effect (Hernández-Ruiz et al., 2004). In order to widen their inquiries, Hernández-Ruiz et al. (2005) indicated that melatonin acts as a growth-stimulating compound in four kinds of monocot species, including wheat (*Triticum aestivum* L.), oat (*Avena sativa* L.), barley (*Hordeum vulgare* L.), and canary grass (*Phalaris canariensis* L.). In particular, the role of melatonin as a growth promoter is extended to coleoptiles of canary grass, wheat, barley, and oat, in which it shows a relative IAA activity of between 10% and 55%. In addition, melatonin is seen to have an important inhibitory growth effect on roots, similar to that of IAA. Quantitation by liquid chromatography with electrochemical detection and identification by tandem mass spectrometry of melatonin and IAA in etiolated coleoptiles of the monocots assayed showed that both compounds are present in similar levels in these tissues. These results point to the coexistence of auxin and melatonin in tissues and raise the possibility of their co-participation as auxinic hormones in some physiological actions in plants. Based on the functional relationship between melatonin and IAA as reported in previous studies, melatonin may be able to interfere with typical auxin functions such as the regulation of morphogenesis *in vitro*, the promotion of coleoptile growth, and the inhibition of root elongation. Murch et al. (2001) applied serotonin and melatonin to a culture medium. In subsequent research, six inhibitors of auxin and indoleamine action or transport—2,3,5-triiodobenzoic acid, p-chlorophenoxyisobutyric acid, p-chlorophenylalanine, D-amphetamine, fluoxetine, and methylphenidate—were included in a culture medium in the presence or absence of IAA. Interestingly, significant reductions in *de novo* root regeneration were found to correspond with decreases in the pool of both IAA and melatonin. An increase in the endogenous concentration of melatonin was correlated with an increase in *de novo* root formation, and increased serotonin levels corresponded to increased shoot formation on the explants. These results provide powerful evidence that a balance of the endogenous concentration of serotonin and melatonin may modulate plant morphogenesis *in vitro*.

The degree of root branching impacts the efficiency of water uptake, nutrient acquisition, and anchorage by plants. A well-developed root system is essential for the vegetative growth and fruit development of plants. In recent years, the plant root system has emerged as a central focus of research in many laboratories across the world; and researchers have successively reported the effects of serotonin and melatonin on root architecture in different crops found in agricultural production systems. The systematic research for melatonin in root development was initiated by Arnao et al. (2007), who investigated the possible effect of melatonin on the regeneration of lateral and adventitious roots in etiolated hypocotyls of *Lupinus albus* L. compared with the effect of IAA. In this study, 6-day-old derooted lupin hypocotyls immersed in several melatonin or IAA concentrations were used to induce roots. Arnao et al. observed that both melatonin and IAA induced the appearance of root primordia from pericycle cells, modifying the pattern of distribution of adventitious or lateral roots, the time course, the number and length of adventitious roots, and the number of lateral roots (Figure 13.2). Melatonin and IAA were detected and quantified in lupin primary roots, where both molecules were present in similar concentrations. The researchers concluded that this demonstrated the physiological effect of exogenous melatonin as a root promoter, its action being similar to that of IAA. From a similar viewpoint, Chen et al. (2009) observed that 0.1 μM melatonin had a stimulatory effect on root growth, while 100 μM was inhibitory. Furthermore, the stimulatory effect was only detectable in young (2-day-old) seedlings. Older (4-day-old) seedlings appeared to be less susceptible to both the stimulatory and the inhibitory effects of melatonin. Exogenous application of 0.1 μM melatonin also raised the endogenous levels of free IAA in roots, while higher concentrations had no significant effect. The specific mechanism that causes exogenous

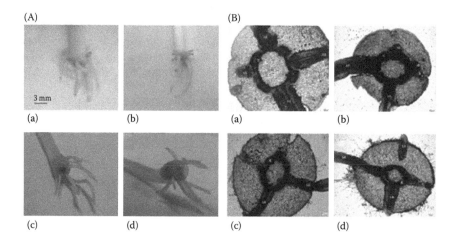

(A)   (B)

(a)   (b)   (a)   (b)

(c)   (d)   (c)   (d)

FIGURE 13.2 Images of the root regeneration (A) and transverse sections (B) induced by melatonin or IAA in 6-day-old etiolated lupin after 8 days of treatment: (a) adventitious root induction by 10 μM melatonin; (b) idem by 1 μM IAA; (c) lateral root induction by 1 μM melatonin; (d) idem by 1 μM IAA. (From Arnao et al., *Journal of Pineal Research*, 42, 147–152, 2007.)

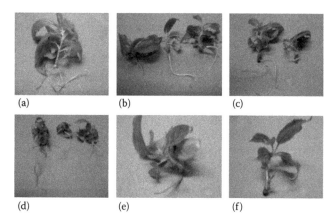

(a)   (b)   (c)

(d)   (e)   (f)

FIGURE 13.3 Formation of roots in the rootstock PHL-C: (a) control; (b) 0.05 μM melatonin; (c) 0.5 μM melatonin; (d) 1 μM melatonin; (e) 5 μM melatonin; (f) 10 μM melatonin. (From Sarropoulou et al., *Plant Physiology and Biochemistry*, 61, 162–168, 2012a.)

melatonin to increase the amount of free IAA in roots, paired with a stimulation of root growth, remains to be uncovered. As a potential hormone, melatonin, irrespective of its concentration, had a negative effect on the number of roots (Sarropoulou et al., 2012a,b). However, the application of 0.5 μM melatonin significantly increased the root length, while 1 μM melatonin increased the root length 2.5 times and the fresh weight of the roots 4 times, in comparison to the control. Although 0.05 μM melatonin increased rooting by 11.11%, 5 μM melatonin had a significant reduction on the number and fresh weight of roots and the rooting percentage (Figure 13.3). Progressively more data support the idea that a primary function of melatonin in plants is to exhibit effects similar to auxin. This suggestion is also substantiated by Sarropoulou et al.'s (2012a,b) observation that root regeneration was significantly regulated by melatonin and auxin in sweet cherry rootstock (*Prunus cerasus* L.) plants. Their data show that melatonin has a promotional effect on rooting at low concentrations but an inhibitory effect on growth at high concentrations. In the absence of auxin, 1 μM melatonin induced an auxinic response in terms of the number and length of roots, but 10 μM melatonin was inhibitory to rooting in all the tested rootstocks. The final conclusion of this experiment was that exogenously applied melatonin acted as a rooting promoter and its action was similar to that of IAA.

Other studies have pointed out that melatonin regulates plant root system architecture likely acting independently of auxin signaling. Pelagio-Flores et al. (2012) found that melatonin modulated the root system architecture by stimulating lateral and adventitious root formation but minimally affected primary root growth or root hair development. They used the auxin-responsive marker constructs *DR5:uidA*, *BA3:uidA*, and *HS:AXR3NT-GUS* to investigate the auxin activity of melatonin in *Arabidopsis* roots. To their surprise, melatonin neither activated auxin-inducible gene expression nor induced the degradation of *HS:AXR3NT-GUS*, indicating that root developmental changes elicited by melatonin were independent of auxin signaling. In recent years, several differential screening techniques, such as RNA deep-sequencing technology (RNA-seq), have made it possible to characterize genes that are differentially expressed after melatonin treatments. Recently, Zhang et al. (2014) used RNA-seq technology to rapidly identify and analyze the effects of melatonin on cucumber lateral root formation. The RNA-seq analysis generated 17,920 genes, which provided abundant data for the analysis of melatonin-mediated lateral root formation. These genes were significantly enriched in 57 KEGG pathways and 16 GO terms. Based on their expression pattern, several transcription factor families might play important roles in lateral root formation processes. A number of genes related to cell wall formation, carbohydrate metabolic processes, oxidation–reduction processes, and catalytic activity also showed different expression patterns as a result of melatonin treatments. Based on the preceding studies, we think that serotonin and melatonin could be used in agricultural production: for example, to promote root elongation in some vegetable crops with weaker root systems (e.g., Cucurbitaceae, including cucumber, watermelon, and melon) or in some crops that are grown in droughty, saline–alkaline, water logged, or sticky soils; or to promote the regeneration of adventitious roots and survival in some cutting garden plants; and so on (Figure 13.4).

More recently, Wei et al. (2014) showed that exogenous spraying of melatonin could significantly enhance growth and yield in soybean plants. This study showed that melatonin-treated plants grew bigger than control seedlings (Figure 13.5a,b). After harvest, yield-related traits were measured: melatonin-treated plants produced more pods, more seeds, and more yield than control plants (Figure 13.5c–e). The results suggest that melatonin improves plant growth and soybean production under field conditions. However, a similar but negative study indicated enhanced seedling growth but low grain yield in melatonin-rich transgenic (MRT) rice (Byeon and Back, 2014). The average biomass in the transgenic lines was two times higher than that of the wild-type lines, suggesting that melatonin increased seedling vigor in the transgenic rice compared with the wild-type rice (Figure 13.6). Regrettably, the grain weight of the MRT rice decreased by 33% on average (Figure 13.6). So, the mechanism by which melatonin affects crop production still needs to be clarified, and the rational utilization of melatonin should also be considered carefully.

As we know, melatonin is a natural health supplement in organisms. Is it possible that melatonin can be used in the postharvest ripening of horticultural crops? Sun et al. (2014) observed that exogenous melatonin treatment significantly increased lycopene levels compared with a control (Figure 13.7a,c). Meanwhile, the key genes involved in fruit color development, including phytoene synthase1 (*PSY1*) and carotenoid isomerase (*CRTISO*), showed a twofold increase in expression levels (Figure 13.7b). In addition, the application of 50 µM melatonin enhanced fruit softening, increased water-soluble pectin, and

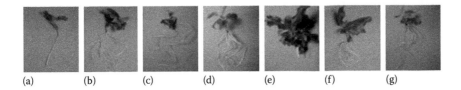

(a)   (b)   (c)   (d)   (e)   (f)   (g)

FIGURE 13.4   (a) Formation of roots in the rootstock Gisela 6 in the absence of melatonin and auxin (control); (b) promotion of rooting in the presence of 1 µM melatonin compared with control; (c) rooting with IAA (5.71 µM); (d) increase in the number of roots per rooted explant in the treatment IAA plus 0.1 µM melatonin compared with IAA alone; (e) significant inhibition of rooting in the treatment IAA plus 10 µM melatonin; (f) rooting with IBA (4.92 µM); (g) elongation of roots with 10 µM melatonin plus IBA (4.92 µM) in comparison with IBA without melatonin. Color figures only from rootstock Gisela 6 are included. (From Sarropoulou et al., *Journal of Pineal Research*, 52, 38–46, 2012b.)

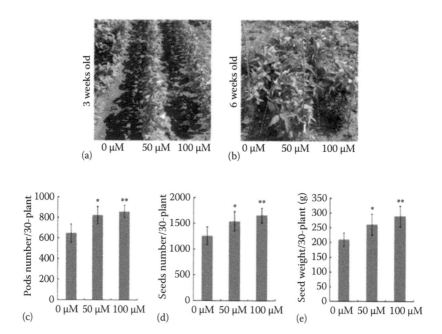

FIGURE 13.5 Melatonin effects on soybeans grown under field conditions: (a) comparison of 3-week-old plants grown in the field; (b) phenotypes of 6-week-old plants; (c) comparison of pod numbers from 30 plants; (d) seed numbers from 30 plants; (e) seed weight from 30 plants. (From Byeon et al., *Journal of Pineal Research*, 56, 408–414, 2014.)

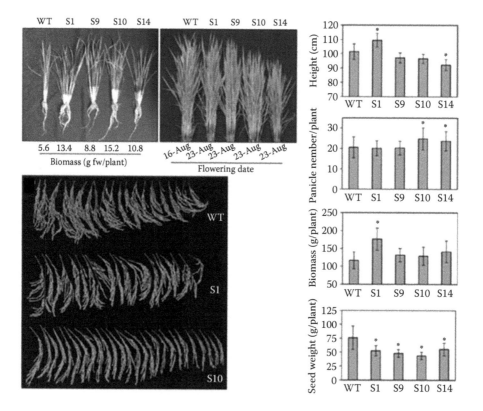

FIGURE 13.6 Plant growth and crop yield of wild-type (WT) rice and melatonin-rich transgenic (S1–S14) plants. (From Wei et al., *Journal of Experimental Botany*, 66, 695–707, 2014.)

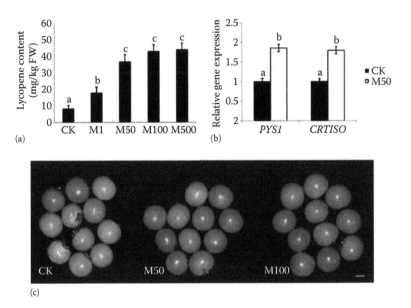

(a)   (b)   (c)

FIGURE 13.7 Melatonin's effect on tomato ripening and pigment accumulation. CK: fruits pretreated with water; M1–M500: fruits pretreated with 1, 50, 100, 500 μM melatonin separately. (a) Lycopene concentration at 17 days after treatment; (b) real-time PCR analysis of PSY1 and CRTISO expression levels in fruit at 17 days after treatment; (c) phenotype of tomato at 17 days after treatment. (From Sun et al., *Journal of Experimental Botany*, 66, 657–668, 2014.)

decreased protopectin, which might be attributed to increased ethylene (ET) biosynthesis and perception by melatonin.

Leaf senescence is a highly regulated, ordered series of events involving the cessation of photosynthesis, the disintegration of chloroplasts, the breaking down of leaf proteins, the loss of chlorophyll, and the removal of amino acids. As we know, some stresses including drought, high temperatures, low temperatures, and disease cause leaf senescence, which usually reduces agricultural production. Melatonin has an important anti-aging role in plant physiology. Wang et al. (2013a) tested the effects of long-term melatonin exposure on metabolic status and protein degradation during natural leaf senescence in apple trees. They found that leaves from treated plants had significantly higher photosynthetic activity, chlorophyll concentrations, and levels of photosynthetic end products when compared with the control. The plants also exhibited better preservation of their nitrogen, total soluble protein, and Rubisco protein concentrations than the control. The slower process of protein degradation might be a result of melatonin-linked inhibition on the expression of autophagy-related apple genes. Melatonin keeps plant leaves vibrant not only in normal conditions, but also in detrimental or stressed conditions. Recent studies have indicated that melatonin has a delaying effect on leaf senescence under dark and long-term slight water deficit conditions (Figure 13.8) (Arnao et al., 2009; Wang et al., 2012, 2013b). However, the relationship between melatonin and aging and the overall changes and regulation of proteome profiling by long-term melatonin exposure during leaf senescence is not well understood. Wang et al. (2014a,b) extended the investigation of the mechanism of melatonin against leaf senescence by proteomics analysis. The proteomics data showed that 622 and 309 proteins, respectively, were altered by senescence, and melatonin involved in major metabolic processes exhibited hydrolase activity and was mainly located in the plastids. These proteins were classified into several senescence-related functional categories, including degradation of macromolecules, redox and stress responses, transport, photosynthesis, development, and other regulatory proteins. They found that melatonin treatment led to the downregulation of proteins that are normally upregulated during senescence. The melatonin-related delay in senescence might have occurred due to the altering of proteins involved in associated processes. In view of the anti-senescent effects of melatonin, we consider exogenous melatonin to be useful as a healthy preservative for fresh vegetables, fruits, cutting flowers, and so on.

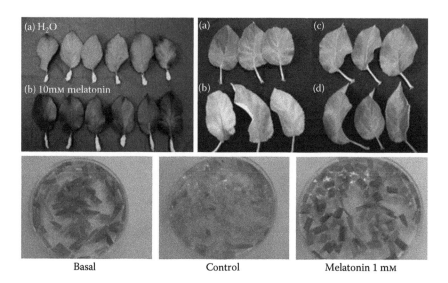

FIGURE 13.8 (Left) Effects of melatonin on phenotype of detached, fully mature apple leaves at day 12 after dark treatment: (a) leaves treated with double-distilled water as control; (b) leaves treated with melatonin. (From Wang et al., *Journal of Pineal Research*, 53, 11–20, 2012.) (Right) Effects of melatonin on leaf yellowing after 90 days of treatment: (a) well-watered control; (b) drought control; (c) melatonin-applied/well-watered treatment; (d) melatonin-applied/drought treatment. (From Wang et al., *Journal of Pineal Research*, 54, 292–302, 2013b.) (Bottom) Effects of melatonin treatment with respect to control on senescence of barley leaf sections after 48 h in the dark. (From Arnao et al., *Journal of Pineal Research*, 46, 58–63, 2009.)

In addition, Park et al. (2013c) have suggested that the induction of melatonin during flower development is regulated by the transcriptional control of its biosynthesis genes and that melatonin may participate in flower development. Melatonin and serotonin were found at the highest levels in the least developed flower buds of *Datura metel* L. with decreasing concentrations as the flower buds matured. In the developing fruit, melatonin was present at relatively stable, high concentrations for the first 10 days after anthesis. After 10–15 days, the ovule had grown to a sufficient size for excision and analysis, and melatonin was found to be at its highest concentrations in the developing ovule, with minimal concentrations of neuroindoles in the fleshy fruit. Together, these data indicate that melatonin may play a role in protecting the reproductive tissues during flower and seed formation in plant species.

Application Values of Serotonin and Melatonin in Overcoming Abiotic Stress in Plants

Temperature stress, of both high and low temperatures, is one of the most destructive environmental stresses and considerably restricts the efficiency and quality of products growing all over the world. Temperature stress brings about a wide range of physiological, morphological, and anatomical disruptions in plants. The major negative effect of temperature stress is that it causes cell death usually associated with membrane dysfunction, chloroplast disintegration, enzyme inactivation, and chromatin fragmentation (Turk et al., 2014). To date, at least eight reports have indicated the role of melatonin in alleviating temperature stress tolerance in plants (Figure 13.9). Cold stress mitigation by melatonin was first reported in carrot suspension cells (*Daucus carota* L.) by Lei et al. (2004). Pretreatment with 43–86 nM melatonin significantly attenuated cold-induced apoptosis in carrot suspension cells. The antiapoptotic effect of melatonin was initially thought to be a result of its antioxidant actions. However, ROS generation remained unaffected by melatonin treatment, suggesting that melatonin's protective role is not related to its role as a direct ROS scavenger. At the same time, notable increases in putrescine and spermidine levels were observed in melatonin-treated cells, which may be responsible for the alleviation

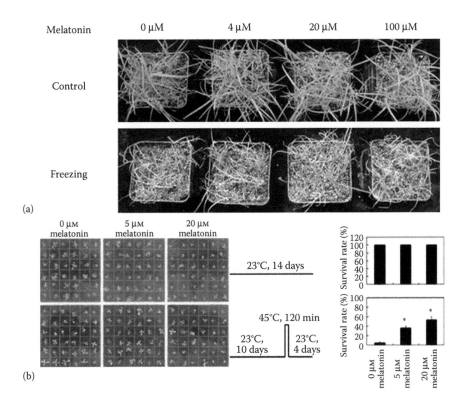

FIGURE 13.9   (a) Effects of melatonin on phenotype of bermudagrass (*Cynodon dactylon* L.) under control and freezing treatments. (From Shi et al., *Journal of Experimental Botany*, 66, 681–694, 2015a.) (b) Effects of melatonin on phenotype of *Arabidopsis* under control and heat treatments. (From Shi et al., *Journal of Pineal Research*, 58, 335–342, 2015b.)

of cold-induced apoptosis. Melatonin protected the membrane structures of cucumber seeds against peroxidation during chilling, but excessive melatonin levels provoked oxidative changes in proteins (Posmyk et al., 2009). In addition, melatonin can improve plant resistance to cold stress in wheat (Turk et al., 2014) and bermudagrass (Shi et al., 2015a) seedlings by directly scavenging ROS and by modulating redox balance and other defense mechanisms. There is still a lack of information clearly explaining the role of melatonin in plant physiology. This molecule acts multidirectionally and is usually aligned with other compounds. To aid in the understanding of the role of melatonin in alleviating cold stress, Bajwa et al. (2014) investigated the effects of melatonin treatment on the expression of cold-related genes in *Arabidopsis*. Melatonin upregulated the expression of C-repeat-binding factors/drought response element–binding factors (*CBFs/DREBs*), a cold-responsive gene (*COR15a*), a transcription factor involved in freezing and drought-stress tolerance (*CAMTA1*), and transcription activators of ROS-related antioxidant genes (*ZAT10* and *ZAT12*) following cold stress. The upregulation of cold-signaling genes by melatonin may stimulate the biosynthesis of cold-protecting compounds and contribute to the increased growth of plants treated with exogenous melatonin under cold stress. Recently, it has been revealed that endogenous melatonin levels are significantly induced by cold stress in *Arabidopsis*. In addition, a cysteine2/histidine2-type zinc finger transcription factor (*ZAT6*)-activated CBF pathway was suggested to be involved in melatonin-mediated freezing stress response in *Arabidopsis* (Shi and Chan, 2014). Though melatonin incorporated into a priming medium significantly reversed the inhibitory effects of high temperatures on *Phacelia tanacetifolia* seed germination, the concrete roles of melatonin in promoting heat tolerance are still unclear and underdiscussed. Exhilaratingly, Shi et al. (2015b) recently found that the endogenous melatonin levels in *Arabidopsis* leaves were significantly induced by heat stress treatment, and exogenous melatonin treatment conferred improved thermotolerance in *Arabidopsis*. The transcript levels of class-A1 heat-shock factors (*HSFA1s*), which serve as the master regulators of heat stress responses, were significantly upregulated by heat stress and exogenous melatonin treatment in

Arabidopsis. Notably, exogenous melatonin largely alleviated heat stress in *HSFA1* knockout mutants, and *HSFA1*-activated transcripts of heat-responsive genes (heat stress–associated 32 [*HSA32*], heat-shock proteins 90 [*HSP90*], and 101 [*HSP101*]) might be attributed to melatonin-mediated thermotolerance. Taken together, this study provided a direct link between melatonin and thermotolerance in plants.

Under water deficit conditions, ROS form and rapidly accumulate in plants. Free radicals can damage plants by oxidizing photosynthetic pigments, membrane lipids, proteins, and nucleic acids. A comprehensive investigation was carried out to determine the changes that occurred in osmosis-stressed cucumber (*Cucumis sativus* L.) in response to melatonin treatment. Zhang et al. (2013) observed that the ultrastructure of chloroplasts in osmosis-stressed cucumber leaves was maintained after melatonin treatment. The antioxidant levels and activities of the ROS scavenging enzymes were also increased by melatonin. Exposure to drought stress was observed to decrease plant growth rate and chlorophyll content compared with a control in both bermudagrass and soybean (Figure 13.10a). Melatonin application to drought-stressed plants significantly improved these growth parameters. These results suggest that melatonin, as an important signal, is involved in drought stress tolerance in the plant kingdom. Melatonin treatment also helped preserve the internal lamellar system of chloroplasts and alleviated the ultrastructural damage induced by drought stress (Figure 13.10b). This ameliorating effect may be ascribed to the enhanced activity of antioxidant enzymes, increased levels of nonenzymatic antioxidants, and increased amount of osmoprotectants (Meng et al., 2014). In addition, the application of melatonin to wine grapes is effective in reducing drought stress by regulating stoma characteristics (Figure 13.10c).

Salinity–alkalinity has greatly limited crop production in semiarid and arid regions. There are 831 million ha of soils affected by excessive salinity–alkalinity in the world. Of this, 434 million ha are sodic soils, compared with 397 million ha of saline soils (Gong et al., 2014a). Osmotic-induced water deficiency, ion-induced nutrient imbalance, and membrane injury are generally considered to be involved in salinity–alkalinity stress on plants (Gong et al., 2014b). In addition, the high pH has a negative effect on plants under alkaline conditions (Gong et al., 2014b). Recently, melatonin, as well as serotonin, has emerged in the literature on plant response and adaption to salinity–alkalinity stress in several plant species, including *Malus hupehensis* (Li et al., 2012), *Helianthus annuus* (Mukherjee et al., 2014),

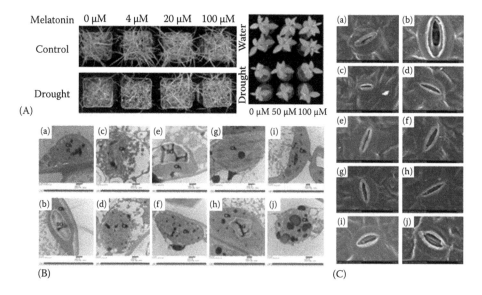

FIGURE 13.10   (a) Effects of melatonin on phenotype of bermudagrass and soybean under control and drought treatments. (From Shi et al., *Journal of Experimental Botany*, 66, 681–694, 2015a; Wei et al., *Journal of Experimental Botany*, 66, 695–707, 2014.) (b, c) Effects of melatonin treatment on ultrastructure of chloroplasts and stomatal characteristics in grape leaves under drought stress induced by 10% PEG. a, c, e, g, and i (b: top row; c: left column) and b, d, f, h, and j (b: bottom row; c: right column) represent young leaves and mature leaves, respectively, from subgroups, including those grown under control, 10% PEG, 10% PEG and 50 nM melatonin, 10% PEG and 100 nM melatonin, 10% PEG and 200 nM melatonin. (From Meng et al., *Journal of Pineal Research*, 57, 200–212, 2014.)

Cucumis sativus L. (Zhang et al., 2014), *Citrus aurantium* L. (Kostopoulou et al., 2015), and *Solanum lycopersicum* L. (Liu et al., 2015). The major role of melatonin on improving salinity–alkalinity stress tolerance depends on scavenging excess ROS (Li et al., 2012; Kostopoulou et al., 2015). Melatonin also alleviated salinity stress by affecting abscisic acid (ABA) and gibberellin acid (GA) biosynthesis and catabolism during cucumber seed germination (Zhang et al., 2014). Compared with NaCl treatment, melatonin significantly upregulated ABA catabolism genes and downregulated ABA biosynthesis genes, resulting in a rapid decrease of ABA content during the early stages of germination. At the same time, melatonin positively upregulated GA biosynthesis genes, contributing to a significant increase in GA content. However, how does salt stress influence the metabolism of serotonin and melatonin in plants? Mukherjee et al. (2014) have indicated that exogenous serotonin and melatonin treatments lead to variable effects on hypocotyl elongation and root growth under salt stress. Salt stress for 48 h increased endogenous serotonin and melatonin content in roots and cotyledons, thus indicating their involvement in salt-induced long-distance signaling from roots to cotyledons. Salt stress–induced accumulation of serotonin and melatonin exhibited differential distribution in the vascular bundles and cortex in the differentiating zones of the primary roots, suggesting their compartmentalization in the growing region of roots (Figure 13.11).

Serotonin and melatonin accumulation in oil body–rich cells of salt-treated seedling cotyledons correlates with the longer retention of oil bodies in cotyledons. These results indicate the possible role of serotonin and melatonin in regulating root growth during salt stress in sunflowers. The effects of exogenous serotonin and melatonin treatments on sunflower seedlings grown in the absence or presence of 120 mM NaCl substantiates their role in seedling growth. Auxin and serotonin biosynthesis are coupled to the common precursor tryptophan. Salt stress–induced root growth inhibition thus pertains to the partial impairment of auxin functions caused by increased serotonin biosynthesis. In seedling cotyledons, salt stress modulates the activity of N-acetylserotonin O-methyltransferase, the enzyme responsible for melatonin biosynthesis from N-acetylserotonin.

In addition, we have conducted some studies about the effects of exogenous melatonin on plant growth, reactive oxygen metabolism, and ion homeostasis in alkaline-treated tomato seedlings (Liu et al., 2015). Exogenous melatonin increased the tolerance of tomato plants to sodic alkaline stress, and the effect

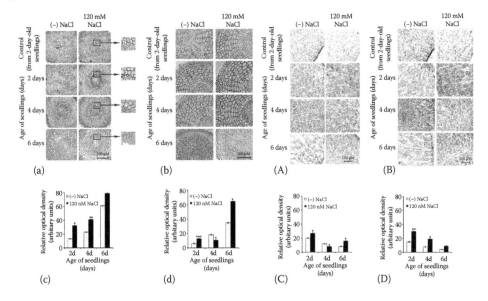

FIGURE 13.11 (Left) Immunohistochemical localization of (a) serotonin and (b) melatonin accumulation in roots of seedlings grown in the absence and presence of NaCl; (c) and (d) represent quantification of serotonin and melatonin, respectively, in various treatments. (Right) Immunohistochemical localization of (A) serotonin and (B) melatonin accumulation in cotyledons of seedlings grown in the absence and presence of NaCl; (C) and (D) represent quantification of serotonin and melatonin, respectively, in various treatments. (From Mukherjee et al., *Physiologia Plantarum*, 152, 714–728, 2014.)

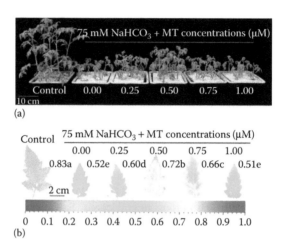

(a)

(b)

FIGURE 13.12 Effects of melatonin (MT) on phenotype (a), and maximum photochemistry efficiency of PSII (b) of tomato seedlings under 75 mM NaHCO₃ stress. (From Liu et al., *Scientia Horticulturae*, 181, 18–25, 2015.)

depended on the concentration. In the experiment, treatment with 0.5 μM melatonin showed the strongest promotional effect on tomato plant growth under alkaline stress and reduced oxidative stress by inducing the activity of antioxidant enzymes and the accumulation of ascorbic acid and glutathione, which could be partly responsible for protecting chlorophyll from degradation and maintaining higher Fv/Fm (Figure 13.12). Furthermore, we found that melatonin treatment decreased Na^+ content and increased K^+ content in tomato leaves under alkaline stress, which showed that ion homeostasis maintenance by melatonin also played an important role in enhancing tomato tolerance to alkaline stress.

It has been suggested that melatonin improves the redox state of cells, decreasing ROS and reactive nitrogen species (RNS) levels and stabilizing biological membranes, as it does in animal cells (Arnao and Hernández-Ruiz, 2014). To examine whether melatonin-rich plants can defend themselves against oxidative stress, Park et al. (2013b) subjected MRT rice plants to the singlet oxygen–generating herbicide butafenacil. Both MRT and transgenic control rice seeds germinated and grew equally well in continuous dark on half-strength Murashige and Skoog (MS) medium containing 0.1 μM butafenacil. However, after transferring the seedlings to light, the control seedlings rapidly necrotized, whereas the MRT seedlings showed resistant phenotypes. Seven-day-old MRT seedlings treated with 0.1 μM butafenacil were resistant to the herbicide and contained high chlorophyll levels and low malondialdehyde and hydrogen peroxide levels compared with the control seedlings. As they did before the herbicide treatment, the MRT plants also produced much more melatonin after the herbicide treatment than the control seedlings. In addition, the MRT plants exhibited higher superoxide dismutase and catalase activities before and after the herbicide treatment compared with the control seedlings. This result indicates that melatonin scavenges ROS efficiently *in vivo* in the transgenic plants, leading to oxidative stress resistance. Heavy-metal stress usually leads to oxidative damage.

When red cabbage (*Brassica oleracea rubrum*) seedlings pretreated with melatonin were subjected to a high level of Cu^{2+} stress, seed germination, seedling weight, and lipid peroxidation of membrane were less affected by the toxic copper ion compared with untreated plants (Posmyk et al., 2008). Interestingly, the Cd-induced synthesis of melatonin coincided with the increased expression of melatonin biosynthetic genes including tryptophan decarboxylase (*TDC*), tryptamine 5-hydroxylase (*T5H*), and N-acetylserotonin methyltransferase (AsSNMT) (Byeon et al., 2015). However, the expression of serotonin N-acetyltransferase (*SNAT*), the penultimate gene in melatonin biosynthesis, was downregulated, suggesting that melatonin synthesis was counter-regulated by *SNAT*. Notably, the induction of melatonin biosynthetic gene expression was coupled with the induction of four melatonin 2-hydroxylase (*M2H*) genes involved in melatonin degradation, which suggests that genes for melatonin synthesis and degradation are coordinately regulated. The induced *M2H* gene expression was correlated with enhanced M2H enzyme activity. Three of the M2H proteins were localized to the cytoplasm and one M2H protein was localized to chloroplasts, indicating that melatonin degradation occurs both in the cytoplasm and in chloroplasts. The biological activity of 2-hydroxymelatonin in the induction of the plant defense gene expression was 50% less than that of melatonin, which indicates that 2-hydroxymelatonin may be a metabolite of melatonin. Overall, melatonin synthesis occurs in parallel with melatonin degradation in both chloroplasts and cytoplasm, and the resulting melatonin metabolite 2-hydroxymelatonin also acts as a signaling molecule for defense gene induction.

In the most complete study to date of melatonin-mediated genetic functional analysis (Weeda et al., 2014), mRNA-seq technology was used to analyze *Arabidopsis* plants subjected to a 16-h 100 pM (low) and 1 mM (high) melatonin treatment, and the expression profiles were analyzed to identify differentially expressed genes. The 100 pM melatonin treatment significantly affected the expression of 81 genes, with 51 being downregulated and 30 upregulated. However, the 1 mM melatonin treatment significantly altered the expression of 1308 genes, with 566 being upregulated and 742 downregulated. Not all the genes that showed altered expression when subjected to low melatonin were affected by high melatonin, suggesting that melatonin plays different roles in plant growth and development regulation at low and high concentrations. Furthermore, many of the genes whose expression were altered by melatonin were involved in plant stress defense: transcript levels for many stress receptors, mitogen-activated protein kinases, and stress-associated calcium signals were upregulated. Most of the transcription factors identified were also involved in plant stress defense and in the redox network. Interestingly, chlorophyllase and PaO, which are involved in chlorophyll degradation, were both downregulated, confirming preliminary studies on the ability of melatonin to delay senescence (Arnao and Hernández-Ruiz, 2009). Furthermore, cell death–associated genes were mostly downregulated by the 100 pM melatonin treatment. In total, 183 genes involved in hormone signaling were identified in this study (Weeda et al., 2014). Genes in the ABA, ET, salicylic acid (SA), and jasmonic acid pathways were upregulated, whereas genes related to auxin responses, transport, homeostasis, and signaling, and those associated with cell wall synthesis, were mostly downregulated. Two 1-aminocyclopropane-1-carboxylate (ACC) synthases were also upregulated in response to melatonin, suggesting that melatonin may induce ET biosynthesis. One of these two ACC synthases is also auxin inducible. None of the auxin biosynthesis pathway genes were significantly altered by the melatonin treatment, with the exception of an IAA–amino synthase that conjugates amino acids to IAA. In conclusion, the results indicate that during plant growth and development melatonin plays important physiological roles in many biological processes including abiotic stress defense, photosynthesis, cell wall modification, and redox homeostasis (Weeda et al., 2014).

Potentiality of Serotonin and Melatonin in Controlling Plant Diseases and Insect Pests

Indications that serotonin and melatonin are involved in signaling defense responses during plant–pathogen interactions have been documented in some experiments during recent years. The challenge of a pathogen very often leads to the induction of the hypersensitive response (HR). There is some evidence that melatonin plays a key signaling role during HR, next to the accumulation of ROS and SA. Exogenously applied melatonin could improve the resistance of apple to Marssonina apple leaves blotch (*Diplocarpon mali*) (Figure 13.13). When plants were pretreated with melatonin, resistance was increased in their leaves, indicating the potential role for melatonin in modulating levels of ROS, as well the activities of antioxidant enzymes and pathogenesis-related proteins during these plant–pathogen interactions. Pretreatment enabled plants to maintain intracellular ROS concentrations at steady-state levels and enhance the activities of plant defense-related enzymes, possibly improving disease resistance. Defense responses and disease resistance via melatonin is a typical example of activating the expression of pathogenesis-related (PR) genes (Yin et al., 2012). The application of a 10 µM concentration of melatonin on *Arabidopsis* and tobacco leaves induced various PR genes, as well as a series of defense genes activated by SA and ET, two key factors involved in plant defense response, compared with mock-treated leaves (Lee et al., 2014). The induction of these defense-related genes in melatonin-treated *Arabidopsis* matched an increase in resistance against the bacterium by suppressing its multiplication

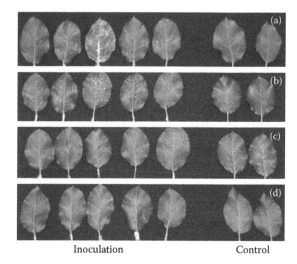

Inoculation   Control

FIGURE 13.13   Effect of melatonin concentration on apple leaf phenotype at 20 dpi. Fully mature leaves were collected from plants pretreated with (a) water only and no exogenous melatonin, (b) 0.05 mM melatonin, (c) 0.1 mM melatonin, or (d) 0.5 mM melatonin. (From Yin et al., *Journal of Pineal Research*, 54, 426–434, 2012.)

about tenfold, relative to the mock-treated *Arabidopsis*. Like melatonin, N-acetylserotonin also plays a role in inducing a series of defense genes, although serotonin does not. Furthermore, melatonin-induced PR genes were almost completely or partially suppressed in the *NPR1*, *EIN2*, and *MPK6 Arabidopsis* mutants, indicative of SA and ET dependency in melatonin-induced plant defense signaling. Lee et al. (2015) discovered that the serotonin N-acetyltransferase (*SNAT*) knockout mutant lines, which had 50% less melatonin than the wild-type lines, exhibited susceptibility to pathogen infection that coincided with decreased induction of defense genes, including *PR1*, *ICS1*, and *PDF1.2*. Because melatonin acts upstream of SA synthesis, the reduced melatonin levels in the *SNAT* mutant lines led to decreased SA levels compared with the wild-type lines, suggesting that the increased pathogen susceptibility of the *SNAT* mutant lines could be attributed to decreased SA levels and subsequent attenuation of defense gene induction. Exogenous melatonin treatment failed to induce defense gene expression in *nahG Arabidopsis* plants, but restored the induction of defense gene expression in the *SNAT* mutant lines. In addition, melatonin caused the translocation of the *NPR1* (nonexpressor of PR1) protein from the cytoplasm into the nucleus, indicating that melatonin-elicited pathogen resistance in response to avirulent pathogen attack is SA dependent in *Arabidopsis*. Because melatonin is safe and beneficial to animals and humans, exogenous pretreatment might represent a promising cultivation strategy to protect plants against pathogen infection in agricultural systems.

Perspectives of Serotonin and Melatonin in Nutritional and Agricultural Science

Melatonin is a well-known molecule that possesses many beneficial effects on human health. The beneficial effects of melatonin as a potent antioxidant and its potential to improve human health have been extensively reviewed (Paradies et al., 2010), and it has been suggested for clinical use in a variety of pathologies, including neurodegenerative diseases, heart disease, metabolic disorders, tumors, and accidental nuclear radiation (Tan et al., 2012). The significantly high levels of melatonin in popular beverages and foods could prove beneficial for some of these conditions. As we know, melatonin exists in almost all plants and plant products tested (Table 13.2). Based on this, it is probably not surprising that melatonin has also been discovered in diets that are based on plant products. Many agricultural products are a natural source of melatonin. Thus, how to improve the levels of melatonin in agricultural crops has become the focus of some interest recently.

TABLE 13.2

Data on the Presence and Content of Melatonin in Some Edible Plant Organs

Family	Common Name	Species Name	Melatonin Content (pg g^{-1} FW or DW)
Actinidiaceae	Kiwi fruit	*Actinidia deliciosa*	24.4
Amaranthaceae	Beetroot	*Beta vulgaris* L.	2
Araceae	Taro	*Colocasia esculenta* L.	54.6
Basellaceae	Indian spinach	*Basella alba* L.	38.7
Asparagaceae	Asparagus	*Asparagus officinalis* L.	9.5
Brassicaceae	Cabbage	*Brassica oleracea capitata* L.	107.4
	White radish	*Raphanus sativus* L.	657.2
	Chinese cabbage	*Brassica rapa* L.	112.5
	Black mustard	*Brassica nigra* L.	129,000 DW
	White mustard	*Brassica hirta* L.	189,000 DW
Bromeliaceae	Pineapple	*Ananas comosus* L.	36.2
Asteraceae	Shungiku	*Chrysanthemum coronarium* L.	416.8
	Butterbur (fuki)	*Petasites japonicus* Maxim.	49.5
	Milk thistle seed	*Silybum marianum* L.	2,000 DW
Cucurbitaceae	Cucumber fruit	*Cucumis sativus* L.	24.6
	Cucumber seed	*Cucumis sativus* L.	11,000
Fabaceae	Alfalfa seed	*Medicago sativa* L.	16,000 DW
	Fenugreek seed	*Trigonella foenum-graecum* L.	43,000 DW
	Lupin seed	*Lupinus albus* L.	3,830
Juglandaceae	Walnut	*Juglans regia* L.	3,500 DW
Poaceae	Rice seed	*Oryza sativa* L.	1,006
	Barley seed	*Hordeum vulgare* L.	378.1
	Sweet corn	*Zea mays* L.	1,366
	Oat seed	*Avena sativa* L.	1,796
	Tall fescue	*Festuca arundinacea* Schreb.	5,288
Papaveraceae	Poppy seed	*Papaver somniferum* L.	6,000 DW
Liliaceae	Onion	*Allium cepa* L.	31.5
	Welsh onion	*Allium fistulosum* L.	85.7
Lythraceae	Pomegranate	*Punica granatum* L.	540–5,500
Musacea	Banana	*Musa acuminata* Colla	0.46–8.9
Oleracea	Olive oil	*Olea europaea* L.	50–119 pg mL^{-1}
Rosaceae	Apple	*Malus domestica* Borkh.	47.6
	Strawberry	*Fragaria x ananassa* Duch.	12.4
	Almond seed	*Prunus amygdalus* Batsch	39,000 DW
	Tart cherries	*Prunus cerasus* L.	1,000–19,500
	Sweet cherries	*Prunus avium* L.	6–224
Rubiaceae	Robusta	*Coffea canephora* Pierr.	5,800 000 DW
	Arabica	*Coffea arabica* L.	6,800 000 DW
Rutaceae	Orange juice	*Citrus × sinensis* L.	150
Umbelliferae	Carrot	*Daucus carota* Hoffm.	55.3
	Anise seed	*Pimpinella anisum* L.	7,000 DW
	Coriander seed	*Coriandrum sativum* L.	7,000 DW
	Celery seed	*Apium graveolens* L.	7,000 DW
	Linseed (flax)	*Linum usitatissimum* L.	12,000 DW
	Fennel seed	*Foeniculum vulgare* L.	28,000 DW
	Sunflower seed	*Helianthus annuus* L.	29,000 DW
Solanaceae	Tomato fruit	*Solanum lycopersicum* L.	2,500
	Currant tomato	*Solanum pimpinellifolium* L.	0.1
	Green bell pepper	*Capsicum annuum* L.	521.4 DW
	Orange bell pepper	*Capsicum annuum* L.	581.1 DW
	Red bell pepper	*Capsicum annuum* L.	179.5 DW
	Wolf berry (goji)	*Lycium barbarum* L.	103,000 DW

(Continued)

TABLE 13.2 (CONTINUED)

Data on the Presence and Content of Melatonin in Some Edible Plant Organs

Family	Common Name	Species Name	Melatonin Content (pg g^{-1} FW or DW)
Vitaceae	Grapevine	*Vitis vinifera* L.	5–965
Zingiberaceae	Cardamom seed	*Elettaria cardamomum* L.	15,000 DW
	Curcuma	*Curcuma aeruginosa* Roxb.	120,000 DW
	Ginger	*Zingiber officinale* Rose	583.7

Source: Arnao, *Advances in Botany*, 2014.

Cherries are one such fruit, as they are rich in melatonin. In order to understand the biological roles of melatonin in cherry fruit, melatonin synthesis was systematically monitored over a 24-h period both during development and in ripe cherries in two cultivars, *Prunus avium* L. cv. Hongdeng and *Prunus avium* L. cv. Rainier (Zhao et al., 2013). It was found that both darkness and oxidative stress induced melatonin synthesis, which led to dual melatonin synthetic peaks during a 24-h period. The high levels of malondialdehyde induced by high temperatures and high-intensity light exposure were directly related to upregulated melatonin production. It has been speculated that a primary function of melatonin in cherry fruits is as an antioxidant to protect the cherry from oxidative stress. Importantly, plant tryptophan decarboxylase gene (*TDC*) was identified in cherry fruits. Zhao et al. (2013) showed that *TDC* expression is positively related to melatonin production in cherries, which provides additional information to suggest that *TDC* is a rate-limiting enzyme of melatonin synthesis in plants. As we know, cultivar and cultivated methods usually affect the quality of agricultural production. So, Riga et al. (2014) evaluated the effect of cultivar and solar radiation on the melatonin content of pepper and tomato fruits. The melatonin content of the red pepper fruits ranged from 31 to 93 ng g^{-1} (dry weight). The melatonin content of the tomatoes ranged from 7.5 to 250 ng g^{-1} (dry weight). These data indicated that select suitable cultivars are important for food nutrition in agricultural production. Interestingly, the researchers measured the effect of ripeness on melatonin content and identified one group of pepper cultivars in which the melatonin content increased as the fruit ripened and another in which it decreased. Under shade conditions, the melatonin content in most tomato cultivars tended to increase (up to 135%), whereas that of most pepper cultivars decreased (to 64%). Overall, the results also demonstrated that the melatonin content of the fruits was not related to carbon fluxes from leaves. Korkmaz et al. (2014) determined the presence of melatonin in different organs (leaves, roots, fruits, and seeds) of two pepper cultivars and its variation during various growth stages (germination, seedling, flowering, and harvest). They found very high concentrations of melatonin in seedlings at the cotyledon stage, with a progressive decline as the plants matured. An abundance of melatonin was also found to accumulate in mature seeds, and to varying concentrations in fruits, leaves, and roots depending on the developmental stage. The effects of sample preparation, cultivar, leaf age, and tea processing on the melatonin content of mulberry (*Morus* spp.) leaves were investigated by Pothinuch and Tongchitpakdee (2011). A human study has shown that individuals who consume vegetables rich in melatonin increase their urinary melatonin metabolite excretion, which is consistent with an elevated level of melatonin in the blood (Oba et al., 2008). In these individuals, their daily melatonin consumption derived from vegetables was calculated to be 1.288 μg. This amount is much less than what individuals would obtain from eating corn or rice. The impact of melatonin consumed from these popular crops currently remains unknown. Over the course of a lifetime, the regular consumption of melatonin-rich foods may improve the general health of consumers.

Transgenic Modification of Serotonin and Melatonin Metabolism in Plants: Outcomes, Problems, and Perspectives

Several investigators have studied, via transgenic modification, the functions of the relative genes involved in the melatonin biosynthetic pathway in plants. To generate melatonin-rich rice plants, Byeon et al. (2014) first independently overexpressed three tryptophan decarboxylase isogenes in the rice genome. Melatonin levels were altered in the transgenic lines through overexpression of *TDC1*, *TDC2*, and *TDC3*; *TDC3* transgenic seed had melatonin concentrations 31 times higher than those of wild-type seeds. In *TDC3* transgenic seedlings, however, only a doubling of melatonin content occurred over wild-type levels. Thus, a seed-specific accumulation of melatonin appears to occur in *TDC3* transgenic lines. In addition to increased melatonin content, *TDC3* transgenic lines also had enhanced levels of melatonin intermediates, including 5-hydroxytryptophan, tryptamine, serotonin, and N-acetylserotonin. In contrast, expression levels of melatonin biosynthetic mRNA did not increase in *TDC3* transgenic lines, indicating that increases in melatonin and its intermediates in these lines are attributable exclusively to overexpression of the *TDC3* gene.

Okazaki et al. (2009) isolated cDNA-coded arylalkylamine N-acetyltransferase (*CrAANAT*), a possible limiting enzyme for melatonin biosynthesis, from *Chlamydomonas reinhardtii*. The predicted amino acid sequence of *CrAANAT* shares 39% homology to *AANAT* from *Ostreococcus tauri* and lacks cyclic adenosine monophosphate (cAMP)-dependent protein kinase phosphorylation sites in the N- and C-terminal regions that are conserved in vertebrates. Transgenic plants constitutively expressing *CrAANAT* were produced using a model cultivar of tomato (Micro-Tom). The transgenic Micro-Tom exhibited higher melatonin content compared with the wild type, suggesting that melatonin was synthesized from serotonin via N-acetylserotonin in plants. Moreover, the MRT Micro-Tom can be used to elucidate the role of melatonin in plant development. Tryptamine 5-hydroxylase (*T5H*) is the second enzyme in melatonin biosynthesis, catalyzing tryptamine into serotonin. Transgenic rice plants, in which the expression of endogenous *T5H* was either overexpressed or repressed, were examined for alteration in melatonin biosynthesis. Unexpectedly, the overexpression genotypes showed reduced levels of melatonin, while the repression genotypes had elevated levels with an average four-fold increase (Park et al., 2013d). With regard to melatonin intermediates, tryptamine and serotonin levels decreased, but tryptophan and N-acetylserotonin were unaltered in the overexpression genotypes compared with the wild type. In contrast, tryptamine levels in the repression genotypes were seven times higher than the wild type. In addition, tryptophan and 5-hydroxytryptophan were present at higher levels in the repression genotypes than in both the wild-type and the overexpression genotypes. The enhanced melatonin synthesis in the repression genotypes was closely associated with a transcriptional increase in *TDC1*. When these rice plants were challenged by oxidative stressors such as herbicides, much higher melatonin synthesis was also observed in the repression genotypes than in either the wild-type or overexpression genotypes. These results suggest that the tryptamine increase through the suppression of *T5H* plays an important signaling role in triggering melatonin biosynthesis in rice, although the exact role of tryptamine remains to be discovered. Whether both animals and plants use a similar biosynthetic pathway of melatonin remains to be seen. An interesting study was performed to examine the phenotype of transgenic Micro-Tom tomato plants overexpressing the homologous sheep *oAANAT* and *oHIOMT* genes responsible for the last two steps of melatonin synthesis (Wang et al., 2014a,b). The *oAANAT* transgenic plants had higher melatonin levels and lower IAA contents than the control due to the competition for tryptophan, the same precursor for both melatonin and IAA. Therefore, the *oAANAT* lines lost their "apical dominance," inferring that melatonin likely lacks auxin activity (Figure 13.14a). The significantly higher melatonin content in *oHIOMT* lines than in *oAANAT* lines provides new proof for the importance of AsSNMT in plant melatonin synthesis. In addition, the enhanced drought tolerance of *oHIOMT* lines will also make an important contribution to plant engineering (Figure 13.14b).

Kang et al. (2010) generated transgenic rice plants via expression of the human serotonin N-acetyltransferase (*SNA*) gene under a constitutive ubiquitin promoter using *Agrobacterium tumefaciens*–mediated gene transformation. They investigated the role of *SNA* in the biosynthesis of melatonin and the physiological role of melatonin in rice plants. High *SNA*-specific enzyme activities were observed in the transgenic rice plants, whereas the wild type revealed a trace level of *SNA* enzyme activity. The functional expression of the *SNA* protein was closely associated with the elevated synthesis of N-acetylserotonin and melatonin in the transgenic rice plants. Experiments using both exogenous treatment of serotonin and senescent detached leaves, which contain a pool of serotonin, significantly enhanced melatonin biosynthesis, indicating that endogenous serotonin levels play a bottleneck role in the pathway of melatonin biosynthesis. Finally, the transgenic rice seedlings with high levels of melatonin showed elevated chlorophyll synthesis during cold stress, suggesting a role for melatonin in cold

(a)

(b)

FIGURE 13.14   (a) Observations of "branching" phenotype in two oAANAT lines. (From Wang et al., *Journal of Pineal Research*, 57, 291–307, 2014b.) (b) The wilting of leaves of three lines and wild type (WT) after drought treatment for 20 days and the phenotype of transgenic lines and wild type after rewatering for 48 h. (From Wang et al., *Journal of Pineal Research*, 56, 134–142, 2014a.)

stress resistance. Zhang et al. (2012) have observed transgenic *Nicotiana sylvestris* plants expressing an arylalkylamine N-acetyltransferase (*AANAT*) gene and a hydroxyindole-O-methyltransferase (*HIOMT*) gene using *A. tumefaciens*–mediated transformation. Both *AANAT* and *HIOMT* are key enzymes in melatonin synthesis. The content of melatonin was significantly higher in the transgenic plants than in the nontransgenic plants. The highest melatonin content of transgenic plant leaves reached 50.4 μg g^{-1} (dry weight), while almost no melatonin could be detected in the nontransformed plants. To investigate the effects of the expression of the *AANAT* and *HIOMT* genes on melatonin function in plants, isolated protoplasts of *N. sylvestris* were exposed to ultraviolet (UV)-B radiation for different durations. DNA damage was evaluated by single-cell gel electrophoresis, which showed that the tail DNA percentage value in transgenic protoplasts was lower than that in the nontransformed protoplasts in the range of 0–30 s of UV-B radiation. DNA damage caused by UV-B was therefore reduced in transgenic *N. sylvestris* plants. All these studies indicate that genetically engineering melatonin's synthetic enzymes is a feasible method to elevate endogenous melatonin production in plants. Moreover, the utility of melatonin in agricultural production is a new frontier for melatonin research and deserves further investigation.

Conclusion

Serotonin and melatonin are pleiotropic molecules with multiple physiological functions in many plants. The multiple functions of serotonin and melatonin may be a consequence of their biologically active metabolites, as well as their receptor-mediated or receptor-independent actions. Serotonin and melatonin have emerged in the literature as animal hormones for nearly five decades. However, increasing evidence from recent studies indicates that serotonin and melatonin also act as constitutive and important molecules in plants. Several studies indicate that the primary functions of serotonin and melatonin in plants are as ROS scavengers under stress conditions. Besides functioning as an antioxidant to protect against various forms of stresses, serotonin and melatonin have acquired other functions in plants as well.

The concentration of melatonin in plants is significantly higher than that in animals. As a powerful antioxidant, the high levels of melatonin in plants are also beneficial to animals, including humans, who consume them. Melatonin levels in fruits, vegetables, beverages, and several major agriculture crops are sufficiently high to raise blood melatonin levels after their consumption. Nearly six billion people all over the world depend on these products as their major food source, so the potentially beneficial effects of melatonin consumed in these products is obvious.

We hypothesized that the application of serotonin and especially melatonin in agricultural systems may significantly enhance stress tolerance and improve crop production, according to the data available and preliminary observations. The reasons for these are: (1) melatonin increases the germination rate of seeds and simulates the physiological function of auxin signaling; (2) melatonin stimulates the development of the root system and promotes root regeneration; (3) melatonin preserves chloroplast against oxidative stress and accelerates the rate of photosynthesis in plants; and (4) as an antioxidant, melatonin enhances the tolerance of crops against drought, heat, cold, salinity–alkalinity, chemical pollutants, and other environmental stresses. Genetic manipulations to elevate melatonin production have proved feasible. Several transgenic plants have successfully expressed their transfected genes of enzymes for melatonin synthesis and have elevated the melatonin concentrations compared with their wild counterparts. The application of melatonin in agriculture is a new frontier to be explored.

Acknowledgments

This work was supported by the National Natural Science Foundation of China (No. 31501779) and the Excellent Young Scientist Foundation of Shandong Province (BS2014NY005).

REFERENCES

Arnao MB. Phytomelatonin: Discovery, content, and role in plants. *Advances in Botany*. 2014, e815769(2).

Arnao MB, Hernández-Rui J. The physiological function of melatonin in plants. *Plant Signaling and Behavior*. 2006; 1: 89–95.

Arnao MB, Hernández-Ruiz J. Melatonin promotes adventitious and lateral root regeneration in etiolated hypocotyls of *Lupinus albus* L. *Journal of Pineal Research*. 2007; 42: 147–152.

Arnao MB, Hernández-Ruiz J. Protective effect of melatonin against chlorophyll degradation during the senescence of barley leaves. *Journal of Pineal Research*. 2009; 46: 58–63.

Arnao MB, Hernández-Ruiz J. Melatonin: Plant growth regulator and/or biostimulator during stress? *Trends in Plant Science*. 2014; 19: 789–797.

Bajwa VS, Shukla MR, Sherif SM, Murch SJ, Saxena PK. Role of melatonin in alleviating cold stress in *Arabidopsis thaliana*. *Journal of Pineal Research*. 2014; 56: 238–245.

Byeon Y, Back K. An increase in melatonin in transgenic rice causes pleiotropic phenotypes, including enhanced seedling growth, delayed flowering, and low grain yield. *Journal of Pineal Research*. 2014; 56: 408–414.

Byeon Y, Lee HY, Hwang OJ, Lee HJ, Lee K, Back K. Coordinated regulation of melatonin synthesis and degradation genes in rice leaves in response to cadmium treatment. *Journal of Pineal Research*. 2015; 58: 470–478.

Byeon Y, Park S, Lee HY, Kim Y, Back K. Elevated production of melatonin in transgenic rice seeds expressing rice tryptophan decarboxylase. *Journal of Pineal Research*. 2014; 56: 275–282.

Chen Q, Qi W, Reiter RJ, Wei W, Wang B. Exogenously applied melatonin stimulates root growth and raises endogenous indoleacetic acid in roots of etiolated seedlings of *Brassica juncea*. *Journal of Plant Physiology*. 2009; 166: 324–328.

Gong B, Li X, Bloszies S, Wen D, Sun S, Wei M, Li Y, Yang F, Shi Q, Wang X. Sodic alkaline stress mitigation by interaction of nitric oxide and polyamines involves antioxidants and physiological strategies in *Solanum lycopersicum*. *Free Radical Biology and Medicine*. 2014b; 71: 36–48.

Gong B, Li X, Vanden Langenberg KM, Wen D, Sun S, Wei M, Li Y, Yang F, Shi Q, Wang X. Overexpression of S-adenosyl-L-methionine synthetase increased tomato tolerance to alkali stress through polyamine metabolism. *Plant Biotechnology Journal.* 2014a; 12: 694–708.

Hardeland R, Poeggeler B. Non-vertebrate melatonin. *Journal of Pineal Research.* 2003; 34: 233–241.

Hernández-Ruiz J, Cano A, Arnao MB. Melatonin: A growth-stimulating compound present in lupin tissues. *Planta.* 2004; 220: 140–144.

Hernández-Ruiz J, Cano A, Arnao MB. Melatonin acts as a growth-stimulating compound in some monocot species. *Journal of Pineal Research.* 2005; 39: 137–142.

Kang K, Kim S, Park S, Back K. Senescence-induced serotonin biosynthesis and its role in delaying senescence in tice leaves. *Plant Physiology.* 2009; 150: 1380–1393.

Kang K, Lee K, Park S, Kim YS, Back K. Enhanced production of melatonin by ectopic overexpression of human serotonin N-acetyltransferase plays a role in cold resistance in transgenic rice seedlings. *Journal of Pineal Research.* 2010; 49: 176–182.

Korkmaz A, Değer O, Cuci Y. Profiling the melatonin content in organs of the pepper plant during different growth stages. *Scientia Horticulturae.* 2014; 172: 242–247.

Kostopoulou Z, Therios I, Roumeliotis E, Kanellis AK, Molassiotis A. Melatonin combined with ascorbic acid provides salt adaptation in *Citrus aurantium* L. seedlings. *Plant Physiology and Biochemistry.* 2015; 86: 155–165.

Lee HY, Byeon Y, Back K. Melatonin as a signal molecule triggering defense responses against pathogen attack in *Arabidopsis* and tobacco. *Journal of Pineal Research.* 2014; 57: 262–268.

Lee HY, Byeon Y, Tan DX, Reiter RJ, Back K. *Arabidopsis* serotonin N-acetyltransferase knockout mutant plants exhibit decreased melatonin and salicylic acid levels resulting in susceptibility to an avirulent pathogen. *Journal of Pineal Research.* 2015; 58: 291–299.

Lei XY, Zhu RY, Zhang GY, Dai YR. Attenuation of cold-induced apoptosis by exogenous melatonin in carrot suspension cells: The possible involvement of polyamines. *Journal of Pineal Research.* 2004; 36: 126–131.

Li C, Wang P, Wei Z, Liang D, Liu C, Yin L, Jia D, Fu M, Ma F. The mitigation effects of exogenous melatonin on salinity-induced stress in *Malus hupehensis. Journal of Pineal Research.* 2012; 53: 298–306.

Liu N, Jin Z, Wang S, Gong B, Wen D, Wang X, Wei M, Shi Q. Sodic alkaline stress mitigation with exogenous melatonin involves reactive oxygen metabolism and ion homeostasis in tomato. *Scientia Horticulturae.* 2015; 181: 18–25.

Meng JF, Xu TF, Wang ZZ, Fang YL, Xi ZM, Zhang ZW. The ameliorative effects of exogenous melatonin on grape cuttings under water-deficient stress: Antioxidant metabolites, leaf anatomy, and chloroplast morphology. *Journal of Pineal Research.* 2014; 57: 200–212.

Mukherjee S, David A, Yadav S, Baluška F, Bhatla SC. Salt stress-induced seedling growth inhibition coincides with differential distribution of serotonin and melatonin in sunflower seedling roots and cotyledons. *Physiologia Plantarum.* 2014; 152: 714–728.

Murch SJ, Campbell SS, Saxena PK. The role of serotonin and melatonin in plant morphogenesis: Regulation of auxin-induced root organogenesis in in vitro-cultured explants of St. John's wort (*Hypericum perforatum* L.). *In Vitro Cellular and Developmental Biology-Plant.* 2001; 37: 786–793.

Murch SJ, Saxena PK. Melatonin: A potential regulator of plant growth and development? *In Vitro Cellular and Developmental Biology—Plant.* 2002; 38: 531–536.

Oba S, Nakamura K, Sahashi Y, Hattori A, Nagata C. Consumption of vegetables alters morning urinary 6-sulfatoxymelatonin concentration. *Journal of Pineal Research.* 2008; 45, 17–23.

Okazaki M, Higuchi K, Hanawa Y, Shiraiwa Y, Ezura H. Cloning and characterization of a *Chlamydomonas reinhardtii* cDNA arylalkylamine N-acetyltransferase and its use in the genetic engineering of melatonin content in the Micro-Tom tomato. *Journal of Pineal Research.* 2009; 46: 373–382.

Paradies G, Petrosillo G, Paradies V, Reiter RJ, Ruggiero FM. Melatonin, cardiolipin and mitochondrial bioenergetics in health and disease. *Journal of Pineal Research.* 2010; 48: 297–310.

Park WJ. Melatonin as an endogenous plant regulatory signal: Debates and perspectives. *Journal of Pineal Research.* 2011; 54: 143–149.

Park S, Lee K, Back K. Tryptamine 5-hydroxylase-deficient Sekiguchi rice induces synthesis of 5-hydroxytryptophan and N-acetyltryptamine but decreases melatonin biosynthesis during senescence process of detached leaves. *Journal of Pineal Research.* 2012; 52: 211–216.

Park S, Byeon Y, Kim YS, Back K. Kinetic analysis of purified recombinant rice N-acetylserotonin methyl-transferase and peak melatonin production in etiolated rice shoots. *Journal of Pineal Research*. 2013a; 54: 139–144.

Park S, Lee D, Jang H, Byeon Y, Kim Y, Back K. Melatonin-rich transgenic rice plants exhibit resistance to herbicide-induced oxidative stress. *Journal of Pineal Research*. 2013b; 54: 258–263.

Park S, Le TNN, Byeon Y, Kim YS, Back K. Transient induction of melatonin biosynthesis in rice (*Oryza sativa* L.) during the reproductive stage. *Journal of Pineal Research*. 2013c; 55: 40–45.

Park S, Byeon Y, Back K. Transcriptional suppression of tryptamine 5-hydroxylase, a terminal serotonin biosynthetic gene, induces melatonin biosynthesis in rice (*Oryza sativa* L.). *Journal of Pineal Research*. 2013d; 55: 131–137.

Pelagio-Flores R, Muñoz-Parra E, Ortiz-Castro R, López-Bucio J. Melatonin regulates *Arabidopsis* root system architecture likely acting independently of auxin signaling. *Journal of Pineal Research*. 2012; 53: 279–288.

Posmyk MM, Bałabusta M, Wieczorek M, Sliwinska E, Janas KM. Melatonin applied to cucumber (*Cucumis sativus* L.) seeds improves germination during chilling stress. *Journal of Pineal Research*. 2009; 46: 214–223.

Posmyk MM, Janas KM. Melatonin in plants. *Acta Physiologiae Plantarum*. 2009; 31: 1–11.

Posmyk MM, Kuran H, Marciniak K, Janas KM. Presowing seed treatment with melatonin protects red cabbage seedlings against toxic copper ion concentrations. *Journal of Pineal Research*. 2008; 45: 24–31.

Pothinuch P, Tongchitpakdee S. Melatonin contents in mulberry (*Morus* spp.) leaves: Effects of sample preparation, cultivar, leaf age and tea processing. *Food Chemistry*. 2011; 128: 415–419.

Ramakrishna A. Indoleamines in edible plants: Role in human health effects. In A. Catalá, Ed. *Indoleamines: Sources, Role in Biological Processes and Health Effects*. Hauppauge, NY: Nova, 2015; 279.

Ramakrishna A, Giridhar P, Jobin M, Paulose CS, Ravishankar GA. Indoleamines and calcium enhance somatic embryogenesis in cultured tissues of *Coffea canephora* P ex Fr. *Plant Cell, Tissue and Organ Culture*. 2012a; 108: 267–278.

Ramakrishna A, Giridhar P, Ravishankar GA. Indoleamines and calcium channels influence morphogenesis in in vitro cultures of *Mimosa pudica* L. *Plant Signaling and Behavior*. 2009; 12: 1136–1141.

Ramakrishna A, Giridhar P, Ravishankar GA. Phytoserotonin: A review. *Plant Signaling and Behavior*. 2011; 6: 800–809.

Ramakrishna A, Giridhar P, Udaya SK, Ravishankar GA. Melatonin and serotonin profiles in beans of *Coffea sps*. *Journal of Pineal Research*. 2012b; 52: 470–476.

Ramakrishna A, Giridhar P, Udaya SK, Ravishankar GA. Endogenous profiles of indoleamines: Serotonin and melatonin in different tissues of *Coffea canephora* P ex Fr. as analyzed by HPLC and LC-MS-ESI. *Acta Physiologiae Plantarum*. 2012c; 34: 393–396.

Riga P, Medina S, García-Flores LA, Gil-Izquierdo Á. Melatonin content of pepper and tomato fruits: Effects of cultivar and solar radiation. *Food Chemistry*. 2014; 156: 347–352.

Sarropoulou V, Dimassi-Theriou K, Therios I, Koukourikou-Petridou M. Melatonin enhances root regeneration, photosynthetic pigments, biomass, total carbohydrates and proline content in the cherry rootstock PHL-C (*Prunus avium* × *Prunus cerasus*). *Plant Physiology and Biochemistry*. 2012a; 61: 162–168.

Sarropoulou VN, Therios IN, Dimassi-Theriou KN. Melatonin promotes adventitious root regeneration in *in vitro* shoot tip explants of the commercial sweet cherry rootstocks CAB-6P (*Prunus cerasus* L.), Gisela 6 (*P. cerasus* × P. canescens), and MxM 60 (*P. avium* × P. mahaleb). *Journal of Pineal Research*. 2012b; 52: 38–46.

Shi H, Chan Z. The cysteine2/histidine2-type transcription factor ZINC FINGER OF *ARABIDOPSIS THALIANA* 6-activated C-REPEAT-BINDING FACTOR pathway is essential for melatonin-mediated freezing stress resistance in *Arabidopsis*. *Journal of Pineal Research*. 2014; 57: 185–191.

Shi H, Jiang C, Ye T, Tan DX, Reiter RJ, Zhang H, Liu R, Chan Z. Comparative physiological, metabolomic, and transcriptomic analyses reveal mechanisms of improved abiotic stress resistance in bermudagrass [*Cynodon dactylon* (L). Pers.] by exogenous melatonin. *Journal of Experimental Botany*. 2015a; 66: 681–694.

Shi H, Tan DX, Reiter RJ, Ye T, Yang F, Chan Z. Melatonin induces class A1 heat-shock factors (HSFA1s) and their possible involvement of thermotolerance in *Arabidopsis*. *Journal of Pineal Research*. 2015b; 58: 335–342.

Sun Q, Zhang N, Wang J, Zhang H, Li D, Shi J, Li R, Weeda S, Zhao B, Ren S, Guo YD. Melatonin promotes ripening and improves quality of tomato fruit during postharvest life. *Journal of Experimental Botany.* 2014; 66: 657–668.

Tan DX, Hardeland R, Manchester LC, Korkmaz A, Ma S, Rosales-Corral S, Reiter RJ. Functional roles of melatonin in plants, and perspectives in nutritional and agricultural science. *Journal of Experimental Botany.* 2012; 63: 577–597.

Turk H, Erdal S, Genisel M, Atici O, Demir Y, Yanmis D. The regulatory effect of melatonin on physiological, biochemical and molecular parameters in cold-stressed wheat seedlings. *Plant Growth Regulation.* 2014; 74: 139–152.

Vantassel DL, Li JA, O'Neill SD. Melatonin-identification of a potential dark signal in plants. *Plant Physiology.* 1993; 102: 117–117.

Wang L, Zhao Y, Reiter RJ, He C, Liu G, Lei Q, Zuo B, Zheng XD, Li Q, Kong J. Changes in melatonin levels in transgenic "Micro-Tom" tomato overexpressing ovine AANAT and ovine HIOMT genes. *Journal of Pineal Research.* 2014a; 56: 134–142.

Wang P, Sun X, Chang C, Feng F, Liang D, Cheng L, Ma F. Delay in leaf senescence of *Malus hupehensis* by long-term melatonin application is associated with its regulation of metabolic status and protein degradation. *Journal of Pineal Research.* 2013; 55: 424–434.

Wang P, Sun X, Li C, Wei Z, Liang D, Ma F. Long-term exogenous application of melatonin delays drought-induced leaf senescence in apple. *Journal of Pineal Research.* 2013b; 54: 292–302.

Wang P, Sun X, Xie Y, Li M, Chen W, Zhang S, Liang D, Ma F. Melatonin regulates proteomic changes during leaf senescence in *Malus hupehensis. Journal of Pineal Research.* 2014b; 57: 291–307.

Wang P, Yin L, Liang D, Li C, Ma F, Yue Z. Delayed senescence of apple leaves by exogenous melatonin treatment: Toward regulating the ascorbate-glutathione cycle. *Journal of Pineal Research.* 2012; 53: 11–20.

Weeda S, Zhang N, Zhao X, Ndip G, Guo Y, Buck GA, Fu C, Ren S. *Arabidopsis* transcriptome analysis reveals key roles of melatonin in plant defense systems. *Plos One.* 2014; 9: e93462.

Wei W, Li QT, Chu YN, Reiter RJ, Yu XM, Zhu DH, Zhang WK, Ma B, Lin Q, Zhang JS, Chen SY. Melatonin enhances plant growth and abiotic stress tolerance in soybean plants. *Journal of Experimental Botany.* 2014; 66: 695–707.

Yin L, Wang P, Li M, Ke X, Li C, Liang D, Wu S, Ma X, Li C, Zuo Y, Ma F. Exogenous melatonin improves Malus resistance to Marssonina apple blotch. *Journal of Pineal Research.* 2012; 54: 426–434.

Zhang HJ, Zhang N, Yang RC, Wang L, Sun QQ, Li DB, Cao YY, Weeda S, Zhang B, Ren S, Guo YD. Melatonin promotes seed germination under high salinity by regulating antioxidant systems, ABA and GA4 interaction in cucumber (*Cucumis sativus* L.). *Journal of Pineal Research.* 2014; 57: 269–279.

Zhang L, Jia J, Xu Y, Wang Y, Hao J, Li T. Production of transgenic *Nicotiana sylvestris* plants expressing melatonin synthetase genes and their effect on UV-B-induced DNA damage. *In Vitro Cellular and Developmental Biology—Plant.* 2012; 48: 275–282.

Zhang N, Zhao B, Zhang HJ, Weeda S, Yang C, Yang ZC, Ren S, Guo YD. Melatonin promotes water-stress tolerance, lateral root formation, and seed germination in cucumber (*Cucumis sativus* L.). *Journal of Pineal Research.* 2013; 54: 15–23.

Zhao Y, Tan D, Lei Q, Chen H, Wang L, Li Q, Gao Y, Kong J. Melatonin and its potential biological functions in the fruits of sweet cherry. *Journal of Pineal Research.* 2013; 55: 79–88.

Section III

Medicinal Plants: Occurrence and Efficacy in Humans

14

Melatonin in Traditional Herbal Medicine

Jeffrey Johns and Tanit Padumanonda
Khon Kaen University
Khon Kaen, Thailand

CONTENTS

ABSTRACT This chapter reviews the literature on Chinese traditional medicines that have been shown to contain melatonin, and discusses the possible therapeutic links between melatonin and the indications and treatments in which these remedies are used. The feasible uses of these herbal products as possible sources of natural melatonin are discussed. Melatonin in a traditional Thai sleeping remedy is reported, as well as melatonin in Thai herbs with anti-inflammatory, antibacterial, and glucose control applications. The melatonin contents of a number of Thai herbs, fruits, and vegetables used in traditional Thai medicine is reported for the first time. Finally, the potential of herbal melatonin as a source of melatonin is discussed, with consideration of the probable doses needed to achieve an equivalent to endogenous nighttime levels as an alternative to synthetic melatonin for use in jet lag, sleep disorders, headaches, etc. The need for studies on the effects of storage and herbal medicine preparations on melatonin stability is highlighted.

KEY WORDS: *traditional medicinal herbs, melatonin, sleep, inflammation, antibacterial, glucose control.*

Abbreviations

AChE:	acetylcholinesterase
ELISA:	enzyme-linked immunosorbent assay
FRAP:	ferric-reducing antioxidant power
GAE:	gallic acid equivalents
GME:	geissoschizine methy ether
GC–MS:	gas chromatography–mass spectroscopy
HPLC:	high-performance liquid chromatography
NMU:	N-methyl-N-nitrosourea

RIA: radioimmunoassay
TP: total phenol content

Introduction

The World Health Organization (WHO) defines traditional medicine as "the sum total of the knowledge, skills, and practices based on the theories, beliefs, and experiences indigenous to different cultures, whether explicable or not, used in the maintenance of health as well as in the prevention, diagnosis, improvement or treatment of physical and mental illness" (WHO, 2000). The philosophy of traditional medicine is to create wholeness and harmony within a person with various combinations of herbs rather than focus on the active compounds in the herbs. In this chapter we review the literature on traditional medicines that have been shown to contain melatonin, and discuss the possible therapeutic links between melatonin and the indications and treatments in which these remedies are used. We also look at the feasible use of these herbal products as possible sources of natural melatonin. The online database The Plant List (2013) was used to confirm and rationalize the use of different species names found in the literature.

First Reports of Melatonin in Plants

Melatonin was first reported in edible plants of the Cramineae family by Hattori et al. (1995). Radioimmunoassay (RIA) and high-performance liquid chromatography (HPLC) with fluorescence detection–mass spectroscopy showed levels of melatonin of 1 ng/g in rice tissue, 0.4 ng/g in barley, 1.4 ng/g in sweet corn (maize), and 1.8 ng/g in oat. This was a significant finding, as maize and rice are the two most consumed staple foods worldwide. Japanese Ashitaba (family Umberlliferae) and Japanese radish (family Cruciferae) also showed high melatonin levels (0.62 and 0.66 ng/g tissue, respectively), along with cabbage (*Brassica oleracea*) at 0.11 ng/g wet weight and ginger (*Zingiber officinale*) at 0.58 ng/g wet weight. This study showed that feeding chicks a diet containing plant products rich in melatonin produced levels of melatonin in their blood measureable by RIA. Likewise, melatonin extracted from plants inhibited binding of [^{125}I]iodomelatonin to rabbit brain. Thus, melatonin ingested in foodstuffs was shown to enter the blood and be capable of binding to melatonin-binding sites in the brains of mammals.

At about the same time, Dubbels et al. (1995) investigated melatonin in nine edible plants by RIA and HPLC, and found concentrations ranging from 0 to 0.86 ng melatonin/mg protein. The presence of melatonin was verified by gas chromatography–mass spectroscopy (GS–MS). Levels of 2 to nearly 200 ng/g dry weight in seeds of 15 edible plants have also been reported, the highest being white mustard (*Brassica hirta*), with 189 ng/g dry weight (Manchester et al., 2000). Moreover, melatonin profiles have been reported in coffee beans (Ramakrishna et al., 2012a, 2012b).

Traditional Medicinal Herbs with Reportedly High Levels of Melatonin

Levels of melatonin in herbs are much higher than in edible plants and seeds. Melatonin was identified at levels of 1.75 and 4.39 µg/g in dried leaves and flowers, respectively, of St John's wort (*Hypericum perforatum*) (Murch et al., 1997). St John's wort has long been used as a sedative for the treatment of depression and anxiety, irritability, and nervous tension (Brown et al., 2001) at doses of 2–4 g. It has found significant use for mild to moderate depression, especially in children and adolescents in some countries (Fegert et al., 2006). These doses can provide between 3.5 and 17.5 µg of melatonin (depending on bioavailability), which is within the lower pharmacological range of melatonin when used as a drug for sleep purposes. Melatonin is well recognized for its sedative properties (Cardinali et al., 2012; Marseglia et al., 2015). It has been shown to significantly reduce the risk of depressive symptoms in women with breast cancer during a three-month period after surgery (Hansen et al., 2014) and provided significant improvements in depression and sleep over time (Serfaty et al., 2010) in randomized double-blind placebo-controlled trials. Its effect on mood disorders has been reviewed by Lanfumey et al.

(2013). These effects could be due to other constituents in St John's wort, however, or a synergy between active components, including melatonin.

To date and to the best of our knowledge, the highest reported level from traditional medicinal melatonin sources is 7.11 µg/g dry weight in *Scutellaria baicalensis* Georgi (Huang-qin, Baikal skullcap) (Murch et al., 1997). Using HPLC-electrochemical detection (HPLC-ECD), validated by LC–MS/MS and RIA, the Murch group also quantified melatonin at 0.04 ng/g (9 nmol/g) to 0.19 µg/g (44–362 nmol/g) dry weight in *in vitro* propagated shoots from 26 distinct germplasm lines, a nearly 5000-fold difference in concentration (Murch et al., 2004), and significantly lower than in their previous study. Huang-qin is one of the 50 fundamental herbs used in traditional Chinese medicine and is usually used as the dry root. In addition to melatonin, it contains a large number of flavonoid compounds (Tang and Eisenbrand, 1992; Yang et al., 2002). It is commonly used for fevers, neurological disorders, cancer and inflammatory diseases, respiratory infections including hay fever, and gastrointestinal infections (Cole et al., 2008), as well as liver problems including viral hepatitis and jaundice, among others.

Melatonin at 1.37 to 2.45 µg/g dry weight in various dried leaves of *Tanacetum parthenium* (L.) Sch. Bip. (feverfew) has been reported (Murch et al., 1997). Melatonin has also been found at around 1.1 µg/g dry weight from hot water extraction and 1.8–2.1 µg/g dry weight from 50% methanolic extraction by HPLC and ELISA (Ansari et al., 2010). The efficacy and safety of feverfew has been extensively reviewed (Ernst et al., 2000). Feverfew is commonly used to prevent migraine headaches, arthritis, and digestive problems, though earlier reviews did not find sufficient scientific evidence to support anything beyond a placebo effect (Pittler and Ernst, 2004; Pareek et al., 2011). Since the 2004 Cochrane review, however, new evidence supporting the use of feverfew-based remedies for migraine has been emerging (Pfaffenrath et al., 2002; Shrivastava et al., 2006; Cady et al., 2011). The status of melatonin for preventing or treating migraine is in a similar state, with some positive results (Gagnier, 2001; Peres et al., 2004, 2006). Melatonin appears to be a significant contributor to the pathogenesis of migraine (Toglia, 2001), and migraine has been linked to disrupted sleep and circadian rhythm (Gagnier, 2001). One study has also shown lower levels of melatonin than normal in migraine patients (Burns et al., 2007). Thus, the administration of exogenous melatonin may normalize the circadian cycle, helping to relieve migraines. Melatonin and melatonin-containing traditional medicines such as feverfew may have an important therapeutic role for the treatment of migraine (Matharu et al., 2004).

The melatonin content in more than 100 Chinese medicinal herbs was investigated using solid-phase extraction followed by HPLC (Chen et al., 2003). Sixty-four herbs contained melatonin in excess of 10 ng/g dry mass and 10 contained more than 1000 ng/g dry mass (Table 14.1). These levels are generally significantly higher than previously found in other plant products (Hattori et al., 1995).

A rich source of natural melatonin is *chantui* (cicada molting; *Periostracum cicadae*). This commonly used ingredient in Chinese medicine formulation is not actually a herb, but the molted skin of cicada, and consists of about 50% chitin and 50% proteins as well as small amounts of amino acids, lipids, minerals, and wax. Its use was first mentioned by Tao Hongjing, a physician, naturalist, and the most eminent Daoist of his time, in his *materia medica*, *Mingyi bielu* (c. AD 500). Chantui has subsequently been included in the *materia medica* of China and is an official component of the modern pharmacopoeia of the PRC (Chang and But, 1986). Usually a decoction of 6–15 g is prepared and taken as an infusion. This could provide up to 22–55 µg of melatonin. Chantui has indications for tetanus (using a complex decoction that includes cicada at 30 g per day or cicada powder used as a single ingredient at 9–15 g per dose, three times daily), corneal opacity (topically or subconjunctivally by injection), allergic skin diseases such as urticaria (3 g cicada powder with 50 g rice wine), Bell's palsy (chantui with powder of gypsum, swallowed down with heated wine), influenza fever (combinations of chantui and silkworm [*baijiang-can*]), acute nephritis (a decoction with cicada and spirodela [*fuping*]), and cataracts (daily administration of 9 g chantui) (Chang and But, 1986).

Viola philippica Cav. (*zi hua di ding*; purpleflower violet) is reported to contain 2.4 µg/g melatonin (Chen et al., 2003). It is used in doses of 9–15 g and is indicated for hot swellings or sores; red, swollen eyes, throat, and/or ears; mumps; sores and abscesses of the head and/or back; and venomous snake bites (Liu et al., 2005). Its properties have been attributed to the antibacterial activities of long-chain carboxylic acids (Xie et al., 2004), but this herb may be a useful source of natural melatonin. Interestingly, a

TABLE 14.1

Melatonin Contents of Chinese Herbs

Name	ng/g	Scientific Name	Name	ng/g	Scientific Name
Chantui	3771	*Periostracum cicadae*	Kushen	190	*Sophora flavescens* Ait.
Diding	2368	*Viola philippica* Cav.	Dansnen	187	*Salvia miltiorrhiza* Bge
Gouteng	2460	*Uncaria rhynchophylla*	Qinjiu	180	*Gentiana macrophylla* Pall.
Shiya tea leaf	2120	*Babreum coscluea*	Huangqin	178	*Scutellaria amoena* C.H. Wright
Sangye	1510	*Morus alba L* (Leaf)	Jinqiancao	169	*Desmodium styracifolium* Merr
Huangbo	1235	*Phellodendron amurense* Rupr.	Sanqi	169	*Panax notoginsneg* Burk
Sangbaipi	1110	*Mori albae* (Cortex)	Yimucao	169	*Leonurus japonicus* Houtt.
Yinyanghuo	1105	*Epimedium brevicornum* Maxim	Juhua	160	*Dendranthema morifolium*
Huanglian	1008	*Coptis chinensis* Franch	Zicao	158	*Arnebia euchroma*
Dahuang	1078	*Rheum palmatum* L.	Gegen	150	*Pueraria lobata* Willd
Yuanzhi	850	*Polygala tenuifolia* Willd.	Dazao	146	*Ziziphus jujuba* Mill.
Shanyurou	821	*Coruns officinalis* Sieb.	Yejiaoteng	143	*Caulis Polygonam multiflorum* Thunb.
Longdacao	780	*Gentiana scabra* Bge.	Shuanhua	140	*Lonicera japonica* Thunb.
Luxiancao	750	*Pirola decorata* H.	Erzhu	120	*Curcuma aeruginosa* Roxb.
Danghui	698	*Angelica sinensis* Oliv.	Gancao	112	*Glycyrrhiza uralensis* Fisch
Sangjisheng	648	*Taxillus chinensis* DC	Dilong	97	*Pheretimaaspergillum*
Fuling	585	*Poria cocos* Schw. Wolf.	Shengdi	97	*Rehmannia glutinosa*
Gouqi	530	*Lycium barbarum* L.	Wuweizi	86	*Schisondra chinensis*
Bihu	523	*Gekko japonicus* (Dumeril et Bibron)	Qinghao	84	*Artemisis annua* L.
Luhui	516	*Aloe vela* L.	Banlangen	79	*Isatis indigotica* Fort.
Chuanxinlian	511	*Andrographis paniculats* Burm.	Fangfeng	60	*Saposhmikovia divaricata*
Duzhong	497	*Eucommia ulmoides* Oliv.	Zhuye	55	Not available
Laifuzi	485	*Raphanus sativus* L.	Shidagonglao	52	*Mahonia bealei* (Fort.)
Dingxiang	446	*Syzygium aromaticum* L.	Lianqiao	45	*Forsythia suspensa* (Thunb.)
Fupenzi	387	*Rubus chingii* Hu	Huangjing	45	*Polygonatum sibiricum* Delar.
Xuanshen	342	*Scrophularia ningpoensis* Hemsl.	Damzjuye	38	*Lophartherum gracile* Brongn.
Huoxiang	302	*Agastaches rugosa*	Xiakucao	34	*Prunella vulgaris* L.
Banzhilian	257	*Lobelia chinesis* Lour.	Baijing	32	*Herba Patriniae scabiosaefoliae*
Suanzhaoren	256	*Ziziphus jujuba* Mill.	Duhuo	31	*Angelica biserrata*
Wugong	248	*Scolopendra subspinipes mutilang.*	Rouchongrong	28	*Cistanche desericola* Y.
Jiangcan	227	*Bombyx batryticatus*	Chenpi	25	*Citrus reticulata* Blanco
Maidong	198	*Ophiopogon japonicus*	Zhizi	12	*Galdenia jasminoides* Ellis

Source: Adapted from Chen et al., *Life Sci* 73(1), 19–26, 2003.

related species, *Viola odorata* L., has been found to contain negligible amounts of melatonin (Ansari et al., 2010).

Uncaria rhynchophylla (*diào gōu téng*; cat's claw) is used at doses of 10–15 g for convulsions and seizures (as an antiepileptic), headaches, red swollen eyes/throat, toothache, skin disorders, hypertension, and for the treatment of agitation in elderly persons. It contains a variety of indole alkaloid structures, most notably rhynchophylline, and a potent drug-like alkaloid, geissoschizine methyl ether (GME). It appears to have neuroprotective effects (Shi et al., 2003; Ho et al., 2014; Ndagijimana et al., 2013) on glial cells, which support the neurons from activating in response to inflammation. This anti-inflammatory effect in the brain appears to underlie its antiepileptic properties. This herb is reported to contain 2.5 µg/g melatonin (Chen et al., 2003). Melatonin, also an indole compound, is widely accepted to have powerful anti-inflammatory (Mauriz et al., 2013), antioxidant and neuroprotective (Carpentieri et al., 2012), analgesic (Ambriz-Tututi et al., 2009), and sedative properties (Johnson et al., 2002; Marseglia et al., 2015), and may contribute to the therapeutic action of this herb.

Shiya tea leaf (*Babreum coscluea*) was found to contain 2.1 µg/g dry weight melatonin (Chen et al., 2003), and although there is little in the literature about its use as a traditional medicine, it contains a significant amount of melatonin.

The melatonin content of *Tripleurospermum (Chrysanthemum) disciforme* (C.A. Mey) Schultz Bip. (Asteraceae) was determined as 3.07 and 2.916 µg/g by HPLC and ELISA, respectively (Ansari et al., 2010). The extract showed promising anti-ulcer effects on pylorus ligated (Shay) rats, resulting in significant reductions in ulcer area and ulcer index, and for the latter the range of reduction was 21.8%–39.1% (Minaiyan et al., 2006). Extracts of *T. disciforme* were shown to inhibit acetylcholinesterase (AChE) in a dose-dependent manner (Mandegary et al., 2014). Extracts also showed analgesic effects in rat (Parvini, 2007) and inhibited carrageenan-induced edema in rat (Bakhtiarian, 2007).

Morus alba L., from the family Moraceae, is a short-lived, fast-growing, small- to medium-sized mulberry tree, cultivated to feed silkworms used in the commercial production of silk, and as food for livestock during dry seasons. It is also widely consumed worldwide as a tea. This herb has been extensively reported both in the traditional medicinal literature and the modern literature, with more than 350 entries in PubMed by 2015. Readers are referred to the recent review by Devi et al. (2013). Reported levels of melatonin vary from 1.5 µg/g dry leaf (Chen et al., 2003) to 0.04–0.28 µg/g dry leaf (Pothinuch and Tongchitpakdee, 2011), both by solid-phase extraction purification/HPLC. The melatonin contents of all cultivars tested were highest in the tip of the leaves, followed by that in the young leaves, whereas the lowest was found in the old leaves. Heat treatment during tea processing decreased the melatonin content in mulberry leaves by approximately 87% when compared with that of the fresh leaves (Pothinuch and Tongchitpakdee, 2011). A recent study (Natić et al., 2015) identified 14 hydroxycinnamic acid esters, 13 flavonol glycosides, and 14 anthocyanins in *Morus alba* fruit extracts; total phenolic content ranged from 43.84 to 326.29 mg GAE/100 g frozen fruit and ferric-reducing antioxidant power (FRAP) (0.03–38.45 µM ascorbic acid), indicating potentially potent antioxidant activity. Zou et al. (2012) also identified four phenolic compounds ranging from 2.3 to 4.2 mg/g dry weight, while the mean total phenol (TP) content of the six cultivars ranged from 30.4 to 44.7 GAE mg/g dry weight.

The dry leaves of white mulberry (5–10 g) have various indications in Chinese medicine for the treatment of fever, dizziness, and blurred vision (Ou Ming, 1999; Pharmacopoeia Commission of the Ministry of Public Health, 1995). Furthermore, dry fruits of white mulberry are also used to treat vertigo, tinnitus, and palpitations, as a sleeping aid, and for hair rejuvenation (Zhang, 1990). In human studies, *Morus alba* exhibited antidiabetic and antioxidant effects with a fruit polyphenol–enhanced extract (Wang et al., 2013, 2014), anti-inflammatory and anti-cancer activity of the root bark in cell lines (Eo et al., 2014), potent inhibition of mammary carcinogenesis induced by N-methyl-N-nitrosourea (NMU) in Sprague Dawley rats (Harshavardhan et al., 2012), and antidepressant-like effects of an ethyl acetate soluble fraction of the root bark in rats (Lim et al., 2014). *Morus alba* has significant antioxidant properties (Yang et al., 2014), and contains several novel alkaloids (Wang et al., 2014) and flavanes (Yang et al., 2010).

As well as herbs, various seeds have been shown to contain melatonin. Beans are extensively consumed as macrobiotics and dietary supplements (Penny et al., 2002), and many have indications in Chinese herbal medicine. Soybean is the most important cereal, and contains a great deal of protein, fat, carbohydrate, carotene, vitamins B1, B2, and B12, nicotinic acid, isoflavonoids, and saponins. In traditional Chinese medicine, soybean is used for the treatment of postpartum convulsive disease, convulsion, and lockjaw, and for relieving drug toxicity (Zhang, 1990). Mung bean (*Vigna radiate* L. R. Wilczek) seeds are said to be a traditional cure for paralysis, rheumatism, coughs, fever, and liver ailments (Ou Ming, 1999; Wang et al., 2009) In addition, the study of Wu et al. (2001) showed that the mung bean and black gram can improve liver injury caused by acetaminophen, with a comparable result to silymarin, a hepatoprotective agent. It is also well publicized that legumes normally have high protein contents (Shalaby, 1997).

Wine made from grapes (*Vitis vinifera*) is often taken medicinally, and can contain high levels of melatonin, from 25 to 428 ng/g, as measured by HPLC and ELISA (Iriti et al., 2006; Iriti and Faoro, 2006). By comparison, red wine contains between 0.2 and 5.8 ng/g resveratrol, depending on the grape variety (Gu et al., 1999). Coffee (*Coffea canephora* and *Coffea arabica*), although not a traditional medicine, also exhibits high levels of melatonin, containing 5.8–12.5 (green) and 8.0–9.6 (roasted) μg/g dry weight, as measured by ELISA (Ramakrishna et al., 2012a).

Melatonin in a Traditional Thai Sleeping Remedy

The rationale for the selection of herbs in our group is based on the traditional formulation and plant chemotaxonomy. The correlation of melatonin content with treatment using herbal sleeping aids is the main focus of this section. Thailand has longstanding knowledge of traditional herbal healing passed from generation to generation. In one particular formulation, there are seven edible plants recommend as sleeping aids: black pepper (*Piper nigrum* L.), Burmese grape (*Baccaurea ramiflora* Lour.), hummingbird tree/scarlet wisteria (*Sesbania glandiflora* [L.] Desv.), bitter gourd (*Moringa charantia*), foetid cassia (*Senna tora* [L.] Roxb.), drumstick tree (*Moringa oleifera* Lam.), and *Sesbania sesban* (L.). Formulations of these herbs are included in a Thai medical textbook called *Tumra Paetsart Sonkrau: Textbooks of Thai Traditional Medicine* (Phitsanuprasātwēt, 1992) as sleeping aids. This textbook is a compilation of teachings from the Thai Traditional Medicine and Ayurvedh Vidhayalai (Jevaka Komarapaj) Colleges, elaborating the content with scientific explanations, and using contemporary language so that it may be easily understood by lay people and students. The textbook indicates that the consumption of the cooked mature leaves of the seven plants with rice is believed to increase blood tonic and induce sleep. These seven herbs were analyzed for their melatonin contents using ultrasonication, solid-phase extraction, and HPLC and ELISA analysis. The limit of detection was 2.4 ng/g dry weight. The results showed that some of these edible herb extracts contained significant amounts of melatonin, particularly black pepper (1093 ng/g), but also Burmese grape (43.2 ng/g), hummingbird tree/scarlet wisteria (26.3 ng/g), bitter gourd (21.4 ng/g), foetid cassia (10.5 ng/g), and *Sesbania sesban* (8.7 ng/g), all dry sample weight (Padumanonda et al., 2014). Melatonin was not detected in the drumstick tree extract. The high content of melatonin found in the black pepper is within the range of the highest amounts previously reported for dried herbs (Chen et al., 2003), and suggests it may be very effective as a traditional medicine, especially when used in combination with other herbs with sedative properties. This explains the traditional formulation of a combination of seven herbs for sedative properties. The leaves of black pepper are rarely consumed as a food, unlike the other parts, such as the seed, which is a very well-known spice. The high melatonin content in black pepper leaves is promising for the future development of this overlooked part of the plant as a health food supplement.

Melatonin in Thai Herbs with Inflammation, Antibacterial, and Glucose Control Applications

Ethanolic extracts of 34 Thai herbs, fruits, and vegetables that are used in traditional Thai medicine for inflammation, antibacterial, and glucose control were screened for melatonin content. Melatonin content was quantified using HPLC with a fluorescent detector. The quantification limit was about 0.5 ng/mL (0.1 ng/g extract). Extract yields varied from 0.01% (*Momordica charantia* L.) to 3.7% (*Allium cepa*) (wet weight) and 0.3% (*Oryza sativa* L.) to 22% (*Allium cepa*) (dry weight). Confirmation of melatonin in the herbs was shown by ELISA analysis of the extracts. Ten of the extracts showed quantifiable amounts of melatonin (Table 14.2).

TABLE 14.2

Melatonin Contents of Thai Herbs, Fruits, and Vegetables Used in Traditional Thai Medicine

Common Name	Scientific Name	Family	Part	% Yield of Extract	Melatonin ng/g Dry Weight	Melatonin ng/g Wet Weight
Gingerspan	*Zingiber officinale* Roscoe	*Zingiberaceae*	Fresh rhizome	0.60	6.9	0.041
Tumeric	*Curcuma longa* L.	*Zingiberaceae*	Powdered root	8.24	6.3	0.52
Carcum	*Catymbium peciosum* (Wendl.) Holtt.	*Zingiberaceae*	Fresh rhizome	2.6	7.0	0.18
Indian gooseberry	*Phyllanthus emblica* L.	*Euphobiaceae*	Branch	6.4	146.1	9.35
Shikakai	*Acacia concinna* (Willd.) DC.	*Mimosoideae*	Fruit	28.2	4.5	1.27
Siamese rough bush, khoi, toothbrush tree	*Streblus asper* Lour.	*Moraceae*		5.9	139.6	8.24
Vallarai	*Centella asiatica* (L.) Urb.	*Umbelliferae*	Leaf, stem	1.9	3.8	0.073
Long coriander, cilantro, *pak chi farang*	*Eryngium foetidum* Linn.	*Umbelliferae*	Leaf, stem	1.60	0.5	0.0079
Tomato	*Lycopersicon esculentum* Mill.	*Solanaceae*	Fruit	0.090	4.7	0.0042
Pink mempat, *tiu yang*, *tiu dang*	*Cratoxylum formosum* (Kurz) Gogel (Jack.) Dyer spp. *Pruniflorum* (Kurz) Gogel	*Guttiferae*	Leaf	3.78	191.0	7.22

Lycopersicon esculentum (tomato) contained 0.0042 ng/g wet weight of melatonin, lower than in previous studies: 0.0322 ng/g by RIA (Hattori et al., 1995) and 0.302 ng/g by GC–MS (Badria, 2002). *Zingiber officinale* (ginger) contained 0.041 ng/g melatonin, also lower than that reported by Hattori et al. (1995) of 0.538 ng/g by RIA. *Phyllanthus emblica* and *Streblus asper* exhibited melatonin content of 9.4 ng/g and 8240 pg/g wet weight, respectively, similar to the high levels previously reported for some herbs—for example, 7.1 ng/g for *Huang qin* and 4.39 ng/g for *Hypericum perforatum* (St John's wort flowers) (Murch et al., 1997), and 12.9 ng/g (leaves) for *Ficus carica* (Zohara et al., 2011). Surprisingly, melatonin was not detected by HPLC or ELISA in the Thai sample of *Morus alba*. Some Thai herbal medicines show potential for providing supplemental dietary melatonin.

Potential of Herbal Melatonin

Although there is great potential for herbal medicine as a source of melatonin, there are very few studies of the bioavailability of melatonin from plant materials, showing that an update is indeed needed. Reiter et al. (2005) showed that consumption of melatonin-containing walnuts (*Juglans regia* L.) influenced levels of melatonin and the total antioxidant capacity of blood in rats. Serum melatonin concentrations were 38.0 ± 4.3 pg/mL serum for walnut-fed animals compared with only 11.5 ± 1.9 for chow-fed animals, significantly different at $p < 0.01$. The "total antioxidant power" of the serum estimated by TEAC and FRAP showed significant ($p < 0.01$) increases after walnut consumption. Studies that investigated the bioavailability of melatonin after the consumption of fruit in humans showed significantly increased circulating melatonin levels and improved FRAP and ORAC antioxidant capacities of plasma; these studies are described in Chapter 22 of this book. The bioavailability of melatonin in humans from herbal sources has not yet been reported.

Melatonin is endogenously produced by the pineal gland, but is also consumed in edible plants. Because physiological concentrations of melatonin in the blood are known to correlate with the total antioxidant capacity of the serum, consuming foodstuffs containing melatonin may be helpful in lowering oxidative stress (Reiter et al., 2001; Reiter and Tan, 2002) and preventing age-related degenerative diseases.

Many medicinal herbs contain in excess of 1 µg melatonin/g dry mass. Considering that the healthy adult endogenous production of melatonin is about 28.8 µg/day (4.6 µg/h for males and 2.8 µg/h for females; Macchi et al., 2004), one can envisage that a human would need to consume about 3–5 g of dry herb to achieve the equivalent of 1 h of endogenous production, assuming complete absorption. This would bring melatonin levels to around normal in patients with low levels—for example, the elderly (Benot et al., 1999)—or those exposed to light at night—for example, shift workers (Reiter et al., 2007; Dumont and Paquet, 2014). Thus, it does not seem unreasonable to be able to complement endogenous levels though the intake of herbal medicines.

Conclusion

Herbal formulation should be able to provide doses in the microgram range, and extracts may be able to provide much higher levels, equivalent to pharmacological doses of pure drug of 0.1–20 mg per person/night, as normally administered for jet lag, sleep disorders, and headaches. Further trials on the human bioavailability of herbal melatonin and its effect on diseases are needed to explore the potential of herbal melatonin in traditional medicine. Little is also known about the effects of storage and herbal medicine preparations on melatonin stability, so further study is needed.

Acknowledgment

Special thanks is given to Professor Bung-orn Sripanidkulchai of the Center of Research and Development in Herbal Health, Faculty of Pharmaceutical Sciences, Khon Kaen University, for providing some of the plant materials described in this study.

REFERENCES

Ambriz-Tututi M, Rocha-González HI, Cruz SL, Granados-Soto V. Melatonin: A hormone that modulates pain. *Life Sci* 2009;84(15–16):489–498.

Ansari M, Rafiee K, Yasa N, Vardasbi S, Naimi SM, Nowrouzi A. Measurement of melatonin in alcoholic and hot water extracts of *Tanacetum parthenium*, *Tripleurospermum disciforme* and *Viola odorata*. *Daru* 2010;18(3):173–178.

Badria FA. Melatonin, serotonin, and tryptamine in some egyptian food and medicinal plants. *J Med Food* 2002;5(3):153–157.

Bakhtiarian A. Inhibition of carrageenan-induced edema by tripleurospermum disciforme extract in rats. *Pak J Biol Sci* 2007;10:2237–2240.

Benot S, Goberna R, Reiter RJ, Garcia-Mauriño S, Osuna C, Guerrero JM. Physiological levels of melatonin contribute to the antioxidant capacity of human serum. *J Pineal Res* 1999;27(1):59–64.

Brown RP, Gerbarg PL. Herbs and nutrients in the treatment of depression, anxiety, insomnia, migraine, and obesity. *J Psychiatr Pract* 2001;7(2):75–91.

Burns B, Watkins L, Goadsby PJ. Treatment of medically intractable cluster headache by occipital nerve stimulation: Long-term follow-up of eight patients. *Lancet* 2007;369(9567):1099–1106.

Cady RK, Goldstein J, Nett R, Mitchell R, Beach ME, Browning R. A double-blind placebo-controlled pilot study of sublingual feverfew and ginger (LipiGesic™ M) in the treatment of migraine. *Headache* 2011;51(7):1078–1086.

Cardinali DP, Srinivasan V, Brzezinski A, Brown GM. Melatonin and its analogs in insomnia and depression. *J Pineal Res* 2012;52(4):365–375.

Carpentieri A, Díaz de Barboza G, Areco V, Peralta López M, Tolosa de Talamoni N. New perspectives in melatonin uses. *Pharmacol Res* 2012;65(4):437–444.

Chang H-M, But PP-H, eds. *Pharmacology and Applications of Chinese Materia Medica*, Vol. 2. Singapore: World Scientific, 1986.

Chen G, Huo Y, Tan DX, Liang Z, Zhang W, Zhang Y. Melatonin in Chinese medicinal herbs. *Life Sci* 2003;73(1):19–26.

Cole IB, Cao J, Alan AR, Saxena PK, Murch SJ. Comparisons of *Scutellaria baicalensis*, *Scutellaria lateriflora* and *Scutellaria racemosa*: Genome size, antioxidant potential and phytochemistry. *Planta Med* 2008;74(4):474–481.

Devi B, Sharma N, Kumar D, Jeet K. *Morus alba* Linn: A phytopharmacological review. *Intl J Pharm Pharmaceut Sci* 2013;5(2):14–18.

Dubbels R, Reiter RJ, Klenke E, Goebel A, Schnakenberg E, Ehlers C, Schiwara HW, Schloot W. Melatonin in edible plants identified by radioimmunoassay and high performance liquid chromatography–mass spectroscopy. *J Pineal Res* 1995;18(1):28–31.

Dumont M, Paquet J. Progressive decrease of melatonin production over consecutive days of simulated night work. *Chronobiol Int* 2014;31(10):1231–1238.

Eo HJ, Park JH, Park GH, Lee MH, Lee JR, Koo JS, Jeong JB. Anti-inflammatory and anti-cancer activity of mulberry (*Morus alba* L.) root bark. *BMC Complement Altern Med* 2014;14:200.

Ernst E, Pittler MH. The efficacy and safety of feverfew (*Tanacetum parthenium* L.): An update of a systematic review. *Public Health Nutr* 2000;3(4A):509–514.

Fegert JM, Kölch M, Zito JM, Glaeske G, Janhsen K. Antidepressant use in children and adolescents in Germany. *J Child Adolesc Psychopharmacol* 2006;16(1–2):197–206.

Gagnier JJ. The therapeutic potential of melatonin in migraines and other headache types. *Altern Med Rev* 2001;6(4):383–389.

Gu X, Creasy L, Kester A, Zeece M. Capillary electrophoretic determination of resveratrol in wines. *J Agric Food Chem* 1999;47(8):3223–3227.

Hansen MV, Andersen LT, Madsen MT, Hageman I, Rasmussen LS, Bokmand S, Rosenberg J, Gögenur I. Effect of melatonin on depressive symptoms and anxiety in patients undergoing breast cancer surgery: A randomized, double-blind, placebo-controlled trial. *Breast Cancer Res Treat* 2014;145(3):683–695.

Harshavardhan G, Vinay KK, Manohar RE, Rajesh KJ, Arihara SKG, Veera Reddy Y, Akbar MD. Chemomodulatory influence of melatonin rich *Morus alba* L. on N-methyl-N-nitrosourea (NMU) induced mammary carcinoma in Sprague-Dawley rats. *Res J Pharm Biol Chem Sci* 2012;3(4):994–1000.

Hattori A, Migitaka H, Iigo M, Itoh M, Yamamoto K, Ohtani-Kaneko R, Hara M, Suzuki T, Reiter RJ. Identification of melatonin in plant seeds and its effects on plasma levels and binding to melatonin receptors in vertebrates. *Biochem Mol Biol Int* 1995;35:627–634.

Ho TY, Tang NY, Hsiang CY, Hsieh CL. Uncaria rhynchophylla and rhynchophylline improved kainic acid-induced epileptic seizures via IL-1β and brain-derived neurotrophic factor. *Phytomedicine* 2014;21(6):893–900.

Iriti M, Faoro F. Grape phytochemicals: A bouquet of old and new nutraceuticals for human health. *Med Hypotheses* 2006;67(4);833–838.

Iriti M, Rossoni M, Faoro F. Melatonin content in grape: Myth or panacea? *J Sci Food Agr* 2006;86(10):1432–1438.

Johnson K, Page A, Williams H, Wassemer E, Whitehouse W. The use of melatonin as an alternative to sedation in uncooperative children undergoing an MRI examination. *Clin Radiol* 2002;57(6):502–506.

Lanfumey L, Mongeau R, Hamon M. Biological rhythms and melatonin in mood disorders and their treatments. *Pharmacol Ther* 2013;138(2):176–184.

Lim DW, Kim YT, Park JH, Baek NI, Han D. Antidepressant-like effects of the ethyl acetate soluble fraction of the root bark of *Morus alba* on the immobility behavior of rats in the forced swim test. *Molecules* 2014;19(6):7981–7989.

Liu C, Tseng A, Yang S. *Chinese Herbal Medicine Modern Applications of Traditional Formulas*. Boca Raton: CRC Press, 2005.

Macchi MM, Bruce JN. Human pineal physiology and functional significance of melatonin. *Front Neuroendocrinol* 2004;25(3–4):177–195.

Manchester LC, Tan DX, Reiter RJ, Park W, Monis K, Qi W. High levels of melatonin in the seeds of edible plants: Possible function in germ tissue protection. *Life Sci* 2000;67(25):3023–3029.

Mandegary A, Soodi M, Sharififar F, Ahmadi S. Anticholinesterase, antioxidant, and neuroprotective effects of *Tripleurospermum disciforme* and *Dracocephalum multicaule*. *J Ayurveda Integr Med* 2014;5(3):162–166.

Marseglia L, D'Angelo G, Manti S, Aversa S, Arrigo T, Reiter RJ, Gitto E. Analgesic, anxiolytic and anaesthetic effects of melatonin: New potential uses in pediatrics. *Int J Mol Sci* 2015;16(1):1209–1220.

Matharu MS, Bartsch T, Ward N, Frackowiak RS, Weiner R, Goadsby PJ. Central neuromodulation in chronic migraine patients with suboccipital stimulators: A PET study. *Brain* 2004;127:220–230.

Mauriz JL, Collado PS, Veneroso C, Reiter RJ, González-Gallego J. A review of the molecular aspects of melatonin's anti-inflammatory actions: Recent insights and new perspectives. *J Pineal Res* 2013;54(1):1–14.

Minaiyan M, Ghassemi-Dehkordi N, Mohammadzadeh B. Anti-ulcer effect of *Tripleurospermum disciforme* (C.A. Mey) Shultz Bip. on pylorus ligated (Shay) rats. *Res Pharm Sci* 2006;1:15–21.

Murch SJ, Simmons SB, Saxena PX. Melatonin in feverfew and other medicinal plants. *Lancet* 1997;350:1598–1599.

Murch SJ, Vasantha Rupasinghe HP, Goodenowe D, Saxena PK. A metabolomic analysis of medicinal diversity in Huang-qin (*Scutellaria baicalensis* Georgi) genotypes: Discovery of novel compounds. *Plant Cell Rep* 2004;23:419–425.

Natić MM, Dabić DČ, Papetti A, Fotirić Akšić MM, Ognjanov V, Ljubojević M, Tešić Ž. Analysis and characterisation of phytochemicals in mulberry (*Morus alba* L.) fruits grown in Vojvodina, North Serbia. *Food Chem* 2015;171:128–36.

Ndagijimana A, Wang X, Pan G, Zhang F, Feng H, Olaleye O. A review on indole alkaloids isolated from *Uncaria rhynchophylla* and their pharmacological studies. *Fitoterapia* 2013;86:35–47.

Ou Ming. *Regular Chinese Medicine Handbook*. Taipei: Wang Wen She, 1999.

Padumanonda T, Johns J, Sangkasat A, Tiyaworanant S. Determination of melatonin content in traditional Thai herbal remedies used as sleeping aids. *Daru* 2014;22(1):1–5.

Pareek A, Suthar M, Rathore GS, Bansal V. Feverfew (*Tanacetum parthenium* L.): A systematic review. *Pharmacogn Rev* 2011;5(9):103–110.

Parvini S. The study of analgesic effects and acute toxicity of *Tripleurospermum disciforme* in rats by formalin test. *Toxicol Mech Methods* 2007;17:575–580.

Penny M, Hecker KD, Bonanome A, Coval SM, Binkoski AE, Hilpert KF, Griel AE, Etherton TD. Bioactive compounds in foods: Their role in the prevention of cardiovascular disease and cancer. *Amer J Med* 2002;113(9)Supp.2:71–88.

Peres MF, Masruha MR, Zukerman E, Moreira-Filho CA, Cavalheiro EA. Potential therapeutic use of melatonin in migraine and other headache disorders. *Expert Opin Investig Drugs* 2006;15(4):367–375.

Peres MF, Zukerman E, da Cunha Tanuri F, Moreira FR, Cipolla-Neto J. Melatonin, 3 mg, is effective for migraine prevention. *Neurology* 2004;63(4):757.

Pfaffenrath V, Diener HC, Fischer M, Friede M, Henneicke-von Zepelin HH. The efficacy and safety of *Tanacetum parthenium* (feverfew) in migraine prophylaxis: A double-blind, multicentre, randomized placebo-controlled dose-response study. *Cephalalgia* 2002;22(7):523–532.

Pharmacopoeia Commission of the Ministry of Public Health. *A Coloured Atlas of the Chinese Material Medica Specified in the Pharmacopoeia of the People's Republic of China* (1995 Edition). Guangzhou: Guangdong Science & Technology Press, 1996.

Phitsanuprasātwēt P. *Tumra Paetsart Sonkhrau Chabub Anurak: Textbooks of Thai Traditional Medicine*. Bangkok: Sarm Charoen Panich, 1992.

Pittler MH, Ernst E. Feverfew for preventing migraine. *Cochrane Database Syst Rev* 2004;(1):CD002286.

Plant List, The. Version 1.1, released September 2013. http://www.theplantlist.org. Last accessed 27 Jan 2015.

Pothinuch P, Tongchitpakdee S. Melatonin contents in mulberry (*Morus* spp.) leaves: Effects of sample preparation, cultivar, leaf age and tea processing. *Food Chem* 2011;128:415–419.

Ramakrishna A, Giridhar P, Sankar KU, Ravishankar GA. Melatonin and serotonin profiles in beans of *Coffea* species. *J Pineal Res* 2012a;52:470–476.

Ramakrishna A, Giridhar P, Sankar KU, Ravishankar GA. Endogenous profiles of indoleamines: Serotonin and melatonin in different tissues of *Coffea canephora* P ex Fr. as analyzed by HPLC and LC-MS-ESI. *Acta Physiol Plant* 2012b;34:393–396.

Reiter RJ, Manchester LC, Tan DX. Melatonin in walnuts: Influence on levels of melatonin and total antioxidant capacity of blood. *Nutrition* 2005;21(9):920–924.

Reiter RJ, Tan DX. Melatonin: An antioxidant in edible plants. *Ann NY Acad Sci* 2002;957:341–344.

Reiter, RJ, Tan DX, Burkhardt S, Manchester LC. Melatonin in plants. *Nutr Rev* 2001;59(9):286–290.

Reiter RJ, Tan DX, Korkmaz A, Erren TC, Piekarski C, Tamura H, Manchester LC. Light at night, chronodisruption, melatonin suppression, and cancer risk: A review. *Crit Rev Oncog* 2007;13(4):303–328.

Serfaty MA, Osborne D, Buszewicz MJ, Blizard R, Raven PW. A randomized double-blind placebo-controlled trial of treatment as usual plus exogenous slow-release melatonin (6 mg) or placebo for sleep disturbance and depressed mood. *Int Clin Psychopharmacol* 2010;25(3):132–142.

Shalaby AR. Significance of biogenic amines in food safety and human health. *Food Res Int* 1997;29(7):675–690.

Shi JS, Yu JX, Chen XP, Xu RX. Pharmacological actions of Uncaria alkaloids, rhynchophylline and isorhynchophylline. *Acta Pharmacol Sin* 2003;24(2):97–101.

Shrivastava R, Pechadre JC, John GW. *Tanacetum parthenium* and *Salix alba* (Mig-RL) combination in migraine prophylaxis: A prospective, open-label study. *Clin Drug Investig* 2006;26(5):287–296.

Tang W, Eisenbrand G. *Chinese Drugs of Plant Origin: Chemistry, Pharmacology, and Use in Traditional and Modern Medicine*. Berlin: Springer-Verlag, 1992.

Toglia JU. Melatonin: A significant contributor to the pathogenesis of migraine. *Med Hypotheses* 2001;57(4):432–434.

Wang X, Kang J, Wang HQ, Liu C, Li BM, Chen RY. Three new alkaloids from the fruits of *Morus alba*. *J Asian Nat Prod Res* 2014;16(5):453–458.

Wang Y, Wang Y, Hao J, Li Q, Jia J. Defend effects of melatonin on mung bean UV-B irradiation. *Acta Photonica Sinica* 2009;38:2629–2633.

Wang Y, Xiang L, Wang C, Tang C, He X. Antidiabetic and antioxidant effects and phytochemicals of mulberry fruit (*Morus alba* L.) polyphenol enhanced extract. *PLoS One* 2013;8(7):e71144.

WHO. *General Guidelines for Methodologies on Research and Evaluation of Traditional Medicine*. Geneva: World Health Organization, 2000.

Wu SJ, Wang JS, Lin CC, Chang CH. Evaluation of hepatoprotective activity of legumes. *Phytomedicine* 2001;8(3):213–219.

Xie C, Kokubun T, Houghton PJ, Simmonds MS. Antibacterial activity of the Chinese traditional medicine, Zi Hua Di Ding. *Phytother Res* 2004;18(6):497–500.

Yang LX, Liu D, Feng XF, Cui SL, Yang JY, Tang XJ, et al. Determination of flavone for *Scutellaria baicalensis* from different areas by HPLC. (Article in Chinese.) *Zhongguo Zhong Yao Za Zhi* 2002;27(3):166–170.

Yang Y, Tan YX, Chen RY, Kang J. The latest review on the polyphenols and their bioactivities of Chinese Morus plants. *J Asian Nat Prod Res* 2014;16(6):690–702.

Yang Y, Zhang T, Xiao L, Chen RY. Two novel flavanes from the leaves of *Morus alba* L. *J Asian Nat Prod Res* 2010;12(3):194–198.

Zhang E, ed. *Chinese Medicated Diet*. Shanghai: Shanghai College of Traditional Chinese Medicine, Shanghai University, China, 1990.

Zohara R, Izhakib I, Koplovichb A, Ben-Shlomoa R. Phytomelatonin in the leaves and fruits of wild perennial plants. *Phytochem Lett* 2011;4(3):222–226.

Zou Y, Liao S, Shen W, Liu F, Tang C, Chen CY, Sun Y. Phenolics and antioxidant activity of mulberry leaves depend on cultivar and harvest month in Southern China. *Int J Mol Sci* 2012;13(12):16544–16553.

15

Association of Medicinal Plants and Melatonin in Human Health

Eduardo Luzía França, Cristina Filomena Justo, Michelangelo
Bauwelz Gonzatti, and Adenilda Cristina Honorio-França
Federal University of Mato Grosso
Mato Grosso, Brazil

CONTENTS

ABSTRACT   Melatonin hormone (N-acetyl-5-methoxytryptamine) is produced by the pineal gland and plays an important protective role for humans. Many of the benefits of melatonin and its metabolites are related to their antioxidant, anti-inflammatory, and prooxidative effects. Melatonin has been shown to act on human phagocytes as well as rat splenic macrophages, and may be considered a potent immunomodulator at physiological concentrations. It can stimulate natural immunity, which is an important defense against microbial infections. This molecule is also present in different plants and is denominated *phytomelatonin*. Its presence in edible and medicinal plants calls for attention due to the possible effect of exogenous melatonin on human physiology, related to sleep and other functions correlated to circadian rhythm. Phytomelatonin concentrations vary from the order of pictograms to micrograms per gram, detected by different techniques. Many melatonin-rich plants in the diet are Eurasian in origin, such as tart cherry, red wine grape, and oat, but some are from America, such as tomato, pineapple, and corn. Investigations into the traditional knowledge of tea infusions and other homemade remedies in the induction of sleep have revealed the presence of phytomelatonin in medicinal plants from China and some other countries. Some native medicinal plants in Brazil might also have phytomelatonin. These plants belong to a rich flora, in many cases endemic to the biomes of the Amazon, Cerrado, and Mata Atlântica, that are now under pressure from deforestation. Previous studies have shown that several melatonin-rich medicinal plants have the ability to induce immune system activation and display beneficial effects against diseases. This chapter attempts to evaluate the potential presence of phytomelatonin in Brazilian flora and its relation to medicinal use, and to evidence the main effects of this hormone's associated plants on human health.

KEY WORDS: *melatonin, medicinal plants, phytomelatonin, immunomodulatin, Brazilian flora.*

Introduction

Melatonin (N-acetyl-5-methoxytryptamine) is a substance with many diverse functions in the human body. It is a molecule characteristic of indoleamines, which can affect sleep, humor, body temperature, retinal activity, and sexual behavior, among other physiological manifestations. These aspects are mainly modulated by the regulation of the body's circadian rhythm by melatonin (Reiter 2003; Fuller et al. 2006; Maronde and Stehle 2007; Slominski et al. 2008; Jan et al. 2009; Zmijewski et al. 2009; Hardeland 2012; Arnao 2014).

The action of melatonin was first detected from extracts of cattle pineal gland in 1958, being identified as present in the human body around one year later (Lerner et al. 1958, 1959). Initially, it was thought to be synthesized only in the pineal gland, but recent studies have demonstrated that it can be synthesized in different kinds of tissue (Kvetnoy 1999; Stefulj et al. 2001).

Melatonin and its metabolites are associated with their antioxidant effects, which inhibit the damage caused by reactive species through various mechanisms. They act directly on free radicals and indirectly by stimulating the production and/or activation of intracellular antioxidant enzymes (Tan et al. 2007).

This hormone plays an important immunomodulatory role in controlling infectious diseases (França et al. 2009a), and is reported to have a stimulatory effect on human colostrum macrophages, the activation of cellular defense mechanisms such as superoxide formation and phagocytosis, and intracellular calcium release (Hara et al. 2013; Honorio-França et al. 2013).

Moreover, plants which are now reported to be rich in melatonin have long been used for medicinal purposes (Violante et al. 2009). Advances in ethnopharmacology have led to investigations into several bioactive compounds and their functional attributes. Many plants rich in melatonin also exhibit immunomodulatory effects, antimicrobial properties that need to be further explored for their therapeutic potential (Reinaque et al. 2012; Pessoa et al. 2014).

Thus, the search for antimicrobial properties in extracts of these plants and their association with endogenous substances has intensified, requiring further studies directed at the development of new substances, as well as the development of new approaches to the treatment of diseases (Cunico et al. 2004). Active principles in native plants have been the subject of increasing research due to the large investment in alternative medicines and because of their therapeutic actions.

Physiological Processes of Melatonin

Melatonin is involved in many physiological processes related to sleep and circadian rhythm, and its dysfunction may contribute to many diseases or conditions, such as insomnia, anxiety, and mood disorders (Comai and Gobbi 2014). Melatonin is an important regulator of the total oxidative status of the cell and body and seems to have a number of means of protection against oxidative stress. This molecule can act as a direct antioxidant, scavenging reactive species of oxygen and nitrogen, and also works indirectly when stimulating antioxidant enzymes (Reiter et al. 2003).

The effectiveness of melatonin in the improvement of sleep quality is already being studied, but more conclusive trials are still to be run (Costello et al. 2014). Melatonin has been demonstrated to decrease sleep onset latency, increase total sleep time, and improve sleep quality (Ferracioli-Oda et al. 2013). The current pharmacological treatment for sleep disorders is mainly based on benzodiazepine receptor agonists, but they are known to present several side effects, including addiction, withdrawal symptoms, and drug resistance (Wheatley 2005). In the past few decades, there has been increased interest in the use of complementary and alternative medicines as a natural form of treatment, focusing on reducing the amount of side effects and the cost of medications currently used to treat numerous types of anxiety and sleep disorders (Gooneratne 2008).

The activity of melatonin in the brain seems to be mostly related to two high-affinity G protein–coupled receptors, MT1 and MT2. Ramelteon is an MT1/MT2 receptor agonist that demonstrates sleep-promoting effects and has been used as a treatment for insomnia, indicating that melatonin compounds facilitate the induction of sleep. The action of ramelteon on both subsets of receptors makes it difficult to discern which receptor is involved in its hypnotic and sedative properties. With the use of antibodies that target MT2 receptors, it has been demonstrated that most of the neurons bearing this subset of receptor are GABAergic, indicating that a selective MT2 agonist might have a better pharmacological profile than the classical benzodiazepines (Comai and Gobbi 2014).

Melatonin has also been associated with memory formation. A facilitating influence in short-term memory formation was observed in an olfactory social memory test and this effect was blocked with the use of the MT1 receptor antagonist luzindole, suggesting the involvement of endogenous melatonin (Argyriou et al. 1998). Melatonin also demonstrated improvements in rats with learning and memory impairment caused by induced hyperhomocysteinemia, and it is possible that melatonin has a protective effect against oxidative stress caused by homocysteine neurotoxicity, which can involve the production of reactive species of oxygen (Baydas et al. 2005).

Melatonin and Medicinal Plants

Herbal medicine as an alternative treatment for anxiety, mood, and sleep disorders has been widely used in many different cultures, and a variety of plants and compounds obtained from herbs are used in folk medicine due to their medicinal properties (Lakhan and Vieira 2010; Sarris et al. 2011). Besides having fewer side effects and being more cost-effective, some extracts from single plants contain numerous potentially psychoactive components, some of which may present synergistic effects and improve outcomes (Sarris et al. 2011). Lots of plants have shown similar effects on anxiety and sleep-related disorders, suggesting a relation to melatonin content or the presence of molecules that may be related to melatonin secretion or affect indirectly the physiological process mediated by this hormone (Miyasaka et al. 2007; Shin et al. 2013).

A few plants presenting sedative and anxiolytic effects have already been described and analyzed in clinical trials, but there is still a need for more research to better elucidate many plants' potential and mechanisms of action (Lakhan and Vieira 2010). Kava-kava (*Piper methysticum* G. Forster, Piperaceae) and valerian (*Valeriana officinalis* L., Caprifoliaceae) are two examples of plants that have been studied and used in many settings as alternative medication for anxiety and sleep disorders. A few reviews have summarized their pharmacology, efficacy, and adverse effects from an analysis of numerous randomized and placebo-controlled trials (Wheatley 2005; Antoniades et al. 2012).

Most plants used in popular medicine, especially in underdeveloped countries, have yet to be assessed for melatonin content, but some of the medicinal effects of these plants could be related to physiological processes regulated by melatonin. A variety of herbs in Brazilian popular medicine have recently been studied for their medicinal properties. The leaves of *Cymbopogon citratus* Stapf. are used in the form of tea for their possible anxiolytic, hypnotic, and anticonvulsant properties. The effect of the essential oil of this species was evidenced in mice, increasing sleeping time and suggesting the regularization of circadian rhythm (Blanco et al. 2009).

A Brazilian native species of passion fruit, *Passiflora actinia* Hooker (*maracujá-do-mato*), has also been evaluated for its behavioral effects in mice. In this study, carried out in Brazil, the sedative and anxiolytic effects of the plant's leaves were blocked by a GABA-A-benzodiazepine receptor antagonist, suggesting the involvement of a GABA-A system in this effect (Lolli et al. 2007).

The Verbenaceae *Lippia alba* (Mill.) is used in Brazilian popular medicine mainly for its sedative and hypnotic properties. The central nervous system activity of the plant extract was tested in mice in a Brazilian study. Different extracts were analyzed for their flavonoid content in comparison with their significant sedative and myorelaxant effects (Zétola et al. 2002).

Citrus aurantium L. and *Citrus limon* (L.) Burn. are two species commonly used in Brazilian popular medicine as an alternative treatment for insomnia and anxiety (Carvalho-Freitas and Costa 2002; Lopes-Campêlo et al. 2011). Researchers in Brazil analyzed the potential effects of *C. aurantium* L. in mice, suggesting a depressive action on the central nervous system and elucidating its potential use as an anticonvulsant in the treatment of epilepsy as well as its anxiolytic and sedative actions (Carvalho-Freitas and Costa 2002).

The neurobehavioral effect of the essential oil obtained from *C. limon* was analyzed using mice in an elevated plus maze anxiety test and compared with the effects of currently used medication. The study demonstrated sedative, anxiolytic, and antidepressant effects, and suggested that these effects might involve action on benzodiazepine-type receptors (Lopes-Campêlo et al. 2011).

Many studies on the identification and quantification of melatonin presence in edible and medicinal plants have already been done (Badria 2002; Chen et al. 2003; Reiter and Tan 2002). Melatonin seems to be present in a variety of plant species, some of which are commonly present in our diet (Huang and Mazza 2011). Melatonin has a low molecular weight and can easily cross lipid membranes; this is important to the evaluation of melatonin consumed in the diet because it is absorbed in the gut and significantly alters circulating levels of the indoleamine (Reiter and Tan 2002).

Studies have shown the presence of melatonin and serotonin in different parts of plants, including leaves, roots, flowers, fruits, and seeds (Paredes et al. 2009). It is reportedly extremely difficult to assess and measure melatonin content in plants, this being the main aspect to have slowed down research in this field (Arnao and Hernández-Ruiz 2007). Also, melatonin can have a direct antioxidant effect, without the mediation of membrane receptors, thus reacting quickly with reactive oxygen species (Tan et al. 2007).

Hypericum perforatum, commonly known as St John's wort, is evidently effective in the treatment of depression (Sarris et al. 2011). Clinical trials and meta-analysis have demonstrated its efficacy and tolerability in the treatment of mild to moderate depression, supported by a comparison with the outcomes of other antidepressants (Kasper et al. 2010). *H. perforatum* has been described as having significant melatonin content in its leaves (Murch and Saxena 2006; Reiter and Tan 2002), and this could be related to some of the outcomes observed in its medicinal effects. An analysis of brain neurotransmitter receptor binding in rats after the administration of *H. perforatum* showed a significant decrease in the binding of a dopamine receptor associated with the anxiolytic effect and an upregulation of serotonin and benzodiazepine receptors (Kumar et al. 2002).

A report on the antidepressant mechanisms of *H. perforatum* suggested that the presence of hyperforin and hypericin is responsible for most of the antidepressant effects via nonspecific presynaptic inhibition of neurotransmitter reuptake in the brain (Mennini and Gobbi 2004). Another study evaluated the sedative and anxiolytic effects of *H. perforatum* devoid of hyperforin on mice in an anxiety test. The results showed positive antidepressant effects, suggesting that the observed effects were not attributed to this substance (Coleta et al. 2001). This plant species has also been shown to have free radical scavenging activities and to protect against lipid peroxidation (Silva et al. 2005).

H. perforatum, *Tanacetum parthenium*, and *Scutellaria baicalensis* were some of the first plants to be characterized as containing melatonin in a study done in 1997 on plants traditionally used for the treatment of neurological disorders, with the amount of melatonin ranging from 0.09 to 7.11 μg g^{-1} among the three species (Murch et al. 1997). *T. parthenium* is commonly known as *feverfew*, receiving this name due to its potentially anti-inflammatory effects. The plant has also been shown to have anticancer activity, and is suggested to be effective in the treatment of migraine headaches and rheumatoid arthritis (Pareek et al. 2011).

In regard to its antioxidant activity, *T. parthenium* alcoholic extract possessed strong free radical scavenging activity and moderate Fe^{2+}-chelating capacity (Wu et al. 2006). This species demonstrated immunomodulatory properties, regulating gene expression profiles in human monocytic THP-1 cells, and mediating cellular metabolism, migration, and cytokine production (Chen and Cheng 2009). The molecular components responsible for the observed antioxidant and immunomodulatory effects are not completely understood, and have been attributed to other components rather than melatonin. On the other hand, melatonin has already been characterized as an antioxidant and as a potential modulator for immune cells (França et al. 2010); thus, the melatonin content in feverfew plants could be related to the presented effects.

Herbs containing high levels of melatonin have been used traditionally in the treatment of neurological problems and disorders involving oxidative metabolism, acting in the scavenging of free radicals. *T. parthenium*, *Tripleurospermum disciforme*, and *Viola odorata* are three plant species that have a long history in traditional medicine for their sedative and anti-inflammatory properties, and are also used to treat stress-related conditions. Different methods for the identification and quantification of melatonin in their flowers and leaves have been used, and their high melatonin content might be evidenced by the physiological effects in humans (Ansari et al. 2010).

Balaton and Montmorency tart cherries (*Prunus cerasus*) have also been reported as having high melatonin content in comparison with normal levels of the indoleamine in the blood. In the study by Burkhardt et al. (2001), Montmorency cherries contained on average 13.46 ng g^{-1} of melatonin, presenting almost six times more melatonin than Balaton cherries, which had a mean of 2.06 ng g^{-1}. In a randomized double-blind, placebo-controlled study, the effect on sleep quality was assessed, demonstrating improvement in sleep duration and quality in healthy subjects. This study also demonstrated an increase in urinary melatonin after dietary intake of concentrated tart cherry juice by analysis of the metabolite 6-sulfatoxymelatonin (Howatson et al. 2011). The consumption of cherries has also shown potentially protective benefits in regard to cancer and cardiovascular and inflammatory diseases, demonstrating high antioxidant activity supported by its melatonin and vitamin C content (McCune et al. 2011).

Cranberries (*Vaccinium* spp., Ericaceae) have demonstrated antioxidant activity and are suggested to have protective effects against cancer and cardiovascular and neurodegenerative diseases. It has been proposed that the components of the fruit might counteract the effects of oxidative stress and damage, and these properties have been mainly attributed to the presence of flavonoids, especially anthocyanins (Neto 2007; Kähkönen et al. 2001). A comparison between three species of cranberry in relation to their melatonin content confirmed their free radical scavenging capacity and demonstrated that this could be partially related to the presence of melatonin and serotonin, though there was not a very strong correlation. The commercially cultivated large cranberry *V. macrocarpon* Ait. showed a significantly higher amount of melatonin in comparison with the moderate amounts of two smaller species, *V. oxycoccos* L. and *V. vitis-idaea* L. Serotonin content seemed to be constant among the three species, but all of them tested positive for the presence of indoleamines (Brown et al. 2012).

Chen et al. (2003) screened 108 herbs widely used in Chinese medicine for the presence of melatonin. The concentration of melatonin measured using advanced high-performance liquid chromatography (HPLC) were found to vary among species from 10 to 3000 ng g^{-1}. From the total plants analyzed, 64 of them contained an amount of melatonin greater than 10 ng g^{-1} of dry mass and 43 had more than 100 ng g^{-1}. The majority of herbs that presented a significantly higher amount of melatonin are traditionally used to retard aging and to treat diseases related to free radical production and oxidative stress. The species *Epimedium brevicornum* Maxim (Berberidaceae) and the root cortex of *Morus alba* L. (Moraceae), for example, contained 1105 and 1110 ng g^{-1}, respectively, and are commonly used to slow the aging process. The leaf of *M. alba* is believed to have a beneficial effect on irradiation damage caused by ultraviolet (UV) light or radiotherapy, and also presented a significant amount of melatonin (1510 ng g^{-1}) (Chen et al. 2003).

Seven Thai herbal remedies that are used in traditional medicine as sleeping aids were screened for their melatonin content using HPLC with fluorescent detection as the most sensitive method; enzyme-linked immunosorbent assay (ELISA) was used as confirmation. Six of the seven plants tested positive for melatonin content and *Piper nigrum* L. (Piperaceae), known as black pepper, showed the highest amount of the indoleamine (1092.7 ng g^{-1} dry sample weight). The other herbs that contained melatonin were from Fabaceae (*Sesbania grandifora* [L.] Desv., *S. sesban* [L.] Merr., and *Senna tora* [L.] Roxb.) and from Cucurbitaceae (*Momordica charantia* L.), and presented a modest amount in comparison with *P. nigrum*, ranging from 8.7 to 43.2 ng g^{-1} (Padumanonda et al. 2014).

Serotonin is a precursor to the formation of melatonin, and both indoleamines are synthesized in a similar pathway from the amino acid L-tryptophan (Murch et al. 2000). A lot of plants were identified for the presence of serotonin, and it is possible that the intake of serotonin might increase its concentration in the blood, leading to a greater synthesis of endogenous melatonin. Serotonin has been identified in a variety of edible plants very common in human diets, such as lettuce, strawberry, green onion, spinach, hot pepper, and tomato (Ly et al. 2008).

The consumption of the fruit or juice of pineapple, orange, and banana demonstrated an increase in serum levels of melatonin, indicating that the presence of this indoleamine or its precursors can directly affect concentration in the blood. The study also correlated this variation with the antioxidant effects observed for those fruits (Sae-Teaw et al. 2013). The consumption of orange, banana, and apple demonstrated a protective effect on neurotoxicity induced by oxidative stress, and a decrease in the risk of neurodegenerative disorders (Heo et al. 2008), properties that might be somehow related to the melatonin and serotonin content of these edible plants.

Fernández-Pachõn et al. (2014) demonstrated that the concentration of melatonin increases during the fermentation of orange juice. This synthesis seems to be related to tryptophan content, as the levels of tryptophan were inversely correlated with melatonin levels, and the melatonin would appear earlier in the process with the addition of tryptophan before fermentation. The presence of melatonin is also described in species of grapes, and is related to the antioxidant effects of this fruit (Vitalini et al. 2011). A similar process happens during the fermentation of grapes in the production of wine: yeast fermentation can contribute to the synthesis of melatonin from tryptophans and increase the amount of exogenous melatonin present in wine samples (Gomez et al. 2012; Rodriguez-Naranjo et al. 2011).

Phytomelatonin

The concentration of melatonin in different plants is variable and significant, but in general it is present in very low concentrations, so the analytical methods must be sensitive and the extraction methods must be chosen wisely to guarantee complete recovery and accurate measurements (Huang and Mazza 2011).

The first articles to mention the presence of melatonin in plants were published in 1993 and the following years (Tassel and O'Neill 1993; Tassel et al. 1995; Hattori et al. 1995), although there were previous reports of melatonin presence in dinoflagellates (Poeggeler et al. 1989, 1991).

Phytomelatonin has a structure similar to auxin because they are both derived from tryptophan, and more than 100 plants have already had their melatonin content measured in different parts of the plants (Arnao 2014); concentrations vary from picograms to micrograms per gram of tissue. These differences do not seem to have a taxonomical correlation (Hardeland 2015; Ramakrishna et al. 2009, 2012a,b,c; Ramakrishna 2015). In general, seeds present higher contents than vegetative parts (Manchester et al. 2000; Reiter and Tan 2002; Hardeland et al. 2007; Arnao 2014), although this varies depending on the cultivar, place of harvesting, physiological state, and method of extraction (Janas and Posmyk 2013).

Some physiological functions in the plants have been attributed to phytomelatonin—for example, the regulation of root development—suggesting cooperation with auxin in biological processes in *Arabidopsis* (Koyama et al. 2013) and *Hypericum* (Murch et al. 2001); on the other hand, other studies have shown that melatonin does not mimic the response to auxin (Arnao 2014).

The greater concentration of phytomelatonin at nighttime periods led to the hypothesis that it could also have a function in the perception of circadian rhythm in plants (Kolar and Machackova 1994; Kolar et al. 1999), but this hypothesis was subsequently disregarded (Afreen et al. 2000; Tettamanti et al. 2000; Tan et al. 2007). It has also been proposed that melatonin has an antioxidative role in plant tissue (Reiter and Tan 2002; Reiter et al. 2005; Hardeland et al. 2007). Besides that, the response of plants differs in relation to concentration. It is possible that there is more than one type of cellular receptor for phytomelatonin, but other types have not yet been identified (Hardeland 2015); therefore, lots of questions remain unanswered in respect to phytomelatonin, indicating that new studies might have promising results (Arnao 2014; Arnao and Hernández-Ruiz 2007; Paredes et al. 2009). Weeda et al. (2014) demonstrated that melatonin could activate a diverse range of genes related to stress signaling in plants.

Melatonin is easily absorbed in the intestinal tract (Reiter et al. 2007) and it is preserved in dried plant material (Hardeland and Poeggeler 2003; Hardeland 2008). It has been suggested that the intake of plants with elevated levels of melatonin can alter the concentration of the compound in the blood (Dubbels et al. 1995). Hardeland (2015) argued that the use of transgenic plants does not induce a greater concentration of melatonin, having observed the production of toxic derivatives from the metabolic pathway of serotonin and melatonin synthesis in these organisms.

The abundance of melatonin-rich plants found in regions of greater altitude might be related to protection against UV radiation damage, which is more intense in this kind of ecosystem (Caniato et al. 2003; Kolar and Machackova 2005; Tan et al. 2012). It is possible that there are many plants in Brazil with high melatonin contents, because in most of the country the rate of UV radiation is elevated (Corrêa et al. 2003).

Beyond that, ethnobotanical studies are important tools in the discovery of new active ingredients (Elisabetsky 2003; Elisabetsky and Souza 2004) and they reveal that teas of a diversity of plants are reported as calmative, sedative and used for insomnia (Agra et al. 2007, 2008, Di Stasi and Hiruma-Lima 2002, Guarim-Neto and de Morais 2003, Lorenzi and Abreu-Matos 2008). Clinical trials have tried to evaluate the efficacy of phytotherapics used in traditional and folk medicine for the treatment of insomnia (Feyzabadi et al. 2014). The preparation of infusions with hot water is enough to extract a significant amount of melatonin (Ansari et al. 2010); homemade preparations are therefore efficient for this extraction, and in general an infusion or decoction is prepared to be taken at night (Agra et al. 2007, 2008).

Different authors have evaluated the melatonin content of plants used in traditional medicine in their countries, frequently observing concentrations superior to 1.0 $\mu g\ g^{-1}$ of dried material (Padumanonda et al. 2014; Chen et al. 2003). In Brazil, in a syncretism of knowledge, native plants are generally used as well as plants that have been introduced (Elisabetski and Shanley 1994; Elisabetsky 2003; Elisabetsky and Souza 2004), because the popular culture in the country is a summation of knowledge inherited from the native indigenous population, from Africans that were brought over as slaves, and from Europeans of different nationalities that colonized the country (Mello 1980).

Only a few species native to Brazil were analyzed for their melatonin content (Arnao 2014; Hardeland 2008), but some are closely related to other analyzed species and the results seem to be promising (Hardeland 2008). A variety of species of *Hypericum* occur in Brazil (Slusarski et al. 2007) and they are close to *H. perforatum* L. (Hypericaceae), which presents great levels of phytomelatonin (Arnao 2014; Murch et al. 2001; Barnes et al. 2001). Among the species of *Hypericum* native to Brazil, there are no studies on the characterization of their phytomelatonin content, and only *H. connatum* Lam. has been reported in ethnobotanical studies as an alternative treatment for oral lesions and with potentially protective effects against the herpes simplex virus (Fritz et al. 2007).

There is enormous floral diversity in the Amazon, but only a small fraction has been tested for its pharmacological activity (Elisabetski and Shanley 1994). Due to the high cost of allopathic medications and the difficulty in accessing medical care in the region, medicinal plants are generally utilized by populations with low incomes and poor education. Knowledge about medicinal plants is usually carried by the women, and the information is orally transmitted to the next generation (Costa and Mitja 2010). Certain communities utilize more than 100 species of medicinal plants. Some of those species are described as antidepressant and analgesic; there is no mention of their direct usage in the treatment of insomnia, but research indicates that a lot of the herbs used as "tonic for the nerves" have antidepressant properties (Elisabetski and Shanley 1994).

In the Caatinga region, the rural population is very poor and has a tradition of using medicinal herbs from the area; some of them are used to help with sleep, as they are thought to have a sedative effect (Agra et al. 2007, 2008). The region is very likely to suffer from a lack of rain, presenting an adapted xerophytic vegetation (Almeida 2006).

The pharmacopoeia of the region known as Cerrado is diverse and has over 200 plants; however, only *Duguetia furfuracea* (Anonaceae), *Pterodon emarginatus* (Fabaceae), and *Guapia tomentosa* (Nyctaginaceae) are mentioned as having sedative effects (Vieira and Martins 2000). Some other species have been identified by Maroni et al. (2006) and are mentioned in Table 15.1.

The Pantanal region is subject to periodical overflow, frequently isolating the population. The regional culture values local solutions, and that includes knowledge of medicinal plants and homemade remedies. Only three plants are mentioned as having soothing properties in ethnobotanical studies (Guarim-Neto and de Morais 2003).

In communities of the Mata Atlântica region in the south of Brazil, the use of exotic plants such as *T. parthenium* L. (Asteraceae) is predominant. Among species native from the region, *Aloysia gratissima* (Gillies and Hook) Tronc. (Verbenaceae) stands out, as it has already been studied in relation to its neuroprotective effects and as an antidepressant (Zeni 2011). *Aloysia triphylla* (L'Hér.) Britton (Verbenaceae) is native to the south of the country and some surrounding countries (Brant et al. 2009), and presents significant amounts of melatonin (Reiter et al. 2001).

TABLE 15.1

Medicinal Plants Cited in Ethnobotanical Studies in Brazil to Be Explored for Phytomelatonin

Botanical Family Scientific Name	Vernacular Name(s)	Used Parts	Ethnobotanical Registry Region	Confirmed Activity	Active Constituent	Bibliographic Source
Annonaceae						
Annona muricata L.	Graviola, araticum-de-paca, nona	Fruit juice, infusion of dried leaves	AMA	ANL, SED, CTX	Acetogenin, cyanohydrins, terpenes, tannins, steroids, cardiac glycosides, cyclopeptides	Di Stasi and Hiruma-Lima 2002; Mass 2015; Gajalakshmi et al. 2012; Rodrigues et al. 2008
Duguetia furfuracea (A. St.-Hil.)	Beladona, sofre-do-rim-quem-quer, araticum bravo, ata-de-lobo	Decoction of root	CER	CPR	Tannins, flavonoids, alkaloids	Vieira and Martins 2000; Maas et al. 2015; Fernandes-Lima et al. 2014; Guarim-Neto and Morais 2003
Apocynaceae						
Calotropis procera (Ait.) W.T. Aiton	Algodão-de-seda, algodão-da-praia, flor-de-seda, flor-de-cera	Decoction of leaves	CAA	ANT, ANL, PRT	Cardiac glycosides	Agra et al. 2007; Sousa-Rangel and Trindade-Nascimento 2011; Koch et al. 2015; Lorenzi and Matos 2008
Skytanthus hancorniifolius (A. DC.) Miers	Leiteiro	Infusion of leaves, flowers, and stem bark	CAA	—	No phytochemical data	Agra et al. 2008; Araújo et al. 2008; Koch et al. 2015; Albuquerque et al. 2007
Thevetia peruviana (Pers.) K. Schum.	Castanha-da-índia, chapéu-de-Napoleão, fava-elétrica, noz-de-cobra, coração de Jesus	Seeds	AMA	POI, BAC	Alkaloids, cardiac glycosides, flavonoids, tannins, fatty acids	Di Stasi and Hiruma-Lima 2002; Koch et al 2015
Aristolochiaceae						
Aristolochia esperanzae Kuntze.	Cipó-mil-homens, papo de peru do cerrado, jarrinha	Infusion of root	CER	BAC	Flavonoids, diterpenes, sesquiterpenoids	Maroni et al. 2006; Lorenzi and Matos 2008; Pacheco 2010
Asteraceae						
Achyrocline satureioides (Lam.) DC.	Macela-do-campo, camomila nacional	Infusion of inflorescences and leaves	FLO	ANL, ANT, REL, ANV	Flavonoids, monoterpenes, sesquiterpenes	De Paula et al. 2009; Loeuille and Monge 2014; Lorenzi and Matos 2008
Egletes viscosa (L.) Less.	Macela-do-campo, macela-da-terra, chá-da-lagoa	Inflorescences and leaves	CAA	ANV, REL, ANT	Flavonoids, saponins, terpenoids	Agra et al. 2007, 2008; Araujo et al. 2008; Borges and Teles 2014; Lorenzi and Matos 2008
Mikania glomerata Spreng.	Guaco, coração de Jesus, erva-de-sapo	Decoction of leaves	FLO, CAA	BRO	Coumarin, diterpenes, lactones, sesquiterpenes	De Paula et al. 2009; Ritter et al. 2015; Lorenzi and Matos 2008; Santana et al. 2014

Species	Common names	Part used			Compounds	References
Porophyllum ruderale (Jacq.) Cassini	Arnica, couve cravinho, arruda de galinha, erva de veado, cravo de urubu	Leaves	CAA	ANT	Essential oils, tannins	Agra et al. 2007; 2008, Fonseca et al. 2007; Nakajima 2015a
Tagetes minuta L.	Cravo bravo	Infusion of flowers and leaves	CAA	GAB	Essential oils	Agra et al. 2007, 2008; Rodrigues et al. 2008; Nakajima 2015b
Bignoniaceae						
Anemopaegma arvense (Vell.) Stellfeld ex de Souza	Catuaba, alecrim do campo, marapuama, caramuru	Decoction of stem bark and xylopodium	CER	ADP, APH, AOX	Alkaloids, tannins, resins	Maroni et al. 2006; Lorenzi and Matos 2008; Silva et al. 2012
Tynanthus fasciculatus Miers	Cipó-cravo	Infusion of stem and leaves	FLO	SPE, ANL, AOX	Tannins, saponins, flavonoids, cardiac glycosides	De Paula et al. 2009; Lohmann 2015; Melo et al. 2010; Carvalho et al. 2009; Carvalho-Freitas and Costa 2002
Convolvulaceae						
Merremia dissecta (Jacq.) Hallier f.	Jitirana	Infusion of leaves	CAA	—	—	Agra et al. 2007, 2008
Fabaceae						
Anadenanthera colubrina var. *cebil* (Griseb.) Altschul	Angico	Decoction or syrup of stem bark	CAA	HAL	Tannins, alkaloids, phytosteroids, flavonoids, triterpenoids, phenolic compounds	Agra et al. 2007; Lorenzi and Matos 2008
Caesalpinia microphylla Mart. ex G. Don	Arranca-estribo	Decoction of stem bark	CAA	—	—	Agra et al. 2007, 2008
Erythrina velutina Willd.	Mulungu, corticeira	Infusion of stem bark	CAA, FLO	SNC, ACO, ANX	Alkaloids	Agra et al. 2007, 2008; De Paula et al. 2009; Lorenzi and Matos 2008; Rodrigues et al. 2008
Mimosa acutistipula (Mart.) Benth.	Jurema-preta	Stem bark	CAA	—	—	Agra et al. 2007, 2008
Mimosa verrucosa Benth.	Jurema	Stem bark	CAA	—	—	Agra et al. 2007, 2008
Pterodon emarginatus Vog. (= *Pterodon polygalaeflorus* Benth.)	Sucupira	Seeds	CER	—	Essential oils, isoflavones, diterpenes	Vieira and Martins 2000; Lorenzi and Matos 2008
Senna obtusifolia (L.) H.S. Irwin & Barneby	Mata-pasto, fedegoso	Decoction of leaves	CAA	PSO, ANF, BAC	Anthraquinones, naphthalene compounds, galactomannans	Agra et al. 2007, 2008; Lorenzi and Matos 2008
Lamiaceae						
Leonotis nepetifolia (L.) R.Br.	Cordão-de-são-francisco, rubim-de-bola, emenda-nervos	Decoction of leaves and flowers	CAA	REL	Diterpenes, coumarin, sesquiterpene lactones, flavonoids, triterpenoids, caffeine	Agra et al. 2007, 2008; Lorenzi and Matos 2008

(Continued)

TABLE 15.1 (CONTINUED)

Medicinal Plants Cited in Ethnobotanical Studies in Brazil to Be Explored for Phytomelatonin

Botanical Family Scientific Name	Vernacular Name(s)	Used Parts	Ethnobotanical Registry Region	Confirmed Activity	Active Constituent	Bibliographic Source
Vitex cymosa Bert.	Tarumã-do-brejo, azeitona-do-mato	Infusion of stem bark and leaves	PAN	BAC, ANF	Iridoids, essential oils	Guarim-Neto and Morais 2003; Santos et al. 2001; Fonseca et al. 2006
Vitex gardneriana Schauer	Jaramataia, tamanqueira	Infusion of leaves	CAA	SED	Terpenoids, iridoids, saponins, sugars, flavonoids, phenylpropane glycoside	Agra et al. 2007; Sá-Barreto et al. 2008; Sá-Barreto 2004
Lythraceae						
Cuphea cartagenensis (Jacq.) J.F. Macbr.	Sete-sangrias, guaxuma	Whole plant	CER	AOX, ANX	Essential oils, tannins, pigments, mucilage, saponins, flavonoids	Maroni et al. 2006; Lorenzi and Matos 2008; Lorenzo 2000
Malvaceae						
Hibiscus rosa-sinensis L.	pampoela, firmeza-dos-homens, rosa paulista, rosa louca	Infusion of flowers	AMA, ATL	HAL, HGL	Phenolic acids [339], fatty acid methyl esters, anthocyanins	Di Stasi and Hiruma-Lima 2002
Nyctaginaceae						
Guapira tomentosa (Casar.) Lund. (= *Pisonia tomentosa* Casar.)	NI	Folha	CER	—	—	Vieira and Martins 2000
Orchidaceae						
Vanilla palmarum Lindley	Baunilha, bonilha	Seeds	PAN	—	—	Guarim-Neto and Morais 2003
Passifloraceae						
Passiflora alata	Maracujá	Decoction of leaves and fruit juice	CER	ANL, SED, SNC, ACO	Alkaloids, cyanogenic glycosides, flavonoids	Di Stasi and Hiruma-Lima 2002; Vieira and Martins 2000; Lorenzi and Matos 2008
Passiflora amethystina J.C. Mikan	Maracujá	Flower, leaves, fruit peel	FLO	—	—	De Paula et al. 2009
Passiflora coccinea Abl.	Maracujá-do-mato, maracujá-poranga	Infusion of leaves, fruit juice	AMA, ATL	—	Cyanogenic glycosides	Di Stasi and Hiruma-Lima 2002
Passiflora edulis Sims	Maracujá	Fruit juice, decoction of leaves	CAA	SED, SNC	Carotenoids, monoterpenoids, alkaloids, cyanogenic glycosides, flavonoids	Agra et al. 2007, 2008; Di Stasi and Hiruma-Lima 2002; Lorenzi and Matos 2008
Passiflora incarnata	Maracujá-guaçu	Decoction of leaves	PAM	SED, REL, AOX	Alkaloids, flavonoids, cyanogenic glycosides	Di Stasi and Hiruma-Lima 2002; Lorenzi and Matos 2008; Rodrigues et al. 2008

Species	Common name	Preparation	Biome	Activity	Compounds	References
Passiflora macrocarpa Mart.	Maracujá gigante	Infusion of leaves, fruit juice	AMA	—	—	Di Stasi and Hiruma-Lima 2002
Piperaceae						
Ottonia leptostachya Kunth	Jaborandi, falso jaborandi	Infusion or decoction of whole plant	CAA	—	—	Agra et al. 2007, 2008; Mello 1980
Plumbaginaceae						
Plumbago scandens L.	Caataia, Louco, jasmim-azul, erva-do-diabo	Decoction or infusion of whole plant	CAA	ANF	Naphtoquinones, flavonoids	Agra et al. 2007, 2008; Lorenzi and Matos 2008
Rosaceae						
Rubus brasiliensis Mart	Amora-brasileira	—	CER	ANS, HYP, REL, ACO	—	Maroni et al. 2006; Nogueira et al. 1998a,b; Nogueira and Vassilieff 2000
Rubiaceae						
Alibertia edulis (L.C. Rich) A. Rich D.C.	Marmelada-bola	Leaves	PAN	—	—	Guarim-Neto and Morais 2003
Palicourea coriacea (Cham.) K. Schum	Douradinha	NI	CER	BAC	Alkaloids	Guarim-Neto and Morais 2003; Vieira and Martins 2000; Kato et al. 2006; Nascimento et al. 2006
Palicourea rigida H.B.K.	douradão, gritadeira	Decoction of Leaves and root	CER	ANX	Flavonoids, alkaloids, iridoid glicosides	Vieira and Martins 2000; Rosa et al. 2010; Rodrigues et al. 2008
Solanaceae						
Brunfelsia uniflora (Pohl) D. Don	Manacá, romeu-e-julieta, geretaca	Infusion or decoction of root and stem bark	CAA	ANL, ANT, HAL	Alkaloids	Agra et al. 2007, 2008; Lorenzi and Matos 2008
Physalis angulata L. (*Physalis pubescens*)	Camapu, bate-testa, juá	Decoction of whole plant	CAA, AMA	ANV, IMU, ADI	Flavonoids, Alkaloids, phytosterols	Agra et al. 2007, 2008; Lorenzi and Matos 2008; Giorgetti and Negri 2011; Hassan and Ghoneim 2013
Solanum americanum Mill.	Aguiraquia, erva-moura, Maria preta, pimenta de cachorro, caraxixá	Decoction of leaves	CAA, CER	POI, ANF	Glicoalkaloids, acetylcholine, saponins	Agra et al. 2007; Maroni et al. 2006; Lorenzi and Matos 2008
Solanum licocarpum A. St.-Hill.	Lobeira, capoeira-branca, jurubebão, baba-de-boi	Decoction of leaves	CER	ANT, ADI	Alkaloids	Maroni et al. 2006; Lorenzi and Matos 2008

(Continued)

TABLE 15.1 (CONTINUED)

Medicinal Plants Cited in Ethnobotanical Studies in Brazil to Be Explored for Phytomelatonin

Botanical Family Scientific Name	Vernacular Name(s)	Used Parts	Ethnobotanical Registry Region	Confirmed Activity	Active Constituent	Bibliographic Source
Verbenaceae						
Aloysia gratissima (Gill et Hook) Troncoso	Erva-de nossa-senhora, garupa, mimo-do-Brasil	Decoction of shoot	PAM	BAC, ANF, AOX	Flavonoids, sesquiterpenes, triterpenes, ferulic acid	Zeni et al. 2011
Aloysia triphylla (L'Hér.) Britton	Erva-luíza, cidró-pessegueiro	—	—	BAC	Essential oils	Lorenzi and Matos 2008
Lippia alba (Mill.) N.E. Br.	Erva cidreira brasileira, sálvia-brava, falsa-melissa	Infusion of leaves and flowers	FLO, CAA	ANL, CAL, ACO, ANX, HYP	Essential oils, alkaloids, saponins, sterols, flavonoids, prenylated naphtoquinones, iridoids, terpenoids, phenolic acids	De Paula et al. 2009; Lorenzi and Matos 2008; Rodrigues et al. 2008

Note: Ethnobotanical registry region: AMA = Amazonian Forest, ATL = Atlantic Forest, CAA = Caatinga, CER = Cerrado, FLO = Floresta Estacional Semidecidual, PAM = Pampas, PAN = Pantanal, RES = Restinga, ROC = rocky fields. NI = not informed. Confirmed activity in scientific research: ACO = anticonvulsant, ADI = antidiabetic, ADP = adaptogens, ANF = antifungal, ANL = analgesic, ANT = anti-inflammatory, ANV = antiviral, ANX = anxiolytic, AOX = antioxidant, APH = aphrodisiac, BAC = bacteriostatic or bactericidal action, BRO = bronchodilator action, CAL = calming, CPR = cytoprotective activity, CTX = cytotoxicity, GAB = action on GABA modulation, HGL = hypoglycemic action, HYP = hypnotic, INS = insomnia treatment, IMU = immunostimulant, POI = poison, PRT = proteolytic action, PSO = psoriasis treatment, REL = muscle relaxant, SED = sedative, SNC = nonspecific depressant activity of central nervous system, SPE = sperm production stimulus.

Some plants used in traditional medicine as sleeping aids have already been characterized for their melatonin content (Padumanonda et al. 2014), and melatonin seems to be more prevalent among medicinal plants in general (Chen et al. 2003). Rodrigues et al. (2008) summarized the medicinal properties of Brazilian plants used in popular medicine for their hypnotic, tranquilizing, and anxiolytic effects. Most of the plants presented have not yet been assessed for melatonin content, but the presented properties suggest that some of these plants might contain melatonin as a bioactive compound or affect physiological processes that are regulated by this hormone.

Numerous plants presented in Table 15.1 have other uses besides inducing sleep, and some plants have multiple indications (Agra et al. 2007, 2008), suggesting the presence of more than one bioactive compound. Also, in distant locations the same species might have different indications, revealing cultural diversity (Di Stasi and Hiruma-Lima 2002). However, it is noticeable that many communities predominantly use medicinal plants that were introduced into the ecosystem, brought from different people and from different regions (Coutinho et al. 2002), indicating cultural gaps (Medeiros 2013).

Some of these plants have toxic effects and can represent a risk to human health (Fritz et al. 2007), especially if ingested in frequent or high dosages; as an example, *Annona muricata* has been reported as presenting the degeneration of nerve cells *in vitro* (Di Stasi and Hiruma-Lima 2002).

A few plants have narcotic or hallucinogenic effects (Fritz et al. 2007), perhaps related to the presence of N,N-dimethyltryptamine (DMT), though none of the studies mention melatonin and DMT occurring concomitantly. It is possible that these plants also contain melatonin, because the precursor for both molecules is tryptophan and they are synthesized for similar metabolic pathways (Tan et al. 2014). Moreover, the presence of melatonin and serotonin was verified in the flowers and fruits of *Datura metel* L. (Solanaceae), among other examples (Murch et al. 2009; Mukherjee et al. 2014). DMT can cause psychedelic or hallucinogenic effects, being found in diverse species from different botanical families. Some of these plants are used in the preparation of hallucinogenic drinks that are used for religious purposes (Gable 2007; Pires et al. 2010; Barker et al. 2012).

Table 15.1 presents a collection of data on plants that have been identified in ethnobotanical studies in different regions of Brazil, with popular use as an alternative treatment for insomnia, and as having a calming and sedative effect. For some of them, it was possible to find information on the active ingredients already identified, but for most of them the presence of melatonin has not yet been assessed. Also mentioned is the part of the plant that is utilized in the preparations and the pharmacological activities that have already been confirmed by scientific studies. The inclusion of nonnative plants has been avoided; however, some have been able to adapt after introduction and have become spontaneous in areas with anthropic presence, sometimes being invasive (Sousa-Rangel and Trindade-Nascimento 2011).

Immunomodulatory Effects of Melatonin and Medicinal Plants

Melatonin may be considered a potent immunomodulatory agent in both animals and humans (Maestroni 2001; Srinivasan et al. 2005; Pandi-Perumal et al. 2008; Honorio-França et al. 2013). Melatonin has immunomodulatory effects partially due to its specific actions on melatonin receptors in immunocompetent cells (Maestroni et al. 2002). For example, melatonin was found to enhance cytokine production in a human lymphocytic cell line (Garcia-Maurino et al. 2000). In physiological concentrations, melatonin can also stimulate the phagocytic activity of macrophages in animals (Pawlak et al. 2005).

Neurohormonal control is very important to modulate immunomodulatory effects. Melatonin has beneficial free radical scavenging actions beyond its stimulatory effects on diverse cytosolic antioxidant enzyme systems (Klepac et al. 2005; Sudnikovich et al. 2007; Pandi-Perumal et al. 2008). Melatonin strongly stimulates the immune cells and previous studies have reported that melatonin stimulated cellular oxidative metabolism and the functional activity of phagocytes in animal models (França et al. 2009a; Honorio-França et al. 2009) and in humans (Morceli et al. 2013; Hara et al. 2013; Honorio-França et al. 2013).

Melatonin is not a conventional hormone because it displays both receptor-mediated and receptor-independent actions. Therefore, regardless of whether the cells possess indoleamine receptors or not, all cells in the body are a target for melatonin. This molecule interacts with membrane and nuclear receptors.

Using receptor-mediated and receptor-independent mechanisms, melatonin seems to be involved in a number of physiological and metabolic processes. Melatonin can affect the levels of 3',5'-cyclic adenosine monophosphate (cAMP) and intracellular Ca^{2+}, and its effect seems to be transmitted through the Gq-coupled membrane receptor acting on phospholipase C (PLC) and inositol triphosphate (IP3) (Ramracheya et al. 2008).

When melatonin interacts with its receptors, it activates the latter signal to the second messenger, cAMP or IP3/calcium (Ca2þ), changing intracellular concentrations of either cAMP or calcium. Studies have demonstrated via the pretreatment of phagocytes with TMB-8 (an intracellular calcium inhibitor) that the mediation of phagocytic activity by melatonin is calcium dependent, because melatonin induced the release of intracellular calcium by these cells (Morceli et al. 2013).

At physiological concentrations, the pineal hormone melatonin can stimulate natural immunity, an important defense with anti-infectious and microbicidal actions. On the other hand, previous studies have shown that several medicinal plants have the ability to induce immune system activation and display beneficial effects against disease (Scherer et al. 2011; Reinaque et al. 2012; Côrtes et al. 2013; França et al. 2014). The immunomodulatory effects of plants and their isolated constituents with endogen substances have attracted worldwide attention as an alternative to traditional treatment methods.

In Brazilian ethnopharmacology, a popular mixture of seven plants has proved to be useful for immune system enhancement as well as increasing resistance to the development of malignant cells. This herbal mixture includes extracts from the following plants: *Attalea speciosa Marti, Handreanthus impetiginosus (Mart.). Arctum lappa L., Rosa centifolia L., Mayteneus ilicifolia Mart., Gymnanthemum amygdalinum* and *Thuja occidentalis L.* It has been demonstrated that this herbal mixture has the ability to activate oxidative mechanisms (Corrêa et al. 2006, Reinaque et al. 2012) in phagocytosis and to have microbicidal activity (França et al. 2010).

Melatonin and these medicinal plants have been shown to activate cellular oxidative metabolism and exert important immunomodulatory effects on phagocytes (França et al. 2010; Ormonde et al. 2015). The association of melatonin with the herbal mixture has also been shown to have a potent immunostimulatory effect on the functional activity of phagocytes. This study reported that melatonin presented immunostimulatory effects that may be potentiated by the herbal mixture, and confirmed an additive effect between endogenous peptides and plants with medicinal activities. The combination of melatonin and the herbal mixture increased the functional activity of phagocytes (França et al. 2010).

The functional activity of phagocytes modulated by the association of melatonin and the herbal mixture was able to activate cellular oxidative metabolism, because the treatment of phagocytes by superoxide dismutase (SOD) reduced the microbicidal activity of cells stimulated by both the melatonin and the herbal mixture. This result suggests that the association between melatonin and the herbal mixture exerts important immunomodulatory effects on the functional activity of phagocytes, and this interaction may represent an alternative mechanism of defense against infection.

A study with *M. charantia* extract in combination with melatonin also demonstrated increased oxidative activation of phagocytes against bacteria. In the presence of the plant extract, there was an increase in the microbicidal activity of phagocytes, but the hormone melatonin reduced the microbicidal activity modulated by the extract on the phagocytes. This study suggests that the immunomodulatory activity of extract of *M. charantia* varies with the presence of the hormone melatonin (Ormonde et al. 2015).

It is possible that many plants associated with melatonin can stimulate immune cells, and so are promising in the treatment of diseases. These plants are a source of traditional medicines, with chemical substances other than melatonin that have potentially useful properties for the modulation of the immune system. Figure 15.1 shows the possible immunomodulatory effects of interactions between plants and melatonin.

Effects of Melatonin and Plants in the Pathologies

The benefits of melatonin and its metabolites are related to their antioxidant and anti-inflammatory properties (Reiter et al. 2008; Morceli et al. 2013) and their prooxidant effects (França-Botelho et al. 2011). Melatonin also affects glucose regulation in humans (Peschke et al. 2012).

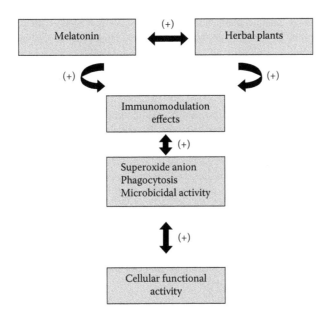

FIGURE 15.1 Immunomodulatory effects of interactions between herbal plants and melatonin.

Diabetic patients have lower diurnal serum melatonin levels and more pancreatic melatonin receptors (Peschke 2008). *Diabetes mellitus* is a metabolic disease characterized by elevated blood glucose levels. It results from the absence of, or inadequate levels of, pancreatic insulin secretion, with or without concurrent impairment of insulin action.

The role of melatonin in preventing or delaying diabetes onset, however, is not well established, because studies showing the beneficial effects of melatonin have been conducted after the development of the clinical manifestation of diabetes (Kadhim et al. 2006). In addition, the actions of melatonin on endocrine pancreas physiology, including a probable reduction in incidences of diabetes, are not well described (Peschke 2008).

High concentrations of melatonin have been reported in colostrum of diabetic mothers (Morceli et al. 2013). This hormone, contained in hyperglycemic maternal milk, can contribute to protecting the gastrointestinal mucosa of newborns. The action of melatonin in the newborn's intestine may be associated with the prevention of gastrointestinal mucosa ulceration by antioxidant action, the reduction of hydrochloric acid secretion, and immune system stimulation, fostering epithelial regeneration and increased microcirculation, especially in infants of diabetic mothers.

A previous study showed an interactive effect of glucose metabolism and melatonin on phagocytes. In phagocytes from normoglycemic patients, melatonin likely increases cellular functional activity. In diabetic patients, because hyperglycemia modifies the functional activity of phagocytes, melatonin effects are likely limited to anti-inflammatory processes, with low activity (França et al. 2014).

Otherwise, the therapeutic properties of compounds isolated from plants offer an important strategy. In the literature, the immunomodulatory effects of *M. charantia* on phagocytes suggest that this plant may be very useful, having therapeutic diversity. A study has evaluated the effects on blood viscosity in diabetic patients of *M. charantia* adsorbed in microsphere. Treatment with a nanofraction extract of *M. charantia* was associated with a significant reduction in the viscosity of hyperglycemic whole blood.

The development of plant-based drugs in nanoscale doses presents a number of advantages, including enhancements of solubility and bioavailability, protection from toxicity, the enhancement of pharmacological activity, the enhancement of stability, the improvement of tissue macrophage distribution, sustained delivery, and protection from physical and chemical degradation. High glucose levels impair a series of physiological processes, including blood viscosity. However, *M. charantia* was effective in the reduction of blood viscosity and presented immunostimulatory effects in diabetic patients. The effects of this plant in association with melatonin in patients have not yet been reported.

The effects of plants and melatonin on blood viscosity were investigated in patients with malaria. Malaria is a major infectious disease and constitutes a serious health problem worldwide, with approximately 300–500 million cases annually and almost 1 million deaths (Murray et al. 2012). The disease is caused by protozoa of the genus *Plasmodium*; it is endemic in several countries and considered to be re-emerging. *P. vivax* is the most prevalent species and is responsible for most cases of malaria (Gething et al. 2012).

In malaria, the development of intracellular parasites is accompanied by a series of structural, biological, chemical, and functional changes in erythrocytes, as evidenced by changes in the cell membrane, which is responsible for the clinical and pathological symptoms. Changes in the adhesion and rheological properties of erythrocytes are very important because these traits are directly linked to the increased destruction of these cells, leading to anemia and the sequestration of cells (Mohandas and An 2012). Due to erythrocyte tropism and their modifications to the structure of the infected cells, *Plasmodium* spp. infections can play an important role in blood flow, particularly in the vascular system. It is known that blood rheology can be influenced by the viscosity of plasma, erythrocytes, and the deformation of erythrocyte aggregation (Piagnerelli et al. 2003), which are blood alterations largely caused by infections such as malaria. The blood from individuals infected with *P. falciparum* shows changes in the conformational and rheological properties of blood cells, resulting in the obstruction of capillary vessels and hampering the flow of erythrocytes (Fedosov et al. 2011).

Rheological alterations play an important role in the pathogenesis of malaria (Figure 15.2), and changes in viscosity vary according to their shear rates (Figure 15.3).

Melatonin and *M. charantia* have been shown to modify the viscosity of blood from patients with malarial infection (Figures 15.4 and 15.5). A study with *Orbignya phalerata* Mart. extract in combination with melatonin also showed that it modified the viscosity of the blood of these patients (Figure 15.6). These studies suggest that the association of melatonin with the plant is linked to a significant reduction in the viscosity of blood of patients with malaria. Rheological parameters are used in cardiovascular research, reporting the detection of patients who are prone to cardiovascular disease, assessing hydration status in patients prior to surgery or catheterization, and monitoring blood viscosity at low temperatures, such as in bypass surgery. Figure 15.6 shows the possible effects of interactions between plants and melatonin in diseases. These data may represent an alternative to the development of new therapies toward reducing the risk of cardiovascular disorders resulting from this pathology.

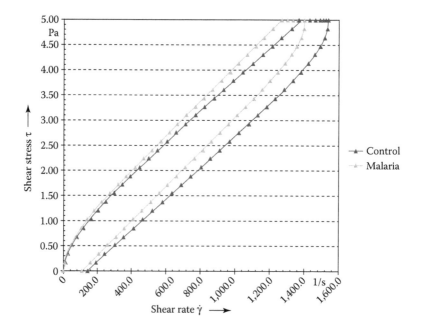

FIGURE 15.2 Flow curve of whole blood from individuals infected or not with malaria.

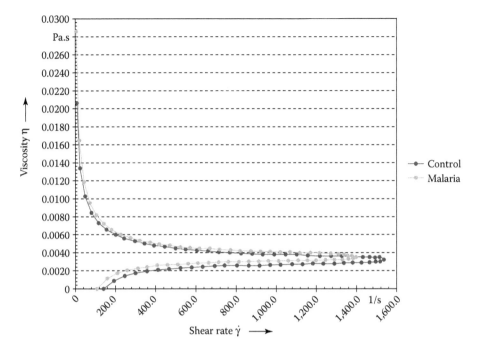

FIGURE 15.3 Viscosity curve of whole blood from individuals infected with malaria or not (Control).

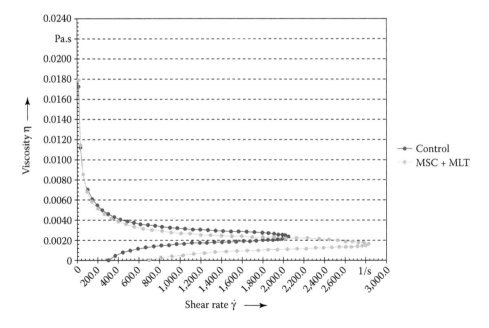

FIGURE 15.4 Viscosity curve of whole blood from individuals infected with malaria or not (Control) after treatment with melatonin (MLT) plus *Momordica charantia* (*M. charantia*) (MSC).

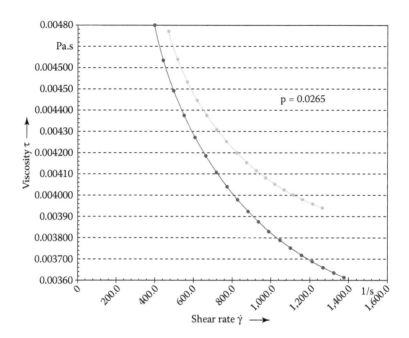

FIGURE 15.5 Amplification of viscosity curve of whole blood from individuals infected with malaria or not (Control) after treatment with melatonin (MLT) plus *Momordica charantia*.

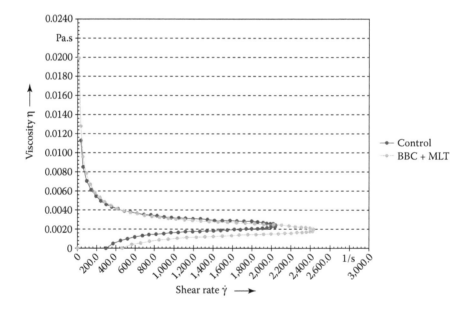

FIGURE 15.6 Viscosity curve of whole blood from individuals infected with malaria or not (Control) after treatment with melatonin (MLT) plus *Orbignya phalerata* Mart (*O. phalerata*) (BBC).

Conclusion

Considering the numerous plants with therapeutic potential and the effects of melatonin hormone, their simultaneous use can improve the pharmacological properties of active plant extracts in relation to conventional administration. These associations are capable of altering the pharmacokinetics and biodistribution of the active plant, reducing the cytotoxicity exhibited by many extracts. These characteristics favor the development of new herbal products to increase the therapeutic arsenal. In recent years, the increasing demand for new therapeutics has resulted in more efficient alternatives with substantial therapeutic levels and minimal collateral effects. Conversely, medicinal plants and melatonin still have great relevance because of the need to obtain drugs for the treatment of both emerging and re-emerging diseases.

Acknowledgments

The authors are very grateful to Dr Laercio Wandeley dos Santos for his helpful suggestions, Evelly Caroline Botelho de Miranda and Renato Queiroz Ferraz for their rheological technical assistance, and Valeria Conde Correa of the Environmental Biodiversity Center, "EcoCerrado" Reserve, Araxá, MG, Brazil, for her herbal technical assistance. This work was supported by the CNPq (Conselho Nacional de Desenvolvimento Científico e Tecnológico), Brazil, Fundação de Apoio a Pesquisa de Mato Grosso (FAPEMAT), Brazil, and Coordenação de Aperfeiçoamento de Pessoal de Nível Superior (CAPES), Brazil.

REFERENCES

Afreen, F., S.M. Zobayed, and T. Kozai. 2000. Melatonin in *Glycyrrhiza uralensis*: Response of plant roots to spectral quality of light and UV-B radiation. *Journal of Pineal Research* 41: 108–115.

Agra, M. de F., P.F. de Freitas, and J.M. Barbosa-Filho. 2007. Synopsis of the plants known as medicinal and poisonous in Northeast of Brazil. *Brazilian Journal of Pharmacognosy* 17(1): 114–140.

Agra, M. de F., K. Nurit-Silva, I.J.L.D. Basílio, P.F. de Freitas, and J.M. Barbosa-Filho. 2008. Survey of medicinal plants used in the region Northeast of Brazil. *Brazilian Journal of Pharmacognosy* 18(3): 472–508.

Albuquerque, U.P., P.M. de Medeiros, A.L.S. de Almeida, et al. 2007. Medicinal plants of the *caatinga* (semi-arid) vegetation of NE Brazil: A quantitative approach. *Journal of Ethnopharmacology* 114: 325–354.

Almeida, C. de F.C., E.L.C. de Amorim, U.P. de Albuquerque, M.B.S. Maia. 2006. Medicinal plants popularly used in the Xingó region: A semi-arid location in Northeastern Brazil. *Journal of Ethnobiology and Ethnomedicine* 2: 15–21.

Ansari, M., K. Rafiee, N. Yasa, S. Vardasbi, S.M. Naimi, and A. Nowrouzi. 2010. Measurement of melatonin in alcoholic and hot water extracts of *Tanacetum parthenium*, *Tripleurospermum disciforme* and *Viola odorata*. *DARU Journal of Pharmaceutical Sciences* 18(3): 173–178.

Antoniades, J., K. Jones, C. Hassed, and L. Piterman. 2012. Sleep… naturally: A review of the efficacy of herbal remedies for managing insomnia. *Alternative and Complementary Therapies* 18(3): 136–140.

Araújo, A.A.S., L.R. Bonjardim, E.M. Mota, et al. 2008. Antinociceptive activity and toxicological study of aqueous extract of *Egletes viscosa* Less. (Asteraceae). *Brazilian Journal of Pharmaceutical Sciences* 44(4): 707–715.

Araujo, D.S.D. de, C.F.C. de Sá, J. Fontella-Pereira, et al. 2009. Área de proteção ambiental de Massambaba, Rio de Janeiro: Caracterização fitofisionômica e florística. *Rodriguésia* 60(1): 67–96.

Argyriou, A., H. Prast, and A. Philippu. 1998. Melatonin facilitates short-term memory. *European Journal of Pharmacology* 349(2–3): 159–162.

Arnao, M.B. 2014. Phytomelatonin: Discovery, content, and role in plants. *Advances in Botany* 2014: 11.

Arnao, M.B., and J. Hernández-Ruiz. 2007. Melatonin in plants: More studies are necessary. *Plant Signaling and Behavior* 2(5): 381–382.

Badria, F.A. 2002. Melatonin, serotonin, and tryptamine in some Egyptian food and medicinal plants. *Journal of Medicinal Food* 5(3): 153–157.

Barker, S.A., E.H. McIlhennya, and R. Strassman. 2012. A critical review of reports of endogenous psychedelic N, N-dimethyltryptamines in humans: 1955–2010. *Drug Testing and Analysis* 4(7–8): 617–635.

Barnes, J., L.A. Anderson, and J.D. Phillipson. 2001. St John's wort (*Hypericum perforatum* L.): A review of its chemistry, pharmacology and clinical properties. *Journal of Pharmacy and Pharmacology* 53: 583–600.

Baydas, G., M. Özer, A. Yasar, M. Tuzcu, and S.T. Koz. 2005. Melatonin improves learning and memory performances impaired by hyperhomocysteinemia in rats. *Brain Research* 1046(1–2): 187–194.

Blanco, M.M., C.A. Costa, A.O. Freire, J.G. Santos Jr, and M. Costa. 2009. Neurobehavioral effect of essential oil of *Cymbopogon citratus* in mice. *Phytomedicine* 16(2–3): 265–270.

Borges, R.A.X., and A.M. Teles. 2014. Egletes. In *Lista de espécies da flora do Brasil*. Jardim Botânico do Rio de Janeiro. http://reflora.jbrj.gov.br/jabot/floradobrasil/FB16094 (accessed April 13, 2015).

Brant, R. da S., J.E.B.P. Pinto, S.K.V. Bertolucci, A. da Silva, and C.J.B. Albuquerque. 2009. Teores do óleo essencial de Cidrão [Aloysia triphylla (L'Hérit) Britton (Verbanaceae)] em diferentes horários de colheita e processamento pós-colheita. *Ciências agrotécnicas* 33: 2065–2068.

Brown, P.N, C.E. Turi, P.R. Shipley, and S.J. Murch. 2012. Comparison of large (*Vaccinum macrocarpon* Ait.) and small (*Vaccinium oxycoccos* L., *Vaccinum vitis-ideae* L.) cranberry in British Columbia by phytochemical determination, antioxidant potential, and metabolomic profiling with chemometric analysis. *Planta Medica* 78: 630–640.

Burkhardt, S., D.-X. Tan, L.C. Manchester, R. Hardeland, and R.J. Reiter. 2001. Detection and quantification of the antioxidant melatonin in Montmorency and Balaton tart cherries (*Prunus cerasus*). *Journal of Agricultural and Food Chemistry* 49(10): 4898–4902.

Caniato, R., R. Filippini, A. Piovan, L. Puricelli, A. Borsarini, and E.M. Cappelletti. 2003. Melatonin in plants. In G. Allegri, C.V.L. Costa, E. Ragazzi, H. Steinhart, and L. Varesio (eds), *Advances in Experimental Medicine and Biology, Developments in Tryptophan and Serotonin Metabolism*, vol. 527, pp. 593–597. New York: Springer.

Carvalho-Freitas, M.I.R., and M. Costa. 2002. Anxiolytic and sedative effects of extracts and essential oil from *Citrus aurantium* L. *Biological & Pharmaceutical Bulletin* 25(12): 1629–1633.

Carvalho, C.A., S.L.P. Matta, F.C.S.A. Melo, et al. 2009. Cipó-cravo (Tynnanthus fasciculatus Miers—Bignoniaceae): Estudo fitoquímico e toxicológico envolvendo Artemia salina. *Revista Eletrônica de Farmácia* [S.l.], 6(1). https://revistas.ufg.emnuvens.com.br/REF/article/view/5861/4561 (accessed May 23, 2016).

Chen, C.F., and C.H. Cheng. 2009. Regulation of cellular metabolism and cytokines by the medicinal herb feverfew in the human monocytic THP-1 cells. *Evidence-Based Complementary and Alternative Medicine* 6(1): 91–98.

Chen, G., Y. Huo, D.-X. Tan, Z. Liang, W. Zhang, and Y. Zhang. 2003. Melatonin in Chinese medicinal herbs. *Life Science* 73(1): 19–26.

Coleta, M., M.G. Campos, M.D. Cotrim, and A. Proença-da-Cunha. 2001. Comparative evaluation of *Melissa officinalis* L., *Tilia europaea* L., *Passiflora edulis* Sims. and *Hypericum perforatum* L. in the elevated plus maze anxiety test. *Pharmacopsychiatry* 34(Suppl. 1): S20–S21.

Comai, S., and G. Gobbi. 2014. Unveiling the role of melatonin MT_2 receptors in sleep, anxiety and other neuropsychiatric diseases: A novel target in psychopharmacology. *Journal of Psychiatry and Neuroscience* 39(1): 6–21.

Corrêa, M.P., P. Dubuisson, and A. Plana-Fattori. 2003. An overview of the ultraviolet index and the skin cancer cases in Brazil. *Photochemistry and Photobiology* 78(1): 49–54.

Corrêa, V.S.C., J.C. Maynié, E.L. França, and A.C. Honorio-França. 2006. Activity of phagocytes in the presence of the "Mais Vida" (more life) herbal remedy. *Brazilian Journal of Medicinal Plants* 8: 26–32.

Côrtes, M.A., E.L. França, A.P.B. Reinaque, E.F. Scherer, and A.C. Honorio-França. 2013. Immunomodulation of human blood phagocytes by *Strychnos pseudoquina* ST. HILL adsorbed to polyethylene glycol (PEG). *Polímeros* 23(3): 402–409.

Costa, J.R., and D. Mitja. 2010. Uso dos recursos vegetais por agricultores familiares de Manacapuru (AM). *Acta Amazonica* 40(1): 49–58.

Costello, R.B., C.V. Lentino, C.C. Boyd, et al. 2014. The effectiveness of melatonin for promoting healthy sleep: A rapid evidence assessment of the literature. *Nutrition Journal* 13(1): 106–122.

Coutinho, D.F., L.M.A. Travassos, and F.M.M. do Amaral. 2002. Ethnobotanical study of medicinal plants used by the Indian communities from Maranhão, Brazil. *Visão acadêmica, Curitiba* 3(1): 7–12.

Cunico, M.M., J.L.S. Carvalho, V.A. Kerber, C.E.K. Higaskino, S.C. Cruz Almeida, M.D. Miguel, and O.G. Miguel. 2004. Antimicrobial activity of crude ethanolic extracts of *Ottonia martiana Miq.* (Piperaceae) roots and aerial parts. *Brazilian Journal of Pharmacognosy* 14(2): 97–103.

De-Paula, M.C.Z., M. Bolson, E.L. Cardozo-Junior, and S.M. Hefler. 2009. Ethnopharmacological survey and ransom germplasm in forest remnants of semi-deciduous forest in western Paraná, Brazil. *Pontifical Catholic University of Paraná, Technical report*. http://www.iap.pr.gov.br/arquivos/File/Pesquisa%20em%20UCs/resultados%20de%20pesquisa/132_09_Maria_Cristina_Zborowski_de_Paula.pdf (accessed March 30, 2015).

Di Stasi, L.C., and C.A. Hiruma-Lima. 2002. *Plantas medicinais na Amazônia e na Mata Atlântica*, 2nd ed. *Revista e ampliada*. São Paulo: Editora Unesp.

Dubbels, R., R.J. Reiter, E. Klenke, et al. 1995. Melatonin in edible plants identified by radioimmunoassay and by high performance liquid chromatography–mass spectrometry. *Journal of Pineal Research* 18(1): 28–31.

Elisabetsky, E. 2003. Etnofarmacologia. *Ciencia e cultura* 55(3): 35–36.

Elisabetsky, E., and P. Shanley. 1994. Ethnopharmacology in the Brazilian Amazon. *Pharmacology and Therapeutics* 64: 201–214.

Elisabetsky, E., and G.C. Souza. 2004. Etnofarmacologia como ferramenta na busca de substâncias ativas. In C.M.O. Simões, E.P. Schenkel, G. Gosmann, J.C.P. Mello, L.A. Mentz, and P.R. Petrovick (eds), *Farmacognosia: Da planta ao medicamento*, 5th edn, pp. 107–122. Porto Alegre/Florianópolis: Ed. da UFRGS/UFSC.

Fedosov, D.A., B. Caswell, and G.E. Karniadakis. 2011. Wall shear stress-based model for adhesive dynamics of red blood cells in malaria. *Biophysical Journal* 100: 2084–2093.

Fernandes-Lima, C.N., T. Francelino-Valero, N. Figueiredo-Leite, et al. 2014. Ação protetora de *Duguetia furfuracea* (A. St.-Hil.) Saff. contra a toxicidade do cloreto de mercúrio em *Escherichia coli*. *Revista Cubana de plantas medicinales* 19(3): 179–188.

Fernández-Pachõn, M.S., S. Medina, G. Herrero-Martín, I. Cerrillo, G. Berná, B. Escudero-Lõpez, F. Ferreres, F. Martín, M.C. García-Parrilla, and A. Gil-Izquierdo. 2014. Alcoholic fermentation induces melatonin synthesis in orange juice. *Journal of Pineal Research* 56(1): 31–38.

Ferracioli-Oda, E., A. Qawasmi, and M.H. Bloch. 2013. Meta-analysis: Melatonin for the treatment of primary sleep disorders. *PLoS ONE* 8(5): e 63773.

Feyzabadi, Z., F. Jafari, S.H. Kamali, et al. 2014. Efficacy of *Viola odorata* in treatment of chronic insomnia. *Iran Red Crescent Medicine Journal* 16(12): e17511.

Fonseca, E.N., A. Figer, D.T. Furtado, et al. 2006. Análise química e atividade antimicrobiana do óleo essencial dos frutos de *Vitex cymosa* Bertero. *Brazilian Journal of Medicinal Plants* 8(4): 87–89.

Fonseca, M.C.M., V.W.D. Casali, and L.C.A. Barbosa. 2007. Influência da época e do horário de colheita nos teores de óleo essencial e de taninos em couve-cravinho (*Porophyllum ruderale*) (Jacq.) Cassini. *Brazilian Journal of Medicinal Plants* 9(2): 75–79.

França, E.L., N.D. Feliciano, K.A. Silva, C.K.B. Ferrari, and A.C. Honorio-França. 2009a. Modulatory role of melatonin on superoxide release by spleen macrophages isolated from alloxan-induced diabetic rats. *Bratislava Medical Journal* 7: 163–173.

França, E.L., J.C. Maynie, V.C. Correa, U.C.R. Pereira, C. Batalini, C.K.B. Ferrari, and A.C. Honorio-França. 2010. Immunomodulation effects of herbal plants plus melatonin on human blood phagocytes. *International Journal of Phytomedicine* 2: 354–362.

França, E.L., A. Pereira Jr, S.L. Oliveira, and A.C. Honorio-França. 2009b. Chronoimmunomodulation of melatonin on bactericidal activity of human blood phagocytes. *International Journal of Microbiology* 6: 1–15.

França E.L., E.B. Ribeiro, E.F. Scherer, D.G. Cantarini, R.S. Pessôa, F.L. França, and A.C. Honorio-França. 2014. Effects of *Momordica charantia* L. on the blood rheological properties in diabetic patients. *BioMed Research International* 2014(840379): 1–8.

França-Botelho, A.C., J.L. França, F.M. Oliveira, et al. 2011. Melatonin reduces the severity of experimental amoebiasis. *Parasite & Vectors* 4: 62.

Fritz, D., C.R. Venturi, S. Cargnin, et al. 2007. Herpes virus inhibitory substances from *Hypericum connatum* Lam., a plant used in southern Brazil to treat oral lesions. *Journal of Ethnopharmacology* 113: 517–520.

Fuller, P.M., J.J. Gooley, and C.B. Saper. 2006. Neurobiology of the sleep–wake cycle: Sleep architecture, circadian regulation, and regulatory feedback. *Journal of Biological Rhythms* 21(6): 482–493.

Gable, R.S. 2007. Risk assessment of ritual use of oral dimethyltryptamine (DMT) and harmala alkaloids. *Addiction* 102(1): 24–34.

Gajalakshmi, S., S. Vijayalakshmi, and V.R. Devi. 2012. Phytochemical and pharmacological properties of *Annona muricata*: A review. *International Journal of Pharmacy and Pharmaceutical Sciences* 4(2): 3–6.

Garcia-Maurino, S., D. Pozo, J.R. Calvo, and J.M. Guerrero. 2000. Correlation between nuclear melatonin receptor expression and enhanced cytokine production in human lymphocytic and monocytic cells lines. *Journal of Pineal Research* 29: 129–137.

Gething, P.W., I.R. Elyazar, and C.L. Moyes. 2012. A long neglected world malaria map: *Plasmodium vivax* endemicity in 2010. *PLoS Neglected Tropical Diseases* 6: e1814.

Giorgetti, M., and G. Negri. 2011. Plants from Solanaceae family with possible anxiolytic effect reported on 19th century's Brazilian medical journal. *Revista Brasileira de farmacognosia* 21(4): 772–780.

Gomez, F.J.V., J. Raba, S. Cerutti, and M.F. Silva. 2012. Monitoring melatonin and its isomer in *Vitis vinifera* Cv. Malbec by UHPLC-MS/MS from grape to bottle. *Journal of Pineal Research* 52(3): 349–355.

Gooneratne, N.S. 2008. Complementary and alternative medicine for sleep disturbances in older adults. *Clinics in Geriatric Medicine* 24(1): 121–138.

Guarim-Neto, G., and R.G. de Morais. 2003. Recursos medicinais de espécies do cerrado de Mato Grosso: Um estudo bibliográfico. *Acta Botanica Brasilica* 17(4): 561–584.

Hara, C.C.P., A.C. Honório-França, D.L.G. Fagundes, P.C.L. Guimarães, and E.L. França. 2013. Melatonin nanoparticles adsorbed to polyethylene glycol microspheres as activators of human colostrum macrophages. *Journal of Nanomaterials* 2: 1–8.

Hardeland, R. 2008. Plants: Sources of melatonin. In R.R. Watson and V.R. Preedy (eds), *Botanical Medicine in Clinical Practice*, pp. 752–761. Wallingford, UK: CABI.

Hardeland, R. 2012. Melatonin in aging and disease: Multiple consequences of reduced secretion, options and limits of treatment. *Aging and Disease* 3(2): 194–225.

Hardeland, R. 2015. Melatonin in plants and other phototrophs: Advances and gaps concerning the diversity of functions. *Journal of Experimental Botany* 66(3): 627–646.

Hardeland, R., S.R. Pandi-Perumal, and B. Poeggeler. 2007. Melatonin in plants: Focus on a vertebrate night hormone with cytoprotective properties. *Functional Plant Science and Biotechnology* 1: 32–45.

Hardeland, R., and B. Poeggeler. 2003. Non-vertebrate melatonin. *Journal of Pineal Research* 34: 233–241.

Hassan, A.I., and M.A.M. Ghoneim. 2013. A possible inhibitory effect of physalis (*Physalis pubescens* L.) on diabetes in male rats. *World Applied Sciences Journal* 21(5): 681–688.

Hattori, A., H. Migitaka, M. Iigo, et al. 1995. Identification of melatonin in plants and its effects on plasma melatonin levels and binding to melatonin receptors in vertebrates. *Biochemistry and Molecular Biology International* 35(3): 627–634.

Heo, H.J., S.J. Choi, S.G. Choi, D.H. Shin, J.M. Lee, and C.Y. Lee. 2008. Effects of banana, orange, and apple on oxidative stress-induced neurotoxicity in PC12 cells. *Journal of Food Science* 73(2): H28–H32.

Honorio-França A.C., C.C.P. Hara, J.V.J. Ormonde, G.N. Triches, and E.L. França. 2013. Human colostrum melatonin exhibits a day–night variation and modulates the activity of colostral phagocytes. *Journal of Applied Biomedicine* 11: 153–162.

Honorio-França, A.C., K.A. Silva, N.D. Feliciano, I.M.P. Calderon, M.V.C. Rudge, and E.L. França. 2009. Melatonin effects on macrophages in diabetic rats and the maternal hyperglycemic implications for newborn rats. *International Journal of Diabetis Metabolism* 17: 87–92.

Howatson, G., P.G. Bell, J. Tallent, B. Middleton, M.P. McHugh, and J. Ellis. 2011. Effect of tart cherry juice (*Prunus cesarus*) on melatonin levels and enhanced sleep quality. *European Journal of Nutrition* 51: 909–916.

Huang, X., and G. Mazza. 2011. Application of LC and LC-MS to the analysis of melatonin and serotonin in edible plants. *Critical Reviews in Food Science and Nutrition* 51: 269–284.

Jan, J.E., R.J. Reiter, M.B. Wasdell, and M. Bax. 2009. The role of the thalamus in sleep, pineal melatonin production, and circadian rhythm sleep disorders. *Journal of Pineal Research* 46(1): 1–7.

Janas, K.M., and M.M. Posmyk. 2013. Melatonin, an underestimated natural substance with great potential for agricultural application. *Acta Physiologiae Plantarum* 35(12): 3285–3292.

Kadhim, H.M., S.H. Ismail, K.I. Hussein, et al. 2006. Effects of melatonin and zinc on lipid profile and renal function in type 2 diabetic patients poorly controlled with metformin. *Journal of Pineal Reserch* 41: 189–193.

Kähkönen, M.P., A.I. Hopia, and M. Heinonen. 2001. Berry phenolics and their antioxidant activity. *Journal of Agricultural and Food Chemistry* 49(8): 4076–4082.

Kasper, S., F. Caraci, B. Forti, F. Drago, and E. Aguglia. 2010. Efficacy and tolerability of *Hypericum* extract for the treatment of mild to moderate depression. *European Neuropsychopharmacology* 20(11): 747–765.

Kato, L., C.M.A. de Oliveira, C.A. do Nascimento, and L.M. Lião. 2006. Atividade antimicrobiana de *Palicourea coriacea* (Cham.) K. Schum. Paper presented at the 29th annual meeting of the Brazilian Chemistry Society. https://sec.sbq.org.br/cd29ra/resumos/T0224-1.pdf.

Klepac, N., Z. Rudes, and R. Klepac. 2005. Effects of melatonin on plasma oxidative stress in rats with streptozotocin induced diabetes. *Biomedicine and Pharmacotherapy* 60: 32–35.

Koch, I., A. Rapini, A.O. Simões, L.S. Kinoshita, A.P. Spina, and A.C.D. Castello. 2015. Apocynaceae. In *Lista de espécies da flora do Brasil*. Jardim Botânico do Rio de Janeiro. http://floradobrasil.jbrj.gov.br/jabot/floradobrasil/FB16093 (accessed April 13, 2015).

Kolar, J., C.H. Johnson, and I. Machackova. 1999. Presence and possible role of melatonin in a short-day flowering plant, *Chenopodium rubrum*. *Advances in Experimental Medicine and Biology* 460: 391–393.

Kolar, J., and I. Machackova. 1994. Melatonin: Does it regulate rhythmicity and photoperiodism also in higher plants? *Flower Newsletter* 17: 53–54.

Kolar, J., and I. Machackova. 2005. Melatonin in higher plants: Occurrence and possible functions. *Journal of Pineal Research* 39: 333–341.

Koyama, F.C., T.L. Carvalho, E. Alves, H.B. da Silva, M.F. de Azevedo, A.S. Hemerly, and C.R. Garcia. 2013. The structurally related auxin and melatonin tryptophan-derivatives and their roles in *Arabidopsis thaliana* and in the human malaria parasite *Plasmodium falciparum*. *Journal of Eukaryotic Microbiology* 60(6): 646–651.

Kumar, V., V.K. Khanna, P.K. Seth, P.N. Singh, and S.K. Bhattacharya. 2002. Brain neurotransmitter receptor binding and nootropic studies on Indian *Hypericum perforatum* Linn. *Phytotherapy Research* 16(3): 210–216.

Kvetnoy, I.M. 1999. Extrapineal melatonin: Location and role within diffuse neuroendocrine system. *Histochemical Journal* 31(1): 1–12.

Lakhan, S.E., and K.F. Vieira. 2010. Nutritional and herbal supplements for anxiety and anxiety-related disorders: Systematic review. *Nutrition Journal* 9(1): 42–55.

Lerner, A.B., J.D. Case, W. Mori, and M.R. Wright. 1959. Melatonin in peripheral nerve. *Nature* 183(4678): 1821.

Lerner, A.B., J.D. Case, Y. Takahashi, T.H. Lee, and W. Mori. 1958. Isolation of melatonin, the pineal gland factor that lightens melanocytes. *Journal of the American Chemical Society* 80(10): 2587–2592.

Loeuille, B., and M. Monge. 2014. Achyrocline. In *Lista de espécies da flora do Brasil*. Jardim Botânico do Rio de Janeiro. http://floradobrasil.jbrj.gov.br/jabot/floradobrasil/FB102953 (accessed April 13, 2015).

Lohmann, L.G. 2015. Bignoniaceae. In *Lista de Espécies da Flora do Brasil*. Jardim Botânico do Rio de Janeiro. http://floradobrasil.jbrj.gov.br/jabot/floradobrasil/FB112409 (accessed April 13, 2015).

Lolli, L.F., C.M. Sato, C.V. Romanini, L.B. Villas-Boas, C.M. Santos, and R.M. Oliveira. 2007. Possible involvement of GABA A-benzodiazepine receptor in the anxiolytic-like effect induced by *Passiflora actinia* extracts in mice. *Journal of Ethnopharmacology* 111: 308–314.

Lopes-Campêlo, L.M., C. Gonçalves-e-Sá, A.A. Almeida, et al. 2011. Sedative, anxiolytic and antidepressant activities of *Citrus limon* (Burn.) essential oil in mice. *Pharmazie* 66(8): 623–627.

Lorenzi, H. and F.J. Abreu-Matos. 2008. *Plantas medicinais no Brasil*, 2nd ed. Nova Odessa, Brazil: Instituto Plantarum.

Lorenzo, M.A. 2000. Estudo do efeito do tipo ansiolítico da Cuphea carthagenensis (JACQ.) J. F. MACBR. (sete-sangrias) em camundongos. MA thesis, Universidade Federal de Santa Catarina.

Ly, D., K. Kang, J.Y. Choi, A. Ishihara, K. Back, and S.G. Lee. 2008. HPLC analysis of serotonin, tryptamine, tyramine and the hydroxycinnamic acid amides of serotonin and tyramine in food vegetables. *Journal of Medicinal Food* 11(2): 385–389.

Maas, P., A. Lobão, and H. Rainer. 2015. Annonaceae. In *Lista de espécies da flora do Brasil*. Jardim Botânico do Rio de Janeiro. http://floradobrasil.jbrj.gov.br/jabot/floradobrasil/FB110219 (accessed April 13, 2015).

Maestroni, G.J. 2001. The immunotherapeutic potential of melatonin. *Expert Opinion on Investigational Drugs* 10: 467–476.

Maestroni, G.J., A. Sulli, C. Pizzorni, B. Villaggio, and M. Cutolo. 2002. Melatonin in rheumatoid arthritis: Synovial macrophages show melatonin receptors. *Annals of the New York Academy of Sciences* 966: 271–275.

Manchester, L.C., D.-X. Tan, R.J. Reiter, W. Park, K. Monis, and W. Qi. 2000. High levels of melatonin in the seeds of edible plants: Possible function in germ tissue protection. *Life Sciences* 67(25): 3023–3029.

Maronde, E., and J.H. Stehle. 2007. The mammalian pineal gland: Known facts, unknown facets. *Trends in Endocrinology & Metabolism* 18(4): 142–149.

Maroni, B.C., L.C. Di-Stasi, and S.R. Machado. 2006. *Plantas medicinais do cerrado de Botucatu*. São Paulo, Brazil: Editora UNESP.

McCune, L.M., C. Kubota, N.R. Stendell-Hollis, and C.A. Thomson 2011. Cherries and health: A review. *Critical Reviews in Food Science and Nutrition* 51(1): 1–12.

Medeiros, P.M. de. 2013. Why is change feared? Exotic species in traditional pharmacopoeias. *Ethnobiology and Conservation* 2(3): 1–5.

Melo, F.C., S.L. Matta, T.A. Paula, M.L. Gomes, and L.C. Oliveira. 2010. The effects of *Tynanthus fasciculatus* (Bignoniaceae) infusion on testicular parenchyma of adult Wistar rats. *Biological Research* 43(4): 445–450.

Mello, J.F. de. 1980. Plants in traditional medicine in Brazil. *Journal of Ethnopharmacology* 2: 49–55.

Mennini, T., and M. Gobbi. 2004. The antidepressant mechanism of *Hypericum perforatum*. *Life Sciences* 75(9): 1021–1027.

Miyasaka, L.S., A.N. Atallah, and B.G.O. Soares. 2007. Passiflora for anxiety disorder. *Cochrane Database of Systematic Reviews* 1: CD004518.

Mohandas, N., and X. An. 2012. Malaria and human red blood cells. *Medical Microbiology and Immunology* 201: 593–598.

Morceli G., A.C. Honorio-França, D.L.G. Fagundes, I.M.P. Calderon, and E.L. Franca. 2013. Antioxidant effect of melatonin on the functional activity of colostral phagocytes in diabetic women. *PloS One* 8: e56915.

Mukherjee, S., A. David, S. Yadav, F. Baluška, and S.C. Bhatla. 2014. Salt stress–induced seedling growth inhibition coincides with differential distribution of serotonin and melatonin in sunflower seedling roots and cotyledons. *Physiologia Plantarum* 152(4): 714–728.

Murch, S.J, A.R. Alan, J. Cao, and P.K. Saxena. 2009. Melatonin and serotonin in flowers and fruits of *Datura metel* L. *Journal of Pineal Research* 47(3): 277–283.

Murch, S.J., S.S. Campbell, and P.K. Saxena. 2001. The role of serotonin and melatonin in plant morphogenesis: Regulation of auxin-induced root organogenesis in *in vitro*-cultured explants of St John's wort (*Hypericum perforatum* L.). *In Vitro Cellular and Developmental Biology: Plant* 37(6): 786–793.

Murch, S.J., S. KrishnaRaj, and P.K. Saxena. 2000. Tryptophan is a precursor for melatonin and serotonin biosynthesis *in vitro* regenerated St. John's wort (*Hypericum perforatum* L. cv. Anthos) plants. *Plant Cell Reports* 19(7): 698–704.

Murch, S.J., and P.K. Saxena. 2006. A melatonin-rich germplasm line of St John's wort (*Hypericum perforatum* L.). *Journal of Pineal Research* 41(3): 284–287.

Murch, S.J., C.B. Simmons, and P.K. Saxena. 1997. Melatonin in feverfew and other medicinal plants. *Lancet* 350(9091): 1598–1599.

Murray, C.J., L.C. Rosenfeld, S.S. Lim, et al. 2012 Global malaria mortality between 1980 and 2010: A systematic analysis. *Lancet* 379: 413–431.

Nakajima, J.N. 2015a. Porophyllum. In *Lista de espécies da flora do Brasil*. Jardim Botânico do Rio de Janeiro. http://floradobrasil.jbrj.gov.br/jabot/floradobrasil/FB16259 (accessed April 13, 2015).

Nakajima, J.N. 2015b. Tagetes. In *Lista de espécies da flora do Brasil*. Jardim Botânico do Rio de Janeiro. http://floradobrasil.jbrj.gov.br/reflora/floradobrasil/FB16340 (accessed April 13, 2015).

Nascimento, C.A., M.S. Gomes, L.M. Lião, et al. 2006. Alkaloids from *Palicourea coriacea* (Cham.) K. Schum. *Z. Naturforsch* 61b: 1443–1446.

Neto, C.C. 2007. Cranberry and blueberry: Evidence for protective effects against cancer and vascular diseases. *Molecular Nutrition and Food Research* 51(6): 652–664.

Nogueira, E., G.J.M Rosa, M. Haraguchi, and V.S. Vassilieff. 1998a. Anxiolytic effect of *Rubus brasilensis* in rats and mice. *Journal of Ethnopharmacology* 61: 111–117.

Nogueira E., G.J.M. Rosa, and V.S. Vassilieff. 1998b. Involvement of GABAA-benzodiazepine receptor in the anxiolytic effect induced by hexanic fraction of Rubus brasilensis. *Journal of Ethnopharmacology* 61: 119–126.

Nogueira, E., and V.S. Vassilieff. 2000. Hypnotic anticonvulsant and muscle relaxing of *Rubus brasiliensis*: Involvement of GABAA system. *Journal of Ethnopharmacology* 70: 275–280.

Ormonde J.V.S., R.T.S. Fernandes, G.T. Nunes, E.F. Scherer, A.C. Honorio-França, and E.L. França. 2015. *Momordica charantia* L. plus melatonin present an effective immunomodulatory effect to eliminated enteropathogenic *Escherichia coli* and *Streptococcus pneumonia*. *World Journal of Pharmacy and Pharmaceutical Sciences* 4(3): 58–75.

Pacheco, A.G., T.M. Silva, R.M. Manfrini, et al. 2010. Chemical study and antibacterial activity of stem of Aristolochia esperanzae Kuntze (*Aristolochiaceae*). *Química Nova* 33(8): 1649–1652.

Padumanonda, T., J. Johns, A. Sangkasat, and S. Tiyaworanant. 2014. Determination of melatonin content in traditional Thai herbal remedies used as sleeping aids. *DARU Journal of Pharmaceutical Sciences* 22(1): 6.

Pandi-Perumal, S.R., I. Trakht, V. Srinivasan, et al. 2008. Physiological effects of melatonin: Role of melato-nin receptors and signal transduction pathways. *Progress in Neurobiology* 85: 335–353.

Paredes, S.D., A. Korkmaz, L.C. Manchester, D.-X. Tan, and R.J. Reiter. 2009. Phytomelatonin: A review. *Journal of Experimental Botany* 60(1): 57–69.

Pareek, A., M. Suthar, G.S. Rathore, and V. Bansal. 2011. Feverfew (*Tanacetum parthenium* L.): A systematic review. *Pharmacognosy Reviews* 5(9): 103–110.

Pawlak, J., J. Singh, R.W. Lea, and K. Skwarlo-Sonta. 2005. Effect of melatonin on phagocytic activity and intracellular free calcium concentration in testicular macrophages from normal and streptozotocin-induced diabetic rats. *Molecular and Cellular Biochemistry* 275: 207–213.

Peschke, E. 2008. Melatonin, endocrine pancreas and diabetes. *Journal of Pineal Research* 44: 26–40.

Peschke, E., K. Hofmann, K. Pönicke, D. Wedekind, and E. Mühlbauer. 2012. Catecholamines are the key for explaining the biological relevance of insulin–melatonin antagonisms in type 1 and type 2 diabetes. *Journal of Pineal Research* 52: 389–396.

Pessoa, R.S., E.L. França, E.B. Ribeiro, N.G. Abud, and A.C. Honorio. 2014. Microemulsion of babassu oil as a natural product to improve human immune system function. *Drug Design, Development and Therapy* 9: 21–31.

Piagnerelli, M., K.Z. Boudjeltia, M. Vanhaeverbeek, and J.L. Vincent. 2003. Red blood cell rheology in sepsis. *Intensive Care Medicine* 29: 1052–1061.

Pires, A.P.S., C.D.R. Oliveira, and M. Yonamine. 2010. Ayahuasca: Uma revisão dos aspectos farmacológicos e toxicológicos. *Journal of Basic and Applied Pharmaceutical Sciences* 31(1): 15–23.

Poeggeler, B., I. Balzer, J. Fischer, G. Behrmann, and R. Hardeland. 1989. A role of melatonin in dinoflagel-lates? *Acta Endocrinologica* (Copenhagen) 120(Suppl. 1): 97.

Poeggeler, B., I. Balzer, R. Hardeland, and A. Lerchl. 1991. Pineal hormone melatonin oscillates also in the dinoflagellate *Gonyaulax polyedra*. *Naturwissenschaften* 78: 268–269.

Quintans, L.J., Jr, J.R.G.S. Almeida, J.T. Lima, et al. 2008. Plants with anticonvulsant properties: A review. *Revista Brasileira de farmacognosia* 18: 798–819.

Ramakrishna, A. 2015. Indoleamines in edible plants: Role in human health effects. In Angel Catalá (ed.), *Indoleamines: Sources, Role in Biological Processes and Health Effects*, p. 279. Biochemistry Research Trends. Hauppauge, NY: Nova.

Ramakrishna, A., P. Giridhar, M. Jobin, C.S. Paulose, and G.A. Ravishankar. 2012a. Indoleamines and cal-cium enhance somatic embryogenesis in cultured tissues of *Coffea canephora* P ex Fr. *Plant Cell, Tissue and Organ Culture* 108: 267–278.

Ramakrishna, A, P. Giridhar, and G.A. Ravishankar. 2009. Indoleamines and calcium channels influ-ence morphogenesis in *in vitro* cultures of *Mimosa pudica* L. *Plant Signaling & Behavior* 12: 1136–1141.

Ramakrishna, A., P. Giridhar, K.U. Sankar, and G.A. Ravishankar. 2012b. Melatonin and serotonin profiles in beans of *Coffea* sps. *Journal of Pineal Research* 52: 470–476.

Ramakrishna, A., P. Giridhar, K.U. Sankar, and G.A. Ravishankar. 2012c. Endogenous profiles of indoleam-ines: Serotonin and melatonin in different tissues of *Coffea canephora* P ex Fr. as analyzed by HPLC and LC-MS-ESI. *Acta Physiologia Plantarum* 34: 393–396.

Ramracheya, R.D., D.S. Muller, P.E. Squires, H. Brereton, D. Sugden. 2008. Function and expression of mela-tonin receptors on human pancreatic islets. *Journal of Pineal Research* 44: 273–279.

Reinaque A.P.B., E.L. França, E.F. Scherer, M.A. Côrtes, F.J.D. Souto, and A.C. Honorio-França. 2012. Natural material adsorbed onto a polymer to enhance immune function. *Drug Design, Development and Therapy* 6: 209–216.

Reiter, R.J. 2003. Melatonin: Clinical relevance. *Best Practice & Research: Clinical Endocrinology & Metabolism* 17(2): 273–285.

Reiter, R.J., L.C. Manchester, and D.-X. Tan. 2005. Melatonin in walnuts: Influence on levels of melatonin and total antioxidant capacity of blood. *Nutrition* 21: 920–924.

Reiter, R.J., and D.-X. Tan. 2002. Melatonin: An antioxidant in edible plants. *Annals of the New York Academy of Sciences* 957: 341–344.

Reiter, R.J., D.-X. Tan, S. Burkhardt, and L.C. Manchester. 2001. Melatonin in plants. *Nutrition Reviews* 59(9): 286–290.

Reiter, R.J., D.-X. Tan, M.J. Jou, et al. 2008. Biogenic amines in the reduction of oxidative stress: Melatonin and its metabolites. *Neuro Endocrinology Letters* 29: 391–398.

Reiter, R.J., D.-X. Tan, L. Manchester, et al. 2007. Melatonin in edible plants (phytomelatonin): Identification, concentrations, bioavailability and proposed functions, *World Review of Nutrition and Dietetics* 97: 211–230.

Reiter, R.J., D.-X. Tan, J.C. Mayo, R.M. Sainz, J. Leon, and Z. Czarnocki. 2003. Melatonin as an antioxidant: Biochemical mechanisms and pathophysiological implications in humans. *Acta Biochimica Polonica* 50(4): 1129–1146.

Ritter, M.R., R.M. Liro, N. Roque, J. Nakajima, F.O. Souza-Buturi, and C.T. Oliveira. 2015. Mikania. In *Lista de espécies da flora do Brasil*. Jardim Botânico do Rio de Janeiro. http://floradobrasil.jbrj.gov.br/jabot/floradobrasil/FB5344 (accessed April 13, 2015).

Rodrigues, E., R. Tabach, J.C.F. Galduróz, G. Negri. 2008. Plants with possible anxiolytic and/or hypnotic effects indicated by three Brazilian cultures: Indians, Afro-Brazilians and River-Dwellers. *Studies in Natural Products Chemistry* 35: 549–595.

Rodriguez-Naranjo, M.I., A. Gil-Izquierdo, A.M. Troncoso, E. Cantos-Villar, and M.C. Garcia-Parrilla. 2011. Melatonin is synthesised by yeast during alcoholic fermentation in wines. *Food Chemistry* 126(4): 1608–1613.

Rosa, E.A. da, B.C. e Silva, F.M. da Silva, et al. 2010. Flavonoides e atividade antioxidante em Palicourea rigida Kunth, Rubiaceae. *Revista Brasileira de farmacognosia* 20(4): 484–488.

Sá-Barreto, L.C.L. 2004. Estudo farmacognóstico e determinação da atividade biológica de *Vitex gardneriana* Schauer (Verbenaceae). MA thesis, Universidade Federal de Pernambuco.

Sá-Barreto, L.C.L., M.S.S. Cunha-Filho, I.A. Souza, M.C. Fraga, and H.S. Xavier. 2008. Avaliação preliminar da atividade biológica e toxicidade aguda de *Vitex gardneriana* Schauer. *Latin American Journal of Pharmacy* 27(6): 909–913.

Sae-Teaw, M., J. Johns, N.P. Johns, and S. Subongkot. 2013. Serum melatonin levels and antioxidant capacities after consumption of pineapple, orange, or banana by healthy male volunteers. *Journal of Pineal Research* 55(1): 58–64.

Santana, L.C.L.R., M.R.M. Brito, G.L.S. Oliveira, et al. 2014. *Mikania glomerata*: Phytochemical, pharmacological, and neurochemical study. *Evidence-Based Complementary and Alternative Medicine* 2014:11.

Santos, T.C. dos, J. Schripsema, F.D. Monache, and S. G Leitão. 2001. Iridoids from *Vitex cymosa*. *Journal of the Brazilian Chemical Society*, 12(6): 763–766.

Sarris, J., A. Panossian, I. Schweitzer, C. Stough, and A. Scholey. 2011. Herbal medicine for depression, anxiety and insomnia: A review of psychopharmacology and clinical evidence. *European Neuropsychopharmacology* 21(12): 841–860.

Scherer, E.F., A.C. Honorio-França, C.C.P. de Hara, A.P.B. Reinaque, M.A. Côrtes, and E.L. França 2011. Immunomodulatory effects of poly(ethylene glycol) microspheres adsorbed with nanofractions of *Momordica charantia* L. on diabetic human blood phagocytes. *Science of Advanced Materials* 3(5): 687–694.

Shin, J.C., H.-Y. Jung, A. Harikishored, et al. 2013. The flavonoid myricetin reduces nocturnal melatonin levels in the blood through the inhibition of serotonin N-acetyltransferase. *Biochemical and Biophysical Research Communications* 440(2): 312–316.

Silva, B.A., F. Ferreres, J.O. Malva, and A.C.P. Dias. 2005. Phytochemical and antioxidant characterization of *Hypericum perforatum* alcoholic extracts. *Food Chemistry* 90(1–2): 157–167.

Silva, C.V. da, F.M. Borges, and E.S. Velozo. 2012. Phytochemistry of some Brazilian plants with aphrodisiac activity. In V. Rao (ed.), *Phytochemicals: A Global Perspective of Their Role in Nutrition and Health*, Chapter 15, Rijeka: Intech.

Slominski, A., D.J. Tobin, M.A. Zmijewski, J. Wortsman, and R. Paus. 2008. Melatonin in the skin: Synthesis, metabolism and functions. *Trends in Endocrinology & Metabolism* 19(1): 17–24.

Slusarski, S.R., A.C. Cervi, and O.A. Guimarães. 2007. Estudo taxonômico das espécies nativas de *Hypericum* L. (Hypericaceae) no Estado do Paraná, Brasil. *Acta Botanica Brasilica* 21(1): 163–184.

Sousa-Rangel, E. de, and M. Trindade-Nascimento. 2011. Ocorrência de *Calotropis procera* (Ait.) R. Br. (Apocynaceae) como espécie invasora de restinga. *Acta Botanica Brasilica* 25(3): 657–663.

Srinivasan, V., G.J. Maestroni, D.P. Cardinalli, et al. 2005. Melatonin, immune function and aging. *Immunity & Ageing* 2: 17–49.

Stefulj, J., M. Hörtner, M. Ghosh, et al. 2001. Gene expression of the key enzymes of melatonin synthesis in extrapineal tissues of the rat. *Journal of Pineal Research* 30(4): 243–247.

Sudnikovich, E.J., Y.Z. Maksimchik, S.V. Zabrodskaya, et al. 2007. Melatonin attenuates metabolic disorders due to streptozotocin-induced diabetes in rats. *European Journal of Pharmacology* 569: 180–187.

Tan, D.-X., R. Hardeland, L.C. Manchester, et al. 2012. Functional roles of melatonin in plants, and perspectives in nutritional and agricultural science. *Journal of Experimental Botany* 63(2): 577–597.

Tan, D.-X., L.C. Manchester, P. Di Mascio, G.R. Martinez, F.M. Prado, and R.J. Reiter. 2007. Novel rhythms of N1-acetyl-N2-formyl-5-methoxykynuramine and its precursor melatonin in water hyacinth: Importance for phytoremediation. *The FASEB Journal* 21: 1724–1729.

Tan, D.X., X. Zheng, J. Kong, et al. 2014. Fundamental issues related to the origin of melatonin and melatonin isomers during evolution: Relation to their biological functions. *International Journal of Molecular Science* 15: 15858–15890.

Tassel, D. van, and S. O'Neill. 1993. Melatonin: Identification of a potential dark signal in plants. *Plant Physiology* 102(Suppl. 1): 659.

Tassel, D. van, N. Roberts, and S. O'Neill. 1995. Melatonin from higher plants: Isolation and identification of N-acetyl-5-methoxytryptamine. *Plant Physiology* 108(Suppl. 2): 101.

Tettamanti, C., B. Cerabolini, P. Gerola, and A. Conti. 2000. Melatonin identification in medicinal plants. *Acta Phytotherapeutica* 3: 137–144.

Vieira, R.F., and M.V.M. Martins. 2000. Recursos genéticos de plantas medicinais do cerrado: Uma compilação de dados. *Brazilian Journal of Medicinal Plants* 3(1): 13–36.

Violante, I.M.P., Souza, I.M., Venturini, C.L., Ramalho, A.F., Santos, R.A.N., and Ferrari, M. 2009. *In vitro* sunscreen activity evaluation of plants extracts from Mato Grosso cerrado. *Brazilian Journal of Pharmacognosy* 19: 452–457.

Vitalini, S., C. Gardana, A. Zanzotto, P. Simonetti, F. Faoro, G. Fico, and M. Iriti. 2011. The presence of melatonin in grapevine (*Vitis Vinifera* L.) berry tissues. *Journal of Pineal Research* 51(3): 331–337.

Weeda S., N. Zhang, X. Zhao, et al. 2014. Arabidopsis transcriptome analysis reveals key roles of melatonin in plant defense systems. *PLoS One* 9(3): e93462.

Wheatley, D. 2005. Medicinal plants for insomnia: A review of their pharmacology, efficacy and tolerability. *Journal of Psychopharmacology* 19(4): 414–421.

Wu, C., F. Chen, X. Wang, et al. 2006. Antioxidant constituents in feverfew (*Tanacetum parthenium*) extract and their chromatographic quantification. *Food Chemistry* 96(2): 220–227.

Zeni, A.L.B. 2011. Phytochemistry, toxicology, neuroprotective and antidepressant-like effects of aqueous extract of *Aloysia gratissima* (Gill et Hook) Troncoso (erva santa) study. PhD thesis, Federal University of Santa Catarina, Florianópolis, Brazil.

Zétola, M., T.C.M. de Lima, D. Sonaglio, et al. 2002. CNS activities of liquid and spray-dried extracts from *Lippia alba*: Verbenaceae (Brazilian false melissa). *Journal of Ethnopharmacology* 82(2–3): 207–215.

Zmijewski, M.A., T.W. Sweatman, and A.T. Slominski. 2009. The melatonin-producing system is fully functional in retinal pigment epithelium (ARPE-19). *Molecular and Cellular Endocrinology* 307(1–2): 211–216.

16

Overview of the Occurrence of Melatonin in Medicinal Plants with Special Reference to Crataeva nurvala *Buch-Ham: Isolation, Purification, and Characterization*

Atanu Bhattacharjee, Shastry Chakrakodi Shashidhara, and Santanu Saha
NGSM Institute of Pharmaceutical Sciences
Karnataka, India

CONTENTS

ABSTRACT Melatonin (N-acetyl-5-methoxytryptamine), an indole compound of tryptophan precursor, is ubiquitous in the plant kingdom. The objectives of the current study were to isolate, purify, and characterize a bioactive tryptamine derivative from Varuna (*Crataeva nurvala* Buch-Ham), a well-explored traditional Indian medicinal plant with historical evidence of efficacy in the treatment of neurological and antioxidant deficiency–related disorders. In this study, a chloroform fraction of *C. nurvala* stem bark was analyzed using column chromatography and thin-layer chromatography (TLC), which lead to the isolation of melatonin, a biogenic indoleamine. The structure of the isolated compound was determined by various spectroscopic analyses, such as ultraviolet (UV), infrared (IR), carbon nuclear magnetic resonance (^{13}C NMR), proton nuclear magnetic resonance (^{1}H NMR), two-dimensional correlated spectroscopy (2-D COSY), and mass spectroscopy. Mass spectroscopy of the isolated compound showed a parent molecular ion (M$^+$) peak at m/z 233.2 g/mol, corresponding to the molecular formula $C_{13}H_{16}N_2O_2$. In the ^{1}H NMR spectrum, singlet (s) at δ_H 3.79 corresponds to 3 H of -OCH$_3$, multiplet (m) at δ_H 6.7, 7.0, 7.7 denotes 4 Ar. protons, singlet (s) at δ_H 1.9 is assigned to 3 H of -CH$_3$CO, triplet (t) at δ_H 2.8 corresponds to 2 H of N-CH$_2$, multiplet (m) at δ_H 3.27 denotes 2 H of indolyl CH$_2$, singlet (s) at δ_H 10.41 is assigned to 1 H of NH indole, and singlet (s) at δ_H 8.01 corresponds to 1 H, NH of secondary amide. ^{13}C NMR showed the presence of a total of 13 carbon atoms. Based on physical and spectral characteristics, the isolated compound was identified and reported for the first time as N-acetyl-5-methoxytryptamine (melatonin) from *C. nurvala*. With this background, the chapter is designed to provide an overview of phytomelatonin, specifically on its distribution and occurrence, isolation, and structural elucidation.

KEY WORDS: Crataeva nurvala, *tryptophan, chromatography, isolation, spectral analysis, phytomelatonin.*

Abbreviations

^{13}C NMR:	carbon nuclear magnetic resonance
^{1}H NMR:	proton nuclear magnetic resonance
2-D COSY:	two-dimensional correlated spectroscopy
EI:	electron ionization
ELISA:	enzyme-linked immunosorbent assay
GC–MS:	gas chromatography–mass spectroscopy
HPLC:	high-performance liquid chromatography
HPLC–MS:	high-performance liquid chromatography–mass spectroscopy
IAA:	indole-3-acetic acid
LR–EI–MS:	low resolution–electron ionization–mass spectroscopy
RIA:	radioimmunoassay
ROS:	reactive oxygen species
TLC:	thin-layer chromatography

Introduction

Melatonin is an "old and well-known friend" in human and animal physiology but "newfangled" to plant physiology (Reiter et al., 2011). Melatonin was first isolated from bovine pineal gland and identified as N-acetyl-5-methoxytryptamine by Lerner and coworkers in 1958 (Lerner et al., 1958). It was named melatonin due to its ability to blanch the skin in certain fish, reptiles, and amphibians (Chava et al., 2012). In mammals, melatonin plays a key role in regulating circadian rhythm, with the highest levels during the scotophase and baseline levels during the photoperiod (Sahna et al., 2005; Ramakrishna, 2015). This molecule has been linked to a number of physiological and pathological functions, including the prevention of ischemia–reperfusion damage, chronic pain relief, immunity enhancement, antibacterial and oncostatic action, the treatment of neurological disorders, and anti-inflammation and antioxidation (Nitulescu-Arsene et al., 2009; Carrillo-Vico et al., 2013; Fatma et al., 2013; Bhavini et al., 2009; Russel et al., 2010). Melatonin passively diffuses into the cerebrospinal fluid and bloodstream, where it shows maximum effectiveness (Ayushi et al., 2007). Melatonin preserves mitochondrial homeostasis, increases gene expression for antioxidant enzymes, and thereby acts as a powerful antioxidant. Hence, it is extremely beneficial in neurodegenerative disorders such as Alzheimer's and Parkinson's disease, whose pathogenesis is associated with the cytotoxic effect of reactive oxygen species (ROS) (Hardeland, 2005; Jian-zhi and Ze-fen, 2006; Venkatramanujam, 2011).

For a long time, this neurohormone was portrayed as being synthesized only in the pineal gland of vertebrates (Ebels and Tommel, 1972). The existence of melatonin in plants was independently identified for the first time by Dubbels and coworkers and Hattori and coworkers in 1995 (Dubbels et al., 1995; Hattori et al., 1995). Since then, the search for plant-derived melatonin—that is, *phytomelatonin*—has become one of the fastest emerging fields of research in plant physiology. During the last two decades, the universal presence of melatonin in plants has been supported by numerous scientific evidence. Tryptophan is an essential amino acid, and since animals lack the ability to synthesize it, they must obtain it from other natural sources (Marino and Hernandez-Ruiz, 2007).

In plants, tryptophan provides precursors for melatonin along with the hormone auxin, phytoalexins, glucosinolates, alkaloids, and indoleamines (Bandurski et al., 1995). Melatonin has been identified in diverse organisms including prokaryotes, eukaryotes, fungi, algae, and higher plants (Rudiger and Burkhard, 2003). Based on its ubiquitous distribution and multidirectional activity, melatonin is recommended as one of the most versatile biological signals in nature (Hardeland et al., 1996). Indeed, recent research suggests that this classical indole derivative is both synthesized in and taken up by plants (Marino and Josefa, 2006). The role of phytomelatonin as an antioxidant, a free radical scavenger, and a growth promoter is clearly supported by the experimental outcomes (Russel et al., 2014). Studies suggest that the production of indole compounds is augmented under high ultraviolet (UV) radiation and thus provide substantial evidence for the role of phytomelatonin as a free radical scavenger, thereby protecting plants against oxidative stress and reducing the damage of macromolecules in a manner similar to that in animals (Katerova et al., 2012; Dun-xian, 2007). It plays a key role in the regulation of plant reproductive physiology and the defense of plant cells against apoptosis induced by unfavorable environmental conditions (Dun-xian, 2015). Several physiological roles of phytomelatonin, including possible roles in flowering, maintaining circadian rhythms, and photoperiodicity, and as a growth regulator, have been identified. Melatonin content varies in different plant organs or tissues and seems to be more profuse in aromatic plants, and in leaves more than seeds (Krystyna and Małgorzata, 2013; Hernandez-Ruiz et al., 2005). It shows auxin-like activity and thus regulates the growth of roots, shoots, and explants, activating seed germination and rhizogenesis (lateral and adventitious roots), and delaying induced leaf senescence (Katarzyna et al., 2014). Recently, a possible role in rhizogenesis in lupin has also been proposed (Hernandez-Ruiz et al., 2004, 2008).

Although the presence of melatonin in plants is a universal phenomenon, still very little information regarding its occurrence outside angiosperms (with the exception of micro- and macroalgae and other photoautotrophic microorganisms) has been reported (Kolar et al., 2001; Marcello et al., 2012). This is largely attributed to inadequate detection methods and a lack of experimental protocols to investigate the biochemical and molecular aspects of phytomelatonin. However, during the last few years, certain methodological protocols regarding extraction, isolation, and quantification have been successfully designed and optimized with sufficient complexity to obtain quick, reliable results on phytomelatonin content. The complete biosynthetic pathways and enzymatic involvement in phytomelatonin production are yet to be explored; studies with radioisotope tracer techniques have revealed tryptophan as a common precursor for both serotonin and melatonin, as well as for indole-3-acetic acid (IAA) (Marino, 2014; Russel et al., 2007; Van Tassel et al., 1995). It has been reported that plants may be able to absorb melatonin from the soil in which they grow. Evidence also indicates the involvement of melatonin in chlorophyll preservation, thereby promoting photosynthesis (Kolar and Machackova, 2005). Transgenic plants with high levels of melatonin may play a significant role in increasing crop production and improving the general health of humans (Amit and Vinod, 2014). Besides discussing interesting data on phytomelatonin, this chapter is constructed with the objective to deepen our understanding of the isolation and structural elucidation of melatonin in plants.

Occurrence and Distribution

The occurrence of melatonin has been identified in more than 140 different aromatic and medicinal plants and plants edible by humans (Jan and Ivana, 2005). Several sophisticated analytical techniques have been developed to detect the presence of melatonin in plant tissues. Among them, radioimmunoassay (RIA), enzyme-linked immunosorbent assay (ELISA), high-performance liquid chromatography (HPLC), and gas chromatography–mass spectrophotometry (GC–MS) are considered the most reliable (John et al., 2011; Kolar, 2003; Marcello et al., 2012; Tettamanti et al., 2000). The ubiquitous distribution of melatonin is observed in different parts of plants, viz. leaves, roots, fruits, and seeds (Pandi-Perumal et al., 2006). It has been reported that crops belonging to the family Graminae (e.g., rice, barley, sweet corn, oat) contain high amounts of melatonin (Dun-xian et al., 2007). GC–MS analysis showed that banana contains melatonin at a concentration of 0.655 ng/g, but HPLC–MS suggested significantly higher levels of melatonin (1 ng/g of plant tissue) (Badria, 2002). Melatonin was also found in fruits such as strawberries, kiwis, pineapples, apples, and grapes, as well as tart cherries and tomatoes (Burkhardt et al., 2001; Russel, 2001). RIA showed the presence of melatonin in both white and black mustard seeds (189 and 123 ng/g of plant tissue, respectively; Manchester et al., 2000). The presence of melatonin was also identified in both green and roasted beans of *Coffea canephora* and *Coffea arabica* at concentrations of 5.8 ± 0.8 and 8.0 ± 0.9 µg/g dry weight, respectively (Ramakrishna et al., 2012a,b). Brief information regarding the distribution of melatonin in plants according to the families is summarized in Table 16.1.

TABLE 16.1

Occurrence and Distribution of Melatonin in Edible and Medicinally Important Plants

Family	Common Name	Scientific Name	Method of Detection	Amount of Melatonin (pg/g)	References
Actinidiaceae	Kiwi fruit	*Actinidia deliciosa* Liang-Ferg.	ELISA	24.4	(Hattori et al., 1995)
Amaranthaceae	Beetroot	*Beta vulgaris* L.	RIA	2	(Dubbels et al., 1995)
Araceae	Taro	*Colocasia esculenta* L.	ELISA	54.6	(Hattori et al., 1995)
Asparagaceae	Asparagus	*Asparagus officinalis* L.	ELISA	9.5	(Hattori et al.
		Asparagus racemosus L.	ELISA	10	(Hattori et al., 1995)
Asteraceae	Feverfew	*Tanacetum parthenium* L.	HPLC–UV	1,300–7,000 ng/g	(Ansari et al., 2010)
		Tripleurospermum disciforme Schultz. Bip.	HPLC–UV	1,305.8 ng/g (in hot water extract)	(Ansari et al., 2010)
			ELISA	3,073.3 ng/g (in 50% methanol extract)	
				1,112.0 ng/g (in hot water extract)	
				2,096.2 ng/g (in 50% methanol extract)	
	Shungiku	*Chrysanthemum coronarium* L.	ELISA	416.8	(Hattori et al., 1995)
	Butterbur (fuki)	*Petasites japonicus* Maxim	ELISA	49.5	(Hattori et al., 1995)
	Milk thistle seed	*Silybum marianum* L.	ELISA	2,000	(Manchester et al., 2000)
Berberidaceae	Barren wort	*Epimedium brevicornum* M.	ELISA	1,105 ng/g	(Sergio et al., 2009)
Basellaceae	Indian spinach	*Basella alba* L.	ELISA	38.7	(Hattori et al., 1995)
Brassicaceae	Cabbage	*Brassica oleracea* L.	ELISA	107.4	(Hattori et al., 1995)
	White radish	*Raphanus sativus* L.	ELISA	657.2	(Hattori et al., 1995)
	Chinese cabbage	*Brassica rapa* L.	ELISA	485,000	(Hattori et al., 1995)
	Black, white mustard seed	*Brassica nigra* L., *Brassica hirta* L.	ELISA	112.5, 129,000	(Manchester et al., 2000)
Bromeliaceae	Pineapple	*Ananas comosus* L.	ELISA	36.2	(Hattori et al., 1995)
Cucurbitaceae	Cucumber fruit	*Cucumis sativus* L.	HPLC	24.6	(Hattori et al., 1995)
Fabaceae	Alfalfa seed	*Medicago sativa* L.	HPLC–UV	16,000	(Manchester et al., 2000)

Family	Common name	Scientific name	Method	Melatonin content	References
	Fenugreek seed	*Trigonella foenum-graecum* L.	HPLC–UV	43,000	(Manchester et al., 2000)
	Lupin seed	*Lupinus albus* L.	HPLC–UV	3,830	(Hernandez-Ruiz et al., 2004)
	Tora	Senna tora	ELISA	—	(Chen et al., 2003)
	Hummingbird tree	*Sesbania glandiflora* L.	HPLC–UV	10.5	(Chen et al., 2003)
			ELISA	43.7	(Chen et al., 2003)
			HPLC–UV	26.3	
Hypericaceae	St John's wort	*Hypericum perforatum* L.	Leaf	1,750 ng/g	(Murch et al., 2006)
			Flower	2,400–4,000 ng/g	(Russel and Manchester, 2005)
Juglandaceae	Walnut	*Juglans regia* L.	ELISA	3,500	(Manchester et al., 2000)
Papaveraceae	Poppy seed	*Papaver somniferum* L.	RIA	6,000	(Chen et al., 2003)
Phyllanthaceae	Burmese grape	*Baccaurea ramiflora* L.	ELISA	76.7	
			HPLC–UV	43.2	
Poaceae	Rice seed	*Oryza sativa* L.	ELISA	1,006	(Hattori et al., 1995)
	Barley seed	*Hordeum vulgare* L.	ELISA	378.1	(Hattori et al., 1995)
	Sweet corn	*Zea mays* L.	ELISA	580	(Hattori et al., 1995)
	Oat seed	*Avena sativa* L.	ELISA	1,366	(Hattori et al., 1995)
Lamiaceae		*Scutellaria baicalensis* L.	ELISA	2,000–7,000 ng/g	(Van Tassel et al., 1995)
Liliaceae	Onion	*Allium cepa* L.	RIA	31.5	(Hattori et al., 1995)
	Welsh onion	*Allium fistulosum* L.	RIA	85.7	(Hattori et al., 1995)
Lythraceae	Pomegranate	*Punica granatum* L.	HPLC–MS	540–5,500	(Mena et al., 2012)
Moraceae	White mulberry	*Morus alba* M.	—	1,510 ng/g	(Tanit et al., 2014)
Musacea	Banana	*Musa acuminata* Colla.	GC–MS	0.46	(Marino, 2014)
Oleracea	Olive oil	*Olea europaea* L.	ELISA	50–119 pg/mL	(De la Puerta et al., 2007)
Polygonaceae	Chinese rhubarb	*Rheum palmatum* L.	ELISA	1,078 ng/g	(Marino, 2014)
Ranunculaceae	Chinese goldthread	*Coptis chinensis* F.	ELISA	1,008 ng/g	(Marino, 2014)
Rosaceae	Apple	*Malus domestica* Borkh.	ELISA	47.6	(Hattori et al., 1995)
	Strawberry	*Fragaria ananassa* Duch.	ELISA	12.4	(Hattori et al., 1995)
	Almond seed	*Prunus amygdalus* Batsch.	ELISA	1,400–11,260	(Marino et al., 2007)
Rubiaceae	Coffee beans	*Coffea arabica* L.	ELISA	5.8 µg/g	(Hattori et al., 1995)
	Gambir vine	*Uncaria rhynchophylla*	ELISA	2,460 ng/g	(Hattori et al., 1995)

(Continued)

TABLE 16.1 (CONTINUED)

Occurrence and Distribution of Melatonin in Edible and Medicinally Important Plants

Family	Common Name	Scientific Name	Method of Detection	Amount of Melatonin (pg/g)	References
Rutaceae	Orange juice	*Citrus × sinensis* L.	HPLC–UV	150	(Sae-Teaw et al., 2012)
	Amur cork tree	*Phellodendron amurense*	ELISA	1,235 ng/g	(Hattori et al., 1995)
Solanaceae	Tomato fruit	*Solanum lycopersicum* L.	HPLC–UV	32.2	(Marino, 2014)
	Silver leaf nightshade fruit	*Solanum elaeagnifolium* Cav.	HPLC	7,895	(Marino, 2014)
	Black nightshade fruit	*Solanum nigrum* L.	HPLC	323	(Marino, 2014)
	Tobacco leaf	*Nicotiana tabacum* L.	HPLC	50	(Marino, 2014)
	Devil's trumpet flower	*Datura metel* L.	HPLC	1,500	(Marino, 2014)
Umbelliferae	Carrot	*Daucus carota* Hoffm.	ELISA	55.3	(Hattori et al., 1995)
	Anise seed	*Pimpinella anisum* L.	ELISA	7,000	(Hattori et al., 1995)
	Coriander seed	*Coriandrum sativum* L.	ELISA	7,000	(Manchester et al., 2000)
	Fennel seed	*Foeniculum vulgare* L.	ELISA	28,000	(Manchester et al., 2000)
	Sunflower seed	*Helianthus annuus* L.	ELISA	29,000	(Manchester et al., 2000)
Violaceae		*Viola philipica* Cav.	ELISA	2,368 ng/g	(Ansari et al., 2010)
Vitaceae	Grapevine	*Vitis vinifera* L.	ELISA	5–965	(Manchester et al., 2000)
Zingiberaceae	Cardamom seed	*Elettaria cardamomum* L.	HPLC–MS	15,000	(Marino, 2014)
	Curcuma	*Curcuma aeruginosa* Roxb.	GC–MS	120,000	(Marino, 2014)
	Ginger	*Zingiber officinale* Rose	HPLC–MS	583.7	(Marino, 2014)
	Piper	*Piper nigrum* L.	ELISA	865 ng/g	(Marino, 2014)
			HPLC–UV	1,092.7 ng/g	

Isolation and Structural Elucidation

The majority of herbs containing high levels of melatonin have traditionally been used to treat neurological disorders associated with the generation of free radicals, which might be linked to its potent antioxidant activity (Kazutaka et al., 2013). Recently, high levels of melatonin were identified in *Crataeva nurvala* Buch-Ham (Capparidaceae), a well-explored traditional Indian medicinal plant used to treat various ailments—in particular, urolithiasis and neurological disorders (Shiddamallayya et al., 2010; Amod et al., 2005).

   C. nurvala stem bark was extracted by cold maceration with ethanol and concentrated through rotary flash evaporator. The concentrated ethanolic extract was defatted with petroleum ether and fractionated with chloroform (Parvin et al., 2011). A detailed flowchart of the method of extraction and fractionation is given in Figure 16.1.

   A chloroform fraction (10 g) was purified over silica gel (60–120 #) column chromatography using chloroform–ethyl acetate eluent with increasing order of polarity (the gradient elution technique). The progress of separation was monitored by TLC (silica gel G 60 F_{254} plates, Merck). The eluent was collected into 12 different fractions. Fraction IV was rechromatographed and eluted with chloroform and ethyl acetate (7:3), resulting in a crude, amorphous yellowish-white substance, which was crystallized with methanol. The crystals were further analyzed spectrophotometrically to elucidate the structure. TLC chromatogram developed with methanol and chloroform (2:8) was homogeneous with R_f 0.58. Further, qualitative analysis suggested an alkaloid nature of the compound. The melting point was observed at 117°C (Atanu et al., 2014).

Structural Elucidation

UV spectroscopic analysis (Shimadzu UV-1700 Pharmac-spec UV-Vis spectrophotometer [Japan]) revealed the λ_{max} at 278.0 nm. IR spectrum (Alpha-Bruker IR spectrophotometer) showed the presence of –NH (amide) str. at 3726 cm^{-1}, –NH–indole str. at 3432 cm^{-1}, –CH$_3$/–CH$_2$ str. at 2945 cm^{-1}, C=O

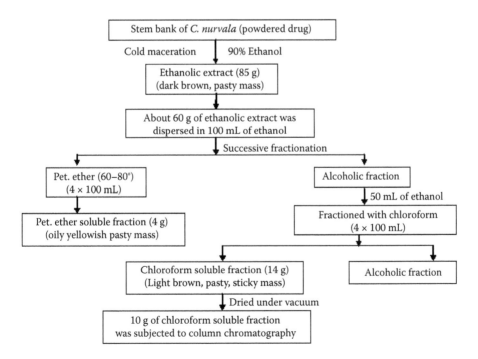

FIGURE 16.1 Extraction and fractionation of *C. nurvala* Buch-Ham. (Adapted from Atanu et al., *Am J Phytomed Clin Therap.* 2(3), 301–9, 2014.)

TABLE 16.2

^{1}H NMR Data of Melatonin: Chemical Shifts in δ_{ppm} Relative to the
Internal Standard DMSO

δ (ppm)	Spin Multiplicity	Integration	Comment
3.79	s	3 H	–OCH$_3$
6.7, 7.0, 7.7	m	4 H	Ar. protons
1.9	s	3 H	–CH$_3$CO
2.8	t	2 H	–N–CH$_2$
3.27	m	2 H	indolyl–CH$_2$
10.41	s	1 H	–NH gr. of indole
8.01	s	1 H	–NH gr. of sec. amide

Source: Data from Atanu et al., *Am J Phytomed Clin Therap.* 2(3), 301–9, 2014.
Note: s = singlet, m = multiplet, t = triplet.

str. at 1644 cm^{-1}, Ar. C=C str. at 1400 cm^{-1}, C–O str. at 1216 cm^{-1}, and C–N (amine) str. at 1071 cm^{-1} (Atanu et al., 2014).

The proton nuclear magnetic resonance (^{1}H NMR) spectrum (Bruker Advance II 400 NMR spectro-photometer [Karlsruhe, Germany]) of the isolated compound dissolved in dimethyl sulfoxide (DMSO) showed a sharp singlet at δ_H 3.79, which indicated the presence of the Ar.–OCH$_3$ group. Aromatic protons were also accounted for by the multiplet signals δ_H 6.7, 7.0, and 7.7. Singlet signals appearing at δ_H 1.9 accounted for 3H of the aliphatic –COCH$_3$. The triplet signal appearing at δ_H 2.8 indicated the protons of N–CH$_2$. The multiplet appearing at δ_H 3.27 accounting for 2H was assigned to indolyl–CH$_2$ of the alkaloid. Singlet signals appearing at δ_H 10.41 corresponded to 1 H, the –NH group of the indole nucleus (Atanu et al., 2014). The singlet appearing at δ_H 8.01 accounted for 1 H, the –NH group of the secondary amide (Table 16.2 and Figures 16.2 through 16.4). Moreover, in a two-dimensional correlated spectroscopy (2-D COSY) NMR spectrum, the cross peak attributed between the protons at δ 7.0 ppm and the multiplet near δ 3.0 ppm corresponded to the coupling of the indole –NH and the H-8 of the aromatic ring of melatonin (Figure 16.5) (Ackermann et al., 2006).

Carbon nuclear magnetic resonance (^{13}C NMR) spectrum revealed the presence of a total 13 carbon atoms; δ_{ppm} 152.92 at C-5 suggested the presence of the methoxy group (–OCH$_3$), whereas δ_{ppm} 55.20 at C-15 corresponded to the Ar.–OCH$_3$ group. Indolyl–CH$_2$ peak at C-10 was observed at δ_{ppm} 29.07, and δ_{ppm} 25.03 at C-11 revealed the presence of the aliphatic N–CH$_2$ group. Further, δ_{ppm} 22.60 at C-14 denoted the presence of the -CH$_3$CO group (Figures 16.6 and 16.7) (Atanu et al., 2014).

Low-resolution and high-resolution electron ionization MS (LR–EI–MS and HR–EI–MS) were recorded by the Waters Q-TOF (Micromass, Altrincham, UK) and a mass spectrometer connected with a GC system HP 6890 series (Hewlett Packard, Palo Alto, CA). LR–EI–MS (resolution power 1500) and HR–EI–MS (resolution power 8000, 10% resolution valley definition) were performed under the following experimental conditions: electron beam energy 70 eV, source temperature 210°C, source pressure 10^{-7} Torr, trap current 250 μA, emission current 2.3 μA, accelerating voltage 8.0 kV. Accurate mass measurements (±10 ppm) were determined by HR–EI–MS using perfluoro-kerosene (PFK) as the internal standard. Gas-chromatographic conditions were: injector temperature 290°C, column AT^{TM}-5 (Alltech, Deerfield, FL), film thickness 0.25 μm, length 30 m, ID 0.25 mm, carrier gas (helium) flow 1.0 mL/min, isotherm at 120°C (5 min), ramp 120–240°C (20°C/min), isotherm 240°C (9 min).

Mass spectroscopy revealed a molecular ion peak at m/z = 233.2 g/mol and base peak at m/z = 255.1 g/mol. These assignments revealed the molecular formula of the isolated compound to be C$_{13}$H$_{16}$N$_2$O$_2$. By comparing infrared (IR), time of light mass spectroscopy electrospray ionization (TOF MS ES), ^{1}H, and ^{13}C NMR data with the existing literature, the isolated compound was assigned as N-acetyl-5-methoxytryptamine (melatonin) (Table 16.3 and Figures 16.8 through 16.10) (Pasquale et al., 2003).

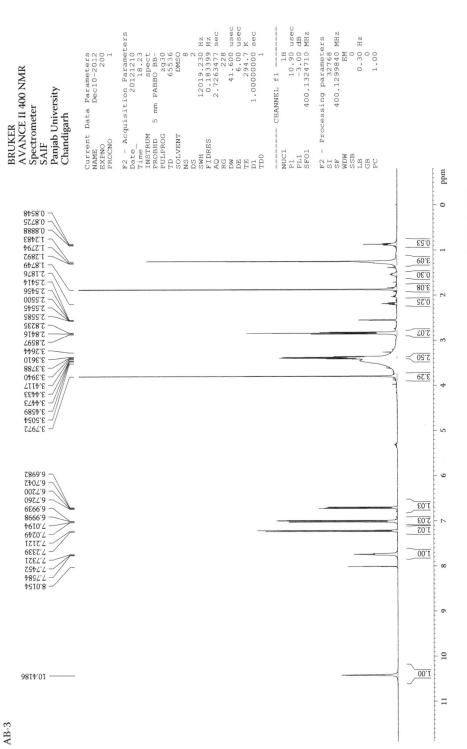

FIGURE 16.2 ^{1}H NMR spectra of melatonin (δ ppm=0–11). (Adapted from Atanu et al., *Am J Phytomed Clin Therap.* 2(3), 301–9, 2014.)

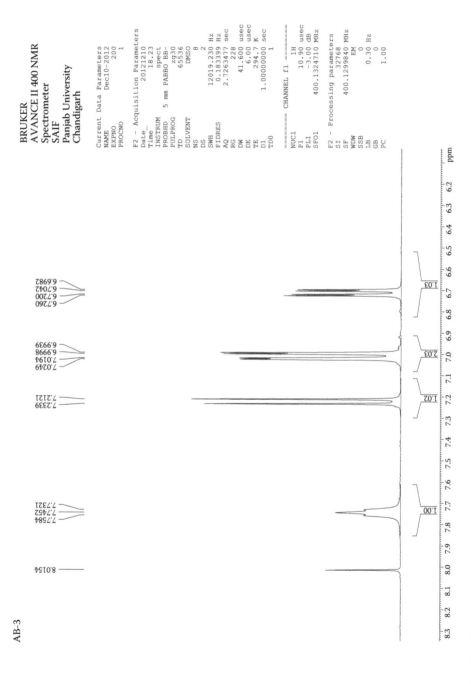

FIGURE 16.3 ¹H NMR spectra of melatonin (δ ppm = 6–9).

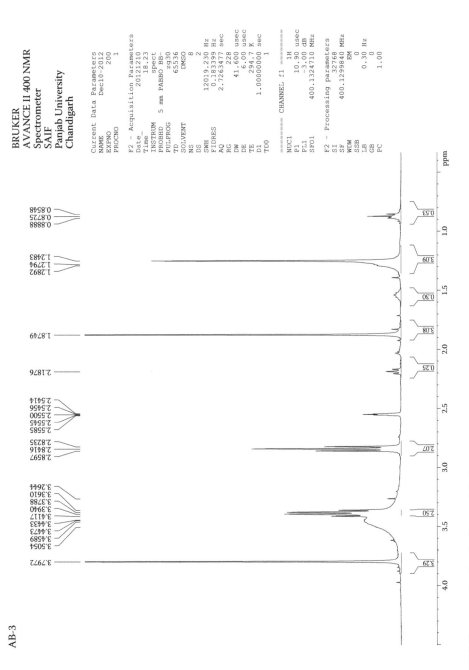

FIGURE 16.4 ¹H NMR spectra of melatonin (δ ppm = 0.5–5).

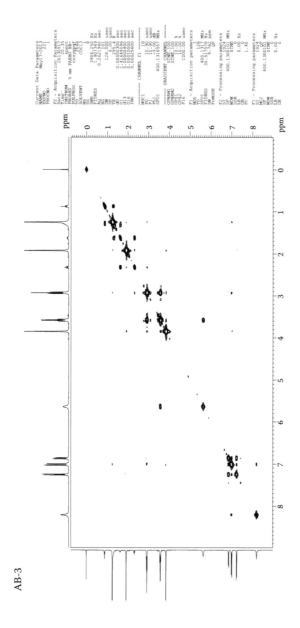

FIGURE 16.5 The 2-D COSY ¹H-nuclear magnetic resonance (CDCl₃) spectrum of melatonin. (Adapted from Atanu et al., *Am J Phytomed Clin Therap.* 2(3), 301–9, 2014.)

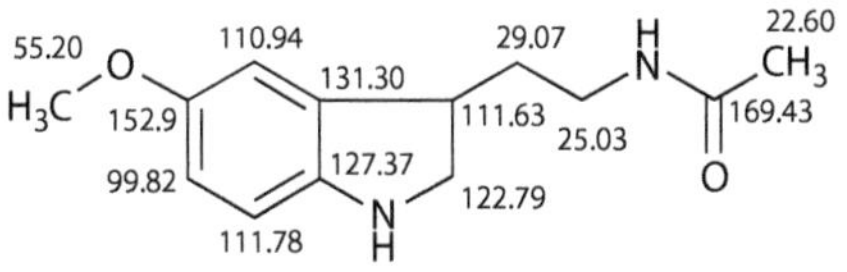

FIGURE 16.6 ^{13}C-NMR chemical shifts of melatonin. (Adapted from Atanu et al., *Am J Phytomed Clin Therap.* 2(3), 301–9, 2014.)

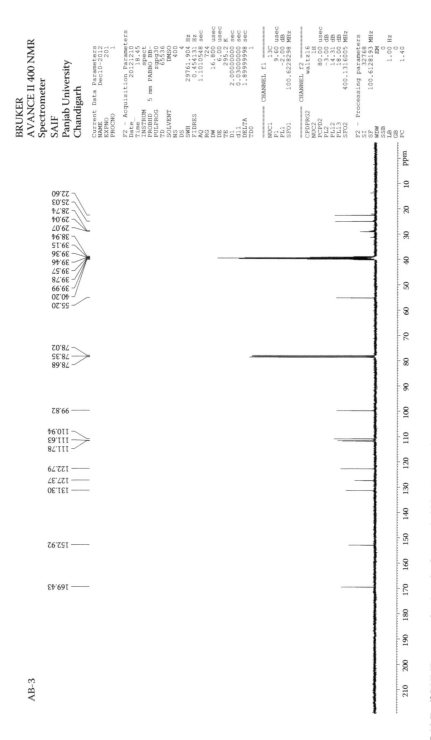

FIGURE 16.7 ^{13}C NMR spectra of melatonin (δ ppm = 0–200). (Adapted from Atanu et al., *Am J Phytomed Clin Therap.* 2(3), 301–9, 2014.)

TABLE 16.3

Mass Fragmentation Pattern of Melatonin

EIMS (m/z) (%)	233.2 [M$^+$] (8%)
Relative intensity	174.1 (20%), 255.1 (100%), 255.8 (78%), 256.1 (25%), 271.1 (9%)

Source: Data from Atanu et al., *Am J Phytomed Clin Therap.* 2(3), 301–9, 2014.

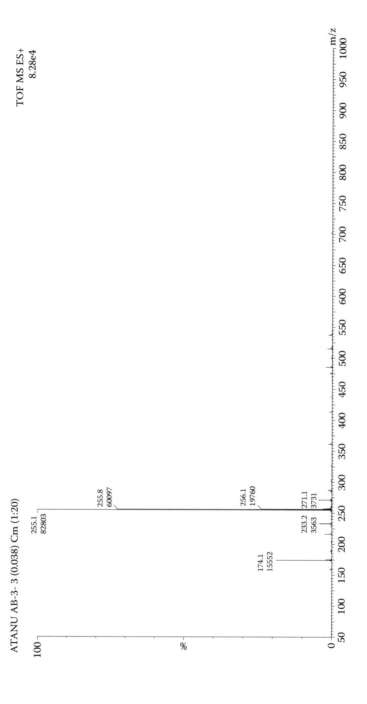

FIGURE 16.8 High resolution–electron ionization–mass spectrum (70 eV) of melatonin (m/z: 0–1000). (Adapted from Atanu et al., *Am J Phytomed Clin Therap.* 2(3), 301–9, 2014.)

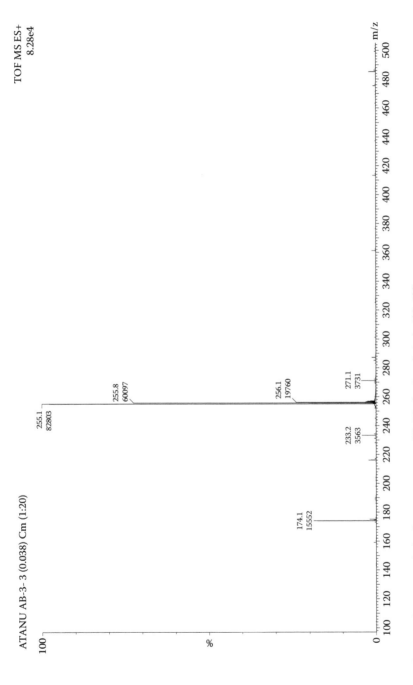

FIGURE 16.9 High resolution–electron ionization–mass spectrum (70 eV) of melatonin (m/z: 100–500).

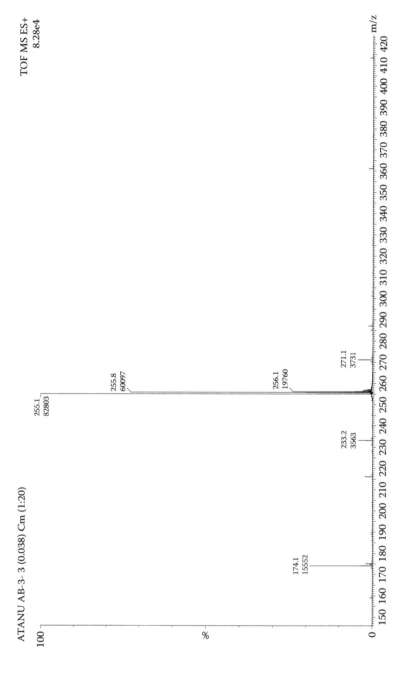

FIGURE 16.10 High resolution–electron ionization–mass spectrum (70 eV) of melatonin (m/z: 100–450).

Conclusion

The possibility of introducing melatonin-rich plant foods or food supplements is due to its immense health benefits, particularly against neurodegenerative disorders such as Alzheimer's. Studies have revealed that oral doses of melatonin of up to 1 g/day produce no adverse effects in humans. In addition, melatonin is easily absorbed via the gastrointestinal tract. So, melatonin as a nutraceutical seems to have a promising future in the promotion of a healthier life (Charanjit et al., 2008; Jemima et al., 2011). In this regard, the identification of melatonin in *C. nurvala* may prove to be extremely beneficial, and new preclinical aspects will be explored to find its utility against neurological disorders associated with the generation of free radicals, such as Alzheimer's (Chandana et al., 2014). In addition, melatonin may act as a suitable marker compound for the standardization and validation of commercial extracts and herbal preparations containing *C. nurvala*.

Acknowledgment

The authors acknowledge the financial support of Nitte University, Mangalore, India, for the research work.

REFERENCES

Ackermann K, Bux R, Rub U, Korf H, Kanert G, Stehle JH. Characterization of human melatonin synthesis using autoptic pineal tissue. *Endocrin* 2006;23:55–9.

Amit KT, Vinod K. Melatonin: An integral signal for daily and seasonal timing. *Ind J Exp Biol* 2014;52:425–37.

Amod PK, Kellaway LA, Girish JK. Herbal complement inhibitors in the treatment of neuroinflammation. *Ann NY Acad Sci* 2005;1056:413–29.

Ansari M, Rafiee K, Yasa N, Vardasbi S, Naimi SM, Nowrouzi A. Measurement of melatonin in alcoholic and hot water extracts of *Tanacetum parthenium*, *Tripleurospermum disciforme* and *Viola odorata*. *Daru* 2010;18(3):173–78.

Atanu B, Shastry CS, Santanu S. Isolation, purification and structural elucidation of N-acetyl-5- methoxy-tryptamine (melatonin) from *Crataeva nurvala* Buch-Ham stem bark. *Am J Phytomed Clin Therap.* 2014;2(3):301–9.

Ayushi J, Maheep B. Melatonin: A "magic biomolecule." *Ann Neurosci* 2007;14(4): 1–5.

Badria FA. Melatonin, serotonin, and tryptamine in some Egyptian food and medicinal plants. *J Med Food* 2002;5:153–57.

Bandurski RS, Cohen JD, Slovin J, Reinecke DM. Auxin biosynthesis and metabolism. In: Davies PJ, ed. *Plant Hormones: Physiology, Biochemistry and Molecular Biology*, pp. 39–65. Dordrecht, the Netherlands: Kluwer Academic, 1995.

Bhavini B, Muhammad UF, Archit B. The therapeutic potential of melatonin in neurological disorders. *Recent Pat Endocr Metab Immune Drug Discov* 2009;3:60–64.

Burkhardt S, Tan DX, Manchester LC, Hardeland R, Reiter RJ. Detection and quantification of the anti-oxidant melatonin in Montmorency and Balaton tart cherries (*Prunus cerasus*). *J Agri Food Chem* 2001;49:4898–902.

Carrillo-Vico A, Patricia JL, Alvarez-Sánchez N, Rodríguez-Rodríguez A, Guerrero JM. Melatonin: Buffering the immune system. *Int J Mol Sci* 2013;14:8638–83.

Chandana H, Somenath G, Amaresh KS. Phyto-melatonin: A novel therapeutic aspect of melatonin in nature's way. In: V. Srinivasan, G. Gobbi, S.D. Shillcutt, S. Suzen (eds) *Melatonin: Therapeutic Value and Neuroprotection*, pp. 515–520. Boca Raton, FL: CRC Press, 2014.

Charanjit K, Sivakumar V, Ling EA. Melatonin and its therapeutic potential in neuroprotection. *CNS Agents in Med Chem.* 2008;8:260–66.

Chava VK, Sirisha K. Melatonin: A novel indolamine in oral health and disease. *Int J Dent* 2012;2012:1–9.

Chen G, Huo Y, Tan DX, Liang Z, Zhang W, Zhang Y. Melatonin in Chinese medicinal herbs. *Life Sci* 2003;73:19–26.

De la Puerta C, Carrascosa-Salmoral MP, Garcia-Luna PP. Melatonin is a phytochemical in olive oil. *Food Chem* 2007;104(2):609–12.

Dubbels R, Reiter RJ, Klenke E, Goebel A, Schnakenberg E, Ehlers C, et al. Melatonin in edible plants identified by radioimmunoassay and by high performance liquid chromatography–mass spectrometry. *J Pineal Res* 1995;18:28–31.

Dun-xian T. Melatonin and plants. *J Exp Bot* 2015;66(3):625–26.

Dun-xian T, Lucien CM, Pat H, Russel JR. Phytoremediative capacity of plants enriched with melatonin. *Plant Signal Behav* 2007;2(6):514–16.

Dun-xian T, Rudiger H, Lucien CM, Ahmet K, Shuran M, Sergio RC, et al. Functional roles of melatonin in plants, and perspectives in nutritional and agricultural science. *J Exp Botany* 2012;63(2):577–97.

Ebels MGMB, Tommel DKJ. Separation of pineal extracts on Sephadex G-10. *Anal Biochem* 1972;50:234–44.

Fatma PK, Alper K, Arzu UT, Arzu BY. Antibacterial and antitumor activities of melatonin. *Spatula DD* 2013;3(2):33–39.

Hardeland R. Antioxidative protection by melatonin: Multiplicity of mechanisms from radical detoxification to radical avoidance. *Endocrine* 2005;27(2):119–30.

Hardeland R, Fuhrberg B. Ubiquitous melatonin: Presence and effects in unicells, plants and animals. *Trends Comp Biochem Physiol* 1996;2:25–45.

Hattori A, Migitaka H, Iigo M, Itoh M, Yamamoto K, Ohtani-Kaneko R, et al. Identification of melatonin in plants and its effects on plasma melatonin levels and binding to melatonin receptors in vertebrates. *Biochem Mol Bio Int* 1995;35:627–34.

Hernandez-Ruiz J, Arnao MB. Distribution of melatonin in different zones of lupin and barley plants at different ages in the presence and absence of light. *J Agric Food Chem* 2008;56:10567–73.

Hernandez-Ruiz J, Cano A, Arnao MB. Melatonin: A growth-stimulating compound present in lupin tissues. *Planta* 2004;220:140–44.

Hernandez-Ruiz J, Cano A, Arnao MB. Melatonin acts as a growth-stimulating compound in some monocot species. *J Pineal Res* 2005;39:137–42.

Jan K, Ivana M. Melatonin in higher plants: Occurrence and possible functions. *J Pineal Res* 2005;39:333–41.

Jemima J, Bhattacharjee P, Singhal RS. Melatonin: A review on the lesser known potential nutraceutical. *IJPSR* 2011;2(8):1975–87.

Jian-zhi W, Ze-fen W. Role of melatonin in Alzheimer-like neurodegeneration. *Acta Pharmacol Sin* 2006;27(1):41–9.

John TY, Christian B, Miriam W, David WP, Leo Meerts W, Michael S. Rotationally resolved electronic spectroscopy of biomolecules in the gas phase melatonin. *J Mol Spectroscopy* 2011;268:115–22.

Katarzyna S, Rafał S, Krystyna MJ. Involvement of melatonin applied to *Vigna radiata* L. seeds in plant response to chilling stress. *Cent Eur J Biol* 2014;9(11):1117–26.

Katerova Z, Todorova D, Tasheva K, Sergiev I. Influence of ultraviolet radiation on plant secondary metabolite production. *Gen Plant Physiol* 2012;2(3–4):113–44.

Kazutaka S, Meaghan S, Borlongan CV. Melatonin-based therapeutics for neuroprotection in stroke. *Int J Mol Sci* 2013;14:8924–47.

Kolar J, Johnson CH, Machackova I. Exogenously applied melatonin affects flowering of the short-day plant *Chenopodium rubrum*. *Physiol Plant* 2003;118:605–12.

Kolar J, Machackova I. Occurrence and possible function of melatonin in plants: A review. *Endocytobiosis Cell Res* 2001;14:75–84.

Kolar J, Machackova I. Melatonin in higher plants: Occurrence and possible functions. *J Pineal Res* 2005;39:333–41.

Krystyna MJ, Małgorzata MP. Melatonin, an underestimated natural substance with great potential for agricultural application. *Acta Physiol Plant* 2013;35:3285–92.

Lerner AB, Case JD, Takahashi Y, Lee TH, Mori W. Isolation of melatonin, a pineal factor that lightens melanocytes. *J Am Soc* 1958;80:2587.

Manchester LC, Tan DX, Reiter RJ, Park W, Monis K, Qi W. High levels of melatonin in the seeds of edible plants: Possible function in germ tissue protection. *Life Sci* 2000;67(25):3023–29.

Marcello I, Mara R, Franco F. Melatonin content in grape: Myth or panacea? *J Sci Food Agric* 2006;86:1432–38.

Marcello I, Sara V, Mara R, Franco F. Occurrence and analysis of melatonin in food plants. In: L.M.L. Nollett and F. Toldra (eds) *Handbook of Analysis of Active Compounds in Functional Foods*, pp. 651–64. Boca Raton, FL: CRC Press, 2012:651–58.

Marino BA. Phytomelatonin: Discovery, content, and role in plants. *Adv Bot* 2014;2014:1–11.

Marino BA, Hernandez-Ruiz J. Melatonin in plants: More studies are necessary. *Plant Sig Behav* 2007;2(5):381–82.

Marino BA, Josefa HZ. The physiological function of melatonin in plants. *Plant Signal Behav* 2006;1(3):89–95.

Mena P, Gil-Izquierdo A, Moreno DA, Marti N, Garcia-Viguera C. Assessment of the melatonin production in pomegranate wines. *LWT: Food Sci Tech* 2012;47(1):13–18.

Nitulescu-Arsene AL, Niculina M, Cristea A, Manuela Dragoi C. Experimental research on mice regarding the implication of melatonin in pain management. *Farmacia* 2009;57(2):223–28.

Pandi-Perumal SR, Srinivasan V, Maestroni GJM, Cardinali DP, Poeggeler B, Hardeland R. Melatonin: Nature's most versatile biological signal? *FEBS J* 2006;5:2813–38.

Parvin S, Md. Abdul K, Md. Abdul M, Ekramul H, Md. Ashik M, Mir Imam IW. Triterpenoids and phytosteroids from stem bark of *Crataeva nurvala* Buch-Ham. *J Appl Pharm Sci* 2011;1(9):47–50.

Pasquale A, Giuseppe A, David B, Leopoldo C, Felice F, Maria CN, et al. Melatonin: Structural characterization of its non-enzymatic mono-oxygenate metabolite. *J Pineal Res* 2003;35:269–75.

Ramakrishna A. Indoleamines in edible plants: Role in human health effects. In: A. Catalá (ed.) *Indoleamines: Sources, Role in Biological Processes and Health Effects.* pp. 235–46. Hauppauge, NY: Nova, 2015.

Ramakrishna A, Giridhar P, Udaya Sankar K, Ravishankar GA. Melatonin and serotonin profiles in beans of *Coffea* sps. *J Pineal Res* 2012a;52:470–76.

Ramakrishna A, Giridhar P, Udaya Sankar K, Ravishankar GA. Endogenous profiles of indoleamines: Serotonin and melatonin in different tissues of *Coffea canephora* P ex Fr. as analyzed by HPLC and LC–MS–ESI. *Acta Physiol Plant* 2012b;34:393–96.

Reiter RJ, Coto-Montes A, Boga JA, Fuentes-Broto L, Rosales-Corral S, Tan DX. Melatonin: New applications in clinical and veterinary medicine, plant physiology and industry. *Neuro Endocrinol Lett* 2011;32(5):575–87.

Rudiger H, Burkhard P. Non-vertebrate melatonin. *J Pineal Res* 2003;34:233–41.

Russel JR. Melatonin in plants. *Nutr Rev* 2001;59(9):286–90.

Russel JR, Dun-xian T, Annia G. Melatonin reduces lipid peroxidation and membrane viscosity. *Front Physiol* 2014;5:1–4.

Russel JR, Dun-xian T, Pilar Terron M, Luis JF, Zbigniew C. Melatonin and its metabolites: New findings regarding their production and their radical scavenging actions. *Acta Biochem Polonica* 2007;54(1):1–9.

Russel JR, Reyes-Gonzales M, Fuentes-Broto L, Dun-Xian T. Melatonin reduces oxidative catastrophe in neurons and glia. *Act Nerv Super Rediviva* 2010;52(2):93–103.

Sae-Teaw M, Johns J, Johns NP, Subongkot S. Serum melatonin levels and antioxidant capacities after consumption of pineapple, orange, or banana by healthy male volunteers. *J Pineal Res* 2012;55(1):58–64.

Sahna E, Parlakpinar H, Turkoz Y, Acet A. Protective effects of melatonin on myocardial ischemia: Reperfusion induced infarct size and oxidative changes. *Physiol Res* 2005;54:491–95.

Sergio DP, Ahmet K, Lucien CM, Dun-Xian T, Russel JR. Phytomelatonin: A review. *J Exp Bot* 2009;60(1):57–69.

Shiddamallayya N, Azara Y, Gopakumar K. Hundred common forest medicinal plants of Karnataka in primary healthcare. *Ind J Trad Knowl* 2010;9(1):90–95.

Tanit P, Jeffrey J, Autcharaporn S, Suppachai T. Determination of melatonin content in traditional Thai herbal remedies used as sleeping aids. *Daru J Pharm Sci* 2014;22(6):2–5.

Tettamanti C, Cerabolini B, Gerola P, Conti A. Melatonin identification in medicinal plants. *Acta Phytother* 2000;3:137–44.

Van Tassel DL, Roberts N, O'Neill S. Melatonin from higher plants: Isolation and identification of N-acetyl-5-methoxytryptamine. *Plant Physiol* 1995;108(2):101–12.

Venkatramanujam S. Therapeutic potential of melatonin and its analogs in Parkinson's disease: Focus on sleep and neuroprotection. *Ther Adv Neurol Disord* 2011;4(5):297–317.

Section IV

Food: Occurrence and Dietary Implications

17

Melatonin in Plant-Based Food: Implications for Human Health

Gaia Favero, Rita Rezzani, and Luigi Fabrizio Rodella
University of Brescia
Brescia, Italy

Lorenzo Nardo
University of California
San Francisco, USA

Vitor Antunes Oliveira
Federal University of Santa Maria
Santa Maria, RS, Brazil

CONTENTS

ABSTRACT The utilization of dietary compounds to target, treat, and even prevent certain diseases has become an area of great and fundamental interest. The Mediterranean diet, rich in fruits and vegetables, is associated with a low incidence of chronic degenerative disorders common to Western populations, and compelling evidence points to a reduced risk of cancer and cardiovascular and neuro-degenerative diseases in Mediterranean countries compared with other industrialized countries. These observations may be related to beneficial compounds in the diet, including antioxidants, minerals, fiber, and many biologically active molecules. A new diet element has recently contributed to improve the phytochemical diversity of healthy diets and their beneficial potential effects: melatonin. Melatonin was first identified in plants in 1995, and its medicinal and nutritional relevance to both animals and humans and its functions in plants have acquired diffuse interest. In particular, melatonin has been identified in a very large number of plant species. So far, it has been detected and quantified in the roots, shoots, leaves, flowers, fruits, and seeds of a considerable variety of spermatophyte species that are relevant to humans as food and medicinal plants. When plant products are eaten, melatonin is absorbed through the gut and influences the levels of indoleamine in the blood as well as improving antioxidant status. The concentration of melatonin in edible plants varies widely and, to date, it is difficult to calculate how a specific diet may positively influence human health. This chapter will summarize and update current knowledge about melatonin in plant foods and medicinal herbs, and provide an overview of the major implications of the consumption of plant foods containing melatonin for the maintenance of a healthy lifestyle and protection against diseases.

KEY WORDS: *melatonin, food, beverages, medicinal herbs.*

Abbreviations

aMT6s: 6-sulfatoxymelatonin
ASMT: acetylserotonin O-methyltransferase
AANAT: aralkylamine N-acetyltransferase
CoQ10:   coenzyme Q10
CYP: cytochrome P450
DO: designation of origin
NAS: N-acetylserotonin
SNAT: serotonin N-acetyltransferase
TDC: tryptophan decarboxylase

Introduction

Epidemiological, experimental, and clinical trials have pointed to a positive correlation between lifestyle and dietary factors as they relate to degenerative disorders such as metabolic syndrome, cancer, and cardiovascular diseases. Dietary patterns high in refined starches, sugar, saturated fatty acids, and trans-fatty acids, and low in natural antioxidants and fiber from fruits, vegetables, and whole grains have been found to predispose susceptible people to metabolic syndrome and cardiovascular diseases (Giugliano et al., 2006). Moreover, a high caloric diet has been characterized as a powerful mechanism in the development and prognosis of some types of cancer; in fact, in colorectal cancer, it has been proposed that dietary factors contribute to more than 70% of cases (Daniel and Tollefsbol, 2015; Pericleous et al., 2013). On the other hand, a reduction of caloric consumption in humans has been linked to reduced risks of diabetes and cardiovascular diseases (Cruzen and Colman, 2009) and increased longevity and health span (Daniel and Tollefsbol, 2015).

Because diet can influence epigenetic changes, utilizing dietary compounds to target, treat, and even prevent certain diseases has become an area of great and fundamental interest. Epigenetic mechanisms that protect against cancer and aging are observed after the consumption of healthy foods such as soy, fruits, vegetables, and green tea (Daniel and Tollefsbol, 2015; Hardy and Tollefsbol, 2011).

The Mediterranean diet, rich in fruits and vegetables, is associated with a low incidence of the chronic degenerative disorders common to Western populations and compelling evidence points out the reduced risk of cancer and cardiovascular and neurodegenerative diseases in Mediterranean countries compared with other industrialized countries (Iriti et al., 2010; La Vecchia and Bosetti, 2006; Visioli et al., 2004). The healthy properties of the Mediterranean diet have been attributed to the phytochemical diversity of this traditional dietary style (Iriti and Varoni, 2015).

The health benefits of eating fruits and vegetables are based on evidence provided by numerous studies that have reported their inverse relationship with the onset of various diseases (Garcia-Moreno et al., 2013; Maderuelo-Fernandez et al., 2015), including different forms of cancer (Bamia et al., 2013; Couto et al., 2011), inflammatory stages (Urpi-Sarda et al., 2012), cardiovascular diseases (Maderuelo-Fernandez et al., 2015), Alzheimer's disease (Rosales-Corral et al., 2012), depression (Cardinali et al., 2012), metabolic syndrome (Kitagawa et al., 2012; Koziróg et al., 2011), and osteoporosis (Garcia-Moreno et al., 2013; Kotlarczyk et al., 2012). These observations may be related to beneficial compounds in fruits and vegetables, including antioxidants (e.g., vitamin C, vitamin E, carotenoids, flavonoids, and anthocyanins; Speciale et al., 2011), minerals, fiber, and many biologically active molecules (Lampe, 1999; Johns et al., 2013). A new element has recently contributed to improving the phytochemical diversity of a correct diet and, possibly, its healthy potential: melatonin. Tryptophan, an essential amino acid that can be obtained only from the diet, is the unique precursor of the indoleamine melatonin, mainly known for its property to regulate the circadian cycle (Iriti and Varoni, 2015; Iriti et al., 2012; Johns et al., 2013; Reiter et al., 2011). It is known that melatonin has regulatory effects on body weight in a high-fat diet–induced obese rat model (Prunet-Marcassus et al., 2003; Ríos-Lugo et al., 2010, 2015) and it may prevent obesity-induced complications such as cardiovascular alteration (Agabiti-Rosei et al., 2014; Favero et al., 2013; Stacchiotti et al., 2014).

Melatonin was first found in plants in 1995 (Hattori et al., 1995), and since then studies have focused primarily on its medical and nutritional relevance to both animals and humans, with little attention given to the potential physiological and biological roles in the plants themselves. Over the last few years, however, identifying the functions of melatonin in plants has been of fundamental interest.

Interestingly, age-associated reductions in endogenous melatonin secretion may alter energy regulation in middle age, resulting in elevated body weight and adiposity, and their associated detrimental metabolic consequences (Korkmaz et al., 2009). Moreover, night shift workers have lower levels of melatonin because of their nightly light exposure (Burch et al., 2005) and a body mass index significantly higher and inversely associated with their levels of 6-sulfatoxymelatonin (aMT6s) (Cocco et al., 2005), melatonin's major metabolite, secreted in urine, which reflects the peak of melatonin production (Dopfel et al., 2007). In addition, lower nocturnal urinary aMT6s level has been associated with the use of diffuse medications such as beta blockers, calcium channel blockers, and psychotropics (Davis et al., 2001).

To date, it is widely recognized that one's diet can no longer be considered simply for basic nutrition, but is an important factor for human health, too. In fact, a healthy lifestyle with a dietary intake of fruit and vegetables is important in order to maintain good health (Maderuelo-Fernandez et al., 2015; Moore et al., 2015; Vitaglione et al., 2015). This chapter will summarize and update current knowledge about melatonin in plant-based foods and medicinal herbs, and provide an overview of the major implications of the consumption of plant foods containing melatonin for the maintenance of a healthy lifestyle and protection against diseases.

Melatonin in Plants

Outside the animal kingdom, melatonin was first discovered in the photosynthesizing unicellular planktonic dinoflagellate alga *Lingulodinium polyedrum* (Stein) J.D. Dodge (*Gonyaulax polyedra* Stein) (Balzer and Hardeland, 1991; Hardeland et al., 1996). Since then, melatonin has been identified in a large number of plant species, including edible plants and medicinal herbs (Parades et al., 2009; Park et al., 2012; Tan et al., 2012a; Ramakrishna et al., 2009, 2011a, 2012a,b,c; Reiter et al., 2013). So far, melatonin has been detected and quantified in the roots, shoots, leaves, flowers, fruits, and seeds of a considerable variety of spermatophyte species (Burkhardt et al., 2001; Dubbels et al., 1995; Hattori et al., 1995; Iriti and Varoni, 2015; Manchester et al., 2000; Paredes et al., 2009; Reiter et al., 2001; Zhao et al., 2013). In flowering plants, this indoleamine was found in more than 100 species belonging to at least 19 different families and in both mono- and dicotyledons (Iriti and Varoni, 2015; Paredes et al., 2009; Reiter et al., 2001; Riga et al., 2014; Tan et al., 2012a), which are relevant to humans as food and medicinal plants. The concentration of melatonin varies extremely in different plants and in plant parts (Reiter et al., 2001, 2013; Ramakrishna et al., 2011a, 2012a,b,c), and some plants have been genetically engineered to produce more melatonin (Okazaki et al., 2010; Reiter et al., 2013).

When plant products are eaten, melatonin is absorbed through the gut and influences the levels of indoleamine in the blood as well as improving antioxidant status (Reiter et al., 2013). As ingredients in a healthy diet, many fruits and vegetables, including cherries, grapes, bananas, strawberries, pineapples, and tomatoes, provide melatonin (Murch et al., 2010; Zhao et al., 2013).

In plants, tryptophan is first decarboxylated to tryptamine by tryptophan decarboxylase (TDC), and then tryptamine is hydroxylated via tryptamine 5-hydroxylase, a cytochrome P450 (CYP) monooxygenase (Hardeland, 2015; Park et al., 2012; Tan et al., 2012a), to form serotonin. Thus, TDC serves as a rate-limiting enzyme for melatonin synthesis in plants and its gene has been identified in several plant species

(Zhao et al., 2013); however, its significance in plants remains to be amplified. In fact, transgenic rice plants overexpressing TDC resulted not only in considerable increases in tryptamine and serotonin, but also in the formation of a toxic serotonin dimer, which caused low fertility, stunted growth, and resulted in a dark brown phenotype (Hardeland, 2015; Kanjanaphachoat et al., 2012). Because of the reduced grain yield, but especially with regard to the neurotoxicity of several indole dimers formed in oxidative reactions (Behrends et al., 2004), it actually may not be recommendable to elevate melatonin levels by overexpressing TDCs in edible crops (Hardeland, 2015).

Finally, serotonin is N-acetylated to N-acetylserotonin (NAS) in plants and dinoflagellates by, presumably different, serotonin N-acetyltransferases (SNATs). NAS is subsequently O-methylated by enzymes usually termed acetylserotonin O-methyltransferases (ASMTs). All enzymes of the melatonin biosynthetic pathway have recently been cloned and characterized (Byeon et al., 2014; Kang et al., 2011, 2013). The SNAT from rice does not exhibit homology to the vertebrate paralog aralkylamine N-acetyltransferase (AANAT) (Kang et al., 2013), but it is similar to a homolog from cyanobacteria (Byeon et al., 2013; Hardeland, 2015), which is assumed to represent the phylogenetic origin of the plant SNAT (Byeon et al., 2013, 2014; Hardeland, 2015; Park et al., 2014). However, so far, knowledge is mostly restricted to rice, and pertinent data from other plants is still missing.

The gene expression of melatonin synthetic enzymes, particularly SNAT, is significantly upregulated in response to stresses such as elevated temperature (Byeon and Back, 2014; Park et al., 2014), and, in contrast to ASMT, SNAT seems to play a pivotal role in melatonin biosynthesis, as ectopic overexpression leads to melatonin overexpression (Park et al., 2014; Wang et al., 2014). In addition, plant SNAT proteins, such as those in loblolly pine, showed thermophilic patterns, with an optimal temperature of 55°C. In particular, rice and cyanobacteria SNAT proteins showed high catalytic activities even at 70°C (Byeon et al., 2013), but the SNAT lost its enzyme activity at 70°C, indicating that plant SNATs have different thermophilic patterns, depending on the species. Moreover, in rice, ASMT is localized in the cytosol (Byeon et al., 2014), and the three different isoforms (ASMT1, ASMT2, and ASMT3) are expressed at various levels (Park et al., 2013). ASMT from rice differed from the respective mammalian enzymes by a lack of substrate inhibition (Hardeland, 2015; Park et al., 2013).

Chloroplasts, a specialized organelle for photosynthesis, are believed to be a primary site of melatonin synthesis. A large quantity of reactive oxygen species are produced during photosynthesis, and locally generated melatonin would provide on-site protection against these toxic oxidative stresses. In fact, melatonin has been observed to preserve chlorophyll and improve the photosynthetic efficiency of chloroplasts in plants under stress (Byeon et al., 2015).

To date, studies on the melatonin metabolism of plants have focused mainly on biosynthetic pathways, and the catabolic processes have not been investigated with regard to those known from vertebrates (Hardeland, 2015; Hardeland and Poeggeler, 2012). In the mammalian liver, the primary enzymes for melatonin breakdown are CYP isoforms, which form, in turn by monooxygenase activities, 6-hydroxymelatonin (Tan et al., 2014a). Plants possess a high number of CYP isoforms (Zhong et al., 2002) and enzymes for hydroxylation and dealkylation that may be involved in melatonin catabolism in plants, but this has not yet been investigated (Hardeland, 2015).

Melatonin is synthesized and accumulated to varying degrees in almost all plant tissues (Byeon et al., 2014; Hardeland, 2015; Ramakrishna and Ravishankar, 2013; Reiter et al., 2011); in fact, it has been identified in many vegetables, fruits, seeds, edible oils, coffees, teas, wines, and beers (Badria et al., 2002; Fernández-Pachón et al., 2014; Ramakrishna et al., 2012a,b,c; Tan et al., 2014a). However, in most plant species, including tomatoes and rice, tissue concentrations reach only a few nanograms per gram of fresh mass (Byeon et al., 2014; Kang et al., 2010; Okazaki and Ezura, 2009; Okazaki et al., 2009). On the other hand, melatonin levels are much higher in plants than they are in vertebrates. High levels of melatonin presumably are due to the fact that plants, unlike animals, cannot actively avoid environmental hazards, because they are immobile.

Many studies are conducted from a primarily medicinal or nutraceutical point of view with the aim of identifying plants containing particularly high levels of melatonin. Among these, species with astonishingly high concentrations have been identified, not rarely in the range of micrograms per gram, first in St John's wort (Hardeland, 2015; Murch and Saxena, 2006), later in numerous Chinese herbs (Chen et al., 2003), and in alpine as well as Mediterranean plants (Hardeland, 2015; Iriti and Varoni, 2015).

The extremely broad range of melatonin concentrations detected in the various plant species and in different plant parts leads to the indisputable conclusion that melatonin must have different functions in various plants (Hardeland, 2015).

Seeds from a further 15 edible plants have been found to have melatonin (Dopfel et al., 2007; Manchester et al., 2000), ranging between 2 and 190 ng/g dry weight of seed; the highest levels were found in black (*Brassica nigra*) and white mustard (*B. hirta*), with 189 and 129 ng/g dry weight of seed, respectively. Other seeds contained melatonin at various lower levels, including almond (39 ng/g dry seed), sunflower (29 ng/g dry seed), fennel (28 ng/g dry seed), green cardamom (15 ng/g dry seed) (Manchester et al., 2000), celery (7 ng/g), and wolf berry (103 ng/g) (Reiter et al., 2001).

Unfortunately, some gaps exist in the current knowledge about melatonin in plants and plant-based foods. The first concerns the distribution of melatonin in tissues, although the relative fractions of melatonin in the cytoplast, the vacuole, and the apoplast have not yet been determined in any plant (Hardeland, 2015). Moreover, melatonin may be taken up by lipid stores (Hardeland, 2015), in accordance with the high levels of melatonin found in oily seeds (Manchester et al., 2000; Reiter and Tan, 2002). In fact, the uptake of melatonin into oil bodies has recently been demonstrated in cotyledons of sunflower (*Helianthus annuus*) seedlings (Mukherjee et al., 2014). It is important to remember that the main enzymes of mammal melatonin synthesis, AANAT and hydroxyindole-O-methyltransferase, have also been detected in the gastrointestinal mucosa, supporting its synthesis from tryptophan present in food (Carpentieri et al., 2012; Stefulj et al., 2001). The function of mammal gut melatonin is not yet entirely understood because the intestine is not only the source but also the sink of melatonin brought by the circulation (Carpentieri et al., 2012; Poeggeler et al., 2005).

It would be beneficial to know whether other specific enzymes are involved in plants producing much higher quantities of melatonin. Interest in metabolic engineering procedures aimed at improving melatonin levels in food plants has increased, partly due to the beneficial effects of melatonin on human health.

Melatonin in Food

Melatonin is widely used as a food supplement in several countries, including the United States. However, in most countries this important indoleamine is categorized as a drug and is often unavailable. Alternative natural sources through the consumption of food rich in melatonin are therefore of interest.

It is widely recognized that the dietary intake of fruit and vegetables is important to staying healthy. The concentration of melatonin in edible plants varies widely and, to date, it is difficult to calculate how a specific diet may positively influence human health (Reiter et al., 2001). Melatonin is available in edible foodstuffs, melatonin-rich diets increase circulating concentrations of the indoleamine, and dietary-derived melatonin is an adequate contributor to the total antioxidative capability of the blood (Reiter et al., 2001).

Melatonin exists in almost all plants and their products; based on this, it is probably not surprising that melatonin has also been discovered in beverages derived from plant products. In fact, melatonin has been identified in the most popular beverages all over the world: coffee, tea, wine, and beer (Garcia-Moreno et al., 2013; Iriti et al., 2010; Maldonado et al., 2009; Ramakrishna et al., 2012a,b,c; Stege et al., 2010; Tan et al., 2012a). However, lower recoveries of this indoleamine were observed in oily foods such as walnut, and it was not detected in potatoes (Kocadağlı et al., 2014). Moreover, yeast-fermented foods contain several times more melatonin isomers than those fermented by bacteria. Melatonin synthesis by yeast during alcoholic fermentation has been observed and evaluated in white and red wines (Rodriguez-Naranjo et al., 2012; Vitalini et al., 2013), beer (Garcia-Moreno et al., 2013; Iriti et al., 2010), and orange and pomegranate juices (Fernández-Pachón et al., 2014). In-depth studies are necessary to better understand the effects of food-processing conditions on the formation and degradation of melatonin and its isomers (Kocadağlı et al., 2014).

The possibility of modulating the circulating levels of melatonin in mammals (about 200 pg/mL at the maximum nighttime peak and less than 10 pg/mL during the day; Bonnefont-Rousselot and Collin, 2010) through the intake of plant-based foods represents an exciting and valuable research topic.

TABLE 17.1

Melatonin in Edible Plant-Based Foods

Plant-Based Foods	Melatonin Content
Apple	160 ng/g
Argan oil	10–305 ng/kg
Banana	500 ng/kg
Black olive	5.3 pg/g
Bread crumb	350–380 pg/g
Bread crust	150 pg/g
Ginger	600 pg/g
Grape exocarp	5–96 pg/g
Green coffee bean	40–45 pg/g
Mango	700–750 pg/g
Modena balsamic vinegar (DO)	0.1 ng/mL
Olive oil	70–120 pg/mL
Orange	150 pg/g
Papaya	250 pg/g
Pineapple	35–40 pg/g
Pistachio	230 µg/g
Pomegranate	180 ng/g
Potato	0
Red pepper	31–93.4 ng/g
Rice	1000–1050 pg/g
Strawberry	8.46–136.6 pg/g
Tomato	4.11–114.52 ng/g

TABLE 17.2

Melatonin in Edible Plant Beverages

Plant Beverage	Melatonin Content
Beer	51.8–169.7 pg/mL
Fermented orange juice	21.9 ng/mL
Pomegranate juice	8.78 ng/mL
Red wine	0.005–0.9 ng/mL

Tryptophan, the precursor of melatonin, cannot be produced by humans and must therefore be part of their diet. Thus, foods high in tryptophan, such as milk, poultry, fish, sesame seeds, beans, lentils, rice, and certain nuts, may be associated with variations in melatonin concentrations. Few studies have explored the influence of these and other nutritional factors on melatonin concentrations (Schernhammer et al., 2009).

Also, how plant-based foods are eaten should be taken into consideration, as different food forms might affect the uptake of melatonin by the body (Johns et al., 2013). Further studies are needed to better understand the importance of a healthy diet rich in food containing melatonin.

In the following paragraphs we will summarize the many pieces of the puzzle to identify melatonin-containing foods and plants and their impact on health. Tables 17.1 and 17.2 summarize the melatonin content of the main edible plant-based foods and plant beverages, respectively.

Grapes and Wines

Melatonin has been widely described in grapes and wines (Iriti and Varoni, 2015; Kocadağlı et al., 2014; Murch et al., 2010; Rodriguez-Naranjo et al., 2013). Scientific evidence shows that alcoholic fermentation is crucial for its synthesis. Indeed, *Saccharomyces cerevisiae* produces not only bioactive melatonin, but also other deleterious substances, such as biogenic amines, during fermentation (Rodriguez-Naranjo et al., 2013).

The grapevine berry consists of three major types of tissue: exocarp, flesh, and seed. For wine making, exocarp has greater importance, because the polyphenols are more easily extracted from the skins than from the seeds, even though the latter represent 1%–6% of berry weight and are an abundant source of polyphenols, mainly proanthocyanidins (Cadot et al., 2006; Vitalini et al., 2011). During Stage I of grape berry development, the berry grows initially by cell division and later by cell expansion; also during this stage, all seed structures are differentiated and the seed approaches its full size. During Stage II, berry growth slows, whereas organic acid concentration reaches its highest level. The seed accumulates reserves and the embryo develops with a concomitant hardening of the seed coat. Stage III is the ripening phase, when sugars and aroma components rapidly accumulate and organic acids decline. The beginning of this stage is characterized by berry softening and the accumulation of red pigments: anthocyanins, a class of flavonoidic polyphenols. The seed is almost entirely developed and its color begins to change from an initial green to a dark brown at harvest, because of polyphenol oxidation in the seed coat (Cadot et al., 2006; Kennedy et al., 2000; Vitalini et al., 2011).

The melatonin content of berry exocarp tissues ranges from 5 to 96 pg/g (Iriti et al., 2010). In particular, in the berry exocarp of different Italian and French wine grapes, the highest melatonin concentrations were detected in Nebbiolo and Croatina varieties (0.9 and 0.8 ng/g respectively), whereas the lowest concentration occurred in the Cabernet Franc cultivar (0.005 ng/g) (Iriti et al., 2010). Differences in the treatment or chemical processing of the products may explain the diverse concentrations measured. It was also shown that, in the whole berry, melatonin decreases with ripening (Murch et al., 2010; Vitalini et al., 2011).

Melatonin was also detected in grape seed, suggesting a primary role of this indoleamine in the grape antioxidant defense system, the germ being particularly vulnerable to oxidative damage (Manchester et al., 2000). Owing to its amphipathic nature and antioxidant activity (Hardeland, 2015; Hardeland and Pandi-Perumal, 2005; Kocadağlı et al., 2014; Reiter et al., 2013; Vitalini et al., 2011; Zhao et al., 2011), melatonin is able to permeate all seed tissues, particularly those rich in storage lipids and membranes, by virtue of both its lipophilic and hydrophilic properties; it is thus able to reach the subcellular compartments where reactive oxygen species are endogenously produced (Reiter et al., 2009).

Many endogenous and exogenous factors may noticeably influence melatonin levels in grapes and their products; these factors include the genetic traits of the cultivars and their geographic origin, the berry tissues/plant organs analyzed, differences between thin- and thick-skinned grapes, phenological stages, day/night fluctuations, pathogen (mainly fungal) infections, agrochemical treatments, agrometeorological conditions, environmental stresses (altitude, ultraviolet radiation, and high light irradiance), and wine-making procedures (Boccalandro et al., 2011; Hardeland et al., 2009; Iriti, 2009; Iriti et al., 2010; Iriti and Varoni, 2015; Paredes et al., 2009; Reiter et al., 2007; Rodriguez-Naranjo et al., 2011; Vitalini et al., 2011).

Importantly, Rodriguez-Naranjo and colleagues (2011) and Gomez and colleagues (2012) showed increased levels of melatonin in wines where tryptophan was added during the fermentation process. However, Tan and colleagues (2012b) pointed to melatonin production without the presence of tryptophan and observed that the production of melatonin and isomers was independent of exogenous tryptophan during wine fermentation, suggesting that yeasts can still synthesize this indoleamine without tryptophan in the medium (Fernández-Pachón et al., 2014).

Interestingly, red wines showed a higher melatonin content than white wines (Mercolini et al., 2008; Stege et al., 2010) or distillate beverages such as whiskey, gin, vodka, and rum (Iriti et al., 2010). On the other hand, the presence of melatonin was also reported in other grapevine products: balsamic vinegars of Modena designation of origin (DO) (0.1 ng/mL), grape juice (0.5 ng/mL), and Albana grappa (0.3 ng/mL) (Mercolini et al., 2012; Vitalini et al., 2013).

The consumption of melatonin-containing foods increases circulating melatonin levels, which, in turn, correlate with the total antioxidant potential in humans and animals (Hattori et al., 1995; Reiter et al., 2011; Sae-Teaw et al., 2013; Tan et al., 2014a,b). Lamont and colleagues (2011) suggested that the beneficial effects of red wine consumption on cardiovascular health might be associated with its melatonin levels rather than the presence of resveratrol, which is considered a fundamental beneficial ingredient in wine (Kitada and Koya, 2013; Penumathsa and Maulik, 2009; Rodella et al., 2008; Tan et al., 2014b). Moreover, Tan and colleagues (2012a) discussed the possibility that ethanol formed during alcoholic fermentation could be a stressor factor for *Saccharomyces* to synthesize melatonin (Rodriguez-Naranjo et al., 2013).

In addition, drinking a glass of red wine (125 mL) does not decrease the salivary antiradical capacity in humans. This suggests that the prooxidant effects of ethanol may be counteracted by melatonin (0.23 ng/mL) and polyphenols present (Varoni et al., 2013). Red wine melatonin and polyphenol extract effectively enhance salivary antioxidant capacity; these wine components may stimulate, by unknown mechanism(s), the powerful endogenous antioxidant defenses of the oral cavity. This observation is just a starting point for further evaluations of the chronic red wine assumption effects at the level of the oral cavity in subjects with altered redox status (patients with periodontal disease or other oxidative stress–related disorders).

During the process of fermentation, which is associated with gradually increased alcohol levels in wine, the degree of melatonin only rises gradually. Considerable amounts of melatonin isomers are generated during the process of wine making (Rodriguez-Naranjo et al., 2011; Tan et al., 2012b). However, the levels of melatonin isomers increase markedly with time toward the end of the fermentation process. This indicates that the inducibility of melatonin isomers owing to environmental stressors, particularly in yeast, is much greater than that of melatonin per se (Tan et al., 2012a).

Interestingly, three melatonin isomers were detected in different grape products: Italian mono- and polyvarietal red, white, and dessert wines from different geographical areas, Modena balsamic vinegars, and grape juices. In particular, isomer 1 and isomer 2 were the most recurring and abundant in all wines and vinegars assayed (Iriti and Varoni, 2015; Vitalini et al., 2013).

The amount of melatonin isomers formed during wine fermentation is not related to melatonin concentration (Gardana et al., 2014; Tan et al., 2012b); this suggests that melatonin isomers probably are not derived from the biosynthetic pathway of melatonin, but rather from a different pathway (Gardana et al., 2014). Certainly, it would be of great interest to ascertain the nutritional significance of these isomers, comparing their biological activity with that of melatonin (Iriti and Varoni, 2015).

Beer

Melatonin has been reported in different commercial beers (Garcia-Moreno et al., 2013; Iriti et al., 2010). In particular, melatonin was measured at concentrations ranging from 51.8 to 169.7 pg/mL, and the highest melatonin concentration was in beers with higher alcoholic content, likely related to the solubility of melatonin in alcohol (Maldonado et al., 2009). Also, beer was found to contain melatonin isomers (14.3 ng/mL). The brewery industry should investigate new methods to extract alcohol from beer without the removal of melatonin, or perhaps alcohol-free beers could be enriched with the melatonin that is initially lost. The amphiphilic properties of melatonin could explain why it is still present in dealcoholized beer (Maldonado et al., 2009; Garcia-Moreno et al., 2013).

Beer is a fermented beverage of low alcoholic graduation originating from the fermentation of cereals and contains about 400 compounds extracted from raw materials or generated during the process of elaboration, such as ethanol, amino acids, minerals, vitamins, carbohydrates, polyphenols, aromatics compounds, and other important elements for life (Garcia-Moreno et al., 2013). Thus, a craft beer elaborated with natural carbonation in a bottle is much more nutritive and healthy than an industrial beer filtered and artificially carbonated. In addition, industrial beers have a series of chemical additives that are not carried in craft beers, such as foam stabilizers (E-405 and E-224) and preservatives (sulfites, citric acid, ascorbic acid, etc.), with unknown effects on the body (Feick et al., 2007; Garcia-Moreno et al., 2013).

A possible source of melatonin in beer is barley, although *S. cerevisiae* is able to produce melatonin in the presence of tryptophan (Garcia-Moreno et al., 2013; Hernández-Ruiz et al., 2008; Iriti et al., 2010). During the process of beer fermentation, the alcohol concentration gradually increases until the levels become toxic to yeasts. As a defense mechanism, these organisms produce additional melatonin, which functions as an antioxidant to protect against oxidative stress caused by the alcohol (Tan et al., 2012b,

2014b). Recently, this was also seen in fermented orange juice (a new alcoholic beverage) (Fernández-Pachón et al., 2014; Tan et al., 2014b).

Some studies have shown that beer contains other antioxidants, such as B vitamin complex, citric acid, silicic acid, and resveratrol, that might contribute to increasing the antioxidant status of human serum (González-Muñoz et al., 2008a,b; Gorinstein et al., 2007). In fact, beer appears to be an example of a functional drink, with varied components contributing to its overall beneficial impact on the organism compared with alcohol abuse or abstinence (Romeo et al., 2007; Tousoulis et al., 2008), and some of the beneficial effects of beer are produced by melatonin. In particular, Maldonado and colleagues (2009) showed that both melatonin levels and total antioxidant capability of healthy volunteers were increased 45 min after drinking beer (330 mL for women and 660 mL for men), indicating that moderate beer consumption might protect organisms from overall oxidative stress. It also would be extremely interesting to measure the effects of beer on sleep induction and to correlate these effects with the melatonin and alcohol content of each beer. These data would provide a scientific explanation to the common assumption that beer has a sleep-inducing effect (Molfino et al., 2010).

Among the yeast-fermented samples, besides beer, there is also bread, which contains relatively higher amounts of melatonin isomers compared with other food samples (Kocadağlı et al., 2014). The formation of melatonin isomers in bread during fermentation might contribute to its nutritional value. The concentration of melatonin isomers in crumb and crust was found to be 15.7 and 0.4 ng/g, respectively. It is thought that temperatures exceeding 100°C degrade melatonin in bread by approximately 95%. On the other hand, as the fermentation time increases, the degree of degradation decreases. This must be due to the accumulation of carbon dioxide and ethanol, leading to a sponge-like texture in oven spring, which limits heat transfer into the loaf during baking. Given the worldwide consumption of bread, it is relevant to investigate the production of melatonin during fermentation and its degradation during baking (Yilmaz et al., 2014). The melatonin isomer content of non-fermented dough was estimated at 4.02 ng/g. This initial concentration might come from both the yeast culture and the flour, since flour also contains a natural microbial load. The concentration of isomers in dough increased up to 16.71 ng/g during fermentation (Yilmaz et al., 2014).

In addition, lower concentrations of melatonin isomer have been observed in probiotic yogurt (0.9 ng/g) and kefir (0.6 ng/g) (Kocadağlı et al., 2014).

Coffee

Coffee is one of the most important and widely consumed beverages in the world. From a commercial point of view, only two coffee species are cultivated extensively: *Coffea arabica* (Arabica) and *Coffea canephora* (Robusta) (Cagliani et al., 2013; Perrois et al., 2015; Ramakrishna et al., 2012a,b). Many beneficial health properties are attributed to coffee drinks (Ramakrishna et al., 2012a,b,c; Sato et al., 2011; Sirota et al., 2015). The presence of biogenic amines in coffee is interesting because of their physiological actions as neurological mediators in humans (Cano-Marquina et al., 2013; Nakaso et al., 2008). Furthermore, they are used as chemical markers to differentiate between Arabica and Robusta coffees (Casal et al., 2004).

Ramakrishna and colleagues (2012a,b) detected significant levels of serotonin and melatonin in the green and roasted beans of both *Coffea* species. The presence of serotonin and melatonin in coffee beans might have a beneficial influence on consumer health (Ramakrishna et al., 2012a,b; Tan et al., 2012a), as with melatonin in walnuts, which contributes to protection against cardiovascular damage (Feldman et al., 2002; Reiter et al., 2005; Ros et al., 2004).

Caffeine is metabolized in the liver by CYP enzymes (Ursing et al., 2002, 2003; Tanaka et al., 1998) and it is known that these enzymes might also be involved in the metabolism of human melatonin. The same enzyme metabolizes both caffeine and melatonin; therefore, drinking two ordinary cups of coffee in the evening can augment nocturnal serum melatonin levels (Ursing et al., 2003). On the other hand, coffee consumption interferes with sleep quantity and quality and decreases aMT6s excretion, so individuals who suffer from sleep abnormalities should avoid caffeinated coffee during the evening hours (Shilo et al., 2002). It is known how caffeine induces wakefulness by blocking adenosine receptors, but it is not clearly understood how caffeine interferes with melatonin levels (MacKenzie et al., 2007; Shilo et al., 2002; Urising et al., 2003; Wright et al., 2000).

Several epidemiological studies provide solid evidence of the preventative effect of coffee consumption against prostate cancer (Nilsson et al., 2010; Wilson et al., 2011; Yu et al., 2011). It has been estimated that the incidence of prostate cancer decreases by 30% and 60% in subjects who drink three or six cups, respectively, of coffee every day. The active ingredients accounting for this protection have not yet been identified; however, it is not caffeine per se, since decaffeinated coffee did not jeopardize the protective effects of coffee drinking against prostate cancer (Tan et al., 2012a). Melatonin might contribute to the beneficial antitumoral action of coffee due to its known cancer-preventive effects (Bizzarri et al., 2013; Proietti et al., 2013; Wang et al., 2015).

Oils

Melatonin was also detected in olive oil, within the range of 70 to 120 pg/mL. Comparing refined olive oils with DO extra-virgin olive oils, as well as comparing DO oils with others, revealed different concentrations of the indoleamine (e.g., 71 pg/mL in DO Bajo Aragon and 119 pg/mL in DO Baena). Diversities in the heat treatment or chemical processing of the products might explain the different measured concentrations (Iriti and Varoni, 2015). Interestingly, black olives contained 5.3 pg/g of melatonin (Iriti and Varoni, 2015; Kocadağlı et al., 2014).

Melatonin levels in argan oil are higher than in extra-virgin olive oil and higher still in refined linseed oil and virgin soybean oil. Virgin argan oil possesses one of the highest antioxidant capacities among edible oils and represents a rich dietary source of coenzyme Q10 (CoQ10) (López et al., 2013; Venegas et al., 2011). In addition to CoQ10 and melatonin, virgin argan oil contains other antioxidants that contribute to its strong antioxidant capacity, such as tocopherols and polyphenols (López et al., 2013). Consequently, the content of CoQ10 and melatonin might also be essential for future applications of virgin argan oil in the food and cosmetic industries (Venegas et al., 2011).

Fruits

Melatonin is also present as metabolites in edible foods such as tomatoes, strawberries, cherries, nuts, and cereals (Burkhardt et al., 2001; Iriti et al., 2010; Mercolini et al., 2012; Stürtz et al., 2011). Given that melatonin is absorbed when melatonin-containing foods are eaten (Hattori et al., 1995), the intake of these foods could prevent the reduction of melatonin plasma concentration that occurs with age (Cardinali et al., 2008) or in some degenerative disorders, such as metabolic syndrome (Nduhirabandi et al., 2012).

The average amounts of melatonin in 100 g of fruit tissue were: 47 ng in banana, 25 ng in tomato, 9 ng in cucumber, and 0.9 ng in beetroot. Melatonin was not detected in potato. The quantity of serotonin exhibited a strong parallelism with the amount of melatonin present: 28 µg/g in banana, 12 µg/g in tomato, and none in potato (Badria, 2002). Sae-Teaw and colleagues (2013) observed that the consumption of fresh fruits (orange, banana, and pineapple) significantly increased levels of melatonin in human blood, and this increased concentration was positively correlated with the total serum antioxidant capacity of the subjects. It has also been demonstrated that the concentration of urinary aMT6s is augmented significantly after the consumption of fruits. In addition, in premenopausal women, urinary excretion of aMT6s was considered a biomarker of fruit intake and, of greater importance, it was inversely correlated with breast cancer risk (Nagata et al., 2005; Oba et al., 2008).

It has been suggested that under intense radiation, larger amounts of plant melatonin are consumed, because of its role as a free radical scavenger, with respect to the quantity produced or imported from leaves, thus leading to low melatonin contents in fruit. In contrast, under low levels of radiation, less melatonin is consumed and the content in fruits is therefore higher (Riga et al., 2014). The decrease in melatonin content observed in fruit exposed to intense light may be a regulatory response to maintain a high level of melatonin in leaves exposed to light-stress conditions and thereby to provide effective protection against free radicals derived from photosynthetic processes, so that the allocation of melatonin to a sink organ like fruit is reduced (Arnao and Hernández-Ruiz, 2013; Reiter et al., 2011; Riga et al., 2014).

Bananas are one of the most important fruits in world trade, second only to citrus plants (Pérez-Pérez et al., 2006). Badria (2002) observed that bananas contain a nonnegligible concentration of melatonin and observed variability in indoleamine content in different parts and at different times of the fruit's development. The highest concentration of this indoleamine is in the peel, and the concentration reached maximum level when the fruit was overripe.

Melatonin has also been found at high levels in walnuts (*Juglans regia* L.), known for their beneficial effects in reducing the risk of cardiovascular disease development (Feldman, 2002; Reiter et al., 2005; Ros et al., 2004). The health benefits of walnuts are usually attributed to their high content of ω-3 fatty acids (Amaral et al., 2003; Kris-Etherton et al., 1999) and, less so, vitamin E (Maguire et al., 2004). In addition, walnuts have other components that may be beneficial for health, including arginine, folate, fiber, tannins, polyphenols, and melatonin (Anderson et al., 2001; Muthaiyah et al., 2014; Reiter et al., 2005; Vanden Heuvel et al., 2012). Melatonin may contribute to the beneficial cardiovascular actions of walnuts due to its important cardioprotective effects (Favero et al., 2014; Reiter et al., 2003, 2005; Rezzani et al., 2013; Rodella et al., 2013), suggesting that the increase in blood melatonin levels after walnut consumption increases the protection of the heart against oxidative damage, which is the basis of a variety of deteriorative cardiac conditions. Interestingly, melatonin has been shown to synergize with other antioxidants found in walnuts (Reiter et al., 2005).

Recently, unexpectedly high levels of melatonin were identified in pistachio. Oladi and colleagues (2014) analyzed the kernels of four different pistachio varieties—Ahmad Aghaei, Akbari, Kalle Qouchi, and Fandoghi—and reported that the levels of melatonin in the pistachio nut reached 230 µg/g.

Sweet cherry is a highly valuable fruit that is rich in tryptophan, serotonin, and melatonin (Garridoa et al., 2012; Zhao et al., 2013). The consumption of Jerte Valley cherries by humans enhances mood, increases 5-hydroxyindoleacetic acid, and reduces cortisol levels in urine (Garridoa et al., 2012,2013; Zhao et al., 2013); similar results were obtained after ingesting Jerte Valley sweet cherry products, Japanese plums, and tropical fruits (including oranges) (Garrido et al., 2010; González-Flores et al., 2012; Johns et al., 2013). Interestingly, Zhao and colleagues (2013), for the first time, monitored melatonin synthesis throughout the entire period of cherry development and in two types of cherry cultivar. In particular, they observed a rise in melatonin production, probably for mitigating reactive oxygen species formation, during embryo development and the gradual lignification of the endocarp, and during the dark period, leading to dual melatonin synthetic peaks over a 24 h period. Importantly, the gene TDC was identified in cherry fruits, providing additional information to suggest that TDC is a rate-limiting enzyme of melatonin synthesis in plants. In addition, ingestion of Jerte Valley cherry products or tart cherry juice concentrate increased the urinary aMT6s levels and improved sleep duration and quality (Garrido et al., 2013; Howatson et al., 2012; Pigeon et al., 2010).

Melatonin has also been detected in strawberries (*Fragaria magna*)—in particular, in four different varieties, in a low concentration (12.4–136.6 pg/g) (Badria, 2002; Stürtz et al., 2011). The Festival and Primoris varieties presented the highest melatonin content (11.26 and 8.46 ng/g, respectively). The amount of melatonin in dietary supplements ranges from 0.5 to 5 mg, and the weight of strawberries required to reach this dose is too high: 44–442 kg (Stürtz et al., 2011).

In the study by Fernández-Pachón and colleagues (2014), the tryptophan concentration of orange juice was 13.8 mg/L before fermentation (day 0), while at the end of the fermentation process (day 15), tryptophan content reached a minimal value of 3.19 mg/L. Additionally, melatonin concentration underwent a significant increase in the pellet after alcoholic fermentation from 0.66 ng/mL (day 0) to 1.45 ng/mL (day 7), until maximal value at day 15 (21.9 ng/mL). These observations suggest that the increase in melatonin content during alcoholic fermentation is mainly due to the increase in the soluble fraction (a major portion of the melatonin content was present in the supernatant, but small amounts were identified in the pellet fraction). The consumption of fermented orange juice could induce higher beneficial effects than orange juice alone, and the difference between both beverages would be caused by the higher melatonin content. It is important to underline that melatonin levels in fermented orange juice are also higher than those found in fresh fruits (Fernández-Pachón et al., 2014).

The increase of melatonin content during fermentation was also observed by Mena and colleagues (2012) in pomegranate juice, in which it reached concentrations of about 8.78 ng/mL. The differences among fermented products can be explained by the basal level of melatonin before fermentation onset,

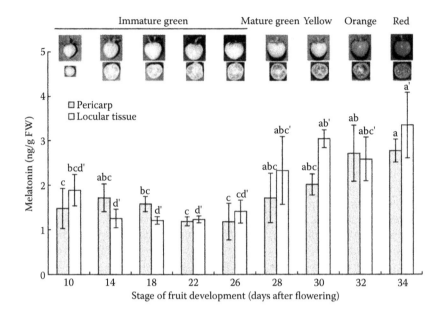

FIGURE 17.1 Melatonin concentrations in pericarps (filled bar) and locular tissues (open bar) at different stage of tomato development. (From Okazaki and Ezura, *J Pineal Res* 46, 338–343, 2009. License no. 3591230947857.)

growth phase, and type of yeast, which determine the concentration of melatonin in the final product (Rodriguez-Naranjo et al., 2012).

Melatonin is present also in other vegetables, such as tomatoes. Tomato (*Solanum lycopersicum* L.) was one of the first edible plants in which melatonin was identified (Dubbels et al., 1995; Kocadağlı et al., 2014). Melatonin concentration in tomatoes varies during the developmental stage (Okazaki and Ezura, 2009; Sun et al., 2015), but it has been detected in all tomato parts (Okazaki and Ezura, 2009; Okazaki et al., 2009; Pape and Lüning, 2006) (Figure 17.1). Leaves and stems show the most significant differences in melatonin content, while root melatonin content differs very little with growth conditions (remaining between 7 and 10 ng/g fresh weight).

Melatonin was first detected, in 1995, in the wild tomato species *S. pimpinellifolium* (currant tomato), which is highly sensitive to ozone and has much lower melatonin content (Dubbels et al., 1995; Okazaki and Ezura, 2009). It was also reported that melatonin accumulates in mature tomato fruits and seeds, and the melatonin content in the pericarp of tomato fruit increases from the mature green stage to the red stage (Okazaki and Ezura, 2009). However, the role of melatonin in tomato fruit ripening is not well understood (Sun et al., 2015). Interestingly, tomato fruit may biosynthesize melatonin: recent evidence has suggested that mitochondria and chloroplasts have the capacity to synthesize melatonin *in situ* (Tan et al., 2013). Melatonin concentration ranges from 4.11 to 114.52 ng/g (fresh weight) and has been identified in 11 tomato varieties, with Marbone and RAF varieties presenting the highest melatonin concentrations (114.52 and 50.1 ng/g, respectively), whereas Lucinda and Catalina varieties show the lowest melatonin concentrations (4.45 and 4.11 ng/g, respectively) (Stürtz et al., 2011).

Sun et al. (2015) suggest that melatonin may influence ethylene to regulate fruit textural changes and that exogenous melatonin could promote tomato fruit ripening during postharvest life. In addition, the data of Arnao and Hernández-Ruiz (2013) clearly indicate that plants cultivated indoors under moderate conditions have a lower melatonin content than those cultivated outdoors in more variable conditions. Tomato plants cultivated *in vitro* had the lowest melatonin content, while plants cultivated in pots and in controlled conditions (in a growth chamber) had an intermediate melatonin content. The leaves of plants cultivated in outdoor conditions had roughly 10 times more melatonin than the leaves of plants cultivated *in vitro* and 7 times more than the leaves of chamber-grown plants. In stem, melatonin levels are lower than in leaves, but the differences with respect to the growth cultivation mode are also revealing, following the same trend observed in leaves.

Tomato is not only an important vegetable crop in an economic sense, but it also has an important role as an experimental model plant for the study of fruit development (Mueller et al., 2005; Okazaki and Ezura, 2009). The miniature tomato cultivar Micro-Tom has attracted attention for its ability to grow at high densities (up to 1357 plants/m²), its short life cycle (70–90 days from sowing to fruit ripening), and its ability to be transformed at frequencies of up to 80% through *Agrobacterium*-mediated transformation of cotyledons (Mueller et al., 2005). Hence, various functional genomic resources for Micro-Tom have been created (Matsukura et al., 2008). In Micro-Tom, melatonin was found to accumulate abundantly in seeds and to varying concentrations in fruits and leaves, depending on the developmental stage, strongly suggesting that melatonin acts during these developmental processes (Okazaki and Ezura, 2009).

The melatonin content in red pepper fruits ranged from 31.0 to 93.4 ng/g (dry weight) (Riga et al., 2014). Two types of pepper cultivars are known in which the melatonin content increases concomitantly with the degree of maturation (Barranca and F26) and another two cultivars where its content decreases (Velero and Derio), demonstrating that the effect of the stage of ripeness on melatonin content in pepper is not straightforward and heavily depends on the genotype (Riga et al., 2014).

At the very least, melatonin has also been identified in apple, cucumber, cabbage, and beetroot (Schernhammer et al., 2009), and in five species of green algae that are commonly eaten in Asia (Pape and Lüning, 2006; Ramakrishna et al., 2011b; Tal et al., 2011). Further studies should be performed in order to elucidate whether the melatonin content of fruit is derived only from leaves or is also biosynthesized in the fruit, in order to identify the factors that affect melatonin transport from leaves and/or roots and to clarify the physiological functions of melatonin in fruit (Riga et al., 2014). Subsequent intervention studies are necessary to evaluate the health effects of fruit consumption and to verify whether any of the benefits observed may be, at least in part, due to melatonin and its isomers, possibly acting synergically with other bioactive compounds such as flavones, polyphenols, and carotenoids.

Rice

Cultivated rice (*Oryza sativa*), the foremost food for more than half of the world's population (Muthayya et al., 2014a; de Pee, 2014), contains naturally bioactive compounds, including melatonin (Setyaningsih et al., 2015). The melatonin contents are noticeably different between most pigmented and nonpigmented rice samples. The highest recorded concentration of melatonin in pigmented rice is 207.79 ± 3.18 µg/kg (red rice), and this is followed by black rice (182.04 ± 2.79 µg/kg) and black glutinous rice (73.81 ± 1.13 µg/kg). Glutinous rice mainly differs from nonglutinous rice in that it has low levels of, or almost no, amylose in its starch, but it has high levels of amylopectin (Setyaningsih et al., 2015). In addition, plant samples with high amylose contents exhibit more antioxidant activity than their glutinous genotypes (Li et al., 2007). In nonpigmented rice, melatonin content is reduced by as much as 33% in comparison with the whole and polished grains: whole grains have a melatonin content ranging from 42.95 ± 0.64 to 47.83 ± 0.12 µg/kg, whereas polished grains range from 28.33 ± 0.61 to 31.99 ± 0.31 µg/kg (Setyaningsih et al., 2015).

A cluster of pigmented rice grains has a higher concentration of melatonin, as rice contains a higher level of vitamin B (Setyaningsih et al., 2015). It is recognized that niacin (vitamin B3) prevents the breakdown of the melatonin precursor, whereas pyridoxine (vitamin B6), a coenzyme in the one-carbon metabolism pathways, is involved in the synthesis of melatonin (Fardet, 2010).

Melatonin-rich transgenic rice plants showed enhanced seedling growth but delayed flowering compared with the wild type. All transgenic lines studied had a lower grain yield compared with the wild type. The main reason for the yield penalty was a reduction in the number of spikelets per panicle (Byeon and Back, 2014). These data clearly indicate that melatonin has a profound effect on rice reproduction, especially for spikelet formation in the panicle; whether the reduced number of spikelets per panicle from high melatonin levels is associated with the mitotic arrest function is currently unknown (Byeon and Back, 2014). So far, spikelet number in rice inflorescences is thought to be regulated by various genes, including cytokinin oxidase (Ashikari et al., 2005) and heat-shock protein (Dong et al., 2013). However, nothing is known about the regulation of these genes by melatonin, and the effects of endogenous melatonin increase or exogenous melatonin treatment in rice still need to be clarified.

Byeon et al. (2014) demonstrated for the first time that overexpression of the TDC3 isogene in rice plants, rather than overexpression of TDC1 and TDC2, shows normal phenotype and induces the accumulation of melatonin and melatonin intermediates such as serotonin and NAS.

Thus, the biotechnological development of melatonin-rich crops is a worthwhile pursuit that will potentially improve the health of consumers. However, the health benefits of melatonin-rich plants are contingent upon melatonin being absorbed in sufficient quantities after the foods are consumed (Byeon and Back, 2014).

Melatonin has also been discovered in widely consumed crops besides rice, including corn, wheat, barley, ginger, and oats (Hattori et al., 1995; Hernández-Ruiz and Arnao, 2008), and its concentration in these crops varies from several nanograms to several thousand picograms per gram (Tan et al., 2012a).

Melatonin in Medicinal Herbs

In the last decade, many plant species that are active in human health have been found to contain melatonin and serotonin (Hardeland, 2015; Kolár and Machackova, 2005; Reiter and Tan, 2002; Reiter et al., 2011); when these plants are consumed, they have important effects in humans. Melatonin in plant-based medicines can affect health and may have an impact on several chronic diseases (Murch et al., 2009). Plant mutants/variants with higher levels of melatonin may be highly desirable for both basic research and commercial product development, but have yet to be further investigated.

Very high melatonin levels have been measured in some plants that have been used as herbal medicines for centuries (Chen et al., 2003), so it is conceivable that some of the known benefits of the regular use of these herbs may be due to their high melatonin concentrations (Chen et al., 2003; Hattori et al., 1995).

A large number of Chinese medicinal herbs were analyzed to determine whether these commonly used herbs contain melatonin, and it was identified in the majority of these herbs. Of the 108 herbs screened, 64 contained melatonin in excess of 10 ng/g (dry weight) and 43 had in excess of 100 ng/g (dry weight). Interestingly, the majority of the herbs containing the highest levels of melatonin are traditionally used to retard aging and to treat diseases characterized by oxidative stress. For example, *yinyanghuo* (*Epimedium brevicornum* Maxim) had a melatonin concentration of 1105 ng/g, and *sangbaipi* (*Mori Albae Cortex*) showed a melatonin level of 1110 ng/g; these herbs are typically used to slow the aging process. *Sangye* (leaf of *Morus alba* L.) had a melatonin concentration of 1510 ng/g, and is believed to have a beneficial effect on irradiation damage induced by UV radiation or radiotherapy (Chen et al., 2003).

Datura metel L., commonly known as Devil's Trumpet or Angel's Trumpet, is a narcotic plant with a long history of use as a medicinal herb. In Ayurvedic medicine, the seeds of *D. metel* are used to treat skin rashes, ulcers, bronchitis, jaundice, and diabetes (Murch et al., 2009), whereas, in Brazil, the seeds are prepared as a sedative beverage and the flowers are dried and smoked as cigarettes (Agra et al., 2007). Moreover, traditional Chinese medicine uses the flowers to treat skin inflammation and psoriasis (Yang et al., 2010). Murch and colleagues (2009) observed the presence of melatonin and serotonin in developing flower buds and ovules of *D. metel*. Interestingly, only the young buds and the early stages of ovule and fruit development showed high melatonin and serotonin content, while the mature buds or fruit had low levels of both compounds. Moreover, exposing the buds to cold resulted in elevated levels of both melatonin and serotonin, confirming that melatonin in plants is stress inducible.

St John's wort (*Hypericum perforatum*), which has been used for several ailments such as depression and apoptosis of cancer cells, including superficial bladder, prostate, and mammary carcinomas (Agostinis et al., 2002; Blank et al., 2001; Colasanti et al., 2000), has also been reported to contain melatonin (Murch and Saxena, 2006). St John's wort, besides melatonin, contains a variety of other active compounds, including flavonoids, xanthones, proanthocyanidins, hyperforin, pseudohyperforin, and hypericin, with antidepressant and anticancer activities (Müller, 2003; Murch and Saxena, 2006; Rodríguez-Landa and Contreras, 2003). St John's wort has also been shown to inhibit the reuptake of several synaptosomal neurotransmitters, including serotonin, noradrenalin, dopamine, and gamma-aminobutyric acid (Müller, 2003), indicating different mechanisms of action and a possible synergy (Murch and Saxena, 2006).

The selected stable, melatonin-rich germplasm line of St John's wort should facilitate studies on the elucidation of the role of melatonin in plant metabolism and may provide tissues to gain new insights into the interactions between plants and human health (Murch and Saxena, 2006).

Melatonin was also identified in extract of feverfew (*Tanacetum parthenium*), Chamomilla (*Tripleurospermum disciforme*), and Viola (*Viola odorata*), plants with a long history in traditional Iranian medicine for the treatment of migraine, cancer (Ansari et al., 2010), and menstrual cramps (Avallone et al., 2000), and have also been used as sedatives (Parvini et al., 2007) and as antimicrobial and anti-inflammatory agents (Ansari et al., 2010). There are several different varieties of feverfew, and melatonin was found at significantly higher levels in the green leaf variety (Murch et al., 1997).

Drinking water extract or boiled plant parts is the most common method of using medicinal plants in household and traditional medicine, so Ansari and colleagues (2010) compared the melatonin contents in boiled and alcoholic extracts and showed that water fraction of both the heated and boiled plant flowers contained a considerable amount of melatonin. However, the level of this indoleamine in non-heat-treated food products is much higher than in heated food products (Pothinuch and Tongchitpakdee, 2011). In addition, a melatonin loss of 15% and 30% was observed during freeze- and oven-drying, respectively, though melatonin has been reported to be stable at low temperatures.

Mulberry (*Morus* spp.) leaves are generally used as a food source for silkworms in silk production in Thailand, since they contain high levels of nutritional factors, especially proteins and amino acids (Pothinuch and Tongchitpakdee, 2011), but they also contain melatonin. The fresh leaves are also traditionally processed into tea using two different methods (blanching and nonblanching). In the blanching method, mulberry leaves undergo blanching before kneading and drying. Mulberry leaf tea obtained from this method is usually green in color (Mulberry green tea). Another method for mulberry tea production involves kneading and drying without blanching and so the finished product is normally black in color (Mulberry black tea). There are no significant differences in the melatonin content of either mulberry green tea or mulberry black tea; in particular, the melatonin content of the mulberry green tea and the mulberry black tea was 46.5 and 40.6 ng/g (dry weight), respectively (Pothinuch and Tongchitpakdee, 2011). Mulberry leaf tea has been reported to have antioxidant activity (Pothinuch and Tongchitpakdee, 2011; Wanyo et al., 2010) and to prevent *Diabetes mellitus* (Hunyadi et al., 2013; Naowaboot et al., 2012).

Mulberry leaf extracts have also been shown to have an anti-obesity effect (Zhong et al., 2006) as well as alleviating insomnia, and this last effect could be attributed mainly to melatonin. The melatonin concentrations in Chinese green tea and black tea have also been measured: melatonin concentrations in Shiya tea (green tea) (Chen et al., 2003) produced in the Guangxi province of China were about 2.12 µg/g, whereas in Chinese Longjing tea (green tea) and Wulong tea (black tea) they were several hundred nanograms per gram (Tan et al., 2012a).

Tryptophan, on the other hand, is contained in the medicinal plants feverfew, St John's wort, and Aloe Vera (Murch et al., 1997; Reiter and Tan, 2002), and their consumption increases the circulating melatonin supply in mammals. Table 17.3 summarizes the melatonin content of these medicinal herbs.

Identifying melatonin in traditional medicinal herbs provides justification for their "clinical" applications, particularly for those containing the highest levels of melatonin (Chen et al., 2003). In fact, Chinese medical herbs customarily prescribed to retard the effects of ageing and combat diseases associated with free radical damage tend to have the highest melatonin contents (Pandi-Perumal et al., 2006).

TABLE 17.3

Melatonin in Medicinal Herbs

Medicinal Herb	Melatonin Content
Feverfew	90–4500 ng/g
Mulberry tea	40.6–46.5 ng/g
St John's wort	2500 ng/g

Conclusion

Melatonin from plants "enters" animals via their diets, so melatonin may be considered a new alternative functional food element for health-conscious consumers. In addition, the application of melatonin in agriculture is a valuable frontier to be explored.

The observations and data summarized in this chapter have created a new area of melatonin research. However, the health benefits attributed to a (plant) food or beverage do not depend on a single compound present in it, but on a combination of phytochemicals. Moreover, the amount of melatonin found in a plant may not be enough to have a pharmacological effect, but should be sufficient for health promotion and to raise melatonin to the physiological level.

Based on the data available, it is hypothesized that melatonin application in agriculture may significantly improve crop production. The reasons for this are (1) as an antioxidant, melatonin enhances the tolerance of crops against heat, cold, chemical pollutants, and other environmental insults; (2) it increases the germination rate of seeds under environmental stresses; (3) it preserves chlorophyll against oxidative stress; (4) it accelerates the rate of photosynthesis in plants; and (5) it stimulates the development of the root system and promotes root regeneration (Tan et al., 2012a).

Actually, compelling evidence indicates that increased consumption of foods containing nutritive and nonnutritive compounds with, at least, antioxidant properties may contribute to improved quality of life by delaying the onset of degenerative diseases and reducing the risk of their beginning. In this context, wine, tea, fruits, and olive oil have received much attention, because they are particularly rich in natural bioactive elements.

Further studies are needed to determine the importance of melatonin in human health, and future studies should examine the associations between diet and melatonin in other populations, particularly those in Asia, as their diets may be richer in melatonin content.

REFERENCES

Agabiti-Rosei C, De Ciuceis C, Rossini C, Porteri E, Rodella LF, Withers SB, Heagerty AM, et al. Anticontractile activity of perivascular fat in obese mice and the effect of long-term treatment with melatonin. *J Hypertens* 2014;32:1264–1274.

Agostinis P, Vantieghem A, Merlevede W, de Witte PA. Hypericin in cancer treatment: More light on the way. *Int J Biochem Cell Biol* 2002;34:221–241.

Agra MF, Baracho GS, Nurit K, Basílio IJ, Coelho VP. Medicinal and poisonous diversity of the flora of "Cariri Paraibano," Brazil. *J Ethnopharmacol* 2007;111:383–395.

Amaral JS, Casal S, Pereira JA, Seabra RM, Oliveira BP. Determination of sterol and fatty acid compositions, oxidative stability, and nutritional value of six walnut (*Juglans regia* L.) cultivars grown in Portugal. *J Agric Food Chem* 2003;51:7698–7702.

Anderson KJ, Teuber SS, Gobeille A, Cremin P, Waterhouse AL, Steinberg FM. Walnut polyphenolics inhibit *in vitro* human plasma and LDL oxidation. *J Nutr* 2001;131:2837–2842.

Ansari M, Rafiee K, Yasa N, Vardasbi S, Naimi SM, Nowrouzi A. Measurement of melatonin in alcoholic and hot water extracts of *Tanacetum parthenium*, *Tripleurospermum disciforme* and *Viola odorata*. *Daru* 2010;18:173–178.

Arnao MB, Hernández-Ruiz J. Growth conditions influence the melatonin content of tomato plants. *Food Chem* 2013;138:1212–1214.

Ashikari M, Sakakibara H, Lin S, Yamamoto T, Takashi T, Nishimura A, Angeles ER, Qian Q, Kitano H, Matsuoka M. Cytokinin oxidase regulates rice grain production. *Science* 2005;309:741–745.

Avallone R, Zanoli P, Puia G, Kleinschnitz M, Schreier P, Baraldi M. Pharmacological profile of apigenin, a flavonoid isolated from *Matricaria chamomilla*. *Biochem Pharmacol* 2000;59:1387–1394.

Badria FA. Melatonin, serotonin, and tryptamine in some Egyptian food and medicinal plants. *J Med Food* 2002;5:153–157.

Balzer I, Hardeland R. Photoperiodism and effects of indoleamines in a unicellular alga, *Gonyaulax polyedra*. *Science* 1991;253:795–797.

Bamia C, Lagiou P, Buckland G, Grioni S, Agnoli C, Taylor AJ, Dahm CC, et al. Mediterranean diet and colorectal cancer risk: Results from a European cohort. *Eur J Epidemiol* 2013;28:317–328.

Behrends A, Hardeland R, Ness H, Grube S, Poeggeler B, Haldar C. Photocatalytic actions of the pesticide metabolite 2-hydroxyquinoxaline: Destruction of antioxidant vitamins and biogenic amines: Implications of organic redox cycling. *Redox Rep* 2004;9:279–288.

Bizzarri M, Proietti S, Cucina A, Reiter RJ. Molecular mechanisms of the pro-apoptotic actions of melatonin in cancer: A review. *Expert Opin Ther Targets* 2013;17:1483–1496.

Blank M, Mandel M, Hazan S, Keisari Y, Lavie G. Anti-cancer activities of hypericin in the dark. *Photochem Photobiol* 2001;74:120–125.

Boccalandro HE, González CV, Wunderlin DA, Silva MF. Melatonin levels, determined by LC-ESI-MS/MS, fluctuate during the day/night cycle in *Vitis vinifera* cv Malbec: Evidence of its antioxidant role in fruits. *J Pineal Res* 2011;51:226–232.

Bonnefont-Rousselot D, Collin F. Melatonin: Action as antioxidant and potential applications in human disease and aging. *Toxicology* 2010;278:55–67.

Burch JB, Yost MG, Johnson W, Allen E. Melatonin, sleep, and shift work adaptation. *J Occup Environ Med* 2005;47:893–901.

Burkhardt S, Tan DX, Manchester LC, Hardeland R, Reiter RJ. Detection and quantification of the antioxidant melatonin in Montmorency and Balaton tart cherries (*Prunus cerasus*). *J Agric Food Chem* 2001;49:4898–4902.

Byeon Y, Back K. An increase in melatonin in transgenic rice causes pleiotropic phenotypes, including enhanced seedling growth, delayed flowering, and low grain yield. *J Pineal Res* 2014;56:408–414.

Byeon Y, Lee HY, Hwang OJ, Lee HJ, Lee K, Back K. Coordinated regulation of melatonin synthesis and degradation genes in rice leaves in response to cadmium treatment. *J Pineal Res* 2015;58:470–478.

Byeon Y, Lee HY, Lee K, Park S, Back K. Cellular localization and kinetics of the rice melatonin biosynthetic enzymes SNAT and ASMT. *J Pineal Res* 2014;56:107–114.

Byeon Y, Lee K, Park YI, Park S, Back K. Molecular cloning and functional analysis of serotonin N-acetyltransferase from the cyanobacterium *Synechocystis* sp. PCC 6803. *J Pineal Res* 2013;55:371–376.

Cadot Y, Miñana-Castelló MT, Chevalier M. Anatomical, histological, and histochemical changes in grape seeds from *Vitis vinifera L.* cv Cabernet franc during fruit development. *J Agric Food Chem* 2006;54:9206–15.

Cagliani LR, Pellegrino G, Giugno G, Consonni R. Quantification of *Coffea arabica* and *Coffea canephora* var. robusta in roasted and ground coffee blends. *Talanta* 2013;106:169–173.

Cano-Marquina A, Tarín JJ, Cano A. The impact of coffee on health. *Maturitas* 2013;75:7–21.

Cardinali DP, Esquifino AI, Srinivasan V, Pandi-Perumal SR. Melatonin and the immune system in aging. *Neuroimmunomodulation* 2008;15:272–278.

Cardinali DP, Srinivasan V, Brzezinski A, Brown GM. Melatonin and its analogs in insomnia and depression. *J Pineal Res* 2012;52:365–375.

Carpentieri A, Díaz de Barboza G, Areco V, Peralta López M, Tolosa de Talamoni N. New perspectives in melatonin uses. *Pharmacol Res* 2012;65:437–444.

Casal S, Mendes E, Alves MR, Alves RC, Beatriz M, Oliveira PP, Ferreira MA. Free and conjugated biogenic amines in green and roasted coffee beans. *J Agric Food Chem* 2004;52:6188–6192.

Chen G, Huo Y, Tan DX, Liang Z, Zhang W, Zhang Y. Melatonin in Chinese medicinal herbs. *Life Sci* 2003;73:19–26.

Cocco P, Cocco ME, Paghi L, Avataneo G, Salis A, Meloni M, Atzeri S, Broccia G, Ennas MG, Erren TC, Reiter RJ. Urinary 6-sulfatoxymelatonin excretion in humans during domestic exposure to 50 hertz electromagnetic fields. *Neuro Endocrinol Lett* 2005;26:136–142.

Colasanti A, Kisslinger A, Liuzzi R, Quarto M, Riccio P, Roberti G, Tramontano D, Villani F. Hypericin photosensitization of tumor and metastatic cell lines of human prostate. *J Photochem Photobiol B* 2000;54:103–107.

Couto E, Boffetta P, Lagiou P, Ferrari P, Buckland G, Overvad K, Dahm CC, et al. Mediterranean dietary pattern and cancer risk in the EPIC cohort. *Br J Cancer* 2011;104:1493–1499.

Cruzen C, Colman RJ. Effects of caloric restriction on cardiovascular aging in non-human primates and humans. *Clin Geriatr Med* 2009;25:733–743.

Daniel M, Tollefsbol TO. Epigenetic linkage of aging, cancer and nutrition. *J Exp Biol* 2015;218:59–70.

Davis S, Kaune WT, Mirick DK, Chen C, Stevens RG. Residential magnetic fields, light-at-night, and nocturnal urinary 6-sulfatoxymelatonin concentration in women. *Am J Epidemiol* 2001;154:591–600.

Dong X, Wang X, Zhang L, Yang Z, Xin X, Wu S, Sun C, Liu J, Yang J, Luo X. Identification and characterization of OsEBS, a gene involved in enhanced plant biomass and spikelet number in rice. *Plant
Biotechnol J* 2013;11:1044–1057.

Dopfel RP, Schulmeister K, Schernhammer ES. Nutritional and lifestyle correlates of the cancer-protective
hormone melatonin. *Cancer Detect Prev* 2007;31:140–148.

Dubbels R, Reiter RJ, Klenke E, Goebel A, Schnakenberg E, Ehlers C, Schiwara HW, Schloot W. Melatonin
in edible plants identified by radioimmunoassay and by high performance liquid chromatography–mass
spectrometry. *J Pineal Res* 1995;18:28–31.

Fardet A. New hypotheses for the health-protective mechanisms of whole-grain cereals: What is beyond fibre?
Nutr Res Rev 2010;23:65–134.

Favero G, Lonati C, Giugno L, Castrezzati S, Rodella LF, Rezzani R. Obesity-related dysfunction of the aorta
and prevention by melatonin treatment in ob/ob mice. *Acta Histochem* 2013;115:783–788.

Favero G, Rodella LF, Reiter RJ, Rezzani R. Melatonin and its atheroprotective effects: A review. *Mol Cell
Endocrinol* 2014;382:926–937.

Feick P, Gerloff A, Singer MV. Effect of non-alcoholic compounds of alcoholic drinks on the pancreas.
Pancreatology 2007;7:124–130.

Feldman EB. The scientific evidence for a beneficial health relationship between walnuts and coronary heart
disease. *J Nutr* 2002;132:1062S–1101S.

Fernández-Pachón MS, Medina S, Herrero-Martín G, Cerrillo I, Berná G, Escudero-López B, Ferreres F,
Martín F, García-Parrilla MC, Gil-Izquierdo A. Alcoholic fermentation induces melatonin synthesis in
orange juice. *J Pineal Res* 2014;56:31–38.

Garcia-Moreno H, Calvo JR, Maldonado MD. High levels of melatonin generated during the brewing process.
J Pineal Res 2013;55:26–30.

Gardana C, Iriti M, Stuknytė M, De Noni I, Simonetti P. "Melatonin isomer" in wine is not an isomer of the
melatonin but tryptophan-ethylester. *J Pineal Res* 2014;57:435–441.

Garrido M, Espino J, González-Gómez D, Lozano M, Barriga C, Paredes SD, Rodríguez AB. The consumption of a Jerte Valley cherry product in humans enhances mood, and increases 5-hydroxyindoleacetic
acid but reduces cortisol levels in urine. *Exp Gerontol* 2012;47:573–580.

Garrido M, González-Gómez D, Lozano M, Barriga C, Paredes SD, Rodríguez AB. A Jerte valley cherry product provides beneficial effects on sleep quality. Influence on aging. *J Nutr Health Aging* 2013;17:553–560.

Garrido M, Paredes SD, Cubero J, Lozano M, Toribio-Delgado AF, Muñoz JL, Reiter RJ, Barriga C, Rodríguez
AB. Jerte Valley cherry-enriched diets improve nocturnal rest and increase 6-sulfatoxymelatonin and
total antioxidant capacity in the urine of middle-aged and elderly humans. *J Gerontol A Biol Sci Med
Sci* 2010;65:909–914.

Giugliano D, Ceriello A, Esposito K. The effects of diet on inflammation: Emphasis on the metabolic syndrome. *J Am Coll Cardiol* 2006;48:677–685.

Gomez FJ, Raba J, Cerutti S, Silva MF. Monitoring melatonin and its isomer in *Vitis vinifera* cv. Malbec by
UHPLC-MS/MS from grape to bottle. *J Pineal Res* 2012;52:349–355.

González-Flores D, Gamero E, Garrido M, Ramírez R, Moreno D, Delgado J, Valdés E, Barriga C, Rodríguez
AB, Paredes SD. Urinary 6-sulfatoxymelatonin and total antioxidant capacity increase after the intake
of a grape juice cv. Tempranillo stabilized with HHP. *Food Funct* 2012;3:34–39.

Gonzalez-Muñoz MJ, Meseguer I, Sanchez-Reus MI, Schultz A, Olivero R, Benedí J, Sánchez-Muniz FJ. Beer
consumption reduces cerebral oxidation caused by aluminum toxicity by normalizing gene expression of
tumor necrotic factor alpha and several antioxidant enzymes. *Food Chem Toxicol* 2008a;46:1111–1118.

González-Muñoz MJ, Peña A, Meseguer I. Role of beer as a possible protective factor in preventing Alzheimer's
disease. *Food Chem Toxicol* 2008b;46:49–56.

Gorinstein S, Caspi A, Libman I, Leontowicz H, Leontowicz M, Tashma Z, Katrich E, Jastrzebski Z, Trakhtenberg
S. Bioactivity of beer and its influence on human metabolism. *Int J Food Sci Nutr.* 2007;58:94–107.

Hardeland R. Melatonin in plants and other phototrophs: Advances and gaps concerning the diversity of functions. *J Exp Bot* 2015;66:627–646.

Hardeland R, Behrmann G, Fuhrberg B, Poeggeler B, Burkhardt S, Uria H, Obst B. Evolutionary aspects of
indoleamines as radical scavengers: Presence and photocatalytic turnover of indoleamines in a unicell,
Gonyaulax polyedra. *Adv Exp Med Biol* 1996;398:279–284.

Hardeland R, Pandi-Perumal SR. Melatonin, a potent agent in antioxidative defense: Actions as a natural food constituent, gastrointestinal factor, drug and prodrug. *Nutr Metab (Lond)* 2005;2:22.

Hardeland R, Poeggeler B. Melatonin and synthetic melatonergic agonists: Actions and metabolism in the central nervous system. *Cent Nerv Syst Agents Med Chem* 2012;12:189–216.

Hardeland R, Tan DX, Reiter RJ. Kynuramines, metabolites of melatonin and other indoles: The resurrection of an almost forgotten class of biogenic amines. *J Pineal Res* 2009;47:109–126.

Hardy TM, Tollefsbol TO. Epigenetic diet: Impact on the epigenome and cancer. *Epigenomics* 2011;3:503–518.

Hattori A, Migitaka H, Iigo M, Itoh M, Yamamoto K, Ohtani-Kaneko R, Hara M, Suzuki T, Reiter RJ. Identification of melatonin in plants and its effects on plasma melatonin levels and binding to melatonin receptors in vertebrates. *Biochem Mol Biol Int* 1995;35:627–634.

Hernández-Ruiz J, Arnao MB. Distribution of melatonin in different zones of lupin and barley plants at different ages in the presence and absence of light. *J Agric Food Chem* 2008;56:10567–10573.

Howatson G, Bell PG, Tallent J, Middleton B, McHugh MP, Ellis J. Effect of tart cherry juice (*Prunus cerasus*) on melatonin levels and enhanced sleep quality. *Eur J Nutr* 2012;51:909–916.

Hunyadi A, Veres K, Danko B, Kele Z, Weber E, Hetenyi A, Zupko I, Hsieh TJ. *In vitro* anti-diabetic activity and chemical characterization of an apolar fraction of *Morus alba* leaf water extract. *Phytother Res* 2013;27:847–851.

Iriti M. Melatonin in grape, not just a myth, maybe a panacea. *J Pineal Res* 2009;46:353.

Iriti M, Varoni EM, Vitalini S. Melatonin in traditional Mediterranean diets. *J Pineal Res* 2010;49:101–105.

Iriti M, Varoni EM. Melatonin in Mediterranean diet: A new perspective. *J Sci Food Agric* 2015;95:2355–2559.

Iriti M, Vitalini S. Health-promoting effects of traditional Mediterranean diet: A review. *Pol J Food Nutr Sci* 2012;62:71–76.

Johns NP, Johns J, Porasuphatana S, Plaimee P, Sae-Teaw M. Dietary intake of melatonin from tropical fruit altered urinary excretion of 6-sulfatoxymelatonin in healthy volunteers. *J Agric Food Chem* 2013;61:913–919.

Kang K, Kong K, Park S, Natsagdorj U, Kim YS, Back K. Molecular cloning of a plant N-acetylserotonin methyltransferase and its expression characteristics in rice. *J Pineal Res* 2011;50:304–309.

Kang K, Lee K, Park S, Byeon Y, Back K. Molecular cloning of rice serotonin N-acetyltransferase, the penultimate gene in plant melatonin biosynthesis. *J Pineal Res* 2013;55:7–13.

Kang K, Lee K, Park S, Kim YS, Back K. Enhanced production of melatonin by ectopic overexpression of human serotonin N-acetyltransferase plays a role in cold resistance in transgenic rice seedlings. *J Pineal Res* 2010;49:176–182.

Kanjanaphachoat P, Wei BY, Lo SF, Wang IW, Wang CS, Yu SM, Yen ML, Chiu SH, Lai CC, Chen LJ. Serotonin accumulation in transgenic rice by over-expressing tryptophan decarboxylase results in a dark brown phenotype and stunted growth. *Plant Mol Biol* 2012;78:525–543.

Kennedy JA, Matthews MA, Waterhouse AL. Changes in grape seed polyphenols during fruit ripening. *Phytochemistry* 2000;55:77–85.

Kitada M, Koya D. Renal protective effects of resveratrol. *Oxid Med Cell Longev* 2013;2013:568093.

Kitagawa A, Ohta Y, Ohashi K. Melatonin improves metabolic syndrome induced by high fructose intake in rats. *J Pineal Res* 2012;52:403–413.

Kocadağlı T, Yılmaz C, Gökmen V. Determination of melatonin and its isomer in foods by liquid chromatography tandem mass spectrometry. *Food Chem* 2014;153:151–156.

Kolář J, Macháčková I. Melatonin in higher plants: Occurrence and possible functions. *J Pineal Res* 2005;39:333–341.

Korkmaz A, Topal T, Tan DX, Reiter RJ. Role of melatonin in metabolic regulation. *Rev Endocr Metab Disord* 2009;10:261–270.

Kotlarczyk MP, Lassila HC, O'Neil CK, D'Amico F, Enderby LT, Witt-Enderby PA, Balk JL. Melatonin osteoporosis prevention study (MOPS): A randomized, double-blind, placebo-controlled study examining the effects of melatonin on bone health and quality of life in perimenopausal women. *J Pineal Res* 2012;52:414–426.

Koziróg M, Poliwczak AR, Duchnowicz P, Koter-Michalak M, Sikora J, Broncel M. Melatonin treatment improves blood pressure, lipid profile, and parameters of oxidative stress in patients with metabolic syndrome. *J Pineal Res* 2011;50:261–266.

Kris-Etherton PM, Yu-Poth S, Sabaté J, Ratcliffe HE, Zhao G, Etherton TD. Nuts and their bioactive constituents: Effects on serum lipids and other factors that affect disease risk. *Am J Clin Nutr* 1999;70:504S–511S.

La Vecchia C, Bosetti C. Diet and cancer risk in Mediterranean countries: Open issues. *Public Health Nutr* 2006;9:1077–1082.

Lamont KT, Somers S, Lacerda L, Opie LH, Lecour S. Is red wine a SAFE sip away from cardioprotection? Mechanisms involved in resveratrol- and melatonin-induced cardioprotection. *J Pineal Res* 2011;50:374–380.

Lampe JW. Health effects of vegetables and fruit: Assessing mechanisms of action in human experimental studies. *Am J Clin Nutr* 1999;70:475S–490S.

Li W, Wei CV, White PJ, Beta T. High-amylose corn exhibits better antioxidant activity than typical and waxy genotypes. *J Agric Food Chem* 2007;55:291–298.

López LC, Cabrera-Vique C, Venegas C, García-Corzo L, Luna-Sánchez M, Acuña-Castroviejo D, Escames G. Argan oil-contained antioxidants for human mitochondria. *Nat Prod Commun* 2013;8:47–50.

MacKenzie T, Comi R, Sluss P, Keisari R, Manwar S, Kim J, Larson R, Baron JA. Metabolic and hormonal effects of caffeine: Randomized, double-blind, placebo-controlled crossover trial. *Metabolism* 2007;56:1694–1698.

Maderuelo-Fernandez JA, Recio-Rodríguez JI, Patino-Alonso MC, Pérez-Arechaederra D, Rodriguez-Sanchez E, Gomez-Marcos MA, García-Ortiz L. Effectiveness of interventions applicable to primary health care settings to promote Mediterranean diet or healthy eating adherence in adults: A systematic review. *Prev Med* 2015;76 Suppl:S39–S55.

Maguire LS, O'Sullivan SM, Galvin K, O'Connor TP, O'Brien NM. Fatty acid profile, tocopherol, squalene and phytosterol content of walnuts, almonds, peanuts, hazelnuts and the macadamia nut. *Int J Food Sci Nutr* 2004;55:171–178.

Maldonado MD, Moreno H, Calvo JR. Melatonin present in beer contributes to increase the levels of melatonin and antioxidant capacity of the human serum. *Clin Nutr* 2009;28:188–191.

Manchester LC, Tan DX, Reiter RJ, Park W, Monis K, Qi W. High levels of melatonin in the seeds of edible plants: Possible function in germ tissue protection. *Life Sci* 2000;67:3023–3029.

Matsukura C, Aoki K, Fukuda N, Mizoguchi T, Asamizu E, Saito T, Shibata D, Ezura H. Comprehensive resources for tomato functional genomics based on the miniature model tomato Micro-Tom. *Curr Genomics* 2008;9:436–443.

Mena P, Gil-Izquierdo A, Moreno DA, Martí N, García-Viguera C. Assessment of the melatonin production in pomegranate wines. *LWT: Food Sci Technol* 2012;47:13–18.

Mercolini L, Addolorata Saracino M, Bugamelli F, Ferranti A, Malaguti M, Hrelia S, Raggi MA. HPLC-F analysis of melatonin and resveratrol isomers in wine using an SPE procedure. *J Sep Sci* 2008;31:1007–1014.

Mercolini L, Mandrioli R, Raggi MA. Content of melatonin and other antioxidants in grape-related foodstuffs: Measurement using a MEPS-HPLC-F method. *J Pineal Res* 2012;53:21–28.

Molfino A, Laviano A, Rossi Fanelli F. Sleep-inducing effect of beer: A melatonin- or alcohol-mediated effect? *Clin Nutr* 2010;29:272.

Moore LL, Singer MR, Bradlee ML, Daniels SR. Adolescent dietary intakes predict cardiometabolic risk clustering. *Eur J Nutr* 2016;55:461–468.

Mueller LA, Solow TH, Taylor N, Skwarecki B, Buels R, Binns J, Lin C, et al. The SOL Genomics Network: A comparative resource for Solanaceae biology and beyond. *Plant Physiol* 2005;138:1310–1317.

Mukherjee S, David A, Yadav S, Baluška F, Bhatla SC. Salt stress-induced seedling growth inhibition coincides with differential distribution of serotonin and melatonin in sunflower seedling roots and cotyledons. *Physiol Plant* 2014;152:714–728.

Müller WE. Current St John's wort research from mode of action to clinical efficacy. *Pharmacol Res* 2003;47:101–109.

Murch SJ, Alan AR, Cao J, Saxena PK. Melatonin and serotonin in flowers and fruits of *Datura metel* L. *J Pineal Res* 2009;47:277–283.

Murch SJ, Hall BA, Le CH, Saxena PK. Changes in the levels of indoleamine phytochemicals during véraison and ripening of wine grapes. *J Pineal Res* 2010;49:95–100.

Murch SJ, Saxena PK. A melatonin-rich germplasm line of St John's wort (*Hypericum perforatum* L.). *J Pineal Res* 2006;41:284–287.

Murch SJ, Simmons CB, Saxena PK. Melatonin in feverfew and other medicinal plants. *Lancet* 1997;350:1598–1599.

Muthaiyah B, Essa MM, Lee M, Chauhan V, Kaur K, Chauhan A. Dietary supplementation of walnuts improves memory deficits and learning skills in transgenic mouse model of Alzheimer's disease. *J Alzheimer's Dis* 2014;42:1397–1405.

Muthayya S, Sugimoto JD, Montgomery S, Maberly GF. An overview of global rice production, supply, trade, and consumption. *Ann NY Acad Sci* 2014;1324:7–14.

Nagata C, Matsubara T, Fujita H, Nagao Y, Shibuya C, Kashiki Y, Shimizu H. Associations of mammographic density with dietary factors in Japanese women. *Cancer Epidemiol Biomarkers Prev* 2005;14:2877–2880.

Nakaso K, Ito S, Nakashima K. Caffeine activates the PI3K/Akt pathway and prevents apoptotic cell death in a Parkinson's disease model of SH-SY5Y cells. *Neurosci Lett* 2008;432:146–150.

Naowaboot J, Pannangpetch P, Kukongviriyapan V, Prawan A, Kukongviriyapan U, Itharat A. Mulberry leaf extract stimulates glucose uptake and GLUT4 translocation in rat adipocytes. *Am J Chin Med* 2012;40:163–175.

Nduhirabandi F, du Toit EF, Lochner A. Melatonin and the metabolic syndrome: A tool for effective therapy in obesity-associated abnormalities? *Acta Physiol (Oxf)* 2012;205:209–223.

Nilsson LM, Johansson I, Lenner P, Lindahl B, Van Guelpen B. Consumption of filtered and boiled coffee and the risk of incident cancer: A prospective cohort study. *Cancer Causes Control* 2010;21:1533–1544.

Oba S, Nakamura K, Sahashi Y, Hattori A, Nagata C. Consumption of vegetables alters morning urinary 6-sulfatoxymelatonin concentration. *J Pineal Res* 2008;45:17–23.

Okazaki M, Ezura H. Profiling of melatonin in the model tomato (*Solanum lycopersicum* L.) cultivar Micro-Tom. *J Pineal Res* 2009;46:338–343.

Okazaki M, Higuchi K, Aouini A, Ezura H. Lowering intercellular melatonin levels by transgenic analysis of indoleamine 2,3-dioxygenase from rice in tomato plants. *J Pineal Res* 2010;49:239–247.

Okazaki M, Higuchi K, Hanawa Y, Shiraiwa Y, Ezura H. Cloning and characterization of a *Chlamydomonas reinhardtii* cDNA arylalkylamine N-acetyltransferase and its use in the genetic engineering of melatonin content in the Micro-Tom tomato. *J Pineal Res* 2009;46:373–382.

Oladi E, Mohamadi M, Shamspur T, Mostafavi A. Spectrofluorimetric determination of melatonin in kernels of four different Pistacia varieties after ultrasound-assisted solid-liquid extraction. *Spectrochim Acta A Mol Biomol Spectrosc* 2014;132:326–329.

Pandi-Perumal SR, Srinivasan V, Maestroni GJ, Cardinali DP, Poeggeler B, Hardeland R. Melatonin: Nature's most versatile biological signal? *FEBS J* 2006;273:2813–2838.

Pape C, Lüning K. Quantification of melatonin in phototrophic organisms. *J Pineal Res* 2006;41:157–65.

Paredes SD, Korkmaz A, Manchester LC, Tan DX, Reiter RJ. Phytomelatonin: A review. *J Exp Bot* 2009;60:57–69.

Park S, Byeon Y, Back K. Functional analyses of three ASMT gene family members in rice plants. *J Pineal Res* 2013;55:409–415.

Park S, Byeon Y, Lee HY, Kim YS, Ahn T, Back K. Cloning and characterization of a serotonin N-acetyltransferase from a gymnosperm, loblolly pine (*Pinus taeda*). *J Pineal Res* 2014;57:348–355.

Park S, Lee K, Kim YS, Back K. Tryptamine 5-hydroxylase–deficient Sekiguchi rice induces synthesis of 5-hydroxytryptophan and N-acetyltryptamine but decreases melatonin biosynthesis during senescence process of detached leaves. *J Pineal Res* 2012;52:211–216.

Parvini S, Hosseini MJ, Bakhtiarian A. The study of analgesic effects and acute toxicity of *Tripleurospermum disciforme* in rats by formalin test. *Toxicol Mech Methods* 2007;17:575–580.

Pee S de. Proposing nutrients and nutrient levels for rice fortification. *Ann NY Acad Sci* 2014;1324:55–66.

Penumathsa SV, Maulik N. Resveratrol: A promising agent in promoting cardioprotection against coronary heart disease. *Can J Physiol Pharmacol* 2009;87:275–286.

Pérez-Pérez EM, Rodríguez-Malaver AJ, Padilla N, Medina-Ramírez G, Dávila J. Antioxidant capacity of crude extracts from clones of banana and plane species. *J Med Food* 2006;9:517–523.

Pericleous M, Mandair D, Caplin ME. Diet and supplements and their impact on colorectal cancer. *J Gastrointest Oncol* 2013;4:409–423.

Perrois C, Strickler SR, Mathieu G, Lepelley M, Bedon L, Michaux S, Husson J, Mueller L, Privat I. Differential regulation of caffeine metabolism in *Coffea arabica* (Arabica) and *Coffea canephora* (Robusta). *Planta* 2015;241:179–191.

Pigeon WR, Carr M, Gorman C, Perlis ML. Effects of a tart cherry juice beverage on the sleep of older adults with insomnia: A pilot study. *J Med Food* 2010;13:579–83.

Poeggeler B, Cornélissen G, Huether G, Hardeland R, Józsa R, Zeman M, Stebelova K, et al. Chronomics affirm extending scope of lead in phase of duodenal vs. pineal circadian melatonin rhythms. *Biomed Pharmacother* 2005;59 Suppl 1:S220–S224.

Pothinuch P, Tongchitpakdee S. Melatonin contents in mulberry (*Morus* spp.) leaves: Effects of sample preparation, cultivar, leaf age and tea processing. *Food Chem* 2011;128:415–419.

Proietti S, Cucina A, Reiter RJ, Bizzarri M. Molecular mechanisms of melatonin's inhibitory actions on breast cancers. *Cell Mol Life Sci* 2013;70:2139–2157.

Prunet-Marcassus B, Desbazeille M, Bros A, Louche K, Delagrange P, Renard P, Casteilla L, Pénicaud L. Melatonin reduces body weight gain in Sprague Dawley rats with diet-induced obesity. *Endocrinology* 2003;144:5347–5352.

Ramakrishna A, Dayananda C, Giridhar P, Rajasekaran T, Ravishankar GA. Photoperiod influences endogenous indoleamines in cultured green alga *Dunaliella bardawil*. *Ind J Exp Biol* 2011b;49(3):234–240.

Ramakrishna A, Giridhar P, Jobin M, Paulose CS, Ravishankar GA. Indoleamines and calcium enhance somatic embryogenesis in cultured tissues of *Coffea canephora* P ex Fr. *Plant Cell Tiss Org Cult* 2012c;108:267–278.

Ramakrishna A, Giridhar P, Ravishankar GA. Indoleamines and calcium channels influence morphogenesis in *in vitro* cultures of *Mimosa pudica* L. *Plant Sig Behav* 2009;12:1136–1141.

Ramakrishna A, Giridhar P, Ravishankar GA. Phytoserotonin. *Plant Signal Behav* 2011a;6:800–809.

Ramakrishna A, Giridhar P, Sankar KU, Ravishankar GA. Melatonin and serotonin profiles in beans of *Coffea* species. *J Pineal Res* 2012b;52:470–6.

Ramakrishna A, Giridhar P, Sankar KU, Ravishankar GA. Endogenous profiles of indoleamines: Serotonin and melatonin in different tissues of *Coffea canephora* P ex Fr. as analyzed by HPLC and LC-MS-ESI. *Acta Physiol Plant* 2012a;34:393–396.

Ramakrishna A, Ravishankar GA. Role of plant metabolites in abiotic stress tolerance under changing climatic conditions with special reference to secondary compounds. In N Tuteja and SS Gill, eds. *Climate Change and Abiotic Stress Tolerance* (Chapter 26). Wiley, Weinheim, Germany, 2013.

Reiter RJ, Coto-Montes A, Boga JA, Fuentes-Broto L, Rosales-Corral S, Tan DX. Melatonin: New applications in clinical and veterinary medicine, plant physiology and industry. *Neuro Endocrinol Lett* 2011;32:575–587.

Reiter RJ, Manchester LC, Tan DX. Melatonin in walnuts: Influence on levels of melatonin and total antioxidant capacity of blood. *Nutrition* 2005;21:920–924.

Reiter RJ, Paredes SD, Manchester LC, Tan DX. Reducing oxidative/nitrosative stress: A newly-discovered genre for melatonin. *Crit Rev Biochem Mol Biol* 2009;44:175–200.

Reiter RJ, Tan DX. Melatonin: A novel protective agent against oxidative injury of the ischemic/reperfused heart. *Cardiovasc Res* 2003;58:10–19.

Reiter RJ, Tan DX. Melatonin: An antioxidant in edible plants. *Ann NY Acad Sci* 2002;957:341–344.

Reiter RJ, Tan DX, Burkhardt S, Manchester LC. Melatonin in plants. *Nutr Rev* 2001;59:286–290.

Reiter RJ, Tan DX, Manchester LC, Pilar Terron M, Flores LJ, Koppisepi S. Medical implications of melatonin: Receptor-mediated and receptor-independent actions. *Adv Med Sci* 2007;52:11–28.

Reiter RJ, Tan DX, Rosales-Corral S, Manchester LC. The universal nature, unequal distribution and antioxidant functions of melatonin and its derivatives. *Mini Rev Med Chem* 2013;13:373–384.

Rezzani R, Favero G, Stacchiotti A, Rodella LF. Endothelial and vascular smooth muscle cell dysfunction mediated by cyclophylin A and the atheroprotective effects of melatonin. *Life Sci* 2013;92:875–882.

Riga P, Medina S, García-Flores LA, Gil-Izquierdo Á. Melatonin content of pepper and tomato fruits: Effects of cultivar and solar radiation. *Food Chem* 2014;156:347–352.

Ríos-Lugo MJ, Cano P, Jiménez-Ortega V, Fernández-Mateos MP, Scacchi PA, Cardinali DP, Esquifino AI. Melatonin effect on plasma adiponectin, leptin, insulin, glucose, triglycerides and cholesterol in normal and high fat-fed rats. *J Pineal Res* 2010;49:342–348.

Ríos-Lugo MJ, Jiménez-Ortega V, Cano-Barquilla P, Mateos PF, Spinedi EJ, Cardinali DP, Esquifino AI. Melatonin counteracts changes in hypothalamic gene expression of signals regulating feeding behavior in high-fat fed rats. *Horm Mol Biol Clin Investig* 2015;21:175–183.

Rodella LF, Favero G, Foglio E, Rossini C, Castrezzati S, Lonati C, Rezzani R. Vascular endothelial cells and dysfunctions: Role of melatonin. *Front Biosci (Elite Ed)* 2013;5:119–129.

Rodella LF, Vanella L, Peterson SJ, Drummond G, Rezzani R, Falck JR, Abraham NG. Heme oxygenase-derived carbon monoxide restores vascular function in type 1 diabetes. *Drug Metab Lett* 2008;2:290–300.

Rodríguez-Landa JF, Contreras CM. A review of clinical and experimental observations about antidepressant actions and side effects produced by *Hypericum perforatum* extracts. *Phytomedicine* 2003;10:688–699.

Rodriguez-Naranjo MI, Gil-Izquierdo A, Troncoso AM, Cantos-Villar E, Garcia-Parrilla MC. Melatonin is synthesised by yeast during alcoholic fermentation in wines. *Food Chem* 2011;126:1608–1613.

Rodriguez-Naranjo MI, Ordóñez JL, Callejón RM, Cantos-Villar E, Garcia-Parrilla MC. Melatonin is formed during winemaking at safe levels of biogenic amines. *Food Chem Toxicol* 2013;57:140–146.

Rodriguez-Naranjo MI, Torija MJ, Mas A, Cantos-Villar E, Garcia-Parrilla MeC. Production of melatonin by *Saccharomyces* strains under growth and fermentation conditions. *J Pineal Res* 2012;53:219–224.

Romeo J, Wärnberg J, Nova E, Díaz LE, Gómez-Martinez S, Marcos A. Moderate alcohol consumption and the immune system: A review. *Br J Nutr* 2007;98 Suppl 1:S111–S115.

Ros E, Núñez I, Pérez-Heras A, Serra M, Gilabert R, Casals E, Deulofeu R. A walnut diet improves endothelial function in hypercholesterolemic subjects: A randomized crossover trial. *Circulation* 2004;109:1609–1614.

Rosales-Corral SA, Acuña-Castroviejo D, Coto-Montes A, Boga JA, Manchester LC, Fuentes-Broto L, Korkmaz A, Ma S, Tan DX, Reiter RJ. Alzheimer's disease: Pathological mechanisms and the beneficial role of melatonin. *J Pineal Res* 2012;52:167–202.

Sae-Teaw M, Johns J, Johns NP, Subongkot S. Serum melatonin levels and antioxidant capacities after consumption of pineapple, orange, or banana by healthy male volunteers. *J Pineal Res* 2013;55:58–64.

Sato Y, Itagaki S, Kurokawa T, Ogura J, Kobayashi M, Hirano T, Sugawara M, Iseki K. *In vitro* and *in vivo* antioxidant properties of chlorogenic acid and caffeic acid. *Int J Pharm* 2011;403:136–138.

Schernhammer ES, Feskanich D, Niu C, Dopfel R, Holmes MD, Hankinson SE. Dietary correlates of urinary 6-sulfatoxymelatonin concentrations in the Nurses' Health Study cohorts. *Am J Clin Nutr* 2009;90:975–985.

Setyaningsih W, Saputro IE, Barbero GF, Palma M, García Barroso C. Determination of melatonin in rice (*Oryza sativa*) grains by pressurized liquid extraction. *J Agric Food Chem* 2015. [Epub ahead of print].

Shilo L, Sabbah H, Hadari R, Kovatz S, Weinberg U, Dolev S, Dagan Y, Shenkman L. The effects of coffee consumption on sleep and melatonin secretion. *Sleep Med* 2002;3:271–273.

Sirota R, Gibson D, Kohen R. The role of the catecholic and the electrophilic moieties of caffeic acid in Nrf2/Keap1 pathway activation in ovarian carcinoma cell lines. *Redox Biol* 2015;4:48–59.

Speciale A, Chirafisi J, Saija A, Cimino F. Nutritional antioxidants and adaptive cell responses: An update. *Curr Mol Med* 2011;11:770–789.

Stacchiotti A, Favero G, Giugno L, Lavazza A, Reiter RJ, Rodella LF, Rezzani R. Mitochondrial and metabolic dysfunction in renal convoluted tubules of obese mice: Protective role of melatonin. *PLoS One* 2014;9:e111141.

Stefulj J, Hörtner M, Ghosh M, Schauenstein K, Rinner I, Wölfler A, Semmler J, Liebmann PM. Gene expression of the key enzymes of melatonin synthesis in extrapineal tissues of the rat. *J Pineal Res* 2001;30:243–247.

Stege PW, Sombra LL, Messina G, Martinez LD, Silva MF. Determination of melatonin in wine and plant extracts by capillary electrochromatography with immobilized carboxylic multi-walled carbon nanotubes as stationary phase. *Electrophoresis* 2010;31:2242–2248.

Stürtz M, Cerezo AB, Cantos-Villar E, Garcia-Parrilla MC. Determination of the melatonin content of different varieties of tomatoes (*Lycopersicon esculentum*) and strawberries (*Fragaria ananassa*). *Food Chem* 2011;127:1329–1334.

Sun Q, Zhang N, Wang J, Zhang H, Li D, Shi J, Li R, Weeda S, Zhao B, Ren S, Guo YD. Melatonin promotes ripening and improves quality of tomato fruit during postharvest life. *J Exp Bot* 2015;66:657–668.

Tal O, Haim A, Harel O, Gerchman Y. Melatonin as an antioxidant and its semi-lunar rhythm in green macroalga *Ulva* sp. *J Exp Bot* 2011;62:1903–1910.

Tan DX, Hardeland R, Manchester LC, Korkmaz A, Ma S, Rosales-Corral S, Reiter RJ. Functional roles of melatonin in plants, and perspectives in nutritional and agricultural science. *J Exp Bot* 2012a;63:577–597.

Tan DX, Hardeland R, Manchester LC, Paredes SD, Korkmaz A, Sainz RM, Mayo JC, Fuentes-Broto L, Reiter RJ. The changing biological roles of melatonin during evolution: From an antioxidant to signals of darkness, sexual selection and fitness. *Biol Rev Camb Philos Soc* 2010;85:607–623.

Tan DX, Hardeland R, Manchester LC, Rosales-Corral S, Coto-Montes A, Boga JA, Reiter RJ. Emergence of naturally occurring melatonin isomers and their proposed nomenclature. *J Pineal Res* 2012b;53:113–121.

Tan DX, Manchester LC, Liu X, Rosales-Corral SA, Acuna-Castroviejo D, Reiter RJ. Mitochondria and chloroplasts as the original sites of melatonin synthesis: A hypothesis related to melatonin's primary function and evolution in eukaryotes. *J Pineal Res* 2013;54:127–138.

Tan DX, Zanghi BM, Manchester LC, Reiter RJ. Melatonin identified in meats and other food stuffs: Potentially nutritional impact. *J Pineal Res* 2014b;57:213–218.

Tan DX, Zheng X, Kong J, Manchester LC, Hardeland R, Kim SJ, Xu X, Reiter RJ. Fundamental issues related to the origin of melatonin and melatonin isomers during evolution: Relation to their biological functions. *Int J Mol Sci* 2014a;15:15858–15890.

Tanaka E. Clinical importance of non-genetic and genetic cytochrome P450 function tests in liver disease. *J Clin Pharm Ther* 1998;23:161–170.

Tousoulis D, Ntarladimas I, Antoniades C, Vasiliadou C, Tentolouris C, Papageorgiou N, Latsios G, Stefanadis C. Acute effects of different alcoholic beverages on vascular endothelium, inflammatory markers and thrombosis fibrinolysis system. *Clin Nutr* 2008;27:594–600.

Urpi-Sarda M, Casas R, Chiva-Blanch G, Romero-Mamani ES, Valderas-Martínez P, Salas-Salvadó J, Covas MI, et al. The Mediterranean diet pattern and its main components are associated with lower plasma concentrations of tumor necrosis factor receptor 60 in patients at high risk for cardiovascular disease. *J Nutr* 2012;142:1019–1025.

Ursing C, Härtter S, von Bahr C, Tybring G, Bertilsson L, Röjdmark S. Does hepatic metabolism of melatonin affect the endogenous serum melatonin level in man? *J Endocrinol Invest* 2002;25:459–462.

Ursing C, Wikner J, Brismar K, Röjdmark S. Caffeine raises the serum melatonin level in healthy subjects: An indication of melatonin metabolism by cytochrome P450(CYP)1A2. *J Endocrinol Invest* 2003;26:403–406.

Vanden Heuvel JP, Belda BJ, Hannon DB, Kris-Etherton PM, Grieger JA, Zhang J, Thompson JT. Mechanistic examination of walnuts in prevention of breast cancer. *Nutr Cancer* 2012;64:1078–1086.

Varoni EM, Vitalini S, Contino D, Lodi G, Simonetti P, Gardana C, Sardella A, Carrassi A, Iriti M. Effects of red wine intake on human salivary antiradical capacity and total polyphenol content. *Food Chem Toxicol* 2013;58:289–294.

Venegas C, Cabrera-Vique C, García-Corzo L, Escames G, Acuña-Castroviejo D, López LC. Determination of coenzyme Q10, coenzyme Q9, and melatonin contents in virgin argan oils: Comparison with other edible vegetable oils. *J Agric Food Chem* 2011;59:12102–12108.

Visioli F, Grande S, Bogani P, Galli C. The role of antioxidants in the Mediterranean diets: Focus on cancer. *Eur J Cancer Prev* 2004;13:337–343.

Vitaglione P, Mennella I, Ferracane R, Rivellese AA, Giacco R, Ercolini D, Gibbons SM, et al. Whole-grain wheat consumption reduces inflammation in a randomized controlled trial on overweight and obese subjects with unhealthy dietary and lifestyle behaviors: Role of polyphenols bound to cereal dietary fiber. *Am J Clin Nutr* 2015;101:251–261.

Vitalini S, Gardana C, Simonetti P, Fico G, Iriti M. Melatonin, melatonin isomers and stilbenes in Italian traditional grape products and their antiradical capacity. *J Pineal Res* 2013;54:322–333.

Vitalini S, Gardana C, Zanzotto A, Fico G, Faoro F, Simonetti P, Iriti M. From vineyard to glass: Agrochemicals enhance the melatonin and total polyphenol contents and antiradical activity of red wines. *J Pineal Res* 2011;51:278–285.

Wang L, Zhao Y, Reiter RJ, He C, Liu G, Lei Q, Zuo B, Zheng XD, Li Q, Kong J. Changes in melatonin levels in transgenic "Micro-Tom" tomato overexpressing ovine AANAT and ovine HIOMT genes. *J Pineal Res* 2014;56:134–142.

Wang Z, Dabrosin C, Yin X, Fuster MM, Arreola A, Rathmell WK, Generali D, et al. Broad targeting of angiogenesis for cancer prevention and therapy. *Semin Cancer Biol* 2015;35 Suppl:S224–243.

Wanyo P, Siriamornpun S, Meeso N. Improvement of quality and antioxidant properties of dried mulberry leaves with combined far-infrared radiation and air convection in Thai tea process. *Food Bioprod Process* 2010;89:22–30.

Wilson KM, Kasperzyk JL, Rider JR, Kenfield S, van Dam RM, Stampfer MJ, Giovannucci E, Mucci LA. Coffee consumption and prostate cancer risk and progression in the health professionals follow-up study. *J Natl Cancer Inst* 2011;103:876–884.

Wright KP, Myers BL, Plenzler SC, Drake CL, Badia P. Acute effects of bright light and caffeine on nighttime melatonin and temperature levels in women taking and not taking oral contraceptives. *Brain Res* 2000;873:310–317.

Yang BY, Xia YG, Wang QH, Dou DQ, Kuang HX. Baimantuoluosides D–G, four new withanolide glucosides from the flower of *Datura metel* L. *Arch Pharm Res* 2010;33:1143–1148.

Yılmaz C, Kocadağlı T, Gökmen V. Formation of melatonin and its isomer during bread dough fermentation and effect of baking. *J Agric Food Chem* 2014;62:2900–2905.

Yu X, Bao Z, Zou J, Dong J. Coffee consumption and risk of cancers: A meta-analysis of cohort studies. *BMC Cancer* 2011;11:96.

Zhao Y, Qi LW, Wang WM, Saxena PK, Liu CZ. Melatonin improves the survival of cryopreserved callus of *Rhodiola crenulata*. *J Pineal Res* 2011;50:83–88.

Zhao Y, Tan DX, Lei Q, Chen H, Wang L, Li QT, Gao Y, Kong J. Melatonin and its potential biological functions in the fruits of sweet cherry. *J Pineal Res* 2013;55:79–88.

Zhong L, Furne JK, Levitt MD. An extract of black, green, and mulberry teas causes malabsorption of carbohydrate but not of triacylglycerol in healthy volunteers. *Am J Clin Nutr* 2006;84:551–555.

Zhong L, Wang K, Tan J, Li W, Li S. Putative cytochrome P450 genes in rice genome (*Oryza sativa* L. ssp. indica) and their EST evidence. *Sci China C Life Sci* 2002;45:512–517.

18

Intake of Tryptophan-Rich Foods, Light Exposure, and Melatonin Secretion

Yumi Fukuda and Takeshi Morita
Department of Environmental Science, Fukuoka Women's University
Fukuoka, Japan

CONTENTS

ABSTRACT   Melatonin is a hormone secreted mainly by the pineal gland in the brain, and its secretion is controlled by the suprachiasmatic nucleus (SCN) in the hypothalamus, the central biological clock responsible for circadian rhythms. Daily exposure to light resets the phase of the clock in the SCN. Tryptophan is the precursor of melatonin and its intake increases the availability of melatonin in the body. In this chapter, the sequence of the metabolism of tryptophan is described and the relation between tryptophan, melatonin, and human health is discussed. One study that investigated effects of tryptophan intake and light exposure on sleep and melatonin secretion is introduced and discussed with regard to the importance of daytime light exposure for melatonin secretion at night and the existence of a combined effect of light and tryptophan intake. The effects of the timing of pure tryptophan intake on mood, melatonin secretion, and sleep require further study.

KEY WORDS: *tryptophan, melatonin, light exposure.*

Abbreviations

ALAC: alpha-lactalbumin whey protein
AUC: area under curve
AANAT: arylalkylamine N-acetyltransferase
DLMO: dim-light melatonin onset
EPH: egg protein hydrolysate
DIN: German Institute for Standardization (Deutsches Institut für Normung)
LNAA: large neural amino acid
LAN: light at night
5-HT: serotonin
SCN: suprachiasmatic nucleus
IARC: International Agency for Research on Cancer
TRP: tryptophan

Melatonin and Human Health

Melatonin (N-acetyl-5-methoxytryptamine) is a hormone secreted mainly by the pineal gland in the brain, while other sources of melatonin have also been reported in some species (Vakkuri et al., 1985; Huether et al., 1992b; Yaga et al., 1993; Liu et al., 2007; Maldonado et al., 2010). In mammals, the secretion from the pineal grand is controlled by the suprachiasmatic nucleus (SCN) in the hypothalamus, the central biological clock responsible for circadian rhythms. Daily exposure to light resets the phase of the clock in the SCN (Saper et al., 2005) through photoreceptors in the retina. Retinal ganglion cells containing melanopsin were found early in the twenty-first century and drew attention to an important photoreceptor for the photic entrainment of circadian rhythms (Berson et al., 2002; Hattar et al., 2002; Ruby et al., 2002; Panda et al., 2002). Melatonin secretion usually shows a peak at midnight and is suppressed by exposure to light at night (LAN). It is often used as an indicator of sleep quality and phase advances or delays in the sleep–wake cycle (Claustrat et al., 2005; Zisapel, 2007). The temporal relationship between the temperature nadir, sleepiness, and the peak of melatonin excretion has been shown experimentally (Akerstedt et al., 1979). A field study showed that nocturnal melatonin secretion depends on the amount of light exposure during the daytime and nighttime (Morita et al., 2002). Put simply, bright-light exposure in the daytime increases melatonin secretion at night (Hashimoto et al., 1997; Park and Tokura, 1999), whereas, in contrast, nighttime exposure suppresses melatonin secretion (Lewy et al., 1980; Aoki et al., 1998). Moreover, it has been reported that the amount of melatonin suppression depends not only on the light intensity but also its spectrum (Brainard et al., 2001; Thapan et al., 2001). Melatonin suppression is most sensitive to light with short wavelengths (about 460 nm, in the blue part of the spectrum) and this is close to the peak sensitivity (about 480 nm) of melanopsin-expressing ganglion cells to light. Based on the action spectrum of melatonin, the German Institute for Standardization (Deutsches Institut für Normung; DIN) and the Lighting Research Center in the United States (Rea et al., 2010) have proposed methods to calculate the impact of light on melatonin suppression based on the spectral distribution of the light and the spectral sensitivity of the photoreceptors. However, further research and validation are still required to reach the international standard.

The International Agency for Research on Cancer (IARC) has classified shift work (which disrupts circadian rhythms) as a probable carcinogen (Group 2A). Indeed, night/rotating shift workers have higher risks of breast (Davis et al., 2001; Schernhammer et al., 2001), colorectal (Schernhammer et al., 2003), prostate (Kubo et al., 2006), and endometrial (Viswanathan et al., 2007) cancers, and suppression of pineal melatonin production by LAN is considered to be one of the causes (Blask et al., 2005). However, despite the fact that the rhythm of melatonin production is still functional in blind people (Lockley et al., 1997), breast cancer risk is actually decreased in blind women (Verkasalo et al., 1999). Although the mechanism of tumorigenesis has not been fully elucidated, LAN causes circadian disruption of hormonal rhythms, and this might stimulate tumorigenesis. In the clinical field, doctors and researchers use melatonin to improve sleep quality to treat insomnia and depression and to reduce jet lag, and it is recommended by some clinicians as a preventive agent for breast cancer (Mundey et al., 2005; Srinivasan et al., 2006; Lemoine et al., 2007; Waterhouse et al., 2007; Sanchez-Barcelo et al., 2012; Kostoglou-Athanassiou, 2013). However, the concentrations of plasma melatonin are age dependent (Watson, 2012), and the effects of LAN on individuals differ according to their light history (Hébert et al., 2002), race (Higuchi et al., 2007), and ocular conditions (Brainard et al., 1997; Tanaka et al., 2010). Therefore, light effects on human health will differ between individuals, and such individual differences should be considered when people are advised to take melatonin.

Tryptophan Intake and Metabolism

Tryptophan (TRP; L-alpha-aminoindole-3-propionic acid) is one of the essential amino acids and is contained in proteins sourced from milk, eggs, meat, grain, and beans. Tryptophan is metabolized in mammals through three pathways: (1) over 95% of ingested tryptophan is degraded to niacin, pyruvate, and acetyl-CoA through kynurenine formation; (2) 2%–3% of tryptophan is used for deamination and decarboxylation to yield indoleacetic acid; and (3) 1%–2% is hydroxylated and decarboxylated to generate serotonin (5-HT; 5-hydroxytryptamine) and melatonin (Yao et al., 2011). In the last pathway, only a very small amount of tryptophan is necessary, but this yields the important bioactive compounds serotonin and melatonin. Serum melatonin concentrations show a dose-dependent rise after tryptophan intake in rats, chickens (Huether et al., 1992b), and humans (Huether et al., 1992a). Serotonin synthesis is also increased by tryptophan intake in humans (Eccleston et al., 1970). Serotonin is one of the monoamine neurotransmitters and is assumed to be involved with mental stability. The daily intake of tryptophan in healthy adults is recommended to be 4 mg/day/kg body weight (Firk and Markus, 2009). Tryptophan can cross the blood–brain barrier by competing with other large neural amino acids (LNAAs) and is then transformed into serotonin and melatonin in the brain. The biosynthesis rate of melatonin is limited by the activity of arylalkylamine N-acetyltransferase (AANAT) (Ganguly et al., 2002). Zawilska et al. (2007) reported that AANAT activity is affected by light; its activity declines in the photoperiod (when melatonin secretion also decreases) and increases in the scotoperiod (when melatonin secretion also increases). Therefore, tryptophan intake and the timing of light exposure must be considered together if the effect of tryptophan upon melatonin secretion and sleep quality is to be maximized.

Tryptophan Intake and Mood and Mental Activity

Joseph and Kennett (1983) reported that high-stress situations increased serotonin in rat brain. Therefore, serotonin is assumed to be taken up by the brain to cope with stress and to prevent stress-induced moods and cognitive deterioration. Some experiments have been performed with the aim of examining the effects of tryptophan intake in the morning on serotonin synthesis, which improves mood and mental activity. Markus et al. (2008) examined the effects of tryptophan intake (from different sources) at breakfast on the plasma tryptophan/large neutral amino acids (TRP/LNAA) ratio and mood. They showed that intake of a tryptophan-rich source from hydrolyzed protein had significantly greater effects on the plasma TRP/LNAA ratio and improvement of mood than did pure tryptophan and alpha-lactalbumin whey protein (ALAC). In addition, they then measured depressive mood and perceptual-motor and vigilance performance in subjects (who had high or low chronic stress vulnerabilities) under stress after they had ingested tryptophan-rich egg protein hydrolysate (EPH) in the morning (Markus et al., 2010). The results revealed that EPH improved the depressive mood (i.e., decreased depression) in all subjects and also improved perceptual-motor and vigilance performance in those subjects who had low chronic stress vulnerability.

Effects of Tryptophan Intake and Light Exposure on Subsequent Melatonin Secretion and Sleep

Hudson et al. (2005) reported that using protein as a source of tryptophan before sleep improved sleep quality at night by amounts that were comparable with the intake of pharmaceutical-grade tryptophan. Nakade et al. (2012) reported the relationship between tryptophan intake and sleep quality (in children aged 2–6) in a survey of daily breakfast composition, morning light exposure, and sleep. They found that tryptophan intake at breakfast coupled with morning light exposure was associated with higher melatonin secretion and easier onset of sleep the following night. There is also evidence that tryptophan intake in the morning and light exposure at night affect sleep; Wada et al. (2013) reported that tryptophan-rich breakfasts and low color-temperature light sources at night increased saliva melatonin concentration in university soccer club members. Nevertheless, these investigations were carried out without regulation of several aspects of the individuals' lifestyle, including their overall diet and daytime light exposure, and do not provide quantitative analysis of these effects.

The quantitative effects of altered tryptophan intake and light exposure were first studied by Fukushige et al. (2014). They investigated the effects of tryptophan intake and light exposure on sleep and melatonin secretion by modifying tryptophan ingestion at breakfast and light exposure during the daytime, and measuring sleep quality and melatonin secretion at night. Male university students were randomly divided into four groups and studied for four days: Poor*Dim ($n = 10$), tryptophan-poor breakfast (55 mg/meal) in the morning and dim-light environment (<50 lx) during the daytime; Rich*Dim ($n = 7$), tryptophan-rich breakfast (476 mg/meal) and dim-light environment; Poor*Bright ($n = 9$), tryptophan-poor breakfast and bright-light environment (>5000 lx); and Rich*Bright ($n = 7$), tryptophan-rich breakfast and bright-light environment. Melatonin concentrations were analyzed from saliva samples collected once every hour between 1800 h and 2400 h, and the areas under the curve (AUC) of the melatonin profiles on the first and fourth experimental days were compared. Figure 18.1 shows the AUCs in the four groups. Changes in melatonin AUC between days 1 and 4 were significantly different with light exposure (three-way ANOVA; the interaction between day × light, $p < .01$), while no significant differences were found in the AUCs associated with tryptophan intake. Significant differences were also found between Rich*Bright versus Rich*Dim and Rich*Bright versus Poor*Dim on day 4 (multiple comparisons by Kruskal–Wallis test, $p < .05$), while no significant differences were found between the four groups on day 1 (baseline). In the Poor*Dim and Rich*Dim groups, the AUCs of saliva melatonin concentrations showed a decrease of about 30% on day 4 compared with day 1; however, the Poor*Bright and Rich*Bright groups showed increases of melatonin concentration, by about 55% and 80%, respectively. The dim-light melatonin onset (DLMO) was also considered for phase shifts of melatonin secretion (Table 18.1), as the DLMO is believed to be the most accurate and useful marker for assessing phase delays and advances of circadian rhythms (Lewy et al., 1999; Pullman et al., 2012). The DLMOs between days 1 and 4 were significantly different with light exposure and tended to be different with tryptophan intake (three-way ANOVA; day × light × TRP, $p < .05$; day × light, $p < .01$; day × TRP, $p = .09$). The DLMOs (days 1–4) were significantly advanced in both groups given bright-light exposure, but were delayed in those given dim-light exposure, significantly so in the Rich*Dim group (multiple comparisons by Tukey test, Poor*Dim

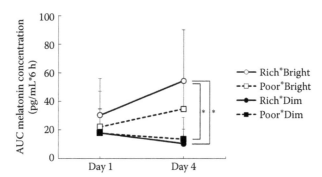

FIGURE 18.1 The area under curve (AUC) of saliva melatonin concentrations during the evenings of days 1 and 4. Mean AUCs and SDs are shown on days 1 and 4 in each group. The dim-light conditions are expressed as a filled square or circle, and the bright-light conditions as an open square or circle. The mean AUCs on days 1 and 4 under the tryptophan-poor conditions are connected with dashed lines, and the tryptophan-rich conditions with straight lines. Statistical significance is shown by asterisks (* $p < .05$).

TABLE 18.1

DLMOs of the Four Groups on Days 1 and 4

Group	DLMO[a]		
	Day 1 (h:min)	Day 4 (h:min)	Day 1–Day 4 (min)
Poor*Dim	21:53 (54)	22:12 (58)	−19 (13)
Rich*Dim	21:30 (60)	22:36 (47)	−66 (14)
Poor*Bright	21:23 (64)	19:57 (75)	86 (32)
Rich*Bright	21:38 (60)	20:07 (52)	90 (53)

Note: DLMO was defined at the absolute thresholds of 3 pg/mL.

[a] Mean time of day and SDs are shown for DLMO on days 1 and 4.

vs. Poor*Bright, Poor*Dim vs. Rich*Bright, Rich*Dim vs. Poor*Bright, Rich*Dim vs. Rich*Bright, all $p < .01$).

This study by Fukushige et al. (2014) confirms that bright-light exposure during the daytime is an important contributor to raised melatonin levels in the evening, as has previously been found by others (Hashimoto et al., 1997; Park and Tokura, 1999; Mishima et al., 2001; Takasu et al., 2006). In addition, the study suggests that, even though the contribution of bright-light exposure to melatonin secretion is more important than that of tryptophan intake, a combined effect of tryptophan intake and bright-light exposure might exist and promote melatonin secretion. On the other hand, there were no significant differences with regard to sleep efficiency and sleep latency (as assessed from the acti-graphs) regarding the effects of tryptophan intake and light exposure. However, subjective feelings of sleep were improved after consuming tryptophan-rich food and being exposed to light during daytime, although not statistically significantly so. Silber and Schmitt (2010) suggested that, in healthy adults, tryptophan intake during the daytime had a relaxing and calming effect, whereas taking it at night might have only a minimal effect on the following sleep. Since the relaxing and calming effects of tryptophan were not investigated by Fukushige et al. (2014), the effects of the timing of tryptophan intake on mood and sleep require further study. In addition, even though, in the study of Fukushige et al. (2014), tryptophan intake was controlled, other nutrients and calorie intake were not controlled. The effects of pure tryptophan or tryptophan-rich protein hydrolysate on melatonin secretion and sleep also need to be investigated.

Summary

Tryptophan is the precursor of the important bioactive compounds serotonin and melatonin. Tryptophan intake increases the availability of serotonin and melatonin in the body. Recently, one study has revealed that the contribution of bright-light exposure to melatonin secretion is more important than that of tryptophan intake. However, it also indicates that a combined effect of these factors might exist. The effects of the timing of pure tryptophan intake upon mood, melatonin secretion, and sleep require further study.

REFERENCES

Akerstedt T, Fröberg JE, Friberg Y, Wetterberg L. Melatonin excretion, body temperature and subjective arousal during 64 hours of sleep deprivation. *Psychoneuroendocrinology*. 1979;4:219–225.

Aoki H, Yamada N, Ozeki Y, Yamane H, Kato N. Minimum light intensity required to suppress nocturnal melatonin concentration in human saliva. *Neurosci Lett*. 1998;252:91–94.

Berson DM, Dunn FA, Takao M. Phototransduction by retinal ganglion cells that set the circadian clock. *Science*. 2002;295:1070–1073.

Blask DE, Brainard GC, Dauchy RT, Hanifin JP, Davidson LK, Krause JA, Sauer LA, Rivera-Bermudez MA, Dubocovich ML, Jasser SA, Lynch DT, Rollag MD, Zalatan F. Melatonin-depleted blood from premenopausal women exposed to light at night stimulates growth of human breast cancer xenografts in nude rats. *Cancer Res*. 2005;65:11174–11184.

Brainard GC, Hanifin JP, Greeson JM, Byrne B, Glickman G, Gerner E, Rollag MD. Action spectrum for melatonin regulation in humans: Evidence for a novel circadian photoreceptor. *J Neurosci*. 2001;21:6405–6412.

Brainard GC, Rollag MD, Hanifin JP. Photic regulation of melatonin in humans: Ocular and neural signal transduction. *J Biol Rhythms*. 1997;12:537–546.

Claustrat B, Brun J, Chazot G. The basic physiology and pathophysiology of melatonin. *Sleep Med Rev*. 2005;9:11–24.

Davis S, Mirick DK, Stevens RG. Night shift work, light at night, and risk of breast cancer. *J Natl Cancer Inst*. 2001;93:1557–1562.

Eccleston D, Ashcroft GW, Crawford TB. Effect of tryptophan administration on 5HIAA in cerebrospinal fluid in man. *J Neurol Neurosurg Psychiatry*. 1970;33:269–272.

Firk C, Markus CR. Mood and cortisol responses following tryptophan-rich hydrolyzed protein and acute stress in healthy subjects with high and low cognitive reactivity to depression. *Clin Nutr.* 2009;28:266–271.

Fukushige H, Fukuda Y, Tanaka M, Inami K, Wada K, Tsumura Y, Kondo M, Harada T, Wakamura T, Morita T. Effects of tryptophan-rich breakfast and light exposure during the daytime on melatonin secretion at night. *J Physiol Anthropol.* 2014;33:33.

Ganguly S, Coon SL, Klein DC. Control of melatonin synthesis in the mammalian pineal gland: The critical role of serotonin acetylation. *Cell Tissue Res.* 2002;309:127–137.

Hashimoto S, Kohsaka M, Nakamura K, Honma H, Honma S, Honma K. Midday exposure to bright light changes the circadian organization of plasma melatonin rhythm in humans. *Neurosci Lett.* 1997;221:89–92.

Hattar S, Liao HW, Takao M, Berson DM, Yau KW. Melanopsin-containing retinal ganglion cells: Architecture, projections, and intrinsic photosensitivity. *Science.* 2002;295:1065–1070.

Hébert M, Martin SK, Lee C, Eastman CI. The effects of prior light history on the suppression of melatonin by light in humans. *J Pineal Res.* 2002;33:198–203.

Higuchi S, Motohashi Y, Ishibashi K, Maeda T. Less exposure to daily ambient light in winter increases sensitivity of melatonin to light suppression. *Chronobiol Int.* 2007;24:31–43.

Hudson C, Hudson SP, Hecht T, MacKenzie J. Protein source tryptophan versus pharmaceutical grade tryptophan as an efficacious treatment for chronic insomnia. *Nutr Neurosci.* 2005;8:121–127.

Huether G, Hajak G, Reimer A, Poeggeler B, Blömer M, Rodenbeck A, Rüther E. The metabolic fate of infused L-tryptophan in men: Possible clinical implications of the accumulation of circulating tryptophan and tryptophan metabolites. *Psychopharmacology.* 1992a;109:422–432.

Huether G, Poeggeler B, Reimer A, George A. Effect of tryptophan administration on circulating melatonin levels in chicks and rats: Evidence for stimulation of melatonin synthesis and release in the gastrointestinal tract. *Life Sci.* 1992b;51:945–953.

Joseph MH, Kennett GA. Stress-induced release of 5-HT in the hippocampus and its dependence on increased tryptophan availability: An *in vivo* electrochemical study. *Brain Res.* 1983;270:251–257.

Kostoglou-Athanassiou I. Therapeutic applications of melatonin. *Ther Adv Endocrinol Metab.* 2013;4:13–24.

Kubo T, Ozasa K, Mikami K, Wakai K, Fujino Y, Watanabe Y, Miki T, Nakao M, Hayashi K, Suzuki K, Mori M, et al. Prospective cohort study of the risk of prostate cancer among rotating-shift workers: Findings from the Japan collaborative cohort study. *Am J Epidemiol.* 2006;164:549–555.

Lemoine P, Nir T, Laudon M, Zisapel N. Prolonged-release melatonin improves sleep quality and morning alertness in insomnia patients aged 55 years and older and has no withdrawal effects. *J Sleep Res.* 2007;16:372–380.

Lewy AJ, Cutler NL, Sack RL. The endogenous melatonin profile as a marker for circadian phase position. *J Biol Rhythms.* 1999;14:227–236.

Lewy AJ, Wehr TA, Goodwin FK, Newsome DA, Markey SP. Light suppresses melatonin secretion in humans. *Science.* 1980;210:1267–1269.

Liu YJ, Zhuang J, Zhu HY, Shen YX, Tan ZL, Zhou JN. Cultured rat cortical astrocytes synthesize melatonin: Absence of a diurnal rhythm. *J Pineal Res.* 2007;43:232–238.

Lockley SW, Skene DJ, Arendt J, Tabandeh H, Bird AC, Defrance R. Relationship between melatonin rhythms and visual loss in the blind. *J Clin Endocrinol Metab.* 1997;82:3763–3770.

Maldonado MD, Mora-Santos M, Naji L, Carrascosa-Salmoral MP, Naranjo MC, Calvo JR. Evidence of melatonin synthesis and release by mast cells: Possible modulatory role on inflammation. *Pharmacol Res.* 2010;62:282–287.

Markus CR, Firk C, Gerhardt C, Kloek J, Smolders GF. Effect of different tryptophan sources on amino acids availability to the brain and mood in healthy volunteers. *Psychopharmacology.* 2008;201:107–114.

Markus CR, Verschoor E, Firk C, Kloek J. Gerhardt CC. Effect of tryptophan-rich egg protein hydrolysate on brain tryptophan availability, stress and performance. *Clin Nutr.* 2010;29:610–616.

Mishima K, Okawa M, Shimizu T, Hishikawa Y. Diminished melatonin secretion in the elderly caused by insufficient environmental illumination. *J Clin Endocrinol Metab.* 2001;86:129–134.

Morita T, Koikawa R, Ono K, Terada Y, Hyun K, Tokura H. Influence of the amount of light received during the day and night times on the circadian rhythm of melatonin secretion in women living diurnally. *Biol Rhythm Res.* 2002;33:271–277.

Mundey K, Benloucif S, Harsanyi K, Dubocovich ML, Zee PC. Phase-dependent treatment of delayed sleep phase syndrome with melatonin. *Sleep.* 2005;28:1271–1278.

Nakade M, Akimitsu O, Wada K, Krejci M, Noji T, Taniwaki N, Takeuchi H, Harada T. Can breakfast tryptophan and vitamin B6 intake and morning exposure to sunlight promote morning-typology in young children aged 2 to 6 years? *J Physiol Anthropol.* 2012;31:11.

Panda S, Sato TK, Castrucci AM, Rollag MD, DeGrip WJ, Hogenesch JB, Provencio I, Kay SA. Melanopsin (Opn4) requirement for normal light-induced circadian phase-shifting. *Science.* 2002;298:2213–2216.

Park SJ, Tokura H. Bright light exposure during the daytime affects circadian rhythms of urinary melatonin and salivary immunoglobulin A. *Chronobiol Int.* 1999;16:359–371.

Pullman RE, Roepke SE, Duffy JF. Laboratory validation of an in-home method for assessing circadian phase using dim light melatonin onset (DLMO). *Sleep Med.* 2012;13:703–706.

Rea MS, Figueiro MG, Bierman A, Bullough JD. Circadian light. *J Circadian Rhythms.* 2010;8:2.

Ruby NF, Brennan TJ, Xie X, Cao V, Franken P, Heller HC, O'Hara BF. Role of melanopsin in circadian responses to light. *Science.* 2002;298:2211–2213.

Sanchez-Barcelo EJ, Mediavilla MD, Alonso-Gonzalez C, Reiter RJ. Melatonin uses in oncology: Breast cancer prevention and reduction of the side effects of chemotherapy and radiation. *Expert Opin Investig Drugs.* 2012;21:819–831.

Saper CB, Lu J, Chou TC, Gooley J. The hypothalamic integrator for circadian rhythms. *Trends Neurosci.* 2005;28:152–157.

Schernhammer ES, Laden F, Speizer FE, Willett WC, Hunter DJ, Kawachi I, Colditz GA. Rotating night shifts and risk of breast cancer in women participating in the nurses' health study. *J Natl Cancer Inst.* 2001;93:1563–1568.

Schernhammer ES, Laden F, Speizer FE, Willett WC, Hunter DJ, Kawachi I, Fuchs CS, Colditz GA. Night-shift work and risk of colorectal cancer in the nurses' health study. *J Natl Cancer Inst.* 2003;95:825–828.

Silber BY, Schmitt JA. Effects of tryptophan loading on human cognition, mood, and sleep. *Neurosci Biobehav Rev.* 2010;34:387–407.

Srinivasan V, Smits M, Spence W, Lowe AD, Kayumov L, Pandi-Perumal SR, Parry B, Cardinali DP. Melatonin in mood disorders. *World J Biol Psychiatry.* 2006;7:138–151.

Takasu NN, Hashimoto S, Yamanaka Y, Tanahashi Y, Yamazaki A, Honma S, Honma K. Repeated exposures to daytime bright light increase nocturnal melatonin rise and maintain circadian phase in young subjects under fixed sleep schedule. *Am J Physiol Regul Integr Comp Physiol.* 2006;291:R1799–1807.

Tanaka M, Hosoe K, Hamada T, Morita T. Change in sleep state of the elderly before and after cataract surgery. *J Physiol Anthropol.* 2010;29:219–224.

Thapan K, Arendt J, Skene DJ. An action spectrum for melatonin suppression: Evidence for a novel non-rod, non-cone photoreceptor system in humans. *J Physiol.* 2001;535:261–267.

Vakkuri O, Rintamäki H, Leppäluoto J. Plasma and tissue concentrations of melatonin after midnight light exposure and pinealectomy in the pigeon. *J Endocrinol.* 1985;105:263–268.

Verkasalo PK, Pukkala E, Stevens RG, Ojamo M, Rudanko SL. Inverse association between breast cancer incidence and degree of visual impairment in Finland. *Br J Cancer.* 1999;80:1459–1460.

Viswanathan AN, Hankinson SE, Schernhammer ES. Night shift work and the risk of endometrial cancer. *Cancer Res.* 2007;67:10618–10622.

Wada K, Yata S, Akimitsu O, Krejci M, Noji T, Nakade M, Takeuchi H, Harada T. A tryptophan-rich breakfast and exposure to light with low color temperature at night improve sleep and salivary melatonin level in Japanese students. *J Circadian Rhythms.* 2013;11:4.

Waterhouse J, Reilly T, Atkinson G, Edwards B. Jet lag: Trends and coping strategies. *Lancet.* 2007;369:1117–1129.

Watson RR. *Melatonin in the Promotion of Health.* Boca Raton, FL: CRC Press. 2012.

Yaga K, Reiter RJ, Richardson BA. Tryptophan loading increases daytime serum melatonin levels in intact and pinealectomized rats. *Life Sci.* 1993;52:1231–1238.

Yao K, Fang J, Yin YL, Feng ZM, Tang ZR, Wu G. Tryptophan metabolism in animals: Important roles in nutrition and health. *Front Biosci.* 2011;3:286–297.

Zawilska JB, Lorenc A, Berezínska M, Vivien-Roels B, Pévet P, Skene DJ. Photoperiod-dependent changes in melatonin synthesis in the turkey pineal gland and retina. *Poult Sci.* 2007;86:1397–1405.

Zisapel N. Sleep and sleep disturbances: Biological basis and clinical implications. *Cell Mol Life Sci.* 2007;64:1174–1186.

19

Melatonin in Grapes and Wine: A Bioactive Phytochemical

Marcello Iriti and Elena Maria Varoni
Milan State University
Milan, Italy

CONTENTS

ABSTRACT   Among the food plants, the presence of melatonin in grapes (*Vitis vinifera* L.) deserves particular attention because of the production of wine, an alcoholic beverage of economic relevance and with putative healthy effects. Furthermore, a number of melatonin isomers have been detected in grape products, too. In this chapter, we discuss the topic of melatonin in grapes and wine, focusing on the contribution of grapes and yeasts to the final melatonin levels in wine and on the oral bioavailability of dietary melatonin.

KEY WORDS: *grapevine, red wine, melatonin isomer, bioavailability, yeasts.*

Introduction

Although melatonin is commonly portrayed as an animal neurohormone, it now seems to be a ubiquitous molecule in the living world (Tan et al., 2010). In particular, this indoleamine has been significantly detected in many medicinal and food plants (Parades et al., 2009; Tan et al., 2012a; Reiter et al., 2013; Ramakrishna et al., 2009, 2012a,b) and, consequently, it now can be considered a dietary component, even if its daily intake is very difficult to estimate (Iriti et al., 2010). Furthermore, bioavailability of dietary melatonin has been demonstrated both in animals and in humans (Reiter et al., 2005; Maldonado et al., 2009; Sae-Teaw et al., 2013). In this chapter, we discuss the topic of melatonin and its isomers in grapes and wine (Table 19.1), focusing on the contribution of grapes and yeasts to the final melatonin levels in wine.

Melatonin in Grapes and Wine

The issue of melatonin in grape products began less than a decade ago, when it was detected, for the first time, in berry skin of Italian and French grapevine varieties (*Vitis vinifera* L. cv. Barbera, Croatina, Cabernet Sauvignon, Cabernet Franc, Marzemino, Nebbiolo, Sangiovese, and Merlot) grown

TABLE 19.1

Melatonin Content in Grapes and Wine

Grapes	Melatonin (ng/g)	References
Nebbiolo, Croatina, Barbera, Sangiovese, Marzemino, Cabernet Sauvignon, Merlot, Cabernet Franc (skin, Italy)	0.005–0.96	Iriti et al., 2006
Malbec, Cabernet Sauvignon, Chardonnay (skin, Argentina)	0.6–1.2	Stege et al., 2010
Merlot (whole berry, Canada)	100,000–150,000	Murch et al., 2010
Merlot (skin, Italy)	9.3–17.5	Vitalini et al., 2011a
Merlot (seed, Italy)	3.5–10	Vitalini et al., 2011a
Merlot (flesh, Italy)	0.2–3.9	Vitalini et al., 2011a
Malbec (skin, Argentina)	9–159	Boccalandro et al., 2011
Malbec (skin, Argentina)	120–160	Gomez et al., 2012
Albana, Sangiovese (whole berry, Italy)	1.2, 1.5	Mercolini et al., 2012
Malbec (skin, Argentina)	440	Gomez et al., 2013
Wine	**Melatonin (ng/mL)**	**References**
Albana, Sangiovese, Trebbiano (Italy)	0.6, 0.4	Mercolini et al., 2008, 2012
Chardonnay, Malbec, Cabernet Sauvignon (Argentina)	0.16–0.32	Stege et al., 2010
Groppello, Merlot (Italy)	8.1, 5.2	Vitalini et al., 2011b
Cabernet Sauvignon, Merlot, Syrah, Tempranillo, Tintilla de Rota, Petit Verdot, Prieto Picudo, Palomino Fino (Spain)	5.1–420	Rodriuez-Naranjo et al., 2011a,b

in northwestern Italy, with levels ranging from 0.005 to 0.96 ng/g (Iriti et al., 2006). Similar results (from 0.6 to 1.2 ng/g) were reported by Stege and colleagues in the same tissue of Malbec, Cabernet Sauvignon, and Chardonnay varieties cultivated in Argentina (Stege et al., 2010). In the whole berry (i.e., skin, flesh, and seeds analyzed together) of Merlot cultivar grown in Canada, much higher melatonin concentrations were measured (between 100 and 150 μg/g), depending on the phenological stage (Murch et al., 2010). In Merlot berry skin, we found 17.5 and 9.3 ng/g of melatonin at pre-*véraison* and *véraison*, respectively (Vitalini et al., 2011a). Conversely, the transition from pre-*véraison* to *véraison* raised the melatonin content both in Merlot berry seeds (from 3.5 to 10 ng/g) and flesh (from 0.2 to 3.9 ng/g) (Vitalini et al., 2011a). In berry skin of the Malbec variety cultivated in Argentina, melatonin concentration showed similar values during the night (around 10 ng/g), reaching a strong peak in the morning (159 ng/g) (Boccalandro et al., 2011). Melatonin levels of about 1.2–1.5 ng/g were measured in the whole berry of Albana and Sangiovese cultivars grown near Bologna, Italy (Mercolini et al., 2012).

As expected, further studies ascertained the occurrence of melatonin in wine. Mercolini and colleagues (Mercolini et al., 2008, 2012) detected it at 0.5 ng/mL in Sangiovese red wine and 0.6 and 0.4 ng/mL in Albana and Trebbiano white wines, respectively. Stege and coworkers (Stege et al., 2010) reported melatonin at 0.16, 0.24, and 0.32 ng/mL in Chardonnay, Malbec, and Cabernet Sauvignon wines, respectively. Our results showed that the levels of melatonin in Groppello and Merlot wines varied between 5.2 and 8.1 ng/mL, depending on agrochemical treatments (Vitalini et al., 2011b), although Rodriguez-Naranjo and colleagues (Rodriguez-Naranjo et al., 2011a,b) measured much higher melatonin concentrations, up to 130 and 420 ng/mL in racked wines.

Melatonin Isomers

The occurrence of melatonin isomers in nature represents an emerging topic. Isomers can be classified according to the position of the two side chains present in the melatonin structure, the methoxy (M) group at position 5 and the N-acetylaminoethyl (A) group at position 3. Hypothetically, either one of these two side chains can be relocated to any one of the seven positions in the indole nucleus of melatonin to form isomers (Tan et al., 2013). In particular, different isomers were found in grape products, including red wine, even if their chemical structure has not yet been identified (Rodriguez-Naranjo et al., 2011b; Gomez et al., 2012; Vitalini et al., 2013).

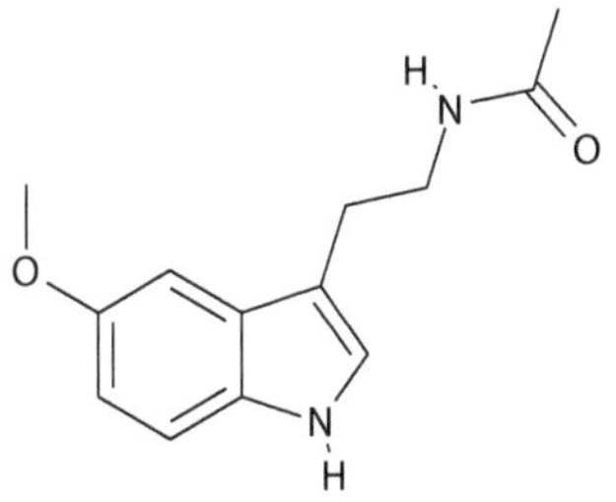

FIGURE 19.1 Chemical structure of (1) melatonin and (2) tryptophan-ethylester.

Interesting findings were reported by Gomez and coworkers (Gomez et al., 2012). They found melatonin in grape (cv. Malbec) berry skin within the range of 120 to 160 ng/g, whereas the indoleamine was not detected in experimental wines (produced from the same grapes), where a melatonin isomer was reported (Gomez et al., 2012). The same authors determined the levels of melatonin and its isomer in grape products. They detected melatonin in berry skin (440 ng/g), but not in commercial and experimental wines, whereas melatonin isomer was measured in wine (from 60 to 211 ng/mL) and not in berry skin (Gomez et al., 2013). Interestingly, in must, neither melatonin (as previously reported by Rodriguez-Naranjo et al., 2011b) nor its isomer were determined (Gomez et al., 2013). Very recently, in an attempt to determine the conformation of the most abundant (putative) melatonin isomer detected in red wine, we have identified it as tryptophan-ethylester, a compound with the same molecular weight as melatonin (Figure 19.1) (Gardana et al., 2014).

In particular, the concentrations of tryptophan-ethylester and melatonin in wine were 84 and 3 ng/mL, respectively (Gardana et al., 2014). In our opinion, a unique pattern does not exist: some wines only contain melatonin, some contain isomer(s) alone, others contain both melatonin and its isomer(s) or neither.

However, the contribution of grapes to melatonin in wine has not yet been entirely elucidated; therefore, a pivotal role of yeasts and, possibly, of bacteria in the production of melatonin and its isomer(s) in wine should be taken into account. In a pioneering paper, Sprenger and colleagues (Sprenger et al., 1999) demonstrated that, in *Saccharomyces cerevisiae*, melatonin is synthesized and metabolized to other 5-methoxylated indoles (5-methoxytryptamine and 5-methoxytryptophol). In other yeast species (i.e., *S. uvarum* and *S. cerevisiae* var. *bayanus*), melatonin production in synthetic grape must depends on growth conditions and medium, including tryptophan concentrations (Rodriguez-Naranjo et al., 2012).

The ability of yeasts to enrich fermented foods and beverages other than wine with indoleamines is corroborated by a number of studies. During the brewing process, yeasts and malted barley are the major contributors of melatonin in beer (Garcia-Moreno et al., 2013). Although melatonin was found not to be present in pomegranate juices, it was significantly detected in pomegranate wines, ranging from 0.54 to 5.50 ng/mL (Mena et al., 2012). During the alcoholic fermentation of orange juice, melatonin significantly increased from day 0 (3.15 ng/mL) until day 15 (21.80 ng/mL), whereas its isomer remained stable, ranging from 11.59 to 14.18 ng/mL (Fernández-Pachón et al., 2014). Although the formation of melatonin was not significant during dough fermentation, the content of its isomer was 4.02 ng/g in nonfermented dough and increased up to 16.71 ng/g at the end of fermentation. Furthermore, compared with their corresponding dough, both the crumb and crust parts of baked bread contained significantly lower amounts of melatonin isomer, which, therefore, seems not to be stable at elevated temperatures (Yılmaz et al., 2014). The same authors also determined melatonin isomer in a variety of products, with the highest levels in fermented food—that is, red wine (170.7 ng/mL), beer (14.3 ng/mL), and bread crumb (15.7 ng/mL) (Kogadağli et al., 2014).

Oral Bioavailability of Melatonin

The discovery of melatonin—a vertebrate neurohormone—in grapes greatly changed food and nutrition sciences. From the human perspective, an efficient uptake of melatonin from food sources should be expected to influence its circulating levels, basically very low (~200 pg/mL at the maximum nighttime peak and lower than 10 pg/mL during the day; Bonnefont-Rousselot and Collin, 2010). In fact, in humans, the intake of foodstuffs containing melatonin may contribute to increase both the serum levels of the indoleamine and the urinary concentrations of its main metabolite, 6-sulfatoxymelatonin (Maldonado et al., 2009; Garrido et al., 2010; González-Flores et al., 2012; Sae-Teaw et al., 2013; Johns et al., 2013). Regarding wine, a plethora of beneficial biological and pharmacological activities has been ascribed, in past decades, to grape and wine polyphenols, mostly based on preclinical *in vitro/in vivo* studies (Iriti and Faoro, 2009). From this viewpoint, we previously hypothesized that the additive and/or synergistic effects of melatonin in wine may maximize the healthy properties attributed to polyphenols (Iriti et al., 2010).

Conclusion

Both endogenous and exogenous factors may influence the biosynthesis and accumulation of melatonin in grapevine tissues and organs: genetic traits and varietal differences, phenological stages of development, day–night cycles and light exposure, and agricultural practices. Probably, as with other secondary plant metabolites, many other environmental parameters may regulate melatonin production in grapevine, including climatic and edaphic conditions, plant–microbe interactions, and agrochemical treatments. Likewise, the techniques and procedures of wine making and the involvement of yeasts may also influence the levels of this indoleamine (and its isomers) in wine. In conclusion, on the strength both of our experience and the available literature, we can state that, in general, the levels of melatonin are near 1 ng/g and 0.5 ng/mL in grapevine berry skin and wine, respectively.

REFERENCES

Boccalandro HE, Gonzáles CV, Wunderlin DA, Silva MF. Melatonin levels, determined by LC–ESI–MS/MS, deeply fluctuate during the day in *Vitis vinifera* cv Malbec: Evidences for its antioxidant role in fruits. *J Pineal Res* 2011;51:226–232.

Bonnefont-Rousselot D, Collin, F. Melatonin: Action as antioxidant and potential applications in human disease and aging. *Toxicology* 2010;278:55–67.

Fernández-Pachón MS, Medina S, Herrero-Martín G, Cerrillo I, Berna G, Escudero-López B, Ferreres F, Martín F, Garcia-Parrilla MDC, Gil-Izquierdo A. Alcoholic fermentation induces melatonin synthesis in orange juice. *J Pineal Res* 2014;56:31–38.

Garcia-Moreno H, Calvo JR, Maldonado MD. High levels of melatonin generated during the brewing process. *J Pineal Res* 2013;55:26–30.

Gardana C, Iriti M, Stuknytė M, De Noni I, Simonetti P. "Melatonin isomer" in wine is not an isomer of the melatonin but tryptophan-ethylester. *J Pineal Res* 2014;57:435–441.

Garrido M, Paredes SD, Cubero J, Lozano M, Toribio-Delgado AF, Munoz JL, Reiter RJ, Barriga C, Rodriguez AB. Jerte Valley cherry–enriched diets improve nocturnal rest and increase 6-sulfatoxymelatonin and total antioxidant capacity in the urine of middle-aged and elderly humans. *J Gerontol A Biol Sci Med Sci* 2010;65:909–914.

Gomez FJV, Hernández IG, Martinez LD, Silva MF, Cerutti S. Analytical tools for elucidating the biological role of melatonin in plants by LC–MS/MS. *Electrophoresis* 2013;34(12):1749–1756.

Gomez FJV, Raba J, Cerutti S, Silva MF. Monitoring melatonin and its isomer in *Vitis vinifera* cv. Malbec by UHPLC–MS/MS from grape to bottle. *J Pineal Res* 2012;52(3):349–355.

González-Flores D, Gamero E, Garrido M, Ramírez R, Moreno D, Delgado J, Valdés E, Barriga C, Rodríguez AB, Paredes SD. Urinary 6-sulfatoxymelatonin and total antioxidant capacity increase after the intake of a grape juice cv. Tempranillo stabilized with HHP. *Food Func* 2012;3:34–39.

Iriti M, Faoro F. Bioactivity of grape chemicals for human health. *Nat Prod Commun* 2009;4:1–24.

Iriti M, Rossoni M, Faoro F. Melatonin content in grape: Myth or panacea? *J Sci Food Agr* 2006;86:1432–1438.

Iriti M, Varoni EM, Vitalini S. Melatonin in Mediterranean diets. *J Pineal Res* 2010:49;101–105.

Johns NP, Johns J, Porasuphattana S, Plaimee P, Sae-Teaw M. Dietary intake of melatonin from tropical fruit altered urinary excretion of 6-sulfatoxymelatonin in healthy volunteers. *J Agr Food Chem* 2013;61:913–919.

Kocadağlı T, Yılmaz C, Gökmen V. Determination of melatonin and its isomer in foods by liquid chromatography tandem mass spectrometry. *Food Chem* 2014;153:151–156.

Maldonado MD, Moreno H, Calvo JR. Melatonin present in beer contributes to increase the levels of melatonin and antioxidant capacity of the human serum. *Clin Nutr* 2009;28:188–191.

Mena P, Gil-Izquierdo Á, Moreno DA, Martí N, García-Viguera C. Assessment of the melatonin production in pomegranate wines. *LWT-Food Sci Technol* 2012;47:13–18.

Mercolini L, Addolorata Saracino M, Bugamelli F, Ferranti A, Malaguti M, Hrelia S, Raggi MA. HPLC–F analysis of melatonin and resveratrol isomers in wine using an SPE procedure. *J Sep Sci* 2008;31:1007–1114.

Mercolini L, Mandrioli R, Raggi MA. Content of melatonin and other antioxidants in grape related foodstuffs: Measurement using a MEPS–HPLC–F method. *J Pineal Res* 2012;55:21–28.

Murch SJ, Hall BA, Le CH, Saxena PK. Changes in the levels of indoleamine phytochemicals during *véraison* and ripening of wine grapes. *J Pineal Res* 2010;49:95–100.

Paredes SD, Korkmaz A, Manchester LC, Tan DX, Reiter RJ. Phytomelatonin: A review. *J Exp Bot* 2009;60:57–69.

Ramakrishna A, Giridhar P, Ravishankar GA. Indoleamines and calcium channels influence morphogenesis in *in vitro* cultures of *Mimosa pudica* L. *Plant Sign Behav* 2009;12:1136–1141.

Ramakrishna A, Giridhar P, Udaya Sankar K, Ravishankar GA. Melatonin and serotonin profiles in beans of *Coffea* sps. *J Pineal Res* 2012;52:470–476.

Ramakrishna A, Giridhar P, Udaya Sankar K, Ravishankar GA. Endogenous profiles of indoleamines: Serotonin and melatonin in different tissues of *Coffea canephora* P ex Fr. as analyzed by HPLC and LC–MS–ESI. *Acta Physiol Plant* 2012a;34:393–396.

Reiter R.J., Manchester L.C., Tan D.X. Melatonin in walnuts: Influence on levels of melatonin and total antioxidant capacity of blood. *Nutrition*, 2005;21:920–924.

Reiter RJ, Tan DX, Rosales-Corral S, Manchester LC. The universal nature, unequal distribution and antioxidant functions of melatonin and its derivatives. *Mini-Rev Med Chem* 2013;13:373–384.

Rodriguez-Naranjo MI, Gil-Izquierdo A, Troncoso AM, Cantos-Villar E, Garcia-Parrilla MC. Melatonin is synthesized by yeast during alcoholic fermentation in wines. *Food Chem* 2011a;126:1608–1613.

Rodriguez-Naranjo MI, Gil-Izquierdo A, Troncoso AM, Cantos E, Garcia-Parrilla MC. Melatonin: A new bioactive compound in wine. *J Food Compos Anal* 2011b;24:603–608.

Rodriguez-Naranjo MI, Torija MJ, Mas A, Cantos-Villar E, Garcia-Parrilla MDC. Production of melatonin by *Saccharomyces* strains under growth and fermentation conditions. *J Pineal Res* 2012;53:219–224.

Sae-Teaw M, Johns J, Johns NP, Subongkot S. Serum melatonin levels and antioxidant capacities after consumption of pineapple, orange, or banana by healthy male volunteers. *J Pineal Res* 2013;55(1):58–64.

Sprenger J, Hardeland R, Fuhrberg B, Han SZ. Melatonin and other 5-methoxylated Indoles in yeast: Presence in high concentrations and dependence on tryptophan availability. *Cytologia* 1999;64:209–213.

Stege PW, Sombra LL, Messina G, Martinez LD, Silva MD. Determination of melatonin in wine and plant extracts by capillary electrochromatography with immobilized carboxylic multi-walled carbon nanotubes as stationary phase. *Electrophoresis* 2010;31:2242–2248.

Tan D-X, Hardeland R, Manchester LC, Paredes SD, Korkmaz A, Sainz RM, Reiter RJ. The changing biological roles of melatonin during evolution: From an antioxidant to signals of darkness, sexual selection and fitness. *Biol Rev* 2010;85(3):607–623.

Tan D-X, Hardeland R, Manchester LC, Rosales-Corral S, Coto-Montes A, Boga JA, Reiter RJ. Emergence of naturally occurring melatonin isomers and their proposed nomenclature. *J Pineal Res* 2012;53:113–121.

Tan DX, Hardeland R, Manchester LC, Korkmaz A, Ma S, Rosales-Corral S, Reiter RJ. Functional roles of melatonin in plants, and perspectives in nutritional and agricultural science. *J Exp Bot* 2012a;63:577–597.

Vitalini S, Gardana C, Simonetti P, Fico G, Iriti M. Melatonin, melatonin isomers and stilbenes in Italian traditional grape products and their antiradical capacity. *J Pineal Res* 2013;54:322–333.

Vitalini S, Gardana C, Zanzotto A, Fico G, Faoro F, Simonetti P, Iriti M. From vineyard to glass: Agrochemicals enhance the melatonin and total polyphenol contents and antiradical activity of red wines. *J Pineal Res* 2011b;51:278–285.

Vitalini S, Gardana C, Zanzotto A, Simonetti P, Faoro F, Fico G, Iriti M. The presence of melatonin in grapevine (*Vitis vinifera* L.) berry tissues. *J Pineal Res* 2011a;51:331–337.

Yılmaz C, Kocadağlı T, Gökmen V. Formation of melatonin and its isomer during bread dough fermentation and effect of baking. *J Agr Food Chem* 2014;62:2900–2905.

20

Metabolism of Serotonin and Melatonin in Yeast

Albert Mas, Maria Jesus Torija, and Gemma Beltran
Universitat Rovira i Virgili
Tarragona, Spain

Jose Manuel Guillamon
Instituto de Agroquímica y Tecnología de los Alimentos (CSIC)
Valencia, Spain

Ana B. Cerezo, Ana M. Troncoso, and M. Carmen Garcia-Parrilla
Universidad de Sevilla
Seville, Spain

CONTENTS

ABSTRACT   The fermentation of grapes into wine results in the increase of several compounds that are seldom present in the primary substrate. Among them, melatonin has been described in wines and also in beers as a result of the metabolic activity of yeast. Serotonin is supposed to be a precursor of melatonin in yeast, although it has not been detected in the final products. In the present review, recent findings on the yeast metabolism of aromatic amino acids are presented and discussed.

KEY WORDS: *tryptophan, wine, beer, fermented beverages, antioxidants.*

Introduction

The biotransformation of grapes into wine is a process where microorganisms, primarily yeast, convert sugar into ethanol. This takes place in water solution and is accompanied by flavor and a pleasant aroma. Yeast uses the nutrients present in the medium for growth, producing a range of metabolites that contribute to the complexity of this fermented beverage.

The grape is very complex, with a variety of compounds ranging from mainstream (sugars) to very small but important quantities, with nutritional (vitamins, minerals, and polyphenols) and organoleptic (flavor and precursors) roles. Grapes contain high sugar levels, present in equimolar concentrations of glucose and fructose between 170 and 280 g/L. Additionally, there is a strong imbalance with the nitrogen fraction, with concentrations three orders of magnitude lower (between 70 and 600 mg/L). Grape must contain a variety of nitrogen compounds, among which the most important for the yeast are amino acids, ammonium ion, and small peptides. These nitrogen compounds, excluding proline, constitute what is called *yeast assimilable nitrogen*. Nitrogen affects yeast cells in two aspects: biomass production during fermentation and the stimulation of the use of sugars (Varela et al., 2004). Therefore, nitrogen regulates the rate and the end of fermentation. In fact, the lack of nitrogen has been indicated as one of the main reasons for stuck or sluggish fermentations (Bisson, 1999; Taillandier et al., 2007). The nitrogen content also affects other pathways in yeast, in particular, through the redox status of the cells, which affects the production of ethanol and other metabolites such as glycerol, acetic acid, and succinic acid (Albers et al., 1996; Camarasa et al., 2003; Radler, 1993). Finally, the main wine yeast, *Saccharomyces cerevisiae*, produces different concentrations of aromatic compounds according to the fermentation conditions, including the quality and quantity of the nitrogen sources (Gutierrez et al., 2013). A range of volatile compounds such as acetate and ethyl esters, higher alcohols, volatile fatty acids, and carbonyls, which are the main compounds contributing to secondary or fermenting wine flavor, are mainly synthesized as metabolites derived from the metabolism of nitrogen (Gutierrez et al., 2013; Henschke and Jiranek, 1993; Rapp and Versini, 1996; Swiegers et al., 2005). Thus, nitrogen availability modulates the organoleptic quality and the taste of wine (Bell and Henschke, 2005).

The presence of nitrogen in any of these chemical forms is highly variable, depending on various factors, including grape variety, degree of ripeness, soil, climate characteristics, and various technological aspects, viz., type of vinification, pressing, and so on (Ribereau-Gayon et al., 2006). The current context of global warming, which results in overripened grapes, has two very direct effects on the composition of the must: increased sugars to ferment and decreased nitrogen content. These low nitrogen concentrations increase the risk of stuck or sluggish fermentations and even lower organoleptic quality. The widespread response by winemakers is currently the addition of nitrogen to must. However, the lack of knowledge of the real nitrogen needs of each strain could result in excessive concentrations of this nutrient in the final wine, which would have negative consequences. The most relevant of these are the microbial instability of wines due to the nitrogen availability for the proliferation of other microorganisms or the synthesis of unhealthy substances, such as ethyl carbamate formed by yeast or biogenic amines, due to lactic acid bacteria using this residual nitrogen during malolactic fermentation. Therefore, there is a need to optimize the use of the nitrogen by wine yeast, leaving very limited amounts of amino acids left.

Metabolism of Melatonin and Serotonin in Yeast

Although the aromatic amino acids are in low concentrations, their impact is important; they are the precursors of higher alcohols through the Ehrlich pathway. However, there are other metabolites derived from these aromatic amino acids that are worth taking into account, as they could be bioactive molecules with interesting properties. They have been only very recently described, and their metabolic pathways, regulation, coding genes, and so on, are still under research. One of them is melatonin, which has been recently detected in wine, and its presence has been associated with the activity of the yeast involved in the fermentation process (Arevalo-Villena et al., 2010; Rodriguez-Naranjo et al., 2012). Originally, melatonin was seen as a unique product of the pineal gland of vertebrates and was called a *neurohormone*. However, in the last two decades, it has been identified in a wide range of invertebrates, plants, bacteria, and fungi. Therefore, today melatonin is considered a ubiquitous molecule in most living organisms (Tan et al., 2012). Although little information is available on melatonin biosynthesis in organisms other than vertebrates, it seems to share the synthetic route and enzymes described in vertebrates, in the case of yeast. This synthesis route is very simple, with four enzymes involved in the conversion of tryptophan to serotonin and N-acetylserotonin intermediates and finally to melatonin.

Except for the earlier work of Sprenger et al. (1999), who demonstrated the link between *S. cerevisiae* and melatonin production, recent studies report the detection of melatonin in wines (Gomez et al., 2012; Rodriguez-Naranjo et al., 2011a,b) and beer (Maldonado et al., 2009). These studies also describe its presence in grapes and other tissues of the vine, which could create doubt about whether its origin is in the substrate or if it is a fermentation by-product (Iriti et al., 2006; Murch et al., 2010; Rodriguez-Naranjo et al., 2012). However, all the references indicate that most melatonin is produced during fermentation, being absent in the initial grape must (Brzezinski, 1997; Rodriguez-Naranjo et al., 2012). The description of melatonin in wine has linked its formation with yeast metabolism (Arevalo-Villena et al., 2010; Rodriguez-Naranjo et al., 2012), although the number of references in this case is still rare, indicating the need to pursue this subject further. In addition, all previous studies have focused exclusively on melatonin production by *Saccharomyces* yeasts, without considering the presence and metabolic activities of non-*Saccharomyces* wine yeast, significantly present in grapes and at the beginning of alcoholic fermentation, and therefore its possible relation to the production of melatonin. So far, there is no evidence on the production of serotonin by *S. cerevisiae*, although serotonin was some years ago found to be synthesized by yeast in response to ultraviolet (UV) radiation (Strakhovskaia et al., 1983). In fact, the enzymatic step to produce serotonin is rate limiting in yeast (Park et al., 2008). Serotonin is found in wines at levels ranging from 2 to 23 mg/L, mainly as a result of malolactic fermentation, and significantly higher serotonin levels were observed when *Lactobacillus plantarum* was used (Landete et al., 2007; Manfroi et al., 2009).

Possible Roles of Melatonin and Serotonin in Yeast

Although the functions of melatonin are clear in mammals and animals (Boccalandro et al., 2011), mainly being related to the regulatory mechanisms involved in circadian rhythms (Eelderink-Chen et al., 2010), the role of melatonin in yeast and other microorganisms seems to be far from being understood. Indeed, although the presence of circadian rhythms in yeast has been determined (Merrow and Raven, 2010), this seems to be far from an independent daily rhythm regulated in response to the light as it has been described in multicellular microorganisms. Instead, the response in yeast is induced by temperature changes only after several generations in chemostats, and appears to be related to the primary nitrogen metabolism, particularly the expression of transporter genes of some nitrogen compounds. Thus, although melatonin is a ubiquitous molecule, its function in microorganisms is unknown. However, it has to be emphasized that in the organisms where it has been studied, it exerts potent regulatory functions.

Melatonin can present up to nine isomers (Diamantini et al., 1998), including melatonin itself, because of the different pattern substitutions of the groups (N-acetyl-(2-aminoethyl)) and methoxy in the indolic ring. An isomer of melatonin was detected in wine that showed the transition from 233 to 216 with shorter retention time compared to standard melatonin (Rodriguez-Naranjo et al., 2011b). Indeed, the MS fragmentation ions of melatonin were different from those of the isomers found in wine (Rodriguez-Naranjo et al., 2011a). Both melatonin and its isomer are present in different wine varieties, with Jaen Tinto showing the highest amount of melatonin isomer (21.9 ng/mL).

This finding was confirmed later by Gomez et al. (2012), who described the isomer in Malbec wine and its formation during the fermentation step. Recently, Kocadagli et al. (2014) detected the highest amount of melatonin isomer in yeast-fermented products. Up to now, there is just one commercially available standard of an isomer of melatonin, N-acetyl-3-(2-aminoethyl)-6-methoxyindole, which was not the compound detected in wine (Rodriguez-Naranjo et al., 2011a). However, it is not possible to determine the position of the methoxy group in the indolic ring by mass spectrometry. Further analysis by nuclear magnetic resonance (NMR) spectroscopy is required to elucidate its structure.

The synthesis of serotonin in yeast as a final product has not yet been described, so the possible effects are still not known. Only some effects in *S. cerevisiae* have been described, using the pharmacological presentation of sertraline, indicating that only a small proportion (15%) is internalized and that this provokes the start of autophagy (Chen et al., 2012).

Conclusion

The metabolism of the aromatic amino acids in yeasts can produce a broad array of molecules that could be relevant from different aspects related to both yeast regulation and human health. The activities of these compounds as neurohormones and antioxidants open a new scenario of applications from nutritional supplements to functional foods. However, the role of these compounds in yeast is still far from being completely understood. Thus, it is still beyond our capabilities to modulate their production and their appearance in fermented food. Further research in the field of yeast metabolism in relation to the presence of aromatic amino acids would provide the theoretical basis for a broad array of applications in modern nutrition.

Acknowledgments

This work was supported by the Ministry of Economy and Competitiveness, Spain (grant no. AGL2013-47300-C3).

REFERENCES

Albers, E., Larsson, C., Liden, G., Niklasson, C., and Gustafsson, L. 1996. Influence of the nitrogen source on *Saccharomyces cerevisiae* anaerobic growth and product formation. *Applied and Environmental Microbiology*, 62, 3187–3195.

Arevalo-Villena, M., Bartowsky, E. J., Capone, D., and Sefton, M. A. 2010. Production of indole by wine-associated microorganisms under oenological conditions. *Food Microbiology*, 27, 685–690.

Bell, S. J., and Henschke, P. A. 2005. Implications of nitrogen nutrition for grapes, fermentation and wine. *Australian Journal of Grape and Wine Research*, 11, 242–295.

Bisson, L. F. 1999. Stuck and sluggish fermentations. *American Journal of Enology and Viticulture*, 50, 107–119.

Boccalandro, H. E., González, C. V., Wunderlin, D. A., and Silva, M. F. 2011. Melatonin levels, determined by LC–ESI–MS/MS, fluctuate during the day/night cycle in *Vitis vinifera* cv Malbec: Evidence of its antioxidant role in fruits. *Journal of Pineal Research*, 51, 226–232.

Brzezinski, A. 1997. Melatonin in humans. *New England Journal of Medicine*, 336, 186–195.

Camarasa, C., Grivet, J. P., and Dequin, S. 2003. Investigation by ^{13}C-NMR and tricarboxylic acid (TCA) deletion mutant analysis of pathways of succinate formation in Saccharomyces cerevisiae during anaerobic fermentation. *Microbiology*, 149, 2669–2678.

Chen, J., Korostyshevsky, D., Lee, S., and Perlstein, E. O. 2012. Accumulation of an antidepressant in vesiculogenic membranes of yeast cells triggers autophagy. *PLoS One*, 7, e34024.

Diamantini, G., Tarzia, G., Spadoni, G., D'Alpaos, M., and Traldi, P. 1998. Metastable ion studies in the characterization of melatonin isomers. *Rapid Communications in Mass Spectrometry*, 153, 1538–1542.

Eelderink-Chen, Z., Mazzotta, G., Sturre, M., Bosman, J., Roenneberg, T., and Merrow, M. 2010. A circadian clock in *Saccharomyces cerevisiae*. *PNAS*, 107, 2043–2047.

Gomez, F. J. V., Raba, J., Cerutti, S., and Silva, M. F. 2012. Monitoring melatonin and its isomer in *Vitis vinifera* cv Malbec by UHPLC-MS/MS from grape to bottle. *Journal of Pineal Research*, 52, 349–355.

Gutierrez, A., Beltran, G., Warringer, J., and Guillamon, J. M. 2013. Genetic basis of variations in nitrogen source utilization in four wine commercial yeast strains. *PLoS One*, 8, E67166.

Henschke, P. A., and Jiranek, V. 1993. Yeasts-metabolism of nitrogen compounds. In G. H. Fleet (ed.), *Wine Microbiology and Biotechnology*, pp. 77–164. Singapore: Harwood Academic.

Iriti, M., Rossoni, M., and Franco, F. 2006. Melatonin content in grape: Myth or panacea? *Journal of the Science of Food and Agriculture*, 86, 1432–1438.

Kocadagli, T., Yilmaz, C., and Gokmen, V. 2014. Determination of melatonin and its isomer in foods by liquid chromatography tandem mass spectrometry. *Food Chemistry*, 153, 151–156.

Landete, J. M., Ferrer, S., and Pardo, I. 2007. Biogenic amine production by lactic acid bacteria, acetic bacteria and yeast isolated from wine, *Food Control*, 18, 1569–1574.

Maldonado, M. D., Moreno, H., and Calvo, J. R. 2009. Melatonin present in beer contributes to increase the levels of melatonin and antioxidant capacity of the human serum. *Clinical Nutrition*, 28, 188–191.

Manfroi, L., Silva, P. H. A., Rizzon, L. A., Sabaini, P. S., and Gloria, M. B. A. 2009. Influence of alcoholic and malolactic starter cultures on bioactive amines in Merlot wines. *Food Chemistry*, 116, 208–213.

Merrow M., and Raven, M. 2010. Finding time: A daily clock in yeast. *Cell Cycle*, 9, 1671–1672.

Murch, S. J., Hall, B. A., Le, C. H., and Saxena, P. K. 2010. Changes in the levels of indoleamine phytochemicals during *véraison* and ripening of wine grapes. *Journal of Pineal Research*, 49, 95–100.

Park, M., Kang, K., Park S., et al., 2008. Expression of serotonin derivative synthetic genes on a single self-processing polypeptide and the production of serotonin derivatives in microbes. *Applied Microbiology and Biotechnology*, 81, 43–49.

Radler, F. 1993. Yeasts: Metabolism of organic acids. In G. H. Fleet (ed.), *Wine Microbiology and Biotechnology*, pp. 165–182. Singapore: Harwood Academic.

Rapp, A. and Versini, G. 1996. Influence of nitrogen on compounds in grapes on aroma compounds in wines. *Journal International des Sciences de la Vigne et du Vin*, 51, 193–203.

Ribereau-Gayon, P., Dubourdieu, D., Doneche, B., and Lonvaud, A. 2006. *Handbook of Enology*, 2nd edn, vol. 1. New York: Wiley.

Rodriguez-Naranjo, M. I., Gil-Izquierdo, A., Troncoso, A. M., Cantos-Villar, E., and Garcia-Parrilla, M. C. 2011a. Melatonin is synthesised by yeast during alcoholic fermentation in wines. *Food Chemistry*, 126, 1608–1613.

Rodriguez-Naranjo, M. I., Gil-Izquierdo, A., Troncoso, A. M., Cantos-Villar, E., and Garcia-Parrilla, M. C. 2011b. Melatonin: A new bioactive compound present in wine. *Journal of Food Composition and Analysis*, 24, 603–608.

Rodriguez–Naranjo, M. I., Torija, M. J., Mas, A., Cantos-Villar, E., and Garcia-Parrilla, M. C. 2012. Production of melatonin by *Saccharomyces* strains under growth and fermentation conditions. *Journal of Pineal Research*, 53, 219–224.

Sprenger, J., Hardeland, R., Fuhrberg, B., and Han, S. Z. 1999. Melatonin and other 5-methoxylated indoles in yeast: Presence in high concentrations and dependence on tryptophan availability. *Cytologia*, 64, 209–213.

Strakhovskaia, M. G., Serdalina, A. M., and Fraikin, G. I. 1983. Effect of the photo-induced synthesis of serotonin on the photoreactivation of *Saccharomyces cerevisiae* yeasts. *Nauchnye doklady vysshei shkoly. Biologicheskie Nauki*, 3, 25–28.

Swiegers, J. H., Bartowsky, E. J., Henschke, P. A., and Pretorius, I. S. 2005. Yeast and bacterial modulation of wine aroma and flavour. *Australian Journal of Grape and Wine Research*, 11, 139–173.

Taillandier, P., Portugal, F. R., Fuster, A., and Strehaiano, P. 2007. Effect of ammonium concentration on alcoholic fermentation kinetics by wine yeasts for high sugar content. *Food Microbiology*, 24, 95–100.

Tan, D. X., Hardeland, R., Manchester, L. C., et al., 2012. Functional roles of melatonin in plants, and perspectives in nutritional and agricultural science. *Journal of Experimental Botany*, 63, 577–597.

Varela, C., Pizarro, F., and Agosin, E. 2004. Biomass content governs fermentation rate in nitrogen-deficient wine musts. *Applied and Environmental Microbiology*, 70, 3392–3400.

21

Role of Dietary Serotonin and Melatonin in Human Nutrition

Paramita Bhattacharjee and Probir Kumar Ghosh
Jadavpur University
Kolkata, India

Manvi Vernekar and Rekha S. Singhal
Institute of Chemical Technology
Mumbai, India

CONTENTS

ABSTRACT   This chapter presents nutritional sources of tryptophan, serotonin, and melatonin in plants and animals, and the biological and therapeutic roles of these biomolecules in human health, with special emphasis on the role of melatonin as an antioxidant. Among processed foods, cherry-based beverages, coffee, and chocolate are the chief sources of serotonin, while fermented foods and the Mediterranean diet are good sources of melatonin. The uses of tryptophan and melatonin as food supplements are also discussed, as well as their bioavailability, stability, and safety aspects.

KEY WORDS: *tryptophan, serotonin, melatonin, sources, processed foods, stability, food supplements, bioavailability, biological role, safety.*

Abbreviations

ACTH: adrenocorticotropin
ATP: adenosine triphosphate
BA: biogenic amine
CNS: central nervous system
CRH: corticotropin-releasing hormone
CRPP: carbohydrate-rich protein-poor (diet)
GI: gastrointestinal
GSH: glutathione
5-HT: 5-hydroxytryptamine
5-HTP: 5-hydroxytryptophan
LNAA: large neutral amino acids
MAO: monoamine oxidase
MI: melatonin isomer
NK: natural killer (cells)
NO: nitric oxide
QPM: quality protein maize
RIA: radioimmunoassay
RDI: recommended dietary intake
SAD: seasonal affective disorder
SLM: solid-liquid microparticles
Trolox: 6-hydroxy-2,5,7,8-tetramethylchroman-2-carboxylic acid

Introduction

Serotonin (5-hydroxytryptamine) is an indoleamine monoamine neurotransmitter and is mainly found in the enterochromaffin cells of the gastrointestinal tract (GI tract), in the serotonergic neurons of the central nervous system (CNS), and in blood platelets of animals. In humans, approximately 90% of serotonin is located in the enterochromaffin cells in the GI tract, where it regulates intestinal movements (Berger et al., 2009).

Melatonin (5-methoxy-N-acetyltryptamine), the "hormone of darkness," is an indole hormone produced in the pineal gland of the brain and controls several functions in our daily lives (Utiger, 1992). Secretion of melatonin commences three months after birth, before which it is primarily supplied through the mother's milk or cow's milk. It is an excellent antioxidant, is beneficial for the immunological system, enhances resistance to infection and diseases, has inhibitory activity on some cancer forms, and induces beneficial effects on neuronal disorders. Several characteristics of melatonin, principally its direct, non-receptor-mediated free-radical scavenging activity, distinguish it from classic hormones (Stege et al., 2010; Rodriguez-Naranjo et al., 2011; Karunanithi et al., 2014). Melatonin is a ubiquitous molecule that plays a decisive role in the development of the brain and body, the regulation of circadian and seasonal rhythms, the sleep–wake process, reproduction, and retinal function. In 2011, the European

Food Safety Authority (EFSA) accepted the health claims related to melatonin and its alleviation of subjective feelings of jet lag.

In the mammalian brain, tryptophan, an essential amino acid, is the precursor of both serotonin and melatonin. In the pineal gland, L-tryptophan is first converted to 5-hydroxytryptophan (5-HTP), the immediate nutrient precursor of the neurotransmitter serotonin 5-hydroxytryptamine (5-HT). In the pineal gland, serotonin is methylated and acetylated to melatonin (Utiger, 1992; Rodriguez et al., 1994). The food sources of serotonin and melatonin are elaborated in the following sections.

Nutritional Sources of Tryptophan, Serotonin, and Melatonin

Plant-Based Sources

Pumpkin and squash seeds are some of the richest sources of tryptophan (576 mg/100 g). Whole oats, wheat bran, wheat germ, and brown rice are also good sources of tryptophan. Butternut is one of the richest sources of serotonin (398 μg/g tissue). Appreciable amounts of serotonin are also reported in plantain, apricots, cherries, peaches, and Chinese plums (García-Moreno et al., 1983; Garrido et al., 2010; Gónzalez-Flores et al., 2011; Huang and Mazza, 2011; Ramakrishna et al., 2011). Seeds of the plant *Griffonia simplicifolia* cultivated on the west coast of Africa are also reportedly a good source of serotonin (Pathak et al., 2010).

Melatonin was initially thought to be exclusively produced in animals. It was found outside the animal kingdom for the first time in the dinoflagellate algae *Lingulodinium polyedrum*. Since then, melatonin has come to be regarded as a ubiquitous molecule, although its presence is not clearly established in some important taxa, particularly in archaea, mosses, ferns, gymnosperms, sponges, annelids, chelicerates, and echinoderms (Hardeland and Pandi-Perumal, 2005). Melatonin was first identified in edible plants by Hattori et al. (1995). Since then, melatonin has been found in many plants and in different parts of plants, such as the roots of *Lupinus albus* (Hernandez-Ruiz et al., 2004), sunflower, mustard, and walnut seeds (Manchester et al., 2000), tomato fruit (Van-Tassel and O'Neill, 2001), coffee (Ramakrishna et al., 2012a,b), grape skin (Iriti et al., 2008), and the rind of tart cherries (Burkhardt et al., 2001) at concentrations usually ranging from picograms to nanograms per gram of tissue (Van-Tassel and O'Neill, 2001; Ramakrishna, 2015). Walnut is the first common tree nut in which melatonin has been studied from a nutritional perspective (Reiter et al., 2005). Mushrooms are one of the major sources of melatonin in addition to tryptophan and serotonin (Muszyńska and Sułkowska-Ziaja, 2015). The levels of tryptophan, serotonin, and melatonin present in various plants and plant-based foods are presented in Table 21.1.

Animal-Based Sources

Lamb, beef, and pork are a few of the major sources of tryptophan (415 mg/100 g). Other tryptophan-rich sources include chicken, tuna, shellfish, and crab (USDA, 2015) (Table 21.2). Salmon is one of the major sources of melatonin (3.7 ng/g). Other sources such as chicken, lamb, halibut, snapper, mackerel, lobster, octopus, clams, prawns, and oysters contain appreciable quantities of serotonin (Iriti and Varoni, 2015; USDA, 2015). Melatonin is also present in fungi, especially in yeasts (Hardeland and Pandi-Perumal, 2005). These natural sources of serotonin and melatonin have several biological and therapeutic roles in humans, which are elaborated in the subsequent sections.

Biological and Therapeutic Functions of Serotonin and Melatonin in Human Health

The various biological and therapeutic roles of serotonin and melatonin in human health are as follows.

Maintenance Role of Serotonin in Human Health

Serotonin regulates practically every type of behavior, such as emotional, motor, cognitive, and autonomic, and appears to be involved in a wide variety of physiological functions and behaviors, such as

TABLE 21.1

L-Tryptophan, Serotonin, and Melatonin in Various Plant-Based Sources and Foods

Sl. No.	Plant/Food Source	L-Tryptophan Content	Method of Detection
1	Mushroom (*Basidiomycota* sp.)	0.01 ± 0.00 − 4.47 ± 0.01 mg/100 g dry weight (DW)	HPLC
2	Oat bran	335 mg/100 g	-do-
3	Pumpkin and squash seeds	576 mg/100 g	-do-

Sl. No.	Plant/Food Source	Serotonin Content	Method of Detection
4	Mushroom (*Basidiomycota* sp.)	0.52 ± 0.01 − 29.61 ± 0.17 mg/100 g DW	HPLC
5	Hazelnut	3.4 ± 0.1 mg/kg tissue	-do-
6	Banana	11.5 ± 0.4 mg/kg tissue	-do-
7	Sweet cherry	8.5 − 37.6 ng/100 g fresh weight (FW)	-do-
8	Pomegranate and strawberry	8–12 µg/g	-do-
9	Plantain	30.0 ± 7.5 µg/g tissue	Radioenzymatic analysis and thin-layer chromatography (TLC)
10	Pineapple	17.0 ± 5.1 µg/g tissue	-do-
11	Banana	15.0 ± 2.4 µg/g tissue	-do-
12	Plums	4.7 ± 0.80 µg/g tissue	-do-
13	Tomatoes	30.2 ± 0.6 µg/g tissue	-do-
14	Avocados (Haas variety) (California)	1.6 ± 0.4 µg/g tissue	-do-
15	Avocados (Fuerte variety) (California)	1.5 ± 0.21 µg/g tissue	-do-
16	Avocados (Booth variety) (California)	0.2 ± 0.04 µg/g tissue	-do-
17	Dates	1.3 µg/g tissue	-do-
18	Grapefruit	0.9 µg/g tissue	-do-
19	Cantaloupe	0.9 µg/g tissue	-do-
20	Honeydew melon	0.6 µg/g tissue	-do-
21	Olives (black)	0.2 µg/g tissue	-do-
22	Broccoli	0.2 µg/g tissue	-do-
23	Figs	0.2 µg/g tissue	-do-
24	Eggplant	0.2 µg/g tissue	-do-
25	Spinach	0.1 µg/g tissue	-do-
26	Cauliflower	0.1 µg/g tissue	-do-
27	Kiwifruit	3 µg/g tissue	-do-
28	Butternut	398 ± 90 µg/g tissue	-do-
29	Black walnut	304 ± 46 µg/g tissue	-do-
30	English walnut	87 ± 20 µg/g tissue	-do-
31	Shagbark	143 ± 23 µg/g tissue	-do-
32	Mocernut	67 ± 13 µg/g tissue	-do-
33	Sweet pignut	25 ± 8 µg/g tissue	-do-

	Plant/Food Source	Melatonin Content	Method of Detection
34	Ground coffee beans	6.15 ± 0.10 µg/g	LC–UV
35	Cocoa	5.2 ± 0.1 µg/g	-do-
36	Cow's milk	14.45 pg/mL	LC–MS/MS
37	Toned milk	18.41 pg/mL	-do-
38	Human milk	15.92 pg/mL	-do-
39	Mushroom (*Basidiomycota* sp.)	0.14 ± 0.01–1.29 ± 0.07 mg/100 g DW	HPLC
40	Sweet cherry	0.6 ± 0.0–22.4 ± 1.3 ng/100 g FW	-do-
41	Corn	187.8 ng/100 g	-do-
42	Rice	149.8 ng/100 g	-do-
43	Barley	87.3 ng/100 g	-do-

TABLE 21.1 (CONTINUED)

L-Tryptophan, Serotonin, and Melatonin in Various Plant-Based Sources and Foods

Sl. No.	Plant/Food Source	L-tryptophan Content	Method of Detection
44	Ginger	142.3 ng/100 g	-do-
45	Mushroom (*Basidiomycetes* sp.)	0.71 ± 0.02–4.40 ± 0.05 mg/100 g DW	-do-
46	White mustard	189 ng/g dry seed	-do-
47	Black mustard	129 ng/g dry seed	-do-
48	Sunflower seeds	29 ng/g dry seed	-do-
49	Poppy seeds	6 ng/g dry seed	-do-
50	Tall fescue	5288.1 ± 368.3 pg/g tissue	RIA
51	Red radish	0.6 ng/g	-do-
52	Japanese radish	0.6 ng/g	-do-
53	Ginger	583.7 ± 50.3 pg/g tissue	-do-
54	Sweet corn	1366.1 ± 465.1 pg/g tissue	-do-
55	Barley	378.1 ± 25.8 pg/g tissue	-do-
56	Rice	1006.0 ± 58.5 pg/g tissue	-do-
57	Oat	1796.1 ± 43.3 pg/g tissue	-do-
58	Onion	31.5 ± 4.8 pg/g tissue	-do-
59	Welsh onion	85.7 ± 8.0 pg/g tissue	-do-
60	Asparagus	9.5 ± 3.2 pg/g tissue	-do-
61	Pineapple	36.2 ± 8 pg/g tissue	-do-
62	Tomato	0.5 ng/g	RIA and HPLC–MS
63	St John's wort (leaf)	1750 ng/g DW	Not reported

Source: 1, 4, 40, 45–46: Muszyńska and Sułkowska-Ziaja, *Food Chemistry* 132, 455–459, 2012; 2–3: USDA, Composition of Foods Raw, Processed, Prepared, USDA National Nutrient Database for Standard Reference, Release 27, US Department of Agriculture, Agricultural Research Service, Beltsville Human Nutrition Research Center, Beltsville, Nutrient Data Laboratory, Maryland, 2015; 5–6: Lavizzari et al., *Journal of Chromatography A* 1129, 67–72, 2006; 7, 39: González-Gómez et al., *European Food Research and Technology* 229, 223–229, 2009; 8, 9–33: Feldman and Lee, *American Journal of Clinical Nutrition* 42(4), 639–643, 1985; 34–35: Restuccia et al., *Food Additives and Contaminants: Part A* 32(7), 1156–1163, 2015b; 36–38: Karunanithi et al., *Journal of Food Science and Technology* 51(4), 805–812, 2014; 41–44: Badria, *Journal of Medicinal Food* 5, 153–157, 2002; 47–49: Manchester et al., *Life Sciences* 67(25), 3023–3029, 2000; 50–61: Hattori et al., *International Journal of Biochemistry and Molecular Biology* 35, 627–634, 1995; 62: Dubbels et al., *Journal of Pineal Research* 18, 28–31, 1995; 63: Murch et al., *The Lancet* 350, 1597–1599, 1997.

eating, circadian rhythmicity, and neuroendocrine functions (Frazer and Hensler, 1999). Serotonin also participates in the hypothalamic control of pituitary secretion, particularly in the regulation of adrenocorticotropin (ACTH), prolactin, and growth hormone secretion, and a direct synaptic connection is observed between serotonergic terminals and corticotropin-releasing hormone (CRH), containing neurons in the paraventricular nucleus of the hypothalamus (Frazer and Hensler 1999). Serotonin also has effects on numerous diseases, including depression, anxiety, social phobia, schizophrenia, and obsessive compulsive and panic disorders, in addition to migraine, hypertension, pulmonary hypertension, eating disorders, vomiting, and irritable bowel syndrome (Pauwels, 2003).

Maintenance Role of Melatonin in Human Health

Melatonin regulates the circannual cycle required to regulate growth patterns, mating behavior, and certain movements such as migration, metabolism, and other physiological functions (Morera and Abreu, 2006). The rhythmic production of melatonin is controlled by light and it is regarded as an endogenous circadian cycle synchronizer. It influences sleep in mammals by thermoregulatory function, since the nocturnal fall in body temperature coincides with the peak production of endogenous melatonin (Claustrat et al., 2005). Melatonin also lowers the risks of myocardial infarction and stroke, since such attacks are more common in the early morning, when the levels of melatonin fall significantly (Macchi

TABLE 21.2

Reported Levels of Melatonin in Animal Sources

Sl. No.	Food Source	Level of Tryptophan	Method of Analysis
1	Lamb, beef, and pork	415 mg/100 g	HPLC
2	Chicken	404 mg/100 g	-do-
3	Tuna	335 mg/100 g	-do-
4	Shellfish and crab	330 mg/100 g	-do-
		Level of Melatonin	
5	Salmon	3.7 ng/g	LC–MS
6	Chicken	2.3 ng/g	-do-
7	Lamb	1.6 ng/g	-do-

Source: 1–4: USDA, Composition of Foods Raw, Processed, Prepared, USDA National Nutrient Database for Standard Reference, Release 27, US Department of Agriculture, Agricultural Research Service, Beltsville Human Nutrition Research Center, Beltsville, Nutrient Data Laboratory, Maryland, 2015; 5–7: Iriti and Varoni, *Journal of Applied Biomedicine* 12(4), 193–202, 2014.

and Bruce, 2004; Claustrat et al., 2005). Melatonin controls the photoperiodic information that modulates reproductive activity and also controls puberty (Macchi and Bruce, 2004). It is also known to contribute to the release, regulation, and proliferation of T, beta, and NK cells, monocytes, thymocytes, cytokines, met-enkephalin and other immune lipids, anti-apoptosis, and anti-tumor activities (Claustrat et al., 2005). It also alters the activity of pancreatic cells, which is essential for preventing diabetes (Karthikeyan et al., 2014).

Preventive/Therapeutic Role of Melatonin in Human Health

Melatonin has preventive roles in immunity-enhancing activity (Haldar and Ahmad, 2010), oncostatic and antitumor activity (Claustrat et al., 2005), and anticancer activity (Anisimov et al., 2010). It controls neuronal disorders such as senile dementia, Alzheimer's, epilepsy, and fibromyalgia (Macchi and Bruce, 2004; Claustrat et al., 2005), aids in the prevention of type-2 diabetes (Karthikeyan et al., 2014), reduces hypertension (Houston, 2013), controls body weight (Gündüz, 2014), slows ageing (Anisimov et al., 2006), and also controls irritable bowel syndrome (Chang, 2014). Melatonin binds with Ca^{2+}-calmodulin in cells and regulates many calcium-dependent cellular functions (Kolar and Machackova, 2005). Winter-type seasonal affective disorder (SAD) is characterized by recurrent depression during the short photoperiod, and its remission during summer is directly associated with photoperiodic variation. It has been hypothesized that the change in phase of melatonin secretion plays a major role in the pathogenesis of SAD; melatonin is therefore widely recommended for therapy (Claustrat et al., 2005).

Melatonin as an Antioxidant

Melatonin is a free radical scavenger and has been reported to significantly lower oxidative stress. It is a stimulator of antioxidative enzymes, enhances the efficiency of mitochondrial oxidative phosphorylation, and reduces electron leakage, and thereby augments the efficiency of other antioxidants (Reiter et al., 2003).

Melatonin as a Direct Radical Scavenger

Melatonin is an endogenous free radical scavenger, as it detoxifies a variety of free radicals and reactive oxygen intermediates such as hydroxyl radical, hydrogen peroxide, peroxynitrite anion, singlet oxygen, and nitric oxide. It is five times more efficient than glutathione (GSH) in neutralizing hydroxyl radical and fifteen times more effective than the exogenous scavenger mannitol, and twice as efficient in scavenging lipid peroxide radicals as vitamin E. Therefore, melatonin could have a potential role as a food antioxidant (Poeggeler et al., 2002). Melatonin directly detoxifies H_2O_2 in living organisms. One important function of melatonin may be complementary in function to catalase and glutathione peroxidase in keeping intracellular H_2O_2 concentrations at steady-state levels (Tan et al., 2000).

Role of Melatonin in Activation of Antioxidant Enzymes

Melatonin acts as a signaling molecule at the cellular level and upregulates a number of antioxidant enzymes. This has been demonstrated for GSH peroxidase, superoxide dismutase, catalase, and GSH reductase. In some tissues, Cu/Zn- and/or Mn-superoxide dismutases and, rarely, catalase are also upregulated. Melatonin's free radical scavenging ability is well known and is widely documented, while the antioxidant enzyme activation by the indoleamines and the mechanisms underlying melatonin-regulated processes have not yet been clearly elucidated. The activation of enzymes by melatonin could be a consequence of antioxidant enzyme mRNA synthesis. Although there are studies that have established melatonin as an inducer for gene expression, no clear evidence exists to date (Tomas-Zapico and Coto-Montes, 2005).

Melatonin in Synergism with Other Antioxidants

An important aspect of the antioxidant action of melatonin is its interaction with classic antioxidants. This is of profound significance from a nutritional point of view, since melatonin spares the antioxidant vitamins for their biological functions. Melatonin is reported to potentiate the effects of ascorbate, Trolox, reduced GSH, and NADH. These findings suggest that melatonin exhibits synergistic rather than additive effects with the same. Its interaction with other antioxidants is by the redox-based regeneration of antioxidants transiently consumed. This may, in fact, be of practical importance, since melatonin is also shown to prevent the decrease in hepatic ascorbate and alpha-tocopherol levels *in vivo* under conditions of long-lasting experimental oxidative stress induced by a high-cholesterol diet (Hardeland and Pandi-Perumal, 2005).

Tryptophan, Serotonin, and Melatonin in Processed Foods

Tryptophan-Enriched Cereals as Sources of Serotonin and Melatonin

Since humans cannot synthesize tryptophan, it can only be obtained from the diet (Richard et al., 2009). Studies have indicated that carbohydrate-rich meals promote serotonin production. As blood glucose levels rise, insulin is released. This then enables muscle tissues to take up most amino acids except for tryptophan, which is bound to albumin in the blood. As a result, the ratio of tryptophan relative to other amino acids in the blood increases, which enables tryptophan to bind to transporters, enter the brain in large amounts, and stimulate serotonin synthesis (Wurtman and Wurtman, 1995).

The consumption of cereals with higher contents of tryptophan increases sleep efficiency, actual sleep time, immobile time, and decreases total nocturnal activity, the sleep fragmentation index, and sleep latency. Thus cereals enriched with tryptophan may be useful for age-related alterations in the sleep–wake cycle (Bravo et al., 2013).

A carbohydrate-rich protein-poor (CRPP) diet reportedly increases the ratio of plasma tryptophan to the sum of the other *large neutral amino acids* (the TRP/LNAA ratio), giving tryptophan an advantage in the competition for access to the brain (Markus et al., 1998). Recently, alpha-lactalbumin was shown to increase the plasma TRP/LNAA ratio by 46%–48% and to have mood-improving effects in highly stress-vulnerable subjects under acute stress (Markus et al., 2002). Certain serotonin-friendly whole foods also inhibit adipogenesis or fat cell production. Wheat products, due to their high carbohydrate content, may increase serotonin levels (Wurtman and Wurtman, 1995).

Cereals enriched with high levels of tryptophan (60 mg tryptophan in 30 g cereals per serving) compared to standard cereal (22.5 mg tryptophan in 30 g cereals per serving) can be used as a chrono-nutritional aid to correct age-related sleep–wake cycle alterations. Tryptophan absorbed at breakfast is very important for keeping children to a morning-type diurnal rhythm and maintaining a high quality of sleep, presumably through the metabolism of tryptophan to serotonin in the daytime and further to melatonin at night. Serotonin and melatonin dietary interventions have been reported to enhance sleep in elite athletes (Hardeland and Pandi-Perumal, 2005).

Cherry Fruit–Based Nutraceutical Products as Sources of Serotonin and Melatonin

Melatonin is known to exhibit immunomodulatory actions that are mainly mediated through the modulation of cytokine production via binding to specific receptors expressed by different immune cells. A nutraceutical beverage formulated using Jerte Valley sweet cherries, which contain substantial amounts of melatonin, exhibited anti-inflammatory properties. This also reaffirms that melatonin acts in synergism with other phytonutrients—mainly antioxidants such as phenolic acids, anthocyanins, and carotenoids (Delgado et al., 2012a). Consumption of cherry nutraceutical products helps to counteract decreases in melatonin and serotonin in circulating blood and therefore combats increases in oxidative stress accompanied with aging (Delgado et al., 2012b).

Coffee as a Source of Serotonin and Melatonin

Serotonin (5-HT) is one of the important biogenic amines (BAs) present in coffee. Because of the relevance of indoleamines as bioactive molecules with implications for food, nutritional sciences, and human health, it is of interest to explore their levels in coffee, an important universal beverage (Ramakrishna et al., 2012a,b). BA content in roasted ground coffee (13.30–88.85 µg/g) is higher than in brewed coffee (< 9.88 µg/g). Also, the BAs in ground coffee decrease with increasing degrees of roasting. The level of BAs in coffee largely depends on the brew preparation technique. Beverages prepared by espresso, capsule, and pod machines have the lowest BA content, owing to the thermal and physical stress on ground coffee in these methods. Mocha has the highest BA content owing to lower pressure and longer brewing time, which allows longer coffee–water contact, allowing better extraction of these compounds from ground coffee (Restuccia et al., 2015a). Both melatonin and 5-HT were detected in green coffee beans (5.8 ± 0.8 and 10.5 ± 0.6 µg/g dry weight [DW], respectively), in roasted beans of *Coffea canephora* (8.0 ± 0.9 and 7.3 ± 0.5 µg/g [DW], respectively) and also in brewed coffee (3.9 ± 0.2 µg/50 mL and 7.3 ± 0.6 µg/50 mL, respectively). In *C. arabica*, levels of 5-HT were higher in green beans (12.5 ± 0.8 µg/g DW) compared to that in roasted beans (8.7 ± 0.4 µg/g DW). On the contrary, in roasted beans, the levels of melatonin were higher (9.6 ± 0.8 µg/g DW) compared to the same in green beans (6.8 ± 0.4 µg/g DW) (Ramakrishna et al., 2012).

Cocoa and Chocolate as Sources of Serotonin

Cocoa contains many BAs, such as serotonin, that promote health. BAs in cocoa range from 5.7–79.9 µg/g depending on the type of cocoa, and processed cocoa products contain more BAs (Restuccia et al., 2015b). Cocoa and chocolate are usually associated with caffeine and other BAs. However, recent studies have shown them to be rich sources of tryptophan and serotonin. Serotonin is also one of the main BAs present in chocolate (2.8 ± 0.2–5.2 ± 0.1 µg/g; Restuccia et al., 2015b). Chocolate with cocoa content between 70% and 85% has 13.27–13.34 µg/g of tryptophan, while 2.93 µg/g of tryptophan is present in chocolate with more than 85% cocoa content (Guillén-Casla et al., 2012).

Miscellaneous Processed Foods as Sources of Serotonin

Roasted soybeans contain around 575 mg/100 g of tryptophan. Alpha-lactalbumin, a minor constituent of milk, improves mood and cognition, presumably owing to increased serotonin (Markus et al., 2002).

Cheeses such as Mozzarella, Parmesan, and Cheddar also contain about 571 mg/100 g of tryptophan. White beans when cooked contain about 115 mg/100 g of tryptophan (USDA, 2015). Six edible mushroom species of Basidiomycota reportedly contain L-tryptophan up to 8.92 mg/100 g DW, post cooking.

Tryptophan as a Source of Serotonin in Indian Traditional Foods

Quality protein maize (QPM) with the opaque-2 gene has twice as much tryptophan and lysine than normal maize. In India, QPM is used to make traditional products to meet the nutritional needs of the vulnerable sections of society (Chaudhary et al., 2014). The GI tract deserves particular attention, not only with respect to serotonin/melatonin uptake, but even more as an extrapineal site of melatonin biosynthesis. In the GI tract, melatonin is present in amounts exceeding those found in the pineal gland several hundred times over, and from where it can be released into the circulation in a postprandial response, especially under the influence of high tryptophan levels.

Germinated Legumes as Source of Melatonin

Germination in legumes leads to an increase in melatonin content and significant antioxidant activity. The highest melatonin content has been reported after six days of germination under 24-hour darkness for legumes such as in lentils and kidney beans. These germinated legume seeds with improved levels of melatonin might play a protective role against free radicals. Thus, these sprouts can be used for combating stress conditions (Aguilera et al., 2014).

Fermented Foods as Sources of Melatonin

In cultures of yeast freshly prepared from commercially available cubes of Baker's yeast, micromolar concentrations of melatonin have been reported, sometimes exceeding 40 µM (Hardeland and Pandi-Perumal, 2005). Therefore, baked food products could also be a source of melatonin.

It has also been reported that melatonin present in beer contributes to the total antioxidative activity of human serum, and moderate beer consumption can be protective toward overall oxidative stress (Dermarderosian et al., 2008; Stege et al., 2010).

The emergence of naturally occurring melatonin isomers (MIs) in fermented foods has opened an exciting new research area since 2012. MIs could possibly increase the performance of yeasts and probiotic bacteria during the processes of fermentation. Therefore, yeasts producing elevated levels of these isomers might have a superior alcohol tolerance and be able to produce higher levels of alcohol. This application could be applied to beer and wine brewing using yeast strains selected for their high isomer levels.

MIs could also find applications in the efficient production of industrial ethanol or for the production of alcohol as a biofuel (Tan et al., 2012).

Mediterranean Diet as a Rich Source of Melatonin

Extra-virgin olive oil has almost double the melatonin content of the most common refined oils. Thus, melatonin may account for the health effects of the Mediterranean diet, in which olive oil is the main source of fat. Besides grapevine and olive products, high levels of this indoleamine have been detected in other typical Mediterranean foods, such as salmon, chicken, lamb, bread, and yogurt (Tables 21.2 and 21.3). It is notable that leaf of purslane (*Portulaca oleracea*), which is the most important source of terrestrial α-linolenic acid in edible wild plants and is a part of the Mediterranean diet, is also rich in melatonin (Puerta et al., 2007).

Epidemiological studies provide evidence to support the concept that a Mediterranean diet has a beneficial influence on diseases associated with oxidative damage, such as coronary heart disease, cancer, and cardiovascular and neurodegenerative diseases (Iriti and Varoni, 2014). It is possible that

TABLE 21.3

Reported Levels of Tryptophan, Serotonin, and Melatonin in Various Processed Foods

Sl. No.	Food Source	Level of Tryptophan	Method of Analysis
1	Chocolate (85% cocoa)	2.93 ± 0.01 µg/g	Capillary liquid chromatography–mass spectrometry (cLC–MS)
2	Chocolate (70% cocoa)	13.27 ± 0.02 µg/g	cLC–MS
3	Mushroom (*Basidiomycota* sp.) (cooked)	1.74 ± 0.04– 9.92 ± 0.20 mg/100 g DW	HPLC

Sl. No.	Food Source	Level of Serotonin	Method of Analysis
4	Coffee brew	0.90 ± 0.01 µg/mL	LC–UV
5	Chocolate (85% cocoa)	2.93 ± 0.01 µg/g	cLC–MS

Sl. No.	Food Source	Level of Melatonin	Method of Analysis
6	Olive oil (virgin)	0.07–0.12 ng/mL	ELISA
7	Olive oil (refined)	0.05–0.08 ng/mL	-do-
8	Sunflower seed oil (refined)	0.05 ng/mL	-do-
9	Wine	0.05–0.08 ng/mL	Capillary electrochromatography
10	Bread crumb	341 pg/g	HPLC
11	Bread crust	138 pg/g	-do-
12	Yogurt	126 pg/g	-do-
13	Roasted soybeans	575 mg/100g	-do-
14	Mozzarella, Parmesan, and Cheddar cheese	571 mg/100g	-do-
15	Mushroom (*Basidiomycota* sp.) (cooked)	0.71 ± 0.02– 4.40 ± 0.05 mg/100 g DW	-do-
16	Chicken cookies and chicken biscuits	1.38 g/100 g	Not reported

Source: 1, 2, 5: Guillén-Casla et al., *Journal of Chromatography A* 1232, 158–165, 2012; 3, 15: Muszyńska and Sułkowska-Ziaja, *Food Chemistry* 132, 455–459, 2012; 4: Restuccia et al., *Food Chemistry* 175, 143–150, 2015a; 6–8: Puerta et al., *Food Chemistry* 104, 609–612, 2007; 9: Stege et al., *Electrophoresis* 31, 2242–2248, 2010; 10–12: Iriti and Varoni, *Journal of Applied Biomedicine* 12(4), 193–202, 2014; 13–14 USDA, Composition of Foods Raw, Processed, Prepared, USDA National Nutrient Database for Standard Reference, Release 27, US Department of Agriculture, Agricultural Research Service, Beltsville Human Nutrition Research Center, Beltsville, Nutrient Data Laboratory, Maryland, 2015; 16: Berwal and Khanna, *Pacific Business Review International* 5(12), 61–70, 2013.

melatonin and other bioactive compounds (carotenoids and glucosinolates) account for the positive effects of Mediterranean dietary habits. This is an additional argument for the adoption of healthy and diversified eating habits rather than the use of dietary supplements (Puerta et al., 2007; Iriti and Varoni, 2014, 2015). Table 21.3 presents the levels of tryptophan, serotonin, and melatonin in various processed foods.

Serotonin and Melatonin as Microparticles or Liposomes

Solid-liquid microparticles (SLMs) as oral pulsatile systems have proved successful in the delivery of melatonin to pediatric populations. Melatonin has been found to be compatible and stable in milk and yogurt, suggesting that SLMs sprinkled into food could be an acceptable means of melatonin administration to children who are unable to swallow capsules or tablets (Albertini et al., 2014). Although similar research on serotonin as liposomes is scarce, the radioprotective effects of liposomes containing serotonin has been reported (Deev et al., 1976).

Stability of Tryptophan and Melatonin in Food Products

The stability of tryptophan in foods is affected by exposure to heat and oxygen during food-processing operations. Thermal degradation of tryptophan in legume proteins during home boiling or pressure cooking is about 5%, while roasting and baking causes a 15% loss of the same. About 22% tryptophan is lost during deep fat frying and the baking of wheat products. Loss of tryptophan also occurs in infant formulations when exposed to 40°C–100°C. Autoclaving rapeseed meal at 121°C for six hours degrades tryptophan by 50% (Friedman and Cuq, 1988).

On the other hand, the amount of melatonin that survives in a food product and is utilized by the body is influenced by the reaction of melatonin with other oxidants and its synergism with other antioxidants (Hardeland and Pandi-Perumal, 2005).

Bioavailability of Serotonin and Melatonin

The bioavailability issues of serotonin and melatonin are intriguing. While serotonin requires elicitor molecules such as ω-3 essential fatty acids, zinc, magnesium, vitamin B, and other enzyme cofactors for its metabolism, melatonin is readily absorbed in the GI tract.

Deficiency of serotonin or its improper metabolism in the body leads to neuronal diseases such as attention deficit hyperactivity disorder (ADHD). This is a matter of serious concern, since in the modern diet of processed foods containing chemical additives, colorings, and preservatives, the serotonin metabolism elicitors are highly compromised (Duff, 2014). Therefore, ω-3 essential fatty acids, zinc, magnesium, and vitamin B have been suggested as food supplements for aiding the bioavailability of serotonin and protecting the body from its deficiency and the consequent adverse neurological implications.

Melatonin, being an amphiphilic molecule, is readily absorbed by the body when taken orally in pure form, and influences the blood plasma concentration. It requires 45 minutes for its absorption into the GI tract and is released to the blood circulation in remarkable amounts with chemical stimuli, particularly in response to tryptophan and other indoleamines. Post absorption, it is a short-lived molecule in the blood with an average life of about 20–40 minutes.

The oral consumption of melatonin-rich foods is reported to increase the serum melatonin concentration. An experiment conducted on rats to monitor blood melatonin concentrations after the consumption of walnuts showed serum melatonin concentrations to increase significantly, which positively correlated with an increase in total antioxidant capacity. Therefore, it could be reasonably concluded that the fluctuations in blood melatonin concentrations can be correlated with the ability of the blood to detoxify toxic free radicals and related reactants in mammals, including humans (Reiter et al., 2003; Maldonado et al., 2009).

Considering that oral melatonin is generally absorbed from the GI tract into the bloodstream in the range 3%–76%, with an average absorption of 15%, the melatonin contents found in grape extracts (from cultivars of Nebbiolo, Croatina, and Cabernet Sauvignon) were up to four times the maximum amount of melatonin found in plasma, which can be regarded as nutraceutically significant (Dermarderosian et al., 2008).

The consumption of tropical fruits too increased the serum melatonin concentration and antioxidant capacity in healthy subjects. In premenopausal women, urinary excretion of 6-sulfatoxy melatonin (a major metabolite of melatonin considered a biomarker of vegetable intake) inversely correlates to breast cancer risk. The intake of 200 mL of grape juice (from Tempranillo cultivar) twice a day for five days significantly increased urinary-6-sulfatoxy melatonin and total antioxidant capacity in young, middle-aged, and elderly healthy individuals. Similar results have been reported after ingesting sweet cherries (Jerte Valley), Japanese plums, and tropical fruits. The administration of red wine (125 mL) did not decrease the salivary antiradical capacity in humans, thus suggesting that the prooxidant effects of ethanol may be counteracted by melatonin and polyphenols (Iriti and Varoni, 2015).

Since melatonin is highly bioavailable, it can also be incorporated into foods as food additives and ingested as food supplements. Serotonin possibly cannot be administered as a food supplement. A tryptophan supplement would be appropriate.

Usage of Tryptophan and Melatonin as Food Supplements

Tryptophan as a Food Supplement

Cubero et al. (2007) reported that tryptophan added to milk could improve sleep duration, efficiency, and latency in new-born infants. Garrido et al. (2013) reported that the consumption of a nutraceutical product made of Jerte Valley cherries twice a day significantly increased urinary 6-sulfatoxy melatonin and actual sleep time and immobility in human participants aged 20–25 years. They concluded that these effects were due to high concentrations of tryptophan in these cherries. In another study, Berwal and Khanna (2013) reported that chicken-incorporated biscuits and cookies would provide a nutritious, convenient, and ready-to-eat food item rich in tryptophan. Cookies with 10% chicken and wheat flour would be useful as nutritional supplements for essential amino acids, chiefly tryptophan, besides lysine and threonine.

Melatonin as a Food Supplement

In view of the strikingly beneficial effects of melatonin as well as its very low toxicity, it is imperative to develop more widespread applications of this molecule as a food supplement. Fortified and/or enriched food products such as milk, bread, and cookies could also be envisaged as food supplements for melatonin. However, no literature reports are available on these aspects of melatonin to date.

Safety of Tryptophan, Serotonin, and Melatonin in Foods and as Food Supplements

Despite the noteworthy benefits of tryptophan and melatonin, their usage as food additives is still a matter of controversy. A naturally occurring compound and constituent of food cannot suffice alone, since processed foods and food supplements will always lead to at most transient pharmacological concentrations in the blood, and the immunomodulatory actions of serotonin/melatonin may not be desired in every case (Hardeland and Pandi-Perumal, 2005).

Oxidation of tryptophan in foods by oxidizing agents or by photooxidation generates toxic compounds. Also, extensive loss of tryptophan occurs when proteins react with oxidized lipids. The carboline compounds are formed when tryptophan-containing food products are heated at high temperatures, such as during deep fat frying. Among these compounds, alpha-carboline and gamma-carboline have mutagenic activities. Carboline compounds also cause hepatogenecity. These compounds are also formed due to Maillard reactions in foods. The browning reactions in foods produce heterocyclic amines, which are genotoxic and carcinogenic. The heterocyclic amines derived from tryptophan induce liver cancer (Friedman and Cuq, 1988). Therefore, food-processing operations should be carried out with caution to avoid the development of these toxic compounds in tryptophan-rich food products. *Serotonin syndrome*, also known as *hyperserotonemia* or *serotonergic syndrome*, is a potentially life-threatening condition where there is an excess of serotonin in the CNS (Frank, 2008). In such cases, serotonin agonists and even foods containing tryptophan must be avoided.

Melatonin could exert unfavorable effects by increasing blood pressure due to the downregulation of NO synthase and NO scavenging by the indoleamine itself or by N^1-acetyl-5-methoxykynuramine. Since circulating melatonin peaks at night, melatonin-rich food and/or food supplements should be consumed at the same time of day in the evening. The usual recommendation of "bedtime" may be insufficient, since in practice this could mean different hours of the day. Melatonin should be taken relatively precisely at the same hour to avoid phase shifts differing in extent and pushing the circadian oscillator back and forth. In patients with rheumatoid arthritis, some symptoms are suspected to be associated with the immunomodulatory actions of melatonin. Melatonin is an easily oxidizable compound; therefore, during food-processing operations, exposure to air should be avoided. All these considerations are important and need more attention for the intake of melatonin through the diet, either from foods or food supplements.

Conclusion

Serotonin and melatonin are important neurotransmitters that have numerous important biological and therapeutic functions. In terms of nutrition, they are interesting both as natural constituents of foods and as food additives. Their usage as food supplements can be recommended only with caution, given the present state of our knowledge, although the risks posed by serotonin and melatonin appear to be remarkably lower than other food additives.

Acknowledgments

The first author acknowledges the cooperation of PhD research scholar Soumi Chakraborty of the Department of Food Technology and Biochemical Engineering, Jadavpur University, Kolkata, India, for her assistance in the literature survey.

REFERENCES

Aguilera, Y., Liébana, R., Herrera, T., et al. 2014. Effect of illumination on the content of melatonin, phenolic compounds, and antioxidant activity during germination of lentils (*Lens culinaris* L.) and kidney beans (*Phaseolus vulgaris* L.). *Journal of Agricultural and Food Chemistry* 62:10736–10743.

Albertini, B., Sabatino, M. D., Melegari, C., Passerini, N. 2014. Formulating SLMs as oral pulsatile system for potential delivery of melatonin to pediatric population. *International Journal of Pharmaceutics* 469:67–79.

Anisimov, V. N., Egormin, P. A., Piskunova, T. S., et al. 2010. Metformin extends life span of HER-2/neu transgenic mice and in combination with melatonin inhibits growth of transplantable tumors *in vivo*. *Cell Cycle* 9(1):188–197.

Anisimov, V. N., Popovich, G. I., Zabezhinski, M. A., Anisimov, S. V., Vesnushkin, G. M., Vinogradova, I. A. 2006. Melatonin as antioxidant, genoprotector and anticarcinogen. *Biochimica et Biophysica Acta* 1757:573–589.

Badria, F. A. 2002. Melatonin, serotonin, and tryptamine in some Egyptian food and medicinal plants. *Journal of Medicinal Food* 5:153–157.

Berger, M., Gray, J. A., Roth, B. L. 2009. The expanded biology of serotonin. *Annual Review of Medicine* 60:355–366.

Berwal, R. K., Khanna, N. 2013. Production of value added chicken meat mince incorporated cookies and their cost economic-benefit ratio. *Pacific Business Review International* 5(12):61–70.

Bravo, R., Matito, S., Cubero, J. 2013. Tryptophan-enriched cereal intake improves nocturnal sleep, melatonin, serotonin, and total antioxidant capacity levels and mood in elderly humans. *Official Journal of the American Aging Association* 35:1277–1285.

Burkhardt, S., Tan, D. X., Manchester, L. C., Hardeland, R., Reiter, R. J. 2001. Detection and quantification of the antioxidant melatonin in Montmorency and Balaton tart cherries (*Prunus cerasus*). *Journal of Agricultural and Food Chemistry* 49:4898–4902.

Chang, F. Y. 2014. Irritable bowel syndrome: The evolution of multi-dimensional looking and multidisciplinary treatments. *World Journal of Gastroenterology* 20(10):2499–2514.

Chaudhary, D. P., Kumar, S., Yadav, O. P. 2014. Nutritive value of maize: Improvements, applications and constraints. In *Maize: Nutrition Dynamics and Novel Uses*, pp. 3–20. Eds Chaudhary, D. P., Kumar, S., Singh, S. New Delhi, India: Springer.

Claustrat, B., Bruna, J., Chazot, G. 2005. The basic physiology and pathophysiology of melatonin. *Sleep Medicine Reviews* 9:11–24.

Cubero, J., Narciso, D., Terrón, M. P., et al. 2007. Chrononutrition applied to formula milks to consolidate infants' sleep/wake cycle. *Neuroendocrinology Letters* 28(4):360–366.

Deev, L. I., Kravtsov, G. M., Kudryashov-Yu, B. 1976. Radioprotective action of liposomes containing serotonin. *Radiobiologiya* 16(2):287–289.

Delgado, J., Terrón, M. P., Garrido, M., et al. 2012a. Jerte Valley cherry-based product modulates serum inflammatory markers in rats and ringdoves. *Journal of Applied Biomedicine* 10:41–50.

Delgado, J., Terrón, M. P., Garrido, M., et al. 2012b. A cherry nutraceutical modulates melatonin, serotonin, corticosterone, and total antioxidant capacity levels: Effect on ageing and chronotype. *Journal of Applied Biomedicine* 10:109–117.

Dermarderosian, A., Liberti, L., Beutler, J. A., et al. 2008. *The Review of Natural Products*, pp. 400–403. St Louis, MO: Facts and Comparisons.

Dubbels, R., Reiter, R. J., Klenke, E., et al. 1995. Melatonin in edible plants identified by radioimmunoassay and by high performance liquid-chromatography-mass spectrometry. *Journal of Pineal Research* 18:28–31.

Duff, J. 2014. Nutrition for ADHD and Autism. In *Clinical Neurotherapy, Application of Techniques for Treatment*, pp. 357–381. Eds Cantor, D. S., and Evans, J. R. Boston, MA: Elsevier Academic Press.

Feldman, J. M., Lee, E. M. 1985. Serotonin content of foods: Effect on urinary excretion of 5-hydroxyindole-acetic acid. *The American Journal of Clinical Nutrition* 42(4):639–643.

Frank, C. 2008. Recognition and treatment of serotonin syndrome. *Canadian Family Physician* 54(7):988–992.

Frazer, A., Hensler, J. G. 1999. Serotonin. In *Basic Neurochemistry: Molecular, Cellular and Medical Aspects*, 6th Edition. Eds Siegel, G.J., Agranoff, B.W., Albers, R.W., et al. Philadelphia, PA: Lippincott-Raven.

Friedman, M., Cuq, J. 1988. Chemistry, analysis, nutritional value, and toxicology of tryptophan in food. A Review. *Journal of Agricultural and Food Chemistry* 36(5):1080–1093.

García-Moreno, C., Rivas-Gonzalo, J. C., Peña-Egido, M. J., Mariné-Font, A. 1983. Improved method for determination and identification of serotonin in foods. *Journal of Association of Official Analytical Chemists* 66(1):115–117.

Garrido, M., González-Gómez, D., Lozano, M., Barriga, C., Paredes, S. D., Rodríguez, A. B. 2013. A Jerte valley cherry product provides beneficial effects on sleep quality: Influence on aging. *The Journal of Nutrition, Health and Aging* 17(6):553–560.

Garrido, M., Paredes, S. D., Cubero, J., et al. 2010. Jerte Valley cherry enriched diets improve nocturnal rest and increase 6-sulfatoxy melatonin and total antioxidant capacity in the urine of middle-aged and elderly humans. *Journals of Gerontology Series A: Biological Sciences and Medical Sciences* 65:909–914.

Gónzalez-Flores, D., Velardo, B., Garrido, M., et al. 2011. Ingestion of Japanese plums (*Prunus salicina Lindl.* cv. Crimson Globe) increases the urinary 6-sulfatoxy melatonin and total antioxidant capacity levels in young, middle-aged and elderly humans: Nutritional and functional characterization of their content. *Journal of Food and Nutrition Research* 50(4):229–236.

González-Gómez, D., Lozano, M., Fernández-León, M. F., Ayuso, M. C., Bernalte, M. J., Rodríguez, A. B. 2009. Detection and quantification of melatonin and serotonin in eight Sweet Cherry cultivars (*Prunus avium* L.). *European Food Research and Technology* 229:223–229.

Guillén-Casla, V., Rosales-Conrado, N., León-González, M. E., Pérez-Arribas, L. V., Polo-Díez, L. M. 2012. Determination of serotonin and its precursors in chocolate samples by capillary liquid chromatography with mass spectrometry detection. *Journal of Chromatography A* 1232:158–165.

Gündüz, B. 2014. Serum leptin profiles, food intake, and body weight in melatonin-implanted Syrian hamsters (*Mesocricetus auratus*) exposed to long and short photoperiods. *Turkish Journal of Biology* 38:185–192.

Haldar, C., Ahmad, R. 2010. Photoimmunomodulation and melatonin. *Journal of Photochemistry and Photobiology B: Biology* 98:107–117.

Hardeland, R., Pandi-Perumal, S. R. 2005. Melatonin, a potent agent in antioxidative defense: Actions as a natural food constituent, gastrointestinal factor, drug and prodrug. *Nutrition and Metabolism* 2:22–37.

Hattori, A., Migitaka, H., Masayake, I., Itoh, M., Yamamoto, K., Ohtani-kaneko, R. 1995. Identification of melatonin in plant seed and its effects on plasma melatonin levels and binding to melatonin receptors in vertebrates. *International Journal of Biochemistry and Molecular Biology* 35:627–634.

Hernandez-Ruiz, J., Cano, A., Arnao, M. B. 2004. Melatonin: A growth-stimulating compound present in lupin tissues. *Planta* 220:140–144.

Houston, M. 2013. Nutrition and nutraceutical supplements for the treatment of hypertension: Part III. *The Journal of Clinical Hypertension* 15(12):931–937.

Huang, X., Mazza, G. 2011. Simultaneous analysis of serotonin, melatonin, piceid and resveratrol in fruits using liquid chromatography tandem mass spectrometry. *Journal of Chromatography A* 1218 (25):3890–3899.

Iriti, M., Rossoni, M., Faoro, F. 2008. Melatonin in grape, not just a myth, maybe a panacea? *Journal of the Science of Food and Agriculture* 86:1432–1438.

Iriti, M., Varoni, E. M. 2014. Cardioprotective effects of moderate red wine consumption: Polyphenols vs. ethanol. *Journal of Applied Biomedicine* 12(4):193–202.

Iriti, M., Varoni, E. M. 2015. Melatonin in Mediterranean diet, a new perspective. *Journal of the Science of Food and Agriculture* 95(12):2355–2359.

Karthikeyan, R., Marimuthu, G., Spence, D. W., Pandi-Perumal, S. R., Salem, A., Brown, M. G. 2014. Should we listen to our clock to prevent type 2 diabetes mellitus? *Diabetes Research and Clinical Practice* 106:182–190.

Karunanithi, D., Radhakrishna, A., Sivaraman, K. P., Biju, V. M. N. 2014. Quantitative determination of melatonin in milk by LC–MS/MS. *Journal of Food Science and Technology* 51(4):805–812.

Kolar, J., Machackova, I. 2005. Melatonin in higher plants: Occurrence and possible functions. *Journal of Pineal Research* 39: 333–341.

Lavizzari, T., Veciana-Nogues, M. T., Bover-Cid, S., Marine-Font, A., Vidal-Carou, M. C. 2006. Improved method for the determination of biogenic amines and polyamines in vegetable products by ion-pair high-performance liquid chromatography. *Journal of Chromatography A* 1129:67–72.

Macchi, M. M., Bruce, J. N. 2004. Human pineal physiology and functional significance of melatonin. *Frontiers in Neuroendocrinology* 25:177–195.

Maldonado, M. D., Moreno, H., Calvo, J. R. 2009. Melatonin present in beer contributes to increase the levels of melatonin and antioxidant capacity of the human serum. *Clinical Nutrition* 28:188–191.

Manchester, L. C., Tan, D. X., Reiter, R. J., Park, W., Monis, K., Qi, W. 2000. High levels of melatonin in the seeds of edible plants: Possible function in germ tissue protection. *Life Sciences* 67(25):3023–3029.

Markus, C. R., Olivier, B., de Haan, E. H. 2002. Whey protein rich in alpha-lactalbumin increases the ratio of plasma tryptophan to the sum of the other large neutral amino acids and improves cognitive performance in stress-vulnerable subjects. *American Journal of Clinical Nutrition* 75:1051–1056.

Markus, C. R., Panhuysen, G., Tuiten, A., Koppeschaar, H., Fekkes, D., Peters, M. L. 1998. Does carbohydrate-rich, protein-poor food prevent a deterioration of mood and cognitive performance of stress-prone subjects when subjected to a stressful task? *Appetite* 31(1):49–65.

Morera, A. L., Abreu, P. 2006. Seasonality of psychopathology and circannual melatonin rhythm. *Journal of Pineal Research* 41:279–283.

Murch, S. J., Simmons, C. B., Saxena, P. K. 1997. Melatonin in feverfew and other medicinal plants. *The Lancet* 350:1597–1599.

Muszyńska, B., Sułkowska-Ziaja, K. 2012. Analysis of indole compounds in edible Basidiomycota species after thermal processing. *Food Chemistry* 132:455–459.

Peters, M. L. 1998. Does carbohydrate-rich, protein-poor food prevent a deterioration of mood and cognitive performance of stress-prone subjects when subjected to a stressful task? *Appetite* 31(1):49–65.

Pathak, S. K., Praveen, T., Jain, N. P., Banweer, J. 2010. A review on *Griffonia simplicifollia*: An ideal herbal antidepressant. *International Journal of Pharmacy and Life Sciences* 1(3):174–181.

Pauwels, P. J. 2003. 5-HT Receptors and their ligands. *Tocris Reviews* 25. https://www.tocris.com/pdfs/5htreview.pdf (accessed on October 16, 2015).

Poeggeler, B., Thuermann, S., Dose, A., Schoenke, M., Burkhardt, S., Hardeland, R. 2002. Melatonin's unique radical scavenging properties: Role of its functional substituents as revealed by its comparison with its structural analogs. *Journal of Pineal Research* 33:20–30.

Puerta, D. C., Carrascosa-Salmoral, M. P., García-Luna, P. P., et al. 2007. Melatonin is a phytochemical in olive oil. *Food Chemistry* 104:609–612.

Ramakrishna, A. 2015. Indoleamines in edible plants: Role in human health effects. In *Indoleamines: Sources, Role in Biological Processes and Health Effects*, p. 279. Ed. Angel Catalá (Biochemistry Research Trends Series). Hauppage, NY: Nova Publishers.

Ramakrishna, A., Giridhar, P., Ravishankar, G. A. Phytoserotonin: A review. *Plant Signaling and Behavior* 2011; 6:800–809.

Ramakrishna, A., Giridhar, P., Udaya Sankar, K., and Ravishankar, G. A. Endogenous profiles of indoleamines: Serotonin and melatonin in different tissues of *Coffea canephora* P ex Fr. as analyzed by HPLC and LC-MS-ESI. *Acta Physiologia Plantarum* 2012a; 34:393–396.

Ramakrishna, A., Giridhar, P., Udaya Sankar, K., and Ravishankar, G. A. Melatonin and serotonin profiles in beans of *Coffea* sps. *Journal of Pineal Research* 2012b; 52:470–476.

Reiter, R. J., Manchester, L. C., Tan, D. X. 2005. Melatonin in walnuts: Influence on levels of melatonin and total antioxidant capacity of blood. *Nutrition* 21:920–924.

Reiter, R. J., Tan, D. X., Mayo, J. C., Sainz, R. M., Leon, J., Czarnocki, Z. 2003. Melatonin as an antioxidant: Biochemical mechanisms and pathophysiological implications in humans. *Acta Biochimica Polonica* 50:1129–1146.

Restuccia, D., Spizzirri, U. G., Parisi, O. I., Cirillo, G., Picci, N., 2015a. Brewing effect on levels of biogenic amines in different coffee samples as determined by LC-UV. *Food Chemistry* 175:143–150.

Restuccia, D., Spizzirri, U. G., Puoci, F., Picci, N. 2015b. Determination of biogenic amine profiles in conventional and organic cocoa-based products. *Food Additives & Contaminants: Part A* 32(7):1156–1163.

Richard, D. M., Dawes, M. A., Mathias, C. W., Acheson, A., Hill-Kapturczak, N., Dougherty, D. M. 2009. L-Tryptophan: Basic metabolic functions, behavioral research and therapeutic indications. *International Journal of Tryptophan Research* 2:45–60.

Rodriguez, I. R., Mazuruk, K., Schoen, T. J., Chader, G. J. 1994. Structural analysis of the human hydroxyindole-O-methyltransferase gene-presence of two distinct promoters. *Journal of Biological Chemistry* 269:31969–31977.

Rodriguez-Naranjo, M. I., Gil-Izquierdo, A., Troncoso, A. M., Cantos-Villar, E., Garcia-Parrilla, M. C. 2011. Melatonin is synthesised by yeast during alcoholic fermentation in wines. *Food Chemistry* 126:1608–1613.

Stege, P. W., Sombra, L. L., Messina, G., Martinez, L. D., Silva, M. F. 2010. Determination of melatonin in wine and plant extracts by capillary electrochromatography with immobilized carboxylic multi-walled carbon nanotubes as stationary phase. *Electrophoresis* 31:2242–2248.

Tan, D., Manchester, L. C., Reiter, R. J., et al. 2000. Melatonin directly scavenges hydrogen peroxide: A potentially new metabolic pathway of melatonin biotransformation. *Free Radical Biology & Medicine* 29(11):1177–1185.

Tan, D. X., Hardeland, R., Manchester, L. C. 2012. Emergence of naturally occurring melatonin isomers and their proposed nomenclature. *Journal of Pineal Research* 53:113–121.

Tomas-Zapico, C., Coto-Montes, A. 2005. A proposed mechanism to explain the stimulatory effect of melatonin on antioxidative enzymes. *Journal of Pineal Research* 39:99–104.

USDA. 2015. Composition of Foods Raw, Processed, Prepared. USDA National Nutrient Database for Standard Reference, Release 27, US Department of Agriculture, Agricultural Research Service, Beltsville Human Nutrition Research Center, Beltsville, Nutrient Data Laboratory, Maryland 20705. http://www.ars.usda.gov/Services/docs.htm?docid=8964 (accessed on September 16, 2015).

Utiger, R. D. 1992. Melatonin: The hormone of darkness. *The New England Journal of Medicine* 327:1377–1379.

Van-Tassel, D. L., O'Neill, S. D. 2001. Putative regulatory molecules in plants: Evaluating melatonin. *Journal of Pineal Research* 31:1–7.

Wurtman, J. J., Wurtman, R. J. 1995. Brain serotonin, carbohydrate-craving, obesity and depression. *Obesity Research* 3(4):477S–480S.

22

Bioavailability of Dietary Phytomelatonin in Animals and Humans

Nutjaree Pratheepawanit Johns and Jeffrey Johns
Khon Kaen University
Khon Kaen, Thailand

CONTENTS

ABSTRACT   Melatonin has been found in a wide variety of plants and plant products and is considered to have a functional role, protecting them against internal and environmental oxidative stressors and to cope with hash environments. Many fruits, vegetables, seeds, grains, herbs, and plant products have been shown to contain high levels of melatonin. These dietary components would be expected to contribute to circulating melatonin levels in human. This chapter reports on selected fruit that are reported to contain substantial amounts of melatonin or are frequently consumed. It reports on the bioavailability of melatonin from plant materials in animals and humans, based on evidence of dietary melatonin being absorbed into the blood or presenting in daytime urine. The best sources of melatonin and the amount of consumption required to achieve doses equivalent to endogenous nighttime levels are postulated. It also considers the interaction of dietary melatonin with drugs and other dietary constituents, particularly those that are substrates for the enzyme cytochrome P4501A2, and the effect of the polymorphism of this enzyme on melatonin uptake from dietary foods. Finally, the role of dietary tryptophan, as the source of precursor for melatonin synthesis, and its synergy with other beneficial phytochemicals—for example, anthocyanins, vitamin C and E, polyphenols, and other antioxidants—is discussed.

KEY WORDS: *phytomelatonin, fruit, plants, bioavailability, diet, drug interactions.*

Abbreviations

aMT6s:	6-sulfatoxymelatonin
CYP1A2:	cytochrome P450 1A2
FRAP:	ferric-reducing antioxidant power
GC–MS:	gas chromatography–mass spectroscopy
HIOMT:	hydroxyindole-O-methyltransferase
HPLC:	high-performance liquid chromatography

NAT: serotonin-N-acetyltransferase
ORAC: oxygen radical antioxidant capacity
TEAC: trolox equivalent antioxidant capacity
UHPLC–QqQ–MS/MS: ultra-high-performance liquid chromatography–triple quadrupole–tandem mass spectrometry

Introduction

Melatonin has been found in a wide variety of plants and plant products (Parades, 2009). In plants, it is considered to have a functional role, protecting them against internal and environmental oxidative stressors and providing a means to cope with harsh environments (Tan et al., 2012; Ramakrishna and Ravishankar 2011, 2013). Food sources rich in melatonin, particularly herbs, grains, seeds, and fruit can be expected to provide dietary melatonin.

To be able to determine the melatonin content of plants, extraction is required, usually followed by purification, and finally quantification. Many different extraction methods have been reported in the literature, including maceration, sonication, and microwave assistance (Setyaningsih et al., 2012) in a large number of different solvent systems and conditions (Arnao and Hernández-Ruiz, 2009). Further purification steps and different analysis methods have been used—for example, solid-phase extraction, immunoassay, high-performance liquid chromatography (HPLC), gas chromatography–mass spectrometry (GC–MS)—resulting in widely varying reports of melatonin contents for different plants (Garcia-Parrilla et al., 2009; Fenga et al., 2014). The variation of melatonin content in each plant also depends on the part or age of the plant (Pothinuch et al., 2011), ripeness (Ozaki and Ezura, 2012), cultivar (Stürtza et al., 2011; Riga et al., 2014), climate, radiation (Riga et al., 2014), and geographical and seasonal variations. It is unsurprising, therefore, that different studies report greatly varying melatonin contents for the same plant or product. Many fruits, vegetables, seeds, grains, herbs, and plant products have been shown to contain high levels of melatonin. These dietary components would be expected to contribute to circulating melatonin levels in humans.

This section describes selected fruit that are reported to contain substantial amounts of melatonin or are frequently consumed, and the evidence that dietary melatonin can be absorbed into the blood—the so-called *bioavailability*—in animals and humans.

Dietary Sources of Melatonin from Plants

Melatonin is being found in an ever-increasing variety of foods. For a comprehensive review of melatonin in fruits or plants, the reader is referred to other studies (Reiter et al., 2007; Parades et al., 2009; Posmyk and Janas, 2009; Tan et al., 2012; Ramakrishna, 2015). Of particular interest is its presence in fruit (which accounts for a substantial fraction of the world's agricultural output), due to its generally wide availability and consumption, and the universally healthy nature of fruit in a regular diet. Fruits also generally contain high levels of water, vitamins, minerals, antioxidants, natural sugars, fiber, and a wide variety of other phytochemicals (Hulme, 1970; USDA, 2014).

Many edible fruits have been shown to contain melatonin (Table 22.1), with levels ranging from about 0.1 ng/g fresh fruit weight for some grapes (Iriti et al., 2006), bananas (Johns et al., 2013), and strawberries (Hattori et al., 1995) to 114.5 ng/g fresh weight for some tomatoes (Stürtza et al., 2001). Melatonin has been detected at quite high levels in fresh-frozen Montmorency tart cherries (13.5 ng/g) and Balaton tart cherries (22.4 ng/g) (*Prunus cerasus*) (Burkhardt, et al., 2001). Neither the orchard of origin nor the time of harvest influenced the amount of melatonin in the fresh cherries. However, melatonin levels of only 0–0.22 ng/g fresh fruit have been reported in eight Spanish sweet-cherry cultivars (*P. avium* L.) (González-Gómez et al., 2009), indicating that there can be great differences between individual species of fruit.

Melatonin in fruit, vegetables, and leaves also depends on the state of ripening and the plant's maturation. The melatonin content in tomato (*Lycopersicon esculentum*) has been widely studied and varies

TABLE 22.1

Melatonin Contents Reported in Selected Fruits

Common Name	Scientific Name	Melatonin/Wet Weight Fresh Fruit	Detection Method	References
Banana	*Musa sapientum*; *Musa ensete*	0.47 ng/g; 0.01 ng/g; 0.66 ng/g	RIA and GC–MS; HPLC–FD, ELISA; GC–MS	Dubbels et al. 1995; Johns et al., 2013; Badria, 2002
Apple	*Malus domestica*	0.05 ng/g; 0.16 ng/g	RIA	Hattori et al., 1995; Badria, 2002
Kiwifruit	*Actinidia chinensis*	0.024 ng/g	RIA	Hattori et al., 1995
Pineapple	*Ananas comosus*	0.04 ng/g; 0.30 ng/g	RIA; HPLC–FD, ELISA	Hattori et al., 1995; Johns et al., 2013
Pomegranate	*Punica granatum*	0.17 ng/g	GC–MS	Badria et al. 2002
Grape (skin)	*Vitis vinifera*	0.01–0.97 ng/g	HPLC–FD, ELISA	Iriti et al., 2006
Orange	*Citrus reticulata*	0.15 ng/g	HPLC–FD, ELISA	Johns et al., 2013
Mango	*Mangifera indica*	0.7 ng/g	HPLC–FD, ELISA	Johns et al., 2013
Papaya	*Carica papaya*	0.24 ng/g	HPLC–FD, ELISA	Johns et al., 2013
Montmorency tart cherry; Balaton tart cherry	*Prunus cerasus*	13.5 ng/g; 20.6 ng/g	HPLC	Burkhardt, et al., 2001
Sweet cherry cultivars	*Prunus avium*	0–0.2 ng/g	LC–MS– ESI	González-Gómez et al., 2009
Strawberry	*Fragaria magna*; *Fragaria ananassa*	0.012 ng/g; 1.4–11.3 ng/g	RIA	Hattori et al., 1995; Stürtza et al., 2001
Fig	*Ficus carica*	3.9 and 12.9 ng/g for whole fruit and leaves	ELISA	Zohar et al., 2011
Tomato	*Lycopersicon esculentum*	4.1–114.5 ng/g	LC–MS	Stürtza et al., 2001

greatly depending on variety and harvest year, ranging from 2.2 pg/g to 114.5 ng/g fresh weight (Dubbels et al., 1995; Hattori et al., 1995; Stürtza et al., 2001; Van Tassel et al., 2001; Badria, 2002; Pape and Lüning, 2006). Varying levels of melatonin were reported in the fruit and leaves of Micro-Tom tomatoes, depending on the developmental stage, suggesting that melatonin is involved in plant maturation (Okazaki and Ezura, 2009). Unlike dry herbs or seeds, the majority of fresh fruits have a high water content and, therefore, levels of melatonin in edible fruits presented as per wet weight are much lower (10–700 pg/g rather than several hundred ng/g wet weight).

Very modest amounts of melatonin (10–113 pg/g wet tissue) were quantified by HPLC analysis in vegetables (Hattori et al., 1995) and by GC–MS analysis (Badria, 2002). High levels of melatonin found in many seeds are thought to be a protectant against UV radiation and oxidative stress during germination and early growth (Tan et al., 2012). Melatonin levels of 2–200 ng/g have been reported in the seeds of 15 edible plants (Manchester et al., 2000).

Melatonin from Food Products

Food products derived from plants show substantial amounts of melatonin, and the food production process can either increase or decrease melatonin content in the product (Table 22.2). Coffee (*Coffea canephora* and *C. arabica*), for example, has been shown to contain significant amounts of melatonin (5,800–12,500 ng/g dry weight in green beans and 8,000–9,600 ng/g in roasted beans) (Ramakrishna et al., 2012a,b). However, brewing the coffee reduced the melatonin to about half these levels.

Iriti et al. (2010) have extensively reviewed the role of melatonin in traditional Mediterranean diets, which are associated with a lower incidence of chronic degenerative disorders. Important Mediterranean

TABLE 22.2

Melatonin Contents Reported in Selected Food Products

Food Product	Scientific Name	Melatonin/Wet Weight	Detection Method	References
Coffee	*Coffea canephora*; *Coffea arabica*	5.8–12.5 (green) and 8.0–9.6 μg/g (roasted) dry weight	ELISA	Ramakrishna et al., 2012
Wines	*Vitis vinifera*	25–428 ng/g	HPLC and ELISA	Iriti et al., 2006; Iriti and Faoro, 2006
Pomegranate wine	*Punica granatum*	0.54–5.50 ng/mL		Mena et al., 2012
Refined and extra virgin Spanish olive oil	*Olea europaea*	53–119 pg/mL	ELISA and immunoprecipitation	Puerta et al., 2007
Beer	—	51.8–169.7 pg/mL	—	Maldonado et al., 2009

foods and food products in which melatonin has been identified include olive oil, wine, and beer. Various refined and extra-virgin Spanish olive oils were found to contain between 650 and 119 pg melatonin/mL (Puerta et al., 2007). Melatonin was also determined in virgin argan oils (Venegas et al., 2011), with extra-virgin olive oil showing higher levels than virgin argan oil.

Several wine cultivars (*Vitis vinifera*) have been reported to contain between 2.4 and 428 pg melatonin/ml (0.005–0.97 ng/g) (Iriti et al., 2006; Iriti and Faoro, 2006). Analysis of eight red wines in different steps of the wine-making process showed between 140 and 277 pg melatonin/mL (Rodriguez-Naranjo et al., 2011). The melatonin content in wines increased during the fermentation process (Sprenger et al., 1999). Levels up to 5.5 ng/mL were detected in pomegranate wine, although, surprisingly, melatonin was not detected in pomegranate fruit itself (Mena et al., 2012). Melatonin from 51.8 to 169.7 pg/mL was found in a wide variety of beers (Maldonado et al., 2009), although there was no correlation with the type of beer (lager, bitter, or stout) or, by inference, barley or hop source. Melatonin content did correlate with beer alcohol level ($r = 0.56$, $p = .045$), perhaps implying, as in the case of wine, production by yeasts during the fermentation process.

Alcoholic fermentation has also been shown to induce melatonin synthesis in orange juice (Fernández-Pachón et al., 2014), with levels increasing (determined by UHPLC–QqQ–MS/MS) from 3.15 ng/mL before fermentation to 21.80 ng/mL by day 15. Tryptophan levels, however, significantly dropped from 13.80 mg/L (day 0) up to 3.19 mg/L (day 15) during fermentation, and melatonin was inversely and significantly correlated with tryptophan ($r = 0.907$).

Bioavailability of Melatonin

Melatonin taken orally is absorbed from the gastrointestinal tract and largely metabolized in the liver. The remainder is distributed into different parts of the body before being excreted mainly as soluble metabolites (6-sulfatoxymelatonin, aMT6s, and some gluconarated products) in the urine. When consumed in food, melatonin is metabolized in a similar fashion, with unmetabolized melatonin entering the blood and being carried to target sites of action, then acting either directly (Galano et al., 2011; Reiter et al., 2013), usually as an antioxidant, or indirectly, through melatonin receptors MT1 and MT2 (Dubocovich et al., 2003; Slominski et al., 2012; Kostoglou-Athanassiou, 2013).

Bioavailability of dietary melatonin from plant sources has been reported in only a few studies. When two-week-old chickens were fed a high-melatonin diet containing 3.5 ng/g, there was a significant increase in plasma melatonin levels (measured using a radioimmunoassay) from 20 to 34 pg/mL after 1.5 hours ($p < 0.01$), compared with controls (8 pg/mL) that ate a low-melatonin diet (less than 100 pg/g) (Hattori et al., 1995). This was the first study to demonstrate that dietary melatonin could cause an increase in circulating melatonin levels in the blood in animals.

In a later study by Reiter et al. (2005), rats fed with walnuts (*Juglans regia* L.) containing 3.5 ± 1.0 ng melatonin/g, showed an increase in serum melatonin concentrations to 38.0 ± 4.3 pg/mL compared with 11.5 ± 1.9 pg/mL for chow-fed only rats ($p < 0.01$). The *total antioxidant power* of the serum measured by trolox equivalent antioxidant capacity (TEAC) and ferric-reducing antioxidant power (FRAP) also increased for the walnut-fed animals. The antioxidant effect, however, could be attributed to the vitamin E content of walnuts as well (Maguire et al., 2004), although the level is very low in walnuts compared with most other nuts (265–436 µg/g oil gamma-tocopherol) (Savage et al., 1999). No correlation was reported between melatonin content and the antioxidant effect measured by TEAC or FRAP.

In humans, melatonin is well absorbed after oral administration, although few studies have considered very low doses, comparable to that from dietary intake. In the study by Fourtillan et al. (2000), 250 µg oral doses of melatonin given to 20 fasting healthy young volunteers resulted in C_{max} concentrations between 155 and 720 pg/mL in the plasma, with females having a level almost three times higher (624 ± 571 pg/mL) than males (244 ± 125 pg/mL). Other pharmacokinetic parameters such as elimination and distribution half-life seem to be constant among subjects. Bioavailability values of oral melatonin reported vary from $8.6\% \pm 3.6\%$ for males and $16.8\% \pm 12.7\%$ for females (Fourtillan et al., 2000) to 22% (Waldhauser et al., 1984) or 33% (Di et al., 1997), both in healthy males.

The absorption of melatonin from plants into the human body was reported in only a few studies. 6-Sulfatoxymelatonin (aMT6s), the primary urinary metabolite of melatonin in humans, is often used as a marker of circulating melatonin in the body. The first-void morning urines of 432 healthy Japanese women (average age of 48 years) showed higher levels of aMT6s that were significantly positively associated with higher vegetable intake. It was not possible to estimate the dietary melatonin intake from vegetables and/or from the entire diet, however, because the melatonin content of all the plants consumed was not known (Nagata et al., 2005). In a further study, healthy Japanese women, aged 24–55, consumed 350 g/day of six selected vegetables (intake ~1288 ng melatonin/day) for 65 days. A control group were requested to avoid the same six vegetables during the study period (intake 5.3 ng melatonin/day). In the melatonin-rich group, urinary aMT6s slightly increased from baseline from 48.1 to 49.6 ng/mg creatinine compared with a decrease in the control group from 55.5 to 50.8 ng/mg creatinine ($p = 0.03$) (Oba et al., 2008). Both of these studies demonstrated that increased consumption of vegetables can raise circulatory melatonin concentrations, determined from aMT6s urinary metabolite.

Tart Montmorency cherries (*P. cerasus*) are reported to contain high levels of melatonin (13.46 ± 1.10 ng/g) (Burkhardt et al., 2001). In a randomized, double-blind, placebo-controlled, crossover design, 20 volunteers who consumed tart cherry juice concentrate for seven days showed significantly elevated total melatonin content ($p < 0.05$) determined by urinary aMT6s, compared with a placebo. Another crossover study of six middle-aged and six elderly people who consumed 200 g of whole cherries (*P. avium* L.) twice a day as lunch and dinner desserts for three days showed a significant increase in urinary aMT6s levels (125%–220%, $p < 0.05$). Total antioxidant capacity of their urine also increased ($p < 0.05$), and additionally, beneficial effects on actual sleep duration, total nocturnal activity, assumed sleep, and immobility were observed (Garrido et al., 2010). The significant rises of urinary aMT6s levels found in these two studies due to cherry intake indicate higher circulating levels, and thus bioavailability, of the melatonin in the fruit.

A crossover study of 30 young healthy volunteers consuming six tropical fruits—banana (Musa sapientum Linn.), pineapple (*Ananus comosus* Merr.), orange (*Citrus reticulata*), papaya (*Carica papaya* L.), mango (*Mangifera indica* Linn.), and *makmao* or Thai berry (*Antidesma thwaiteaianun*)—consumed one at the time, with a one-week washout period between fruits, showed significant increases in urine aMT6s concentrations. Levels increased after consumption of pineapple (266%, $p = 0.004$), banana (180%, $p = 0.001$), and orange (47%, $p = 0.007$). The melatonin content of the fruits was determined by HPLC after methanol extraction and purification by SPE. After correction for recovery rate, the calculated melatonin content of the fruits (wet weight) showed the highest level in mango (699 ± 75 pg/g), followed by pineapple (302 ± 47 pg/g), papaya (241 ± 14 pg/g), orange (150 ± 6 pg/g), and banana (9 ± 0.6 pg/g), respectively (Johns et al., 2012), although the value for banana may be low because of extraction difficulties.

Although aMT6s correlates with melatonin level in the blood (Kovacs et al., 2000), it is not a direct measure of circulating melatonin in the body and has limited implications. Natural secretion of melatonin results in physiological levels of 120–200 pg/ml at night, decreasing to around 10–20 pg/ml

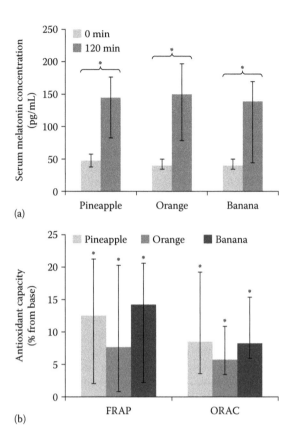

(a)

(b)

FIGURE 22.1 Serum melatonin concentration and antioxidant capacity after consuming pineapple, orange, or banana: (a) comparison of serum melatonin concentration before and 120 minutes (median ± interquartile range) after fruit consumption; (b) percentage change of antioxidant capacity in FRAP and ORAC assays (median ± interquartile range); * significant difference from baseline at $p < 0.05$. (From Sae-teaw et al., *J Pineal Res*, 2013, 55(1), 58–64.)

during the daytime (Brzezinski, 1997). For pharmaceutical levels—for example, a 3 mg oral supplement (Kotlarczyk et al., 2012)—a melatonin range of 940–27,240 pg/mL in plasma is expected (Kovacs et al., 2000). To determine if significant increases in blood levels of melatonin are achievable after consumption of foods containing melatonin, studies collecting and measuring levels in blood are required.

The first evidence of elevated serum melatonin in humans after consumption of food was recently reported in young healthy volunteers who consumed, during the daytime, either juice extracted from 1 kg of orange or pineapple or two whole bananas following a light breakfast, with a one-week washout period between fruits (Sae-teaw et al., 2012). Fruits were selected based on increased urinary aMT6s levels after consumption in a previous study (Johns et al., 2012). The highest serum melatonin concentration (above normal nocturnal levels of about 120 pg/mL plasma) was observed at 120 minutes after fruit consumption, and compared with levels before consumption, significantly increased in pineapple (146.0 vs. 48.0 pg/mL $p = 0.002$), orange (151.0 vs. 39.9 pg/mL, $p = 0.005$), and banana (140.3 vs. 32.4 pg/mL, $p = 0.008$). Serum antioxidant capacity following fruit consumption also significantly increased for both FRAP (7%–14% increase, $p > 0.004$) and oxygen radical antioxidant capacity (ORAC) (6%–9% increase, $p = 0.002$) assays. Surprisingly, there were strong correlations (all $r^2 > 0.984$, $p < 0.01$) of both serum FRAP and ORAC values and melatonin concentrations for all three fruits (Figure 22.1). Selected fruit consumption can elevate serum melatonin up to nocturnal levels and also raise the antioxidant capacity in serum of young healthy volunteers. Although some of the antioxidant effect could be attributed to the polyphenol and vitamin C or E contents of these fruits, the strong correlation with melatonin levels for all three fruits and both antioxidant assays would indicate that melatonin was a significant contributor to the increase antioxidant status of the plasma in those fruits.

Interaction of Dietary Melatonin with Drug and Other Dietary Constituents

Melatonin bioavailability depends not only on the quantity of melatonin in the food source ingested and the quantity consumed, but also on the absorption from the food in the gut, the metabolism in the liver before entering the blood stream, the distribution to tissue, and then the second-stage metabolism, where it is conjugated with sulfate to become water soluble before being excreted in the urine.

In humans, melatonin is principally first-pass metabolized in the liver to 6-hydroxymelatonin by the enzyme cytochrome P450 1A2 (CYP1A2). O-demethylation by CYP2C19 and, to some extent, CYP1A2 represents a minor reaction. CYP1B1, which is not expressed in the liver but has a ubiquitous extrahepatic distribution, has been shown to 6-hydroxylate melatonin. CYP1B1 is likely to metabolize melatonin in tissues that accumulate either melatonin or 6-HMEL, such as intestine and cerebral cortex (Ma et al., 2005).

Drugs that are also substrates for CYP1A2, such as paracetamol, theophylline, and fluvoxamine, would be expected to cause an increase in circulating melatonin levels after consumption of melatonin-containing food, due to direct competition for the enzyme. Strong CYP1A2 inhibitors such as ciprofloxacin, fluoroquinolones, fluvoxamine (Brøsen et al., 1993), and verapamil would have an even greater effect, reducing the amount and activity of CYP1A2. Inducers of CYP1A2 such as smoking (Schrenk et al., 1998), broccoli, Brussels sprouts, and cruciferous vegetables (Lampe et al., 2000), echinacea (Gorski et al., 2004), chargrilled meat (Dorne et al., 2001), cauliflower, insulin (Barnett et al., 1992), nafcillin (a beta-lactam antibiotic) and proton pump inhibitors such as omeprazole (Rost and Roots, 1994), and oral hormone-replacement therapy and contraceptives (Laine et al., 1999) have the opposite effect, decreasing blood levels of absorbed melatonin due to increased expression and activity of the enzyme.

Higher levels of melatonin than expected in the blood, based on normal oral bioavailability, may result after consumption of some foods due to inhibition of CYP1A2 by constituents in the food. Coffee, for example, is an inhibitor of CYP1A2; therefore, drinking coffee increases melatonin taken orally (Ursing et al., 2003). Likewise, mango has been reported to inhibit CYP1A2, CYP 3A1, CYP2C6, CYP2E1, and P-glycoprotein (ABCB1) (Chieli et al., 2009). *In vitro* studies showed that *Mangifera indica* and polyphenols reduce ABCB1/P-glycoprotein activity, while *in vivo* studies showed that mango interacts with midazolam, diclofenac, chlorzoxazone (Rodeiro et al., 2008, 2009; Rodríguez-Fragoso et al., 2011), and verapamil drugs (Chieli et al., 2009). Interactions of polyphenols with the P450 system and possible implications on human therapeutics have been reviewed. Apple is also known to inhibit CYP1A1 and the OATP family (OATP1, OATP3, and NTCP) of enzymes (Dresser et al., 2002; Pohl et al., 2006). These results may explain why *in vivo* bioavailability does not correlate with the melatonin content of the fruit consumed (Sae-teaw et al., 2012; Johns et al., 2012).

The effect of CYP1A2 polymorphism in individuals will also affect the amount of melatonin uptake from dietary foods. The 163C>A single-nucleotide polymorphism encoding the CYP1A2*1F allele (rs762551) of the CYP1A2 gene is a common (40%–50%) polymorphic variant and is associated with functional changes in enzyme activity (higher production and activity); that is, it has the effect of decreasing the enzyme inducibility. This produces individuals who are considered fast metabolizers (163[A;A]) and slow metabolizers (163[A;C] heterozygotes or 163[C;C] homozygotes) of CYP1A2 substrates. Subjects with –163A/A and –163C/A (CYP1A2*1F, rs762551) show lower plasma melatonin concentrations after administration of 6 mg melatonin compared with –163C/C carriers (Hartter et al., 2006).

The frequencies of mutation (CYP1A2*–163A) in Asian groups has also been reported to be high. In Ovambos ($n = 177$), Koreans ($n = 250$), and Mongolians ($n = 153$) they were 0.46, 0.32, and 0.21, respectively (Fujihara et al., 2007). The polymorphisms –2467delT and –163C>A were detected at high genotype frequencies in three healthy Asian populations in Singapore (Chinese [10.5%], Malay [2.6%], and Indian [14.3%], with 0.29, 0.22, and 0.42, respectively) (Lim et al., 2010). 163C/A has a reported frequency of 0.20 in Thai women (Sangrajang et al., 2009). Prawan et al. (2005) observed a 54.1% frequency of CYP1A2*1F/1*F genotype ($n = 233$) in physically healthy Thai subjects. Therefore, despite consuming the same dietary source, levels of melatonin in the circulation of each population can differ.

The effects of dietary tryptophan must also be considered. Melatonin is synthesized from dietary tryptophan, an essential amino acid that cannot be synthesized by the body. Thus, increased levels of melatonin in the body from eating fruit- and vegetable-rich diets may also be due to increased synthesis of melatonin, resulting from increased tryptophan intake. This is unlikely during the daytime, however, as melatonin synthesis is minimal at this time.

Many plants investigated for melatonin also contain appreciable amounts of tryptophan and serotonin, precursors of melatonin synthesis, which may enhance the synthesis of extrapineal melatonin via the biotransformation of serotonin from serotonin-N-acetyltransferase (NAT) by hydroxyindole-O-methyltransferase (HIOMT) enzymes—for example, in the gastrointestinal tract (Garcia-Parrilla et al., 2009). Previous studies report increased serum melatonin concentration after tryptophan product ingestion as a result of gastrointestinal melatonin synthesis (Huether et al., 1992; Huether, 1994; Bubenik, 2002; Sánchez et al., 2008).

Conclusion

Melatonin is available in many types of fruits and related products at widely varying levels, depending on the species, variety, location, and time of harvest. Some studies have now confirmed that consumption of fruit containing melatonin, even though the content may be low, can enhance plasma melatonin levels to around normal nocturnal levels. This is important, as low nocturnal melatonin appears to be associated with many disorders and ageing. Fruits rich in melatonin, in addition to other beneficial phytochemicals, could therefore have positive effects on health. Human bioavailability of melatonin from fruit and the effects of the interaction of melatonin and other constituents of fruit with CYP1A2 need further investigation to determine the best sources of melatonin and the amount of consumption required to achieve therapeutic doses.

It should also be considered that melatonin is present in food with other beneficial phytochemicals—for example, anthocyanins, vitamins C and E, polyphenols, resveratrol (grape), lycopene (tomato), harmanine (passion flower), and so on. They may act synergistically, enhancing the function of melatonin, as in the case of vitamins, and thereby enhancing beneficial health effects (Gitto et al., 2001; López-Burillo et al., 2003).

REFERENCES

Arnao MB, Hernández-Ruiz J. Assessment of different sample processing procedures applied to the determination of melatonin in plants. *Phytochem Anal* 2009;20(1):14–18.

Badria FA. Serotonin, tryptamine and melatonin in some Egyptian food and medicinal plants. *J Med Food* 2002;5:153–157.

Barnett CR, Wilson J, Wolf CR, Flatt PR, Ioannides C. Hyperinsulinaemia causes a preferential increase in hepatic P4501A2 activity. *Biochem Pharmacol* 1992,43:1255–1261.

Brøsen K, Skjelbo E, Rasmussen BB, Poulsen HE, Loft S. Fluvoxamine is a potent inhibitor of cytochrome P4501A2. *Biochem Pharmacol* 1993;45:1211–1214.

Burkhardt S, Tan DX, Manchester LC, Hardeland R, Reiter RJ. Detection and quantification of the antioxidant melatonin in Montmorency and Balaton tart cherries (*Prunus cerasus*). *J Agric Food Chem* 2001;49(10):4898–4902.

Chieli E, Romiti N, Rodeiro I, Garrido G. *In vitro* effects of *Mangifera indica* and polyphenols derived on ABCB1/P-glycoprotein activity. *Food Chem Toxicol* 2009;47(11):2703–2710.

Di WL, Kadva A, Johnston A, Silman R. Variable bioavailability of oral melatonin. *N Engl J Med* 1997;336(14):1028–1029.

Dorne JL, Walton K, Renwick AG. Uncertainty factors for chemical risk assessment: Human variability in the pharmacokinetics of CYP1A2 probe substrates. *Food Chem. Toxicol* 2001;39:681–696.

Dresser GK, Bailey DG, Leake BF, Schwarz UI, Dawson PA, Freeman DJ, Kim RB. Fruit juices inhibit organic anion transporting polypeptide-mediated drug uptake. *Clin Pharmacol Ther* 2002;71:11–20.

Dubbels R, Reiter RJ, Klenke E, Goebel A, Schnakenberg E, Ehlers C, Schiwara HW, Schloot W. Melatonin in edible plants identified by radioimmunoassay and high performance liquid chromatography–mass spectroscopy. *J Pineal Res* 1995;18(1):28–31.

Dubocovich ML, Rivera-Bermudez MA, Gerdin MJ, Masana MI. Molecular pharmacology, regulation and function of mammalian melatonin receptors. *Front Biosci* 2003;8:d1093–d1108.

Feng X, Wanga M, Zhaob Y, Hana P, Daia Y. Melatonin from different fruit sources, functional roles, and analytical methods. *Trends Food Sci Technol* 2014;37(1):21–31.

Fernández-Pachón MS, Medina S, Herrero-Martín G, Cerrillo I, Berná G, Escudero-López B, Ferreres F, Martín F, García-Parrilla MC, Gil-Izquierdo A. Alcoholic fermentation induces melatonin synthesis in orange juice. *J Pineal Res* 2014;56(1):31–38.

Fourtillan JB, Brisson AM, Gobin P, Ingrand I, Decourt JP, Girault J. Bioavailability of melatonin in humans after day-time administration of D(7) melatonin. *Biopharma Drug Dispos* 2000;21:15–22.

Galano A, Tan DX, Reiter RJ. Melatonin as a natural ally against oxidative stress: A physicochemical examination. *J Pineal Res* 2011;51(1):1–16.

Garcia-Parrilla MC, Cantosb E, Troncosoa AM. Analysis of melatonin in foods. *J Food Composition Anal* 2009;22(3):177–183.

Garrido M, Paredes SD, Cubero J, Lozano M, Toribio-Delgado AF, Muñoz JL, Reiter RJ, Barriga C, Rodríguez AB. Jerte Valley cherry–enriched diets improve nocturnal rest and increase 6-sulfatoxymelatonin and total antioxidant capacity in the urine of middle-aged and elderly humans. *J Gerontol A Biol Sci Med Sci* 2010;65:909–914.

Gitto E, Tan DX, Reiter RJ, Karbownik M, Manchester LC, Cuzzocrea S, Fulia F, Barberi I. Individual and synergistic antioxidative actions of melatonin: Studies with vitamin E, vitamin C, glutathione and des-ferrioxamine (desferoxamine) in rat liver homogenates. *J Pharm Pharmacol* 2001;53(10):1393–1401.

González-Gómez D, Lozano M, Fernández-León MF, Ayuso MC, Bernalte MJ, Rodríguez AB. Detection and quantification of melatonin and serotonin in eight Sweet Cherry cultivars (*Prunus avium* L.). *Eur Food Res Technol* 2009;229:223–229.

Gorski JC, Huang SM, Pinto A, Hamman MA, Hilligoss JK, Zaheer NA, Desai M, Miller M, Hall SD. The effect of echinacea (*Echinacea purpurea* root) on cytochrome P450 activity in vivo. *Clin Pharmacol Ther* 2004;75(1):89–100.

Hattori A, Migitaka H, Iigo M, Itoh M, Yamamoto K, Ohtani-Kaneko R, Hara M, Suzuki T, Reiter R J. Identification of melatonin in plant seeds and its effects on plasma levels and binding to melatonin receptors in vertebrates. *Biochem Mol Biol Int* 1995;35:627–634.

Hulme AC (ed.). *The Biochemistry of Fruits and their Products*. New York: Academic Press, 1970.

Johns NP, Porasuphattana S, Johns J, Plaimee P, Sae-teaw M. Quantification of melatonin in tropical fruit and in humans after fruit consumption. *J Agric Food Chem* 2013;61:913–919.

Kostoglou-Athanassiou I. Therapeutic applications of melatonin. *Ther Adv Endocrinol Metab* 2013;4(1):13–24.

Laine K, Palovaara S, Tapanainen P, Manninen P. Plasma tacrine concentrations are significantly increased by concomitant hormone replacement therapy. *Clin. Pharmacol. Therap* 1999;66:602–608.

Lampe JW, King IB, Li S, Grate MT, Barale KV, Chen C, Feng Z, Potter JD. Brassica vegetables increase and apiaceous vegetables decrease cytochrome P450 1A2 activity in humans: Changes in caffeine metabolite ratios in response to controlled vegetable diets. *Carcinogenesis* 2000;21:1157–1162.

López-Burillo S, Tan DX, Mayo JC, Sainz RM, Manchester LC, Reiter RJ. Melatonin, xanthurenic acid, resveratrol, EGCG, vitamin C and alpha-lipoic acid differentially reduce oxidative DNA damage induced by Fenton reagents: A study of their individual and synergistic actions. *J Pineal Res* 2003;34(4):269–277.

Ma X, Idle JR, Krausz KW, Gonzalez FJ. Metabolism of melatonin by human cytochromes p450. *Drug Metab Dispos* 2005;33(4):489–494.

Maguire LS, O'Sullivan SM, Galvin K, O'Connor TP, O'Brien NM. Fatty acid profile, tocopherol, squalene and phytosterol content of walnuts, almonds, peanuts, hazelnuts and the macadamia nut. *Int J Food Sci Nutr* 2004;55:171–178.

Mena P, Gil-Izquierdo A, Moreno DA, Martí N, García-Viguera C. Assessment of the melatonin production in pomegranate wines. *LWT: Food Sci Technol* 2012;47(1):13–18.

Nagata C, Nagao Y, Shibuya C, Kashiki Y, Shimizu H. Association of vegetable intake with urinary 6-sulfatoxymelatonin level. *Cancer Epidemiol Biomarkers Prev* 2005;14(5):1333–1335.

Oba S, Nakamura K, Sahashi Y, et al. Consumption of vegetables alters morning urinary 6-sulfatoxymelatonin concentration. *J Pineal Res* 2008;45:17–23.

Okazaki M, Ezura H. Profiling of melatonin in the model tomato (*Solanum lycopersicum* L.) cultivar Micro-Tom. *J Pineal Res* 2009;46(3):338–343.

Pape C, Lüning K. Quantification of melatonin in phototrophic organisms. *J Pineal Res* 2006;41:157–165.

Pohl C, Will F, Dietrich H, Schrenk D. Cytochrome P450 1A1 expression and activity in Caco-2 cells: Modulation by apple juice extract and certain apple polyphenols. *J Agric Food Chem* 2006;54:10262–10268.

Posmyk MM, Janas KM. Melatonin in plants. *Acta Physiol Planta* 2009;31:1–11.

Pothinuch P, Tongchitpakdee S. Melatonin contents in mulberry (*Morus* spp.) leaves: Effects of sample preparation, cultivar, leaf age and tea processing. *Food Chem* 2011;128:415–419.

Puerta C, Carrascosa-Salmoral MP, García-Luna PP, Lardone PJ, Herrera JL, Fernandez-Montesinos R, Guerrero JM, Pozo D. Melatonin is a phytochemical in olive oil. *Food Chem* 2007;104:609–612.

Ramakrishna A. Indoleamines in edible plants: Role in human health effects. In Catalá A (ed.), *Indoleamines: Sources, Role in Biological Processes and Health Effects,* pp. 235–246. Biochemistry Research Trends. Hauppauge, NY: Nova Publishers, 2015.

Ramakrishna A, Giridhar P, Udaya Sankar K, Ravishankar GA. Melatonin and serotonin profiles in beans of *Coffea* sps. *J Pineal Res* 2012a;52:470–476.

Ramakrishna A, Giridhar P, Udaya Sankar K, Ravishankar GA. Endogenous profiles of indoleamines: Serotonin and melatonin in different tissues of *Coffea canephora* P ex Fr. as analyzed by HPLC and LC-MS-ESI. *Acta Physiol Planta* 2012b;34:393–396.

Ramakrishna A, Ravishankar GA. Influence of abiotic stress signals on secondary metabolites in plants. *Plant Signal Behav* 2011;6:1720–1731.

Ramakrishna A, Ravishankar GA. Role of plant metabolites in abiotic stress tolerance under changing climatic conditions with special reference to secondary compounds. In Tuteja N and Gill SS (eds), *Climate Change and Abiotic Stress Tolerance*, pp. 705–726. Weinheim, Germany: Wiley, 2013.

Reiter RJ, Manchester LC, Tan DX. Melatonin in walnuts: Influence on levels of melatonin and total antioxidant capacity of blood. *Nutrition* 2005;21(9):920–924.

Reiter RJ, Tan DX, Rosales-Corral S, Manchester LC. The universal nature, unequal distribution and antioxidant functions of melatonin and its derivatives. *Mini Rev Med Chem* 2013;13(3):373–384.

Riga P, Medina S, García-Flores LA, Gil-Izquierdo A. Melatonin content of pepper and tomato fruits: Effects of cultivar and solar radiation. *Food Chem* 2014;156:347–352.

Rodríguez-Fragoso L, Martínez-Arismendi JL, Orozco-Bustos D, Reyes-Esparza J, Torres E, Burchiel SW. Potential risks resulting from fruit/vegetable-drug interactions: Effects on drug-metabolizing enzymes and drug transporters. *J Food Sci* 2011;76(4):R112–R124.

Rodeiro I, Donato MT, Jimenez N, Garrido G, Molina-Torres J, Menendez R, Castell JV, Gomez-Lechón MJ. Inhibition of human P450 enzymes by natural extracts used in traditional medicine. *Phytother Res* 2009;23:279–282.

Rodeiro I, Donato MT, Lahoz A, Garrido G, Delgado R, Gómez-Lechón MJ. Interactions of polyphenols with the P450 system: Possible implications on human therapeutics. *Mini Rev Med Chem* 2008;8:97–106.

Rodriguez-Naranjo MI, Gil-Izquierdo A, Troncoso AM, Cantos-Villar E, Garcia-Parrilla MC. Melatonin is synthesised by yeast during alcoholic fermentation in wines. *Food Chem* 2011;126(4):1608–1613.

Rost KL, Brösicke H, Heinemeyer G, Roots I. Specific and dose-dependent enzyme induction by omeprazole in human beings. *Hepatology* 1994;20(5):1204–1212.

Sae-teaw M, Johns J, Johns NP, Subongkot S. Serum melatonin level and antioxidant capacity after consumption of pineapple, orange and banana by healthy volunteers. *J Pineal Res* 2013;55(1):58–64.

Savage GP, Dutta PC, McNeil DL. Fatty acid and tocopherol contents and oxidative stability of walnut oils. *J Am Oil Chem Soc* 1999;76(9):1059–1063.

Setyaningsih W, Palma M, Barroso CG. A new microwave-assisted extraction method for melatonin determination in rice grains. *J Cereal Sci* 2012;56(2):340–346.

Schrenk D, Brockmeier D, Mörike K, Bock M, Eichelbaum M. A distribution study of CYP1A2 phenotypes among smokers and non-smokers in a cohort of healthy Caucasian volunteers. *Eur J Clin Pharmacol* 1998;53:361–367.

Slominski RM, Reiter RJ, Schlabritz-Loutsevitch N, Ostrom RS, Slominski AT. Melatonin membrane receptors in peripheral tissues: Distribution and functions. *Mol Cell Endocrinol* 2012;351(2):152–166.

Stürtza, M, Cerezoa AB, Cantos-Villarb E, Garcia-Parrillaa MC. Determination of the melatonin content of different varieties of tomatoes (*Lycopersicon esculentum*) and strawberries (*Fragaria ananassa*). *Food Chem* 2011;127(3):1329–1334.

Tan DX, Hardeland R, Manchester LC, Korkmaz A, Ma S, Rosales-Corral S, Reiter RJ. Functional roles of melatonin in plants, and perspectives in nutritional and agricultural science. *J Exp Bot* 2012;63(2):577–597.

Ursing C, Wikner J, Brismar K, Röjdmark S. Caffeine raises the serum melatonin level in healthy subjects: An indication of melatonin metabolism by cytochrome P450(CYP)1A2. *J Endocrinol Invest* 2003;26(5):403–406.

USDA. US Department of Agriculture, Agricultural Research Service. 2014. USDA National Nutrient Database for Standard Reference, Release 27. Nutrient Data Laboratory homepage: http://www.ars.usda.gov/ba/bhnrc/ndl (Last accessed 27 Jan 2015).

Van Tassel DL, Roberts N, Lewy, A, O'Neill SD. Melatonin in plant organs. *J Pineal Res* 2001;31(1):8–15.

Waldhauser F, Waldhauser M, Lieberman HR, Deng MH, Lynch HJ, Wurtman RJ. Bioavailability of oral melatonin in humans. *Neuroendocrinology* 1984;39(4):307–313.

Zohara R, Izhakib I, Koplovichb A, Ben-Shlomoa R. Phytomelatonin in the leaves and fruits of wild perennial plants. *Phytochem Lett* 2011;4(3):222–226.

Section V

Human Health: Brain, Behavior, Neurological Disorders, Sleep Disorders, Depression

23

Melatonin, Brain, and Behavior

Adejoke Onaolapo and Olakunle Onaolapo
Ladoke Akintola University of Technology
Ogbomosho, Oyo State, Nigeria

CONTENTS

ABSTRACT Melatonin, an indoleamine found in both plants and animals, is a pleiotropic signaling molecule. In animals, it plays an important role in the maintenance of the circadian rhythm and, to date, debates are still ongoing on its other functions and possible uses. Administered melatonin has effects on many behavioral processes, with animal models demonstrating its anticonvulsant, anxiolytic, antidepressant, and locomotor-suppressant and memory-modulating effects. Presently, melatonin research is looking at new horizons, where its role in the etiology and management of clinical conditions such as schizophrenia and dementia is being deeply examined. In this chapter, the sequence of melatonin metabolism and the relationship between melatonin, brain, and neurobehavior is discussed. The implications of exogenous administration of daytime melatonin and its effect on novelty-induced behaviors, learning, and memory are presented.

KEY WORDS: *melatonin, brain, behavior, chronobiotic, endogenous, exogenous, learning, memory.*

Abbreviations

AANAT: aralkylamine N-acetyltransferase
AFMK: N1-acetyl-N2-formyl-5-methoxykynuramine
SNAT: N-acetyltransferase
CNS: central nervous system
CSF: cerebrospinal fluid
EPM: elevated plus maze
GABA: gamma-aminobutyric acid
5-MIAA: 5-methoxyindole acetic acid
5-MTOL: 5-methoxytryptophol
NMDA: N-methyl-D-aspartate

SCN: suprachiasmatic nucleus
SNAT: Serotonin N-acetyltransferase
TPH: tryptophan hydroxylase

Introduction

Melatonin

Melatonin (N-acetyl-5-methoxytrypamine; Figure 23.1) is an indoleamine, first characterized and isolated from bovine pineal gland in 1958 by American dermatologist Aaron Lerner and his colleagues (Lerner et al., 1959; Arendt, 1988; Commai and Gobbi, 2014; Sugden et al., 2004), when it was noticed to be responsible for the lightening of frog skin through contraction of the epidermal melanophores. This is believed to be related to the mechanism by which some amphibians and reptiles change the color of their skin (Sugden et al., 2004), thus identifying the depigmenting factor that was first described in 1917 by McCord and Allen (Lerner, 1960).

Melatonin is found in humans, animals, bacteria, fungi, algae, and plants, including *Tanacetum parthenium* (feverfew), *Hypericum perforatum* (St John's wort), rice, corn, tomato, grape, edible fruits, olive oil, wine, and beer (Tan et al., 2011; Lamont et al., 2011) Melatonin has been found in a number of tissues and organs that are able to biosynthesize the indoleamine and has been reported to secrete enzymes that catalyze the synthesis of melatonin. These tissues include astrocytes, glial cells, retinal cells, lymphocytes, bone marrow cells, mast cells and epithelial cells, organs like the gut, testes, ovary, placenta, and the skin. In 1939, a reference to the possible functional relationship between melatonin and carbohydrate metabolism was made by the Romanian Constantin Parhon in a short abstract (Parhon, 1939), and debates are still ongoing on melatonin and its possible uses and functions.

Endogenous melatonin is a highly pleiotropic signaling molecule secreted by the pineal gland, predominantly at night; it is primarily known as the signal of darkness (Arendt, 1998). Melatonin is secreted from the pineal gland in young and middle-aged individuals (Haimov et al., 1994; Peirpaoli and Regelson, 1995), and its secretion follows a circadian rhythm with high amplitude, while melatonin secretion from other sites follows a low-amplitude rhythm. In earlier reports, its secretion was believed to decrease with age; more recent studies, however, dispute this. Zeitzer et al. (1999) reported no significant difference in plasma melatonin concentrations in a group of 34 healthy elderly subjects aged between 65–81 years, and 98 healthy drug-free young men aged between 18–30 years. In the Zeitzer study, extensive screening led to the exclusion of subjects on melatonin-lowering drugs like beta blockers and nonsteroidal anti-inflammatory drugs, as well as subjects consuming alcohol, caffeine, or nicotine (Zeitzer et al., 1999). Extrapineal melatonin, however, is thought play a small role in the maintenance of the circadian rhythm, since studies have shown that pinealectomy diminishes this rhythm. It is speculated that melatonin from these sources is used, rather, in the defense against oxidative stress.

Physiology and Pharmacology of Endogenous Melatonin

Melatonin biosynthesis in humans and some other organisms involves four enzymatic steps from the essential dietary amino acid tryptophan. It is synthesized by the pinealocytes in a multistep process, beginning with hydroxylation of aromatic amino acid L-tryptophan to 5-hydroxytryptophan; this is

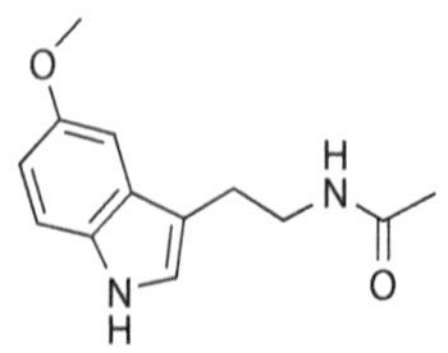

FIGURE 23.1   Structure of melatonin (N-acetyl-5-methoxytrypamine).

catalyzed by tryptophan hydroxylase, an enzyme believed to be key in melatonin synthesis (Schallreuter, 1994). 5-Hydroxytryptophan is then converted to serotonin (5-hydroxytryptamine) by the aromatic amino acid decarboxylase, initially to hydroxytryptophan through hydroxytryptamine and eventually to melatonin. In the dark, aralkylamine N-acetyltransferase (AANAT), a key enzyme in melatonin biosynthesis, is activated, which catalyzes the conversion of serotonin to N-acetylserotonin, which is then converted to melatonin by the final enzyme, acetylserotonin O-methyltransferase (Lovenberg et al., 1967; Iuvone et al., 2005; Norman and Henry, 2012). Aralkylamine N-acetyltransferase is a key regulator of melatonin synthesis from tryptophan, as its gene AANAT is directly influenced by the photoperiod. Tryptophan hydroxylase (TPH), is another enzyme that is important in melatonin biosynthesis, albeit to a lesser extent than AANAT; it controls the availability of serotonin. The final step of melatonin synthesis is the conversion of N-acetylserotonin to melatonin by the enzyme serotonin N-acetyltransferase (SNAT), formerly known as hydroxyindole-O-methyl transferase (HIOMT) (Radomir et al., 2012). Endogenous melatonin upon synthesis by the pineal gland is secreted quickly into the blood and other bodily fluids, including cerebrospinal fluid (CSF) (Rousseau et al., 1999), saliva (Vakkuri et al., 1985), and bile (Tan et al., 1999). Levels of melatonin in the cerebrospinal fluid are usually higher than that seen in the blood. Approximately 50%–75% of melatonin found in the blood is bound reversibly to albumin and alpha-1-acid glycoprotein; it has a half-life estimated to be 30–60 min and first-pass metabolism in the liver results in a clearance rate of 90% (Pardridge and Mietus, 1980; Claustrat et al., 1986).

Melatonin metabolism occurs via three main pathways, which include: (1) the hepatic degradation pathway that produces 6-hydroxymelatonin (Radomir et al., 2012; Hardeland et al., 2011), (2) the indolic pathway that generates 5-methoxyindole acetic acid (5-MIAA) or 5-methoxytryptophol (5-MTOL) (Radomir et al., 2012), and (3) the kynuric pathway that produces N1-acetyl-N2-formyl-5-methoxykynuramine (AFMK) (Tan et al., 2007; Slominski et al., 2008). The most definitive physiological role melatonin is linked to is the regulation of the circadian (Quera and Hartley, 2012) and circannual or seasonal rhythms, and the conveyance of this information to body physiology for the organization of functions that vary with season.

In humans, however, functional relationships between endogenous melatonin rhythms and other physiologic processes remain uncertain. Melatonin also regulates physiologic functions such as reproduction, coat growth and color, appetite, body weight, and sleep (Jolanta et al., 2009; Arendt, 2000). Other functions modulated by melatonin include immune functions, contraction of smooth muscle, body temperature regulation, defense against oxidative stress, balancing of organismal energy metabolism, and retarding the aging process; it has also been described as a putative antihypertensive. Melatonin is a neurohormone with endocrine, paracrine, and autocrine activity (Chun-Qiu et al., 2011). It performs most of these functions by acting as a messenger of the suprachiasmatic nucleus (SCN).

Melatonin and its metabolites are potent antioxidants (Bavithra et al., 2013) that have both direct and indirect antioxidant activity. They act directly by aiding in the reduction of the body's free radical burden and levels of both oxygen and nitrogen species. They are able to extend this antioxidant potential to all subcellular structures due to their highly lipophilic ability; melatonin has the advantage of being soluble in lipids and water, a unique character that enhances its cellular distribution, enabling its passage through the blood–brain barrier (Al-Omary, 2013). Melatonin's indirect antioxidant activity resides in its ability to improve mitochondrial efficiency and stimulate the expression and activation of a number of antioxidants and potentiate their antioxidant activities. Many of melatonin's actions are mediated through the interaction with specific membrane-bound receptors (Radomir et al., 2012; Slominski et al., 2008) such as MT1 and MT2, also known as Mel1a, and Mel1b, respectively (Dubocovich et al., 2010). Both these receptors are members of the G protein–coupled, seven-transmembrane receptor family (Dubocovich et al., 2003). Melatonin also acts through non-receptor-mediated mechanisms when it scavenges reactive nitrogen and oxygen species (Gomez-Moreno et al., 2010).

Exogenous Melatonin

Exogenous melatonin is a powerful *chronobiotic* that aids the synchronicity of the normal circadian rhythms mostly in persons with sleep disorders and/or altered circadian rhythms that can occur in jet lag, night shift work, and various neuropsychiatric disorders (Melatonin Monograph, 2005).

In the United States, the Food and Drug Administration has categorized melatonin as a dietary supplement (Buscemi et al., 2004), and so it is available over the counter in both the United States and Canada; however, in Europe it is considered a *neurohormone* and so a prescription is required for its sale (Guardiola-Lemaître, 1997). Melatonin administered orally is absorbed rapidly and peak serum levels are observed at 60–150 minutes, while peak concentrations after oral dosing are significantly (350–10,000 times) higher than those seen with endogenous melatonin secretion (DeMuro et al., 2000). The range of melatonin bioavailability following an oral dose is 10%–56% (Di et al., 1997) and the half-life of exogenous melatonin is 12–48 minutes (Aldhous et al., 1985). Metabolism of exogenous melatonin is as seen with endogenous melatonin (Melatonin Monograph, 2005).

Exogenous melatonin exerts a hypnotic and sedative effect, especially when administered at doses close to the physiological range of endogenous melatonin, regardless of time of administration (Alternative Medicine Review, 2005). Administration of melatonin has been reported to cause a reduction in core body temperatures (Satoh and Mishima, 2001) through mechanisms that are currently being studied; the relationship between melatonin and the thermoregulatory centers in the hypothalamus is being considered. Other uses of exogenous melatonin include its use in the treatment of insomnias (Haimov et al., 1994; Zhdanova et al., 1995; Dollins et al., 1994). Macfarlane and coworkers in studying the effect of high and low doses of exogenous melatonin on total sleep time and daytime alertness in insomniacs revealed a significant increases in total sleep time, as well as improved daytime alertness and decreases in time needed to fall asleep compared with placebo (MacFarlane et al., 1991). Other studies have also demonstrated an increase in the amount of stage-2 sleep with electroencephalogram (Rajaratnam et al., 2004) and reduction in sleep latency (Pinto et al., 2004).

Melatonin is usually administered at night to mimic endogenous melatonin secretion, though a few studies have tried daytime melatonin administration. Daytime administration of exogenous melatonin has been shown to promote sleep in humans and results in sleep-like brain activity patterns in specific areas of the brain, such as the precuneus and hippocampus (Gorfine and Zisapel, 2007).

Endogenous and Exogenous Melatonin and the Brain

A large part of pineal research has deliberated on the brain's response to melatonin rhythms. This has led to the defining of two main roles of melatonin in humans: the involvement of nocturnal melatonin secretion in initiating and maintaining sleep, and control by the day/night melatonin rhythm of the timing of other 24-h rhythms. Melatonin in the central nervous system (CNS) was believed to originate solely from the pineal gland and the secretion was responsible for melatonin fluctuations in the CSF of mammals. Recent studies, however, have shown that it may not be the only source of melatonin in the CNS (Hardeland and Poeggeler, 2008); there has been documentation of the presence in the brain tissue of rats of messenger RNAs of arylalkylamine N-acetyltransferase (AANAT) and serotonin N-acetyltransferase (SNAT), which are the essential enzymes of melatonin, indicating that other brain regions may possess the machinery for melatonin synthesis. Whether neurons can synthesize melatonin remains to be answered.

Melatonin has multiple receptor-mediated and receptor-independent actions. Its binding sites and messenger RNAs of its receptors have been detected in the hypothalamus, pituitary, retina (Cahill and Besharse, 1990), thalamus, hippocampus (Gorfine and Zisapel, 2007), and neocortex in a variety of mammalian species, humans included (Malpaux et al., 2001). One very important characteristic of melatonin is its ability to readily cross the blood–brain barrier and accumulate at substantial levels in the brain (Al-Omary, 2013). Melatonin exhibits strong neuroprotective effects, especially under the conditions of oxidative stress or neural inflammation (Bromme et al., 2000). It has also been reported to be an important regulator of the sleep–wake cycle; it synchronizes circadian rhythms and organizes several seasonal rhythms (Claustrat et al., 2005; Pevet, 1988). The disruption of these rhythms in humans has been associated with seasonal and nonseasonal affective disorders characterized by disturbances in the secretion pattern of melatonin (Ouakki et al., 2013). In unipolar or bipolar affective disorders, studies have revealed a reduction in circulating levels of melatonin. Similarly, a reduction in nocturnal secretion levels of melatonin is observed in panic disorders, and abnormally high levels of melatonin have been reported in depression (Brown et al., 1985).

Melatonin and Neurobehavior

A number of studies have demonstrated that the administration of melatonin plays a role in many behavioral processes. In animal studies, a number of researchers have demonstrated melatonin's anticonvulsant properties in animal models of epilepsy (De-lima et al., 2005; Yildirim and Marangoz, 2006), although a few have reported proconvulsant effects (Musshoff and Speckmann, 2003; Stewart and Leung, 2005). Several rodent studies have demonstrated the antidepressant and anxiolytic effects of melatonin in animal models of anxiety, using passive avoidance tests, open-field tests, elevated plus maze (EPM) tests, and the light/dark box (Guardiola-Lemaitre et al., 1992; Kopp et al., 1999; Kopp et al., 2000; El Mrabet et al., 2012a; El Mrabet, et al., 2012b).

Melatonin and Exploratory Behavior

Rodent exploratory behavior helps to drive survival because it is important in the search for food, water, and shelter, as animals are able to approach novel environments and objects (Crusio, 2001). Exploration can be defined as an "action that is evoked by novel stimuli and consists of behavioral acts and postures that permit the collection of information about new objects and unfamiliar parts of the environment" (Crusio and van Abeelen, 1986). The behavior of an animal in a novel environment is usually a synergy between its curiosity and fear. Rodent exploratory behavior has been studied widely in behavioral pharmacology using animal models that have been validated. McKinney (1984) defined animal models as experimental preparations developed in a species solely for the purpose of studying phenomena occurring in another species. Kaplan (1973) also suggested that a model may be valid if it has the same structure as the human behavior or pathology being studied; it is believed that whenever a relation exists between two elements of the animal model, a corresponding association may exist among corresponding elements of the human behavior. For an animal model to be suitable for research it has to have predictive validity (pharmacological correlation), face validity (isomorphism), and construct validity (homology and similarity of underlying neurobiological mechanisms) (McKinney and Bunney, 1969; Triet, 1985).

The open-field test is one of the most popular models for measuring exploratory behavior in rodents (Prut and Belzung, 2003). It is a forced exploration test that involves putting an animal into an unfamiliar arena, usually a box from which there is no retreat. The open field measures horizontal and vertical locomotion and grooming (Crusio and van Abeelen, 1986). Both locomotor and rearing activities of rodents are central excitatory behaviors that are indicative of the rodent's explorative ability (Ajayi and Ukponmwan, 1994). A reduction in horizontal locomotion or rearing in rodents points to a central inhibitory or central depressant effect, and an increase in these two parameters is indicative of a central excitatory or stimulatory effect.

Studies on the effect of melatonin on exploratory behavior are few and the results are largely dependent on experimental design, although some consistency seems to exist to a large extent. In studies where melatonin was administered to rodents, either time of administration (usually nighttime) or photoperiodicity (short vs long daylight) were taken into consideration. However, in a study by Onaolapo et al. (2014), the effect of oral daytime administration of two doses of melatonin was tested in the open field and in a Y maze. The results of the study revealed a decrease in grooming, horizontal locomotion, and rearing behavior; while grooming and rearing behavior showed statistically significant differences, horizontal locomotion showed visible differences only when compared to vehicle (Onaolapo et al., 2014). The results corroborate reports that have shown that exogenous melatonin at pharmacological doses, regardless of route of administration, increased blood and brain concentrations of melatonin enough to result in CNS effects (Gorfine and Zisapel, 2007). The study also showed that oral daytime melatonin in mice had a central depressant effect, thereby corroborating the results of other studies by Golombeck et al. (1991) and Maha et al. (2013), who both reported decreases in locomotor activity and rearing following administration of melatonin orally to rats and in adult male golden hamsters following intraperitoneal injection of at least 30 µg melatonin/kg for five days. In these studies, the depressant effect was found to be lowest at midnight and blunted by the administration of a benzodiazepine. Two different routes of administration (oral vs intraperitoneal) were employed and the animal models used were also at variance, and yet melatonin consistently caused locomotor retardation. Kopp et al. (1999), in a study using adult C3H/He mice, administered melatonin beginning in the dark phase following exposure to chronic mild stress procedures; the study also reported decreases in locomotor activity, rearing, and entries into the unfamiliar zone of the free exploratory test. However, when melatonin was administered prenatally to rats that were then tested as adults (Ripple, 2010), an increase in locomotor activity and rearing was seen that was blunted by the administration of luzindole, a melatonin receptor antagonist. In other studies, reductions in locomotor activity and rearing were seen when melatonin was administered prenatally to spontaneously hypertensive rats (Kim et al., 2002), whereas in the study by Brotto et al. (2000), rearing, but not locomotor activity, was decreased following the administration of melatonin to adult rodents.

Grooming is an ancient behavior that is observed in a number of animals (Sachs, 1988; Aldridge and Berridge, 1998). Its primary goal is that of hygiene and body care, although it has many more functions than this, including thermoregulation, chemocommunication, social interaction, and stress reduction (López-Crespo et al., 2009). Grooming is also a measure of emotionality and anxiety-related behavior. Grooming is modulated from a number of brain regions and by neuromediators, hormones, and drugs (Spruijt et al., 1992; Barros et al., 1994). In a number of melatonin studies reviewed (Onaolapo et al., 2014; Maha et al., 2013; Golombeck et al., 1991; Stoessl, 1996), administration of melatonin resulted in a suppression of grooming behavior. Melatonin administration over a 12-week period was also reported to have reduced grooming and other mating behaviors in voles (Ferkin and Kile, 1996), rabbits (Goldman et al., 1993), and hamsters (Fleming et al., 1988; Karp et al., 1990). Maha et al. (2013) reported that administration of melatonin at 10 mg/kg significantly lowered the emotionality of rats in the open-field test, indicated by reductions in grooming frequency (49.1%), defecation (37.9%), and percentage urination occurrence (44.4%) in comparison with a diazinon-treated group. Also reported was a significant reduction in grooming behavior resulting from co-administration of melatonin and diazinon; this effect was a reversal of the central excitatory effect of diazinon.

The anxiolytic effects of melatonin have been buttressed using other behavioral paradigms such as EPM. Ouakki et al. (2013) reported that the dose-dependent anxiolytic effect of melatonin increased significantly with increasing doses of melatonin. Sokolovic et al. (2012) also reported a reduction in microwave radiation–induced anxiety states by administration of melatonin. Administration of melatonin regardless of route of administration significantly reduced anxiety-related behaviors in Wistar rats, Sprague Dawley rats (both normotensive and spontaneously hypertensive rats), and BALB/c and C$_3$H/He mice (Kopp et al., 1999; Krsková et al., 2007; Karakaş et al., 2011). Likewise, melatonin was reported to induce sedative effects against other anxiogenic compounds (Permpoonputtana et al., 2012; Stein et al., 2013).

Numerous reports suggest that melatonin modulates the electrical activity of neurons, although its role in the CNS is still being studied. The reduction in central excitatory behaviors from melatonin administration in adult rodents could be due to a reduction in excitatory input such as that from glutamatergic agents or is probably due to its effects on central dopaminergic activation. The physiological effects of melatonin are believed to involve a decrease in striatal dopaminergic activity via the dopamine D1 and D2 receptors, which in turn inhibit glutamate release (Stewart, 2001).

Gamma-aminobutyric acid (GABA), is the main inhibitory neurotransmitter in the mammalian CNS (Nutt and Malizia, 2001). Alongside its receptors, GABA is involved in the regulation of normal and pathological brain mechanisms, like sleep, epilepsy, memory, emotions (Seighart, 1995), and grooming (Barros et al., 1994). Melatonin can cause a reduction in grooming by an increased inhibitory input such GABAergic. Melatonin is thought to potentiate the effects of GABA via direct interaction with GABA receptors (Ferini-Strambi et al., 1993; Wang et al., 2003). It is known that pelage cleaning in rodents increases with exposure to novelty and this response is modulated by central dopaminergic activation via D1 receptors (Komorowska and Pellis, 2004). Melatonin could also potentiate brain inhibitory transmission via GABAergic synapses. Input from the GABA system has been implicated in the expression of novelty-induced grooming via its GABA-A and GABA-B receptors (Barros et al., 1994).

Melatonin may inhibit calcium influx into the neurons and bind to the calcium–calmodulin complex, inhibiting neuronal nitric oxide synthase activity and causing a decrease in nitric oxide production, resulting in the reduction of the excitatory effect of N-methyl-D-aspartate (NMDA) (Monika et al., 2011;

Muñoz-Hoyos et al., 1998; León et al., 2000) or in some instances specifically inhibiting the NMDA subtype of excitatory glutamate receptors in rodent striatum, and increasing the concentration and receptor affinity of the inhibitory neurotransmitter GABA (Acuña-Castroviejo et al., 1986).

Melatonin, Learning and Memory

Memory is a fundamental mental process, without which daily living becomes difficult, if not impossible, except for simple reflexes and stereotyped behaviors. Thus, learning and memory is one of the most intensively studied subjects in the field of neuroscience. Various approaches have been used to understand the mechanisms underlying this process. Nootropics are pharmaceutical agents, drug supplements, nutraceuticals, or functional foods that improve one or more aspects of memory or mental function; specifically, this could include improvement of spatial working memory, attention, alertness, or motivation. Recent breakthroughs in neuroscience have uncovered a host of melatonin's other vital benefits other than its use as a sleep aid. It has also been seen to inhibit cognitive deficits related to aging (Shen et al., 2001; Sharma and Gupta, 2001), which is especially attributed to its ability to pass the blood–brain barrier and be readily absorbed after oral dosing.

Melatonin has been reported to possess both antidepressant and anxiolytic properties in a number of studies (Barden et al., 2005; Golombek et al., 1993; Micale et al., 2006), and it is also involved in the modulation of complex processes such as learning and memory (Rawashdeh and Maronde, 2012), largely by binding to melatonin receptors MT1 and MT2, which are distributed widely throughout the brain (Morgan et al., 1994). Studies in animal models of learning and memory have reported melatonin's memory-facilitating effects in the novel object recognition test (He et al., 2013) and the olfactory social memory test in rats (Argyriou et al., 1998), enhancement of performance in a verbal association task (Gorfine and Zisapel, 2007), and improvement of memory acquisition in conditions of stress (Rimmele et al., 2009). Further, Shen and coworkers (2001) explored the effects of exogenous melatonin administration on memory impairment induced by amyloid-beta peptides in aged rats; amyloid-beta proteins are known to induce inflammation and brain dysfunction, resulting in amnesia characteristic of Alzheimer's disease. They administered melatonin at increasing doses to both normal and amnesic aged rats for a period of 10 days and deduced using the Morris water maze, shuttle box, and step down tests that melatonin substantially improved impaired learning and memory in amyloid-beta peptide–treated aged rats (Shen et al., 2001). A number of other studies have also reported contrasting results with melatonin. A study by Rawashdeh et al. (2007) using zebrafish reported that the nootropic effects of melatonin were circadian, with better memory performance seen during the day than at night; they concluded that melatonin suppressed nighttime memory formations (Rawashdeh et al., 2007). Yang et al. (2013) also reported that melatonin administration impaired the acquisition but not the expression of contextual fear in rats (Yang et al., 2013). Intraamygdalar melatonin injections have also been reported to impair spatial memory performance in rats (Karakas et al., 2011). More recently, Huang et al. (2014) reported that melatonin (2.5 mg/kg) facilitated the extinction of conditional cued fear without affecting its acquisition or expression, and melatonin facilitates cued fear extinction only when it is present during extinction training (Huang et al., 2014).

The differences in the results of memory in animal models could be attributable to differences in species response (rodents vs amphibians), routes of administration, doses administered, and period of administration. Onaolapo et al. (2014) reported the dose-dependent nootropic effects of daytime melatonin administration in a Y maze, with significant improvement seen at 5 mg/kg and no significant difference seen at 10 mg/kg compared to amnesic (scopolamine) rats. Compared to vehicle treated rats however, the memory response seen with melatonin was significantly lower. With the radial arm maze, however, melatonin significantly improved spatial memory performance at both doses of melatonin (5 and 10 mg/kg) when compared with amnesic rats. Compared with groups of mice administered vehicle, spatial memory improved significantly at 5 mg/kg, and at 10 mg/kg no significant difference was seen (Onaolapo et al., 2014).

Studies in humans have demonstrated the memory-enhancing effects of melatonin (Oxenkrug, 2007). Clinical research on the effect of melatonin on patients with mild cognitive impairment has reported evidence of significant improvements in memory processing (Cardinali, 2010). In another study, melatonin specifically enhanced the recognition memory accuracy of objects encoded under stress; however, 15 minutes after stress, when cortisol levels were highest, retrieval of memories acquired the day before was not influenced by melatonin. Melatonin also did not influence the stress-induced elevation of catecholamine and cortisol levels, which in turn did not correlate with the effects of melatonin on memory. These results may be indicative of a primary action of melatonin on central-nervous stimulus processing under conditions of stress and possibly on memory consolidation (Rimmele et al., 2009).

Conclusion

Exogenously administered melatonin alters behavioral processes in animals, as evidenced by its ability to affect both central excitation and spatial working memory in mice. Presently, beyond the understanding of melatonin's neurobehavioral effects, there are attempts to look at newer areas of interest, including its role in the genesis and management of conditions such as schizophrenia and dementia. These newer areas give a glimpse into future directions in melatonin research.

REFERENCES

Acuña-Castroviejo D, Rosenstein RE, Romeo HE, Cardinali DP. Changes in gamma-aminobutyric acid high affinity binding to cerebral cortex membranes after pinealectomy or melatonin administration to rats. *Neuroendocrinology*, 1986;43:24–31.

Ajayi AA, Ukponmwan OE. Evidence of angiotensin II and endogenous opioid modulation of NIR in the rat. *Afr J Med Med Sci*, 1994;23:287–290.

Aldhous M, Franey C, Wright J, Arendt J. Plasma concentrations of melatonin in man following oral absorption of different preparations. *Br J Clin Pharmacol*, 1985;19:517–521.

Aldridge JW, Berridge KC. Coding of serial order by neostriatal neurons: A natural action approach to movement sequence. *J Neurosci*, 1998;18:2777–2787.

Al-Omary FA. Melatonin: Comprehensive profile. *Profiles Drug Subst Excip Relat Methodol*, 2013;38:159–226.

Arendt J. Melatonin. *Clin Endocrinol (Oxf)*, 1988;29:205–229.

Arendt J. Melatonin and the pineal gland: Influence on mammalian seasonal and circadian physiology. *Rev Reprod*, 1998;3:13–22.

Arendt J. Melatonin, circadian rhythms and sleep. *N Engl J Med*, 2000;343:1114–1116.

Argyriou A, Prast H, Philippu A. Melatonin facilitates short-term memory. *Eur J Pharmacol*, 1998;349(2–3): 159–162.

Barden N, Shink E, Labbé M, Vacher R, Rochford J, Mocaër E. Antidepressant action of agomelatine (S 20098) in a transgenic mouse model. *Prog Neuropsychopharmacol Biol Psychiatry*, 2005;29(6):908–916.

Barros HM, Tannhauser SL, Tannhauser MA. The effects of GABAergic drugs on grooming behaviour in the open field. *Pharmacol Toxicol*, 1994;74:339–344.

Bavithra S, Selvakumar K, Krishnamoorthy G, Venkataraman P, Arunakaran J. Melatonin attenuates polychlorinated biphenyls induced apoptosis in the neuronal cells of cerebral cortex and cerebellum of adult male rats *in vivo*. *Environ Toxicology Pharmacol*, 2013;36:152–163.

Bromme HJ, Morke W, Peschke D, et al. Scavenging effect of melatonin on hydroxyl radicals generated by alloxan. *J Pineal Res*, 2000;29:201–208.

Brotto LA, Barr AM, Gorzalka BB. Sex differences in forced-swim and open field test behaviours after chronic administration of melatonin. *Eur J Pharmacol*, 2000;402:87–93.

Brown R, Kocsis JH, Caroff S, Amsterdam J, Winokur A, Stokes PE, Frazer A. Differences in nocturnal melatonin secretion between melancholic depressed patients and control subjects. *Am J Psychiatry*, 1985;142(7):811–816.

Buscemi N, Vandermeer B, Pandya R, Hooton N, Tjosvold L, Hartling L, Baker G, Vohra S, Klassen T. Melatonin for treatment of sleep disorders. Evidence Report/Technology Assessment No. 108. (Prepared by the University of Alberta Evidence-Based Practice Centre, under Contract No. 290-02-0023.) AHRQ Publication 2004, No. 05-E002-2. Rockville, MD: Agency for Healthcare Research and Quality, US Department of Health and Human Services.

Cahill GM, Besharse JC. Circadian regulation of melatonin in the retina of *Xenopus laevis*: Limitation by serotonin availability. *J Neurochem*, 1990;54:716–719.

Cardinali DP, Furio AM, Brusco LI. Clinical aspects of melatonin intervention in Alzheimer's disease progression. *Curr Neuropharmacol*, 2010;8:218–227.

Chun-Qiu C, Jakub F, Mohammad B, Yong-Yu L, Martin S. Distribution, function and physiological role of melatonin in the lower gut. *World J Gastroenterol*, 2011;17(34):3888–3898.

Claustrat B, Brun J, Chazot G. The basic physiology and pathophysiology of melatonin. *Sleep Med Rev*, 2005;9(1):11–24.

Claustrat B, Brun J, Garry P. A once-repeated study of nocturnal plasma melatonin patterns and sleep recordings in six normal young men. *J Pineal Res*, 1986;3:301–310.

Commai S, Gobbi G. Unveiling the role of melatonin MT2 receptors in sleep, anxiety and other neuropsychiatric diseases: Novel target in psychopharmacology. *J Psychiatry Neurosci*, 2014;39(1):6–21.

Crusio WE. Genetic dissection of mouse exploratory behaviour. *Behav Brain Res*, 2001; 125(1–2):127–132.

Crusio WE, van Abeelen JH. The genetic architecture of behavioural responses to novelty in mice. *Heredity*, 1986;56:55–63.

De-Lima E, Soares Jr JM, del Carmen Sanabria Garrido Y, et al. Effects of pinealectomy and the treatment with melatonin on the temporal lobe epilepsy in rats. *Brain Res*, 2005;1043:24–31.

DeMuro RL, Nafziger AN, Blask DE. The absolute bioavailability of oral melatonin. *J Clin Pharmacol*, 2000;40:781–784.

Di WL, Kadva A, Johnston A, Silman R. Variable bioavailability of oral melatonin. *N Engl J Med*, 1997;336:1028–1029.

Dollins AB, Zhdanova IV, Wurtman RJ. Effect of inducing nocturnal serum melatonin concentrations in daytime on sleep, mood, body temperature, and performance. *Proc Natl Acad Sci USA*, 1994;91:1824–1828.

Dubocovich ML, Delagrange P, Krause DN, Sugden D, Cardinali DP, Olcese J. International union of basic and clinical pharmacology. LXXV nomenclature, classification, and pharmacology of G protein–coupled melatonin receptors. *Pharmacol Rev*, 2010;62:343–380.

Dubocovich ML, Rivera-Bermudez MA, Gerdin MJ, Masana MI. Molecular pharmacology, regulation and function of mammalian melatonin receptors. *Front Biosci*, 2003;8:d1093–d1108.

El Mrabet FZ, Lagbouri I, Mesfioui A, El Hessn A, Ouichou A. The influence of gonadectomy on anxiolytic and antidepressant effects of melatonin in male and female wistar rats: A possible implication of sex hormones. *Neurosci Med*, 2012;3(2):162–173.

El Mrabet FZ, Ouakki S, Mesfioui A, El Hessni A, Ouichou A. Pinealectomy and exogenous melatonin regulate anxiety-like and depressive-like behaviors in male and female wistar rats. *Neurosci Med*, 2012;3:394–403.

Ferini-Strambi L, Zucconi M, Biella G. Effect of melatonin on sleep microstructure: Preliminary results in healthy subjects. *Sleep*, 1993;16:744–747.

Ferkin MH, Kile JR. Melatonin treatment affects the attractiveness of the anogenital area scent in meadow voles (*Microtus pennsylvanicus*). *Horm Behav*, 1996;30(3):227–235.

Fleming AS, Phillips A, Rydall A. Effects of photoperiod, the pineal gland and the gonads on agonistic behavior in female golden hamsters (*Mesocricetus auratus*). *Physiol Behav*, 1988;44:227–234.

Goldman BD, Nelson RJ. Melatonin and seasonality in mammals. In Yu HS, Reiter RJ (eds), *Melatonin: Biosynthesis, Physiological Effects, and Clinical Applications*. Boca Raton, FL: CRC Press, 1993, pp. 225–252.

Golombeck DA, Escolar E, Cardinali DP. Melatonin-induced depression of locomotor activity in hamsters: Time-dependency and inhibition by the central type benzodiazepine antagonist Ro 15-1788. *Physiol Behav*, 1991;49:1091–1097.

Golombeck DA, Martini M, Cardinali DP. Melatonin as an anxiolytic in rats: Time dependence and interaction with the central GABAergic system. *Eur J Pharmacol*, 1993;237(2–3):231–236.

Gomez-Moreno G, Guardia J, Ferrera MJ, Cutando A, Reiter RJ. Melatonin in diseases of the oral cavity. *Oral Dis*, 2010;16:242–247.

Gorfine T, Zisapel N. Melatonin and the human hippocampus, a time dependent interplay. *J Pineal Res*, 2007;43(1):80–86

Guardiola-Lemaître B. Toxicology of melatonin. *J Biol Rhythms*, 1997;12(6):697–706.

Guardiola-Lemaitre B, Lenegre A, Porsolt RD. Combined effects of diazepam and melatonin in two tests for anxiolytic activity in the mouse. *Pharmacol Biochem Behav*, 1992;41(2):405–408.

Haimov I, Laudon M, Zisapel N, et al. The relationship between urinary 6- sulphatoxymelatonin rhythm and insomnia in old age. *Adv Pineal Res*, 1994;8:433–438.

Hardeland R, Cardinali DP, Srinivasan V, Spence DW, Brown GM, Pandi-Perumal SR. Melatonin: A pleiotropic, orchestrating regulator molecule. *Prog Neurobiol*, 2011;93:350–384.

Hardeland R, Poeggeler B. Melatonin beyond its classical functions. *Open Physiol J*, 2008;210–22.

He P, Ouyang X, Zhou S, Yin W, Tang C, Laudon M, Tian S. A novel melatonin agonist Neu-P11 facilitates memory performance and improves cognitive impairment in a rat model of Alzheimer's disease. *Horm Behav*, 2013;64(1):1–7.

Huang F, Zehua Y, Xiaoyan L, Chang-Qi L. Melatonin facilitates extinction, but not acquisition or expression, of conditional cued fear in rats. *BMC Neurosci*, 2014;15:86.

Iuvone PM, Tosini G, Pozdeyev N, Haque R, Klein DC, Chaurasia SS. Circadian clocks, clock networks, arylalkylamin N-acetyltransferase, and melatonin in the retina. *Prog Ret Eye Res*, 2005;24:433–456.

Jolanta BZ, Debra JS, Josephine A. Physiology and pharmacology of melatonin in relation to biological rhythms. *Pharmacol Rep*, 2009;61:383–410.

Kaplan A. The conduct of inquiry. *Methodology for Behavioural Sciences*. Aylesbury, Buckinghamshire: Intertext Books, 1973, pp. 15–22.

Karakaş A, Coşkun H, Kaya A, Kücük A, Gündüz B. The effects of the intraamygdalar melatonin injections on the anxiety like behavior and the spatial memory performance in male Wistar rats. *Behav Brain Res*, 2011;222(1):141–150.

Karp JD, Dixon ME, Powers JB. Photoperiod history, melatonin, and reproductive responses of male Syrian hamsters. *J Pineal Res*, 1990;8:137–152.

Kim CY, Lee BN, Kim JS. Effects of maternal-melatonin treatment on open-field behaviours and hypertensive phenotype in spontaneously hypertensive rats' offspring. *Exp Anim*, 2002;51(1):69–74.

Komorowska J, Pellis S M. Regulatory mechanisms underlying novelty induced grooming in the laboratory rat. *Behav Process*, 2004;67(2):287–293.

Kopp C, Vogel E, Rettori MC, Delagrange P, Guardiola-Lemaître B, Misslin R. Effects of melatonin on neophobic responses in different strains of mice. *Pharmacol Biochem Behav*, 1999;63(4):521–526.

Kopp C, Vogel E, Rettori MC, Delagrange P, Misslin R. Anxiolytic-like properties of melatonin receptor agonists in mice: Involvement of MT1 and/or MT2 receptors in the regulation of emotional responsiveness. *Neuropharmacology*, 2000;39(10):1865–1871.

Krsková L, Vrabcová M, Zeman M. Effect of melatonin on exploration and anxiety in normotensive and hypertensive rats with high activity of renin-angiotensin system. *Neuroendocrinol Lett*, 2007;28,(3):295–301.

Lamont KT, Somers S, Lacerda L et al. Is red wine a SAFE sip away from cardioprotection? Mechanisms involved in resveratrol- and melatonin-induced cardioprotection. *J Pineal Res*, 2011;50(4):374–380.

León J, Macias M, Escames G, Camacho E, Khaldy H, Martin M, Espinosa A, Gallo MA, Acuña-Castroviejo D. Structure-related inhibition of calmodulin-dependent neuronal nitric-oxide synthase activity by melatonin and synthetic kynurenines. *Mol Pharmacol*, 2000;58:967–975.

Lerner AB. Hormonal control of pigmentation. *Annu Rev Med*, 1960;11:187–194.

Lerner AB, Case JD, Heinzelman RV. Structure of melatonin. *J Am Chem Soc*, 1959;81:6084–6092.

López-Crespo GA, Flores P, Sánchez-Santed F, Sánchez-Amate MC. Acute high dose of chlorpyrifos alters performance of rats in the elevated plus-maze and the elevated T-maze. *Neurotoxicology*, 2009;30:1025–1029.

Lovenberg W, Jequier E, Sjoerdsma A. Tryptophan hydroxylation: Measurement in pineal gland, brainstem, and carcinoid tumor. *Science*, 1967;155:217–219.

MacFarlane JG, Cleghorn JM, Brown GM, Streiner DL. The effects of exogenous melatonin on the total sleep time and daytime alertness of chronic insomniacs: A preliminary study. *Biol Psychiatry*, 1991;30:371–376.

Maha AEA, Hebatalla IA, Engy M E. Melatonin protects against diazinon-induced neurobehavioral changes in rats. *Neurochem Res*, 2013;38:2227–2236.

Malpaux B, Migaud M, Tricoire H, Chemineau P. Biology of mammalian photoperiodism and critical role of the pineal gland and melatonin. *J Biol Rhythm*, 2001;16:336–347.

McKinney WT. Animal models of depression: An overview. *Psychiatry Dev*, 1984;2:77–96.

McKinney WT Jr, Bunney WE Jr. Animal model of depression: I. Review of evidence: Implications for research. *Arch Gen Psychiatry*, 1969;21:240–248.

Melatonin monograph, *Alter Med Rev*, 2005;10(4):325–336.

Micale V, Arezzi A, Rampello L, Drago F. Melatonin affects the immobility time of rats in the forced swim test: The role of serotonin neurotransmission. *Eur Neuropsychopharmacol* 2006;16(7):538–545.

Monika B, Elwira G, Marian J, Kinga KB. Melatonin in experimental seizures and epilepsy. *Pharmacol Rep*, 2011;63:1–11.

Morgan PJ, Barrett P, Howell HE, Helliwell R. Melatonin receptors: Localization, molecular pharmacology and physiological significance. *Neurochem Int*, 1994;24(2):M101–M146.

Muñoz-Hoyos A, Molina-Carballo A, Macías M, Rodríguez-Cabezas T, Martín-Medina E, Narbona-López E, Valenzuela-Ruiz A, Acuña-Castroviejo D. Comparison between tryptophan methoxyindole and kynurenine metabolic pathways in normal and preterm neonates and in neonates with acute fetal distress. *Eur J Endocrinol*, 1998;139:89–95.

Musshoff U, Speckmann EJ. Diurnal actions of melatonin on epileptic activity in hippocampal slices of rats. *Life Sci*, 2003; 73:2603–2610.

Norman AW, Henry, HL. *Hormones* (3rd edn). Oxford, UK: Academic Press, 2012, pp. 352–359.

Nutt DJ, Malizia AL. New insights into the role of the GABA(A)-benzodiazepine receptor in psychiatric disorder. *Br J Psychiatry*, 2001;179:390–396.

Onaolapo OJ, Adejoke YO, Akanni AA, Eniafe AL. Central depressant and nootropic effects of daytime melatonin in mice. *Ann Neurosci*, 2014;21(3):90–96.

Ouakki S, Fatima ZEM, Ibtissam L, Aboubaker EH, Abdelhalem M, Paul P, Etienne C, Ali O. Melatonin and diazepam affect anxiety-like and depression-like behavior in Wistar rats: Possible interaction with central GABA neurotransmission. J Behav Brain Sci, 2013;3(7):12.

Oxenkrug GF. Genetic and hormonal regulation of tryptophan kynurenine metabolism: Implications for vascular cognitive impairment, major depressive disorder, and aging. *Ann NY Acad Sci*, 2007;1122:35–49.

Pardridge WM, Mietus LJ. Transport of albumin bound melatonin through the blood-brain barrier. *J Neurochem*, 1980;34:1761–1763.

Parhon CI. Congrès d'Endocrinologie de Bucarest. *Communications, Institutul de Arte Grafice*, 1939;I:187.

Peirpaoli W, Regelson W. *The Melatonin Miracle*. New York, NY: Simon and Schuster, 1995, pp. 24–30.

Permpoonputtana K, Mukda S, Govitrapong P. Effect of melatonin on D-amphetamine-induced neuroglial alterations in postnatal rat hippocampus and prefrontal cortex. *Neurosci Lett*, 2012;524:1–4.

Pevet P. The role of the pineal gland in the photoperiodic control of reproduction in different hamster species. *Reprod Nutr Dev*, 1988;28(2B):443–458.

Pinto LR Jr, Seabra Mde L, Tufik S. Different criteria of sleep latency and the effect of melatonin on sleep consolidation. *Sleep*, 2004;27:1089–1092.

Prut L, Belzung C. The open field as a paradigm to measure the effects of drugs on anxiety-like behaviours: A review. *Eur J Pharmacol*, 2003;463(1–3):3–33.

Quera SMA, Hartley S. Mood disorders, circadian rhythms, melatonin and melatonin agonists. *J Cent Nerv Syst Dis*, 2012;4:15–26.

Radomir MS, Russel JR, Natalia SL, Rennolds SO, Andrzej T. Slominski Melatonin membrane receptors in peripheral tissues: Distribution and functions. *Mol Cell Endocrinol*, 2012;351(2):152–166.

Rajaratnam SM, Middleton B, Stone BM. Melatonin advances the circadian timing of EEG sleep and directly facilitates sleep without altering its duration in extended sleep opportunities in humans. *J Physiol*, 2004;561:339–351.

Rawashdeh O, de Borsetti NH, Roman G, Cahill GM. Melatonin suppresses nighttime memory formation in zebrafish. *Science*, 2007;318(5853):1144–1146.

Rawashdeh O, Maronde E. The hormonal Zeitgeber melatonin: Role as a circadian modulator in memory processing. *Front Mol Neurosci*, 2012;5:27.

Rimmele U, Spillmann M, Bartschi C, Wolf OT, Weber CS, Ehlert U, Wirtz PH. Melatonin improves memory acquisition under stress independent of stress hormone release. *Psychopharmacol (Berl)*, 2009;202(4):663–672.

Ripple J. The effects of altered prenatal melatonin signaling on adult behavior and hippocampal gene expression of the male rat: A circadioneuroendocrine-axis hypothesis of psychopathology. Honor's thesis, Bucknell University, Philadelphia, PA, 2010, Paper 30.

Rousseau A, Petren S, Plannthin J, et al. Serum and cerebrospinal fluid concentrations of melatonin: A pilot study in healthy male volunteers. *J Neural Transm*, 1999;106:883–888.

Sachs BD. The development of grooming and its expression in adult animals. *Ann NY Acad Sci*, 1988;525:1–17.

Satoh K, Mishima K. Hypothermic action of exogenously administered melatonin is dose-dependent in humans. *Clin Neuropharmacol*, 2001;24:334–340.

Schallreuter KU, Wood JM, Pittelkow MR, Gutlich M, Lemke KR, Rodl W, Swanson NN, Hitzemann K, Ziegler I. Regulation of melanin biosynthesis in the human epidermis by tetrahydrobiopterin. *Science*, 1994;263:1444–1446.

Sharma M, Gupta YK. Effect of chronic treatment of melatonin on learning, memory and oxidative deficiencies induced by intracerebroventricular streptozotocin in rats. *Pharmacol Biochem Behav*, 2001;70(2–3):325–331.

Shen YX, Wei W, Yang J, Liu C, Dong C, Xu SY. Improvement of melatonin to the learning and memory impairment induced by amyloid beta-peptide 25–35 in elder rats. *Acta Pharmacol Sin*, 2001;22(9):797–803.

Sieghart W. Structure and pharmacology of gamma-aminobutyric acid receptor subtypes. *Pharmacol Rev*, 1995;47:181–234.

Slominski A., Tobin DJ, Zmijewski MA, Wortsman J, Paus R. Melatonin in the skin: Synthesis, metabolism and functions. *Trends Endocrinol Metab*, 2008;19:17–24.

Sokolovic D, Djordjevic B, Kocic G, Babovic P, Ristic G, Stanojkovic Z, Sokolovic DM, Veljkovic A, Jankovic A, Radovanovic Z. The effect of melatonin on body mass and behaviour of rats during an exposure to microwave radiation from mobile phone. *Bratislavské lekárske listy*, 2012;113(5):265–269.

Spruijt BM, van Hooff JA, Gispen WH, et al. Ethology and neurobiology of grooming behaviour. *Physiol Rev*, 1992;72:825–852.

Stein DJ, Picarel-Blanchot F, Kennedy SH. Efficacy of the novel antidepressant agomelatine for anxiety symptoms in major depression. *Hum Psychopharmacol*, 2013;28:151–159.

Stewart LS. Endogenous melatonin and epileptogenesis: Facts and hypothesis. *Intern J Neurosci*, 2001;107:77–85.

Stewart LS, Leung LS. Hippocampal melatonin receptors modulate seizure threshold. *Epilepsia*, 2005;46:473–480.

Stoessl AJ. Dopamine D1 receptor agonist induced grooming is blocked by the opiod receptor antagonist maloxone. *Eur J Pharmacol*, 1996;259:301–303.

Sugden D, Davidson K, Hough KA, et al. Melatonin, melatonin receptors and melanophores: A moving story. *Pigment Cell Res*. 2004;17(5):454–460.

Tan D, Manchester LC, Reiter RJ. High physiological levels of melatonin in the bile of mammals. *Life Sci*, 1999;65:2523–2529.

Tan DX, Hardeland R, Manchester LC, et al. Functional roles of melatonin in plants, and perspectives in nutritional and agricultural science. *J Exp Bot* 2011;63(2):577–597.

Tan DX, Manchester LC, Terron MP, Flores LJ, Reiter RJ. One molecule, many derivatives: A never-ending interaction of melatonin with reactive oxygen and nitrogen species? *J Pineal Res*, 2007;42:28–42.

Treit D. Animal models for the study of anti-anxiety agents: A review. *Neurosci Biobehav Rev*, 1985;9:203–222.

Vakkuri O. Diurnal rhythm of melatonin in human saliva. *Acta Physiol Scand*, 1985;124:409–412.

Wang F, Li J, Wu C. The GABA(A) receptor mediates the hypnotic activity of melatonin in rats. *Pharmacol Biochem Behav*, 2003;74:573–578.

Yang Z, Li C, Huang F. Melatonin impaired acquisition but not expression of contextual fear in rats. *Neurosci Lett*, 2013;552:10–14.

Yildirim M, Marangoz C. Anticonvulsant effects of melatonin on penicillin-induced epileptiform activity in rats. *Brain Res*, 2006;1099:183–188.

Zeitzer JM, Daniels JE, Duffy JF, et al. Do plasma melatonin concentrations decline with age? *Am J Med*, 1999;107:432–436.

Zhdanova IV, Wurtman RJ, Lynch HJ. Sleep-inducing effects of low doses of melatonin ingested in the evening. *Clin Pharmacol Ther* 1995;57:552–558.

24

5-Hydroxy-tryptamine (5HT)-Related
Therapy in Neurological Diseases

Robert Lalonde
University of Rouen
Mont-Saint-Aignan, France

Catherine Strazielle
University of Lorraine
Vandoeuvre-les-Nancy, France

CONTENTS

ABSTRACT Concentrations of 5HT increased in the cerebellum and brainstem of *Nna1*$^{pcd-1J}$ (*Purkinje cell degeneration*) mutant mice. Increased cerebellar 5HT concentrations were also found in *Grid2*^{Lc} (*Lurcher*), *Grid2*^{ho} (*hot-foot*), *Rora*^{sg} (*staggerer*), and *Girk2*^{Wv} (*Weaver*) mutant mice with cerebellar atrophy. In a similar fashion, increased levels of 5-hydroxyindoleacetic acid (5HIAA), the main 5HT metabolite, were reported in the cerebrospinal fluid of patients with spinocerebellar atrophy. Based on the hypothesis of insufficient neurotransmission, pharmacotherapy with the 5HT precursors 5-hydroxy-tryptophan (5HTP) and tryptophan improved motor coordination in patients or mice with cerebellar atrophy, respectively. 5HT-related therapy has also been successful in other neurological conditions such as akinesia, tardive dyskinesia, and spinal trauma. Tardive dyskinesia can be counteracted by substances that potentiate the 5HT synapse, such as 5HTP, citalopram, fluoxetine, and paroxetine. Likewise, spinal trauma can be counteracted by 5HT and 5HTP as well as 5HT receptor agonists.

KEY WORDS: *serotonin, spinocerebellar atrophy, Parkinson's disease, schizophrenia, tardive dyskinesia, Tourette syndrome, Rett syndrome, Angelman syndrome, amyotrophic lateral sclerosis, spinal cord injury, cerebellum.*

Abbreviations

5HTP:	5-hydroxytryptophan
5HT:	5-hydroxytryptamine
8-OH-DPAT:	8-hydroxy-2-dipropylaminotetralin
ALS:	amyotrophic lateral sclerosis
CSF:	cerebrospinal fluid
GABA:	gamma-aminobutyric acid
mCPP:	meta-chlorophenyl-piperazine
MPTP:	1-methyl-4-phenyl-1,2,3,6-tetrahydropyridine
PCPA:	parachlorophenylalanine

Introduction

Serotonin (5-hydroxytryptamine; 5HT)-related therapy has been initiated in several neurological diseases, including cases of cerebellar degeneration. The main cerebellar afferents are climbing fibers originating from the inferior olive and mossy fibers originating from the pontine, reticular, and spinal regions (Voogd, 1995). The cerebellum is also innervated by 5HT fibers originating from medullary reticular formation and the dorsal raphe (Bishop et al., 1993). 5HT is synthesized from tryptophan, converted to 5-hydroxytryptophan (5HTP) in a reaction catalyzed by tryptophan hydroxylase and in turn converted to 5HT in a reaction catalyzed by aromatic L-amino acid decarboxylase (Cooper et al., 2003). The main metabolite of 5HT is 5-hydroxyindoleacetic acid (5HIAA) in a reaction catalyzed by monoamine oxidase.

Cerebellar Degeneration

Spontaneous Mutant Mice

Levels of 5HT have been examined in several spontaneous mutants with cerebellar degeneration (Table 24.1). In particular, 5HT concentrations (levels relative to brain weight) increased in the cerebellum and brainstem of $Nna1^{pcd-1J}$ (*Purkinje cell degeneration*) mutant mice (Ghetti et al., 1988; Ohsugi et al., 1986). Cerebellar 5HT concentrations likewise increased in $Grid2^{Lc}$ (*Lurcher*) (Reader et al., 1999; Strazielle et al., 1996), $Grid2^{ho}$ (*hot-foot*) (Draski et al., 1994), $Rora^{sg}$ (*staggerer*) (Lalonde and Strazielle, 2006; Ohsugi et al., 1986), and $Girk2^{Wv}$ (*Weaver*) (Ohsugi et al., 1986) mutants, but not the Dst^{dt-J} (*dystonia musculorum*) mutant marked by selective damage to spinocerebellar tracts (Ase et al., 2000). In $Nna1^{pcd-1J}$ (Le Marec et al., 1998) and $Grid2^{Lc}$ (Strazielle et al., 1996) mutants, there was also an increase in cerebellar 5HT reuptake binding. Likewise, $Grid2^{Lc}$ mutants are characterized by increased numbers of cerebellar 5HT fibers (Triarhou and Ghetti, 1991). In addition, 5HT content (levels not relative to brain weight) showed an age-related increase in $Nna1^{pcd-1J}$ mutants, indicating that

TABLE 24.1

Cerebellar 5HT Levels in Spontaneous Mutants with Cerebellar Degeneration

Mutant	5HT levels	References
$Nna1^{pcd-1J}$	↑	Ghetti et al., 1988; Ohsugi et al., 1986
$Grid2^{Lc}$	↑	Reader et al., 1999; Strazielle et al., 1996
$Grid2^{ho}$	↑	Draski et al., 1994
$Rora^{sg}$	↑	Lalonde and Strazielle, 2006; Ohsugi et al., 1986
$Girk2^{Wv}$	↑	Ohsugi et al., 1986

↑ Increased vs. controls.

5HT levels stayed high despite a shrinking cerebellum (Ghetti et al., 1988). The absence of such a result in *Rora^sg* and *Girk2^Wv* (Ohsugi et al., 1986) may be due to their being tested at too young an age. Thus, 5HT fiber innervation of a damaged cerebellum may be elevated or just maintained, but not decreased. A selective increase in 5HT reuptake sites occurred in the deep cerebellar nuclei but not the cerebellar cortex in the *Grid2^Lc* cerebellar mutant (Strazielle et al., 1996) as well as the *Dst^dt-J* spinocerebellar mutant (Ase et al., 2000). In both mutants, elevated binding also occurred in cerebellar-related regions, such as the red nucleus, dorsal raphe, and ventral tegmental areas in *Grid2^Lc* mutants, as well as the vestibular nuclei and ventral tegmental areas in *Dst^dt-J* mutants. In the spinal cord of *Dst^dt-J* mutants, increased 5HT concentrations corresponded to the same total amount of indoleamines in the area when taking into account its reduced size.

The increase in 5HT concentrations, fiber number, and reuptake sites may reflect a compensatory reaction to blocked neurotransmission. Although no change occurred regarding brainstem tryptophan hydroxylase activity in any cerebellar mutant, including *Nna1^pcd-1J*, *Rora^sg*, *Girk2^Wv*, and *Reln^rl* (Ohsugi et al., 1986), 5HT-related therapy has been undertaken based on the hypothesis of insufficient 5HT neurotransmission and positive results have been obtained. Indeed, peripheral injections of the 5HT precursor, L-tryptophan, speeded up movement times of *Grid2^Lc* mutants suspended upside down from a steel bar in the coat-hanger test (Le Marec et al., 2001). In the same mutant and test, buspirone, a 5HT1A receptor agonist, increased latencies before falling. Although buspirone likely decreases 5HT neurotransmission in normal brain due to a predominant action at presynaptic sites, the opposite may occur in brain-damaged animals. On the contrary, buspirone had no effect in *SCA1*/154Q mice on the rotorod test (Nag et al., 2013), perhaps indicating mutation and test selectivity.

Human Spinocerebellar Ataxias

5HT and 5HIAA concentrations were measured postmortem in the cerebellar cortex (Kish et al., 1992a) and striatum (Kish et al., 1992b) of subjects diagnosed with dominantly inherited spinocerebellar ataxias (SCAs). The results showed higher 5HIAA though not 5HT levels than those of age-matched controls in both brain regions, a sign of increased 5HT turnover. No such increase in 5HIAA concentrations occurred in the cerebrospinal fluid of SCA patients of a more varied etiology (Botez and Young, 2001). In contrast, cerebrospinal fluid (CSF) 5HIAA levels decreased in patients with Friedreich's ataxia marked by spinocerebellar atrophy.

Based on the hypothesis of insufficient 5HT neurotransmission in the cerebellum and perhaps cerebellar-related pathways, 5HT-related therapy was initiated in cerebellar patients of various etiologies by Trouillas et al. (1988), who administered the immediate 5HT precursor, 5HTP, and found improved motor control in a double-blind study. Similar ameliorations were reported after the administration of buspirone in open-label (Lou et al., 1995) and double-blind (Trouillas et al., 1997) studies, as found in the murine *Grid2^Lc* cerebellar mutant cited previously (Le Marec et al., 2001). In addition, motor coordination improved after administration of another 5HT1A receptor agonist, tandospirone, in *SCA3* and *SCA6* patients (Takei et al., 2002, 2010). On the contrary, no such effect occurred in *SCA1* and *SCA2* patients (Takei et al., 2010). Likewise, no effect was found with 5HTP in patients with cerebellar cortical atrophy (Manto et al., 1997), or else with buspirone in a double-blind, placebo-controlled, crossover trial (Assadi et al., 2007), or a in single-subject, double-blind, placebo-controlled study (Holroyd-Leduc et al., 2005). In addition, no benefit accrued in *SCA3* patients after the administration of fluoxetine, a 5HT reuptake blocker (Monte et al., 2003). A review based on fourteen clinical trials indicated no effectiveness with any pharmacologic treatment for the speech disorders in hereditary ataxia (Vogel et al., 2014). As in animal models, the differing results are a sign of mutation- and test-specific effects (Gazulla and Modrego, 2008).

Parkinson's Disease, Schizophrenia, and Tardive Dyskinesia

Although dopamine deficiency is the main biochemical deficit in Parkinson's disease, the 5HT system contributes to its symptoms (Beaudoin-Gobert and Sgambato-Faure, 2014; Huot and Fox, 2013; Jenner et al., 1983; Rylander, 2012). This occurs because 5HT-containing neurons originating from dorsal and medial raphe nuclei innervate forebrain dopamine-containing areas (Halliday et al., 1995). In addition to the vital importance of the 5HT transporter in the depressive symptoms seen in patients with Parkinson's disease (Mössner et al., 2000), 5HT reuptake blockers and 5HT receptor agonists or antagonists counteract some of their motor disturbances (Miyawaki et al., 1997). In particular, the 5HT system appears to be involved in akinesia, one of the cardinal symptoms of Parkinson's disease. In rats, the 5HT1A receptor agonist, buspirone, improved the anti-akinesia effect of L-dopa in 6-hydroxydopamine-lesioned rats (Mahmoudi et al., 2011). In addition, the 5HT1B receptor agonist CP93129 reversed the akinesia resulting from intrapallidal injection of the dopamine-depleting agent reserpine, as well as inhibiting GABA release from globus pallidus slices (Chadha et al., 2000). SB242,084, a 5HT2C receptor antagonist, but not 5HT2A (M100,907) or 5HT2A/C (ketanserin) receptor antagonists, reduced haloperidol-induced catalepsy (Creed-Carson et al., 2011).

The 5HT system also appears to be involved in L-dopa-mediated tardive dyskinesia in Parkinson's disease. Postmortem analyses in parkinsonian patients revealed low 5HT levels in dorsomedial but not dorsolateral striatum and no change in the 5HT transporter of either region (Cheshire et al., 2015). In particular, chronic L-dopa administration reduced striatal 5HT concentrations (Riahi et al., 2012) and increased 5HT1A receptor binding (Riahi et al., 2011) in monkeys injected with 1-methyl-4-phenyl-1,2,3,6-tetrahydropyridine (MPTP).

Atypical neuroleptics such as clozapine, quetiapine, and olanzapine used in schizophrenia block not only D2 dopamine receptors but also 5HT2A/2C receptors, a property that contributes to their minimal impact on basal ganglia-related symptoms (Aquino and Lang, 2014; Wolters, 1999), including tardive dyskinesia (Brotchie, 2005). There may be involvement of 5HT2C receptors based on evidence of hypersensitivity to mCPP (meta-chlorophenyl-piperazine), a 5HT2C receptor agonist, after chronic administration of haloperidol in rats (Ikram et al., 2007).

Agents acting on 5HT receptors have been shown to affect dyskinesias in humans and animals. In a single case study, the 5HT1A receptor agonist risperidone reduced dyskinesias caused by chronic haloperidol administration (Suenaga et al., 2000). Meco et al. (2003) investigated the antidyskinetic properties of mirtazapine, another 5HT1A receptor agonist and 5HT2 receptor antagonist, in an open-label study of parkinsonian patients suffering from L-dopa-induced dyskinesias. Mirtazapine was effective in reducing the dyskinesias either alone or combined with the dopamine reuptake blocker amantadine. In rats chronically treated with haloperidol, ketanserin and SB242,084 attenuated vacuous chewing movements, whereas M100,907 had no effect, indicating sensitivity to the 5HT2C receptor. Also in rats, 5HT1A receptor agonists such as buspirone (Azkona et al., 2014; Eskow et al., 2007; Lindenbach et al., 2015) and 8-hydroxy-2-dipropylaminotetralin (8-OH-DPAT) (Lindenbach et al., 2015), 5HT1A/1B receptor agonists such as eltoprazine (Bézard et al., 2013a) and anpirtoline (Bézard et al., 2013b), as well as 5HTP (Tronci et al., 2013) and 5HT reuptake inhibitors, including citalopram (Bishop et al., 2012; Conti et al., 2014; Kuan et al., 2008; Lindenbach et al., 2015), fluoxetine (Nevalainen et al., 2014), and paroxetine (Bishop et al., 2012; Conti et al., 2014), were all effective in reducing L-dopa-induced dyskinesias. 8-OH-DPAT also attenuated dyskinesias induced by D-amphetamine (Smith et al., 2012), a dopamine D1 receptor agonist (Dupre et al., 2013), and dopamine-rich grafts (Shin et al., 2012). Buspirone suppressed the dyskinetic effects of dopamine-rich grafts as well (Shin et al., 2012). Moreover, the dyskinetic effects of D-amphetamine (Smith et al., 2012) and dopamine-rich grafts (Shin et al., 2012) were mitigated after injections of the 5HT1B agonist CP94253. In addition to the 5HT-depleting action of L-DOPA administration (Riahi et al., 2012), these results favor the hypothesis of an antidyskinetic action of substances that facilitate 5HT transmission.

Tourette Syndrome

Only suggestive evidence exists concerning the role of 5HT in Tourette syndrome, marked by tics of the mouth, nose, and eye as well as obsessive compulsive symptoms, probably as a result of interrupted basal ganglia circuitries (Müller-Vahl et al., 2014; Worbe et al., 2015). Dopamine and noradrenaline seem the most important transmitters regarding symptomatology, but 5HT, glutamate, and gamma-aminobutyric acid (GABA) have also been discussed (Paschou et al., 2013; Udvardi et al., 2013). Current therapy most commonly entails the administration of noradrenergic alpha-receptor agonists such as clonidine or antidopaminergic agents (Kurlan, 2014; Roessner et al., 2013), but also combined dopamine-5HT-related pharmacotherapy (Steeves and Fox, 2008).

Whole-blood 5HT and tryptophan levels were measured in patients with Tourette syndrome under medication-free baseline conditions and after acute or chronic treatment with the alpha-2 receptor agonist clonidine (Leckman et al., 1984). Compared with normal controls, Tourette patients had lower baseline tryptophan but not 5HT levels. After an acute dose of clonidine, no difference was observed on either measure. After chronic treatment with clonidine, 5HT but not tryptophan levels were higher when expressed per platelet. In a second study, postmortem tissue from frontal or occipital neocortex of patients with Tourette syndrome was analyzed (Yoon et al., 2007). Although dopamine transporter and D2 dopamine receptors increased in density in the frontal lobe of patients relative to age-matched controls, no such effect occurred on 5HT1A receptors.

5HT reuptake binding was reduced in patients with Tourette syndrome relative to controls on single-photon emission computed tomography (SPECT) scans (Müller-Vahl et al., 2005), and alleles of the 5HT reuptake site have been associated with Tourette syndrome (Moya et al., 2013). The 5HT reuptake blocker fluoxetine counteracted obsessive compulsive symptoms but was without effect on tics in a double-blind, placebo-controlled study (Scahill et al., 1997). Another 5HT reuptake blocker, fluvoxamine, has also been used successfully at least on obsessive compulsive symptoms (Goodman et al., 1997). Ondansetron, a 5HT3 receptor antagonist, reduced tics in a three-week randomized, double-blind, placebo-controlled study (Toren et al., 2005).

Rett Syndrome

Rett syndrome is caused by a deficiency in *MECP2* situated on chromosome X and encoding methyl CpG–binding protein 2 (Amir et al., 1999). The disease starts in early childhood and affects mainly girls, presumably because affected boys die in embryo. Symptoms comprise loss in purposeful hand uses and the appearance of stereotyped ones, together with gait anomalies, autism, seizures, mental retardation, and autonomic dysfunction, as in breathing (Carter et al., 2010; Downs et al., 2010; Goldman and Temudo, 2012; Weng et al., 2011). It has been proposed that brainstem 5HT and noradrenaline neurons are involved in the patients' abnormal locomotion, and dopamine neurons in their stereotypies (Segawa et al., 2001). 5HT may be involved either primarily or through dysfunction of midbrain DA neurons via the pedunculopontine nuclei. *Mecp2*-null mutant mice are liable to deficits in motor coordination (Santos et al., 2010). LP-211, a 5HT7 receptor agonist, improved their motor coordination as well as spatial learning and object recognition (De Filippis et al., 2014, 2015).

5HT transporter binding in the dorsal motor nucleus of the vagus nerve serving preganglionic parasympathetic outflow critical for breathing was higher in subjects with Rett syndrome than that of controls (Paterson et al., 2005). As in human cases, *Mecp2* deficiency in mice disrupted respiratory systems (Viemari et al., 2005). 5HT levels were lower than those of wild type in whole brain (Ide et al., 2005), neocortex (Santos et al., 2010), medulla (Viemari et al., 2005), and cerebellum (Santos et al., 2010). Breathing capacity in *Mecp2*-deficient mice improved after administration of 5HT1A receptor agonists 8-OH-DPAT (Abdala et al., 2010) or WAY100135 (Levitt et al., 2013). Likewise, breathing capacity in patients improved after administration of the 5HT1A receptor agonist buspirone (Andaku et al., 2005). The mouse mutant defect in CO_2 chemosensitivity, which contributes to breathing, was corrected by the 5HT reuptake blocker citalopram (Toward et al., 2013), in line with the hypothesis that breathing troubles are caused by insufficient 5HT transmission.

Angelman Syndrome

Maternal deletion of *UBE3a* causes Angelman syndrome, characterized by mental retardation and disturbances in speech, motor abilities, and emotion (Barry et al., 2005; Beckung et al., 2004; Peters et al., 2004). There is evidence that the *Ube3a* gene product regulates monoamine synthesis in Drosophila brain (Ferdousy et al., 2011). Farook et al. (2012) examined monoamine levels in striatum, ventral midbrain, frontal neocortex, cerebellar cortex, and hippocampus in *Ube3a*-deficient mice. Relative to the wild type, 5HT levels were elevated in neocortex and striatum of *Ube3a*-deficient mice. These results should prompt further investigations into the possible influence of this neurotransmitter on the neurological signs of the mouse model and the patients. *Ube3a*-deficient mice are also characterized by impaired glucocorticoid signaling, high serum corticosterone levels, downregulated parvalbumin-positive inter-neurons in hippocampus and basolateral amygdala, and anxiogenic tendencies, all of these reversed by fluoxetine administration (Godavarthi et al., 2014).

Amyotrophic Lateral Sclerosis

Motoneuron degeneration is the pathological hallmark of amyotrophic lateral sclerosis (ALS) (Ravits et al., 2013). 5HT may contribute to the pathogenesis, a hypothesis based on evidence that regions densely innervated by 5HT projections, such as motoneurons in ventral horn and motor cortex as well as trigeminal motor, facial motor, ambiguus, and hypoglossal nuclei, appear more vulnerable to damage in ALS than regions more sparsely innervated, such as oculomotor, trochlear, and abducens nuclei as well as the cerebellum (Sandyk, 2006). This vulnerability may be mediated by glutamate. 5HT counteracted several effects caused by glutamatergic n-methyl-d-asparate (NMDA) receptors in frontal cortex but also increased the glutamate-induced excitability of spinal motoneurons (Ciranna, 2006).

Several types of data point toward decreased 5HT transmission in ALS. Blood platelet 5HT levels were lower in ALS patients than controls and positively correlated with the patients' life span (Dupuis et al., 2010). In ALS, 5HT projections to spinal motoneurons are liable to degenerate (Dentel et al., 2013). These data are in accord with those indicating lower 5HT and 5HIAA concentrations in the thoracic spinal cord of ALS patients (Sofic et al., 1991). The 5HT/5HIAA ratio increased in the cervical spinal cord of ALS patients relative to controls, indicating reduced 5HT turnover (Bertel et al., 1991). However, 5HT concentrations were higher in ALS subjects in the lateral part of the lumbar ventral horn and lateral white matter. In another study, no change occurred in 5HT or 5HIAA concentrations in the cervical spinal cord (Ohsugi et al., 1987). Nevertheless, 5HT1A receptor densities increased in the spinal cord of patients, perhaps a compensatory response to decreased 5HT levels (Manaker et al., 1988). In an opposite fashion, 5HT1A receptor binding decreased in cortical and raphe nuclei, representing loss of neurons bearing this receptor (Turner et al., 2005).

The functional impacts of altered 5HT neurotransmission on ALS-like animal models have been presented. Unlike normal newborn mice, where flexor-extensor locomotor-like activity was found in lumbar ventral roots of an *in vitro* brainstem spinal cord preparation exposed to NMDA and 5HT, this was rendered more difficult in *SOD1* (superoxide dismutase 1) transgenic mice harboring the G85R mutation (Amendola et al., 2004). But no such effect occurred in *SOD1*/G93A transgenic mice from lumbar ventral roots of the *in vitro* spinal cord preparation (Milan et al., 2014). Moreover, 5HT deple-tion after administration of the tryptophan hydroxylase inhibitor parachlorophenylalanine (PCPA), or 5HT receptor blockade after methysergide, had no impact on the breathing capacity of *SOD1*/G93A rats (Nichols et al., 2014). With respect to *in vivo* experiments, the 5HT precursor 5HTP delayed hindlimb weakness and death in *SOD1*/G93A transgenic mice (Turner et al., 2003). On the contrary, fluoxetine had no effect when administered in juvenile or adult stages and even decreased life span when administered in the neonatal period (Koschnitzky et al., 2014). The 5HT2A receptor antagonist cyproheptadine had no effect.

Despite inconclusive evidence, progressive degeneration of 5HT input may worsen motoneuron degeneration. Conversely, boosting 5HT synaptic function via reuptake inhibitors and 5HT receptor agonists or antagonists may slow down symptoms by preventing excessive glutamatergic transmission on motoneurons.

Spinal Cord Injury

Ghosh and Pearse (2015) and Nardone et al. (2015) reviewed evidence regarding the facilitative effects of 5HT-related substances on limb coordination after spinal cord injury. Facilitating 5HT transmis-sion can promote movements by activating the locomotor central pattern generator below the site of injury. Another method is by transplanting fetal 5HT neurons at the site of injury, supplying 5HT axon regrowth below the lesion. Data on drug-related effects are summarized in Table 24.2. In chronic spinal

TABLE 24.2

Effects of Stimulating 5HT Transmission on Locomotor Abilities after Spinal Cord Injury in Humans and Animals

Drug	Action	References
8-OH-DPAT (5HT1A/7 agonist)	Improved	Antri et al., 2003; Ichiyama et al., 2008; Landry et al., 2006; Lapointe and Guertin, 2008; Musienko et al., 2011; Slawinska et al., 2014
	No change	Brustein and Rossignol, 1999 (cats); Hayashi et al., 2010
Quipazine (5HT2A/C agonist)	Improved	Antri et al., 2002; Brustein and Rossignol, 1999 (cats); Cowley et al., 2015; Landry and Guertin, 2004; Musienko et al., 2011; Slawinska al., 2014
8-OH-DPAT + quipazine	Improved	Antri et al., 2005; Slawinska al., 2014
5HT	Improved	Brustein and Rossignol, 1999
5TP (5HT precursor)	Improved	Brustein and Rossignol, 1999; Hayashi et al., 2010
mCPP (5HT2C receptor agonist)	Improved	Murray et al., 2010
	No change	Hayashi et al., 2010
Fenfluramine (5HT reuptake blocker)	No change	Hayashi et al., 2010
Escitalopram (5HT reuptake blocker)	No change	Thompson and Hornby, 2013 (humans)

Note: Mice or rats were tested unless otherwise indicated.

rats, peripheral injections of 8-OH-DPAT and quipazine, respectively 5HT1A/7 and 5HT2A/C receptor agonists, ameliorated hindlimb coordination on a treadmill (Musienko et al., 2011; Slawinska al, 2014), as did grafting of fetal medulla 5HT neurons below the level of thoracic spinal injury, an effect mediated by the same receptors (Slawinska al., 2013). Intraspinal grafts of embryonic spinal (Kim et al., 1999) or raphe (Majczynski et al., 2005: Ribotta et al., 2000) nuclei and neural stem/progenitor cells (Nishimura et al., 2013) were equally effective. On the contrary, no change occurred in rat open-field locomotion, gait, or footfall errors on a grid after spinal transplant of human 5HT-containing hNT2.19 cells, despite relief from neuropathic pain (Eaton et al., 2011). In confirmation of these reports, Landry et al. (2006), Ichiyama et al. (2008), and Antri et al. (2003) showed improved locomotion after thoracic spinal injury in rats injected peripherally with 8-OH-DPAT, as did Lapointe and Guertin (2008) in mice. Antri et al. (2002), Cowley et al. (2015), and Landry and Guertin (2004) reported the same positive results with quipazine in rats.

The effects of intrathecal 8-OH-DPAT and quipazine as well as 5HT and 5HTP were examined in cats locomoting on a treadmill after thoracic lesions (Brustein and Rossignol, 1999). 5HT, 5HTP, and quipazine improved regular hindlimb stepping characterized by smooth movements with lateral stability and increased step cycle duration. But 8-OH-DPAT did not produce this effect, perhaps because of a nonoptimal dose. When treadmill locomotion was evaluated after moderate or severe thoracic contusion in rats, peripheral 5HTP administration improved weight-supported treadmill locomotion after moderate but not severe contusion, whereas 8-OH-DPAT had no effect, irrespective of lesion severity (Hayashi et al., 2010). Instead, the combined application of 8-OH-DPAT and quipazine was more effective than either substance alone (Antri et al., 2005; Slawinska et al., 2014). Nonoptimal stimulation of these receptors may explain why systemic administration of the 5HT2C receptor agonist metachlorophenylpiperazine (mCPP) or the 5HT reuptake blocker fenfluramine were without effect either (Hayashi et al., 2010). However, in another study (Kim et al., 2001), systemic mCPP promoted recovery of weight-supported steps on a treadmill after thoracic injury in rats. Murray et al. (2010) found that after spinal transection in rats, changes in posttranscriptional editing of 5HT2C receptor mRNA increased expression of 5HT2C receptor isoforms spontaneously active without 5HT, which promoted motor recovery despite debilitating muscle spasms, the latter eliminated by antagonists of constitutively active 5HT2C receptors such as cyproheptadine or SB206553. The 5HT2C receptor was upregulated in dorsal and ventral horns at the lumbar level in severe relative to moderate thoracic contusion in rats (Hayashi et al., 2010). Moreover, 5HT2A receptor immunoreactivity below the lesion increased in motoneuron soma and proximal dendrites after chronic spinal transection in rats (Kong et al., 2010). In humans with spinal cord injury, no change occurred in walking ability in individuals with chronic incomplete spinal cord injury after administration of escitalopram, a 5HT reuptake blocker, or cyproheptadine, a 5HT2A/C receptor antagonist, although the former increased muscle strength, while the latter decreased it (Thompson and Hornby, 2013). Perhaps optimal activation of 5HT1A/7 and 5HT2A/C receptors are necessary to obtain a positive outcome.

Hindlimb tremors were noted as a side effect of 5HTP-induced improvements in rats, a mark of 5HT hypersensitivity (Hayashi et al., 2010). Likewise, supersensitivity to 5HTP occurred on monosynaptic and polysynaptic reflexes in chronic spinal rats (Nagano et al., 1988). The authors proposed that this action is due to a lack of 5HT uptake into 5HT-containing nerve terminals rather than to upregulated 5HT receptors, because the 5HT receptor agonist 5-methoxy-N,N-dimethyltryptamine (5-Me-DMT) had no such effect. The mechanism of 5HTP-related effects on spinal cord lesions was examined by Li et al. (2014). In spinal cord of normal rats, most of the aromatic amino acid decarboxylase is confined to brain stem–derived monoamine fibers, but caudal to transaction, the enzyme was found in blood vessel endothelial cells and pericytes as well as in a novel group of neurons that upregulate it, but without endogenous 5HT synthesis in the spinal cord. However, when 5HTP was applied exogenously *in vitro* or *in vivo*, enzyme-containing vessels and neurons synthesized 5HT, which increased motoneuron activity and muscle spasms (long-lasting reflexes) by acting on SB206553-sensitive 5HT2 receptors located on motoneurons. The blockade of monoamine oxidase increased the sensitivity of long-lasting reflexes to 5HTP more than it increased the sensitivity of motoneurons to 5HT, indicating that 5HT synthesized from aromatic amino acid decarboxylase was mostly metabolized in enzyme-containing neurons and vessels.

In addition to locomotion, the effects of 5HT-related drugs have been examined on breathing rhythms. Zhou and Goshgarian (1999) investigated the effects of 5HTP on crossed phrenic nerve activity after cervical spinal cord hemisection on anesthetized, vagotomized, paralyzed, and artificially ventilated rats. Intravenously administered 5HTP increased crossed phrenic nerve activity, an effect reversed by the nonselective 5HT receptor antagonist methysergide. In the same preparation, 5HTP facilitated respiratory recovery in the phrenic nerve ipsilateral to hemisection, also an effect reversed by methysergide (Zhou and Goshgarian, 2000). Although this recovery was initially accompanied by increased activity in the contralateral phrenic nerve, it persisted after this activity returned to predrug levels. Moreover, episodic spinal serotonin receptor activation elicited long-lasting phrenic motor facilitation (MacFarlane and Mitchell, 2009).

Conclusion

Based on the hypothesis of insufficient 5HT neurotransmission, there exists limited though intriguing evidence of improved motor coordination in patients or mice with cerebellar atrophy after administration of 5HTP, tryptophan, or buspirone. 5HT-related therapy has also been successful in other neurological conditions such as akinesia, tardive dyskinesia, and spinal trauma. Buspirone appears effective in counteracting akinesia but also tardive dyskinesia. The latter is probably mediated at postsynaptic receptors based on findings of improvement of L-dopa-induced dyskinesias with substances that potentiate the 5HT synapse such as 5HTP, citalopram, fluoxetine, and paroxetine. It remains to be determined whether similar improvements may be forthcoming in Tourette, Rett, and Angelman syndromes as well as amyotrophic lateral sclerosis.

REFERENCES

Abdala AP, Dutschmann M, Bissonnette JM, Paton JF. Correction of respiratory disorders in a mouse model of Rett syndrome. *Proc Natl Acad Sci USA* 2010;107:18208–18213.

Amendola J, Verrier B, Roubertoux P, Durand J. Altered sensorimotor development in a transgenic mouse model of amyotrophic lateral sclerosis. *Eur J Neurosci* 2004;20:2822–2826.

Amir RE, Van den Veyver IB, Wan M, Tran CQ, Francke U, Zoghbi HY. Rett syndrome is caused by mutations in X-linked MECP2, encoding methyl-CpG-binding protein 2. *Nat Genet* 1999;23:185–188.

Andaku DK, Mercadante MT, Schwartzman JS. Buspirone in Rett syndrome respiratory dysfunction. *Brain Dev* 2005;27:437–438.

Antri M, Barthe JY, Mouffle C, Orsal D. Long-lasting recovery of locomotor function in chronic spinal rat following chronic combined pharmacological stimulation of serotonergic receptors with 8-OHDPAT and quipazine. *Neurosci Lett* 2005;384:162–167.

Antri M, Mouffle C, Orsal D, Barthe JY. 5-HT1A receptors are involved in short- and long-term processes responsible for 5-HT-induced locomotor function recovery in chronic spinal rat. *Eur J Neurosci* 2003;18:1963–1972.

Antri M, Orsal D, Barthe JY. Locomotor recovery in the chronic spinal rat: Effects of long-term treatment with a 5-HT2 agonist. *Eur J Neurosci* 2002;16:467–476.

Aquino CC, Lang AE. Tardive dyskinesia syndromes: Current concepts. *Parkinsonism Relat Disord* 2014;20(Suppl 1):S113–117.

Ase A, Strazielle C, Hébert C, Botez MI, Lalonde R, Descarries L, Reader TA. The central serotonin system in dystonia musculorum mutant mice: Biochemical, autoradiographic and immunocytochemical data. *Synapse* 2000;37:179–193.

Assadi M, Campellone JV, Janson CG, Veloski JJ, Schwartzman RJ, Leone P. Treatment of spinocerebellar ataxia with buspirone. *J Neurol Sci* 2007;260:143–146.

Azkona G, Sagarduy A, Aristieta A, Vazquez N, Zubillaga V, Ruíz-Ortega JA, Pérez-Navarro E, Ugedo L, Sánchez-Pernaute R. Buspirone anti-dyskinetic effect is correlated with temporal normalization of dysregulated striatal DRD1 signalling in L-DOPA-treated rats. *Neuropharmacology* 2014;79:726–737.

Barry RJ, Leitner RP, Clarke AR, Einfeld SL. Behavioral aspects of Angelman syndrome: A case control study. *Am J Med Genet A* 2005;132:8–12.

Beaudoin-Gobert M, Sgambato-Faure V. Serotonergic pharmacology in animal models: From behavioral disorders to dyskinesia. *Neuropharmacology* 2014;81:15–30.

Beckung E, Steffenburg S, Kyllerman M. Motor impairments, neurological signs, and developmental level in individuals with Angelman syndrome. *Dev Med Child Neurol* 2004;46:239–243.

Bertel O, Malessa S, Sluga E, Hornykiewicz O. Amyotrophic lateral sclerosis: Changes of noradrenergic and serotonergic transmitter systems in the spinal cord. *Brain Res* 1991;566:54–60.

Bézard E, Muñoz A, Tronci E, Pioli EY, Li Q, Porras G, Björklund A, Carta M. Anti-dyskinetic effect of anpirtoline in animal models of L-DOPA-induced dyskinesia. *Neurosci Res* 2013b;77:242–246.

Bézard E, Tronci E, Pioli EY, Li Q, Porras G, Björklund A, Carta M. Study of the antidyskinetic effect of eltoprazine in animal models of levodopa-induced dyskinesia. *Mov Disord* 2013a;28:1088–1096.

Bishop C, George JA, Buchta W, Goldenberg AA, Mohamed M, Dickinson SO, Eissa S, Eskow Jaunarajs KL. Serotonin transporter inhibition attenuates L-DOPA-induced dyskinesia without compromising L-DOPA efficacy in hemiparkinsonian rats. *Eur J Neurosci* 2012;36:2839–2848.

Bishop GA, Kerr CW, Chen YF, King JS. The serotoninergic system in the cerebellum: Origin, ultrastructural relationships, and physiological effects. In P Trouillas, K Fuxe (eds), *Serotonin, the Cerebellum, and Ataxia*. New York: Raven Press, 1993;91–112.

Botez MI, Young SN. Biogenic amine metabolites and thiamine in cerebrospinal fluid in heredo-degenerative ataxias. *Can J Neurol Sci* 2001;28:134–140.

Brotchie JM. Nondopaminergic mechanisms in levodopa-induced dyskinesia. *Mov Disord* 2005;20:919–1031.

Brustein E, Rossignol S. Recovery of locomotion after ventral and ventrolateral spinal lesions in the cat: II. Effects of noradrenergic and serotoninergic drugs. *J Neurophysiol* 1999;81:1513–1530.

Carter P, Downs J, Bebbington A, Williams S, Jacoby P, Kaufmann WE, Leonard H. Stereotypical hand movements in 144 subjects with Rett syndrome from the population-based Australian database. *Mov Disord* 2010;25:282–288.

Chadha A, Sur C, Atack J, Duty S. The 5HT(1B) receptor agonist, CP-93129, inhibits [(3)H]-GABA release from rat globus pallidus slices and reverses akinesia following intrapallidal injection in the reserpine-treated rat. *Br J Pharmacol* 2000;130:1927–1932.

Cheshire P, Ayton S, Bertram KL, et al. Serotonergic markers in Parkinson's disease and levodopa-induced dyskinesias. *Mov Disord* 2015;30:796–804.

Ciranna L. Serotonin as a modulator of glutamate- and GABA-mediated neurotransmission: implications in physiological functions and in pathology. *Curr Neuropharmacol* 2006;4:101–114.

Conti MM, Ostock CY, Lindenbach D, Goldenberg AA, Kampton E, Dell'isola R, Katzman AC, Bishop C. Effects of prolonged selective serotonin reuptake inhibition on the development and expression of L-DOPA-induced dyskinesia in hemi-parkinsonian rats. *Neuropharmacology* 2014;77:1–8.

Cooper JR, Bloom FE, Roth RH. *The Biochemical Basis of Neuropharmacology*. Oxford: Oxford University Press, 2003.

Cowley KC, MacNeil BJ, Chopek JW, Sutherland S, Schmidt BJ. Neurochemical excitation of thoracic propriospinal neurons improves hindlimb stepping in adult rats with spinal cord lesions. *Exp Neurol* 2015;264:174–187.

Creed-Carson M, Oraha A, Nobrega JN. Effects of 5-HT(2A) and 5-HT(2C) receptor antagonists on acute and chronic dyskinetic effects induced by haloperidol in rats. *Behav Brain Res* 2011;219:273–279.

De Filippis B, Chiodi V, Adriani W, Lacivita E, Mallozzi C, Leopoldo M, Domenici MR, Fuso A, Laviola G. Long-lasting beneficial effects of central serotonin receptor 7 stimulation in female mice modeling Rett syndrome. *Front Behav Neurosci* 2015;9:86.

De Filippis B, Nativio P, Fabbri A, Ricceri L, Adriani W, Lacivita E, Leopoldo M, Passarelli F, Fuso A, Laviola G. Pharmacological stimulation of the brain serotonin receptor 7 as a novel therapeutic approach for Rett syndrome. *Neuropsychopharmacology* 2014;39:2506–2518.

Dentel C, Palamiuc L, Henriques A, et al. Degeneration of serotonergic neurons in amyotrophic lateral sclerosis: A link to spasticity. *Brain* 2013;136:483–493.

Downs J, Bebbington A, Jacoby P, Williams AM, Ghosh S, Kaufmann WE, Leonard H. Level of purposeful hand function as a marker of clinical severity in Rett syndrome. *Dev Med Child Neurol* 2010;52:817–823.

Draski LJ, Nash, DJ, Gerhardt GA. CNS monoamine levels and motoric behaviors in the hotfoot ataxic mutant. *Brain Res* 1994;645:69–77.

Dupre KB, Ostock CY, George JA, Eskow Jaunarajs KL, Hueston CM, Bishop C. Effects of 5-HT1A receptor stimulation on D1 receptor agonist-induced striatonigral activity and dyskinesia in hemiparkinsonian rats. *ACS Chem Neurosci* 2013;4:747–760.

Dupuis L, Spreux-Varoquaux O, Bensimon G, Jullien P, Lacomblez L, Salachas F, Bruneteau G, Pradat PF, Loeffler JP, Meininger V. Platelet serotonin level predicts survival in amyotrophic lateral sclerosis. *PLoS One* 2010;5(10):e13346.

Eaton MJ, Widerström-Noga E, Wolfe SQ. Subarachnoid transplant of the human neuronal hNT2.19 serotonergic cell line attenuates behavioral hypersensitivity without affecting motor dysfunction after severe contusive spinal cord injury. *Neurol Res Int* 2011;2011:891605.

Eskow KL, Gupta V, Alam S, Park JY, Bishop C. The partial 5-HT(1A) agonist buspirone reduces the expression and development of L-DOPA-induced dyskinesia in rats and improves L-DOPA efficacy. *Pharmacol Biochem Behav* 2007;87:306–314.

Farook MF, DeCuypere M, Hyland K, Takumi T, LeDoux MS, Reiter LT. Altered serotonin, dopamine and norepinepherine levels in 15q duplication and Angelman syndrome mouse models. *PLoS One* 2012;7(8):e43030.

Ferdousy F, Bodeen W, Summers K, Doherty O, Wright O, Elsisi N, Hilliard G, O'Donnell JM, Reiter LT. Drosophila Ube3a regulates monoamine synthesis by increasing GTP cyclohydrolase I activity via a non-ubiquitin ligase mechanism. *Neurobiol Dis* 2011;41:669–677.

Gazulla J, Modrego P. Buspirone and serotonin in spinocerebellar ataxia. *J Neurol Sci* 2008;268:199–200.

Ghetti B, Perry KW, Fuller RW. Serotonin concentration and turnover in cerebellum and other brain regions of pcd mutant mice. *Brain Res* 1988;458:367–371.

Ghosh M, Pearse DD. The role of the serotonergic system in locomotor recovery after spinal cord injury. *Front Neural Circuits* 2015;8:151.

Godavarthi SK, Sharma A, Jana NR. Reversal of reduced parvalbumin neurons in hippocampus and amygdala of Angelman syndrome model mice by chronic treatment of fluoxetine. *J Neurochem* 2014;130:444–454.

Goldman S, Temudo T. Hand stereotypies distinguish Rett syndrome from autism disorder. *Mov Disord* 2012;27:1060–1062.

Goodman WK, Ward H, Kablinger A, Murphy T. Fluvoxamine in the treatment of obsessive-compulsive disorder and related conditions. *J Clin Psychiatry* 1997;58(Suppl 5):S32–49.

Halliday G, Harding A, Paxinos G. Serotonin and tachykinin systems. In G Paxinos (ed.), *The Rat Nervous System*. New York: Academic Press, 2nd edn, 1995;929–974.

Hayashi Y, Jacob-Vadakot S, Dugan EA, McBride S, Olexa R, Simansky K, Murray M, Shumsky JS. 5-HT precursor loading, but not 5-HT receptor agonists, increases motor function after spinal cord contusion in adult rats. *Exp Neurol* 2010;221:68–78.

Holroyd-Leduc JM, Liu BA, Maki BE, Zecevic A, Herrmann N, Black SE. The role of buspirone for the treatment of cerebellar ataxia in an older individual. *Can J Clin Pharmacol* 2005;12:e218–221.

Huot P, Fox SH. The serotonergic system in motor and non-motor manifestations of Parkinson's disease. *Exp Brain Res* 2013;230:463–476.

Ichiyama RM, Gerasimenko Y, Jindrich DL, Zhong H, Roy RR, Edgerton VR. Dose dependence of the 5-HT agonist quipazine in facilitating spinal stepping in the rat with epidural stimulation. *Neurosci Lett* 2008;438:281–285.

Ide S, Itoh M, Goto Y. Defect in normal developmental increase of the brain biogenic amine concentrations in the mecp2-null mouse. *Neurosci Lett* 2005;386:14–17.

Ikram H, Samad N, Haleem DJ. Neurochemical and behavioral effects of m-CPP in a rat model of tardive dyskinesia. *Pak J Pharm Sci* 2007;20:188–195.

Jenner P, Sheehy M, Marsden CD. Noradrenaline and 5-hydroxytryptamine modulation of brain dopamine function: Implications for the treatment of Parkinson's disease. *Br J Clin Pharmacol* 1983;15(Suppl. 2):S277–289.

Kim D, Adipudi V, Shibayama M, Giszter S, Tessler A, Murray M, Simansky KJ. Direct agonists for serotonin receptors enhance locomotor function in rats that received neural transplants after neonatal spinal transection. *J Neurosci* 1999;19:6213–6224.

Kim D, Murray M, Simansky KJ. The serotonergic 5-HT(2C) agonist m-chlorophenylpiperazine increases weight-supported locomotion without development of tolerance in rats with spinal transections. *Exp Neurol* 2001;169:496–500.

Kish SJ, Robitaille Y, el-Awar M, Clark B, Schut L, Ball MJ, Young LT, Currier R, Shannak K. Striatal monoamine neurotransmitters and metabolites in dominantly inherited olivopontocerebellar atrophy. *Neurology* 1992b;42:1573–1577.

Kish SJ, Robitaille Y, Schut L, el-Awar M, Ball MJ, Shannak K. Normal serotonin but elevated 5-hydroxyindoleacetic acid concentration in cerebellar cortex of patients with dominantly-inherited olivopontocerebellar atrophy. *Neurosci Lett* 1992a;144:84–86.

Kong XY, Wienecke J, Hultborn H, Zhang M. Robust upregulation of serotonin 2A receptors after chronic spinal transection of rats: An immunohistochemical study. *Brain Res* 2010;1320:60–68.

Koschnitzky JE, Quinlan KA, Lukas TJ, Kajtaz E, Kocevar EJ, Mayers WF, Siddique T, Heckman CJ. Effect of fluoxetine on disease progression in a mouse model of ALS. *J Neurophysiol* 2014;111:2164–2176.

Kuan W-L, Zhao JW, Barker RA. The role of anxiety in the development of levodopa-induced dyskinesias in an animal model of Parkinson's disease, and the effect of chronic treatment with the selective serotonin reuptake inhibitor citalopram. *Psychopharmacology* 2008;197:279–293.

Kurlan RM. Treatment of Tourette syndrome. *Neurotherapeutics* 2014;11:161–5.

Lalonde R, Strazielle C. Regional variations of 5HT concentrations in *Rora^sg* (*staggerer*) mutants. *Neurochem Res* 2006;31:921–924.

Landry ES, Guertin PA. Differential effects of 5-HT1 and 5-HT2 receptor agonists on hindlimb movements in paraplegic mice. *Prog Neuropsychopharmacol Biol Psychiatry* 2004;28:1053–1060.

Landry ES, Lapointe NP, Rouillard C, Levesque D, Hedlund PB, Guertin PA. Contribution of spinal 5-HT1A and 5-HT7 receptors to locomotor-like movement induced by 8-OH-DPAT in spinal cord-transected mice. *Eur J Neurosci* 2006;24:535–546.

Lapointe NP, Guertin PA. Synergistic effects of D1/5 and 5-HT1A/7 receptor agonists on locomotor movement induction in complete spinal cord-transected mice. *J Neurophysiol* 2008;100:160–168.

Leckman JF, Anderson GM, Cohen DJ, Ort S, Harcherik DF, Hoder EL, Shaywitz BA. Whole blood serotonin and tryptophan levels in Tourette's disorder: Effects of acute and chronic clonidine treatment. *Life Sci* 1984;35:2497–2503.

Le Marec N, Ase AR, Botez-Marquard T, Marchand L, Reader TA, Lalonde R. Behavioral and biochemical effects of L-tryptophan and buspirone in a model of cerebellar atrophy. *Pharmacol Biochem Behav* 2001;69:333–342.

Le Marec N, Hébert C, Amdiss F, Botez MI, Reader TA. Regional distribution of 5-HT transporters in the brain of wild type and "Purkinje cell degeneration" mutant mice: a quantitative autoradiographic study with [^{3}H]citalopram. *J Chem Neuroanat* 1998;15:155–171.

Levitt ES, Hunnicutt BJ, Knopp SJ, Williams JT, Bissonnette JM. A selective 5-HT1A receptor agonist improves respiration in a mouse model of Rett syndrome. *J Appl Physiol* 2013;115:1626–1633.

Li Y, Li L, Stephens MJ, Zenner D, Murray KC, Winship IR, Vavrek R, Baker GB, Fouad K, Bennett DJ. Synthesis, transport, and metabolism of serotonin formed from exogenously applied 5-HTP after spinal cord injury in rats. *J Neurophysiol* 2014;111:145–163.

Lindenbach D, Palumbo N, Ostock CY, Vilceus N, Conti MM, Bishop C. Side effect profile of 5-HT treatments for Parkinson's disease and L-DOPA-induced dyskinesia in rats. *Br J Pharmacol* 2015;172:119–130.

Lou JS, Goldfarb L, McShane L, Gatev P, Hallett M. Use of buspirone for treatment of cerebellar ataxia: An open-label study. *Arch Neurol* 1995;52:982–988.

MacFarlane PM, Mitchell GS. Episodic spinal serotonin receptor activation elicits long-lasting phrenic motor facilitation by an NADPH oxidase-dependent mechanism. *J Physiol* 2009;587:5469–5481.

Mahmoudi J, Nayebi AM, Samini M, Reyhani-Rad S, Babapour V. Buspirone improves the anti-cataleptic effect of levodopa in 6-hydroxydopamine-lesioned rats. *Pharmacol Rep* 2011;63:908–914.

Majczynski H, Maleszak K, Cabaj A, Slawinska U. Serotonin-related enhancement of recovery of hind limb motor functions in spinal rats after grafting of embryonic raphe nuclei. *J Neurotrauma* 2005;22:590–604.

Manaker S, Caine SB, Winokur A. Alterations in receptors for thyrotropin-releasing hormone, serotonin, and acetylcholine in amyotrophic lateral sclerosis. *Neurology* 1988;38:1464–1474.

Manto M, Hildebrand J, Godaux E, Roland H, Blum S, Jacquy J. Analysis of fast single-joint and multijoint movements in cerebellar cortical atrophy: Failure of L-hydroxytryptophan to improve cerebellar ataxia. *Arch Neurol* 1997;54:1192–1194.

Meco G, Fabrizio E, Di Rezze S, Alessandri A, Pratesi L. Mirtazapine in L-DOPA-induced dyskinesias. *Clin Neuropharmacol* 2003;26:179–181.

Milan L, Barrière G, De Deurwaerdère P, Cazalets JR, Bertrand SS. Monoaminergic control of spinal locomotor networks in SOD1G93A newborn mice. *Front Neural Circuits* 2014;8:77.

Miyawaki E, Meah Y, Koller WC. Serotonin, dopamine, and motor effects in Parkinson's disease. *Clin Neuropharmacol* 1997;20:300–310.

Monte TL, Rieder CR, Tort AB, Rockenback I, Pereira ML, Silveira I, Ferro A, Sequeiros J, Jardim LB. Use of fluoxetine for treatment of Machado-Joseph disease: An open-label study. *Acta Neurol Scand* 2003;107:207–210.

Mössner R, Schmitt A, Syagailo Y, Gerlach M, Riederer P, Lesch KP. The serotonin transporter in Alzheimer's and Parkinson's disease. *J Neural Transm Suppl* 2000;60:345–350.

Moya PR, Wendland JR, Rubenstein LM, Timpano KR, Heiman GA, Tischfield JA, King RA, Andrews AM, Ramamoorthy S, McMahon FJ, Murphy DL. Common and rare alleles of the serotonin transporter gene, *SLC6A4*, associated with Tourette's disorder. *Mov Disord* 2013;28:1263–1270.

Müller-Vahl KR, Grosskreutz J, Prell T, Kaufmann J, Bodammer N, Peschel T. Tics are caused by alterations in prefrontal areas, thalamus and putamen, while changes in the cingulate gyrus reflect secondary compensatory mechanisms. *BMC Neurosci* 2014;15:6.

Müller-Vahl KR, Meyer GJ, Knapp WH, Emrich HM, Gielow P, Brücke T, Berding G. Serotonin transporter binding in Tourette syndrome. *Neurosci Lett* 2005;385:120–125.

Murray KC, Nakae A, Stephens MJ, et al. Recovery of motoneuron and locomotor function after spinal cord injury depends on constitutive activity in 5-HT2C receptors. *Nat Med* 2010;16:694–700.

Musienko P, van den Brand R, Märzendorfer O, Roy RR, Gerasimenko Y, Edgerton VR, Courtine G. Controlling specific locomotor behaviors through multidimensional monoaminergic modulation of spinal circuitries. *J Neurosci* 2011;31:9264–9278.

Nag N, Tarlac V, Storey E. Assessing the efficacy of specific cerebellomodulatory drugs for use as therapy for spinocerebellar ataxia type 1. *Cerebellum* 2013;12:74–82.

Nagano N, Ono H, Ozawa M, Fukuda H. The spinal reflex of chronic spinal rats is supersensitive to 5-HTP but not to TRH or 5-HT agonists. *Eur J Pharmacol* 1988;149:337–344.

Nardone R, Höller Y, Thomschewski A, Höller P, Lochner P, Golaszewski S, Brigo F, Trinka E. Serotonergic transmission after spinal cord injury. *J Neural Transm* 2015;122:279–295.

Nevalainen N, Af Bjerkén S, Gerhardt GA, Strömberg I. Serotonergic nerve fibers in L-DOPA-derived dopamine release and dyskinesia. *Neuroscience* 2014;260:73–86.

Nichols NL, Johnson RA, Satriotomo I, Mitchell GS. Neither serotonin nor adenosine-dependent mechanisms preserve ventilatory capacity in ALS rats. *Respir Physiol Neurobiol* 2014;197:19–28.

Nishimura S, Yasuda A, Iwai H, et al. Time-dependent changes in the microenvironment of injured spinal cord affects the therapeutic potential of neural stem cell transplantation for spinal cord injury. *Mol Brain* 2013;6:3.

Ohsugi K, Adachi K, Ando K. Serotonin metabolism in the CNS in cerebellar ataxic mice. *Experientia* 1986;42:1245–1247.

Ohsugi K, Adachi K, Mukoyama M, Ando K. Lack of change in indoleamine metabolism in spinal cord of patients with amyotrophic lateral sclerosis. *Neurosci Lett* 1987;79:351–354.

Paschou P, Fernandez TV, Sharp F, Heiman GA, Hoekstra PJ. Genetic susceptibility and neurotransmitters in Tourette syndrome. *Int Rev Neurobiol* 2013;112:155–177.

Paterson DS, Thompson EG, Belliveau RA, Antalffy BA, Trachtenberg FL, Armstrong DD, Kinney HC. Serotonin transporter abnormality in the dorsal motor nucleus of the vagus in Rett syndrome: Potential implications for clinical autonomic dysfunction. *J Neuropathol Exp Neurol* 2005;64:1018–1027.

Peters SU, Beaudet AL, Madduri N, Bacino CA. Autism in Angelman syndrome: Implications for autism research. *Clin Genet* 2004;66:530–536.

Ravits J, Appel S, Baloh RH, et al. Deciphering amyotrophic lateral sclerosis: What phenotype, neuropathology and genetics are telling us about pathogenesis. *Amyotroph Lateral Scler Frontotemporal Degener* 2013;14(Suppl 1):S5–18.

Reader TA, Le Marec N, Ase AR, Lalonde R. Effects of L-tryptophan on indoleamines and catecholamines in discrete brain regions of wild type and Lurcher mutant mice. *Neurochem Res* 1999;24:1125–1134.

Riahi G, Morissette M, Lévesque D, Rouillard C, Samadi P, Parent M, Di Paolo T. Effect of chronic L-DOPA treatment on 5-HT(1A) receptors in parkinsonian monkey brain. *Neurochem Int* 2012;61:1160–1171.

Riahi G, Morissette M, Parent M, Di Paolo T. Brain 5-HT(2A) receptors in MPTP monkeys and levodopa-induced dyskinesias. *Eur J Neurosci* 2011;33:1823–1831.

Ribotta MG, Provencher J, Feraboli-Lohnherr D, Rossignol S, Privat A, Orsal D. Activation of locomotion in adult chronic spinal rats is achieved by transplantation of embryonic raphe cells reinnervating a precise lumbar level. *J Neurosci* 2000;20:5144–5152.

Roessner V, Schoenefeld K, Buse J, Bender S, Ehrlich S, Münchau A. Pharmacological treatment of tic disorders and Tourette syndrome. *Neuropharmacology.* 2013;68:143–149.

Rylander D. The serotonin system: A potential target for anti-dyskinetic treatments and biomarker discovery. *Parkinsonism Relat Disord* 2012;18(Suppl 1):S126–128.

Sandyk R. Serotonergic mechanisms in amyotrophic lateral sclerosis. *Int J Neurosci* 2006;116:775–826.

Santos M, Summavielle T, Teixeira-Castro A, Silva-Fernandes A, Duarte-Silva S, Marques F, Martins L, Dierssen M, Oliveira P, Sousa N, Maciel P. Monoamine deficits in the brain of methyl-CpG binding protein 2 null mice suggest the involvement of the cerebral cortex in early stages of Rett syndrome. *Neuroscience* 2010;170:453–467.

Scahill L, Riddle MA, King RA, Hardin MT, Rasmusson A, Makuch RW, Leckman JF. Fluoxetine has no marked effect on tic symptoms in patients with Tourette's syndrome: A double-blind placebo-controlled study. *J Child Adolesc Psychopharmacol* 1997;7:75–85.

Segawa M. Pathophysiology of Rett syndrome from the standpoint of clinical characteristics. *Brain Dev* 2001;23(Suppl 1):S94–98.

Shin E, Tronci C, Carta M. Role of serotonin neurons in L-DOPA- and graft-induced dyskinesia in a rat model of Parkinson's disease. *Parkinsons Dis* 2012;2012:370190.

Slawinska U, Miazga K, Cabaj AM, Leszczynska AN, Majczynski H, Nagy JI, Jordan LM. Grafting of fetal brainstem 5-HT neurons into the sublesional spinal cord of paraplegic rats restores coordinated hindlimb locomotion. *Exp Neurol* 2013;247:572–581.

Slawinska U, Miazga K, Jordan LM. 5-HT2 and 5-HT7 receptor agonists facilitate plantar stepping in chronic spinal rats through actions on different populations of spinal neurons. *Front Neural Circuits* 2014;8:95.

Smith GA, Breger LS, Lane EL, Dunnett SB. Pharmacological modulation of amphetamine-induced dyskenesia in transplanted hemi-parkinsonian rats. *Neuropharmacology* 2012;63:818–828.

Sofic E, Riederer P, Gsell W, Gavranovic M, Schmidtke A, Jellinger K. Biogenic amines and metabolites in spinal cord of patients with Parkinson's disease and amyotrophic lateral sclerosis. *J Neural Transm Park Dis Dement Sect* 1991;3:133–142.

Steeves TD, Fox SH. Neurobiological basis of serotonin-dopamine antagonists in the treatment of Gilles de la Tourette syndrome. *Prog Brain Res* 2008;172:495–513.

Strazielle C, Lalonde R, Riopel L, Botez MI, Reader TA. Regional distribution of the 5-HT innervation in the brain of normal and Lurcher mice as revealed by [^{3}H]citalopram quantitative autoradiography. *J Chem Neuroanat* 1996;10:157–171.

Suenaga T, Tawara Y, Goto S, Kouhata S, Kagaya A, Horiguchi J, Yamanaka Y, Yamawaki S. Risperidone treatment of neuroleptic-induced tardive extrapyramidal symptoms. *Int J Psychiatry Clin Pract* 2000;4:241–243.

Takei A, Hamada S, Homma S, Hamada K, Tashiro K, Hamada T. Difference in the effects of tandospirone on ataxia in various types of spinocerebellar degeneration: An open-label study. *Cerebellum* 2010;9:567–570.

Takei A, Honma S, Kawashima A, Yabe I, Fukazawa T, Hamada K, Hamada T, Tashiro K. Beneficial effects of tandospirone on ataxia of a patient with Machado-Joseph disease. *Psychiatry Clin Neurosci* 2002;56:181–185.

Thompson CK, Hornby TG. Divergent modulation of clinical measures of volitional and reflexive motor behaviors following serotonergic medications in human incomplete spinal cord injury. *J Neurotrauma* 2013;30:498–502.

Toren P, Weizman A, Ratner S, Cohen D, Laor N. Ondansetron treatment in Tourette's disorder: A 3-week, randomized, double-blind, placebo-controlled study. *J Clin Psychiatry* 2005;66:499–503.

Toward MA, Abdala AP, Knopp SJ, Paton JF, Bissonnette JM. Increasing brain serotonin corrects CO_2 chemosensitivity in methyl-CpG-binding protein 2 (*Mecp2*)-deficient mice. *Exp Physiol* 2013;98:842–849.

Triarhou LC, Ghetti B. Serotonin-immunoreactivity in the cerebellum of two neurological mutant mice and the corresponding wild-type genetic stocks. *J Chem Neuroanat* 1991;4:421–428.

Tronci E, Lisci C, Stancampiano R, Fidalgo C, Collu M, Devoto P, Carta M. 5-Hydroxy-tryptophan for the treatment of L-DOPA-induced dyskinesia in the rat Parkinson's disease model. *Neurobiol Dis* 2013;60:108–114.

Trouillas P, Brudon F, Adeleine, P. Improvement of cerebellar ataxia with levorotatory form of 5-hydroxytryptophan: A double-blind study with quantified data processing. *Arch Neurol* 1988;45:1217–1222.

Trouillas P, Xie J, Adeleine P. Buspirone, a 5-hydroxytryptamine1A agonist, is active in cerebellar ataxia: Results of a double-blind drug placebo study in patients with cerebellar cortical atrophy. *Arch Neurol* 1997;54:749–752.

Turner BJ, Lopes EC, Cheema SS. The serotonin precursor 5-hydroxytryptophan delays neuromuscular disease in murine familial amyotrophic lateral sclerosis. *Amyotroph Lateral Scler Other Motor Neuron Disord* 2003;4:171–176.

Turner MR, Rabiner EA, Hammers A, Al-Chalabi A, Grasby PM, Shaw CE, Brooks DJ, Leigh PN. [11C]-WAY100635 PET demonstrates marked 5-HT1A receptor changes in sporadic ALS. *Brain* 2005;128:896–905.

Udvardi PT, Nespoli E, Rizzo F, Hengerer B, Ludolph AG. Nondopaminergic neurotransmission in the pathophysiology of Tourette syndrome. *Int Rev Neurobiol* 2013;112:95–130.

Viemari JC, Roux JC, Tryba AK, et al. *Mecp2* deficiency disrupts norepinephrine and respiratory systems in mice. *J Neurosci* 2005;25:11521–11530.

Vogel AP, Folker J, Poole ML. Treatment for speech disorder in Friedreich ataxia and other hereditary ataxia syndromes. *Cochrane Database Syst Rev* 2014;10:CD008953.

Voogd J. Cerebellum. In G Paxinos (ed.), *The Rat Nervous Stem*. New York: Academic Press, 2nd edn, 1995;309–350.

Weng SM, Bailey ME, Cobb SR. Rett syndrome: From bed to bench. *Pediatr Neonatol* 2011;52:309–316.

Wolters EC. Dopaminomimetic psychosis in Parkinson's disease patients: Diagnosis and treatment. *Neurology* 1999;52(7, Suppl 3):S10–13.

Worbe Y, Marrakchi-Kacem L, Lecomte S, et al. Altered structural connectivity of cortico-striato-pallido-thalamic networks in Gilles de la Tourette syndrome. *Brain* 2015;138:472–482.

Yoon DY, Gause CD, Leckman JF, Singer HS. Frontal dopaminergic abnormality in Tourette syndrome: A postmortem analysis. *J Neurol Sci* 2007;255:50–56.

Zhou SY, Goshgarian HG. 5-Hydroxytryptophan-induced respiratory recovery after cervical spinal cord hemisection in rats. *J Appl Physiol* 2000;89:1528–1536.

Zhou SY, Goshgarian HG. Effects of serotonin on crossed phrenic nerve activity in cervical spinal cord hemisected rats. *Exp Neurol* 1999;160:446–453.

25

Serotonin and Melatonin vis-à-vis Sleep Disorders

Ravindra P. Nagendra
Gadag Institute of Medical Sciences
Karnataka, India

Bindu M. Kutty
National Institute of Mental Health and Neurosciences (NIMHANS)
Karnataka, India

CONTENTS

ABSTRACT   Sleep has fascinated people across the spectrum of society since time immemorial and this continues even to the present day. It was once considered a passive phenomenon that rejuvenated the whole organism. However, a great change in the understanding of sleep began after the discovery of REM sleep; it is now well recognized that sleep is an active process brought about by the complex interaction of various neural networks, neurotransmitters, and neuromodulators. Interaction between serotonin and melatonin, wake and sleep state–promoting agents, respectively, play crucial roles in the initiation and maintenance of each behavioral state at any given instant, and also aid the smooth, precise transition from wake to sleep and vice versa. In addition to its role in sleep, serotonin's association with mood is well studied. Therefore, serotonin forms a common link relating to the sleep disturbances among mood disorders such as depression. Further, serotonin, during sleep, plays an important role in maintaining the patency of the airways, thus linking the malfunction of the serotonin circuitry to obstructive sleep apnea (OSA), especially in the aged. Melatonin, the most versatile chemical in nature, is the best known circadian rhythm generator. It plays a major role in bringing about normal hormonal, temperature, and behavioral variations, such as that of food intake associated with the sleep–wake cycle. Circadian variations in melatonin are subject to great perturbance due to alterations in lifestyle patterns, such as working late at night, shift work, frequent traveling across time zones, and so on, thus increasing the prevalence of circadian sleep disorders. Thus, serotonin and melatonin per se and their interaction play an indispensible role in the development and maintenance of normal adult sleeping patterns, and their alterations can bring about various sleep disorders. In the present chapter, the comprehensive description of the role of serotonin and melatonin in bringing about normal sleep patterns and in various sleep disorders is described.

KEY WORDS: *serotonin, melatonin, circadian rhythm disorders, depression, sleep.*

Abbreviations

ACTH: adrenocorticotrophic hormone
ASPD: advanced sleep phase syndrome
CRH: corticotrophin-releasing hormone
DSPD: delayed sleep phase syndrome
DRN: dorsal raphe nucleus
5HT: 5 hydroxytryptamine
5HTT: 5 hydroxytryptamine transporter
HPA: hypothalamic–pituitary–adrenal (axis)
NREM: non-rapid eye movement
OSA: obstructive sleep apnea
REM: rapid eye movement
SCN: suprachiasmatic nucleus
SIDS: sudden infant death syndrome

Introduction

Sleep is a distinct behavioral state essential for health and survival. It is one of the most cherished behavioral states, in which most healthy individuals spend six to eight hours every day; that is, almost one-third of life is spent sleeping! Up to the late 1940s, sleep was considered a passive phenomenon, and there was little understanding of the brain mechanisms involved in the generation of normal sleep. The landmark discovery by Moruzzi and Magoun (1949) laid the foundations for the subsequently volumi-nous work that followed, establishing the fact that the dynamic interaction of different neuronal groups in the brain is involved in the initiation and maintenance of sleep.

   The homeostatic and circadian mechanisms and their interaction with various wake- and sleep-promoting regions of the brain form the core brain processes in bringing about normal sleep architecture (Pace-Schott and Hobson, 2002). Human sleep architecture normally is a cyclical, alternating pattern of non-rapid eye movement (NREM) and rapid eye movement (REM) sleep (Figure 25.1). Sleep initiates with the NREM state (the lighter sleep stages, S1 and S2), proceeds toward the deep-sleep states (S3 and S4), and then REM sleep. Thus, the NREM–REM sleep cycle continues for three to four such cycles through the night. In the first half of the night, the major part of the sleep is comprised of deep NREM states with a small percentage of REM, whereas, in the later part of the night, REM sleep predomi-nates. The maintenance of this stable sleep architecture is important for quality and refreshing sleep.

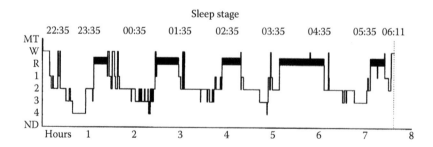

FIGURE 25.1   A normal sleep architecture of a healthy adult. W: wake; R: REM sleep; 1/2/3/4: NREM sleep stages (S1/S2/S3/S4). Note that the deep-sleep state (S3 and S4) is predominant during the first half of the night and REM (R) sleep increases in the second half of the night. (Hypnogram obtained from recordings from the Sleep Research Laboratory, Department of Neurophysiology, National Institute of Mental Health and Neurosciences, Bangalore).

Homeostatic mechanism is the term used to explain the factors and processes associated with increased sleep pressure as a result of extended wake states brought about by sleep deprivation or restriction. The homeostatic drive plays an important role in the initiation of sleep and brings about an increased deep-sleep state in the first half of the night, resulting in dissipated sleep pressure. The circadian mechanism provides the time cue to the physiological system and regulates complex behavior like food intake and temperature, which are associated with sleep. Melatonin forms the indispensible part of the circadian system. Circadian factors play an important role in the timing of sleep, sleep maintenance, and sleep architecture in the later part of the night.

The factors associated with these two mechanisms—that is, homeostatic and circadian—interact with various distinct wake- and sleep-promoting neuronal groups, bringing about the precise regulation of sleep and maintaining a stable sleep architecture. Among the wake-promoting neuronal groups, the serotonergic group (dorsal raphe nucleus) forms an important role, bringing about the wake state. Further, it is of the utmost importance to maintain a stable state either of wake or sleep at any given instance. Therefore, complex and highly synchronized interactions between wake- and sleep-promoting neuronal networks, neurotransmitters, and modulators maintain sleep–wake dynamicity, whereby the transition from one behavioral state to the other is fast, robust, and precise (Clifford et al., 2005). Any alterations in factors influencing sleep initiation, maintenance, and timing of sleep or in state transition, either due to developmental defects, the normal aging process, disease conditions, or altered expression of various associated neurochemical systems, result in the spectrum of sleep disorders. In this chapter, the role of serotonin and melatonin on sleep is discussed comprehensively.

Serotonin: A Predominant Wake-Promoting Neurotransmitter

Our understanding of the role of serotonin in sleep dates back to Brodie et al. (1955), who propounded that decreases in serotonin in the brain induce a "sleep-like" state. Over the years, the pivotal role of serotonin in the sleep–wake mechanism has been established beyond any doubt (Jouvet, 1972). The dorsal raphe nucleus (DRN), the major serotonergic locus that profoundly innervates the forebrain, shows differential activity based on the wake–sleep state. The activity of DRN is high during wake, slows down during NREM sleep, and ceases during REM sleep (Benington et al., 1994). As mentioned earlier, it is the highly precise neural activity of the wake- and sleep-promoting neuronal system that helps in the smooth and robust transitions between wake to sleep and vice versa. Unit activity studies have demonstrated the specific role played by serotonin in inducing state transitions; the serotonergic neurons begin to be activated before waking and hence are important to the process (Takahashi et al., 2008). The activity of serotonergic neurons together with noradrenergic neurons is essential to providing heightened attention- and arousal-promoting mechanisms during wake (Takahashi et al., 2010). However, with extended wakefulness, the serotonin system is known to activate important sleep promoters such as interleukin beta and tumor necrosis factor alpha (Imeri and Opp, 2009). Hence, serotonin's major role is to bring about wake with a smooth state transition; on extended activity, it also promotes sleep. In addition, the role of serotonin in the regulation of mood is well documented. Therefore, it is an important factor associated with alterations in sleep due to mood disorders (e.g., depression) (Peter, 2009). Further, serotonin also plays a major role in orchestrating normal cardiorespiratory events during sleep, with its major role being to maintain the patency of the upper airway during sleep (Baodong et al., 2014). Impaired respiratory events such as sudden infant death syndrome (SIDS), obstructive sleep apnea (OSA) syndrome, or epilepsy are widely attributed either to genetic polymorphism or attenuated drive of the serotonergic system (Baodong et al., 2014; Gordon and George, 2010).

Serotonin: A Common Link between Depression and Sleep

Restricted or disturbed sleep is the most common emerging unhealthy behavior of the modern-day lifestyle. Increased stress and round-the-clock professional and social commitments play a major causative role in dysregulating normal circadian rhythms and thereby mood as well. Insomnia is the most common sleep disorder, wherein the difficulty is sensed either in initiating or maintaining sleep. Acute episodes of insomnia bring down cognitive capability and increases emotional vulnerability, which are recovered by subsequent sleep. However, neurobiological changes due to repeated sleep restrictions or deprivations tend to accumulate, which over time leads to untoward health consequences. Insomnia for longer durations of time increases the risk of developing depression by 39 times when compared with other factors (Ford and Kamerow, 1989). But, interestingly, sleep restriction itself is also known to alleviate the symptoms of depression. Such contrasting relationships between sleep and depression make it interesting and also difficult to understand their relationship.

The interaction of the hypothalamic–pituitary–adrenal (HPA) axis (cholinergic neurotransmitter system) and serotonergic system plays a crucial role in altering sleep in depression. In depression, the HPA axis is hyperactive, which increases the levels of cortisol and corticotrophin-releasing hormone (CRH). Chronically elevated CRH desensitizes the CRH receptors in the pituitary gland, which brings about attenuated pituitary responsiveness to adrenocorticotrophic hormone (ACTH). This attenuated pituitary responsiveness to ACTH is known to reduce the sensitivity of serotonin (5HT1A) receptors, which worsens further if the patient is also sleep deprived (Benca et al., 1992). Hence, in depression, the imbalanced neurochemical state—that is, reduced serotonergic transmission due to increased cholinergic activity—alters the sleep architecture. Emerging evidences suggest that, during the developmental period of life, alterations in REM sleep maturation, which is a serotonin-dependent process, result in sleep-related problems and depression in later life (Kobayashi et al., 2004). Endogenous animal models of depression have provided sufficient data that an altered serotonergic system leads to longer REM sleep periods without much change in other sleep states (Peter, 2009). Clinical data have shown that the sleep architectures of depressive patients are associated with reduced duration of their deep-sleep states, impaired sleep efficiency, increased REM sleep duration, percentage, and density, intermittent awakenings (arousals), thereby reducing the sleep quality (Benca et al., 1992). It is to be noted that reduced deep-sleep states and enhanced REM sleep (due to reduced serotonin levels) form the markers of depression (Giles et al., 1989). In addition to these sleep alterations, reduced serotonergic modulation of prefrontal cortices alters normal executive functions, increases the probability of impulsiveness, diminishes decision-making skills, and increases suicidal ideation (Tsuno et al., 2005). These psychological manifestations in depression and insomnia are attributed to impaired serotonergic function.

Serotonin Plays a Crucial Role in the Development of Respiratory Rhythm and Maintaining Airways during Sleep

During developmental stages, serotonin plays an important role in the formation and functioning of the central neural networks (Kinney et al., 2007) and initiates spontaneous rhythmic activity at the midline raphe (5HT2A). This rhythmic activity propagates and encompasses the pre-Böztinger complex (respiratory rhythm generator) (Thoby-Brisson et al., 2009), thereby initiating the generation of respiratory rhythm. Subsequently, serotonin facilitates respiratory movement generation via 5HT1B receptors acting on phrenic motor neurons (Lindsay and Feldman, 1993). In addition, serotonin neurons located proximally to arteries in the brain stem also detect the level of carbon dioxide and oxygen and, accordingly modulates the respiratory drive and movements, thus bringing about appropriate respiratory adjustments (Feldman et al., 2003). During sleep, serotonin plays a major role in maintaining the patency of the upper airway and the excitatory activity of trigeminal and hypoglossal motor neurons, thereby facilitating comfortable breathing (Rose et al., 1995). Thus, serotonin plays a major role in the development of respiratory rhythm, respiratory movements, and modification of respiration in accordance with the gaseous concentration in the arterial blood during sleep.

In infants, sudden death during sleep (mostly in the prone position), due to reduced oxygen (hypoxia) and increased carbon dioxide (hypercapnia), is termed *sudden infant death syndrome* (SIDS). In SIDS, reduced brain 5HT levels, reduced expression of 5HT1A and 5HT2A receptors, and altered 5HT turnover is reported (Paterson et al., 2009). A genetic defect in the serotonin transporter–promoter gene PHOX2B, which aids in the production of essential enzymes for the synthesis of serotonin that may result in an alteration in the expression of 5HT receptors, has been implicated in SIDS (Rand et al., 2006). This defect in the serotonin system at brain stem level fails to bring about protective respiratory reflexes due to hypoxia and hypercapnia (Paterson et al., 2006a), thus leading to the death of the infant.

During sleep, the upper airways are kept patent through serotonin binding to the 5HT2A (chromosome 13q14-q21) and 5HT2C (chromosome X q24 region) receptors (Sood et al., 2007). The collapse of the airways during sleep brings about obstructive sleep apnea syndrome (OSA; snoring), which is the most common sleep disorder across the globe. OSA is an independent risk factor in the development of cardiovascular disease, diabetes, stroke, cognitive impairment, and so on (Toraldo et al., 2015). The development of OSA is implicated either to the genetic polymorphism of 5HT2A receptors along with 5HT transporter (5HTT) (HuajunXu et al., 2014) or reduced serotonergic drive due to obesity or in the normal aging process (Nakano et al., 2001). Therefore, occurrence of OSA is observed many times more among obese elderly populations. Hypercapnia induced by OSA brings about excitability, thus inducing arousals during sleep. Repeated arousals destabilize the sleep architecture, bringing about reduced efficiency and poor quality of sleep, thus leading to insomnia and excessive daytime sleepiness, which hampers daytime functioning. Therefore, drugs that enhance serotonin have a major role in managing mood and certain sleep disorders.

Melatonin: Its Role in Normal Circadian Rhythm Sleep–Wake Cycles and in Circadian Rhythm Disorders

Melatonin is an evolutionarily highly conserved molecule present in a spectrum of plant and animal species and plays a major role in circadian rhythm regulation. Melatonin is a central time keeper (also called *zeitgeber*) synthesized and secreted by the pineal gland, a central structure in the circadian system. The pineal gland was considered a functionless rudimentary organ until 1958, when Aaron Lerner and colleagues (Lerner et al., 1958) isolated the active pineal substance. They named this compound *melatonin*, the "hormone of darkness," and described its chemical structure as N-acetyl-5-methoxytryptamine. Light is the most dominant powerful environmental source that regulates melatonin synthesis. Photic stimuli through the retina are transmitted to the suprachiasmatic nucleus (SCN), the master circadian clock. SCN in turn conveys the signal to the pineal gland for melatonin synthesis. In this process, noradrenergic fibers of cervical ganglion ensure proper translation of photic information into melatonin synthesis. Therefore, in addition to light, the levels of noradrenalin in these neurons also play an important role in the synthesis of melatonin. It is to be noted that the photoreceptors (melanopsin-containing retinal ganglion cells) that are sensitive to light in the range 446–484 nm (blue light) are primarily involved in stimulating melatonin synthesis (Lewy and Sack, 1989). Therefore, visual blindness doesn't necessarily cause circadian misalignment but for the absence of melanopsin or its connection to the SCN if not intact. However, with the development of cataracts and pupil constriction with age, the transmission of light of shorter wavelengths may be impeded, thereby causing alterations in melatonin rhythm and the ensuing sleep problems.

The most striking features of melatonin are its daily variations that are sensitive to the light–dark cycle and its peak release toward darkness. These circadian variations in melatonin play a major role in bringing about normal daily rhythms in hormonal secretion, temperature variation, the sleep–wake cycle, and so on. The melatonin peak is followed by a drop in core body temperature; these two factors help to induce sleep (Karasek and Winczyk, 2006). This relationship between melatonin and core body temperature follows even during therapeutic administration of melatonin agonists, wherein the core temperature drops down within two to three hours after administration (Verster, 2009). Alterations in normal circadian variations of melatonin, due either to exposure to bright light during the biological nighttime (as in shift work, late-night work, etc.) or to pathological conditions, tends to misalign the circadian rhythm, thereby bringing about various sleep-related problems. The classical circadian rhythm disorders include delayed sleep phase syndrome, advanced sleep phase syndrome, jet lag, shift work, and free-running type, as seen in visually challenged people (Denis and Maria, 2010). However, there circadian disorders can occur due to genetic mutations of circadian genes, medical conditions (insulin resistance, parkinsonism, cirrhosis, renal failure, etc.), medications, and drug abuse as well. A few common circadian sleep disorders are explained briefly here.

Delayed Sleep Phase Syndrome (DSPD)

This is characterized by recurrent difficulty in falling asleep or waking up at a socially acceptable time. There will be a difference of more than two hours to conventional acceptable bedtime and waking time. However, the important characteristic of this disorder is that the circadian rhythm of the person will persistently be delayed, which brings about significant impairment in an individual's social life. DPSD is most commonly observed in adolescents, primarily due to socially enforced alterations in the sleep–wake schedule associated with late-night social activities. The symptoms are exacerbated with decreased light exposure in the morning and bright-light exposure in late evening hours. However, polymorphism of the circadian clock gene is also implicated in developing DPSD. It has been reported that DPSD patients have a diminished ability to compensate for sleep loss, in spite of having a high homeostatic drive.

In managing DPSD, chronotherapy by secluding the timing of sleep will synchronize the circadian rhythm. The sleep schedule is delayed gradually for several hours a day until the target time is reached. After reaching the target time, patients are advised to keep to a regular sleep–wake schedule. In addition to this, bright-light exposure during morning hours with evening melatonin administration is known to realign the circadian rhythm in DPSD patients.

Advanced Sleep Phase Disorder (ASPD)

ASPD is characterized by earlier sleep times than are conventionally socially acceptable—that is, feeling extremely sleepy in the late afternoon or evening hours—and waking earlier than desired (between 2:00 and 5:00 a.m.). This is generally found with aging. Shortened circadian times due to a mutation of the circadian clock gene are one of the causes of ASPD. The most reported complications of this are alcoholism and drug abuse. The usual treatment is exposure to light in the evening so as to suppress melatonin secretion and postpone sleep. Patients are encouraged to take melatonin supplements in the early morning hours so that sleep continues until after sunrise.

Jet Lag Disorders

Jet lag is a temporary misalignment of the circadian clock due to traveling across at least two different time zones. The symptoms, including disturbed sleep, generalized weakness, decreased daytime attention, mood disturbance, and so on, occur within one or two days of travel. The severity of symptoms depends on the number of time zones and the direction of travel. Traveling from east to west is less severe than traveling from west to east. The sleep problems per se with eastward travel include difficulty in falling asleep and waking next day, whereas with westward travel, individuals experience excessive sleepiness in the evening and early morning waking. The treatment of jet lag syndrome is focused on bringing desirable sleep and wake times in accordance with the destination traveled, improving nocturnal sleep, and increasing daytime alertness. However, if the stay in the destination is for two days or less, then measures taken for circadian adaptation will be counterproductive. In the measures to be taken, planned exposure or avoidance to light is advised. Travelers heading east are advised to avoid light exposure in the early morning (enhancing melatonin secretion) and to expose themselves to bright light in late morning and afternoon (decreasing melatonin secretion). In westward travel, travelers are exposed to light during the afternoon and early evening, and aim not to go to sleep until nighttime at their destination. However, with melatonin (0.5–10 mg) administration during the early evening hours well in advance of eastward travel and during the nighttime at the destination, the severity of jet lag syndrome can be minimized.

Shift Work Disorder

This is characterized by a history of excessive sleepiness during the day and insomnia during sleep periods for at least one month due to unconventional work schedules. This will bring about impaired work performance and decreased alertness and cognition, thereby increasing the risk of accidents in the workplace. This is associated with higher incidences of cardiovascular comorbid conditions, metabolic syndromes, endometrial and breast cancer, and so on. Exposure to intense light during night work and avoiding light during the day will help to reduce sleepiness during work and to induce sleep in the day. Exposure to light during night work hours should be 1,000–10,000 lx until two hours prior to the end of the shift. Melatonin consumption before bed will enable good sleep but will not improve alertness during work. Therefore, melatonin and light therapy should be combined in individuals with shift work disorder (Phyllis et al., 2013; Robert et al., 2007a,b).

Conclusion

In addition to these circadian rhythm disorders, melatonin administration in old age aids in preventing many neurodegenerative conditions, insomnia, neuropsychiatric conditions such as depression, schizophrenia, eating disorders, intractable epilepsy, and so on. Interestingly, meditation (Vipassana) is known to enhance endogenous melatonin levels across age groups in experienced practitioners (Ravindra et al., 2013). Apart from its therapeutic potential, melatonin has uses as an antioxidant, an immunity enhancer, and an oncostatic agent; hence, its supplement can have a positive role in maintaining general health and well-being.

REFERENCES

Baodong Q, Zhen S, Yan L, Zaixing Y, Renqian Z. The association of 5-HT2A, 5-HTT, and LEPR polymorphisms with obstructive sleep apnea syndrome: A systematic review and meta-analysis. *PLoS One* 2014;9 (4):e95856.

Benca RM, Obermeyer WH, Thisted RA, Gillin JC. Sleep and psychiatric disorders: A meta-analysis. *Arch Gen Psychiatry* 1992;49(8):651–668; discussion 669–670.

Benington JH, Woudenberg MC, Heller HC. REM-sleep propensity accumulates during 2-h REM-sleep deprivation in the rest period in rats. *Neurosci Lett* 1994;180(1):76–80.

Brodie BB, Pletsher A, Shore A. Evidence that serotonin has a role in brain function. *Science* 1995;122:968.

Clifford BS, Thomas ES, Jun L. Hypothalamic regulation of sleep and circadian rhythms. *Nature* 2005; 237:1257–1263.

Denis M, Maria CL. Circadian rhythm sleep disorders. *Indian J Med Res* 2010;131:141–149.

Feldman JL, Mitchell GS, Nattie EE. Breathing: Rhythmicity, plasticity, chemosensitivity. *Annu Rev Neurosci* 2003;26:239–266.

Ford DE, Kamerow DB. Epidemiologic study of sleep disturbances and psychiatric disorders: An opportunity for prevention? *JAMA* 1989;262(11):1479–1484.

Giles DE, Etzel BA, Reynolds CF. Stability of polysomnographic parameters in unipolar depression: A cross-sectional report. *Biol Psychiatry* 1989;25:807–810.

Gordon FB, George BR. Central serotonin neurons are required for arousal to CO_2. *PNAS* 2010;107(37):16354–16359.

Huajun Xu, Jian Guan, Hongliang Yi, Shankai Yin A. Systematic review and meta-analysis of the association between serotonergic gene polymorphisms and obstructive sleep apnea syndrome. *PLoS One* 2014;9(1):e86460.

Imeri L, Opp MR. How (and why) the immune system makes us sleep. *Nat Rev Neurosci* 2009;10(3): 199–210.

Jouvet M. The role of monoamines and acetylcholine-containing neurons in the regulation of the sleep–waking cycle. *Ergebn Physiol* 1972;64:166–307.

Karasek M, Winczyk K. Melatonin in humans. *J Physiol Pharmacol* 2006;57(Suppl. 5):19–39.

Kinney HC, Belliveau RA, Trachtenberg FL, Rava LA, Paterson DS. The development of the medullary serotonergic system in early human life. *Auton Neurosci* 2007;132:81–102.

Kobayashi T, Good C, Mamiya K, Skinner RD, Garcia-Rill E. Development of REM sleep drive and clinical implications. *J Appl Physiol* 2004;96:735–746.

Lerner AB, Case JD, Heinzelman RV. Structure of melatonin. *J Am Chem Soc* 1959;81:6084–6085.

Lerner AB, Case JD, Takahashi Y, Lee TH, Mori W. Isolation of melatonin, the pineal gland factor that lightens melanocytes. *J Am Chem Soc* 1958;80:2587.

Lewy AJ, Sack RL. The dim light melatonin onset as a marker for circadian phase position. *Chronobiol Int* 1989;6:93–102.

Lindsay AD, Feldman JL. Modulation of respiratory activity of neonatal rat phrenic motoneurones by serotonin. *J Physiol* 1993;461:213–233.

Moruzzi G, Magoun HW. Brainstem reticular formation and activation of the EEG *Electroencephalogr Clin Neurophysiol* 1949;1:455–473.

Nakano H, Magalang UJ, Lee SD, Krasney JA, FarkasGA. Serotonergic modulation of ventilation and upper airway stability in obese zucker rats. *Am J Respir Crit Care Med* 2001;163:1191–1197.

Pace-Schott EF, Hobson AJ. The neurobiology of sleep: Genetics, cellular physiology and subcortical networks. *Nature Rev Neuroscience* 2002;3:591–604.

Paterson DS, Hilaire G, Weese-Mayer DE. Medullary serotonin defects and respiratory dysfunction in sudden infant death syndrome. *Respir Physiol Neurobiol* 2009;168:133–143.

Paterson DS, Thompson EG, Kinney HC. Serotonergic and glutamatergic neurons at the ventral medullary surface of the human infant: Observations relevant to central chemosensitivity in early human life. *Auton Neurosci* 2006a;124:112–124.

Peter BH. The 5-HT7 receptor and disorders of the nervous system: An overview. *Psychopharmacology (Berl)* 2009;206(3):345–354.

Phyllis CZ, Hrayr A, Aleksandar V. Circadian rhythm abnormalities. *Continuum (Minneap Minn)* 2013;19(1):132–147.

Rand CM, Weese-Mayer DE, Zhou L, Maher BS, Cooper ME, Marazita ML, Berry-Kravis EM. Sudden infant death syndrome: Case-control frequency differences in paired like homeoboxPhox2B gene. *Am J Med Genet A* 2006;140(15):1687–1691.

Ravindra PN, Nirmala M, Bindu MK. Meditation its regulatory effect on sleep. *Front Neurol* 2012;3(54):1–3.

Robert LS, Dennis AR, Robert A, Mary AC, Kenneth PW Jr, Michael VV, Irina VZ. Circadian rhythm sleep disorders: Part I. Basic principles, shift work and jet lag disorders; An American Academy of Sleep Medicine review. *Sleep* 2007a;30(11):1460–1483.

Robert LS, Dennis AR, Robert A, Mary AC, Kenneth PW Jr, Michael VV, Irina VZ. Circadian rhythm sleep disorders: Part II. Advanced sleep phase disorder, delayed sleep phase disorder, free-running disorder, and irregular sleep-wake rhythm. An American Academy of Sleep Medicine review. *Sleep* 2007b;30(11):1484–1501.

Rose D, Khater-Boidin J, Toussaint P, Duron B. Central effects of 5-ht on respiratory and hypoglossal activities in the adult cat. *Respir Physiol* 1995;101:59–69.

Sood S, Liu X, Liu H, Horner RL. Genioglossus muscle activity and serotonergic modulation of hypoglossal motor output in obese Zucker rats. *J Appl Physiol* 2007;102:2240–2250.

Takahashi K, Lin JS, Sakai K. Neuronal activity of orexin and non-orexin waking-active neurons during sleep wake–sleep states in the mouse. *Neurosciences* 2008;153:860–870.

Takahashi K, Lin JS, Sakai K. Locus coeruleus neuronal activity during the sleep–waking cycle in mice. *Neurosciences* 2010;169:1115–1126.

Thoby-Brisson M, Karlén M, Wu N, Charnay P, Champagnat J, Fortin G. Genetic identification of an embryonic parafacial oscillator coupling to the preBötzinger complex. *Nat Neurosci* 2009;12(8):1028–1035.

Toraldo DM, DE Nuccio F, DE Benedetto M, Scoditti E. Obstructive sleep apnoea syndrome: A new paradigm by chronic nocturnal intermittent hypoxia and sleep disruption. *Acta Otorhinolaryngol Ital* 2015;35(2):69–74.

Tsuno N, Besset A, Ritchie K. Sleep and depression. *J Clin Psychiatry* 2005;66:1254–1269.

Verster GC. Melatonin and its agonists, circadian rhythms and psychiatry. *Afr J Psychiatry* 2009;12:42–46.

26

Melatonin and Sleep in Humans

Amnon Brzezinski
Hadassah-Hebrew University Medical Center
Jerusalem, Israel

CONTENTS

ABSTRACT   Daytime administration of exogenous melatonin (when it is not present endogenously) promotes sleep in humans and results in sleep-like brain activity patterns at specific areas such as the precuneus and hippocampus. However, existing studies on the hypnotic efficacy of melatonin have been highly heterogeneous in regard to inclusion and exclusion criteria, measures to evaluate insomnia, doses of the medication, and routes of administration. The inconsistent reports about the effectiveness of exogenous melatonin in the treatment of insomnia brought about the development of more potent melatonin analogs with prolonged effects and the design of slow-release melatonin preparations. The melatonergic receptor ramelteon is a selective melatonin-1 (MT1) and melatonin-2 (MT2) receptor agonist with negligible affinity for other neuronal receptors, including gamma-aminobutyric acid and benzodiazepine receptors. It was found effective in increasing total sleep time and sleep efficiency as well as in reducing sleep latency in insomnia patients. The melatonergic antidepressant agomelatine, displaying potent MT1 and MT2 melatonergic agonism and relatively weak serotonin 5HT2C receptor antagonism, reportedly is effective in the treatment of depression-associated insomnia.

KEY WORDS: *melatonin, melatonin agonists, sleep, insomnia, ramelteon, agomelatine.*

Introduction

In humans, the circadian rhythm of melatonin release from the pineal gland is highly synchronized with the habitual hours of sleep, and the daily onset of melatonin secretion is well correlated with the onset of the steepest increase in nocturnal sleepiness ("sleep gate") (Brzezinski, 1997; Shochat, 1998). Serum melatonin levels were reported to be significantly lower (and the time of peak melatonin values delayed) in elderly subjects with insomnia compared with age-matched subjects with no insomnia (Haimov et al., 1995). Exogenous melatonin reportedly induces drowsiness and sleep, and may ameliorate sleep disturbances, including the nocturnal awakenings associated with old age (Zhdanova et al., 1997; 2002). However, existing studies on the hypnotic efficacy of melatonin have been highly heterogeneous in regard to inclusion and exclusion criteria, measures to evaluate insomnia, doses of the medication, and routes of administration. Adding to this complexity, there continues to be considerable controversy over the meaning of the discrepancies that sometimes exist between subjective and objective (polysomnographic) measures of good and bad sleep (Nowell et al., 1997).

Thus, attention has been focused either on the development of more potent melatonin analogs with prolonged effects or on the design of prolonged-release melatonin preparations (Lemoine et al., 2007). The MT1 and MT2 melatonergic receptor ramelteon (Kato et al., 2005; Miyamoto, 2009) was effective in increasing total sleep time and sleep efficiency as well as in reducing sleep latency in insomnia patients (Pandi-Perumal et al., 2009). The melatonergic antidepressant agomelatine, displaying potent MT1 and MT2 melatonergic agonism and relatively weak serotonin 5HT2C receptor antagonism (Millan et al., 2003), was found effective in the treatment of depression-associated insomnia (Srinivasan et al., 2011; Kennedy and Emsley, 2006; Llorca, 2010; Srinivasan et al., 2010). Other melatonergic compounds are currently being developed (Rajaratnam et al. 2009; Zemlan et al., 2005).

Exogenous Melatonin

Melatonin's two well-established physiological effects—promotion of sleep and entrainment of circadian rhythms—are both mediated by two specific receptor proteins in the brain, and not by the gamma-aminobutyric acid (GABA) receptors through which most hypnotic agents act. This difference probably explains why, unlike the GABA-agonist drugs, which are true "sleeping pills," exogenous melatonin does not suppress rapid eye movement (REM) sleep nor, in general, affect the distribution of sleep stages (Zhdanova et al., 2001).

Measurements of melatonin in body fluids in elderly subjects have convincingly demonstrated an age-related impairment of nocturnal pineal melatonin synthesis (Haimov et al., 1994; Hughes and Badia, 1997; Leger et al., 2004). Several studies have shown the importance of melatonin both for the initiation and maintenance of sleep (Cajochen et al., 2003). In all diurnal animals and in human beings, the onset of melatonin secretion coincides with the timing of increase in nocturnal sleep propensity (Lavie, 1997).

In 2005, a meta-analysis (Brzezinski et al., 2005) of 17 studies, involving 284 subjects that satisfied inclusion criteria, demonstrated that melatonin treatment significantly decreased sleep latency and increased sleep efficiency and total sleep duration. The inclusion criteria were that a study include at least six subjects, all adults, be randomized and double-blinded, involve placebo-controlled clinical trials, and use objective measures of sleep evaluation. Studies could utilize crossover or parallel group designs; however, case reports were excluded. Statistical significance was obtained in spite of considerable variations among the studies in melatonin doses and routes of administration, the general health of the subjects, and the measures used to evaluate sleep.

The effects of exogenous melatonin on sleep have been examined under three types of experimental conditions in relation to the onset or offset of endogenous melatonin secretion.

In some studies, the hormone was administered during the daily light period, such that blood melatonin levels would be transiently elevated but would then return to baseline before the initiation of nocturnal melatonin secretion. Such experiments were used to demonstrate that melatonin decreases sleep latency at any time in the afternoon or evening, and that this effect is independent of an action on sleep rhythms (since no treatment can immediately shift the phase of a circadian rhythm by 8–10 hours).

In others, the hormone was given close enough to the onset of darkness for blood melatonin levels to still be elevated when nocturnal melatonin secretion started. The period during which plasma melatonin levels were continuously elevated would thus be prolonged. Such experiments reflected the use of melatonin to decrease sleep latency and maintain continuous sleep in, for example, a shift worker or eastbound world traveler who needed to start sleeping earlier.

In yet others, the hormone was given at the end of the light period to older insomniacs with low nighttime plasma melatonin levels. The intent was to prolong the portion of the night during which their plasma melatonin concentrations would be in the same range as those of noninsomniac young adults.

In all these situations, oral melatonin decreased sleep latency and, when tested, increased sleep duration and sleep efficiency. A 0.3 mg dose was either as effective as or more effective than (18) higher doses, particularly when the hormone was administered for several days. This dose had no effect on body temperature, affirming that, while pharmacological doses can cause hypothermia, melatonin's ability to promote sleep is not mediated by such a change, as had been suggested. The hormone had no consistent effect on sleep architecture (e.g., REM time). Its effects differed from those of most hypnotic drugs, since after receiving melatonin, subjects could readily keep from falling asleep if they chose so, and their cognitive abilities the next morning were unchanged or improved.

In a study of 30 people (Zhdanova et al., 2001) who were fifty years old or older and did or did not suffer from clinically significant insomnia (i.e., sleep efficiencies of 70%–80% in the insomniacs vs 92% in controls), melatonin was found to produce statistically and clinically significant improvements in sleep efficiency among insomniacs.

In yet another meta-analysis (Buscemi et al., 2005) published the same year, the authors found that melatonin decreased sleep onset latency (-11.7 min; 95% confidence interval [CI]: -18.2, -5.2); it was decreased to a greater extent in people with delayed sleep phase syndrome (-38.8 min; 95% CI: -50.3, -27.3; $n = 2$) compared with people with insomnia (-7.2 min; 95% CI: -12.0, -2.4; $n = 12$). The former result appears to be clinically important. However they conclude that melatonin is not effective in treating most primary sleep disorders with short-term use (four weeks or less) but that there was evidence to suggest that melatonin was effective in treating delayed sleep phase syndrome with short-term use.

A meta-analysis published in 2009 (Braam et al., 2009) focused on exogenous melatonin for sleep problems in individuals with intellectual disability. Nine studies (including a total of 183 individuals with intellectual disabilities) showed that melatonin treatment decreased sleep latency by a mean of 34 minutes ($p < .001$), increased total sleep time by a mean of 50 minutes ($p < .001$), and significantly decreased the number of wakes per night ($p < .05$). The authors concluded that melatonin decreases sleep latency and number of wakes per night, and increases total sleep time in individuals with intellectual disabilities.

It should be noted that very recently the European Food Safety Authority (EFSA) has evaluated the available evidence that melatonin can reduce the time it takes for normal sleepers and patients with insomnia to fall asleep. It concluded that there is evidence that "a cause and effect relationship exist between the consumption of melatonin and reduction of sleep onset latency," and that "1 mg of melatonin should be consumed close to bedtime" (EFSA, 2011).

A consensus of the British Association for Psychopharmacology on evidence-based treatment of insomnia, parasomnia, and circadian rhythm sleep disorders concluded that melatonin is the first-choice treatment when a hypnotic is indicated in patients over 55 years (Wilson et al., 2010). More recently it has been reported that exogenous melatonin is a potentially useful therapeutic agent for improving sleep and quality of life in cancer patients (Innominato et al., 2016).

Melatonin Receptor Agonists

Because melatonin has a short half-life (<30 min), its efficacy in promoting and maintaining sleep has not been uniform in the studies undertaken so far. Thus, the need for the development of prolonged-release preparations of melatonin or of melatonin agonists with a longer duration of action on sleep regulatory structures in the brain arose (Turek et al., 2004). In accordance with this idea, slow-release forms of melatonin were developed (Lemoine et al., 2007) (e.g., Circadin®, a 2-mg preparation developed by Neurim, Tel Aviv, Israel, and approved by the European Medicines Agency in 2007). Their efficacy in treatment of sleep disorders in various populations were reported (Garfinkel et al., 2011).

A "sleep-switch" model to describe the regulation of sleep–wakefulness was originally proposed by Saper and his colleagues (Saper et al., 2005a,b). It consists of "flip-flop" reciprocal inhibitions among sleep-associated activities in the ventrolateral preoptic nucleus and wakefulness-associated activities in the locus coeruleus, dorsal raphe, and tuberomammilary nuclei. The suprachiasmatic nucleus (SCN) has an active role both in promoting wakefulness as well as in promoting sleep, and this depends upon a complex neuronal network and a number of neurotransmitters (GABA, glutamate, arginine vasopressin, somatostatin, etc.) (Kalsbeek et al., 2006; Reghunandanan et al., 2006).

The high density of melatonin receptors in the hypothalamic SCN (Reppert et al., 1994; Dubocovich et al., 2010) may suggest that melatonin affects sleep and the sleep–wakefulness cycle by acting on these receptors. Thus, the need arose for the development of melatonin receptor agonists with a longer duration of action, and which were hopefully more potent in affecting sleep quality. The melatonin analogs ramelteon, agomelatine, tasimelteon, and TK-301 are examples of this strategy (Laudon and Frydman-Marom, 2014).

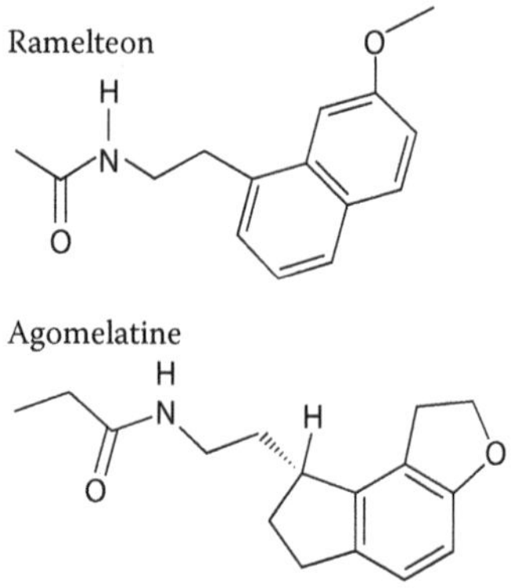

FIGURE 26.1 Chemical structures of ramelteon and agomelatine.

Ramelteon: Ramelteon (Rozerem[R]; Takeda Pharmaceuticals, Osaka, Japan) is a melatonergic hypnotic analog that has been demonstrated to be effective in clinical trials. It is a tricyclic synthetic analog of melatonin with the chemical name (S)-N-[2-(1,6,7,8-tetrahydro-2 H-indeno[5,4-b] furan-8-yl)-ethyl] propionamide. In 2005, ramelteon was approved by the US Food and Drug Administration (FDA) for treatment of insomnia. It is a selective agonist for MT1/MT2 receptors without significant affinity for other receptor sites (Kato et al., 2005; Miyamoto, 2009).

Agomelatine: Disturbances in sleep are prominent features of depression. Antidepressant drugs that are also effective in alleviating sleep disturbances can be of better therapeutic value in treating depressive disorders (Lam, 2006). Agomelatine, a naphthalenic compound chemically designated as N-(2-[7-methoxy-1-naphthalenyl]ethyl)acetamide (Figure 26.1), acts on both MT1 and MT2 melatonergic receptors and also acts as an antagonist to 5HT2C receptors at a concentration three orders of magnitude greater (Millan et al., 2003). Agomelatine has been licensed by the European Medicines Agency (EMA) for the treatment of major depressive disorder (MDD). Agomelatine given before sleep would have an immediate sleep-promoting melatonergic effect that would prevail over its potentially anti-hypnotic 5HT2C antagonism (Millan, 2006).

Conclusion

The data presented clearly indicate that exogenous melatonin and its various analogs promote and maintain sleep. However, there is inconsistency and discrepancy among the large number of reports regarding the degree of efficacy and the clinical significance of these effects. The results of endogenous melatonin's action in insomnia have not been consistent probably because the studies described in existing publications on melatonin's efficacy have utilized different inclusion and exclusion criteria, different outcome measures to evaluate insomnia, different doses of the hormone, and different routes and timing of its administration. There also continues to be considerable controversy over the meaning of the discrepancies that sometimes exist between subjective and objective (polysomnographic) measures of good and bad sleep. Adding to this complexity is melatonin's short half-life and ready metabolism after oral administration of fast-release preparations. Hence, prolonged-release melatonin preparations and melatonin agonists were introduced and have shown good results in treating insomnia.

It is noteworthy that most of the trials that examined the efficacy of melatonin in people with primary sleep disorders were of relatively short trial duration (four to six weeks or less), as were the trials examining the safety of melatonin in this population (three months or less). Therefore, the reported efficacy and safety of melatonin may reflect only the short-term effects of melatonin. Long-term safety and efficacy studies are needed for melatonin and especially for melatonin's agonists, particularly considering the pharmacological activity of their metabolites.

REFERENCES

Braam W, Smits MG, Didden R, Korzilius H, Van Geijlswijk IM, Curfs LM. Exogenous melatonin for sleep problems in individuals with intellectual disability: A meta-analysis. *Dev Med Child Neurol* 2009; 51(5): 340–349.

Brzezinski A. Melatonin in humans. *New Engl J Med* 1997; 336: 186–195.

Brzezinski A, Vangel MG, Wurtman RJ, et al. Effects of exogenous melatonin on sleep: A meta-analysis. *Sleep Med Rev* 2005; 9: 41–50.

Buscemi N, Vandermeer B, Hooten N, et al. The efficacy and safety off exogenous melatonin for primary sleep disorders: A meta-analysis. *J Gen Int Med* 2005; 20: 1151–1158.

Cajochen C, Jewett ME, Dijk DJ. Human circadian melatonin rhythm phase delay during a fixed sleep-wake schedule interspersed with nights of sleep deprivation. *J Pineal Res* 2003; 35: 149–157.

Dubocovich ML, Delagrange P, Krause DN, et al. International union of basic and clinical pharmacology: LXXV. Nomenclature, classification, and pharmacology of G proteincoupled melatonin receptors. *Pharmacol Rev* 2010; 62: 343–380.

European Food Safety Authority. Scientific opinion on the substantiation of a health claim related to melatonin and reduction of sleep onset latency (ID 1698, 1790, 4080) pursuant to Article 13(1) of Regulation (EC) No. 1924/2006. *EFSA J* 2011; 9(6): 2241.

Garfinkel D, Zorin M, Wainstein J, Matas Z, Laudon M, Zisapel N. Efficacy and safety of prolonged-release melatonin in insomnia patients with diabetes: A randomized, double-blind, crossover study. *Diabetes Metab Syndr Obes* 2011; 4: 307–313.

Haimov I, Laudon M, Zisapel N, et al. Sleep disorders and melatonin rhythms in elderly people. *BMJ* 1994; 309: 167.

Haimov I, Lavie P, Laudon M, Herer P, Vigder C, Zisapel N. Melatonin replacement therapy of elderly insomniacs. *Sleep* 1995; 18(7): 598–603.

Hughes RJ, Badia P. Sleep-promoting and hypothermic effects of daytime melatonin administration in humans. *Sleep* 1997; 20: 124–131.

Innominato PF, Lim AS, Palesh O, Clemons M, Trudeau M, Eisen A, Wang C, Kiss A, Pritchard KI, Bjarnason GA. The effect of melatonin on sleep and quality of life in patients with advanced breast cancer. *Support Care Cancer.* 2016; 24(3): 1097–1105.

Kalsbeek A, Perreau-Lenz S, Buijs RM. A network of (autonomic) clock outputs. *Chronobiol Int* 2006; 23: 521–535.

Kato K, Hirai K, Nishiyama K, et al. Neurochemical properties of ramelteon (TAK-375), a selective MT1/MT2 receptor agonist. *Neuropharmacology* 2005; 48: 301–310.

Kennedy SH, Emsley R. Placebo-controlled trial of agomelatine in the treatment of major depressive disorder. *Eur Neuropsychopharmacol* 2006; 16: 93–100.

Lam RW. Sleep disturbances and depression: A challenge for antidepressants. *Int Clin Psychopharmacol* 2006; 21(Suppl. 1): S25–S29.

Laudon M, Frydman-Marom A. Therapeutic effects of melatonin receptor agonists on sleep and comorbid disorders. *Int J Mol Sci* 2014; 15(9): 15924–15950.

Lavie P. Melatonin: Role in gating nocturnal rise in sleep propensity. *J Biol Rhythms* 1997; 12: 657–665.

Leger D, Laudon M, Zisapel N. Nocturnal 6-sulfatoxymelatonin excretion in insomnia and its relation to the response to melatonin replacement therapy. *Am J Med* 2004; 116: 91–95.

Lemoine P, Nir T, Laudon M, Zisapel N. Prolonged-release melatonin improves sleep quality and morning alertness in insomnia patients aged 55 years and older and has no withdrawal effects. *J Sleep Res* 2007; 16: 372–380.

Llorca PM. The antidepressant agomelatine improves the quality of life of depressed patients: Implications for remission. *J Psychopharmacol* 2010; 24: 21–26.

Millan MJ. Multi-target strategies for the improved treatment of depressive states: Conceptual foundations and neuronal substrates, drug discovery and therapeutic application. *Pharmacol Ther* 2006; 110: 135–370.

Millan MJ, Gobert A, Lejeune F, et al. The novel melatonin agonist agomelatine (S20098) is an antagonist at 5-hydroxytryptamine2C receptors, blockade of which enhances the activity of frontocortical dopaminergic and adrenergic pathways. *J Pharmacol Exp Ther* 2003; 306: 954–964.

Miyamoto M. Pharmacology of ramelteon, a selective MT1/MT2 receptor agonist: A novel therapeutic drug for sleep disorders. *CNS Neurosci Ther* 2009; 15: 32–51.

Nowell PD, Mazumdar S, Buysse DJ, Dew MA, Reynolds CF III, Kupfer D. Benzodiazepines and zolpidem for chronic insomnia: A meta-analysis of treatment efficacy. *JAMA* 1997; 278(24): 2170–2177.

Pandi-Perumal SR, Srinivasan V, Spence DW, et al. Ramelteon: A review of its therapeutic potential in sleep disorders. *Adv Ther* 2009; 26: 613–626.

Rajaratnam SM, Polymeropoulos MH, Fisher DM, et al. Melatonin agonist tasimelteon (VEC-162) for transient insomnia after sleep-time shift: Two randomised controlled multicentre trials. *Lancet* 2009; 373: 482–491.

Reghunandanan V, Reghunandanan R. Neurotransmitters of the suprachiasmatic nuclei. *J Circadian Rhythms* 2006; 4: 2.

Reppert SM, Weaver DR, Ebisawa T. Cloning and characterization of a mammalian melatonin receptor that mediates reproductive and circadian responses. *Neuron* 1994; 13: 1177–1185.

Saper CB, Lu J, Chou TC, Gooley J. The hypothalamic integrator for circadian rhythms. *Trends Neurosci* 2005a; 28: 152–157.

Saper CB, Scammell TE, Lu J. Hypothalamic regulation of sleep and circadian rhythms. *Nature* 2005b; 437: 1257–1263.

Shochat T, Haimov I, Lavie P. Melatonin: The key to the gate of sleep. *Ann Med* 1998; 30(1): 109–114.

Srinivasan V, Brzezinski A, Spence DW, et al. Sleep, mood disorders and antidepressants: The melatonergic antidepressant agomelatine offers a new strategy for treatment. *Psychiatria Fennica* 2010; 41: 168–187.

Srinivasan V, Cardinali DP, Pandi-Perumal SR, Brown GM. Melatonin agonists for treatment of sleep and depressive disorders. *J Exp Integrat Med* 2011; 1: 149–158.

Turek FW, Gillette MU. Melatonin, sleep, and circadian rhythms: Rationale for development of specific melatonin agonists. *Sleep Med* 2004; 5: 523–532.

Wilson SJ, Nutt DJ, Alford C, et al. British Association for Psychopharmacology consensus statement on evidence-based treatment of insomnia, parasomnias and circadian rhythm disorders. *J Psychopharmacol* 2010; 24: 1577–1601.

Zemlan FP, Mulchahey JJ, Scharf MB, et al. The efficacy and safety of the melatonin agonist beta-methyl-6-chloromelatonin in primary insomnia: A randomized, placebo-controlled, crossover clinical trial. *J Clin Psychiatry* 2005; 66: 384–390.

Zhdanova IV, Friedman L. Melatonin for treatment of sleep and mood disorders. In Mischolon D, Rosenbaum J (eds), *Natural Medications for Psychiatric Disorders: Considering the Alternatives*, pp. 147–174. Philadelphia, PA: Lippincott Williams & Wilkins, 2002.

Zhdanova IV, Lynch HJ, Wurtman RJ. Melatonin: A sleep-promoting hormone. *Sleep* 1997; 20: 899–907.

Zhdanova IV, Wurtman RJ, Regan MM, et al. Melatonin treatment for age-related insomnia. *J Clin Endocrin Metab* 2001; 86(10): 4727–4730.

27

Relevance of Ayurveda in Management of Sleep Disorders

Rama Jayasundar
All India Institute of Medical Sciences
New Delhi, India

CONTENTS

ABSTRACT   Although sleep, along with diet, is the most powerful factor in regulating health and preventing diseases, it is only in recent times that scientific attention has been drawn to this subject. There is now increasing realization of the physiological and clinical importance of sleep for optimal physical and mental health. While unraveling the molecular mechanisms underlying sleep remains an ongoing challenge, several components involved in the process of sleep have also been identified. For instance, serotonin and melatonin have been found to play important roles in the circadian rhythm of sleep. At the same time, *ayurveda*, the medical system indigenous to India, also deals in detail with the subject of sleep, the related disorders, and their management through therapeutic and nutritional interventions. This chapter provides an overview of the conceptual differences in the understanding of sleep in Western medicine and ayurveda. It also discusses the roles of serotonin and melatonin in the ayurvedic approach to sleep and provides a list of serotonin and melatonin containing medicinal and edible plants used in ayurveda.

KEY WORDS: *ayurveda, sleep, serotonin, melatonin, tridosha, prakruti.*

Abbreviations

GABA: gamma-aminobutyric acid
GIT: gastrointestinal tract
IBS: irritable bowel syndrome
K: *kapha*
P: *pitta*
V: *vata*

Introduction

Human beings spend approximately a third of their lives asleep. There was a time when the importance of sleep was undermined. It was considered a mere necessity and part of one's daily routine, most often taken for granted. It was thus never viewed as a significant area for active research. Now, however, there is increasing realization based on scientific data that sleep, much like eating, is essential for both the physical and mental health of individuals (Swick, 2005; Cirelli and Tononi, 2008; Rial et al., 2007). There is mounting scientific evidence to indicate that sleep not only influences a number of physiological functions, but is also a causative factor for a number of neurological and other diseases (Lautenbacher et al., 2006; Bollinger et al., 2010; Bianchi, 2013; Sharma and Kavuru, 2010). For example, an association between sleep and the onset of obesity has been reported (Bollinger et al., 2010).

On one hand, emerging evidences are reinforcing the importance of sleep history as an integral part of medical diagnosis and clinical research (Kleim et al., 2014). In patient management, sleep, alongside diet and physical activity, is being considered an important component of the prescribed lifestyle package. In fact, clinical applications of sleep as a therapeutic target are being explored, especially in cognition enhancement (Diekelmann, 2014; Sachdeva et al., 2015). At the same time, there is also increasing focus on fundamental research directed toward understanding the molecular mechanisms of sleep, like unraveling the roles of neurotransmitters such as gamma-aminobutyric acid (GABA) and serotonin, and hormones like melatonin in the sleep–wake cycle (Cirelli, 2009; Abbas et al., 2010).

Although sleep as a research topic has gained increasing attention only in recent times in current medical science, one should not sidestep the fact that this rather important physiological process has already been recognized as an essential aspect of life and dealt with in detail in traditional systems of medicine that emphasize preventive health. This is indeed true in the case of *ayurveda*, the medical system indigenous to India, in which subjects of sleep, sleep disorders, and their management through nutritional and therapeutic interventions are dealt with in great detail (Sharma and Dash, 2001; Sharma, 2004; Srikanthamurthy, 2007).

This article explores the possibility of connecting the dots between molecular components like serotonin and melatonin involved in sleep and the ayurvedic parameters understood to play a crucial role in sleep. Before attempting such a connection between ayurveda and the current scientific data obtained in the context of Western medicine, it is essential to understand the different perspectives of these two streams of medicine. This chapter provides a brief overview of their understanding of biological systems and posits sleep in this context.

Current Knowledge and Understanding

Among other molecular components associated with sleep, serotonin and melatonin have been found to play key roles in regulating sleep (Abbas et al., 2010; Van Cauter and Tasali, 2011). Serotonin (neurotransmitter) and melatonin (hormone) are produced in the pineal gland in the brain. Melatonin is synthesized from tryptophan, the immediate precursor, however, being serotonin, which is secreted during the daytime. Melatonin is produced at night and sleep is found to be particularly important for its production.

Circadian rhythm regulates the production of melatonin. The pineal gland responds to light-induced signals and switches off melatonin production during the daytime with levels of melatonin in serum are reported to be lowest during the day and highest at night (Van Cauter and Tasali, 2011). Melatonin plays an important role not only in sleep–wake cycles, but also in a number of other functions such as gastric protection, oxidative stress, and gastrointestinal physiology (Abbas et al., 2010; Van Cauter and Tasali, 2011). While night triggers the release of melatonin from the pineal gland, the gastrointestinal tract (GIT) acts as a source of melatonin during the daytime (Konturek et al., 2011). Melatonin is synthesized in the enterochromaffin cells of the gut and its secretion is considered to be associated with the biological clock related to food intake (Bertrand et al., 2014). Both pineal and GIT melatonin are multifunctional, with some common and function-specific roles (Abbas et al., 2010; Konturek et al., 2011; Bertrand et al., 2014).

While unraveling the molecular mechanisms underlying a biological process remains a fundamental research focus in current medical science, the identified molecular components are also considered, promising therapeutic agents leading to new therapies. Such examples include the therapeutic use of serotonin and melatonin in sleep and mood disorders and irritable bowel syndrome (IBS) (Lu et al., 2005; Pozo et al., 2010). This approach of identifying molecular targets for therapeutic interventions stems from the prevailing perspective of modern biology, which considers life to be organized bottom up. To elaborate, the human system is considered a hierarchy of levels from atom upwards: molecules, organelles, cells, tissues, organs, systems, and so on.

Current understanding views life as a product of complex but coordinated chemical reactions, which can be understood by studying the components involved in these reactions. Although such a reductive approach has provided modern biology and medicine with nuanced molecular-level information about normal and pathophysiological processes, there is also a growing appreciation for a broader systemic approach to understand biological processes (West, 2012). In this context, it will be interesting to understand the viewpoint of holistic medical systems like ayurveda.

Ayurveda: The Art of Healthy Living

Ayurveda has perhaps the longest unbroken chain of clinical experience and health tradition in the world (Mukerjee, 2006). Having been the mainstream medical system in India until colonization, ayurveda carries centuries of observations, experiments, clinical applications, and documentation. It has developed theories based on a string of logical conclusions derived from observational data and has used these theories to handle human complexity (Jayasundar, 2010). The understanding of the human system in ayurveda differs significantly from the way modern medical science perceives it (Jayasundar, 2009, 2012). Viewed as consisting of interdependent physical, physiological, psychological, and spiritual domains, each influencing the other strongly, the human system is understood as a complex but integrated entity in ayurveda (Jayasundar, 2013a). The influence of one domain affects the other, be it in a positive or negative manner. Ayurveda's theories are applicable across these diverse domains.

Tridosha: A Simple Theory to Handle Complexity

Although there are several theories in ayurveda, the guiding paradigm states that every healthy organism is in a state of balance defined by the functional principles of *vata* (V), *pitta* (P), and *kapha* (K) (Sharma and Dash, 2001; Sharma, 2004; Srikanthamurthy, 2007). V, P, and K are individually known as *dosha* and collectively as *tridosha* in Sanskrit, the language of ayurveda. V, P, and K denote movement, metabolic transformation, and growth and support (anabolic transformation), respectively. They also include properties that are physical, physicochemical, and physiological in nature, and which impact the functions associated with V, P, and K. For example, V comprises of properties like dryness and lightness, which affect movement and also have clinical relevance (Pritchard and Fonn, 1995; Bert-Couto et al., 2012). P on the other hand, includes parameters like pH and temperature, which are known to play crucial roles in metabolic transformation (Swainson and Cumming, 1999). Examples of properties involved in K are adhesiveness and viscosity. A more detailed explanation of these and their physiological and clinical relevance can be found in these referred articles: Jayasundar, 2009, 2010, 2012, 2013a.

The parameters associated with V, P, and K are interconnected with one another, making them a network of properties (Jayasundar, 2013a). The interrelations among these properties have inbuilt dynamic feedback loops providing the network with a system of checks and balances. It is pertinent to note that the current systems biology also deals with networks of cellular components like genes, proteins, and metabolites (Cassman et al., 2007). While these associations remain at a cellular level, the network of system properties in ayurveda transcends this level and includes the various planes of structural hierarchy. For example, dryness exists and manifests both at the level of a cell as well as the entire organism (Pritchard and Fonn, 1995; Bert-Couto et al., 2012). V, P, and K in fact encompass physical, physiological, and also mental domains. Ayurveda says that derangement of these very properties results in diseases (Sharma and Dash, 2001; Sharma, 2004; Srikanthamurthy, 2007). Imbalances can lead to sleep disturbances as well. At a pragmatic level, all factors (seasons, activities, dietary ingredients, clinical symptoms, etc.) playing a role in health and disease are also classified in terms of V, P and K. Prevention, diagnosis and therapeutic management involving these factors are strategized using the common platform of V, P and K.

Sleep: An Ayurvedic Perspective

Health in ayurveda reflects the ability of an individual to adapt to stress. Regulated eating and sleeping habits play key roles in aiding this adaptive response. *Nidra* (sleep), according to ayurveda, is a normal physiological process to retain, recover, restore, and maintain a person's health (Sharma and Dash, 2001; Sharma, 2004; Srikanthamurthy, 2007). Ayurveda mentions that optimum conditions for growth or anabolic processes prevail during sleep. The normal physiological process of sleep happens as a consequence of the fatigue of the sense organs, which results in a perceptual disengagement of the senses from the environment. This is a stage when all sense organs have stopped cognizing their objects. At a physical level, sleep is linked to *Kapha dosha*. Sleep and K are directly related: an increase in one will increase the other. Ayurveda strongly advises sleeping at night. The increasing scientific evidence of the positive effects of following circadian rhythms and the negative influence of not doing so supports the ayurvedic viewpoint (Swick, 2005; Cirelli and Tononi, 2008; Rial et al., 2007; Lautenbacher et al., 2006; Bollinger et al., 2010; Bianchi, 2013; Sharma and Kavuru, 2010). Sleep, if suppressed leads to pain in body parts, heaviness of the head, lassitude, exhaustion (even without strain), giddiness, indigestion and diseases of *vata* origin. These clinical observations are consistent with the current research findings on association between sleep deprivation, and pain, immunity and metabolic disturbances (Lautenbacher et al., 2006; Bollinger et al., 2010).

Ayurveda classifies sleep into seven types based on their causative factors (Sharma and Dash, 2001; Srikanthamurthy, 2007): (1) the very nature of night, (2) disease, (3) physical exhaustion, (4) mental exertion, (5) increase of *kapha*, (6) external factors, and (7) the dominance of *tamas*.[*] Of the seven types of sleep, only the natural state of sleeping at night is considered *par excellence* and physiological, facilitating nourishment and protection of the body, while the rest are pathological forms of sleep. While accepting this classification, *Suśruta* provides a physiological classification as well: (1) normal, (2) abnormal, and (3) fatal (*tāmasi*) (Sharma, 2004). Ayurveda also discusses various stages in sleep, the discussion of which would be beyond the scope of this chapter. Ayurveda discusses in detail the role of factors like *prakruti* (bio-psychological phenotype), season, time and disease condition in sleep. *Prakruti* denotes the inherent natural state of physical, physiological and psychological existence of each person and defines their basic tendencies. In the context of sleep, a *vata prakruti* person will have irregular, light and fewer hours of sleep whereas *Kapha prakruti* people are sound sleepers and sleep longer hours.

Diet

A major strength of ayurveda is its in-depth practical knowledge on diet, therapeutic nutrition, and nutraceutical and medicinal plants (Jayasundar, 2013b). Ayurveda provides a number of customized dietary options based on the season, one's *prakruti*, and activities. Information is also provided on dietary ingredients and preparations to help one ease into a deep and restful sleep so that one wakes up with renewed energy and vitality to meet the challenges of the following day.

---

[*] *Tamas* is one of the three qualities used to denote states of mind, the other two being *sattva* and *rajas*. These qualities determine the nature of individuals, their actions, behavior, and attitudes in the objective world in which they live. They reflect individual traits in behavior. To put it simply, *tamas* denotes a state of mind that has inertia, lassitude, and a lack of enthusiasm in activities. *Sattva* reflects a state of mind that is happy, calm, pure in thought, and has no fear. *Rajasic* qualities are midway between *sattva* and *tamas*. They denote passion and desire associated with activities.

Sleep Disorders: Clinical Consequence of Affected Sleep

From the clinical standpoint of ayurveda, disturbances in sleep lead to negative effects which are both short and long term. For example, fatigue, pain and lethargy are short term effects, whereas, the long term effects involve manifestation of diseases like obesity, uneasiness in the chest, obstruction in vessels and gastrointestinal disorders. While deprivation of sleep leads to an increase in *vata* and excess aggravates *kapha* and reduced sleep increases *pitta*. Ayurveda lists the causes, symptoms and therapeutic interventions for each category (Sharma and Dash, 2001; Sharma, 2004; Srikanthamurthy, 2007).

Caraka Samhita makes a special mention of the role of sleep in the onset of obesity (Sharma and Dash, 2001). It is interesting to note similar observations reported in current scientific literature (Lautenbacher et al., 2006; Bollinger et al., 2010; Bianchi, 2013). It is pertinent to mention that while sleep disturbances as a symptom can be indicative of bodily dysfunction, the converse also holds good–disease can cause sleep disturbances. Ayurveda also prescribes sleep as part of therapeutic management in some conditions. Sleep, especially during day time is advised in indigestion, diarrhoea, mental disturbances, hiccough, emaciation, etc. but strictly contraindicated in obesity, disorders of aggravated *kapha*, poisoning and throat disorders. Sleep as a therapeutic intervention is only an emerging concept in Western medicine (Diekelmann, 2014; Sachdeva et al., 2015).

Nutritional Intervention for Sleep Disorders

In line with its focus on molecular understanding, current medical science uses molecules identified in biological processes as agents for both nutritional and therapeutic interventions. For example, melatonin is used both for treating IBS and as a food supplement for addressing oxidative stress (Halson, 2014). Ayurveda, on the other hand, utilizes knowledge of properties involved in the biological processes to target therapeutic and nutritional interventions. For example, lack of sleep increases dryness, indicating increased *vata*. So, dietary ingredients (e.g., milk) and herbs (e.g., *Withania somnifera*) to reduce *vata* would be used to induce sleep.

For those with insufficient sleep, the following are prescribed: milk, sugarcane juice, meat soup, meat of aquatic animals, preparations from jaggery and flour, rice, wine, blackgram, skimmed buttermilk, yoghurt of buffalo milk, and ghee followed by milk processed with a rejuvenating group of herbs such as *Phaseolus trilobus*, *Glycyrrhiza glabra,* and *Teramnus labialis. Vitis vinifera* and sugarcane products are prescribed for consumption at night. Persons with excess sleep should avoid the items enumerated above. They are advised to take food that is dry, pungent, and bitter in taste (Jayasundar, 2013b).

Serotonin, Melatonin, and *Nidra*

From the above description of ayurveda and the parameters it uses to understand a biological system, it is clear there are conceptual differences between ayurveda and current medical science. The latter focuses on structural components involved in a biological process, whereas ayurveda deals with properties involved. The current viewpoint of Western medicine explains complex processes by reduction to its fundamental components. Ayurveda does not look at any process in isolation, but at the same time its framework encompasses structural components as well. As mentioned, the ayurvedic approach typically identifies properties rather than molecules responsible for a process and hence has far less dependence on molecular components. Ayurveda's fundamental approach, thus, differs from that of Western medicine, making one-to-one comparison across these medical systems rather complex. Hence, the direct relevance of serotonin and melatonin to ayurvedic discussions on sleep may be difficult to pinpoint or extrapolate.

That said, there are many points of convergence between the two medical systems in the explanations, clinical observations, and inferences on sleep and related disorders, as evident from this article. These are highlighted in Table 27.1. It is reiterated that the information on sleep in ayurveda stems not

TABLE 27.1

Areas of Convergence in Western Medicine and Ayurveda on Sleep

Current Findings in Western Medicine	Mentioned and Discussed in Ayurveda
Importance of sleep	✓
Optimum conditions for growth during sleep	✓
Association between sleep and circadian cycle	✓
Impact of sleep on:	
Physical and mental health	✓
Physiological functions	✓
Sleep deprivation results in:	
Neurological diseases	✓
Onset of obesity	✓
Disease as a cause of sleep disorders	✓
Neurodegenerative disorders accompanied by changes in sleep patterns	✓
Emerging areas:	
Sleep history as part of medical diagnosis	✓
Sleep as a therapeutic target	✓
Nutritional interventions for inducing sleep	✓
Increasing therapeutic efficacy by following treatment with sleep in certain diseases	✓
Nutritional interventions for inducing sleep	✓
Unraveling the molecular mechanisms of sleep	—

from any recent research but from what has been observed, inferred, and documented over many centuries. In the context of ayurvedic therapeutic and nutritional interventions, the connecting dots between ayurveda and Western medicine could well be the phytochemicals, melatonin and serotonin. These two compounds are found in abundance in plants (Paredes et al., 2009; Ramakrishna et al., 2011; Tan, 2015). Since ayurveda uses plants extensively in both medicine and nutrition, serotonin and melatonin are often consumed by patients.

Table 27.2 lists some of the serotonin- and melatonin-containing medicinal and edible plants used in ayurveda (Odjakova and Hadjiivanova, 1997; Paredes et al., 2009; Murch et al., 2009; Huang and Mazza, 2011; Ramakrishna et al., 2011; Roshchina, 2011; Tan et al., 2012; Arnao, 2014; Servillo et al., 2015; Tan, 2015). Melatonin is found in common food items such as corn, rice, and wheat (Tan et al., 2012). The beneficial effects of serotonin and melatonin consumed through these products, which serve as staple food sources to billions of people, is apparent. Interestingly, many of the food ingredients, such as ginger, garlic, cardamom, fenugreek, and coriander, used in Indian cuisine also contain melatonin and serotonin (Table 27.2). Thus, the daily Indian diet containing some or all of these ingredients could play a prophylactic role in sleep disorders by constant supplementation of these phytochemicals.

Conclusion

In today's stress-filled world and changing lifestyles, sleep-related disorders are recognized as one of the fastest-growing health problems across all age groups. Compromised sleep is now proven to have a negative influence on cognitive and other functions, such as glucose and carbohydrate metabolism, neuroendocrine function, appetite, protein synthesis, and so on (Bianchi, 2013; Sharma and Kavuru, 2010). Although some fundamental questions about sleep are yet to be answered from the perspective of modern science, what is becoming increasingly clear is that balanced sleep is a must for optimal health. Sleep is of fundamental importance for the body and mind to rest, reset, and rejuvenate.

While ongoing research continues to obtain data and draw inferences from them for use in patient management, cognizance should be taken of the already available wealth of information on sleep, the associated disorders, and their management in ayurveda. Interestingly, the therapeutic aspects of sleep is

TABLE 27.2

Serotonin- and Melatonin-Containing Medicinal and Edible
Plants Used in Ayurveda

Plants (Medicinal and Edible) and Food Components Containing:	
Serotonin	**Melatonin**
Mimosa pudica	*Datura metel*
Solanum melongena	*Papaver somniferum*
Mucuna pruriens	*Solanum nigrum*
Datura metel	*Brassica nigra*
Allium cepa	*Allium cepa* (onion)
Capsicum annum	*Capsicum annum*
Solanum lycopersicum	*Allium sativum* (garlic)
Juglans regia (walnut)	*Juglans regia* (walnut)
Citrus genus	*Elettaria cardamomum*
Ananas cosmosus (pineapple)	*Ananas cosmosus* (pineapple)
Carica papaya (papaya)	*Zingiber officinale* (ginger)
Musa paradisiaca (banana)	*Hordeum vulgare*
Musa sapientum (banana)	*Trigonella foenum*
	Oryza sativa (rice)
	Raphanus sativus
	Cucumis sativus
	Punica granatum
	Prunus amygdalus (almond)
	Coriandrum sativum
	Foeniculum vulgare
	Vitis vinifera (grapes)
	corn
	wheat
	oat
	wine

a new frontier being explored in Western medicine (Diekelmann, 2014; Sachdeva et al., 2015), whereas this is already an aspect of ayurveda.

Ayurveda integrates functional and structural information to guide the understanding of physiology. Therefore, redundancy of the focus of Western medicine, namely molecules per se, is inbuilt into the ayurvedic understanding of biological processes. This chapter has explored the differing viewpoints on sleep in ayurveda and current approaches as a means of furthering debate between two different streams of thinking, one focusing on molecular components involved in biological processes and the other on the properties guiding them.

Acknowledgments

I express my appreciation to Dr. Smruthi Jayasundar, University of Cambridge, UK, for her valuable and constructive suggestions.

REFERENCES

Abbas A, Raju J, Milles J, Ramachandran S. A circadian rhythm sleep disorder: Melatonin resets the biological clock. *J R Coll Phys Edinb* 2010; 40: 311–313.
Arnao MB. Phytomelatonin: Discovery, content and role in plants. *Adv Bot* 2014; 2014: 11.

Bert-Couto SA, Couto-Souza PH, Jacobs R, et al. Clinical diagnosis of hyposalivation in hospitalized patients. *J Appl Oral Sci* 2012; 20: 157–161.

Bertrand PP, Polglaze KE, Bertrand RL, Sandow SL, Pozo MJ. Detection of melatonin production from the intestinal epithelium using electrochemical methods. *Curr Pharm Des* 2014; 20: 4802–4806.

Bianchi MT. *Sleep Deprivation and Disease: Effects on the Body, Brain and Behavior.* New York: Springer, 2013.

Bollinger T, Bollinger A, Oster H, Solbach W. Sleep, immunity, and circadian clocks: A mechanistic model. *Gerontology* 2010; 56: 574–580.

Cassman M, Arkin D, Doyle F, Katagiri F, Lauffenburger D, Stokes C. *Systems Biology: International Research and Development.* Dordrecht, the Netherlands: Springer, 2007.

Cirelli C. The genetic and molecular regulation of sleep: From fruit flies to humans. *Nature Rev Neurosci* 2009; 10: 549–560.

Cirelli C, Tononi G. Is sleep essential? *PLoS Biol* 2008; 6(8): e216.

Diekelmann S. Sleep for cognitive enhancement. *Front Syst Neurosci* 2014; 8: 46.

Halson SL. Sleep in elite athletes and nutritional interventions to enhance sleep. *Sports Med* 2014; 44(Suppl. 1): S13–23.

Huang X, Mazza G. Application of LC and LC–MS to the analysis of melatonin and serotonin in edible plants. *Crit Rev Food Sci Nutr* 2011; 51: 269–284.

Jayasundar R. Health and disease: Distinctive approaches of biomedicine and ayurveda. *Leadership Medica* 2009; 15: 6–21.

Jayasundar R. Ayurveda: A distinctive approach to health and disease. *Curr Sci* 2010; 98: 908–914.

Jayasundar R. Contrasting approaches to health and disease: Ayurveda and biomedicine. In Sujata V, Abraham L (eds), *Medicine, State and Society: Indigenous Medicine and Medical Pluralism in Contemporary India*, pp. 37–58. Telangana, India: Orient Blackswan, 2012.

Jayasundar R. Ayurvedic approach to functional foods. In Martirosyan DM (ed.), *Introduction to Functional Food Science*, pp. 454–479 Dallas, TX: Food Science Publisher, 2013a.

Jayasundar R. Quantum logic in ayurveda. In Morandi A, Nambi ANN (eds), *An Integrated Model of Health and Well-Being: Bridging Indian and Western Knowledge*, pp. 115–139. Dordrecht, the Netherlands: Springer, 2013b.

Kleim B, Wilhelm FH, Temp L, Margraf J, Wiederhold BK, Rasch B. Sleep enhances exposure therapy. *Psychol Med* 2014; 44: 1511–1519.

Konturek PC, Brzozowski T, Konturek SJ. Gut clock: Implication of circadian rhythms in the gastrointestinal tract. *J Physiol Pharmacol* 2011; 62: 139–150.

Lautenbacher S, Kundermann B, Krieg JC. Sleep deprivation and pain perception. *Sleep Med Rev* 2006; 10: 357–369.

Lu WZ, Gwee KA, Moochhalla S, Ho KY. Melatonin improves bowel symptoms in female patients with irritable bowel syndrome: A double-blind placebo-controlled study. *Aliment Pharmacol Ther* 2005; 22: 927–934.

Mukerjee GN. *History of Indian Medicine.* Delhi, India: Chaukhamba Sanskrit Pratisthan, 2006.

Murch SJ, Alan AR, Cao J, et al. Melatonin and serotonin in flowers and fruits of *Datura metel* L. *J Pineal Res* 2009; 47: 277–283.

Odjakova M, Hadjiivanova C. Animal neurotransmitter substances in plants: Review. *Bulg J Plant Physiol* 1997; 23: 94–102.

Paredes SD, Korkmaz A, Manchester LC, et al. Phytomelatonin: A review. *J Exptl Bot* 2009; 60: 57–69.

Pozo MJ, Gomez-Pinilla PJ, Camello-Almaraz C, et al. Melatonin, a potential therapeutic agent for smooth muscle-related pathological conditions and aging. *Curr Med Chem* 2010; 17: 4150–4165.

Pritchard N, Fonn D. Dehydration, lens movement and dryness ratings of hydrogel contact lenses. *Ophthalmic Physiol Opt* 1995; 15: 281–286.

Ramakrishna A, Giridhar P, Ravishankar GA. Phytoserotonin: A review. *Plant Signal Behav* 2011; 6: 800–809.

Rial RV, Nicolau MC, Gamundí A, et al. The trivial function of sleep. *Sleep Med Rev* 2007; 11: 311–325.

Roshchina VV. *Neurotransmitters in Plant Life.* Enfield, NH: Science Publishers, 2011.

Sachdeva A, Kumar D, Anand KS. Non-pharmacological cognitive enhancers: Current perspectives. *J Clin Diag Res* 2015; 9: VE01–VE06.

Servillo L, Giovane A, Casale R, et al. Serotonin 5-O-β-glucoside and its N-methylated forms in citrus genus plants. *J Agric Food Chem* 2015; 63: 4220–4227.

Sharma PV (trans.). *Suśruta Samhita*. Varanasi, India: Chaukhamba Visvabharati, 2004.

Sharma RK, Dash B (trans.). *Caraka Samhita*. Varanasi, India: Chaukhamba Sanskrit Series Office, 2001.

Sharma S, Kavuru M. Sleep and metabolism: An overview. *Int J Endocrin* 2010; 2010: 1–12.

Srikanthamurthy KR (trans.). *Ashtanga SaMgraha of Vagbhata*. Varanasi, India: Chaukhamba Orientalia, 2007.

Swainson CP, Cumming AD. Disturbances in water, electrolyte and acid–base balance. In Haslett C, Chilvers ER, Hunter JAA, Boon NA (eds), *Davidson's Principles and Practice of Medicine*, pp. 393–416. New York: Churchill Livingston, 1999.

Swick TJ. The neurology of sleep. *Neurol Clin* 2005; 23: 967–989.

Tan DX. Melatonin and plants. *J Exp Bot.* 2015; 66: 625–626.

Tan DX, Hardeland R, Manchester LC, et al. Functional roles of melatonin in plants, and perspectives in nutritional and agricultural science. *J Exp Bot* 2012; 63: 577–597.

Van Cauter E, Tasali E. Endocrine physiology in relation to sleep and sleep disturbances. In Kryger MH, Roth T, Dement WC (eds), *Principles and Practice of Sleep Medicine*, pp. 291–311. St Louis, MO: Elsevier, 2011.

West GB. The importance of quantitative systemic thinking in medicine. *Lancet* 2012; 379: 1551–1559.

28

Regulation of Serotonin in Depression: Efficacy of Ayurvedic Plants

Rinki Kumari, Aruna Agrawal, and Govind Prasad Dubey
Banaras Hindu University
Varanasi, India

Praveen K. Singh and Gur Prit Inder Singh
Adesh University
Bathinda, India

CONTENTS

ABSTRACT Depression is an etiologically heterogeneous group of chronic psychiatric illnesses associated with significant morbidity, mortality, and disability. Among the several theories behind the pathophysiological basis of depression, imbalances of the monoamine neurotransmitters (monoamine hypothesis), especially serotonin, have been the most extensively investigated. Several antidepressant drugs targeting the serotonin system have been developed (e.g., selective serotonin reuptake inhibitors) with the intended physiological effect to enhance serotonin signaling at the synapse. Although these drugs represent popular and effective treatments for depression, they are also associated with side effects and limitations in their efficacy. Herbal remedies, often employed as part of traditional systems of medicine and generally associated with favorable safety profiles, have the potential to provide effective alternatives to currently employed modern synthetic antidepressants. Ayurveda is one of the oldest systems of medicine in the world and has been practiced for centuries in India. Ayurveda has described the use of several medicinal plants to prevent, treat, and manage numerous diseases. The following chapter discusses a selection of ayurveda medicinal plants with serotonergic properties that are currently used or have the potential to be developed as antidepressants. The plants discussed here include *Areca catechu*, *Withania somnifera*, *Bacopa monnieri*, *Berberis aristata*, *Tinospora cordifolia*, *Curcuma longa*,

Nyctanthes arbor-tristis, Zingiber officinale, and *Ocimum sanctum*. To date, the major scientific evidence signifying the antidepressant properties of these plants comes from animal depression models such as the forced-swim and tail suspension tests coupled with biochemical and/or pharmacological studies, demonstrating that these plants modulate the serotonin system in the brain. Moving forward, standardization of drug products derived from these plants coupled with well-designed clinical studies and further investigations into the molecular mechanisms behind their serotonergic properties will be important to realize their potential to alleviate the global burden of depressive disorders.

KEY WORDS: *depression, serotonin, ayurvedic plants,* Areca catechu, *ashwagandha* (Withania somnifera), *brahmi* (Bacopa monnieri), *daruharidra* (Berberis aristata), *guduchi* (Tinospora cordifolia), *haridra* (Curcuma longa), Nyctanthes arbor-tristis, *sunthi* (Zingiber officinale), *tulsi* (Ocimum sanctum).

Abbreviations

5HT: 5-hydroxytryptamine (serotonin)
BDNF: brain-derived neurotrophic factor
MAO: monoamine oxidase
MAO-A: monoamine oxidase A
MAO-B: monoamine oxidase B
MAOI: MAO inhibitor
SSRI: selective serotonin reuptake inhibitor
SSNRI: selective serotonin and norepinephrine reuptake inhibitor
SNRI: serotonin and norepinephrine reuptake inhibitor
TCA: tricyclic antidepressant

Introduction

Depression is an etiologically heterogeneous group of chronic psychiatric illnesses associated with significant morbidity, mortality, and disability. Depression is clinically characterized by a wide range of symptoms that reflect alternation in cognitive, psychomotor, biological, motivational, behavioral, and emotional processes. These include an inability to concentrate, insomnia, loss of appetite, feelings of extreme sadness, guilt, helplessness and hopelessness, and thoughts of death (Katon and Sullivan, 1990; Belmaker and Agam, 2008). Depression affects the quality of life of many people, is a major cause of suicidal death, and is also considered a significant risk factor for cardiovascular diseases (Baxter et al., 2011; Ferrari et al., 2013).

Depression is a complex mood disorder with many subtypes, multiple etiologies, symptomatology ranging from mild to severe with or without psychotic features, and interactions with other psychiatric and somatic disorders (Merikangas et al., 1994; Bondy, 2002; Østergaard et al., 2011; Sibille and French, 2013; Sharpley and Bitsika, 2013, 2014). Major depressive disorder is one of the most common forms. According to the recent (fifth) edition of the *Diagnostic and Statistical Manual of Mental Disorders,* major depressive disorder consists of at least five symptoms (including one of the first two), involving depressed mood, diminished interest or pleasure, significant increase or decrease in weight or appetite, insomnia or hypersomnia, psychomotor agitation or retardation, fatigue, feelings of worthlessness or excessive or inappropriate guilt, inability to concentrate or indecisiveness, and suicidal thoughts. The symptoms are generally required to be present "most of the day, every day for the last two weeks," "must cause clinically significant distress or impairment in social, occupational, or other areas of functioning," and are "not be due to the direct physiological effects of a substance or a general medical condition" (Sharpley et al., 2014). Major depressive disorder has a high lifetime prevalence rate varying from 3% in Japan to 17% in the United States, with most countries lying within the 8%–12% range. In North America the probability of having a major depressive episode within any year-long period is 5%–8%. Females are at greater risk than males (Andrade et al., 2003; Kessler et al., 2003, 2005).

Several studies indicate that depression is caused by a combination of genetic, biochemical, environmental, and psychological factors, rather than a single cause. The involvement of genetic and environmental factors is highlighted by the fact that some but not all types of depression are familial (Katz et al., 1979; Bierut et al., 1999; Hirschfeld and Weissman, 2002). Over the years, several theories behind the pathophysiological basis of depression have been investigated, including impaired circadian rhythms, dysregulated GABA and glutamate signaling, altered hypothalamic–pituitary–adrenal axis activity, and monoamine deficiencies (Hasler, 2010). The latter, commonly referred to as the *monoamine hypothesis*, has been the most extensively investigated. The monoamine hypothesis gained popularity in the 1960s and implicates imbalances of the monoamine neurotransmitters (dopamine, norepinephrine, and serotonin) in depression (Schildkraut, 1965; Coppen, 1967). Some of the earliest support for this hypothesis comes from the 1950s and 1960s, when the antihypertensive drug reserpine was found to have serotonin-depleting properties and also to induce depressive symptoms, whereas the tuberculosis treatment iproniazid was found to enhance monoamine levels and possess antidepressant properties. Crucially, iproniazid was found to block the effects of reserpine (France et al., 2007; López-Muñoz and Alamo, 2009). Serotonin, in particular, has been the most extensively studied neurotransmitter linked to depression (Coppen, 1967; Bondy, 2002).

Serotonin Synthesis, Signaling, and Degradation

Serotonin, also referred to as 5-hydroxytryptamine (5HT), was first discovered in the enterochromaffin cells of the gut in the 1930s and subsequently isolated and crystallized in the 1940s (Rapport et al., 1948a,b). The presence of serotonin in the nervous system and its role as a neurotransmitter was reported in the 1950s (Twarog and Page, 1953; Bogdanski et al., 1958). Since then, the role of serotonin in numerous processes, including the control of appetite, sleep, learning and memory, temperature, mood, behavior, cardiovascular function, muscle contraction, and endocrine functions, have been identified (Frazer and Hensler, 1999; Berger et al., 2009; Meneses and Liy-Salmeron, 2012).

Serotonin is synthesized using the essential amino acid tryptophan, which first involves the hydroxylation of tryptophan to 5-hydroxytryptophan by tryptophan hydroxylase followed by the decarboxylation of 5-hydroxytryptophan to 5HT (i.e., serotonin) by amino acid decarboxylase (Sumi-Ichinose et al., 1992; Walther and Bader, 2003; Ramakrishna et al., 2011). The serotonergic neurons are primarily found in the raphe nuclei of the brain stem, from where they project to various parts of the brain (Dahlstroem and Fuxe, 1964; Pollak Dorocic et al., 2014). Serotonin is released from the presynaptic terminal into the synaptic cleft, where it binds to serotonin receptors on the pre- and/or postsynaptic cell, thereby modulating numerous processes involved in neurotransmission—for example, neurotransmitter release and synaptic potentiation (Murphy et al., 1998; López-Muñoz and Alamo, 2009). Serotonin receptors have been classified into seven groups (5HT1–7) based on their pharmacological profiles, primary sequences and signal transduction mechanisms, and with the exception of 5HT3 (ligand-gated ion channel) belong to the G protein–coupled receptor super family (Hoyer et al., 1994; Mengod et al., 2006). Indeed, the extensive distribution of serotonergic projections (e.g., forebrain, brain stem, and spinal cord) and serotonin receptors (virtually all parts of the brain) coupled with the diverse inputs received by the serotonergic neurons (e.g., from hypothalamus, cortex, basal ganglia, and midbrain) is in consonance with the fact that this neurotransmitter regulates a range of behavioral and neuropsychological processes (Hornung, 2003; Mengod et al., 2006; Berger et al., 2009; Pollak Dorocic et al., 2014). Excess serotonin in the synaptic cleft is taken back up into the presynaptic cell by serotonin transporters, where it is either broken down by the enzyme monoamine oxidase A (MAO-A) or recycled into vesicles for further signaling (Bondy, 2002).

Considering the evidence linking serotonin to depression, a large number of antidepressant drugs targeting the serotonin system have been developed (Hasler, 2010). MAO, serotonin receptors, and serotonin transporters are all targets for antidepressant drugs. The intended physiological effect of these drugs is to increase the amount of serotonin or enhance serotonin signaling at the synapse. Although these drugs represent popular and effective treatments for depression, they are also associated with side effects and limitations in their efficacy.

The first modern antidepressants consisted of tricyclic antidepressants (TCAs; e.g., imipramine) and MAO inhibitors (MAOIs; e.g., phenelzine) and were first introduced in the 1950–1960s (López-Muñoz and Alamo, 2009). Both acted by enhancing serotonin and/or noradrenalin signaling by preventing the metabolism (MAOIs) or reuptake (TCAs) of these neurotransmitters. However both TCAs and MAOIs are associated with side effects. TCAs were found to inhibit histaminic, cholinergic, and alpha-1-adrenergic receptors, resulting in side effects such as drowsiness and dizziness. MAOIs caused an increase in tyramine levels and thereby was associated with a severe risk of hypertension (Feighner, 1999).

The later generations of antidepressants largely consist of selective serotonin reuptake inhibitors (SSRIs), and serotonin and norepinephrine reuptake inhibitors (SNRI). These include trazodone and nefazodone (serotonin antagonist and reuptake inhibitors), fluoxetine, paroxetine, citalopram and escitalopram (SSRIs), venlafaxine and mirtazapine (SNRI), and the selective SNRI (SSNRI) duloxetine (Gartlehner et al., 2007). Due to their enhanced selectivity, these were generally associated with a better safety profile when compared to the first-generation antidepressants. However, meta-analyses of the clinical studies involving second-generation antidepressants found that 40% of the patients were nonresponders and 50% did not achieve remission; these drugs were also associated with side effects such as nausea, headache, diarrhea, fatigue, dizziness, sweating, tremor, dry mouth, weight gain, and sexual dysfunction (Gartlehner et al., 2007, 2011). The SSRIs fluoxetine, fluvoxamine, and paroxetine are also associated with a substantial risk of drug-to-drug interactions, as they inhibit CYP enzymes (Preskorn et al., 2004). Meta-analyses of clinical studies involving modern antidepressants (e.g., the SSRIs fluoxetine and paroxetine, the SNRI venlafaxine, the 5HT2A receptor antagonist nefazodone, and the TCA imipramine) have also raised questions regarding their effectiveness in patients with mild to moderate symptoms (Khan et al., 2002; Kirsch et al., 2008; Fournier et al., 2010).

Ayurvedic Plants with Antidepressant and Serotonergic Properties

Herbal remedies, often employed as part of traditional systems of medicine and generally associated with favorable safety profiles (owing to their natural sources and history of human use), have the potential to provide effective alternatives to currently employed modern synthetic antidepressants. In recent times, increased efforts have focused on the application of modern scientific technologies to identify pharmacologically active constituents and the mode and mechanism of action of herbal medicines, which is playing a crucial role in promoting the acceptability of these products globally. Several reviews have been published recently that highlight the scientific evidence supporting the antidepressant properties of numerous medicinal plants. Animal behavioral studies (e.g., forced-swim test, tail suspension test), brain neurotransmitter quantification, and pharmacological studies involving the serotonergic system have been routinely employed to investigate the antidepressant properties of the plants (Dhamija et al., 2011; Jawaid et al., 2011; Rajput et al., 2011; Gautam et al., 2013). St John's wort (*Hypericum perforatum*) is a well-known plant with antidepressant properties that have been demonstrated through several clinical trials (Gaster and Holroyd, 2000; Lecrubier et al., 2002; Szegedi et al., 2005; Anghelescu et al., 2006; Rahimi et al., 2009). The St John's wort extract WS 5570 is marketed in Germany for the acute treatment of mild to moderate major depression (Gastpar, 2013). Although the precise mechanism of action of this popular medicinal plant is unknown, *in vitro* and *in vivo* studies have demonstrated that preparations from this plant enhance brain serotonin levels, inhibit serotonin (re)uptake, and upregulate serotonin 5HT2 receptors (Calapai et al., 2001; Kientsch et al., 2001; Butterweck, 2003; Hirano et al., 2004; Schulte-Löbbert et al., 2004). Ayurveda is one of the oldest systems of medicine in the world and has been practiced for centuries in India. Ayurveda has described the use of several medicinal plants to prevent, treat, and manage numerous diseases. The following section discusses a selection of ayurvedic medicinal plants with serotonergic properties that are currently used or have the potential to be developed as antidepressants.

(a)   (b)

FIGURE 28.1 *Areca catechu:* (a) tree (From "Areca catechu 2" by Franz Xaver [Own work], licensed under CC BY-SA 3.0, http://creativecommons.org/licenses/by-sa/3.0, via Wikimedia Commons, https://commons.wikimedia.org/wiki/File:Areca_catechu_2.jpg); (b) fruit (CC0 1.0, https://creativecommons.org/publicdomain/zero/1.0/, via Pixabay, https://pixabay.com/en/areca-nut-palm-seed-areca-palm-241989/).

Areca catechu

A. catechu, commonly known as the betel nut tree, belongs to the family Arecaceae (Palmae) and is grown as a seed crop in many Asian countries, such as India and Malaysia (Figure 28.1). Traditionally, the nut (*supari*) is chewed by locals along with betel leaf (*paan*) due to their psychoactive properties (Jaiswal et al., 2011; Osborne et al., 2011). *A. catechu* nut possesses antiparasitic, anti-inflammatory, antioxidant, analgesic, cholinomimetic, and acetylcholinesterase inhibitory properties, and also promotes digestive function. On the other hand, it has also been associated with oral submucosal cell carcinoma and fibrosis (Gilani et al., 2004; Jaiswal et al., 2011; Peng et al., 2015). *A. catechu* nut contains numerous compounds that possess muscarinic receptor agonist activity, including alkaloids, tannins, flavones, triterpenes, steroids, and fatty acids including the alkaloid arecoline (Yang et al., 2000; Ghelardini et al., 2001; Xie et al., 2004; Peng et al., 2015). In a recent study, treatment with *A. catechu* extracts improved learning and memory in rats (Joshi et al., 2012). *A. catechu* fruit extracts inhibited MAO-A activity (IC50 665 μg/ml) in rat brain homogenates and demonstrated antidepressant activity in rodents (forced-swim and tail suspension tests) (Dar et al., 1997a,b; Dar and Khatoon, 2000; Abbas et al., 2013). *A. catechu* nut ethanol extract and its aqueous fraction also caused an elevation in hippocampal serotonin and noradrenaline levels in rats (Abbas et al., 2013). These preliminary studies encourage further investigation into the antidepressant and serotonergic properties of *A. catechu*.

Ashwagandha (*Withania somnifera*)

Ashwagandha (*W. somnifera*), also known as Indian ginseng, is widely used as part of ayurveda for the treatment of a range of disorders and especially as a *rasayana* (rejuvenator) to promote physical and mental health and healthy ageing (Ven Murthy et al., 2010; Singh et al., 2011) (Figure 28.2). This member of the Solanaceae family possesses anticancer, neuroprotective, antiepileptic, spermatogenic, hepatoprotective, antimicrobial, antioxidant, anti-inflammatory, and anti-arthritic activities (Mishra et al., 2000; Kumar et al., 2015). Ashwagandha has also demonstrated antidepressant, anxiolytic, and adaptogenic properties in a range of rodent behavioral tests, including the open-field test, stress-based models (e.g., cold stress, hot stress, food deprivation, isolation, foot shock), forced-swim test, tail suspension test, learned-helplessness test, elevated open maze, elevated plus maze, and the anti-reserpine test (Tripathi et al., 1998; Rege et al., 1999; Bhattacharya et al., 2000b; Bhattacharya and Muruganandam, 2003; Shah et al., 2006; Gupta and Rana, 2007; Kulkarni et al., 2008b; Maiti et al., 2011; Jayanthi et al., 2012). Results from these studies have found the effects of ashwagandha to be comparable to commonly used antidepressants and anxiolytics such as imipramine (Bhattacharya et al., 2000b; Shah et al., 2006; Maiti et al., 2011), fluoxetine (Shah et al., 2006) diazepam (Gupta and Rana, 2007; Kulkarni et al., 2008b),

FIGURE 28.2 Ashwagandha (*W. somnifera*). (From "An image of Withania somnifera" by Thamizhpparithi Maari (own work), licensed under CC BY-SA 3.0, http://creativecommons.org/licenses/by-sa/3.0, via Wikimedia Commons, https://commons.wikimedia.org/wiki/File:An_image_of_Withania_somnifera.JPG.)

and lorazepam (Bhattacharya et al., 2000b). Ashwagandha contains multiple active constituents and is likely to exert its pharmacological effects via multiple targets, which also include the serotonergic system (Mishra et al., 2000; Kumar et al., 2015). Ashwagandha was found to desensitize the 5HT1A receptor in both control and "depressed" (stress-based model) rats, as determined by reduced expression of some of the components of 8-OH-DPAT (5HT1A full agonist)-induced serotonin syndrome following treatment (eight weeks). Ashwagandha also enhances the wet-dog shake response induced by the serotonin receptor agonists 5-MeO-DMT (5HT1A and 5HT2 agonist) and quipazine, indicating that the 5HT2 receptor may be involved in its antidepressant activity (Tripathi et al., 1998). Shah et al. (2006), using the forced-swim test, conducted a series of experiments investigating the mechanism of action of ashwagandha in rats and found that ashwagandha treatment partially reversed reserpine (vesicular monoamine transporter inhibitor)-induced increases in mean immobility time. However, the results also implicated the adrenergic receptors. Ashwagandha has also been found to potentiate the effects of serotonergic antidepressants such as imipramine or fluoxetine in animal depression models (Shah et al., 2006; Maiti et al., 2011). This relatively limited amount of evidence indicates that the serotonergic system may indeed be involved in the antidepressant activity of ashwagandha. However, more detailed biochemical and pharmacological studies assessing its effects on brain serotonin levels, serotonin reuptake, and serotonin receptor activity are required to provide further evidence on the involvement of the serotonergic system in its antidepressant properties.

Brahmi (*Bacopa monnieri*)

Brahmi (*B. monnieri*) belongs to the family Scrophulariaceae and is a small creeping herb found in marshy grounds throughout India (Figure 28.3). The whole plant, especially the stem and leaves, are used for medicinal purposes. *B. monnieri* has been traditionally used in ayurveda as a *rasayana* (*medhya-rasayana* and *aindra-rasayana*) to rejuvenate the brain and mental health and promote intellect, memory, and longevity (Singh, 2013). Brahmi is widely reputed for its nootropic and cholinergic properties and has therefore attracted attention for its potential to treat neurodegenerative disorders (Joshi and Parle, 2006; Limpeanchob et al., 2008; Peth-Nui et al., 2012; Rao et al., 2012; Rastogi et al., 2012; Le et al., 2013; Neale et al., 2013; Kongkeaw et al., 2014). Several studies have also demonstrated the antidepressant activity of *B. monnieri* in the forced-swim, tail suspension, chronic unpredictable stress, and learned-helplessness models using mice and rats (Sairam et al., 2002; Shen et al., 2009; Chatterjee et al., 2010; Banerjee et al., 2014). Sairam et al. (2002) reported that *B. monnieri* antidepressant activity was comparable to imipramine in the forced-swim and learned-helplessness models. On the other hand,

FIGURE 28.3 Brahmi (*B. monnieri*).

Maiti et al. (2011) only observed small antidepressant effects of *B. monnieri* extracts in the forced-swim and learned-helplessness models, although it did potentiate the effects of imipramine in these tests. In a recent study, Kadali et al. (2014) reported that *B. monnieri* exhibited significant antidepressant activities in the forced-swim and shock-induced depression models but not in the tail suspension test. Potential differences in the chemical composition of the extracts may explain the differences in these studies. Saponins isolated from the whole plant of *B. monnieri*—bacopasides VI–VIII, bacopaside I, bacopaside II, and bacopasaponin C—were all found to possess antidepressant activity in the forced-swim and tail suspension tests using mice (Zhou et al., 2007; Shen et al., 2009). Bacoside-enriched extract containing bacopaside I (5.37%), bacoside A3 (5.59%), bacopaside II (6.9%), bacopasaponin C isomer (7.08%), and bacopasaponin C (4.18%) (the latter four triglycosidic saponins constitute the major component bacoside A) attenuated age-related increases in the duration of immobility in the tail suspension test and declines in serotonin levels in the cortex (Rastogi et al., 2012). Singh et al. (2014) in a recent study evaluated the effects of *B. monnieri* extracts and its constituents (bacopaside I, bacopaside II, bacosaponin C, bacosinne, bacoside A and bacoside A3) on MAO-A and MAO-B activities *in vitro*. Bacopaside I was clearly identified as the more potent inhibitor of both MAO-A (IC50 17.08 μg/ml) and MAO-B (94.22 μg/ml) enzymes when compared with the other constituents or extract but with relatively high IC50s, especially against MAO-B (Singh et al., 2014). Acute and subchronic oral administration of a methanol extract of *B. monnieri* in mice increased brain serotonin levels, although these results were not found to be significant (Rauf et al., 2012). Charles et al. (2011) reported that *B. monnieri* treatment in rats upregulated tryptophan hydroxylase and serotonin transporter expression while also increasing serotonin levels. Rajan et al. (2011) reported that *B. monnieri* extract enhanced hippocampal 5HT3A receptor expression and serotonin and acetylcholine levels, and also attenuated reductions in 5HT3 expression and serotonin and acetylcholine levels induced by a 5HT3 agonist. *In silico* studies also indicate that bacosides (A3/A) may interact and activate tryptophan hydroxylase (Rajathei et al., 2014). *B. monnieri* extract attenuated depression (chronic unpredictable stress)-induced decreases in brain-derived neuro-trophic factor (BDNF) protein and mRNA levels in the hippocampus and frontal cortex; the BDNF and serotonin systems are known to interact at various levels (Martinowich and Lu, 2008; Banerjee et al., 2014). *B. monnieri* also attenuates chronic unpredictable stress-induced decreases in serotonin (and nor-adrenaline) levels in the cortex and hippocampus (and dopamine levels only in the cortex) (Sheikh et al., 2007). These results demonstrate that *B. monnieri* extracts demonstrate antidepressant properties in animal models and can enhance brain serotonin levels by potentially inhibiting MAO activity and enhancing tryptophan hydroxylase and expression (although the exact mechanisms remain unclear). The effectiveness of standardized *B. monnieri* extracts should be explored in well-designed clinical trials to demonstrate clinical utility for the treatment of depression. In a placebo-controlled clinical study involving non-demented elderly subjects (24 completers per group), 12-week treatment with *B. monnieri* whole plant standardized dry extract resulted in an improvement in cognitive performance as well as a reduction in depression and anxiety scores (Calabrese et al., 2008). In a more recent clinical study, a polyherbal formulation consisting of *B. monnieri*, *Dioscorea bulbifera*, and *Hippophae rhamnoides* was also found to improve cognitive performance and lower depression scores in both demented and non-demented elderly subjects (Sadhu et al., 2014).

Daruharidra (*Berberis aristata*)

The shrub *B. aristata*, traditionally known as *daruharidra* or *daruhaldi*, belongs to the family Berberidaceae and is native to the northern Himalayan region (Figure 28.4). This plant has traditionally been used for the treatment of inflammation, jaundice, diarrhea, wounds, skin disease, and eye infec-tions. More recent studies have also demonstrated the hepatoprotective, antidiabetic, anticancer, anti-malarial, antimicrobial, anti-inflammatory and antioxidant properties of this plant (Komal et al., 2011; Potdar et al., 2012; Tamilselvi et al., 2014). Berberine, an isoquinoline alkaloid of the protoberberine type found in *B. aristata*, is gaining increasing attention for its potential to treat a range of neurological and psychiatric disorders, including depression (Kulkarni and Dhir, 2010; Pasrija et al., 2010; Vuddanda et al., 2010). Kulkarni and Dhir (2007) reported that berberine treatment in mice resulted in antidepres-sant effects in the forced-swim and tail suspension tests and also enhanced brain serotonin, noradrena-line, and dopamine levels. Berberine also potentiated the antidepressant effects of subeffective doses of serotonin and/or norepinephrine reuptake inhibitors (imipramine, desipramine, fluoxetine, venlafaxine) and the MAO inhibitor tranylcypromine (Kulkarni and Dhir, 2007). Similarly, Peng et al. (2007) also reported on the antidepressant activity of berberine in forced-swim and tail suspension tests along with

FIGURE 28.4   Daruharidra (*Berberis aristata*). (From "BerberisAculeata" by L. Shyamal (own work), licensed under CC BY-SA 3.0, http://creativecommons.org/licenses/by-sa/3.0, via Wikimedia Commons, https://commons.wikimedia.org/wiki/File:BerberisAculeata.jpg.)

increased serotonin and noradrenaline levels in the hippocampus and frontal cortex. The antidepressant activity of berberine was enhanced by the SNRI desipramine, the SSRI fluoxetine, and the MAO-A inhibitor moclobemide (but not the noradrenaline reuptake inhibitor maprotiline) (Peng et al., 2007). Berberine demonstrated anxiolytic effects in the elevated plus maze test (increased time spent and arm entries in the open arms by mice), which may be linked to the increased monoamine turnover in the brain stem (decreased concentrations of serotonin, noradrenaline, and dopamine, and increased concentrations of the corresponding metabolites 5-hydroxyindoleacetic acid, vanillylmandelic acid, and homovanillic acid) following treatment. Furthermore, in the elevated plus maze test, berberine inhibited the anxiogenic effects of the 5HT1A silent antagonist WAY-100635, the 5HT1A full agonist 8-OH-DPAT, and the 5HT2 agonist DOI, and enhanced the anxiolytic effects of the 5HT1A partial agonist BUS, the 5HT1A antagonist p-MPPI, and the 5HT2 antagonist RIT in the elevated plus maze test. These results potentially implicate decreased postsynaptic 5HT1A and 5HT2 signaling and increased 5HT1A autoreceptor signaling in the anxiolytic effects of berberine (Peng et al., 2004). Berberine has been reported to inhibit MAO-A (IC50 126 µM) and MAO-B (IC50 90 µM) enzymes (Kong et al., 2001; Castillo et al., 2005). On the other hand, berberine has been reported to inhibit tryptophan hydroxylase, decrease serotonin levels, and also increase serotonin transporter mRNA and protein expression and promoter activities in *in vitro* studies involving cell lines (Lee et al., 2001; Hu et al., 2012). Further *in vitro* and *in vivo* pharmacology studies are therefore required to establish the mechanisms of action behind the antidepressant properties of berberine.

Guduchi (*Tinospora cordifolia*)

T. cordifolia, traditionally known as *guduchi* or *amrita*, is a climbing shrub belonging to the family Menispermaceae and is found throughout India (Figure 28.5). This plant has been traditionally used for the treatment of jaundice, fever, skin diseases, and diabetes, and possesses immunomodulatory, antioxidant, anti-inflammatory, anti-hyperglycemic, anti-ulcer, antispasmodic, and antistress properties (Rege et al., 1999; Sinha et al., 2004; Mutalik and Mutalik, 2011). Dhingra et al., reported on the antidepressant properties of *T. cordifolia* (dried stem) extract in the tail suspension test (and forced-swim test), which was reversed by prazosin (alpha-1-adrenoceptor antagonist), sulpiride (dopamine D2-receptor antagonist), p-CPA (serotonin synthesis inhibitor), and baclofen (GABA-B agonist). Treatment also resulted in decreased brain MAO-A and MAO-B activities. Collectively, these results potentially implicate several neurotransmitter systems, including the serotonin system, in the antidepressant properties of this plant (Dhingra and Goyal, 2008). Interestingly, *T. cordifolia* contains berberine, which possesses both antidepressant and serotonergic properties (refer to the section on *B. aristata*) (Singh et al., 2003; Srinivasan et al., 2008).

FIGURE 28.5 Guduchi (*Tinospora cordifolia*). (From "Tinospora cordifolia leaves" by Vinayaraj (own work), licensed under CC BY-SA 3.0, http://creativecommons.org/licenses/by-sa/3.0, via Wikimedia Commons, https://commons.wikimedia.org/wiki/File:Tinospora_cordifolia_leaves.jpg.)

Haridra (*Curcuma longa*)

C. longa, traditionally known as *haridra* or *haldi* in India and commonly known as turmeric, belongs to the family Zingiberaceae and is renowned for its medicinal and culinary applications (Figure 28.6). The rhizome is routinely used as a spice and a herbal medicine in Southern Asia and contains the active alkaloid curcumin, which possesses a range of pharmacological properties including anti-inflammatory, antioxidant, antimicrobial, anticarcinogenic, antidiabetic, and neuroprotective activities (Gupta et al., 2013; Noorafshan and Ashkani-Esfahani, 2013). Traditionally, *C. longa* has been used to treat wounds, digestive problems, and bacterial infections, among several other diseases and disorders (Gupta et al., 2013). Recent studies have also demonstrated the antidepressant activity of this plant (Kulkarni et al., 2009). Yu et al. (2002), using the tail suspension and forced-swim tests, reported on the antidepressant activity in mice of *C. longa* extracts (administered orally), which was of greater potency than fluoxetine. A reduction in brain MAO-A (at lower doses) and MAO-B (at higher doses) was also observed following treatment, thereby indicating a role of the monoaminergic neurotransmitter systems in the antidepressant effects of this plant (Yu et al., 2002). Xia et al. (2007) also reported on the antidepressant activity of *C. longa* extract in mice using the forced-swim test, and demonstrated the ability of *C. longa* extract to attenuate swim stress-induced decreases in serotonin, 5-hydroxyindoleacetic acid, noradrenaline, and dopamine, and increases in serotonin turnover, serum corticotropin-releasing factor, and cortisol levels. Similarly, Xu et al. (2005a) reported on the antidepressant activity of curcumin in mice using the tail suspension and forced-swim tests. Curcumin was also found to decrease brain MAO activity and increase serotonin and noradrenaline levels in the hippocampus and frontal cortex and dopamine levels in the frontal cortex and the striatum (Xu et al., 2005a). Similar results from studies conducted in mice were also reported by Kulkarni et al. (2008a), who further implicated the suppression of MAO activity and the involvement of the serotonin and dopamine (but not noradrenaline) system in the antidepressant activity of curcumin (Kulkarni et al., 2008a). Xu et al. (2005b) also reported on the antidepressant effects of curcumin in the forced-swim test and olfactory bulbectomy models of depression in rats. A reduction in serotonin and noradrenaline levels in the hippocampus, and slight reductions in serotonin, noradrenaline, and dopamine levels in the frontal cortex of olfactory bulbectomized rats were also observed, which were reversed by curcumin treatment (Xu et al., 2005b). Wang et al. (2008) demonstrated that a tryptophan hydroxylase inhibitor (P-chlorophenylalanine) that depletes serotonin also inhibited the antidepressant activity of curcumin in the forced-swim test, thereby implicating the serotonergic system. Results from elaborate pharmacological investigations conducted in this study also implicated the 5HT1 and 5HT2 receptors in the antidepressant activity of curcumin in the forced-swim test. Curcumin can also increase hippocampal neurogenesis in adult mice and chronically stressed rats, which may be linked to the upregulation of 5HT1A receptors and BDNF following treatment (Xu et al., 2007; Kim et al., 2008). This is of significance, as there is a growing amount of evidence supporting the fact that serotonin positively regulates hippocampal neurogenesis and that adult hippocampal neurogenesis is an

(a)   (b)

FIGURE 28.6 Haridra (*Curcuma longa*): (a) plant (From "Curcuma longa (Haldi) Im IMG 2441" by J.M. Garg (own work) licensed under CC BY 3.0, http://creativecommons.org/licenses/by/3.0, via Wikimedia Commons, https://commons. wikimedia.org/wiki/File:Curcuma_longa_(Haldi)_Im_IMG_2441.jpg.); (b) rhizome (From "Curcuma longa roots" by Simon A. Eugster (own work), licensed under CC BY-SA 3.0, http://creativecommons.org/licenses/by-sa/3.0, via Wikimedia Commons, https://commons.wikimedia.org/wiki/File:Curcuma_longa_roots.jpg.)

important mediator of the effects of antidepressants (Mahar et al., 2014). To summarize these reports, *in vivo* studies have demonstrated that curcumin, a pharmacologically active constituent of *C. longa*, possesses antidepressant activity that at least is partially mediated by the serotonin system in the brain. Curcumin can suppress brain MAO activity and enhance serotonin levels while also modulating serotonin receptors (5HT1 and 5HT2). The dopamine and noradrenaline systems may also contribute toward its antidepressant activity. A recent small clinical study has reported on the safety and effectiveness of curcumin for the treatment of major depressive disorder (without concurrent suicidal ideation or other psychotic disorders), and should encourage further clinical studies into the antidepressant effects of this alkaloid and other *C. longa*–derived formulations (Sanmukhani et al., 2014).

Nyctanthes arbor-tristis

N. arbor-tristis (night jasmine), is a small tree or shrub belonging to the family Oleaceae and is native to South Asia (Figure 28.7). It has several traditional names, including *parijatha* in Sanskrit and *harsingar* in Hindi. *N. arbor-tristis* has been traditionally used as part of ayurveda for the treatment of numerous medical conditions, including sciatica, arthritis, fevers, and pain (Rani et al., 2012). Extracts from various parts of the plants possess numerous pharmacological and medicinal properties, including anti-inflammatory (flowers, seeds, leaves, bark, and leaf beta-sitosterol), antioxidant (flowers and leaves), anti-nociceptive (bark, leaves, and leaf beta-sitosterol), antipyretic (leaves), antibacterial (flowers and leaves), larvicidal (flowers), anti-leishmanial, (iridoid-glucosides), immunostimulant (seeds, flowers, and leaves), sedative (flowers and leaves), antihistaminic (leaves), purgative (leaves), antidiabetic (bark) and anti-proliferative (flowers) activities, but also ulcerogenic (leaves) and cytotoxic (flower) activities (Saxena et al., 1984, 1987; Puri et al., 1994; Khatune et al., 2001a, b; Saxena et al., 2002; Ratnasooriya et al., 2005; Das et al., 2006; Kannan, 2010; Suresh et al., 2010; Shukla et al., 2011; Nirmal et al., 2012; Saha et al., 2012; Kakoti et al., 2013; Michael et al., 2013; Pattanayak et al., 2013; Khanapur et al., 2014). Tripathi et al. (2010a,b) recently reported on the antidepressant (using the forced-swim, learned-helplessness, tail suspension, and anti-reserpine tests) as well as anxiolytic (using the elevated zero and plus maze, open-field exploratory behavior, novelty-induced suppressed feeding, and social interaction tests) properties of the hydroalcoholic leaf extract of *N. arbor-tristis* in rats. Das et al. (2008) reported that the water-soluble portion ethanol extracts of the leaves, flowers, and seeds of *N. arbor-tristis* caused a prolongation of onset and duration of sleep in pentobarbital sodium treated mice. These extracts also caused an increase in mice brain serotonin levels (and a decrease in dopamine levels).

FIGURE 28.7 *Nyctanthes arbor-tristis.*

(a)   (b)

FIGURE 28.8 Sunthi (*Zingiber officinale*): (a) plant (From "Zingiber officinale" by Dalgial (own work), licensed under CC BY 3.0, http://creativecommons.org/licenses/by/3.0, via Wikimedia Commons, https://commons.wikimedia.org/wiki/File:Zingiber_officinale.JPG.); (b) rhizome (From "Zingiber officinale 002" by H. Zell (own work), licensed under CC BY-SA 3.0, http://creativecommons.org/licenses/by-sa/3.0, via Wikimedia Commons, https://commons.wikimedia.org/wiki/File:Zingiber_officinale_002.JPG.)

Sunthi (*Zingiber officinale*)

Z. officinale, traditionally referred to as *sunthi* in ayurveda and commonly known as ginger, is a member of the Zingiberaceae family (Figure 28.8). It is a perennial plant with fleshy rhizomes that have several culinary and medicinal applications. In ayurveda, *Z. officinale* has been used to treat digestive disorders and arthritis (Malhotra and Singh, 2003). *Z. officinale* and its constituents also possess antioxidant, anti-inflammatory, antimicrobial, anticancer, antidiabetic, and hepatoprotective properties (Rahmani et al., 2014; Khodaie and Sadeghpoor, 2015). Recent studies have reported on the antidepressant properties of *Z. officinale*, which should encourage further studies. Oral administration of *Z. officinale* rhizome hydroalcoholic extract in rats caused a decrease in the duration of immobility in the forced-swim and tail suspension tests, thereby demonstrating the antidepressant activity of this plant (Singh et al., 2012). *In silico* studies have indicated that the *Z. officinale* constituents gingerol and shogoal can bind to 5HT1A, which warrants further investigation as a mechanism of action behind the antidepressant activity of this plant (Ittiyavirah and Paul, 2013).

Tulsi (*Ocimum sanctum*)

Tulsi (*O. sanctum*), also known as *vishnu priya* and *holy basil*, is a member of the Lamiaceae family and is found in the tropical and subtropical regions of India (Figure 28.9). This herb is widely used in India for religious and medicinal purposes (Kumar et al., 2013). Based on traditional uses and more recent scientific studies, tulsi possesses antimicrobial, hypoglycemic, hypolipidemic, hepatoprotective, immunomodulatory, antioxidant, anticarcinogenic, analgesic, wound healing, anti-inflammatory, antipyretic, antidiarrheal, anti-asthmatic, anti-ulcerogenic, and adaptogenic properties (Gupta et al., 2002; Pandey and Madhuri, 2010). The antidepressant activity of tulsi root extract in mice has also been reported (Maity et al., 2000). Recent studies have also demonstrated the antidepressant (forced-swim and tail suspension tests) and anxiolytic activity of tulsi leaf extract in mice (Pemminati et al., 2010; Tabassum et al., 2010; Chatterjee et al., 2011). Sakina et al. (1990) reported that the tulsi (leaf extract)-induced decreases in duration of immobility were potentiated by the dopamine-2 receptor agonist bromocriptine and blocked by dopamine-2/3 receptor antagonists (haloperidol and sulpiride), indicating the involvement of the dopamine system in the antidepressant activity of tulsi (Sakina et al., 1990). Tulsi attenuates stress-induced changes in brain serotonin levels in rodents (Singh et al., 1991; Ravindran et al., 2005; Samson et al., 2006). Further studies are required to establish the role of the serotonin system in the antidepressant properties of tulsi.

FIGURE 28.9   Tulsi (*Ocimum sanctum*) plant.

Future Perspectives

Depression is a major cause of suffering worldwide and a major social and economic burden (Sartorius, 2001; Ferrari et al., 2013). Several modern synthetic antidepressants have been introduced in the market that have helped in the management of depressive disorders. A large number of these drugs target the serotonin system and include SSRIs, SNRIs, and MAO inhibitors. However, the effectiveness of these drugs is limited in terms of the number of responders as well as the severity of the disease, and virtually all modern antidepressants are associated with a series of side effects (Feighner, 1999; Khan et al., 2002; Preskorn et al., 2004; Gartlehner et al., 2007; Kirsch et al., 2008; Fournier et al., 2010; Gartlehner et al., 2011).

Herbal medicines are encouraging alternatives to currently marketed synthetic antidepressants as they are generally associated with fewer side effects and are often backed by a rich history of human use supporting their safety and efficacy. Ayurveda, one of the oldest systems of medicine in the world, advocates the use of various herbal preparations for the treatment and management of numerous diseases and disorders. In India, plant-based ayurvedic drugs are widely used in both rural and urban India. In this chapter, we have discussed a series of ayurvedic plants with serotonergic and antidepressant properties. Numerous medicinal properties of these plants have been described in ayurveda, and their antidepressant properties have also been demonstrated using modern scientific technologies and methodologies. Indeed, modern science will play a tremendous role toward globally promoting the antidepressant potential of these plants. To date, major scientific evidence demonstrating the antidepressant properties of these plants comes from animal depression models such as the forced-swim and tail suspension tests coupled with biochemical and/or pharmacological studies, demonstrating that these plants modulate the serotonin system in the brain. Further studies are required to elucidate the molecular mechanisms behind their serotonergic properties.

Some pertinent questions to be addressed in these studies include: (1) Can any of these herbal preparations act as MAO-A inhibitors, serotonin reuptake inhibitors, and/or serotonin receptor modulators at therapeutically relevant doses? (2) Do these herbal preparations promote neurogenesis (potentially via their serotonergic properties)? These questions can be addressed by well-designed *in vitro* and *in vivo* pharmacological studies. (3) Do any of these herbal preparations act through alternate pathways to modulate expression levels of serotonin receptors, the serotonin transporter, or MAO? Preliminary experiments monitoring the effects of these plant preparations on mRNA and protein levels of these targets in the brain will help address this. (4) Which molecule(s) in these polymolecular preparations are responsible for the serotonergic and antidepressant properties? Fractionation experiments coupled with pharmacological studies *in vivo* and *in vitro* will help address this. This is indeed a critical aspect of the future work, as identification of the pharmacologically active constituents will help toward formulating standardized antidepressant drug products derived from these plants, which in turn will help ensure batch-to-batch (therapeutic) consistency (Garg et al., 2012). This is indeed a challenging task, as each plant may consist of multiple pharmacologically active and relevant molecules that could target the same or multiple systems. Indeed, some lines of evidence also indicate the involvement of the dopamine and noradrenaline system in the antidepressant activity of some of the plants discussed here. Many of these plants also possess anti-inflammatory properties, which may also contribute toward the antidepressant activity of these plants (Hayley et al., 2005; Haase and Brown, 2014). Some molecules could potentially promote the efficacy of an extract by enhancing the bioavailability of other molecules present in the extract (Efferth and Koch, 2011; Kesarwani and Gupta, 2013). Results from the characterization of pharmacologically active and relevant molecules of these plants and the mechanism of action of these molecules (and extracts) in turn will play an important role in rationalizing the design and/or use of polyherbal formulations with immense therapeutic potential.

Indeed, several polyherbal preparations have been described in ayurveda that contain preparations of the plants discussed here and are currently being evaluated using modern scientific technologies and methodologies. Rasayana Ghana tablets consists of *guduchi* (*T. cordifolia*), *aamalaki* (*Emblica officinalis*), and *gokshura* (*Tribulus terrestris*), and have demonstrated antidepressant activity in mice using a behavior despair test (Deole et al., 2011). Perment consists of *Clitoria ternatea*, *W. somnifera*, *Asparagus racemosus*, and *B. monnieri*, and has demonstrated antidepressant activity while also increasing plasma noradrenaline and serotonin levels in rats subjected to chronic unpredictable mild stress (Ramanatha et al., 2011). EuMil, comprising of standardized extracts of *W. somnifera*, *O. sanctum*, *E. officinalis*, and *A. racemosus*, and Siotone, comprising of *W. somnifera*, *O. sanctum*, *E. officinalis*, and *T. terristris*, have demonstrated antistress and adaptogenic activity in rats subjected to chronic unpredictable mild foot-shock stress (Bhattacharya et al., 2000a; Muruganandam et al., 2002). Mamsyadi Kwatha consists of *jatamansi* (*Nardostachys jatamansi*), *W. somnifera*, and *parasika yavani* (*Hyocymus niger*), and has demonstrated antidepressant activity in the behavioral-despair and anti-reserpine tests (Shreevathsa et al., 2013). NR-ANX-C contains *W. somnifera*, *O. sanctum*, *Camellia sinensis* (green tea), *E. officinalis*, *Terminalia chebula*, *T. bellericans*, and *shilajit*, and reduces ethanol withdrawal–induced anxiety in rats (Mohan et al., 2011). A comprehensive evaluation of the mode and mechanism(s) of action of these polyherbal products (and their individual plant components) using systems biology and *omics* approaches will help assess the role of the serotonin system in their efficacy and go a long way toward promoting these products globally.

Another requirement for the promotion of herbal medicines is the application of well-designed clinical studies to convincingly demonstrate and confirm their safety and efficacy in humans. Placebo-controlled, randomized, multicentric clinical trials using standardized herbal drugs and conducted in different ethnic groups will be critical toward globalizing antidepressants derived from these plants. There are encouraging signs that modern scientific approaches are being utilized to characterize and standardize herbal drugs and further our understanding of their mechanism of action. Well-designed clinical studies involving herbal medicines have also been conducted or are underway to demonstrate and confirm their safety and efficacy. Since the turn of the century, the US Food and Drug Administration (FDA) has approved two botanical (monoherbal) drug products. First, Veregen ointment, which contains sinecatechins derived from green tea leaves, was approved in 2006 for the treatment of genital and perianal warts. Fulyzaq tablets, which contain the active ingredient crofelemer derived from the red sap of *Croton lechleri*, was approved in 2012 for the treatment of noninfectious diarrhea in AIDS patients on antiviral therapy. Several other herbal medicines are currently undergoing clinical development, including

BCM-95, which is a standardized turmeric extract with enhanced bioavailability of curcumin and which has demonstrated antidepressant activity in pilot clinical trials (BCM-95, 2015; Lei et al., 2014; Lopresti et al., 2014; Sanmukhani et al., 2014).

Conclusion

The combination of traditional knowledge and modern science promises to promote the use of herbal medicines to alleviate the burden of diseases and promote healthy living. With the continued application of such approaches over the next few years, society can look forward to utilizing the immense potential of the antidepressant plants discussed here to help alleviate the burden of depressive disorders.

REFERENCES

Abbas G, Naqvi S, Erum S, Ahmed S, Atta-Ur-Rahman, Dar A. Potential antidepressant activity of areca catechu nut via elevation of serotonin and noradrenaline in the hippocampus of rats. *Phyther Res*, 2013;27:39–45.

Andrade L, Caraveo-Anduaga JJ, Berglund P, Bijl RV, de Graaf R, Vollebergh W, Dragomirecka E, et al. The epidemiology of major depressive episodes: Results from the international consortium of psychiatric epidemiology (ICPE) surveys. *Int J Methods Psychiatr Res*, 2003;12:3–21. (Erratum appears in *Int J Methods Psychiatr Res*, 2003;12(3):165).

Anghelescu IG, Kohnen R, Szegedi A, Klement S, Kieser M. Comparison of *Hypericum* extract WS 5570 and paroxetine in ongoing treatment after recovery from an episode of moderate to severe depression: Results from a randomized multicenter study. *Pharmacopsychiatry*, 2006;39:213–219.

Banerjee R, Hazra S, Ghosh AK, Mondal AC. Chronic administration of *Bacopa monniera* increases BDNF protein and mRNA expressions: A study in chronic unpredictable stress induced animal model of depression. *Psychiatry Investig*, 2014;11:297–306.

Baxter AJ, Charlson FJ, Somerville AJ, Whiteford HA. Mental disorders as risk factors: Assessing the evidence for the global burden of disease study. *BMC Med*, 2011;9:134.

BCM-95. Clinical findings on BCM-95. Available from: http://www.bcm95.com/Clinical trials_1.html (accessed May 3, 2015).

Belmaker RH, Agam G. Major depressive disorder. *N Engl J Med*, 2008;358:55–68.

Berger M, Gray JA, Roth BL. The expanded biology of serotonin. *Annu Rev Med*, 2009;60:355–366.

Bhattacharya SK, Bhattacharya A, Chakrabarti A. Adaptogenic activity of Siotone, a polyherbal formulation of Ayurvedic rasayanas. *Indian J Exp Biol*, 2000a;38:119–128.

Bhattacharya SK, Bhattacharya A, Sairam K, Ghosal S. Anxiolytic-antidepressant activity of Withania somnifera glycowithanolides: An experimental study. *Phytomedicine*, 2000b;7:463–469.

Bhattacharya SK, Muruganandam AV. Adaptogenic activity of *Withania somnifera*: An experimental study using a rat model of chronic stress. *Pharmacol Biochem Behav*, 2003;75:547–555.

Bierut LJ, Heath AC, Bucholz KK, Dinwiddie SH, Madden PA, Statham DJ, Dunne MP, Martin NG. Major depressive disorder in a community-based twin sample: Are there different genetic and environmental contributions for men and women? *Arch Gen Psychiatry*, 1999;56:557–563.

Bogdanski DF, Weissbach H, Udenfriend S. Pharmacological studies with the serotonin precursor, 5-hydroxytryptophan. *J Pharmacol Exp Ther*, 1958;122:182–194.

Bondy B. Pathophysiology of depression and mechanisms of treatment. *Dialogues Clin Neurosci*, 2002;4:7–20.

Butterweck V. Mechanism of action of St John's wort in depression: What is known? *CNS Drugs*, 2003;17:539–562.

Calabrese C, Gregory WL, Leo M, Kraemer D, Bone K, Oken B. Effects of a standardized *Bacopa monnieri* extract on cognitive performance, anxiety, and depression in the elderly: A randomized, double-blind, placebo-controlled trial. *J Altern Complement Med*, 2008;14:707–713.

Calapai G, Crupi A, Firenzuoli F, Inferrera G, Squadrito F, Parisi A, De Sarro G, Caputi A. Serotonin, norepinephrine and dopamine involvement in the antidepressant action of *Hypericum perforatum*. *Pharmacopsychiatry*, 2001;34:45–49.

Castillo J, Hung J, Rodriguez M, Bastidas E, Laboren I, Jaimes A. LED fluorescence spectroscopy for direct determination of monoamine oxidase B inactivation. *Anal Biochem*, 2005;343:293–298.

Charles PD, Ambigapathy G, Geraldine P, Akbarsha MA, Rajan KE. *Bacopa monnieri* leaf extract up-regulates tryptophan hydroxylase (TPH2) and serotonin transporter (SERT) expression: Implications in memory formation. *J Ethnopharmacol*, 2011;134:55–61.

Chatterjee M, Verma P, Palit G. Comparative evaluation of *Bacopa monnieri* and *Panax quniquefolium* in experimental anxiety and depressive models in mice. *Indian J Exp Biol*, 2010;48:306–313.

Chatterjee M, Verma P, Maurya R, Palit G. Evaluation of ethanol leaf extract of *Ocimum sanctum* in experimental models of anxiety and depression. *Pharm Biol*, 2011;49:477–483.

Coppen A. The biochemistry of affective disorders. *Br J Psychiatry*, 1967;113:1237–1264.

Dahlstroem A, Fuxe K. Evidence for the existence of monoamine-containing neurons in the central nervous system: I. Demonstration of monoamines in the cell bodies of brain stem neurons. *Acta Physiol Scand Suppl*, 1964;Suppl. 232:1–55.

Dar A, Khatoon S. Antidepressant effects of ethanol extract of Areca catechu in rodents. *Phyther Res*, 1997b;11:174–176.

Dar A, Khatoon S. Behavioral and biochemical studies of dichloromethane fraction from the Areca catechu nut. *Pharmacol Biochem Behav*, 2000;65:1–6.

Dar A, Khatoon S, Rahman G, Atta-Ur-Rahman. Anti-depressant activities of *Areca catechu* fruit extract. *Phytomedicine*, 1997a;4:41–45.

Das S, Basu SP, Sasmal D. Anti-inflammatory activity of the different parts of *Nyctanthes arbortristis* Linn. *Ethiop Pharm J* 2006;24:125–129.

Das S, Sasmal D, Basu SP. Evaluation of CNS depressant activity of different plant parts of *Nyctanthes arbortristis* Linn. *Indian J Pharm Sci*, 2008;70:803–806.

Deole Y, Ashok B, Thakar A, Chavan S, Ravishankar B, Chandola H. Evaluation of anti-depressant and anxiolytic activity of Rasayana Ghana tablet (a compound ayurvedic formulation) in albino mice. *Ayu*, 2011;32:375–379.

Dhamija HK, Parashar B, Singh J. Anti-depression potential of herbal drugs: An overview. *J Chem Pharm Res*, 2011;3:725–735.

Dhingra D, Goyal PK. Evidences for the involvement of monoaminergic and GABAergic systems in antide-pressant-like activity of *Tinospora cordifolia* in mice. *Indian J Pharm Sci*, 2008;70:761–767.

Efferth T, Koch E. Complex interactions between phytochemicals: The multi-target therapeutic concept of phytotherapy. *Curr Drug Targets*, 2011;12:122–132.

Feighner JP. Mechanism of action of antidepressant medications. *J Clin Psychiatry*, 1999;60:4–13.

Ferrari AJ, Charlson FJ, Norman RE, Patten SB, Freedman G, Murray CJL, Vos T, Whiteford HA. Burden of depressive disorders by country, sex, age, and year: Findings from the global burden of disease study 2010. *PLoS Med*, 2013;10:e1001547.

Fournier JC, DeRubeis RJ, Hollon SD, Dimidjian S, Amsterdam JD, Shelton RC, Fawcett J. Antidepressant drug effects and depression severity: A patient-level meta-analysis. *JAMA*, 2010;303:47–53.

France CM, Lysaker PH, Robinson RP. The "chemical imbalance" explanation for depression: Origins, lay endorsement, and clinical implications. *Prof Psychol Res Pract*, 2007;38:411–420.

Frazer A, Hensler JG. Serotonin involvement in physiological function and behavior. In G Siegel, B Agranoff, R Albers, A Et (eds), *Basic Neurochemistry Molecular Cellular and Medical Aspects*, 6th edn. Philadelphia, PA: Lippincott-Raven, 1999.

Garg V, Dhar VJ, Sharma A, Dutt R. Facts about standardization of herbal medicine: A review. *J Chinese Integr Med*, 2012;10:1077–1083.

Gartlehner G, Hansen R, Morgan L, Thaler K, Lux L, Van Noord M, Mager U, et al. Comparative effective-ness of second-generation antidepressants in the pharmacologic treatment of adult depression. *Heal San Fr*, 2007;954.

Gartlehner G, Hansen RA, Morgan LC, Thaler K, Lux L, Van Noord M, Mager U, et al. Comparative benefits and harms of second-generation antidepressants for treating major depressive disorder: An updated meta-analysis. *Ann Intern Med*, 2011;155:772–785.

Gaster B, Holroyd J. St John's wort for depression: A systematic review. *Arch Intern Med*, 2000;160:152–156.

Gastpar M. *Hypericum* extract WS 5570 for depression: An overview. *Int J Psychiatry Clin Pract*, 2013;17:1–7.

Gautam RK, Dixit PK, Mittal S. Herbal sources of antidepressant potential: A review. *Int J Pharm Sci Rev Res*, 2013;18:86–91.

Ghelardini C, Galeotti N, Lelli C, Bartolini A. M1 receptor activation is a requirement for arecoline analgesia. *Farmaco*, 2001;56:383–385.

Gilani AH, Ghayur MN, Saify ZS, Ahmed SP, Choudhary MI, Khalid A. Presence of cholinomimetic and acetylcholinesterase inhibitory constituents in betel nut. *Life Sci*, 2004;75:2377–2389.

Gupta SK, Prakash J, Srivastava S. Validation of traditional claim of Tulsi, *Ocimum sanctum* Linn. as a medicinal plant. *Indian J Exp Biol*, 2002;40:765–773.

Gupta GL, Rana AC. Protective effect of *Withania somnifera* dunal root extract against protracted social isolation induced behavior in rats. *Indian J Physiol Pharmacol*, 2007;51:345–353.

Gupta SC, Sung B, Kim JH, Prasad S, Li S, Aggarwal BB. Multitargeting by turmeric, the golden spice: From kitchen to clinic. *Mol Nutr Food Res*, 2013;57:1510–1528.

Haase J, Brown E. Integrating the monoamine, neurotrophin and cytokine hypotheses of depression: A central role for the serotonin transporter? *Pharmacol Ther*, 2014;147:1–11.

Hasler G. Pathophysiology of depression: Do we have any solid evidence of interest to clinicians? *World Psychiatry*, 2010;9:155–161.

Hayley S, Poulter MO, Merali Z, Anisman H. The pathogenesis of clinical depression: Stressor- and cytokine-induced alterations of neuroplasticity. *Neuroscience*, 2005;135:659–678.

Hirano K, Kato Y, Uchida S, Sugimoto Y, Yamada J, Umegaki K, Yamada S. Effects of oral administration of extracts of *Hypericum perforatum* (St John's wort) on brain serotonin transporter, serotonin uptake and behaviour in mice. *J Pharm Pharmacol*, 2004;56:1589–1595.

Hirschfeld R, Weissman M. Risk factors for major depression and bipolar disorder. In KL Davis, D Charney, JT Coyle, C Nemeroff (eds), *Neuropsychopharmacology: The Fifth Generation of Progress*, 5th edn, pp. 1017–1025. Philadelphia, PA: Lippincott Williams & Wilkins, 2002.

Hornung J-P. The human raphe nuclei and the serotonergic system. *J Chem Neuroanat*, 2003;26:331–343.

Hoyer D, Clarke DE, Fozard JR, Hartig PR, Martin GR, Mylecharane EJ, Saxena PR, Humphrey PP. International Union of Pharmacology classification of receptors for 5-hydroxytryptamine (Serotonin). *Pharmacol Rev*, 1994;46:157–203.

Hu Y, Ehli EA, Hudziak JJ, Davies GE. Berberine and evodiamine influence serotonin transporter (5-HTT) expression via the 5-HTT-linked polymorphic region. *Pharmacogenomics J*, 2012;12:372–378.

Ittiyavirah SP, Paul M. *In silico* docking analysis of constituents of Zingiber officinale as antidepressant. *J Pharmacogn Phyther*, 2013;5:101–105.

Jaiswal P, Kumar P, Singh VK, Singh DK. *Areca catechu* L.: A valuable herbal medicine against different health problems. *Res J Med Plant*, 2011;5:145–152.

Jawaid T, Gupta R, Siddiqui ZA. A review on herbal plants showing antidepressant activity. *Int J Pharm Sci Res*, 2011;2:3051–3060.

Jayanthi M, Prathima C, Huralikuppi J, Suresha R, Murali D. Anti-depressant effects of *Withania somnifera* fat (ashwagandha ghrutha) extract in experimental mice. *Int J Pharma Bio Sci*, 2012;3:33–42.

Joshi M, Gaonkar K, Mangoankar S, Satarkar S. Pharmacological investigation of *Areca catechu* extracts for evaluation of learning, memory and behavior in rats. *Int Curr Pharm J*, 2012;1:128–132.

Joshi H, Parle M. Brahmi rasayana improves learning and memory in mice. *Evid Based Complement Alternat Med*, 2006;3:79–85.

Kadali SLDVRM, Das MC, Rao ASRS, Sri GK. Antidepressant activity of brahmi in albino mice. *J Clin Diagnostic Res*, 2014;8:35–37.

Kakoti BB, Pradhan P, Borah S, Mahato K, Kumar M. Analgesic and anti-inflammatory activities of the methanolic stem bark extract of *Nyctanthes arbor-tristis* Linn. *Biomed Res Int*, 2013;2013:826295.

Kannan M. An immuno-pharmacological investigation of Indian medicinal plant *Nyctanthes arbor-tristis* Linn. *World Appl Sci J*, 2010;11:495–503.

Katon W, Sullivan MD. Depression and chronic medical illness. *J Clin Psychiatry*, 1990;51 Suppl. 3–11; discussion 12–14.

Katz MM, Secunda SK, Hirschfeld RM, Koslow SH. NIMH clinical research branch collaborative program on the psychobiology of depression. *Arch Gen Psychiatry*, 1979;36:765–771.

Kesarwani K, Gupta R. Bioavailability enhancers of herbal origin: An overview. *Asian Pac J Trop Biomed*, 2013;3:253–266.

Kessler RC, Berglund P, Demler O, Jin R, Koretz D, Merikangas KR, Rush AJ, Walters EE, Wang PS. The epidemiology of major depressive disorder: Results from the national comorbidity survey replication (NCS-R). *JAMA*, 2003;289:3095–3105.

Kessler RC, Berglund P, Demler O, Jin R, Merikangas KR, Walters EE. Lifetime prevalence and age-of-onset distributions of DSM-IV disorders in the national comorbidity survey replication. *Arch Gen Psychiatry*, 2005;62:593–602.

Khan A, Leventhal RM, Khan SR, Brown WA. Severity of depression and response to antidepressants and placebo: An analysis of the food and drug administration database. *J Clin Psychopharmacol*, 2002;22:40–45.

Khanapur M, Avadhanula RK, Setty OH. *In vitro* antioxidant, antiproliferative, and phytochemical study in different extracts of *Nyctanthes arbortristis* flowers. *Biomed Res Int*, 2014;2014:291271.

Khatune NA, Haque ME, Mosaddik MA. Laboratory evaluation of *Nyctanthes arbor-tristis* Linn. flower extract and its isolated compound against common filarial vector, *Culex quinquefasciatus* Say (Diptera: Culicidae) larvae. *Pakistan J Biol Sci*, 2001a;4:585–587.

Khatune NA, Mosaddik MA, Haque ME. Antibacterial activity and cytotoxicity of *Nyctanthes arbor-tristis* flowers. *Fitoterapia*, 2001b;72:412–414.

Khodaie L, Sadeghpoor O. Ginger from ancient times to the new outlook. *Jundishapur J Nat Pharm Prod*, 2015;10:e18402.

Kientsch U, Buergi S, Ruedeberg C, Probst S, Honegger UE. St. John's wort extract Ze 117 (*Hypericum perforatum*) inhibits norepinephrine and serotonin uptake into rat brain slices and reduces beta-adrenoceptor numbers on cultured rat brain cells. *Pharmacopsychiatry*, 2001;34:S56–S60.

Kim SJ, Son TG, Park HR, Park M, Kim M-S, Kim HS, Chung HY, Mattson MP, Lee J. Curcumin stimulates proliferation of embryonic neural progenitor cells and neurogenesis in the adult hippocampus. *J Biol Chem*, 2008;283:14497–14505.

Kirsch I, Deacon BJ, Huedo-Medina TB, Scoboria A, Moore TJ, Johnson BT. Initial severity and antidepressant benefits: A meta-analysis of data submitted to the food and drug administration. *PLoS Med*, 2008;5:0260–0268.

Komal S, Ranjan B, Neelam C, Birendra S, Kumar SN. *Berberis aristata*: A review. *Int J Res Ayurveda Pharm*, 2011;2:383–388.

Kong LD, Cheng CHK, Tan RX. Monoamine oxidase inhibitors from rhizoma of *Coptis chinensis*. *Planta Med*, 2001;67:74–76.

Kongkeaw C, Dilokthornsakul P, Thanarangsarit P, Limpeanchob N, Norman Scholfield C. Meta-analysis of randomized controlled trials on cognitive effects of *Bacopa monnieri* extract. *J Ethnopharmacol*, 2014;151:528–535.

Kulkarni SK, Bhutani MK, Bishnoi M. Antidepressant activity of curcumin: Involvement of serotonin and dopamine system. *Psychopharmacology* (Berl), 2008a;201:435–442.

Kulkarni SK, Dhir A. Possible involvement of L-arginine-nitric oxide (NO)-cyclic guanosine monophosphate (cGMP) signaling pathway in the antidepressant activity of berberine chloride. *Eur J Pharmacol*, 2007;569:77–83.

Kulkarni SK, Dhir A. Berberine: A plant alkaloid with therapeutic potential for central nervous system disorders. *Phyther Res*, 2010;24:317–324.

Kulkarni SK, Dhir A, Akula KK. Potentials of curcumin as an antidepressant. *Sci World J*, 2009;9:1233–1241.

Kulkarni SK, Singh K, Bishnoi M. Comparative behavioural profile of newer antianxiety drugs on different mazes. *Indian J Exp Biol*, 2008b;46:633–639.

Kumar V, Dey A, Hadimani M, Marcović T, Emerald M. Chemistry and pharmacology of *Withania somnifera*: An update. *TANG Humanit Med*, 2015;5:1–13.

Kumar A, Rahal A, Chakraborty S, Tiwari R, Latheef SK, Dhama K. *Ocimum sanctum* (Tulsi): A miracle herb and boon to medical science; A Review. *Int J Agron Plant Prod*, 2013;4:1580–1589.

Le XT, Pham HTN, Do PT, Fujiwara H, Tanaka K, Li F, Van Nguyen T, Nguyen KM, Matsumoto K. *Bacopa monnieri* ameliorates memory deficits in olfactory bulbectomized mice: Possible involvement of glutamatergic and cholinergic systems. *Neurochem Res*, 2013;38:2201–2215.

Lecrubier Y, Clerc G, Didi R, Kieser M. Efficacy of St. John's wort extract WS 5570 in major depression: A double-blind, placebo-controlled trial. *Am J Psychiatry*, 2002;159:1361–1366.

Lee MK, Kim E Il, Hur JD, Lee KS, Ro JS. Effects of protoberberine compounds on serotonin content in P815 cells. *Korean J Pharmacogn*, 2001;32:49–54.

Lei X, Chen J, Liu CX, Lin J, Lou J, Shang HC. Status and thoughts of Chinese patent medicines seeking approval in the US market. *Chin J Integr Med*, 2014;20:403–408.

Limpeanchob N, Jaipan S, Rattanakaruna S, Phrompittayarat W, Ingkaninan K. Neuroprotective effect of *Bacopa monnieri* on beta-amyloid-induced cell death in primary cortical culture. *J Ethnopharmacol*, 2008;120:112–117.

López-Muñoz F, Alamo C. Monoaminergic neurotransmission: The history of the discovery of antidepressants from 1950s until today. *Curr Pharm Des*, 2009;15:1563–1586.

Lopresti AL, Maes M, Maker GL, Hood SD, Drummond PD. Curcumin for the treatment of major depression: A randomised, double-blind, placebo-controlled study. *J Affect Disord*, 2014;167:368–375.

Mahar I, Bambico FR, Mechawar N, Nobrega JN. Stress, serotonin, and hippocampal neurogenesis in relation to depression and antidepressant effects. *Neurosci Biobehav Rev*, 2014;38:173–192.

Maiti T, Adhikari A, Bhattacharya K, Biswas S, Debnath PK, Maharana CS. A study on evaluation of antidepressant effect of imipramine adjunct with Aswagandha and Bramhi. *Nepal Med Coll J*, 2011;13:250–253.

Maity TK, Mandal SC, Saha BP, Pal M. Effect of Ocimum sanctum roots extract on swimming performance in mice. *Phyther Res*, 2000;14:120–121.

Malhotra S, Singh AP. Medicinal properties of ginger (*Zingiber officinale* Rosc.). *Nat Prod Radiance*, 2003;2:296–301.

Martinowich K, Lu B. Interaction between BDNF and serotonin: Role in mood disorders. *Neuropsychopharmacology*, 2008;33:73–83.

Meneses A, Liy-Salmeron G. Serotonin and emotion, learning and memory. *Rev Neurosci*, 2012;23:543–553.

Mengod G, Vilaró MT, Cortés R, López-Giménez J, Raurich A, Palacios J. Chemical neuroanatomy of 5-HT receptor subtypes in the mammalian brain. In B Roth (ed.), *The Serotonin Receptors*, pp. 319–364. Totowa, NJ: Humana Press, 2006.

Merikangas KR, Wicki W, Angst J. Heterogeneity of depression: Classification of depressive subtypes by longitudinal course. *Br J Psychiatry*, 1994;164:342–348.

Michael JS, Kalirajan A, Padmalatha C, Singh AJAR. *In vitro* antioxidant evaluation and total phenolics of methanolic leaf extracts of *Nyctanthes arbor-tristis* L. *Chin J Nat Med*, 2013;11:484–487.

Mishra LC, Singh BB, Dagenais S. Scientific basis for the therapeutic use of *Withania somnifera* (ashwagandha): A review. *Altern Med Rev*, 2000;5:334–346.

Mohan L, Rao USC, Gopalakrishna HN, Nair V. Evaluation of the anxiolytic activity of NR-ANX-C (a polyherbal Formulation) in ethanol withdrawal–induced anxiety behavior in rats. *Evid Based Complement Alternat Med*, 2011;2011:327160.

Murphy DL, Andrews AM, Wichems CH, Li Q, Tohda M, Greenberg B. Brain serotonin neurotransmission: An overview and update with an emphasis on serotonin subsystem heterogeneity, multiple receptors, interactions with other neurotransmitter systems, and consequent implications for understanding the actions of serotonergic drugs. *J Clin Psychiatry*, 1998;59 Suppl. 1:4–12.

Muruganandam A, Kumar V, Bhattacharya SK. Effect of poly herbal formulation, EuMil, on chronic stress-induced homeostatic perturbations in rats. *Indian J Exp Biol*, 2002;40:1151–1160.

Mutalik M, Mutalik M. *Tinospora cordifolia*: Role in depression, cognition and memory. *Aust J Med Herbal*, 2011;23:168–173.

Neale C, Camfield D, Reay J, Stough C, Scholey A. Cognitive effects of two nutraceuticals Ginseng and Bacopa benchmarked against modafinil: A review and comparison of effect sizes. *Br J Clin Pharmacol*, 2013;75:728–737.

Nirmal SA, Pal SC, Mandal SC, Patil AN. Analgesic and anti-inflammatory activity of β-sitosterol isolated from *Nyctanthes arbortristis* leaves. *Inflammopharmacology*, 2012;20:219–224.

Noorafshan A, Ashkani-Esfahani S. A review of therapeutic effects of curcumin. *Curr Pharm Des*, 2013;19:2032–2046.

Osborne PG, Chou TS, Shen TW. Characterization of the psychological, physiological and EEG profile of acute betel quid intoxication in naive subjects. *PLoS One*, 2011;6.

Østergaard SD, Jensen SOW, Bech P. The heterogeneity of the depressive syndrome: When numbers get serious. *Acta Psychiatr Scand*, 2011;124:495–496.

Pandey G, Madhuri S. Pharmacological activities of *Ocimum* sanctum (tulsi): A review. *Int J Pharm Sci Rev Res*, 2010;5:61–66.

Pasrija A, Singh R, Katiyar CK. Validated HPLC–UV method for the determination of berberine in raw herb Daruharidra (*Berberis aristata* DC), its extract, and in commercially marketed Ayurvedic dosage forms. *Int J Ayurveda Res*, 2010;1:243–246.

Pattanayak C, Datta PP, Prasad A, Panda P. Evaluation of anti-inflammatory activity of *Nyctanthes arbor-tristis* leaves. *Int J Med Pharm Sci*, 2013;3:18–25.

Pemminati S, Gopalakrishna HN, Alva A, Pai MRSM, Seema Y, Raj V, Pillai D. Antidepressant activity of ethanolic extract of leaves of *Ocimum sanctum* in mice. *J Pharm Res*, 2010;3:624–626.

Peng W, Liu Y-J, Wu N, Sun T, He X-Y, Gao Y-X, Wu C-J. *Areca catechu* L. (Arecaceae): A review of its traditional uses, botany, phytochemistry, pharmacology and toxicology. *J Ethnopharmacol*, 2015;164:340–356.

Peng WH, Lo KL, Lee YH, Hung TH, Lin YC. Berberine produces antidepressant-like effects in the forced swim test and in the tail suspension test in mice. *Life Sci*, 2007;81:933–938.

Peng WH, Wu CR, Chen CS, Chen CF, Leu ZC, Hsieh MT. Anxiolytic effect of berberine on exploratory activity of the mouse in two experimental anxiety models: Interaction with drugs acting at 5-HT receptors. *Life Sci*, 2004;75:2451–2462.

Peth-Nui T, Wattanathorn J, Muchimapura S, Tong-Un T, Piyavhatkul N, Rangseekajee P, Ingkaninan K, Vittaya-Areekul S. Effects of 12-week *Bacopa monnieri* consumption on attention, cognitive processing, working memory, and functions of both cholinergic and monoaminergic systems in healthy elderly volunteers. *Evid Based Complement Alternat Med*, 2012;2012:606424.

Pollak Dorocic I, Fürth D, Xuan Y, Johansson Y, Pozzi L, Silberberg G, Carlén M, Meletis K. A whole-brain atlas of inputs to serotonergic neurons of the dorsal and median raphe nuclei. *Neuron*, 2014;83:663–678.

Potdar D, Hirwani RR, Dhulap S. Phyto-chemical and pharmacological applications of Berberis aristata. *Fitoterapia*, 2012;83:817–830.

Preskorn SH, Ross R, Stanga CY. Selective serotonin reuptake inhibitors. In S Preskorn, J Feighner, C Stanga, R Ross (eds), *Antidepressants Past, Present and Future*, vol. 157, 1st edn, pp. 241–262. Berlin: Springer-Verlag, 2004.

Puri A, Saxena R, Saxena RP, Saxena KC, Srivastava V, Tandon JS. Immunostimulant activity of *Nyctanthes arbor-tristis* L. *J Ethnopharmacol*, 1994;42:31–37.

Rahimi R, Nikfar S, Abdollahi M. Efficacy and tolerability of *Hypericum perforatum* in major depressive disorder in comparison with selective serotonin reuptake inhibitors: A meta-analysis. *Prog Neuropsychopharmacol Biol Psychiatry*, 2009;33:118–127.

Rahmani AH, Shabrmi FMA, Aly SM. Active ingredients of ginger as potential candidates in the prevention and treatment of diseases via modulation of biological activities. *Int J Physiol Pathophysiol Pharmacol*, 2014;6:125–136.

Rajan KE, Singh HK, Parkavi A, Charles PD. Attenuation of 1-(m-chlorophenyl)-biguanide induced hippocampus-dependent memory impairment by a standardised extract of *Bacopa monniera* (BESEB CDRI-08). *Neurochem Res*, 2011;36:2136–2144.

Rajathei DM, Preethi J, Singh HK, Rajan KE. Molecular docking of bacosides with tryptophan hydroxylase: A model to understand the bacosides mechanism. *Nat Products Bioprospect*, 2014;4:251–255.

Rajput MS, Sinha S, Mathur V, Agrawal P. Herbal antidepressants. *Int J Pharm Front Res*, 2011;1:159–169.

Ramakrishna A, Giridhar P, Ravishankar GA. Phytoserotonin: A review. *Plant Sig Behav*, 2011;6:800–809.

Ramanatha M, Balaji B, Justin A, Gopinath N, Vasanthi M, Ramesh RV. Behavioural and neurochemical evaluation of Perment, an herbal formulation in chronic unpredictable mild stress induced depressive model. *Indian J Exp Biol*, 2011;49:269–275.

Rani C, Chawla S, Mangal M, Mangal AK, Kajla S, Dhawan AK. *Nyctanthes arbor-tristis* Linn. (night jasmine): A sacred ornamental plant with immense medicinal potentials. *Indian J Tradit Knowl*, 2012;11:427–435.

Rao RV, Descamps O, John V, Bredesen DE. Ayurvedic medicinal plants for Alzheimer's disease: A review. *Alzheimers Res Ther*, 2012;4:22.

Rapport MM, Green AA, Page IH. Serum vasoconstrictor (Serotonin): IV. Isolation and Characterization. *J Biol Chem*, 1948a;176:1243–1251.

Rapport MM, Green AA, Page IH. Crystalline serotonin. *Science*, 1948b;108:329–330.

Rastogi M, Ojha RP, Prabu PC, Devi BP, Agrawal A, Dubey GP. Prevention of age-associated neurodegeneration and promotion of healthy brain ageing in female Wistar rats by long term use of bacosides. *Biogerontology*, 2012;13:183–195.

Ratnasooriya W, Jayakody J, Hettiarachchi A, Dharmasiri M. Sedative effects of hot flower infusion of *Nyctanthes arbo-tristis* on rats. *Pharm Biol*, 2005;43:140–146.

Rauf K, Subhan F, Abbas M, Haq I ul, Gowhar Ali, Ayaz M. Effect of acute and sub chronic use of *Bacopa monnieri* on dopamine and serotonin turnover in mice whole brain. *African J Pharm Pharmacol*, 2012;6:2767–2774.

Ravindran R, Rathinasamy SD, Samson J, Senthilvelan M. Noise-stress-induced brain neurotransmitter changes and the effect of *Ocimum sanctum* (Linn.) treatment in albino rats. *J Pharmacol Sci*, 2005;98:354–360.

Rege NN, Thatte UM, Dahanukar SA. Adaptogenic properties of six rasayana herbs used in ayurvedic medicine. *Phyther Res*, 1999;13:275–291.

Sadhu A, Upadhyay P, Agrawal A, Ilango K, Karmakar D, Singh GPI, Dubey GP. Management of cognitive determinants in senile dementia of Alzheimer's type: Therapeutic potential of a novel polyherbal drug product. *Clin Drug Investig*, 2014;34:857–869.

Saha RK, Acharya S, Shovon SSH, Apu AS, Roy P. Biochemical investigation and biological evaluation of the methanolic extract of the leaves of *Nyctanthes arbortristis in vitro*. *Asian Pac J Trop Biomed*, 2012;2:S1534–S1541.

Sairam K, Dorababu M, Goel RK, Bhattacharya SK. Antidepressant activity of standardized extract of *Bacopa monnieri* in experimental models of depression in rats. *Phytomedicine*, 2002;9:207–211.

Sakina MR, Dandiya PC, Hamdard ME, Hameed A. Preliminary psychopharmacological evaluation of *Ocimum sanctum* leaf extract. *J Ethnopharmacol*, 1990;28:143–150.

Samson J, Sheela Devi R, Ravindran R, Senthilvelan M. Biogenic amine changes in brain regions and attenuating action of *Ocimum sanctum* in noise exposure. *Pharmacol Biochem Behav*, 2006;83:67–75.

Sanmukhani J, Satodia V, Trivedi J, Patel T, Tiwari D, Panchal B, Goel A, Tripathi CB. Efficacy and safety of curcumin in major depressive disorder: A randomized controlled trial. *Phyther Res*, 2014;28:579–585.

Sartorius N. The economic and social burden of depression. *J Clin Psychiatry*, 2001;62:8–11.

Saxena RS, Gupta B, Lata S. Tranquilizing, antihistaminic and purgative activity of *Nyctanthes arbor-tristis* leaf extract. *J Ethnopharmacol*, 2002;81:321–325.

Saxena RS, Gupta B, Saxena KK, Singh RC, Prasad DN. Study of anti-inflammatory activity in the leaves of *Nyctanthes arbor tristis* Linn: An Indian medicinal plant. *J Ethnopharmacol*, 1984;11:319–330.

Saxena RS, Gupta B, Saxena KK, Srivastava VK, Prasad DN. Analgesic, antipyretic and ulcerogenic activity of *Nyctanthes arbor tristis* leaf extract. *J Ethnopharmacol*, 1987;19:193–200.

Schildkraut JJ. The catecholamine hypothesis of affective disorders: A review of supporting evidence. *Am J Psychiatry*, 1965;122:509–522.

Schulte-Löbbert S, Holoubek G, Müller WE, Schubert-Zsilavecz M, Wurglics M. Comparison of the synaptosomal uptake inhibition of serotonin by St John's wort products. *J Pharm Pharmacol*, 2004;56:813–818.

Shah PC, Trivedi NA, Bhatt JD, Hemavathi KG. Effect of Withania somnifera on forced swimming test induced immobility in mice and its interaction with various drugs. *Indian J Physiol Pharmacol*, 2006;50:409–415.

Sharpley CF, Bitsika V. Differences in neurobiological pathways of four "clinical content" subtypes of depression. *Behav Brain Res*, 2013;256:368–376.

Sharpley CF, Bitsika V. Validity, reliability and prevalence of four "clinical content" subtypes of depression. *Behav Brain Res*, 2014;259:9–15.

Sheikh N, Ahmad A, Siripurapu KB, Kuchibhotla VK, Singh S, Palit G. Effect of *Bacopa monniera* on stress induced changes in plasma corticosterone and brain monoamines in rats. *J Ethnopharmacol*, 2007;111:671–676.

Shen Y-H, Zhou Y, Zhang C, Liu R-H, Su J, Liu X-H, Zhang W-D. Antidepressant effects of methanol extract and fractions of *Bacopa monnieri*. *Pharm Biol*, 2009;47:340–343.

Shreevathsa M, Ravishankar B, Dwivedi R. Anti depressant activity of Mamsyadi Kwatha: An ayurvedic compound formulation. *Ayu*, 2013;34:113–117.

Shukla AK, Patra S, Dubey VK. Deciphering molecular mechanism underlying antileishmanial activity of *Nyctanthes arbortristis*, an Indian medicinal plant. *J Ethnopharmacol*, 2011;134:996–998.

Sibille E, French B. Biological substrates underpinning diagnosis of major depression. *Int J Neuropsychopharmacol*, 2013;16:1893–1909.

Singh HK. Brain enhancing ingredients from Āyurvedic medicine: Quintessential example of *Bacopa monniera*, a narrative review. *Nutrients*, 2013;5:478–497.

Singh N, Bhalla M, de Jager P, Gilca M. An overview on Ashwagandha: A rasayana (rejuvenator) of ayurveda. *African J Tradit Complement Altern Med*, 2011;8:208–213.

Singh N, Misra N, Srivastava AK, Dixit KS, Gupta GP. Effect of anti-stress plants on biochemical changes during stress reaction. *Indian J Pharmacol*, 1991;23:137–142.

Singh SS, Pandey SC, Srivastava S, Gupta VS, Patro B, Ghosh AC. Chemistry and medicinal properties of *Tinospora cordifolia*. *Indian J Pharmacol*, 2003;35:83–91.

Singh R, Ramakrishna R, Bhateria M, Bhatta RS. *In vitro* evaluation of *Bacopa monniera* extract and individual constituents on human recombinant monoamine oxidase enzymes. *Phytother Res*, 2014;28:1419–1422.

Singh RP, Jain R, Mishra R, Tiwari P. Antidepressant activity of hydroalcoholic extract of *Zingiber officinale*. *Int Res J Pharm*, 2012;3:149–151.

Sinha K, Mishra NP, Singh J, Khanuja SPS. *Tinospora cordifolia* (Guduchi), a reservoir plant for therapeutic applications: A review. *Indian J Tradit Knowl*, 2004;3:257–270.

Srinivasan GV, Unnikrishnan KP, Rema Shree AB, Balachandran I. HPLC estimation of berberine in *Tinospora cordifolia* and *Tinospora sinensis*. *Indian J Pharm Sci*, 2008;70:96–99.

Sumi-Ichinose C, Ichinose H, Takahashi E, Hori T, Nagatsu T. Molecular cloning of genomic DNA and chromosomal assignment of the gene for human aromatic L-amino acid decarboxylase, the enzyme for catecholamine and serotonin biosynthesis. *Biochemistry*, 1992;31:2229–2238.

Suresh V, Jaikumar S, Arunachalam G. Anti diabetic activity of ethanol extract of stem bark of *Nyctanthes arbor-tristis* Linn. *Res J Pharm Biol Chem Sci*, 2010;1:311–317.

Szegedi A, Kohnen R, Dienel A, Kieser M. Acute treatment of moderate to severe depression with Hypericum extract WS 5570 (St John's wort): Randomised controlled double blind non-inferiority trial versus paroxetine. *BMJ*, 2005;330:503.

Tabassum I, Siddiqui ZN, Rizvi SJ. Effects of *Ocimum sanctum* and *Camellia sinensis* on stress-induced anxiety and depression in male albino *Rattus norvegicus*. *Indian J Pharmacol*, 2010;42:283–288.

Tamilselvi S, Balasubramani SP, Venkatasubramanian P, Vasanthi NS. A review on the pharmacognosy and pharmacology of the herbals traded as "daruharidra." *Int J Pharma Bio Sci*, 2014;5:556–570.

Tripathi AK, Dey S, Singh RH, Dey PK. Alterations in the sensitivity of 5(th) receptor subtypes following chronic asvagandha treatment in rats. *Anc Sci Life*, 1998;17:169–181.

Tripathi S, Tripathi PK, Singh PN. Antidepressant activity of *Nyctanthes arbor-tristis* leaf extract. *Pharmacologyonline*, 2010a;3:415–422.

Tripathi S, Tripathi PK, Vijayakumar M, Rao C V, Singh PN. Anxiolytic activity of leaf extract of *Nyctanthes arbor-tristis* in experimental rats. *Pharmacologyonline*, 2010b;2:186–193.

Twarog BM, Page IH. Serotonin content of some mammalian tissues and urine and a method for its determination. *Am J Physiol*, 1953;175:157–161.

Ven Murthy MR, Ranjekar PK, Ramassamy C, Deshpande M. Scientific basis for the use of Indian ayurvedic medicinal plants in the treatment of neurodegenerative disorders: Ashwagandha. *Cent Nerv Syst Agents Med Chem*, 2010;10:238–246.

Vuddanda PR, Chakraborty S, Singh S. Berberine: A potential phytochemical with multispectrum therapeutic activities. *Expert Opin Investig Drugs*, 2010;19:1297–1307.

Walther DJ, Bader M. A unique central tryptophan hydroxylase isoform. *Biochem Pharmacol*, 2003;66:1673–1680.

Wang R, Xu Y, Wu HL, Li YB, Li YH, Guo JB, Li XJ. The antidepressant effects of curcumin in the forced swimming test involve 5-HT1 and 5-HT2 receptors. *Eur J Pharmacol*, 2008;578:43–50.

Xia X, Cheng G, Pan Y, Xia ZH, Kong LD. Behavioral, neurochemical and neuroendocrine effects of the ethanolic extract from *Curcuma longa* L. in the mouse forced swimming test. *J Ethnopharmacol*, 2007;110:356–363.

Xie DP, Chen LB, Liu CY, Zhang CL, Liu KJ, Wang PS. Arecoline excites the colonic smooth muscle motility via M3 receptor in rabbits. *Chin J Physiol*, 2004;47:89–94.

Xu Y, Ku B, Cui L, Li X, Barish PA, Foster TC, Ogle WO. Curcumin reverses impaired hippocampal neurogenesis and increases serotonin receptor 1A mRNA and brain-derived neurotrophic factor expression in chronically stressed rats. *Brain Res*, 2007;1162:9–18.

Xu Y, Ku BS, Yao HY, Lin YH, Ma X, Zhang YH, Li XJ. The effects of curcumin on depressive-like behaviors in mice. *Eur J Pharmacol*, 2005a;518:40–46.

Xu Y, Ku BS, Yao HY, Lin YH, Ma X, Zhang YH, Li XJ. Antidepressant effects of curcumin in the forced swim test and olfactory bulbectomy models of depression in rats. *Pharmacol Biochem Behav*, 2005b;82:200–206.

Yang YR, Chang KC, Chen CL, Chiu TH. Arecoline excites rat locus coeruleus neurons by activating the M2-muscarinic receptor. *Chin J Physiol*, 2000;43:23–28.

Yu ZF, Kong LD, Chen Y. Antidepressant activity of aqueous extracts of Curcuma longa in mice. *J Ethnopharmacol*, 2002;83:161–165.

Zhou Y, Shen YH, Zhang C, Su J, Liu RH, Zhang WD. Triterpene saponins from *Bacopa monnieri* and their antidepressant effects in two mice models. *J Nat Prod*, 2007;70:652–655.

29

A Role for the Regulation of the Melatonergic Pathways in Alzheimer's Disease and Other Neurodegenerative and Psychiatric Conditions

George Anderson
CRC Scotland and London
London, UK

Michael Maes
Deakin University
Geelong, Australia

CONTENTS

ABSTRACT When tryptophan is taken up by the body, some is used for the synthesis of serotonin. As well as having neuroregulatory effects, serotonin is also the precursor for the synthesis of N-acetylserotonin (NAS) and melatonin, which form the melatonergic pathways. The melatonergic pathways are associated with many of the changes that are thought to be important in the pathophysiology of Alzheimer's disease, Parkinson's disease, multiple sclerosis, and major depressive disorder. As well as being produced by the pineal gland at night, and thereby contributing to the circadian rhythm, melatonin is also very highly produced in other organs and tissues, including the gut, where it may have an important role in the regulation of gut permeability and thereby with many medical conditions.

Melatonin significantly regulates the immune system, affording protection against immune senescence and the patterning of the immune responses. Recent work on central nervous system (CNS)-associated disorders suggest that altered immune responses are an integral part of their etiology, course, and treatment. Melatonin is also a significant antioxidant and anti-inflammatory as well as optimizing mitochondrial functioning. Melatonin may be produced by many, if not all, human cells, including central glia and immune cells, where autocrine and paracrine effects of melatonin generally decrease the reactivity of these cells. This is important, as glia and immune reactivity are significant pharmaceutical targets.

NAS, the immediate precursor of melatonin, is also a powerful antioxidant. Some NAS effects may be mediated by its ability to mimic brain-derived neurotrophic factor (BDNF), which is commonly decreased in many CNS conditions. Many factors that are associated with benefit in CNS conditions may actually be mediating their benefit by increasing the synthesis of melatonin in a wide array of cells.

This chapter reviews the melatonergic pathway's role in such CNS medical conditions, highlighting its use for a whole host of conditions that are currently very poorly managed.

KEY WORDS: *melatonin, N-acetylserotonin, Alzheimer's disease, Parkinson's disease, multiple sclerosis, major depressive disorder, tryptophan, gut permeability, caffeine, green tea.*

Abbreviations

11B-HSD1:	11β-hydroxysteroid dehydrogenase type 1
AChI:	acetylcholinesterase inhibitor
α7nAChr:	alpha-7 nicotinic acetylcholine receptor
αSyn:	alpha-synuclein
AD:	Alzheimer's disease
AA-NAT:	aralkylamine N-acetyltransferase
AFMK:	N1-acetyl-N2-formyl-5-methoxykynuramine
ALS:	amyotrophic lateral sclerosis
AMK:	N1-acetyl-5-methoxykynuramine
Apo:	apolipoprotein
APP:	amyloid precursor protein
BACE1:	beta-site APP-cleaving enzyme
BAG-1:	Bcl2-associated anthanogene 1
BBB:	blood–brain barrier
BDNF:	brain-derived neurotrophic factor
cAMP:	cyclic adenosine monophosphate
CVD:	cardiovascular disease
CYP:	cytochrome
EGCG:	epigallocatechin-3-gallate
EAE:	experimental autoimmune encephalomyelitis
GBM:	glioblastoma
GSK-3β:	glycogen synthase kinase beta
HO-1:	heme oxygenase 1
HSV:	herpes simplex virus
HMGB1:	high-mobility group box 1
HIOMT:	hydroxyindole-O-methyltransferase
HPA:	hypothalamic–pituitary–adrenal
IDO:	indoleamine 2,3-dioxygenase
IFNγ:	interferon gamma
IL:	interleukin
IPA:	immune–pineal axis

MDD:	major depressive disorder
MMP:	matrix metalloproteinase
MT:	melatonin receptor
miR:	micro-RNA
MCI:	mild cognitive impairment
MAO:	monoamine oxidase
MS:	multiple sclerosis
NK:	natural killer
NAD:	nicotinamide
NAS:	N-acetylserotonin
NE:	norepinephrine
NF-κB:	nuclear factor kappa B
NLRP3:	NOD-like receptor family, pyrin domain–containing 3
NMDA:	N-methyl-D-aspartate
Nrf2:	NF-E2-related factor 2
O&NS:	oxidative and nitrosative stress
PARP1:	poly(ADP-ribose) polymerase 1
PD:	Parkinson's disease
PUFA:	polyunsaturated fatty acid
RAGE:	receptor for advanced glycation end products
ROS:	reactive oxygen species
S1P1r:	sphingosine-1-phosphatase receptor 1
SNP:	single-nucleotide polymorphisms
SubP:	substance P
Th:	T-helper
TLR:	toll-like receptor
TRYCAT:	tryptophan catabolite
TNFα:	tumor necrosis factor alpha
TrKB:	tyrosine kinase receptor B
YY1:	yin yang 1

Introduction

Classically, conceptualizations of neurodegenerative and psychiatric conditions have used changes in neurons and synapses as the basis for the biological underpinnings of conditions such as Alzheimer's disease (AD), multiple sclerosis (MS) and Parkinson's disease (PD), as well as schizophrenia and depression (Anderson and Maes, 2014a). A growing body of data is challenging such neuron-centric conceptualizations, giving greater importance to the immune system, including central astrocytes and microglia (Vincenti et al., 2015), as well as to changes in other organs and tissues, including the gut (Anderson and Maes, 2015a). Such work has highlighted a previously overlooked commonality across such a diverse array of medical conditions, namely the role of pineal and local melatonin synthesis.

There is a wide array of data on the changes in circadian melatonin production across many medical conditions, with pineal melatonin production at night being invariably decreased. However, recent data shows melatonin to be produced by many, if not all, mitochondria-containing cells (Tan et al., 2013), including thyroid cells (Garcia-Marin et al., 2015), immune cells (Muxel et al., 2012; Gupta and Haldar, 2015), and glia (Liu et al., 2007, 2012). Melatonin generally decreases the reactivity of glia and immune cells, suggesting that alterations in melatonin synthesis by these cells may contribute to increases in their reactivity, which current conceptualizations suggest are crucial to a wide range of neurodegenerative and psychiatric conditions.

In this chapter, we look at the role of factors regulating the melatonergic pathways and how this may contribute to the etiology, course, and management of a diverse array of medical conditions. Initially, the melatonergic pathways will be overviewed, before looking at how the melatonergic pathways could be relevant across a host of classically conceived conditions of the central nervous system (CNS).

Melatonergic Pathways Overview

N-acetyl-5-methoxytryptamine (melatonin) is a methoxyindole that has predominantly been studied following its pineal gland efflux during darkness, which is determined by the levels of pineal norepinephrine (NE). Pineal melatonin production and release at night is a significant circadian rhythm driver and modulator. Melatonin is a powerful antioxidant and anti-inflammatory, as well as being an antinociceptive and immune regulator. Importantly, melatonin is also an inducer of endogenous antioxidants, driven by its induction of the transcription factor NF-E2-related factor 2 (Nrf2) (Jumnongprakhon et al., 2015; Hardeland et al., 2011). Further, melatonin enhances neurogenesis (Anderson and Maes, 2014b) and the longevity protein sirtuin-1 (Yang et al., 2015) while also increasing mitochondrial oxidative phosphorylation (Paradies et al., 2015). These are important effects that generally negate the changes associated with neurodegenerative and psychiatric conditions.

N-acetylserotonin (NAS) is the immediate precursor of melatonin. NAS is also a powerful antioxidant as well as an immune and mitochondria regulator. Likewise, NAS also increases neurogenesis, with effects that may be partly derived from its mimicking of brain-derived neurotrophic factor (BDNF). This is due to NAS activating the BDNF receptor tyrosine kinase receptor-B (TrKB) (Jang et al., 2010). It is important to note that the levels of NAS and melatonin are highly dependent on serotonin availability, thereby linking to the decreased serotonin commonly found in major depressive disorder (MDD). Serotonin is enzymatically converted by aralkylamine N-acetyltransferase (AA-NAT) to NAS, with hydroxyindole-O-methyltransferase (HIOMT) (also known as acetylserotonin methyltransferase) then converting NAS to melatonin. Being amphiphilic, melatonin and NAS readily diffuse from the cells in which they are produced, with melatonin often being found to accumulate around intracellular organelles, especially mitochondria (Yu et al., 2016).

Following melatonin metabolism, many of its metabolites also have anti-inflammatory and antioxidant effects as well as modulating the immune system (Galano et al., 2015; Hardeland et al., 2011). Melatonin metabolites include N1-acetyl-5-methoxykynuramine (AMK) and N1-acetyl-N2-formyl-5-methoxykynuramine (AFMK). It is therefore clear that following melatonergic pathway activation many antioxidants are produced, on top of the endogenous antioxidants derived from melatonin induced Nrf2.

Being amphiphilic, melatonin readily crosses into all tissues and organs, including the blood–brain barrier (BBB) as well as generally crossing through cell membranes, where, as indicated, it often accumulates around intracellular organelles, especially mitochondria (Yu et al., 2016). As such, although some of the protection afforded by melatonin is attributable to its activation of the melatonin receptors (MT1 and MT2), many of its effects are melatonin receptor independent (Pappolla et al., 2002). Single nucleotide polymorphisms (SNPs) in melatonergic pathway enzymes, MT1, and MT2 are susceptibility factors for a wide array of medical conditions, including cancer (Deming et al., 2012), MS (Natarajan et al., 2012), bipolar disorder (Anderson et al., 2016; Etain et al., 2012), and depression (Galecki et al., 2010). Many factors can regulate NAS and melatonin synthesis, including via the regulation of pineal NE release, which can be increased by leptin and some 14-3-3 isoforms (Szewczyk-Golec et al., 2015; Beischlag et al., 2016), and decreased by substance P (SubP) and tumor necrosis factor alpha (TNFα).

There is a growing appreciation that the melatonergic pathways are active in many cells, including astrocytes (Liu et al., 2007), skin cells (Liu et al., 2013), macrophages (Muxel et al., 2012), and thyroid cells (Garcia-Marin et al., 2015). It is of note that AA-NAT is also present in neurons (Haghighi et al., 2015; Uz et al., 2002), suggesting that neurons may also be a significant source of autocrine and paracrine melatonin, which may be differentially regulated by a wide array of factors, with consequences for neuronal mitochondrial functioning, reactive oxygen species (ROS) production, and endogenous antioxidant production, among other processes.

A number of factors act to regulate serotonin availability, including regulators of monoamine oxidase (MAO), which break down serotonin; cortisol, which can upregulate tryptophan 2,3-dioxygenase (TDO); and pro-inflammatory cytokine-induced indoleamine 2,3-dioxygenase (IDO), which, like TDO, drives tryptophan to the synthesis of neuroregulatory tryptophan catabolites (TRYCATs) and away from serotonin synthesis. Interestingly, the apolipoprotein ApoE4 allele, which is the major susceptibility gene for AD, significantly regulates astrocytic melatonin production and efflux (Liu et al., 2012). This may be of some importance, given that the ApoE4 allele is a significant genetic determinant of AD risk, as well as being linked to a wide array of other medical conditions, including PD (Mata et al., 2014), MS (Shi et al., 2011), schizoaffective disorder (Hammer et al., 2014), and depression (Skoog et al., 2015). Further investigation is required as to whether factors that can regulate pineal NAS and melatonin production, including NE, cyclic adenosine monophosphate (cAMP), leptin, SubP, cortisol, and specific 14-3-3 dimer isoforms, also regulate the melatonergic pathways in other cell types. Many of these pineal melatonin regulators show changes across a range of medical conditions, including AD, depression, glioblastoma, bipolar disorder, and MS.

Astrocyte, microglia, and immune cell melatonergic pathway activity and regulation, in interaction with the TRYCAT pathways, may be of particular relevance to the modulation of neuronal vulnerability across a host of medical conditions (Morris et al., 2016). Astrocytes may be of particular relevance, given their control of neuronal activity, neurotransmitter levels, and energy provision, as well as their control of vasodilation and therefore oxygen and glucose provision. Recent conceptualizations of astrocytes suggest that they may act as a hub that centrally integrates changes occurring peripherally or systemically, by virtue of their enwrapment of brain microvessels that contribute to the BBB formation (Anderson, 2011). Astrocyte melatonergic pathway regulation may then be a significant treatment target across a host of medical conditions, including neurodegenerative and psychiatric conditions as well as glioblastoma (GBM) (Beischlag et al., 2016; Anderson et al., 2016; Maes and Anderson, 2015; Anderson and Maes, 2014c).

As well as direct effects on central processes that may arise from glia melatonin production and associated alterations in glia reactivity (Ding et al., 2014), there are many other sites where the regulation of local melatonin synthesis is likely to have an impact on neurodegenerative and psychiatric conditions. One such site is the gut, which is generating considerable interest across a wide range of conditions, including MS, the autistic spectrum, depression and PD (Kantarcioglu et al., 2016). A number of gut factors may be relevant across such a diverse array of medical presentations, including variations in gut permeability and the composition of the gut microbiome. Melatonin is very highly synthesized in the gut, up to one-hundred-fold higher than peak pineal melatonin synthesis (Mukherjee and Maitra, 2015). Generally, it is thought that gut melatonin decreases gut permeability, thereby decreasing the influence of gut bacteria and tiny, partially digested fragments of food on systemic immune activation. The gut is therefore an area of intense investigation as to melatonin regulation, including as to whether there are variations in melatonin synthesis by gut bacteria.

Compared with melatonin, NAS is relatively little investigated. Like melatonin, NAS has antioxidant and mitochondria-optimizing effects. However, being a BDNF mimic, NAS is likely to have significantly different effects versus melatonin. As a consequence, factors that act to regulate the NAS/melatonin ratio would also be expected to differentially impact on a wide array of central and systemic processes. The regulation of the NAS/melatonin ratio is potentially of some importance and is in urgent need of further investigation. It is notable that different rat strains show alterations in the NAS/melatonin ratio, including under stress conditions (Oxenkrug and Requintina, 1998). This would suggest the ready biological malleability of the NAS/melatonin ratio, which is likely to be of clinical significance.

The growing body of data showing melatonin synthesis in different cell types has led to the hypothesis that melatonin may be produced by all mitochondria-containing cells, to some degree (Tan et al., 2013). This may be placed in an evolutionary context, given that mitochondria evolved from melatonin-producing bacteria, especially *Rhodospirillum rubrum*, in which melatonin production is circadian (Manchester et al., 1995). This has led to the proposal that over the course of evolution the melatonin biosynthetic capacity has shifted from bacteria/mitochondria synthesis to integration within the nuclear genome, with mitochondria still having at least a residual ability to produce melatonin (Tan et al., 2013). There is currently only limited data to support this, including the mitochondrial localization of MT1 (Wang et al., 2011) and AA-NAT (Kerenyi et al., 1975). This is an important area for future research, given that mitochondrial dysfunction is evident in most, if not all, neurodegenerative and psychiatric conditions (Cassano et al., 2016). As melatonin increases the longevity protein sirtuin 1, which is a powerful mitochondria regulator, melatonin has at least indirect effects in optimizing mitochondria functioning.

Overall, melatonin is a powerful antioxidant, an inducer of endogenous antioxidants, an anti-inflammatory, an optimizer of mitochondria functioning, an immune and glia cell reactivity regulator, and a circadian rhythm regulator. The melatonergic pathway is active in many cell types, tissues, and organs, possibly being expressed in all eukaryotic cells. By regulating glia and immune responses, including via autocrine and paracrine effects, melatonin synthesis is an important pharmaceutical and dietary target across a host of medical conditions, including AD, MS, PD, and depression.

Immune–Pineal Axis (IPA)

Recently, the existence of an IPA has been proposed. The IPA is suggested to be a mechanism whereby pro-inflammatory processes can switch off pineal melatonin synthesis, thereby preventing the immuno-regulatory, generally immune-suppressive, effects of pineal melatonin when an active immune response is required, such as during peripheral infection and inflammation (Markus et al., 2013). Such a switching off of pineal melatonin synthesis will prevent melatonin from decreasing the adhesion of lymphocytes to endothelial cells and vessels, thereby preventing melatonin from suppressing appropriate immune responses under challenge. This is exemplified by data in women following cesarean section (Pontes et al., 2007). In this study, raised TNFα levels following cesarean section switch off pineal melatonin production by acting on pinealocytes (Pontes et al., 2007). Data pertaining to lipopolysaccharide (LPS), acting at toll-like receptor (TLR)4, shows that LPS acts to switch off pineal melatonin by increasing pineal microglia activation, thereby leading to local TNFα synthesis and efflux (da Silveira Cruz-Machado et al., 2012). As such, the IPA hypothesis proposes that pineal melatonin synthesis is a two-way interaction with immune cell activity, with differential consequences under conditions of prolonged systemic immune activation, as in AD, MS, and MDD (Maes et al., 2011). It requires investigation as to how low, but chronic, immune inflammatory activity in AD and in AD risk–related disorders, such as obesity and depression, impact on pineal and local melatonin production.

Melatonin and Alzheimer's Disease

Clinical trials of the utility of melatonin in the treatment of AD have been relatively few, given the plethora of preclinical data showing significant suppressive impacts on most, if not all, of the biological underpinnings of AD (Maes and Anderson, 2015). To some extent this may be due to the relative lack of financial incentive of pharmaceutical companies to fund research involving melatonin, given its ready availability and cheapness. Also, in the United States and many other countries, melatonin is classed as a food, rather than as a drug. As such, patent-driven profit may have contributed to the lack of clinical studies using melatonin. That said, the relatively recent discovery of local melatonin synthesis in brain cells, and their regulation by ApoE4, would suggest that pharmaceutical targeting of local melatonin synthesis, perhaps especially in astrocytes and microglia, should be a significant pharmaceutical company target.

Melatonin decreases amyloid-beta production as well as the levels of hyperphosphorylated tau that underpin intracellular tangles (Maes and Anderson, 2015). These effects of melatonin are mediated via its antioxidant capacity as well as its induction of endogenous antioxidants, which is mediated by melatonin's activation of the PI3K/Akt/Nrf2 signaling pathway (Song et al., 2015), with Nrf2 also regulating the phosphorylation of GSK-3β (O'Connell and Hayes, 2015), leading to a decrease in the phosphorylation of tau (Maqbool et al., 2016). Nrf2 also decreases tau phosphorylation via an increase in autophagy (Jo et al., 2014), with melatonin also mediating some of its effects via the regulation of autophagy (Teng et al., 2015). As such, melatonin has significant impacts on the two classical pathophysiological processes associated with AD.

Melatonin may also have significant impacts on many of the other AD pathophysiological processes. As previously indicated, melatonin is a significant regulator of the α7nAChr (Markus et al., 2010), with the α7nAChr also mediating many of melatonin's effects (Jeong and Park, 2015; Parada et al., 2014). This is likely to be highly relevant to the role of decreased cholinergic activity in AD and its treatment with acetylcholinesterase inhibitors. Also, the α7nAChr is a significant modulator of glia and the immune cell–reactivity threshold, suggesting that a combination of decreased melatonin and ACh will have many impacts on a plethora of processes both centrally and peripherally (Maes and Anderson, 2015). Hung and colleagues have recently shown that the enhanced green fluorescent protein (EGFP)-conjugated microtubule-associated protein 1 light chain 3 (EGFP-LC3), a major regulator of autophagy, mediates its effects in increasing autophagy via the induction of the α7nAChr (Hung et al., 2015). As to whether this would suggest that increasing LC3 is driving an increase in local melatonin synthesis awaits investigation. Given the phase-II trial utility of α7nAChr agonists in AD, it is crucial that the mechanisms by which such α7nAChr agonists interact with endogenous local and pineal melatonin, as well as adjunctive melatonin, are investigated by the pharmaceutical companies.

There is a growing appreciation of the role of the inflammasome in the regulation of pro-inflammatory activity that contributes to neurodegeneration in AD (Marchesi, 2016). The activation of the inflammasome leads to the release of active IL-1β and IL-18, both of which may be associated with an increase in BACE1 and amyloid-beta production (Sutinen et al., 2012; Anderson and Ojala, 2010). IL-1β is also a powerful regulator of BBB permeability, including via effects in astrocytes (Rodrigues and Granger, 2015). The inflammasome may also be a significant determinant of increases in gut permeability (Zaki et al., 2011). As such, the inhibition of inflammasome activation by melatonin, including via its antioxidant effects, is likely to have a wide array of clinically relevant impacts on AD pathophysiology (Volt et al., 2016). Melatonin is a significant inhibitor of the inflammasome, including its effects on mitochondrial functioning (Ortiz et al., 2015). ROS is a significant modulator of inflammasome activity in CNS cells (Alfonso-Loeches et al., 2014), which melatonin will inhibit via its antioxidant capacity as well as its induction of endogenous antioxidants.

IL-18 is also known as *IFNg-inducing factor*, highlighting its powerful induction of IFNg. As IFNg is the major IDO inducer, thereby increasing TRYCATs, including kynurenic acid, which inhibits the α7nAChr, the inhibition of the inflammasome is likely to have negative regulatory impacts on IDO induction and the activation of the neuroregulatory TRYCATs. As such, melatonin will act on the inflammasome to regulate TRYCATs and the wide effects of the α7nAChr in AD. As previously highlighted, by inhibiting the ROS and cytokines that regulate IDO and the TRYCATs, melatonin will increase the availability for serotonin synthesis, therefore increasing its own levels of production and activity.

Melatonin is a significant inhibitor of microglia activation (Ding et al., 2014; Tocharus et al., 2010), including when activation is induced by TLR4 activation (Wong et al., 2010). Likewise, melatonin decreases the reactivity of astrocytes as well as other immune cells (Kong et al., 2008; Muxel et al., 2012; Ren et al., 2015). As such, melatonin is likely to inhibit the role of glia and immune cell reactivity in the etiology and course of AD. Given that stress-induced increases in microglia reactivity/sensitivity may be driven by glucocorticoid induction of high-mobility group box (HMGB)1, with this then mediating its effects via the activation of TLR2/4 (Weber et al., 2015), the ability of melatonin to inhibit microglia activation and to negatively regulate the HPA axis would suggest that it would have protective effects on how stress and microglia reactivity associate with AD. It is of note that melatonin can lower HMGB1 release—for example, in the course of brain ischemia/reperfusion (Kang et al., 2011)—with a decrease in TLR4 activation contributing to decreased levels of fatigue (Lucas et al., 2015), among other symptoms.

Increased gut permeability and variations in gut microbiota are at the cutting edge of research in many medical conditions, including AD (Leblhuber et al., 2015). Melatonin is highly prevalent in the gut, where its levels can be four hundred times that of peak pineal nighttime melatonin (Mukherjee and Maitra, 2015). Melatonin is highly produced in the gut, especially in enterochromaffin cells. As to whether gut melatonin levels are altered in AD or whether different forms of gut microbiota in AD lead to their differential production of melatonin is unknown. Matrix metalloproteinase (MMP)9 is increased in AD and negatively regulates the diversity of gut microbiota (Rodrigues et al., 2012), suggesting that a combination of decreased melatonin and increased MMP9 may contribute to a loss of gut microbiota diversity in AD. It is of note that melatonin can inhibit MMP9, including by direct protein–protein interactions (Rudra et al., 2013). As such, the regulation of the gut–brain axis by melatonin (Anderson and Maes, 2015a) may be another aspect of melatonin effects in AD, perhaps in a subgroup of AD patients.

As previously indicated, an increase in BBB permeability is relevant to the pathophysiology of AD (Dudvarski et al, 2016; Kook et al., 2013). Melatonin is a significant inhibitor of BBB permeability (Chen et al., 2006), as well as being an inhibitor of factors that increase BBB permeability, including IL-1β, MMP9, TLR4 activation, and HMGB1 (Dudvarski et al., 2016; Okuma et al., 2012). As such, the role of BBB permeability in the etiology and course of AD will be significantly inhibited by melatonin, perhaps especially if it is induced locally in the vicinity of BBB permeability changes.

In recent years, a number of studies have suggested that viral infections may play a role in AD, especially herpes simplex virus (HSV)-1 infection (Piacentini et al., 2015; Faldu et al., 2015). Although not an antiviral, melatonin may modulate many viral infections by a number of means (Silvestri and Rossi, 2013), including by increasing the activity of natural killer (NK) cells, which are crucial viral defense regulators (Anderson et al., 2015b). Changes in the activity and expressions of NK cells are known to occur in AD, perhaps especially in the early stages of MCI (Le et al., 2015). As such, the regulation of immune responses to viral infection, including via the regulation of the cytotoxicity of NK cells, may be another means by which melatonin modulates key biological processes in AD.

Overall, the many investigated processes that have been shown to play a role in the pathophysiology of AD can be regulated by melatonin. The melatonergic pathways are also relevant in many other neurodegenerative and psychiatric conditions. We next look at their role in PD.

Parkinson's Disease

PD is a movement disorder, with pathophysiological processes being driven by the loss of dopamine neurons in the substantia nigra. Although the second most common classical neurodegenerative disorder after AD, PD is currently poorly treated, with the primary treatment with L-DOPA tackling the symptoms but having little relevant impact on the underlying pathology, which is defined as the continuing loss of substantia nigra dopamine neurons (Elbaz et al., 2016). As well as being a movement disorder, PD is also associated cognitive loss and wider neurodegenerative processes, which have been widely attributed to the effects of alpha-synuclein (αSyn). However, increases in amyloid-beta are also evident in PD, suggesting that it has more direct overlaps with the processes thought to underpin AD (Shah et al., 2016).

PD is associated with increased oxidative and nitrosative stress (O&NS), TRYCATs, immunoinflammatory activity, and increased levels of depression, paralleling similar changes in AD (Anderson and Maes, 2014). Generally, decreased circadian melatonin occurs in PD in correlation with excessive daytime sleepiness (Videnovic et al., 2014). Decreased melatonin receptor levels are also evident in the substantia nigra of PD patients (Adi et al., 2010). This may be of some relevance, given that melatonin normally induces its own receptors, suggesting that there is a decrease in local melatonin release in the substantia nigra in PD (Anderson and Maes, 2014d).

Alterations in gut functioning are often evident in the early course of PD (Pfeiffer, 2015). This may be due to the effects of αSyn in neurosensory connections to the gut (Klingelhoefer and Reichmann, 2015), with animal models of PD suggesting that αSyn effects may parallel those occurring centrally (Bencsik et al., 2014) and/or that αSyn effects in the gut may contribute to driving central PD symptoms, including via the translocation of αSyn from the gut to the brain (Holmqvist et al., 2014). Wider gut changes are also known to occur in PD, including alterations in the levels and variety of gut microbiota, which correlate with important PD symptoms (Scheperjans et al., 2015). As well as being highly expressed in the gut and a significant modulator of gut permeability, melatonin is also a significant inhibitor of αSyn cytotoxicity and inclusion formation (Ono et al., 2012), therefore having relevance to αSyn effects, both in the gut and centrally.

As well as showing increased levels of immunoinflammatory activity, O&NS, TRYCATs and depression (Anderson and Maes, 2014d), PD is associated with increased levels of HMGB1, which inhibit αSyn inclusion autophagy (Song et al., 2014) and inflammasome activation (Codolo et al., 2013), coupled to significant detrimental interactions with stress (Herrero et al., 2015), as well as some protection afforded by caffeine (Kardani and Roy, in press) and nicotine (Barreto et al., 2015). However, it should be noted that the genetic link of PD with the TRYCAT pathways has still to be fully investigated (Török et al., 2015). As with AD, α7nAChr agonists are a significant treatment target in PD (Quik et al., 2015). The associations of melatonin with all of these processes is as relevant in PD as in AD, with melatonin also having been shown to decrease L-DOPA levels, and associated toxicity, required to manage PD symptoms, as evidenced in PD animal models (Naskar et al., 2015). Melatonin has been used in PD patients predominantly to manage sleep difficulties, with a recent meta-analysis showing its efficacy in sleep regulation in PD (Zhang et al., 2016). However, although having positive benefits in an array of PD models (Su et al., 2015), melatonin has not been extensively investigated as to its role in the management of the etiology and course of more primary PD pathophysiology. Overall, melatonin may be intimately associated with the core features of PD pathophysiology, with treatment efficacy likely to be enhanced by the induction of the melatonergic pathways at local sites, including the gut and immune cells, as well as in the substantia nigra.

Multiple Sclerosis

MS is an immune-mediated condition that is classically associated with demyelination in white matter, leading to a malfunction of neuronal axonal activity. MS is also associated with gray-matter loss that correlates with levels of cognitive deficits. Although difficult to generalize, most investigations of the natural history of MS indicate it to be a two-phase disease: first, arising as focal inflammation with flares; and secondly, with disability progressing independently of focal inflammation (Leray et al., 2016). MS patients show increased levels of pro-inflammatory cytokines, increased O&NS, high levels of comorbid depression, inflammasome activation (Zeis et al., 2015), and high levels of TRYCATs (Watzlawik et al., 2015), coupled to decreased levels of melatonin, perhaps especially when depression is concurrent (Anderson and Rodriguez, 2011).

Melatonin measures in MS have focused on its synthesis and efflux from the pineal gland, driven by evidence showing MS susceptibility genes in tryptophan hydroxylase-2 and melatonin receptors, which correlate with primary and secondary progressive MS (Natarajan et al., 2012). Decreased urinary levels of melatonin are evident in MS patients, with melatonin levels correlating with disease severity (Gholipour et al., 2015); other researchers have also found that decreased circadian melatonin correlates with increased symptom severity, especially increased disability and fatigue scores (Damasceno et al., 2015). In the primary progressive form of MS, clinical case studies show that treatment with melatonin can improve symptomatology (López-González et al., 2015). Other data shows that the treatment of secondary progressive MS patients with melatonin decreases O&NS in red blood cells, as indicated by lowered lipid peroxidation and enhanced endogenous antioxidants (Miller et al., 2013). As such, melatonergic pathways show genetic associations with MS susceptibility, with alterations in pineal melatonin evident in MS and melatonin showing treatment efficacy in this still poorly managed condition.

Some of the efficacy of melatonin in MS is, at least in part, mediated via its induction of sirtuin-1 and antioxidant enzymes, which afford protection against O&NS and pro-inflammatory processes (Emamgholipour et al., 2016). Seasonal variations in levels of melatonin, as with sunlight derived vitamin D, contribute to the seasonality of MS symptomatology (Farez et al., 2015). This seems to be mediated by melatonin suppressing the development of the autoimmune-associated Th17 cells and enhancing the development of anti-inflammatory regulatory T cells and the less damaging Th1 cells (Lee and Cua, 2015).

Animal models of MS also provide support for a melatonin role in its course and treatment. In the experimental autoimmune encephalomyelitis (EAE) animal models of MS, disease activity is controlled by melatonin via its regulation of the ration of effector T cells to regulatory T cells (Álvarez-Sánchez et al., 2015). These authors showed that melatonin decreased levels of Th1 and Th17 pro-inflammatory cytokines, while increasing the Th2 cytokine IL-10 and levels of regulatory T cells (Álvarez-Sánchez et al., 2015). Overall, melatonin has clinical utility in the treatment of MS, which is supported by its immunoregulatory effects in the EAE model.

It should be noted that many melatonin effects in MS may be mediated by its levels of release in local CNS and immune cells, as well as in the gut (Anderson and Rodriguez, 2015c), with changes in gut permeability and gut microbiota evident in MS patients (Miyake et al., 2015).

Major Depressive Disorder

MDD is a commonly diagnosed condition that is also now seen as an integral aspect to the etiology and course of many other disorders of the CNS, including AD, PD, and MS (Anderson and Maes, 2014a, 2014e; Maes et al., 2011), with the newly emerging biological underpinnings of MDD being intimately associated with the pathophysiological changes occurring in these other medical conditions.

Lower melatonin levels are often found in MDD patients, where the melatonin receptor is a susceptibility gene (Galecka et al., 2011). In recurrent MDD, the melatonergic synthesis enzyme is a susceptibility gene (Galecki et al., 2010), indicating a role for decreased melatonin in the etiology and course of the enhanced O&NS and immunoinflammatory activity evident in recurrent MDD patients (Anderson and Maes, 2014e), which are associated with deficits in neurocognitive function (Galecki et al., 2015). In an elderly population sample, higher melatonin levels with were associated with a lower prevalence of cognitive impairment and depressed mood (Obayashi et al, 2015). Much of the protection afforded by melatonin across a range of medical conditions, including depression, is at least partly via its direct effects in mitochondria, where, as previously indicated, it increases mitochondrial complexes and oxidative phosphorylation.

Given the role of melatonin in the genetic susceptibility to depression as well as its role over the course of depression, the regulation of melatonin at different sites forms an important treatment target in MDD. Its anti-inflammatory, antioxidant, anti-nociceptive, and mitochondrial regulatory effects are of particular relevance in the context of immunoinflammatory processes in MDD (Galecka et al., 2011; Kripke et al., 2011). With melatonin impacting on circadian regulation, neurogenesis, gut permeability, sirtuins, HMGB1, inflammasome, and mitochondrial oxidative phosphorylation (Alcocer-Gómez et al., 2016), it is corrective of many of the core biological underpinnings of MDD (Anderson and Maes, 2014e). Via sleep improvements, melatonin counters one of the major reported problems in MDD patients. Via the induction of Bcl2-associated anthanogene 1 (BAG-1), which inhibits the nuclear translocation of the cortisol's glucocorticoid receptor, melatonin inhibits the effects of stress-induced hormones, including the cortisol regulation of MAO (Ou et al., 2006) and stress-induced cortisol-driven gene transcriptions. Stress-driven depression may also require the activation of the inflammasome (Alcocer-Gómez et al., 2016), which melatonin inhibits. Given its inhibition of immunoinflammatory processes and O&NS, melatonin will afford protection against MDD-associated CVD (Opie and Lecour, 2015), thereby also decreasing AD and other neurodegenerative disorders for which CVDs are a risk factor (Claassen, 2015). As well as having systemic applications and associated benefits, the modulation of local glia melatonin production will be a significant treatment target in MDD patients as well as in depression-associated conditions such as MS, PD, and AD.

In a study looking at the combination of buspirone and melatonin in MDD patients, the combination of buspirone and melatonin was of significantly more therapeutic value than buspirone alone, by increasing neurogenesis (Fava et al., 2012) and by decreasing cognitive deficits (Targum et al., 2015). With melatonin decreasing LPS-induced fever (Bruno et al., 2005), it will be interesting to determine whether melatonin, alone or in combination with other antidepressants, has any therapeutic utility in the MDD subgroup that shows increased body temperature (Rausch et al., 2003).

Overall, the melatonergic pathways are intimately associated with MDD via pathways and processes that have significant overlaps with those of AD, PD, and MS. Other subtypes of mood disorders, including bipolar disorder and seasonal affective disorder, are also associated with decreased melatonin and clinical improvements following treatment with melatonin (Anderson et al., 2016; Kaminski-Hartenthaler et al., 2015)

N-acetylserotonin

It should be noted that NAS is the immediate precursor of melatonin, and is released with melatonin. Like melatonin, NAS is amphiphilic and readily crosses cell membranes (Yu et al., 2016). NAS can have effects that are similar and dissimilar to those of melatonin, in part because NAS is a BDNF receptor (TrkB) agonist, and therefore has similar trophic effects to BDNF. Given the significant decreases of BDNF in MDD and AD, as well as in many other CNS disorders, the regulation of NAS is not unimportant. It will also be valuable to investigate the NAS/melatonin ratio in different cell types and tissues. There is some suggestion that the NAS/melatonin ratio may be of particular importance in bipolar disorder (Anderson et al., 2016), suggesting that it could have relevance to MDD and to depression-associated conditions, such as AD, PD, and MS. For example, low folate levels are not uncommon in depression (Petridou et al., 2015), with folate depletion resulting in an increase in the levels of homocysteine, which acts as a marker for decreased S-adenosylmethionine. This may have consequences for melatonin synthesis from NAS, given that S-adenosylmethionine is necessary as a methyl source for melatonin formation from NAS (Axelrod and Weissbach, 1960). As such, decreased folate or increased homocysteine, as evident in MDD (Nabi et al., 2013) and dementia (Miwa et al., 2015), are likely to occur in association with decreased melatonin and a relative increase in the NAS/melatonin ratio. Given the wide-ranging effects of melatonin in a host of different cells and tissues, determination of the NAS/melatonin ratio could be of some importance, with different ratios of NAS/melatonin likely to have differential physiological effects (Álvarez-Diduk et al., 2015).

Melatonin: Role in Pharmaceutical and Nutritional Interventions

In epidemiological studies and physiological experiments, a plethora of pharmaceutical and nutritional factors have been shown to modulate the etiology and course of AD, PD, MS, and MDD, including green tea's epigallocatechin-3-gallate (EGCG) and taurine, as well as MAO inhibitors. In the following sections, we look at some of these factors, suggesting that their impacts on the melatonergic pathways underlie their nonspecific efficacy across a host of medical conditions, including AD, PD, MS, and MDD.

Pharmaceuticals

Serotonin, Tryptophan, and SSRIs

Relatively raised levels of serotonin correlate with lower levels of amyloid-β plaques in AD patients as well as in AD models (Cirrito et al., 2011). This is parsimonious with the efficacy of dietary tryptophan supplementation as the necessary precursor for serotonin synthesis, which also lowers amyloid-beta in rodent AD models (Noristani et al., 2012). Central serotonin receptor levels are decreased in AD patients (Arai et al., 1984; Reynolds et al., 1995), in PD and its models (Deusser et al., 2015), including after L-DOPA treatment (Stansley and Yamamoto, 2015), and in MS, with decreased serotonin classically associated with MDD.

Of note in AD, serotonin or serotonin receptor agonists drive intracellular signaling cascades, which increase soluble amyloid precursor protein (APP)-alpha levels (Roberts et al., 2001) and which parallel melatonin effects (Shukla et al., 2015). Cirrito and colleagues showed that a single SSRI administration to AD murine model mice decreased amyloid-beta levels in the brain interstitial fluid by 25% (Cirrito et al., 2011). This was achieved without any impact on the amyloid-beta clearance rate, suggesting inhibitory impacts on levels of amyloid-beta synthesis.

Serotonergic regulation is therefore a clinically important treatment target in neurodegenerative as well as psychiatric disorders, with serotonin levels being generally decreased in AD, MS (Foley et al., 2014), PD (Bomasang-Layno et al., 2015), and other psychiatric conditions (Anderson et al., 2013). Given the high frequency of an MDD diagnosis in AD, MS, and other psychiatric conditions, SSRIs have been extensively prescribed to these diverse patient groups (Antonsdottir et al., 2015). It still awaits clarification as to whether SSRIs and other serotonin regulators modulate the course of AD, PD, MS, and other CNS disorders. Importantly, it remains to be determined as to whether the putative wider effects of pharmaceutically regulating serotonin is mediated by increasing its availability for NAS and/or melatonin production by glia and immune cells, as well as in the gut and at other sites.

Monoamine Oxidase Inhibitors

Another group of antidepressants, the MAO inhibitors (MAOIs), have also been widely prescribed across a host of medical conditions. MAOIs inhibit the breakdown of monoamines such as serotonin and NE. Consequently, available serotonin levels are increased leading to an enhanced synthesis and efflux of NAS and melatonin, including in astrocytes (Liu et al., 2012).

The two MAO forms, MAO-A and MAO-B, can have differential impacts on neurodegenerative processes (Naoi et al., 2012). MAO-A directly interacts with the AD susceptibility gene presenilin-1 (PSEN1) in the cellular mitochondrial fraction (Wei et al., 2012). It will be interesting to determine any impact on mitochondria melatonin production by MAO-A. In the AD pineal gland, an MAO-A-promoter polymorphism with a variable number of tandem repeats lowers pineal NAS and melatonin production in AD (Wu et al., 2007), while MAO-A inhibition significantly enhances the activation of the pineal melatonergic pathways (Oxenkrug et al., 2007). MAOIs increase levels of ADAM10 in AD platelets (Bianco et al., 2015), which, given that melatonin also increases ADAM10 (Shukla et al., 2015), may suggest that some of the efficacy of MAOIs is mediated by increasing levels of melatonin synthesis.

MAOIs also have beneficial effects on symptomatology in the EAE model of MS (Musgrave et al., 2011). MAOIs have long proven their efficacy in the management of PD. Glucocorticoids raise levels of astrocyte MAO-B (Carlo et al., 1996), although when acutely administered decrease MAO-A levels (Soliman et al., 2012). This seems to reflect the often opposing effects of acute versus chronic cortisol/stress (Ou et al., 2006). This could suggest that the often beneficial effects of acute versus chronic stress may be mediated by increased monoamines, including serotonin, in turn enhancing local NAS and melatonin synthesis and its effects (Anderson and Maes, 2014c). As such, MAO levels are likely to significantly interact with stress processing, including in the etiology and course of neurodegenerative and psychiatric disorders. As a consequence, it is likely that alterations in levels of local melatonergic pathway activity will be modulated by this.

Acetylcholinesterase Inhibitors

Acetylcholinesterase inhibitors (AChIs) are widely utilized in the management of AD, with beneficial effects that involve the activation of the α7nAChr. As such, given the interactions of the α7nAChr and melatonin (Markus et al., 2010), it is not unlikely that variations in pineal and local melatonin synthesis will interact with the effects of AChI. ACh and the α7nAChr are also significant treatment targets for a wide array of CNS disorders, including MS (Li Y et al., 2015), where AChI benefits are at least partly mediated via vagal nerve regulation (Garcia-Oscos et al., 2015). It is not unlikely that AChI also increases the ACh released by T cells and which acts to modulate the gut microbiome–brain axis (Forsythe et al., 2014). Interestingly, subthreshold concentrations of melatonin and an AChI (galantamine) improve AD-like pathology in hippocampal cultures (Buendia et al., in press), perhaps indicative of an interaction of AChI efficacy in AD with levels of local and/or pineal melatonin. Overall, AChI seem likely to interact with melatonin levels in their regulation of α7nAChr levels and activity.

Nutritional Factors

Omega-3 Polyunsaturated Fatty Acids

Omega-3 polyunsaturated fatty acids (omega-3 PUFAs) produce a lower inflammatory cascade versus omega-6 PUFAs, including within the human brain (Chang et al., 2009). Omega-3 PUFAs slow the cognitive decline in MCI and AD patients, with effects being positively correlated with plasma omega-3 levels (Eriksdotter et al., 2015) and more prominent in ApoE4-negative AD patients (Daiello et al., 2015). Omega-3 levels also correlate with brain volume maintenance (Daiello et al., 2015). In microglia, omega-3 PUFAs decrease TNFα levels, while enhancing the phagocytosis of amyloid-β, with this being driven by a higher level of microglia of an M2-like phenotype. Interestingly, this parallels the effects of autocrine melatonin on macrophage activation (Muxel et al., 2012), perhaps indicative of an omega-3

PUFA impact on local melatonin synthesis. In support of this, omega-3 PUFAs lower MAO levels in the rodent prefrontal cortex, suggesting that they act to increasing serotonin availability for NAS and melatonin synthesis (Naveen et al., 2013), while also suggesting that MAO regulation may be one route by which omega-3 PUFAs can counteract chronic cortisol/stress effects (Anderson and Maes, 2014e; Ou et al., 2006; Naveen et al., 2013). If so, omega-3 PUFAs would then act to inhibit the synergistic interactions of chronic stress with the ApoE4 allele in the regulation of cognitive decline in AD, PD, and MS (Anderson and Rodriguez, 2015c; Sheffler et al., 2014). There is indirect support for this in hamsters, where an omega-3 PUFA-deficient diet lowers pineal melatonin production (Lavialle et al., 2008). Fish oil supplementation is effective in reducing the levels of cytokines and O&NS in relapsing-remitting MS patients (Ramirez-Ramirez et al., 2013), as well as having clinical utility in PD (Bousquet et al., 2011) and in MDD, as shown in a recent meta-analysis (Grosso et al., 2014).

Overall, omega-3 PUFAs' efficacy across a range of central disorders may involve the upregulation of the melatonergic pathways.

Curcumin

Curcumin has efficacy across a host of medical conditions, especially in AD models (Yin et al., 2014), with benefits mediated, at least in part, by decreasing O&NS and immunoinflammatory activity (Sun et al., 2016). Curcumin may also have some antimicrobial effects (Naz et al., 2016), which may be relevant to enhanced innate immune activation and levels of circulating LPS in MDD and neurodegenerative conditions. However, much of this research has been carried out in rodents with curcumin administration being intravenous. This is due to the breakdown of curcumin in the digestive tract when ingested orally. Such intravenous injection is not possible in humans, although nasal inhalation may provide utility, especially since curcumin readily crosses the BBB. This is relevant to AD, as curcumin can directly interact with amyloid-beta (McClure et al., 2015). *In vitro*, curcumin, like melatonin, enhances sirtuin-1 (Sun et al., 2014) and the endogenous antioxidant transcription factor Nrf2 (Trujillo et al., 2014), suggesting that it may have anti-ageing as well as mitochondria-regulating effects (Trujillo et al., 2014). Curcumin also significantly modulates a number of immune cells, including regulatory T cells, macrophages, and NK cells (Jagetia and Aggarwal, 2007), as well as decreasing gut permeability and gut inflammation, with the latter being driven by its regulation of the serotoninergic system (Yu et al., 2015). Curcumin therefore has very similar effects to those of melatonin, including the following: gut regulation, possibly via increased gut melatonin synthesis; increased Nrf2 and therefore levels of endogenous antioxidants (Trujillo et al., 2014; Xie et al., 2014); immune regulation and anti-inflammatory effects (Jagetia and Aggarwal, 2007), including increased cytotoxicity of NK cells (Zhang et al., 2007); optimized mitochondrial functioning and oxidative phosphorylation (Trujillo et al., 2014; Jat et al., 2013; Anderson and Maes, 2104e); decreased MAO and peroxide levels, including in astrocytes (Mazzio et al., 1998), which is likely to enhance astrocyte melatonin synthesis (Liu et al., 2012); and lowered effects of chronic stress (Liu D et al., 2014). As with melatonin, curcumin also has antidepressant effects (Lopresti et al., 2014; Anderson and Maes, 2014e), and, given its enhancement of sirtuin-1 (Sun et al., 2014), will similarly have longevity, circadian, and mitochondria-optimizing effects (Trujillo et al., 2014; Zhou et al., 2014; Chang and Guarente, 2013). Not unimportantly, curcumin, like melatonin, also enhances phagocytosis (Mimche et al., 2012; Antoine and Girard, 2014). Curcumin also lowers astrocyte (Seyedzadeh et al., 2014; Anderson and Maes, 2014e) and microglia (Shi et al., 2015; Yang et al., 2014) reactivity, including via the induction of heme oxygenase-1 (HO-1) (Parada et al., 2014), which again parallels the HO-1 induction by melatonin, which is relevant to many of its effects, including viral regulation (Anderson et al., 2015b). It is currently under investigation as to whether curcumin modulates glia and immune cell melatonergic pathways. As to the mechanism(s) underlying such parallels between curcumin and melatonin, it requires investigation as to whether there is a role for curcumin-induced Nrf2 driving an increase in the transcription factor YY1 (René et al., 2010), given that YY1 enhances the transcription of the melatonergic pathways, as evidenced in the retina (Bernard and Voisin, 2008). This will be important to investigate, given that curcumin also has beneficial effects in PD (Cole et al., 2007) and MS (Xie et al., 2011), as well as in AD and MDD (Tizabi et al., 2014).

Green Tea

Green tea consumption lowers the risk of AD, with effects that are, at least in part, determined by the anti-inflammatory effects of green tea's flavonoids, perhaps especially EGCG. EGCG, like melatonin, has beneficial effects across a wide array of medical conditions, including AD (Chesser et al., 2016), PD (Caruana et al., 2015), and MS (Mähler et al., 2015). EGCG also parallels melatonin by having antioxidant (Chen et al., 2012) and anti-inflammatory (Rubio-Perez et al., 2013) effects, as well as optimizing mitochondrial functioning (Schroeder et al., 2009). Interestingly, EGCG, like melatonin, accumulates in mitochondria (Schroeder et al., 2009), where it can also mediate effects via the α7nAChr (Markus et al., 2010; Zhang et al., 2014a). As many of melatonin's effects may be via its induction of the α7nAChr and its interactions with the α7nAChr (Anderson and Maes, 2014e), this could suggest that EGCG modulates the melatonin–α7nAChr interactions. Other EGCG parallels with melatonin include the enhancement of immune cell phagocytic activity (Huang et al., 2013), the lowering of gut permeability (Watson et al., 2004), and a decrease in glia reactivity levels (Álvarez-Pérez et al., 2016; Cai et al., 2014).

Overall, it seems likely that at least of some of EGCG's effects are mediated via its regulation of the melatonergic pathways, which the EGCG inhibition of MAO (Lin et al., 2010) would further suggest.

Coffee and Caffeine

Coffee affords protection against the development of AD (Santos et al., 2010) and PD, in part by its inhibition of adenosine receptors and associated increases in cortical arousal. Some benefits may also be mediated by coffee's regulation of gut microbiota (Scheperjans et al., 2015). Data indicates that coffee may also slow MS progression (D'hooghe et al., 2012), although not all data support such MS-offsetting effects (Massa et al., 2013). However, by keeping cytochrome P450 (CYP)1A2 occupied, coffee lowers human melatonin metabolism (Ursing et al., 2003). By decreasing MAO (Herraiz and Chaparro, 2006), coffee also raises serotonin availability for the synthesis of NAS and melatonin, in part by caffeine's inhibition of the adenosine A2Ar-induced MAO (Raimondi et al., 2000).

In vitro, coffee and caffeine have additive effects with melatonin, including in the prevention of amyloid-beta oligomerization, as well as in lowering levels of tau hyperphosphorylation. Such effects are partly mediated by increasing the PI3K/Akt/pGSK-3b/Nrf2 pathway, thereby increasing endogenous antioxidants and HO-1 (Zhang et al., 2014b). All of these effects mimic those of melatonin. Interestingly, *in vitro* neuronal studies show that coffee/caffeine addition to cells for 12 hours followed by melatonin for 12 hours produces a stronger anti-amyloidogenic effect than either alone, suggesting that the clear demarcation of activity and rest periods, often lost in neurodegenerative conditions, may be an integral part of neuronal aging. It would also be expected in such conditions that the coffee/caffeine induction of CYP1A2 would enhance the availability of, and protection afforded by, melatonin. Coffee also maintains tryptophan levels by decreasing the driving of tryptophan to TRYCATs production, thereby increasing serotonin availability for NAS and melatonin synthesis (Gostner et al., 2015). This requires investigation in different cell types, including glia and immune cells. Coffee also mimics other melatonin effects, including a more beneficial Th1 response, which may inhibit the immune senescence that is associated with age-linked neurodegenerative conditions (Nosáľová et al., 2011); inhibited macrophage inflammatory responses (Kim et al., 2004), as with melatonin (Muxel et al., 2012); and the positive regulation of gut microbiota and a lowering of gut permeability (Nakayama and Oishi, 2013).

Overall, coffee, as with many nutritional factors with proven benefits across a host of diverse medical conditions, may have their efficacy mediated by their regulation of the melatonergic pathways. Other factors with beneficial effects across a host of medical conditions, including taurine, selenium, and walnuts, may also be mediating their effects, in part, via melatonergic pathway regulation.

Conclusions

There is a substantial overlap of key biological underpinnings across a wide array of medical presentations, as exemplified here with AD, PD, MS, and MDD. The specificity of such conditions is likely to be driven by genetic and epigenetic processes that act over a developmental timescale. However, such specificity across these conditions arises in interactions with key processes that may act as information hubs underpinning specific symptomatology. An important aspect of this would seem to be how various medical disorder–specific factors regulate the melatonergic pathways in different cells and tissues, with beneficial dietary and pharmaceutical compounds acting to negate these disorder-specific changes via homeostatic processes driven by the melatonergic pathways. As a consequence, many pharmaceuticals and nutraceuticals may be mediating their effects on the melatonergic pathways, thereby explaining their influence on the etiology, course, and management of a diverse array of medical presentations.

REFERENCES

Alcocer-Gómez E, Ulecia-Morón C, Marín-Aguilar F, Rybkina T, et al. Stress-induced depressive behaviors require a functional NLRP3 inflammasome. *Mol Neurobiol* 2016;53(7):4874–82.

Alfonso-Loeches S, Ureña-Peralta JR, Morillo-Bargues MJ, Oliver-De La Cruz J, et al. Role of mitochondria ROS generation in ethanol-induced NLRP3 inflammasome activation and cell death in astroglial cells. *Front Cell Neurosci* 2014;8:216.

Álvarez-Diduk R, Galano A, Tan DX, Reiter RJ. N-Acetylserotonin and 6-hydroxymelatonin against oxidative stress: Implications for the overall protection exerted by melatonin. *J Phys Chem B* 2015;119(27):8535–43.

Álvarez-Pérez B, Homs J, Bosch-Mola M, Puig T, et al. Epigallocatechin-3-gallate treatment reduces thermal hyperalgesia after spinal cord injury by down-regulating RhoA expression in mice. *Eur J Pain* 2016;20(3):341–52.

Álvarez-Sánchez N, Cruz-Chamorro I, López-González A, Utrilla JC, et al. Melatonin controls experimental autoimmune encephalomyelitis by altering the T effector/regulatory balance. *Brain Behav Immun* 2015;50:101–14.

Anderson G. Neuronal-immune interactions in mediating stress effects in the etiology and course of schizophrenia: Role of the amygdala in developmental co-ordination. *Med Hypotheses* 2011;76:54–60.

Anderson G, Ojala J. Alzheimer's and seizures: Interleukin-18, indoleamine 2,3-dioxygenase and quinolinic acid. *Int J Tryptophan Res* 2010;3:169–73.

Anderson G, Jacob A, Bellivier F, Geoffroy PA. Bipolar disorder: The role of the kynurenine and melatonergic pathways. *Curr Pharm Des* 2016;22(8):987–1012.

Anderson G, Maes M. Neurodegeneration in Parkinson's disease: Interactions of oxidative stress, tryptophan catabolites and depression with mitochondria and sirtuins. *Mol Neurobiol* 2014a;49(2):771–83.

Anderson G, Maes M. Reconceptualizing adult neurogenesis: Role for sphingosine-1-phosphate and fibroblast growth factor-1 in co-ordinating astrocyte-neuronal precursor interactions. *CNS Neurol Disord Drug Targets* 2014b;13(1):126–36.

Anderson G, Maes M. Local melatonin regulates inflammation resolution: A common factor in neurodegenerative, psychiatric and systemic inflammatory disorders. *CNS Neurol Disord Drug Targets* 2014c;13(5):817–27.

Anderson G, Maes M. TRYCAT pathways link peripheral inflammation, nicotine, somatization and depression in the etiology and course of Parkinson's disease. *CNS Neurol Disord Drug Targets* 2014d;13(1):137–49.

Anderson G, Maes M. Oxidative/nitrosative stress and immuno-inflammatory pathways in depression: Treatment implications. *Curr Pharm Des* 2014e;20(23):3812–47.

Anderson G, Maes M. The gut–brain axis: The role of melatonin in linking psychiatric, inflammatory and neurodegenerative conditions. *Adv Integrative Med* 2015a;2(1):31–7.

Anderson G, Maes M, Berk M. Schizophrenia is primed for an increased expression of depression through activation of immuno-inflammatory, oxidative and nitrosative stress, and tryptophan catabolite pathways. *Prog Neuropsychopharmacol Biol Psychiatry* 2013;42:101–14.

Anderson G, Markus RP, Rodriguez M, Maes. The Ebola virus: Melatonin as a readily available treatment option. *J Med Virol* 2015b;87;537–43.

Anderson G, Rodriguez M. Multiple sclerosis, seizures, and antiepileptics: Role of IL-18, IDO, and melatonin. *Eur J Neurol* 2011;18(5):680–5.

Anderson G, Rodriguez M. Multiple sclerosis: The role of melatonin and N-acetylserotonin. *Mult Scler Relat Disord* 2015c;4(2):112–23.

Antoine F, Girard D. Curcumin increases gelatinase activity in human neutrophils by a p38 mitogen-activated protein kinase (MAPK)-independent mechanism. *J Immunotoxicol* 2014;13:1–6.

Antonsdottir IM, Smith J, Keltz M, Porsteinsson AP. Advancements in the treatment of agitation in Alzheimer's disease. *Expert Opin Pharmacother* 2015;16(11):1649–56.

Arai H, Kosaka K, Iizuka R. Changes of biogenic amines and their metabolites in postmortem brains from patients with Alzheimer-type dementia. *J Neurochem* 1984;43:388.

Axelrod J, Weissbach H. Enzymatic O-methylation of N-acetylserotonin to melatonin. *Science* 1960;131(3409):1312.

Barreto GE, Iarkov A, Moran VE. Beneficial effects of nicotine, cotinine and its metabolites as potential agents for Parkinson's disease. *Front Aging Neurosci* 2015;6:340.

Beischlag TV, Anderson G, Mazzoccoli G. Glioma: Tryptophan catabolite and melatoninergic pathways link microRNA, 14-3-3, chromosome 4q35, epigenetic processes and other glioma biochemical changes. *Curr Pharm Des* 2016;22(8):1033–48.

Bencsik A, Muselli L, Leboidre M, Lakhdar L, et al. Early and persistent expression of phosphorylated α-synuclein in the enteric nervous system of A53T mutant human α-synuclein transgenic mice. *J Neuropathol Exp Neurol* 2014;73(12):1144–51.

Bernard M, Voisin P. Photoreceptor-specific expression, light-dependent localization, and transcriptional targets of the zinc-finger protein Yin Yang 1 in the chicken retina. *J Neurochem* 2008;105(3):595–604.

Bianco OA, Manzine PR, Nascimento CM, Vale FA, et al. Serotoninergic antidepressants positively affect platelet ADAM10 expression in patients with Alzheimer's disease. *Int Psychogeriatr* 2015;11:1–6.

Bomasang-Layno E, Fadlon I, Murray AN, Himelhoch S. Antidepressive treatments for Parkinson's disease: A systematic review and meta-analysis. *Parkinsonism Relat Disord* 2015;21(8):833–42.

Bousquet M, Calon F, Cicchetti F. Impact of ω-3 fatty acids in Parkinson's disease. *Ageing Res Rev* 2011;10(4):453–63.

Bruno VA, Scacchi PA, Perez-Lloret S, Esquifino AI, et al. Melatonin treatment counteracts the hyperthermic effect of lipopolysaccharide injection in the Syrian hamster. *Neurosci Lett* 2005;389(3):169–72.

Buendia I, Parada E, Navarro E, León R, et al. Subthreshold concentrations of melatonin and galantamine improves pathological AD-hallmarks in hippocampal organotypic cultures. *Mol Neurobiol*, in press.

Cai J, Jing D, Shi M, Liu Y, et al. Epigallocatechin gallate (EGCG) attenuates infrasound-induced neuronal impairment by inhibiting microglia-mediated inflammation. *J Nutr Biochem* 2014;25(7):716–25.

Carlo P, Violani E, Del Rio M, Olasmaa M, et al. Monoamine oxidase B expression is selectively regulated by dexamethasone in cultured rat astrocytes. *Brain Res* 1996;711(1–2):175–83.

Caruana M, Vassallo N. Tea polyphenols in Parkinson's disease. *Adv Exp Med Biol* 2015;863:117–37.

Cassano T, Pace L, Bedse G, Lavecchia AM, et al. Glutamate and mitochondria: Two prominent players in the oxidative stress-induced neurodegeneration. *Curr Alzheimer Res* 2016;13(2):185–97.

Chang HC, Guarente L. SIRT1 mediates central circadian control in the SCN by a mechanism that decays with aging. *Cell* 2013;153(7):1448–60.

Chang JP, Chen YT, Su KP. Omega-3 polyunsaturated fatty acids (n-3 PUFAs) in cardiovascular diseases (CVDs) and depression: The missing link? *Cardiovasc Psychiatry Neurol* 2009;2009:725310.

Chen F, Jiang L, Shen C, Wan H, et al. Neuroprotective effect of epigallocatechin-3-gallate against N-methyl-D-aspartate-induced excitotoxicity in the adult rat retina. *Acta Ophthalmol* 2012;90(8):e609–15.

Chen HY, Chen TY, Lee MY, Chen ST, et al. Melatonin attenuates the postischemic increase in blood–brain barrier permeability and decreases hemorrhagic transformation of tissue-plasminogen activator therapy following ischemic stroke in mice. *J Pineal Res* 2006;40(3):242–50.

Chesser AS, Ganeshan V, Yang J, Johnson GV. Epigallocatechin-3-gallate enhances clearance of phosphorylated tau in primary neurons. *Nutr Neurosci* 2016;19(1):21–31.

Cirrito JR, Disabato BM, Restivo JL, Verges DK, et al. Serotonin signaling is associated with lower amyloid-β levels and plaques in transgenic mice and humans. *Proc Natl Acad Sci USA* 20116;108(36):14968–73.

Claassen JA. New cardiovascular targets to prevent late onset Alzheimer disease. *Eur J Pharmacol* 2015;763(Pt A):131–4.

Codolo G, Plotegher N, Pozzobon T, Brucale M, et al. Triggering of inflammasome by aggregated α-synuclein, an inflammatory response in synucleinopathies. *PLoS One* 2013;8(1):e55375.

Cole GM, Teter B, Frautschy SA. Neuroprotective effects of curcumin. *Adv Exp Med Biol* 2007;595:197–212.

Daiello LA, Gongvatana A, Dunsiger S, Cohen RA, et al. Alzheimer's disease neuroimaging initiative: Association of fish oil supplement use with preservation of brain volume and cognitive function. *Alzheimers Dement* 2015;11(2):226–35.

Damasceno A, Moraes AS, Farias A, Damasceno BP, et al. Disruption of melatonin circadian rhythm production is related to multiple sclerosis severity: A preliminary study. *J Neurol Sci* 201515;353(1–2):166–8.

da Silveira Cruz-Machado S, Pinato L, Tamura EK, Carvalho-Sousa CE, et al. Glia-pinealocyte network: The paracrine modulation of melatonin synthesis by tumor necrosis factor (TNF). *PLoS One* 2012;7(7):e40142.

Deming SL, Lu W, Beeghly-Fadiel A, Zheng Y, et al. Melatonin pathway genes and breast cancer risk among Chinese women. *Breast Cancer Res Treat* 2012;132(2):693–9.

Deusser J, Schmidt S, Ettle B, Plötz S, et al. Serotonergic dysfunction in the A53T alpha-synuclein mouse model of Parkinson's disease. *J Neurochem* 2015;135(3):589–97.

D'hooghe MB, Haentjens P, Nagels G, De Keyser J. Alcohol, coffee, fish, smoking and disease progression in multiple sclerosis. *Eur J Neurol* 2012;19(4):616–24.

Ding K, Wang H, Xu J, Lu X, et al. Melatonin reduced microglial activation and alleviated neuroinflammation induced neuron degeneration in experimental traumatic brain injury: Possible involvement of mTOR pathway. *Neurochem Int* 2014;76:23–31.

Dudvarski SN, Teodorczyk M, Ploen R, Zipp F, et al. Microglia-blood vessel interactions: A double-edged sword in brain pathologies. *Acta Neuropathol* 2016;131(3):347–63.

Elbaz A, Carcaillon L, Kab S, Moisan F. Epidemiology of Parkinson's disease. *Rev Neurol* (Paris) 2016;172(1):14–26.

Emamgholipour S, Hossein-Nezhad A, Sahraian MA, Askarisadr F, et al. Evidence for possible role of melatonin in reducing oxidative stress in multiple sclerosis through its effect on SIRT1 and antioxidant enzymes. *Life Sci* 2016;145:34–41.

Eriksdotter M, Vedin I, Falahati F, Freund-Levi Y, et al. Plasma fatty acid profiles in relation to cognition and gender in Alzheimer's disease patients during oral omega-3 fatty acid supplementation: The omegAD study. *J Alzheimers Dis* 2015;48(3):805–12.

Etain B, Dumaine A, Bellivier F, Pagan C, et al. Genetic and functional abnormalities of the melatonin biosynthesis pathway in patients with bipolar disorder. *Hum Mol Genet* 201215;21(18):4030–7.

Faldu KG, Shah JS, Patel SS. Anti-viral agents in neurodegenerative disorders: New paradigm for targeting Alzheimer's disease. *Recent Pat Antiinfect Drug Discov* 2015;10(2):76–83.

Farez MF, Mascanfroni ID, Méndez-Huergo SP, Yeste A, et al. Melatonin contributes to the seasonality of multiple sclerosis relapses. *Cell* 2015;162(6):1338–52.

Fava M, Targum SD, Nierenberg AA, Bleicher LS, et al. An exploratory study of combination buspirone and melatonin SR in major depressive disorder (MDD): A possible role for neurogenesis in drug discovery. *J Psychiatr Res* 2012;46(12):1553–63.

Foley P, Lawler A, Chandran S, Mead G. Potential disease-modifying effects of selective serotonin reuptake inhibitors in multiple sclerosis: Systematic review and meta-analysis. *J Neurol Neurosurg Psychiatry* 2014;85(6):709–10.

Forsythe P, Bienenstock J, Kunze WA. Vagal pathways for microbiome–brain–gut axis communication. *Adv Exp Med Biol* 2014;817:115–33.

Galano A, Medina ME, Tan DX, Reiter RJ. Melatonin and its metabolites as copper chelating agents and their role in inhibiting oxidative stress: A physicochemical analysis. *J Pineal Res* 2015;58(1):107–16.

Gałecka E, Szemraj J, Florkowski A, Gałecki P, et al. Single nucleotide polymorphisms and mRNA expression for melatonin MT(2) receptor in depression. *Psychiatry Res* 2011;189(3):472–4.

Gałecki P, Szemraj J, Bartosz G, Bieńkiewicz M, et al. Single-nucleotide polymorphisms and mRNA expression for melatonin synthesis rate-limiting enzyme in recurrent depressive disorder. *J Pineal Res* 2010;48(4):311–7.

Gałecki P, Talarowska M, Anderson G, Berk M, et al. Mechanisms underlying neurocognitive dysfunctions in recurrent major depression. *Med Sci Monit* 2015;21:1535–47.

Garcia-Marin R, Fernandez-Santos JM, Morillo-Bernal J, Gordillo-Martinez F, et al. Melatonin in the thyroid gland: Regulation by thyroid-stimulating hormone and role in thyroglobulin gene expression. *J Physiol Pharmacol* 2015;66(5):643–52.

Garcia-Oscos F, Peña D, Housini M, Cheng D, et al. Vagal nerve stimulation blocks interleukin 6-dependent synaptic hyperexcitability induced by lipopolysaccharide-induced acute stress in the rodent prefrontal cortex. *Brain Behav Immun* 2015;43:149–58.

Gholipour T, Ghazizadeh T, Babapour S, Mansouri B, et al. Decreased urinary level of melatonin as a marker of disease severity in patients with multiple sclerosis. *Iran J Allergy Asthma Immunol* 2015;14(1):91–7.

Gostner JM, Schroecksnadel S, Jenny M, Klein A, et al. Coffee extracts suppress tryptophan breakdown in mitogen-stimulated peripheral blood mononuclear cells. *J Am Coll Nutr* 2015;34(3):212–23.

Grosso G, Pajak A, Marventano S, Castellano S, et al. Role of omega-3 fatty acids in the treatment of depressive disorders: A comprehensive meta-analysis of randomized clinical trials. *PLoS One* 2014;9(5):e96905.

Gupta S, Haldar C. Photoperiodic modulation of local melatonin synthesis and its role in regulation of thymic homeostasis in *Funambulus pennanti*. *Gen Comp Endocrinol* 2015, in press.

Haghighi F, Ge Y, Chen S, Xin Y, et al. Neuronal DNA methylation profiling of blast-related traumatic brain injury. *J Neurotrauma* 2015;32(16):1200–9.

Hammer C, Zerche M, Schneider A, Begemann M, et al. Apolipoprotein E4 carrier status plus circulating anti-NMDAR1 autoantibodies: Association with schizoaffective disorder. *Mol Psychiatry* 2014;19(10):1054–6.

Hardeland R, Cardinali DP, Srinivasan V, Spence DW, et al. Melatonin: A pleiotropic, orchestrating regulator molecule. *Prog Neurobiol* 2011;93(3):350–84.

Herraiz T, Chaparro C. Human monoamine oxidase enzyme inhibition by coffee and beta-carbolines norharman and harman isolated from coffee. *Life Sci* 2006;78(8):795–802.

Herrero MT, Estrada C, Maatouk L, Vyas S. Inflammation in Parkinson's disease: Role of glucocorticoids. *Front Neuroanat* 2015;9:32.

Holmqvist S, Chutna O, Bousset L, Aldrin-Kirk P, et al. Direct evidence of Parkinson pathology spread from the gastrointestinal tract to the brain in rats. *Acta Neuropathol* 2014;128(6):805–20.

Huang AC, Cheng HY, Lin TS, Chen WH, et al. Epigallocatechin gallate (EGCG), influences a murine WEHI-3 leukemia model *in vivo* through enhancing phagocytosis of macrophages and populations of T- and B-cells. *In Vivo* 2013;27(5):627–34.

Hung SY, Huang WP, Liou HC, Fu WM. LC3 overexpression reduces Aβ neurotoxicity through increasing α7nAchR expression and autophagic activity in neurons and mice. *Neuropharmacology* 2015;93:243–51.

Jagetia GC, Aggarwal BB. "Spicing up" of the immune system by curcumin. *J Clin Immunol* 2007;27(1):19–35.

Jang SW, Liu X, Pradoldej S, Tosini G, et al. N-acetylserotonin activates TrkB receptor in a circadian rhythm. *Proc Natl Acad Sci USA* 2010;107(8):3876–81.

Jat D, Parihar P, Kothari SC, Parihar MS. Curcumin reduces oxidative damage by increasing reduced glutathione and preventing membrane permeability transition in isolated brain mitochondria. *Cell Mol Biol (Noisy-le-grand)* 2013;59 Suppl:OL1899–905.

Jeong JK, Park SY. Melatonin regulates the autophagic flux via activation of alpha-7 nicotinic acetylcholine receptors. *J Pineal Res* 2015;59(1):24–37.

Jo C, Gundemir S, Pritchard S, Jin YN, et al. Nrf2 reduces levels of phosphorylated tau protein by inducing autophagy adaptor protein NDP52. *Nat Commun* 2014;5:3496.

Jumnongprakhon P, Govitrapong P, Tocharus C, Pinkaew D, et al. Melatonin protects methamphetamine-induced neuroinflammation through NF-κB and Nrf2 pathways in glioma cell line. *Neurochem Res* 2015;40(7):1448–56.

Kaminski-Hartenthaler A, Nussbaumer B, Forneris CA, Morgan LC, et al. Melatonin and agomelatine for preventing seasonal affective disorder. *Cochrane Database Syst Rev* 2015;11:CD011271.

Kang JW, Koh EJ, Lee SM. Melatonin protects liver against ischemia and reperfusion injury through inhibition of toll-like receptor signaling pathway. *J Pineal Res* 2011;50(4):403–11.

Kantarcioglu AS, Kiraz N, Aydin A. Microbiota–gut–brain axis: Yeast species isolated from stool samples of children with suspected or diagnosed autism spectrum disorders and *in vitro* susceptibility against nystatin and fluconazole. *Mycopathologia* 2016;181(1–2):1–7.

Kardani J, Roy I. Understanding caffeine's role in attenuating the toxicity of α-synuclein aggregates: Implications for risk of Parkinson's Disease. *ACS Chem Neurosci* 2015;16;6(9):1613–25.

Kerenyi NA, Sotonyi P, Somogyi E. Localizing acetylserotonin transferase by electron microscopy. *Histochemistry* 1975;46:77.

Kim JY, Jung KS, Lee KJ, Na HK, et al. The coffee diterpene kahweol suppress the inducible nitric oxide synthase expression in macrophages. *Cancer Lett* 2004;213(2):147–54.

Klingelhoefer L, Reichmann H. Pathogenesis of Parkinson disease: The gut–brain axis and environmental factors. *Nat Rev Neurol* 2015;11(11):625–36.

Kong PJ, Byun JS, Lim SY, Lee JJ, et al. Melatonin induces Akt phosphorylation through melatonin receptor- and PI3K-dependent pathways in primary astrocytes. *Korean J Physiol Pharmacol* 2008;12(2):37–41.

Kook SY, Seok Hong H, Moon M, et al. Disruption of blood–brain barrier in Alzheimer disease pathogenesis. *Tissue Barriers* 20131;1(2):e23993.

Kripke DF, Nievergelt CM, Tranah GJ, Murray SS, et al. Polymorphisms in melatonin synthesis pathways: Possible influences on depression. *J Circadian Rhythms* 2011;9:8.

Lavialle M, Champeil-Potokar G, Alessandri JM, Balasse L, et al. An (n-3) polyunsaturated fatty acid-deficient diet disturbs daily locomotor activity, melatonin rhythm, and striatal dopamine in Syrian hamsters. *J Nutr* 2008;138(9):1719–24.

Leblhuber F, Geisler S, Steiner K, Fuchs D, et al. Elevated fecal calprotectin in patients with Alzheimer's dementia indicates leaky gut. *J Neural Transm* 2015;122(9):1319–22.

Le PA, Bourgade K, Lamoureux J, Frost E, et al. NK Cells are activated in amnestic mild cognitive impairment but not in mild Alzheimer's disease patients. *J Alzheimers Dis* 2015;46(1):93–107.

Lee JS, Cua DJ. Melatonin lulling Th17 cells to sleep. *Cell* 2015;162(6):1212–4.

Leray E, Moreau T, Fromont A, Edan G. Epidemiology of multiple sclerosis. *Rev Neurol (Paris)* 2016;172(1):3–13.

Li Y, Hai S, Zhou Y, Dong BR. Cholinesterase inhibitors for rarer dementias associated with neurological conditions. *Cochrane Database Syst Rev* 2015;3:CD009444.

Lin SM, Wang SW, Ho SC, Tang YL. Protective effect of green tea (–)-epigallocatechin-3-gallate against the monoamine oxidase B enzyme activity increase in adult rat brains. *Nutrition* 2010;26(11–12):1195–200.

Liu D, Wang Z, Gao Z, Xie K, et al. Effects of curcumin on learning and memory deficits, BDNF, and ERK protein expression in rats exposed to chronic unpredictable stress. *Behav Brain Res* 2014;271:116–21.

Liu YJ, Meng FT, Wang LL, Zhang LF, et al. Apolipoprotein E influences melatonin biosynthesis by regulating NAT and MAOA expression in C6 cells. *Neuroscience* 2012;202:58–68.

Liu YJ, Meng FT, Wu L, Zhou JN. Serotoninergic and melatoninergic systems are expressed in mouse embryonic fibroblasts NIH3T3 cells. *Neuro Endocrinol Lett* 2013;34(3):236–40.

Liu YJ, Zhuang J, Zhu HY, Shen YX, et al. Cultured rat cortical astrocytes synthesize melatonin: Absence of a diurnal rhythm. *J Pineal Res* 2007;43(3):232–8.

López-González A, Álvarez-Sánchez N, Lardone PJ, Cruz-Chamorro I, et al. Melatonin treatment improves primary progressive multiple sclerosis: A case report. *J Pineal Res* 2015;58(2):173–7.

Lopresti AL, Maes M, Maker GL, Hood SD, et al. Curcumin for the treatment of major depression: A randomised, double-blind, placebo controlled study. *J Affect Disord* 2014;167:368–75.

Lucas K, Morris G, Anderson G, Maes M. The toll-like receptor radical cycle pathway: A new drug target in immune-related chronic fatigue. *CNS Neurol Disord Drug Targets* 2015;14(7):838–54.

Mackiewicz M, Nikonova EV, Zimmermann JE, Romer MA, et al. Age-related changes in adenosine metabolic enzymes in sleep/wake regulatory areas of the brain. *Neurobiol Aging* 2006;27(2):351–60.

Maes M, Anderson G. Overlapping the tryptophan catabolite (TRYCAT) and melatoninergic pathways in Alzheimer's disease. *Curr Pharm Des* 2016;22(8):1074–85.

Maes M, Kubera M, Obuchowiczwa E, Goehler L, et al. Depression's multiple comorbidities explained by (neuro)inflammatory and oxidative & nitrosative stress pathways. *Neuro Endocrinol Lett* 2011;32(1):7–24.

Mähler A, Steiniger J, Bock M, Klug L, et al. Metabolic response to epigallocatechin-3-gallate in relapsing-remitting multiple sclerosis: A randomized clinical trial. *Am J Clin Nutr* 2015;101(3):487–95.

Manchester LC, Poeggeler B, Alvares FL, Ogden GB, et al. Melatonin immunoreactivity in the photosynthetic prokaryote *Rhodospirillum rubrum*: Implications for an ancient antioxidant system. *Cell Mol Biol Res* 1995;41(5):391–95.

Maqbool M, Mobashir M, Hoda N. Pivotal role of glycogen synthase kinase-3: A therapeutic target for Alzheimer's disease. *Eur J Med Chem* 2016;107:63–81.

Marchesi VT. Gain-of-function somatic mutations contribute to inflammation and blood vessel damage that lead to Alzheimer dementia: A hypothesis. *FASEB J* 2016;30(2):503–06.

Markus RP, Cecon E, Pires-Lapa MA. Immune–pineal axis: Nuclear factor κB (NF-kB) mediates the shift in the melatonin source from pinealocytes to immune competent cells. *Int J Mol Sci* 2013;14(6):10979–97.

Markus RP, Silva CL, Franco DG, Barbosa EM Jr, et al. Is modulation of nicotinic acetylcholine receptors by melatonin relevant for therapy with cholinergic drugs? *Pharmacol Ther* 2010;126(3):251–62.

Massa J, O'Reilly EJ, Munger KL, Ascherio A. Caffeine and alcohol intakes have no association with risk of multiple sclerosis. *Mult Scler* 2013;19(1):53–8.

Mata IF, Leverenz JB, Weintraub D, Trojanowski JQ, et al. APOE, MAPT, and SNCA genes and cognitive performance in Parkinson disease. *JAMA Neurol* 2014;71(11):1405–12.

Mazzio EA, Harris N, Soliman KF. Food constituents attenuate monoamine oxidase activity and peroxide levels in C6 astrocyte cells. *Planta Med* 1998;64(7):603–06.

McClure R, Yanagisawa D, Stec D, Abdollahian D, et al. Inhalable curcumin: Offering the potential for translation to imaging and treatment of Alzheimer's disease. *J Alzheimers Dis* 2015;44(1):283–95.

Miller E, Walczak A, Majsterek I, Kędziora J. Melatonin reduces oxidative stress in the erythrocytes of multiple sclerosis patients with secondary progressive clinical course. *J Neuroimmunol* 201315;257(1–2):97–101.

Mimche PN, Thompson E, Taramelli D, Vivas L. Curcumin enhances non-opsonic phagocytosis of *Plasmodium falciparum* through up-regulation of CD36 surface expression on monocytes/macrophages. *J Antimicrob Chemother* 2012;67(8):1895–904.

Miwa K, Tanaka M, Okazaki S, Yagita Y, et al. Increased total homocysteine levels predict the risk of incident dementia independent of cerebral small-vessel diseases and vascular risk factors. *J Alzheimers Dis* 2015;49(2):503–13.

Miyake S, Kim S, Suda W, Oshima K, et al. Dysbiosis in the gut microbiota of patients with multiple sclerosis, with a striking depletion of species belonging to clostridia XIVa and IV clusters. *PLoS One* 2015;10(9):e0137429.

Morris G, Carvalho A, Anderson G, Galecki P, et al. The many neuroprogressive actions of tryptophan catabolites (TRYCATs) that may be associated with the pathophysiology of neuro-immune disorders. *Curr Pharm Des* 2016;22(8):963–77.

Mukherjee S, Maitra SK. Gut melatonin in vertebrates: Chronobiology and physiology. *Front Endocrinol (Lausanne)* 2015;6:112.

Musgrave T, Benson C, Wong G, Browne I, et al. The MAO inhibitor phenelzine improves functional outcomes in mice with experimental autoimmune encephalomyelitis (EAE). *Brain Behav Immun* 2011;25(8):1677–88.

Muxel SM, Pires-Lapa MA, Monteiro AW, Cecon E, et al. NF-κB drives the synthesis of melatonin in RAW 264.7 macrophages by inducing the transcription of the arylalkylamine-N-acetyltransferase (AA-NAT) gene. *PLoS One* 2012;7(12):e52010.

Nabi H, Bochud M, Glaus J, Lasserre AM, et al. Association of serum homocysteine with major depressive disorder: Results from a large population-based study. *Psychoneuroendocrinology* 2013;38(10):2309–18.

Nakayama T, Oishi K. Influence of coffee (Coffea arabica) and galacto-oligosaccharide consumption on intestinal microbiota and the host responses. *FEMS Microbiol Lett* 2013;343(2):161–68.

Naoi M, Maruyama W, Inaba-Hasegawa K. Type A and B monoamine oxidase in age-related neurodegenerative disorders: Their distinct roles in neuronal death and survival. *Curr Top Med Chem* 2012;12(20):2177–88.

Naskar A, Prabhakar V, Singh R, Dutta D, et al. Melatonin enhances L-DOPA therapeutic effects, helps to reduce its dose, and protects dopaminergic neurons in 1-methyl-4-phenyl-1,2,3,6-tetrahydropyridine-induced parkinsonism in mice. *J Pineal Res* 2015;58(3):262–74.

Natarajan R, Einarsdottir E, Riutta A, Hagman S, et al. Melatonin pathway genes are associated with progressive subtypes and disability status in multiple sclerosis among Finnish patients. *J Neuroimmunol* 2012;250(1–2):106–10.

Naveen S, Siddalingaswamy M, Singsit D, Khanum F. Anti-depressive effect of polyphenols and omega-3 fatty acid from pomegranate peel and flax seed in mice exposed to chronic mild stress. *Psychiatry Clin Neurosci* 2013;67(7):501–08.

Naz RK, Lough ML, Barthelmess EK. Curcumin: A novel non-steroidal contraceptive with antimicrobial properties. *Front Biosci (Elite Ed)* 2016;8:113–28.

Noristani HN, Verkhratsky A, Rodríguez JJ. High tryptophan diet reduces CA1 intraneuronal β-amyloid in the triple transgenic mouse model of Alzheimer's disease. *Aging Cell* 2012;11(5):810–22.

Nosáľová G, Prisenžňáková L, Paulovičová E, Capek P, et al. Antitussive and immunomodulating activities of instant coffee arabinogalactan-protein. *Int J Biol Macromol* 2011;49(4):493–97.

Nunes Oda S, Pereira Rde S. Regression of herpes viral infection symptoms using melatonin and SB-73: Comparison with Acyclovir. *J Pineal Res* 2008;44(4):373–78.

Obayashi K, Saeki K, Iwamoto J, Tone N, et al. Physiological levels of melatonin relate to cognitive function and depressive symptoms: The HEIJO-KYO cohort. *J Clin Endocrinol Metab* 2015;100(8):3090–96.

O'Connell MA, Hayes JD. The Keap1/Nrf2 pathway in health and disease: From the bench to the clinic. *Biochem Soc Trans* 2015;43(4):687–89.

Okuma Y, Liu K, Wake H, Zhang J, et al. Anti-high mobility group box-1 antibody therapy for traumatic brain injury. *Ann Neurol* 2012;72(3):373–84.

Ono K, Mochizuki H, Ikeda T, Nihira T, et al. Effect of melatonin on α-synuclein self-assembly and cytotoxicity. *Neurobiol Aging* 2012;33(9):2172–85.

Opie L, Lecour S. Melatonin, the new partner to aspirin? *Lancet* 2015;385(9970):774.

Ortiz F, Acuña-Castroviejo D, Doerrier C, Dayoub JC, et al. Melatonin blunts the mitochondrial/NLRP3 connection and protects against radiation-induced oral mucositis. *J Pineal Res* 2015;58(1):34–49.

Ou XM, Chen K, Shih JC. Glucocorticoid and androgen activation of monoamine oxidase A is regulated differently by R1 and Sp1. *J Biol Chem* 2006;281(30):21512–25.

Oxenkrug GF, Requintina PJ. The effect of MAO-A inhibition and cold-immobilization stress on N-acetylserotonin and melatonin in SHR and WKY rats. *J Neural Transm Suppl* 1998;52:333–6.

Oxenkrug GF, Sablin SO, Requintina PJ. Effect of methylene blue and related redox dyes on monoamine oxidase activity; rat pineal content of N-acetylserotonin, melatonin, and related indoles; and righting reflex in melatonin-primed frogs. *Ann NY Acad Sci* 2007;1122:245–52.

Pappolla MA, Simovich MJ, Bryant-Thomas T, Chyan YJ, et al. The neuroprotective activities of melatonin against the Alzheimer beta-protein are not mediated by melatonin membrane receptors. *J Pineal Res* 2002;32(3):135–42.

Parada E, Buendia I, León R, Negredo P, et al. Neuroprotective effect of melatonin against ischemia is partially mediated by alpha-7 nicotinic receptor modulation and HO-1 overexpression. *J Pineal Res* 2014;56(2):204–12.

Paradies G, Paradies V, Ruggiero FM, Petrosillo G. Protective role of melatonin in mitochondrial dysfunction and related disorders. *Arch Toxicol* 2015;89(6):923–39.

Petridou ET, Kousoulis AA, Michelakos T, Papathoma P, et al. Folate and B12 serum levels in association with depression in the aged: A systematic review and meta-analysis. *Aging Ment Health* 2015;8:1–9.

Pfeiffer RF. Parkinson's disease and the gut: The wheel is come full circle. *J Parkinsons Dis* 2014;4(4):577–78.

Piacentini R, Li Puma DD, Ripoli C, Elena Marcocci M, et al. Herpes Simplex Virus type-1 infection induces synaptic dysfunction in cultured cortical neurons via GSK-3 activation and intraneuronal amyloid-β protein accumulation. *Sci Rep* 2015;5:15444.

Pontes GN, Cardoso EC, Carneiro-Sampaio MM, Markus RP. Pineal melatonin and the innate immune response: The TNF-alpha increase after cesarean section suppresses nocturnal melatonin production. *J Pineal Res* 2007;43(4):365–71.

Quik M, Zhang D, McGregor M, Bordia T. Alpha7 nicotinic receptors as therapeutic targets for Parkinson's disease. *Biochem Pharmacol* 2015;97(4):399–407.

Raimondi L, Banchelli G, Sgromo L, Pirisino R, et al. Hydrogen peroxide generation by monoamine oxidases in rat white adipocytes: Role on cAMP production. *Eur J Pharmacol* 2000;395(3):177–82.

Ramirez-Ramirez V, Macias-Islas MA, Ortiz GG, Pacheco-Moises F, et al. Efficacy of fish oil on serum of TNF α, IL-1 β, and IL-6 oxidative stress markers in multiple sclerosis treated with interferon beta-1b. *Oxid Med Cell Longev* 2013;2013:709493.

Rausch JL, Johnson ME, Corley KM, Hobby HM, et al. Depressed patients have higher body temperature: 5-HT transporter long promoter region effects. *Neuropsychobiology* 2003;47(3):120–27.

Ren DL, Sun AA, Li YJ, Chen M, et al. Exogenous melatonin inhibits neutrophil migration through suppression of ERK activation. *J Endocrinol* 2015;227(1):49–60.

René C, Lopez E, Claustres M, Taulan M, et al. NF-E2-related factor 2, a key inducer of antioxidant defenses, negatively regulates the CFTR transcription. *Cell Mol Life Sci* 2010;67(13):2297–309.

Reynolds GP, Mason SL, Meldrum A, De Keczer S, et al. 5-Hydroxytryptamine (5-HT)4 receptors in post mortem human brain tissue: Distribution, pharmacology and effects of neurodegenerative diseases. *British J Pharmacol* 1995; 114(5):993–98.

Robert SJ, Zugaza JL, Fischmeister R, Gardier AM, et al. The human serotonin 5-HT4 receptor regulates secretion of non-amyloidogenic precursor protein. *J Biol Chem* 2001;276(48):44881–8.

Rodrigues DM, Sousa AJ, Hawley SP, Vong L, et al. Matrix metalloproteinase 9 contributes to gut microbe homeostasis in a model of infectious colitis. *BMC Microbiol* 2012;12:105.

Rodrigues SF, Granger DN. Blood cells and endothelial barrier function. *Tissue Barriers* 2015;3(1–2):e978720.

Rubio-Perez JM, Morillas-Ruiz JM. Serum cytokine profile in Alzheimer's disease patients after ingestion of an antioxidant beverage. *CNS Neurol Disord Drug Targets* 2013;12(8):1233–41.

Rudra DS, Pal U, Maiti NC, Reiter RJ, et al. Melatonin inhibits matrix metalloproteinase-9 activity by binding to its active site. *J Pineal Res* 2013;54(4):398–405.

Santos C, Costa J, Santos J, Vaz-Carneiro A, et al. Caffeine intake and dementia: Systematic review and meta-analysis. *J Alzheimers Dis* 2010;20(Suppl. 1):S187–204.

Scheperjans F, Aho V, Pereira PA, Koskinen K, et al. Gut microbiota are related to Parkinson's disease and clinical phenotype. *Mov Disord* 2015;30(3):350–8.

Schroeder EK, Kelsey NA, Doyle J, Breed E, et al. Green tea epigallocatechin 3-gallate accumulates in mito-chondria and displays a selective antiapoptotic effect against inducers of mitochondrial oxidative stress in neurons. *Antioxid Redox Signal* 2009;11(3):469–80.

Seyedzadeh MH, Safari Z, Zare A, Gholizadeh Navashenaq J, et al. Study of curcumin immunomodulatory effects on reactive astrocyte cell function. *Int Immunopharmacol* 2014;22(1):230–35.

Shah N, Frey KA, Müller MLTM, Petrou M, et al. Striatal and cortical β-amyloidopathy and cognition in Parkinson's disease. *Mov Disord* 2016;31(1):111–17.

Sheffler J, Moxley J, Sachs-Ericsson N. Stress, race, and APOE: Understanding the interplay of risk factors for changes in cognitive functioning. *Aging Ment Health* 2014;18(6):784–91.

Shi J, Tu JL, Gale SD, Baxter L, et al. APOE ε4 is associated with exacerbation of cognitive decline in patients with multiple sclerosis. *Cogn Behav Neurol* 2011;24(3):128–33.

Shi X, Zheng Z, Li J, Xiao Z, et al. Curcumin inhibits Aβ-induced microglial inflammatory responses *in vitro*: Involvement of ERK1/2 and p38 signaling pathways. *Neurosci Lett* 2015;594:105–10.

Shukla M, Htoo HH, Wintachai P, Hernandez J, et al. Melatonin stimulates the nonamyloidogenic process-ing of βAPP through the positive transcriptional regulation of ADAM10 and ADAM17. *J Pineal Res* 2015;58(2):151–65.

Silvestri M, Rossi GA. Melatonin: Its possible role in the management of viral infections; A brief review. *Ital J Pediatr* 2013;39:61.

Skoog I, Waern M, Duberstein P, Blennow K, et al. A 9-year prospective population-based study on the association between the APOE*E4 allele and late-life depression in Sweden. *Biol Psychiatry* 2015;78(10):730–36.

Soliman A, Udemgba C, Fan I, Xu X, et al. Convergent effects of acute stress and glucocorticoid exposure upon MAO-A in humans. *J Neurosci* 2012;32(48):17120–7.

Song J, Kang SM, Lee KM, Lee JE. The protective effect of melatonin on neural stem cell against LPS-induced inflammation. *Biomed Res Int* 2015;2015:854359.

Song JX, Lu JH, Liu LF, Chen LL, et al. HMGB1 is involved in autophagy inhibition caused by SNCA/α-synuclein overexpression: A process modulated by the natural autophagy inducer corynoxine B. *Autophagy* 2014;10(1):144–54.

Stansley BJ, Yamamoto BK. Behavioral impairments and serotonin reductions in rats after chronic L-DOPA. *Psychopharmacology (Berl)* 2015;232(17):3203–13.

Su LY, Li H, Lv L, Feng YM, et al. Melatonin attenuates MPTP-induced neurotoxicity via preventing CDK5-mediated autophagy and SNCA/α-synuclein aggregation. *Autophagy* 2015;11(10):1745–59.

Sun Q, Jia N, Wang W, Jin H, et al. Activation of SIRT1 by curcumin blocks the neurotoxicity of amyloid-β25–35 in rat cortical neurons. *Biochem Biophys Res Commun* 2014;448(1):89–94.

Sun YP, Gu JF, Tan XB, Wang CF, et al. Curcumin inhibits advanced glycation end product–induced oxidative stress and inflammatory responses in endothelial cell damage via trapping methylglyoxal. *Mol Med Rep* 2016;13(2):1475–86.

Sutinen EM, Pirttilä T, Anderson G, Salminen A, et al. Pro-inflammatory interleukin-18 increases Alzheimer's disease-associated amyloid-β production in human neuron-like cells. *J Neuroinflammation* 2012;9:199.

Szewczyk-Golec K, Woźniak A, Reiter RJ. Inter-relationships of the chronobiotic, melatonin, with leptin and adiponectin: Implications for obesity. *J Pineal Res* 2015;59(3):277–91.

Tan DX, Manchester LC, Liu X, Rosales-Corral SA, et al. Mitochondria and chloroplasts as the original sites of melatonin synthesis: A hypothesis related to melatonin's primary function and evolution in eukary-otes. *J Pineal Res* 2013;54(2):127–38.

Targum SD, Wedel PC, Fava M. Changes in cognitive symptoms after a buspirone–melatonin combination treatment for major depressive disorder. *J Psychiatr Res* 2015;68:392–96.

Teng YC, Tai YI, Huang HJ, Lin AM. Melatonin ameliorates arsenite-induced neurotoxicity: Involvement of autophagy and mitochondria. *Mol Neurobiol* 2015;52(2):1015–22.

Tizabi Y, Hurley LL, Qualls Z, Akinfiresoye L. Relevance of the anti-inflammatory properties of curcumin in neurodegenerative diseases and depression. *Molecules* 2014;19(12):20864–79.

Tocharus J, Khonthun C, Chongthammakun S, Govitrapong P. Melatonin attenuates methamphetamine-induced overexpression of pro-inflammatory cytokines in microglial cell lines. *J Pineal Res* 2010;48(4):347–52.

Török N, Török R, Szolnoki Z, Somogyvári F, et al. The genetic link between Parkinson's disease and the kynurenine pathway is still missing. *Parkinsons Dis* 2015;2015:474135.

Trujillo J, Granados-Castro LF, Zazueta C, Andérica-Romero AC, et al. Mitochondria as a target in the therapeutic properties of curcumin. *Arch Pharm (Weinheim)* 2014;347(12):873–84.

Ursing C, Wikner J, Brismar K, Röjdmark S. Caffeine raises the serum melatonin level in healthy subjects: An indication of melatonin metabolism by cytochrome P450(CYP)1A2. *J Endocrinol Invest* 2003;26(5):403–06.

Uz T, Qu T, Sugaya K, Manev H. Neuronal expression of arylalkylamine N-acetyltransferase (AANAT) mRNA in the rat brain. *Neurosci Res* 2002;42(4):309–16.

Vincenti JE, Murphy L, Grabert K, McColl BW, et al. Defining the microglia response during the time course of chronic neurodegeneration. *J Virol* 2015;90(6):3003–17.

Volt H, García JA, Doerrier C, Díaz-Casado ME, et al. Same molecule but different expression: Aging and sepsis trigger NLRP3 inflammasome activation, a target of melatonin. *J Pineal Res* In 2016;60(2):193–205.

Wang WZ, Fang XH, Stephenson LL, Zhang X, et al. Melatonin attenuates I/R-induced mitochondrial dysfunction in skeletal muscle. *J Surg Res* 2011;171(1):108–13.

Watson JL, Ansari S, Cameron H, Wang A, et al. Green tea polyphenol (-)-epigallocatechin gallate blocks epithelial barrier dysfunction provoked by IFN-gamma but not by IL-4. *Am J Physiol Gastrointest Liver Physiol* 2004;287(5):G954–61.

Watzlawik JO, Wootla B, Rodriguez M. Tryptophan metabolites and their impact on multiple sclerosis progression. *Curr Pharm Des* 2016;22(8):1049–59.

Weber MD, Frank MG, Tracey KJ, Watkins LR, et al. Stress induces the danger-associated molecular pattern HMGB-1 in the hippocampus of male Sprague Dawley rats: A priming stimulus of microglia and the NLRP3 inflammasome. *J Neurosci* 2015;35(1):316–24.

Wei Z, Gabriel GG, Rui L, Cao X, et al. Monoamine oxidase-A physically interacts with presenilin-1(M146V) in the mouse cortex. *J Alzheimers Dis* 2012;28(2):403–22.

Wong CS, Jow GM, Kaizaki A, Fan LW, et al. Melatonin ameliorates brain injury induced by systemic lipopolysaccharide in neonatal rats. *Neuroscience* 2014;267:147–56.

Wu YH, Fischer DF, Swaab, DF. A promoter polymorphism in the monoamine oxidase A gene is associated with the pineal MAOA activity in Alzheimer's disease patients. *Brain Res* 2007;1167:13–19.

Xie L, Li XK, Takahara S. Curcumin has bright prospects for the treatment of multiple sclerosis. *Int Immunopharmacol* 2011;11(3):323–30.

Xie Y, Zhao QY, Li HY, Zhou X, et al. Curcumin ameliorates cognitive deficits heavy ion irradiation–induced learning and memory deficits through enhancing of Nrf2 antioxidant signaling pathways. *Pharmacol Biochem Behav* 2014;126:181–86.

Yang Y, Jiang S, Dong Y, Fan C, et al. Melatonin prevents cell death and mitochondrial dysfunction via a SIRT1-dependent mechanism during ischemic-stroke in mice. *J Pineal Res* 2015;58(1):61–70.

Yang Z, Zhao T, Zou Y, Zhang JH, et al. Curcumin inhibits microglia inflammation and confers neuroprotection in intracerebral hemorrhage. *Immunol Lett* 2014;160(1):89–95.

Yin HL, Wang YL, Li JF, Han B, et al. Effects of curcumin on hippocampal expression of NgR and axonal regeneration in Aβ-induced cognitive disorder rats. *Genet Mol Res* 2014;13(1):2039–47.

Yu H, Dickson EJ, Jung SR, Koh DS, et al. High membrane permeability for melatonin. *J Gen Physiol* 2016;147(1):63–76.

Yu Y, Wu S, Li J, Wang R, et al. The effect of curcumin on the brain–gut axis in rat model of irritable bowel syndrome: Involvement of 5-HT-dependent signaling. *Metab Brain Dis* 2015;30(1):47–55.

Zaki MH, Lamkanfi M, Kanneganti TD. The Nlrp3 inflammasome: Contributions to intestinal homeostasis. *Trends Immunol* 2011;32(4):171–9.

Zeis T, Allaman I, Gentner M, Schroder K, et al. Metabolic gene expression changes in astrocytes in multiple sclerosis cerebral cortex are indicative of immune-mediated signaling. *Brain Behav Immun* 2015;48:313–25.

Zhang HG, Kim H, Liu C, Yu S, et al. Curcumin reverses breast tumor exosomes mediated immune suppression of NK cell tumor cytotoxicity. *Biochim Biophys Acta* 2007;1773(7):1116–23.

Zhang LF, Zhou ZW, Wang ZH, Du YH, et al. Coffee and caffeine potentiate the antiamyloidogenic activity of melatonin via inhibition of Aβ oligomerization and modulation of the tau-mediated pathway in N2a/APP cells. *Drug Des Devel Ther* 2014b;9:241–72.

Zhang W, Chen XY, Su SW, Jia QZ, et al. Exogenous melatonin for sleep disorders in neurodegenerative diseases: A meta-analysis of randomized clinical trials. *Neurol Sci* 2016;37(1):57–65.

Zhang X, Wu M, Lu F, Luo N, et al. Involvement of α7 nAChR signaling cascade in epigallocatechin gallate suppression of β-amyloid-induced apoptotic cortical neuronal insults. *Mol Neurobiol* 2014a;49(1):66–77.

Zhou B, Zhang Y, Zhang F, Xia Y, et al. CLOCK/BMAL1 regulates circadian change of mouse hepatic insulin sensitivity by SIRT1. *Hepatology* 2014;59(6):2196–206.

30

Beneficial Effects of Serotonin and Melatonin on Cancer, Neurodegenerative Diseases, and Related Disorders in Humans

Virginia N. Sarropoulou
Aristotle University of Thessaloniki
Thessaloniki, Greece

CONTENTS

ABSTRACT In this chapter, the function of two neurotransmitters, serotonin and melatonin, and their implications in human health are discussed. Serotonin plays an important role in the regulation of the endogenous circadian clock, depression and insomnia symptoms, migraine, obesity, and alcohol intoxication. In addition, the beneficial effects of melatonin in sleep disorders, epilepsy, autism, diabetes, the aging process, and cardiovascular diseases are elaborated on. The relationship between exogenous melatonin administration and treatment of several diseases including AIDS, rheumatoid arthritis, elderly hypertension, female infertility, and headaches, among others, is discussed. The implications of both serotonin and melatonin in the treatment of neurodegenerative disorders and certain cancer types are presented. The dietary sources of serotonin and melatonin seem to hold promise to meet the human requirements of these molecules of nutraceutical value.

KEY WORDS: *cancer, central nervous system, depression, diabetes, human nutrition, L-tryptophan, melatonin, neurodegenerative diseases, serotonin, sleep disorders.*

Abbreviations

5HT:	serotonin or 5-hydroxytryptamine
5HTP:	5-hydroxytryptophan
13-HODE:	13-hydroxyoctadecadienoic acid
AIDS:	acquired immune deficiency syndrome
ASMT:	acetylserotonin O-methyltransferase
ASD:	autism spectrum disorder
BMI:	body mass index
CO_2:	carbon dioxide
CAT:	catalase
CNS:	central nervous system
cGMP:	current good manufacturing practice
DA:	dopamine
DS:	Down syndrome
EEG:	electroencephalography
ERa:	estrogen receptor
FDA:	food and drug administration
GC–MS:	gas chromatography–mass spectrometry
HPLC:	high-performance liquid chromatography
HPA:	hypothalamic-pituitary-adrenal
MRI:	magnetic resonance imaging
MDA:	malondialdehyde
MDHAR:	monodehydroascorbate reductase
MS:	multiple sclerosis
AFMK:	N1-acetyl-N2-formyl-5-methoxykynuramine
NAA:	neutral amino acid
NOS:	nitrogen species
NIDDM:	non-insulin-dependent diabetes mellitus
OSAS:	obstructive sleep apnea
H_2O_2:	oxygen peroxide
PD:	Parkinson's disease
ROS:	reactive oxygen species
SSRI:	selective serotonin reuptake inhibitor
NaCl:	sodium chloride
SCI:	spinal cord injury
SOD:	superoxide anion dismutase
TAS:	total antioxidative status
T5H:	tryptamine 5-hydrolase
TRP:	tryptophan
TDC:	tryptophan decarboxylase
T2D:	type-2 diabetes
UV-A:	ultraviolet A

UV-B: ultraviolet B
ZnSO$_4$: zinc sulfate

Introduction

Serotonin (5-hydroxytryptamine; 5HT) is a monoamine neurotransmitter that possesses antioxidant and anti-inflammatory activities, as well as antitumor, antibacterial, and antistress potential. In addition, serotonin has a beneficial effect in the treatment of migraines and it may contribute to alcohol intoxication. Furthermore, serotonin plays an important role in the control of appetite, eating behavior, and body weight, thus controlling obesity. What is more, there is a close connection between serotonin and human decision-making, as well as the neurobiology of impulse aggression. Last but not least, serotonin is implicated in the treatment of neurodegenerative disorders such as stroke, Alzheimer's disease, Parkinson's disease (PD), schizophrenia, Down syndrome (DS), and autism, as well as atherosclerosis and atherothrombosis. In humans, melatonin has beneficial effects in the treatment of sleep disorders such as insomnia and obstructive sleep apnea, epilepsy, spinal cord injury, idiopathic scoliosis, autism spectrum disorder, diabetes, melanoma, aging, malaria, post-surgical complications, and newborn asphyxia, as well as certain cancer types and several cardiovascular diseases. Additionally, melatonin regulates core body temperature, has inflammatory properties, and plays a potential role in the menopausal transition in women. What is more, melatonin is used as a therapeutic drug for the treatment of glaucoma, irritable bowel disease, AIDS, rheumatoid arthritis, primary hypertension in elderly patients, female infertility, migraines and cluster headaches, and ulcerative colitis. Finally, among other things, melatonin regulates cholesterol levels in the blood through the intestinal wall and inhibits apoptotic pathways in neurodegenerative disorders such as stroke, Alzheimer's disease, PD, and Huntington's disease, and amyotrophic lateral sclerosis. Both dietary serotonin and melatonin contribute to human nutrition.

Serotonin in Humans

Significance of Tryptophan in Human Nutrition

Aside from its role as one of the limiting essential amino acids in protein metabolism, tryptophan (TRP) serves as the precursor for the synthesis of the neurotransmitters serotonin and tryptamine (TRP) as well as for the synthesis of the antipellagra vitamin nicotinic acid and the epiphyseal hormone melatonin. Through its involvement in manifold pathways, TRP and its metabolites regulate neurobehavioral effects such as appetite, sleeping–waking rhythm, and pain perception. TRP is the only amino acid that binds to serum albumin to a high degree. Its transport through cell membranes is competitively inhibited by large neutral amino acids (NAAs). The TRP/NAA ratio in plasma is essential for TRP availability and thus for serotonin synthesis in the brain. Due to its high TRP concentration, human milk protein provides optimal conditions for the availability of the neurotransmitter serotonin. Low-protein cow's milk–based infant formulas supplemented with alpha-lactalbumin—a whey protein fraction containing 5.8% TRP—present themselves as a new generation of formulas, with an amino acid pattern different from the currently used protein mixtures of adapted formulas, resembling that of human milk to a much higher degree (Heine et al., 1995).

It has long been known that nutritional status can alter brain neurochemistry, especially that involving carbohydrates and the neurotransmitter serotonin in conjunction with various psychological and other disorders, including depression, premenstrual syndrome, sleepiness, impaired perceptual and cognitive function, and seasonal affective disorder, all of which include fatigue as a common symptom (Wurtman and Wurtman, 1989; Fernstrom, 1994; Curzon, 1996).

Serotonin Selectively Modulates Reward Value in Human Decision-Making

The function of serotonin in motivation and choice remains a puzzle. Existing theories propose a diversity of roles that include representing aversive values or prediction errors, behavioral flexibility, delay discounting, and behavioral inhibition (Daw et al., 2002; Doya, 2002; Robbins and Crockett, 2009; Boureau and Dayan, 2011; Cools et al., 2011; Rogers, 2011). Many of these theories appeal to interactions between reward and punishment, with serotonin acting as an opponent to the neuromodulator dopamine (DA), whose involvement in appetitive motivation and choice is rather better established (Schultz et al., 1997; Bayer and Glimcher, 2005).

Role of Serotonin and Dopamine System Interactions in the Neurobiology of Impulsive Aggression and Its Comorbidity with Other Clinical Disorders

Low levels of the neurotransmitter serotonin have been associated with impulsive aggression in both human and animal studies (Linnoila and Virkkunen, 1992). A number of studies indicate that serotonin and DA systems interact closely at a basic neurophysiological level (Daw et al., 2002), and that impaired function of the serotonin system can lead to dysregulation of the DA system (De Simoni et al., 1987). Research indicates that, in general, the neurotransmitter serotonin has an inhibitory action in the brain (Daw et al., 2002) and that it is deeply involved in the regulation of emotion and behavior, including the inhibition of aggression (Davidson et al., 2000). Serotonergic dysfunction has been reliably associated with aggressive behaviors in animals and humans (Van Erp and Miczek, 2000). In humans, a low concentration of 5-hydroxyindoleacetic acid (5HIAA) has been associated with lifetime aggression (Brown et al., 1979), aggression in patients with mental disorders (Virkkunen et al., 1994), violent suicide attempts (Traskman-Bendz et al., 1986), impulsive murder (Lidberg et al., 1985), and recidivism of murderers (Virkkunen et al., 1989). A meta-analytic study (Moore et al., 2002) that examined findings from 20 separate studies found that low levels of serotonin significantly contribute to aggressive behaviors, regardless of the type of crime and mental health problems.

Understanding the brain mechanisms underlying impulsive aggression and identifying the risk factors are important to the prevention and treatment of impulsive aggression. Research has shown that pharmacological interventions such as selective serotonin reuptake inhibitors (SSRIs) or antipsychotics can reduce impulsive aggression by increasing serotonergic activity or decreasing dopaminergic activity (Miczek et al., 2002; Swann, 2003). In addition, aggressive behaviors have been moderated by environmental enrichment, parental treatment, social skills training, and nutritional supplements (August et al., 2002; Gesch et al., 2002; Raine et al., 2003; Caspi et al., 2004).

Roles of Serotonin in Atherothrombosis, Neurodegenerative Disorders, and Related Diseases

There have been no reports of high plasma 5HT levels in patients with stroke at the acute phase. In a study of elderly subjects, plasma 5HT concentrations measured by enzyme immunoassay were significantly higher in patients with vascular dementia caused by stroke or atherosclerotic small-vessel disease compared with age-matched controls (Ban et al., 2007).

In addition, not only does the hypothalamic–pituitary–adrenal (HPA) axis play a critical role in stress-related disorders, but also the brain neuronal systems, including the monoaminergic and in particular the 5HT-containing neuronal systems (Xu et al., 2006; Lanfumey et al., 2008). It is well established that 5HT is a phylogenetically conserved monoaminergic neurotransmitter that is crucial for a number of physiological processes and is dysregulated in several disease states, including depression, anxiety, and schizophrenia (Inoue et al., 1994). The serotonergic system is intensely involved in the pathology and treatment of depression (Mattson et al., 2004). It is also widely accepted that 5HT receptor activation is important for the pharmacotherapeutic effects of antidepressants (Ivy et al., 2003).

Early alterations in serotonin-modulated circuit formation could contribute to complex symptoms in disorders that have a developmental component, such as DS and autism. For example, fetal DS brains exhibit a roughly 40% reduction in frontal cortex serotonin levels compared with unaffected brains (Whittle et al., 2007). This reduction in serotonin levels persists throughout life (Seidl et al., 1999), and pharmacological compounds such as SSRIs have been administered to adult DS patients with some positive effects on cognitive function (Geldmacher et al., 1997), suggesting a role for serotonin dysfunction in DS. Normal developmental changes in serotonin levels are affected in autistic individuals (Chugani et al., 1999), and asymmetries in cortical serotonin synthesis correlate with differences in functions requiring hemispheric specialization, such as language and handedness (Chananda et al., 2005).

Abnormal levels of 5HT in the circulating blood plasma and in intestinal tissue preparations have been associated with diarrhea, dysmotility, and inflammation. Hence, 5HT is also believed to play a significant pathophysiological role in some of the functional gastrointestinal disorders (Gershon, 2003).

Role of Serotonin in Alcohol's Effects on the Brain

Serotonin is an important brain chemical that acts as a neurotransmitter to communicate information among nerve cells. Serotonin's actions have been linked to alcohol's effects on the brain and to alcohol abuse. Alcoholics and experimental animals that consume large quantities of alcohol show evidence of differences in brain serotonin levels compared with nonalcoholics. Both short- and long-term alcohol exposures also affect the serotonin receptors that convert the chemical signal produced by serotonin into functional changes in the signal-receiving cell. Drugs that act on these receptors alter alcohol consumption in both humans and animals. Serotonin, along with other neurotransmitters, may also contribute to alcohol's intoxicating and rewarding effects, and abnormalities in the brain's serotonin system appear to play an important role in the brain processes underlying alcohol abuse (Lovinger, 1999).

Role of Serotonin in Eating Behavior

It has been indicated that serotonin has a suppressive effect on food intake and body weight (Cruzon, 1990). Early pharmacological manipulations identified an inverse relationship between the biogenic amine neurotransmitter serotonin and food intake. More specifically, a selective reduction in serotonin bioavailability was associated with hyperphagia and subsequent weight gain, while diminished food intake was induced by an increase in serotonin efficacy (Heisler et al., 2002). Serotonin, in coordination with the hypothalamus, plays an important role in the central nervous system's (CNS) control of appetite, eating behavior, energy balance, and body weight. It has a special role in the control of carbohydrate intake. It has been observed that a reduction in serotonin levels causes hyperphagia. As a result, carbohydrate intake increases, leading to obesity. Inversely, increased levels of serotonin lead to hypophagia, and as a result, carbohydrate intake decreases. Hence, serotonergic agonists are clinically useful in the treatment of obesity. Obesity (body mass index [BMI] >30) is a risk factor for major causes of death, including cardiovascular disease, numerous cancers, diabetes, and metabolic syndrome, and is linked with markedly diminished life expectancy. The energy regulation of 5HT is mediated in part by 5HT receptors located in various medial hypothalamic nuclei. Along with serotonin, other hormones such as insulin, leptin, and corticosterone are also involved in energy control and regulation. Though large numbers of serotonergic drugs, including SSRIs such as sibutramine and serotonin 5HT2c agonists, are available to treat this deadly disease, these drugs are associated with a large number of side effects (Sharma and Sharma, 2012). Thus, the increasing global prevalence of obesity has renewed interest in the serotonin–hypothalamic regulation of energy balance to find drugs that have maximum pharmacological and minimum toxicological effects.

Serotonin and Migraine

5-Hydroxytryptophan (5HTP) has been shown to have a beneficial effect equivalent to methysergide (a medication for migraine) (Titus et al., 1986). Another double-blind, placebo-controlled study (De Benedittus et al., 1986) showed a significant effect of 5HTP on migraine. In a third study, 5HTP had a beneficial effect on migraine, but propranolol (a drug) did better, and the combination of propranolol and 5HTP did best (Maissen and Ludin, 1991).

Serotonin and Depression

5HTP appears to have equal efficacy to antidepressant medication, but without the drug risks and side effects (Poldinger et al., 1991). Similar studies with depressed children demonstrated equal benefits

(Ryan et al., 1992). Other studies have shown 5HTP to have equal efficacy to antidepressant medication, especially in those who exhibited an anxious, agitated, or irritable mood (Zmilacher et al., 1988).

Serotonin and Sleep

Regarding serotonin, it has been shown that a decrease in daytime levels of serotonin, as a result of low intake of TRP during the day, causes sleep disturbances (Arnulf et al., 2002). Moreover, the neurotransmitter serotonin also plays an important role in the regulation of the endogenous circadian clock. For instance, serotonin raises the proportion of slow-wave sleep (Jouvet, 1999), though it also acts as a wakefulness neurotransmitter.

Health Benefits of Serotonin in Humans

Sweet cherries contain substantial amounts of serotonin, and may have a great number of health benefits if incorporated into a healthy diet (González-Gomez et al., 2009). Such data indicates that serotonin in our diets and plant-based medicines can affect human health and may have an impact on several chronic diseases (Coutts et al., 1986). Serotonin derivatives such as FS and CS have been isolated from safflower seeds and possess antioxidant and anti-inflammatory activities (Hotta et al., 2002), and anti-tumor (Nagatsu et al., 2000), antibacterial (Kumarasamya et al., 2003), and antistress potentials, as well as being involved in reducing depression and anxiety (Yamamotova et al., 2007). Some of the more extensive uses of serotonin include treating patients suffering from Parkinson-like symptoms (Bell and Janzen, 1966) and controlling obesity (Cangiano et al., 1992).

Melatonin in Humans

Role of Melatonin in Humans

Several circadian rhythms, such as the rest–activity cycle, core body temperature, neuronal electrical activity, and locomotor activity, are driven by melatonin (Shibata et al., 1989; Dijk and Cajochen, 1997; Sharma et al., 1999; Krauchi and Wirz-Justice, 2001). Furthermore, melatonin has demonstrated an ability to neutralize oxidative stress in humans across a broad spectrum of conditions, including malaria, post-surgery, newborn asphyxia and Alzheimer's disease, but, more importantly, in vascular disease and atherosclerosis as well (Maharaj et al., 2007). At either physiological or pharmacological concentrations, melatonin appears to be involved in far-flung physiological and pathophysiological processes, including the control of sleep, circadian rhythms, retinal physiology, seasonal reproductive cycles, cancer development and growth, immune activity, antioxidation and free radical scavenging, mitochondrial respiration, cardiovascular function, bone metabolism, intermediary metabolism, and gastrointestinal physiology (Reiter, 2002).

Other physiological and pathophysiological responses modulated by melatonin via melatonin receptor–mediated mechanisms include pituitary hormone release, testosterone production by the testes, cortisol secretion by the adrenal cortex, vascular tone, energy metabolism, fatty acid transport, immune activity, and cancer cell proliferation and tumor growth. Actions of melatonin that appear to involve nuclear receptors include the stimulation of immune cells to produce biologically active substances called *interleukins* and the inhibition of certain types of colon cancer cells (Reiter, 2002; Pandi-Perumal et al., 2006; Blask, 2007).

Melatonin and Related Diseases

In addition to its regulatory role, melatonin has antioxidative capacity, immunomodulatory potency, and also appears to be protective against a variety of cancers, especially breast cancer, although the data are based mostly on observational studies and animal models (Grant et al., 2009; Zawilska et al., 2009; Mediavilla et al., 2010; Peuhkuri et al., 2012). Exogenous melatonin has been used in the treatment of sleep disorders of circadian origin such as jet lag and delayed sleep phase syndrome, and as a complement to other therapeutic drugs for the treatment of numerous diseases, including glaucoma, irritable bowel disease, and certain types of cancers, mainly to either enhance the therapeutic effect of conventional drug therapy or to reduce their toxicity, thus ameliorating the side effects (Buscemi et al., 2005, 2006; Mills et al., 2005; van Geijlswijk et al., 2010; Sanchez-Barcelo et al., 2010; Seely et al., 2012).

Significant increase in total antioxidative status (TAS) levels in the plasma of multiple sclerosis (MS) patients has been observed with melatonin supplementation (Karbownik et al., 2001). In animal models, melatonin has been demonstrated to prevent damage to mitochondrial and nuclear DNA by carcinogens (Karbownik et al., 2001; Tütüncülar et al., 2005; Jou et al., 2007; Greenlee et al., 2009; Nishimura et al., 2010). Few underpowered trials suggest that melatonin may enhance tumor response during the treatment of breast cancer (Anisimov et al., 2003). Melatonin supplementation during chemotherapy holds potential for reducing dose-limiting toxicities (Oaknin-Bendahan et al., 1995).

Melatonin may be beneficial in the treatment of newborn infants with asphyxia, and these beneficial effects may be related both to the antioxidant properties of melatonin and to its ability to increase the efficiency of mitochondrial electron transport (Fulia et al., 2001). Melatonin, via its antioxidant properties, could thus modulate oxidative stress, which may improve organ function and reduce morbidity and mortality, at least as tested in newborns, where the drug was able to decrease oxidative stress induced by surgery (Kucukakin et al., 2009). Melatonin has been investigated in a wide range of diseases, such as heart disease, Alzheimer's disease, AIDS, diabetes, depression, and cancer (Beyer et al., 1998). As an example of the use of melatonin in chronic diseases, it has been shown that melatonin administration in patients with Alzheimer's disease significantly delayed the progression of the disease and decreased brain atrophy, as assessed by MRI (Brusco et al., 1998). Melatonin has been reported to inhibit the intrinsic apoptotic pathways in neurodegenerative diseases, including stroke, Alzheimer's disease, PD, Huntington's disease, and amyotrophic lateral sclerosis (Wang, 2009).

The beneficial antioxidant effects of low doses of melatonin have also been shown in several chronic diseases, such as rheumatoid arthritis (10 mg/day; Forrest et al., 2007), primary essential hypertension in elderly patients (5 mg/day; Kedziora-Kornatowska et al., 2008), type-2 diabetes (T2D) in elderly patients (5 mg/day; Kedziora-Kornatowska et al., 2009), and females suffering infertility (3 mg/day; Tamura et al., 2008). Melatonin attenuates molecular and cellular damage resulting from cardiac ischemia/reperfusion (a massive release of free radicals is involved in the tissue damage following the reperfusion process); anti-inflammatory and antioxidative properties of melatonin also seem to be involved in the protection against vascular disease and atherosclerosis development (Dominguez-Rodriguez et al., 2009). Its protective action in ischemia/reperfusion could also be beneficial in limiting damage following organ transplantation (Fildes et al., 2009). Physiological data suggest that melatonin is an important regulator of both inflammation and motility in the gastrointestinal tract, and some studies in humans suggest that supplemental melatonin may have an ameliorative effect on ulcerative colitis (Terry et al., 2009). A very recent study reported the beneficial action of high doses of melatonin (20 mg/kg) for inhibiting apoptosis and liver damage resulting from oxidative stress in malaria, which could be a novel approach in the treatment of this disease (Srinivasan et al., 2010).

Basic research indicates that melatonin may play a role in modulating the effects of abused drugs such as cocaine (Sircar, 2000; Uz et al., 2003). In exploratory studies, prolonged-release melatonin has shown sleep quality improvement in patients with chronic schizophrenia (Shamir et al., 2000) as well as those with major depressive disorder (Dolberg et al., 1998; Dalton et al., 2000), and treating sleep–wake cycle disorders in children with underlying neurodevelopmental difficulties (Jan et al., 2000; De Leersnyder et al., 2003). Additionally, as an add-on to antihypertensive therapy, prolonged-release melatonin improved blood pressure control in patients with nocturnal hypertension, as shown in a randomized, double-blind, placebo-controlled study (Grossman et al., 2011). Several clinical studies indicate that supplementation with melatonin is an effective preventive treatment for migraines and cluster headaches (Dodick and Capobianco, 2001; Gagnier, 2001). Melatonin presence in the gallbladder has many protective properties, such as converting cholesterol to bile, preventing oxidative stress, and increasing the mobility of gallstones from the gallbladder (Koppisetti et al., 2008). It also decreases the amount of cholesterol produced in the gallbladder by regulating the cholesterol that passes through the intestinal wall. The concentration of melatonin in the bile is two to three times higher than the otherwise very low daytime melatonin levels in the blood across many diurnal mammals, including humans (Tan et al., 2009). Melatonin might improve sleep in people with autism spectrum disorder (ASD) (Braam et al., 2009). Research has shown that children with autism have abnormal melatonin pathways and below-average physiological levels of melatonin (Rossignol and Frye, 2011; Veatch et al., 2015). Melatonin supplementation has been shown to improve sleep duration, sleep onset latency, and nighttime awakenings (Giannotti et al., 2006; Rossignol and Frye, 2011).

Melatonin and Obstructive Sleep Apnea Syndrome (OSAS)

Little is known about melatonin levels in OSAS patients. Wikner et al. (1997) did not find differences between melatonin secretion in OSAS patients before or after continuous positive airway pressure. Ulfberg et al. (1998) showed that OSAS patients presented higher melatonin levels in the afternoon than control subjects who did not snore. Finally, Brzecka et al. (2001) found a correlation between OSAS severity and peak nighttime melatonin value.

Melatonin and Cancer

Melatonin exerts oncostatic activity through several biological mechanisms, including antiproliferative actions, stimulation of anticancer immunity, modulation of oncogen expression, and anti-inflammatory, antioxidant, and antiangiogenic effects. (Mediavilla et al., 2010). It have been shown that the oncostatic actions of melatonin on hormone-dependent mammary tumors are mainly based on its antiestrogenic actions (Hill et al., 2009). Studies performed *in vivo* and *in vitro* have shown that exogenous melatonin exerts oncostatic effects on melanoma cells, especially through its ability to stimulate interleukin-2 production (Cabrera et al., 2010; Srinivasan et al., 2011), and protects cells from ultraviolet A (UV-A) and B (UV-B) actions (Izykowska et al., 2009). The radioprotective action of melatonin is related to its antioxidant properties (Kim et al., 2001).

Several other investigations have assessed the signaling pathways and the expression of melatonin receptors in cells lines of endometrial cancer (Watanabe et al., 2008), glioma (Esposito et al., 2008), renal cancer (Min et al., 2012), leukemia and hematological cancer (Koh et al., 2011; Casado-Zapico et al., 2011), prostate cancer (Tam and Shiu, 2011), gastric adenocarcinoma (Zhang et al., 2012), oral cavity tumors (Cutando et al., 2011), Ewing sarcoma, and bone tumors (Toma et al., 2007; Casado-Zapico et al., 2010), among others. In 100 consecutive patients who were randomized to receive chemotherapy alone or chemotherapy and melatonin, overall tumor regression rate and five-year survival results were significantly higher in patients concomitantly treated with melatonin (Lissoni et al., 2003). It is well documented that melatonin strongly suppresses human prostate tumor cell proliferation in culture, as well as in animal models and in humans (Xi et al., 2001; Shiu et al., 2003; Joo and Yoo, 2009; Jung-Hynes et al., 2011). There is evidence that the administration of melatonin alone or in combination with interleukin-2 in conjunction with chemoradiotherapy and/or supportive care in cancer patients with advanced solid tumors is associated with improved outcomes of tumor regression and survival. Moreover, chemotherapy has been shown to be better tolerated in patients treated with melatonin (Cutando et al., 2012). The results of these studies have provided the basis for the potential utility of melatonin in cancer management.

A new mechanism by which physiological and pharmacological blood levels of melatonin inhibit cancer growth *in vivo* is via a melatonin-induced suppression of tumor linoleic acid (LA) uptake and its metabolism to the important mitogenic signaling molecule 13-hydroxyoctadecadienoic acid (13HODE). Melatonin suppresses cyclic adenosine monophosphate (cAMP) formation and inhibits tumor uptake of LA and its metabolism to 13HODE via a melatonin receptor–mediated mechanism in both tissue-isolated rat hepatoma 7288 CTC and human breast cancer xenografts. cAMP is a tightly regulated second messenger that is critically involved in many intracellular processes (Movsesian, 1999). It has been postulated that in industrialized societies, light at night, by suppressing melatonin production, poses a new risk for the development of breast cancer and, perhaps, other cancers as well. In support of this hypothesis, light during darkness suppresses nocturnal melatonin production and stimulates the LA metabolism and growth of rat hepatoma and human breast cancer xenografts. Nocturnal dietary supplementation with melatonin, at levels contained in a melatonin-rich diet, inhibits rat hepatoma growth via the mechanisms described above. Dietary melatonin supplementation working in concert with the endogenous melatonin signal has potential as a new preventive/therapeutic strategy to optimize the host/cancer balance in favor of host survival and quality of life (Blask et al., 2005b).

Melatonin also produced promising results in patients with cervical cancer, but future research needs to assess the efficacy and safety of melatonin supplements in this specific cancer population (Greenlee et al., 2004). The beneficial effects of melatonin have also been evaluated in breast cancer. Breast cancer diagnoses have peaks in spring and fall, with the authors suggesting that vitamin D in summer and melatonin in winter reduce the breast tumor growth rate in those seasons (Oh et al., 2010; Grant, 2011). A role for melatonin has also been proposed in alleviating the toxicity of anticancer chemotherapy and preventing heart failure (Piasek et al., 2009). However, melatonin requires placebo-controlled trials before being recommended for clinical treatment of muscle wasting and cachexia (Strasser, 2007).

In some neoplastic cells, melatonin acts as a differentiating agent and diminishes their invasive/metastatic potential via alterations in adhesion molecule expression and by supporting the mechanisms responsible for gap-junctional intercellular communication. Additional evidence supports a variety of other biochemical and molecular mechanisms in melatonin's oncostatic action at nocturnal circulating concentrations, including the regulation of estrogen receptor (ERa) expression and transactivation, calcium/calmodulin activity, protein kinase C activity, cytoskeletal architecture and function, intracellular redox status, melatonin receptor–mediated signal transduction cascades, aromatase and telomerase activities, and fatty acid transport and metabolism (Blask et al., 2005b). Melatonin is the first soluble, nocturnal anticancer signal to be identified in humans that directly links the central circadian clock with some of the important mechanisms regulating breast carcinogenesis and possibly other malignancies. These findings also provide the first definitive nexus between the exposure of healthy premenopausal female human subjects to bright white light at night and the enhancement of human breast oncogenesis via disruption (i.e., suppression) of the circadian, oncostatic melatonin signal (Blask et al., 2005a). In breast cancer patients in particular, nocturnal circulating levels of melatonin are negatively correlated with breast cancer ERa content, while tissue levels of melatonin correlate positively with tumor ERa status and negatively with the nuclear grade and proliferative index. These findings suggest that cancer cells elaborate soluble factors that negatively feedback on the mechanisms regulating nocturnal melatonin production (Maestroni and Conti, 1996).

Melatonin and Sleep

Melatonin, which is produced from serotonin, is more involved in sleep quality than in sleep onset (Castro-Silva et al., 2010). Considerable evidence exists for the use of melatonin supplements for sleep problems in long-term care. While melatonin appears to have a modest positive effect on sleep quality among older adults (Pires et al., 2001; Zhdanova et al., 2001; Singer et al., 2003; Kato et al., 2005; Zemlan et al., 2005; Roth et al., 2006; Lemoine et al., 2007; Wade et al., 2007; Catherine et al., 2010), most studies were small in size and included only subjective assessments of sleep quality (Olde Rikkert and Rigaud, 2001), except for actigraphy or polysomnography. However, melatonin caused a statistically significant decrease in sleep onset latency, and an increase in total sleep duration (Brzezinski et al., 2005). In addition, it is inconclusive as to whether melatonin poses risks to long-term care residents due to potential drug interactions (Shimazaki and Martin, 2007; Rondanelli et al., 2011). In hypertensive patients treated with beta blockers, nightly melatonin supplementation significantly improved sleep quality, increased total sleep time, increased sleep efficiency, and decreased sleep onset latency as assessed by polysomnography, without apparent tolerance and without rebound sleep disturbance during withdrawal of melatonin supplementation. These findings may assist in developing countermeasures against sleep disturbances associated with beta blocker therapy (Scheer and Czeisler, 2005). A recent clinical trial also concluded that exogenous melatonin was beneficial in improving sleep duration, quality, and efficiency in healthy volunteers and might be of benefit in managing disturbed sleep (Howatson et al., 2012). Thus, there is considerable evidence to support the effectiveness of melatonin intervention for insomnia in adults, yet this intervention is underutilized. Additional rigorous research is needed prior to making conclusive statements about the safety and risk/benefit of melatonin and recommending an appropriate management strategy for chronic insomnia (Chung and Lee, 2002; Bain, 2006).

In healthy human subjects, the administration of 0.3 or 1 mg of melatonin in the early evening was associated with reduced sleep latency and improved sleep efficiency (Zhdanova et al., 1995; 1996). Another study, which used an electroencephalogram (EEG) to monitor sleep parameters, found no effect of 5 mg melatonin on normal sleep (James et al., 1987). Overall, melatonin has had variable efficacy when studied as a hypnotic agent (Mendelson, 1997). While some studies have reported that exogenous melatonin improves sleep efficiency, this effect is not consistently observed (Mendelson, 1997). The lack of effect on sleep efficiency or total sleep time may not be surprising, given the short half-life of melatonin in the circulation (less than 1 h) (Waldhauser et al., 1984).

Melatonin and Neurological Conditions

Clinical studies have demonstrated abnormalities in melatonin production or release in individuals with ASD (Levy and Hyman, 2008; Melke et al., 2008). One study concluded that unaffected parents of individuals with ASD had lower melatonin levels, and that the deficits were associated with low activity of the ASMT gene, which encodes the enzymes of melatonin synthesis (Melke et al., 2008). Large effect sizes were obtained from randomized trials (Garstang and Wallis, 2006; Andersen et al., 2008). Melatonin was found to decrease sleep latency and increase total sleep time, without any significant effect on the number of nighttime awakenings (Wright et al., 2011). A total 45.6% of individuals with ASD were found to be taking some form of psychotropic agents, including St John's wort and melatonin (Aman et al., 2003).

Melatonin may influence the symptoms of PD and/or the effectiveness of dopaminergic therapy. Preliminary evidence suggests that it may also influence the nonmotor symptoms of PD, such as respiratory, gastrointestinal, mood, and sleep disorders, and orthostatic hypotension. Whenever possible, clinicians should ensure that complementary therapy is used appropriately in PD patients without reducing the benefits of dopaminergic therapy (Zesiewicz and Evatt, 2009). Strategy based on melatonin supplements has been proposed to at risk parents of schizophrenia to eliminate the disorder.

Decreased serum levels of melatonin are suspected to be involved in Alzheimer-like tau hyperphosphorylation (Zhu et al., 2005; Srinivasan et al., 2006; Wang et al., 2013). Thus, melatonin supplementation may retard neurodegenerative changes associated with brain aging (Brusco et al., 1998). A case study revealed that melatonin-treated (6 mg/day) identical twins had less memory loss over 36 months (Jean-Louis et al., 1998).

Melatonin has multiple-cited recommendations for use in epilepsy, but paradoxically often has a proconvulsant effect in addition to potentially serious adverse effects (Pearl et al., 2011). While one study reported a reduction in mean seizure frequency (Peled et al., 2001), a few reported an increase in seizures (Jones et al., 2005). Thus, melatonin appears to have unpredictable effects on seizure frequency, and should be used with caution in patients with epilepsy (Gaby, 2007). Patients should be queried as to the nature of any alternative medicine products they are using, with the view that the use of these products may be reasonable if traditional antiepileptic drug therapy is continued, the potentially adverse effects of the alternative agents are monitored, and the alternative and traditional agents do not conflict (Pearl et al., 2011).

Melatonin supplementation has been shown to lessen spinal cord injury (SCI), but its use has been limited by its side-effect profile (Fee et al., 2010). Furthermore, melatonin deficiency plays a role in the prognosis of idiopathic scoliosis. Therefore, melatonin supplements may prevent the progression of scoliosis, especially in mild cases with less than a 35° curve (Machida et al., 2009).

Melatonin and Diabetes

The use of nutritional supplements by patients with T2D is estimated at somewhere between 8% and 49%. Melatonin as a nutritional supplement was identified as potentially beneficial for T2D treatment or prevention. Health providers should investigate drug–nutritional supplement interactions prior to treatment (Lee and Dugoua, 2011). Another study was conducted to determine the influence of melatonin supplementation on the oxidative stress parameters in elderly NIDDM patients. The findings indicated an improvement in antioxidative defense after melatonin supplementation in the NIDDM individuals and suggested melatonin supplementation as an additional treatment for the control of diabetic complications (Kedziora-Kornatowska et al., 2009).

Interestingly, reductions in plasma melatonin are independently associated with T2D (Radziuk and Pye, 2006). A recent large-scale investigation, a case control study nested within the Nurses' Health Study, adds corroborating evidence demonstrating a greater risk of development of T2D associated with decreased nocturnal melatonin secretion (McMullan et al., 2013). Consequently, this reduction in circulating melatonin, which is associated with significantly higher levels of oxidative stress and reduced antioxidant activity (Kedziora-Kornatowska et al., 2009), is likely a result of increased consumption of melatonin due to hyperglycemia precipitating increased levels of oxidative stress (Tan et al., 2007). The sole prior study that has investigated the effect of melatonin on oxidative stress *in vivo* in T2D was conducted in elderly patients (Kedziora-Kornatowska et al., 2009). Hussain et al. (2006) found that 90 days of supplementation of 10 mg melatonin, in combination with zinc (50 mg/day), significantly decreased fasting plasma glucose (−23%) in individuals with T2D.

Melatonin and the Reproductive System

Melatonin has also been shown to exert beneficial effects on fertility, pregnancy wellness, and embryo development, whose requirements increase during pregnancy (Carlomagno et al., 2011). Associations have been reported between hormonal factors and melanoma; melatonin inhibition increases the risk of melanoma by increasing circulating estrogen levels (Kvaskoff and Weinstein, 2010).

Melatonin has also been shown to be closely linked with reproductive functioning in women (Reiter, 1980; Reiter et al., 2009; Srinivasan et al., 2009; Kotlarczyk et al., 2012), which has led some to speculate that melatonin may play a role in the menopausal transition (Bellipanni et al., 2001; Diaz and Llaneza, 2008). Melatonin levels decrease with age, particularly during the perimenopausal period (Zhou et al., 2003), coincidentally with the appearance of menstrual irregularities, sleep disturbances, bone loss, elevated estrogen, and reduced progesterone levels. Previous studies have shown that women aged 43–49 years taking melatonin have a restoration of their pituitary functions back to juvenile (premenopausal) patterns of regulation (Bellipanni et al., 2001; 2005), a decrease in the number of menstrual cycles, and a normalization of bone markers to resemble the premenopausal state (Kotlarczyk et al., 2012). There are further improvements in various other physical aspects of menopause, including the following: flatulence or gas pains, muscle and joint aches, feeling tired or worn out, difficulty sleeping, aches in the back of the neck or head, decreased physical strength, decreased stamina, lack of energy, dry skin, weight gain, increased facial hair, changes in the appearance, texture, or tone of skin, feeling bloated, lower backache, frequent urination, involuntary urination when laughing or coughing, breast tenderness, vaginal bleeding or spotting, and leg pains or cramps (Kotlarczyk et al., 2012).

Melatonin and Epilepsy

Epilepsy is a chronic disease of the brain characterized by repeated seizures, which originate from hyperexcitation of a certain group of neurons, leading to different transient clinical signs (motor, sensory, autonomic, psychic, or behavioral) and laboratory findings. A major role in the pathogenesis of the disease is played by the formation of the epileptogenic focus, which has an uncontrolled hyperexcitability due to partial prolonged depolarization of cellular membranes. The epileptogenic focus is formed by damaged, but not dead, neurons. There is evidence that the formation of free radicals or the decreased activity of antioxidant systems can cause some forms of epilepsy and can increase the risk of repeated epileptic seizures (Yuksel et al., 2000). The hormone melatonin exerts neuroprotection due to its antioxidant, antiexcitotoxic properties within the central nervous system, and as an adjunct can have a beneficial effect on conditions involving oxidative stress, such as epilepsies (Markov and Kolev, 2009). Kurecki et al. (1995) showed similar results, whereby adding melatonin to the antiepileptic therapy of children led to decreases in sleep disturbances and in the frequency of seizures.

Serotonin and Melatonin in Human Nutrition

Melatonin has been detected in foodstuffs in notable amounts—for example, in tomatoes, olives, cereals (barley, rice), walnuts, strawberry, olive oil, wine, beer, unprocessed cow's milk, and nighttime milk. Caffeine has both stimulatory and inhibitory effects on levels of melatonin (Hattori et al., 1995; Reiter et al., 2005, 2007; Jouan et al., 2006; de la Puerta et al., 2007; Iriti et al., 2010; Castro et al., 2011; Mercolini et al., 2012; Rodriguez-Naranjo et al., 2011; Sturtz et al., 2011; Ramakrishna et al., 2012).

Garrido et al. (2009, 2010) found that sweet cherries from the Jerte Valley (Cáceres, Extremadura, Spain) contained not only high concentrations of anthocyanin pigments and other phenolic compounds (González-Gómez et al., 2009), but also substantial amounts of melatonin, serotonin (González-Gómez et al., 2010), and TRP (Cubero et al., 2010), as recently reported in seven different cultivars of these fruits.

It is well documented that melatonin exhibits circadian rhythm in plasma, saliva, and human milk, among other biological fluids (Almeida et al., 2011; Engler et al., 2012). In breast milk, melatonin shows high levels during dark hours and decreases during the day. In this way, lactating mothers communicate time-of-day information to their babies and coordinate them with zeitgebers. Thereby, mothers prepare babies for sleep through their breast milk. Cubero et al. (2005) demonstrated that the administration of formula milk dissociated into day/night components helps to consolidate the sleep–wake rhythm in babies. On the other hand, some plant extracts are added to increase the sedative effects of formula milk and in this way improve sleep in infants.

High levels of melatonin were identified in fresh green feverfew leaves (*Tanacetum parthenium*), St Johns Wort flowers (*Hypericum perforatum*), and Huang-qin (*Scutellaria baicalensis*) (Murch et al., 1997). Melatonin was also investigated in more than 100 Chinese medicinal herbs as dried powders derived from flowers, seeds, leaves, roots, or stems (Chen et al., 2003). In traditional Thai medicine and for use as a sleep aid, a remedy using the leaves of seven edible herbs showed significant levels of melatonin: black pepper (*Piper nigrum*; 1093 ng/g dry sample weight), Burmese grape (*Baccaurea ramiflora*; 43.2 ng/g), humming bird tree/scarlet wisteria (*Sesbania glandiflora*; 26.3 ng/g), bitter gourd (*Moringa charantia*; 21.4 ng/g), *S. tora* (10.5 ng/g), and *S. sesban* (8.7 ng/g) (Padumanoda et al., 2013). Likewise, other Thai herbs used for inflammation, antibacterial, and glucose control were screened and showed melatonin contents ranging from 0.5 to 146 ng/g dry weight (Johns et al., 2011). High levels of melatonin (2–200 ng/g) were detected in the seeds of 15 edible plants, with the highest concentrations observed in white (*Brassica hirta*) and black mustard (*B. nigra*) seeds (Manchester et al., 2000). None of these plants are primary foodstuffs consumed in appreciable quantities. However, levels of approximately 1 ng/g have been reported in oat, sweet corn, and rice (Hattori et al., 1995), and a variety of Chinese corn seeds and Chinese rice seeds contained amounts of melatonin ranging from 11 to 2034 ng/g and 11 to 264 ng/g dry weight, respectively (Wang et al., 2009).

Melatonin is categorized by the US Food and Drug Administration (FDA) as a dietary supplement. It is sold freely over the counter in both the United States and Canada without any regulation as a pharmaceutical drug (Buscemi et al., 2013). The FDA regulations applying to medications are not applicable to melatonin (Altun and Ugur-Altun, 2007). However, new FDA rules required that by June 2010 all production of dietary supplements must comply with "current good manufacturing practices" (cGMP) and be manufactured with "controls that result in a consistent product free of contamination, with accurate labeling" (FDA, 2009). The industry has also been required to report to the FDA "all serious dietary supplement related adverse events," and the FDA has (within the cGMP guidelines) begun enforcement of that requirement (FDA, 2013).

Conclusion

Neurotransmitters are one of the most profound links between the subjective human experience and the chemical reality of the physical world. Their elegant molecular bends and shapes craft individuals at the most essential level. Although scientists are working ceaselessly to illuminate the roles that these silent messengers play, their roles may be irrevocably intertwined. Serotonin is a paragon of the relationship between human beings and their neurotransmitters. It bridges the gap between the sterile chemical realm and the integrated tapestry of human experience by hovering forever between the extremes of biological function. The idea of using daily exercise to boast serotonin levels is a great discovery that needs to be discussed further. More studies should be conducted to show how daily exercise, along with a good diet, can help raise serotonin levels. Such research may lead to patients learning how to treat themselves without relying on any type of prescription drug. This may also lead to a decrease in the experience of serotonin syndrome. As a factor in restorative sleep, melatonin's benefits extend to neuroprotection and fighting cancer. Its powerful antioxidant effect offers important enhancements to the brain and nervous system, helping protect against age-related damage. Most exciting are melatonin's benefits for relieving anxiety in cancer patients and improving survival from an array of cancers. Finally, migraine sufferers using melatonin may enjoy a vast decline in the frequency and severity of their headaches, leading to a tremendously improved quality of life.

REFERENCES

Almeida EA, di Mascio P, Harumi T, Spence DW, Moscovitch A, Hardeland R, et al. Measurement of melatonin in body fluids: Standards, protocols and procedures. *Childs Nerv Syst* 2011;27:879–891.

Altun A, Ugur-Altun B. Melatonin: Therapeutic and clinical utilization. *Int J Clin Pract* 2007;61:835–845.

Aman MG, Lam KS, Collier-Crespin A. Prevalence and patterns of use of psychoactive medicines among individuals with autism in the Autism Society of Ohio. *J Autism Dev Disord* 2003;33:527–534.

Andersen IM, Kaczmarska J, McGrew SG, et al. Melatonin for insomnia in children with autism spectrum disorders. *J Child Neurol* 2008;23(5):482–485.

Anisimov VN, Alimova IN, Baturin DA, et al. Dose-dependent effect of melatonin on life span and spontaneous tumor incidence in female SHR mice. *Exp Gerontol* 2003;38:449–461.

Arnulf I, Quintin P, Alvarez JC, et al. Mid-morning tryptophan depletion delays REM sleep onset in healthy subjects. *Neuropsychopharmacology* 2002;27:843–851.

August GJ, Hektner JM, Egan EA, Realmuto GM, Bloomquist ML. The early risers longitudinal prevention trial: Examination of 3-year outcomes in aggressive children with intent-to-treat and as intended analyses. *Psychol Add Behav* 2002;16:27–39.

Bain KT. Management of chronic insomnia in elderly persons. *Am J Geriatr Pharmacother* 2006;4:168–192.

Ban Y, Watanabe T, Miyazaki A, et al. Impact of increased plasma serotonin levels and carotid atherosclerosis on vascular dementia. *Atherosclerosis* 2007;195:153–159.

Bayer HM, Glimcher PW. Midbrain dopamine neurons encode a quantitative reward prediction error signal. *Neuron* 2005;47:129–141.

Bell EA, Janzen DH. Medical and ecological considerations of L-DOPA and 5-HTP in seeds. *Nature* 1966;210:529.

Bellipanni G, Bianchi P, Pierpaoli W, Bulian D, Ilyia E. Effects of melatonin in perimenopausal and menopausal women: A randomized and placebo controlled study. *Exp Gerontol* 2001; 36:297–310.

Bellipanni G, Di Marzo F, Blasi F, Di Marzo A. Effects of melatonin in perimenopausal and menopausal women: Our personal experience. *Ann NY Acad Sci* 2005;1057: 393–402.

Beyer CE, Steketee JD, Saphier D. Antioxidant properties of melatonin: An emerging mystery. *Biochem Pharmacol* 1998;56:1265–1272.

Blask DE. Melatonin. *McGraw-Hill Yearbook of Science and Technology*, pp. 142–144. New York: McGraw-Hill, 2007.

Blask DE, Brainard GC, Dauchy RT, et al. Melatonin-depleted blood from premenopausal women exposed to light at night stimulates growth of human breast cancer xenografts in nude rats. *Cancer Res* 2005a;65:11174–11184.

Blask DE, Dauchy RT, Sauer LA. Putting cancer to sleep at night: The neuroendocrine/circadian melatonin signal. *Endocrine* 2005b;27:179–188.

Bouhafs RK, Jarstrand C. Effects of antioxidants on surfactant peroxidation by stimulated human polymorphonuclear leukocytes. *Free Radic Res* 2002;36:727–734.

Boureau YL, Dayan P. Opponency revisited: Competition and cooperation between dopamine and serotonin. *Neuropsychopharmacology* 2011;36:74–97.

Braam W, Smits MG, Didden R, et al. Exogenous melatonin for sleep problems in individuals with intellectual disability: A meta-analysis. *Dev Med Child Neurol* 2009;51:340–349.

Brown GL, Goodwin FK, Ballenger JC, Gover PF. Aggression in humans correlates with cerebrospinal fluid amine metabolites. *Psychiatry Res* 1979;1:131–139.

Brusco LI, Marquez M, Cardinali DP. Monozygotic twins with Alzheimer's disease treated with melatonin: Case report. *J Pineal Res* 1998;25:260–263.

Brzecka A, Piesiak P, Zareba-Bogdał E, Zierkiewicz G, Plamieniak Z. Rhythm of melatonin excretion in obstructive sleep apnoea syndrome. *Pneumonol Alergol Pol* 2001;69:650–654.

Brzezinski A, Vangel MG, Wurtman RJ, et al. Effects of exogenous melatonin on sleep: A meta-analysis. *Sleep Med Rev* 2005;9:41–50.

Buscemi S, Nicolucci A, Mattina A, et al. Association of dietary patterns with insulin resistance and clinically silent carotid atherosclerosis in apparently healthy people. *Eur J Clin Nutr* 2013;67:1284–1290.

Buscemi N, Vandermeer B, Hooton N, et al. The efficacy and safety of exogenous melatonin for primary sleep disorders: A meta-analysis. *J Gen Intern Med* 2005;20:1151–1158.

Buscemi N, Vandermeer B, Hooton N, et al. Efficacy and safety of exogenous melatonin for secondary sleep disorders and sleep disorders accompanying sleep restriction: Meta-analysis. *BMJ* 2006;332:385–393.

Cabrera J, Negrin G, Estevez F, Loro J, Reiter RJ, Quintana J. Melatonin decreases cell proliferation and induces melanogenesis in human melanoma SK-MEL-1 cells. *J Pineal Res* 2010;49:45–54.

Cangiano C, Ceci F, Cascino A, et al. 1992. Eating behavior and adherence to dietary prescriptions in obese adult subjects treated with 5-hydroxytryptophan. *Am J Clin Nutr* 1992;56:863–867.

Carlomagno G, Nordio M, Chiu TT, Unfer V. Contribution of myo-inositol and melatonin to human reproduction. *Eur J Obstet Gynecol Reprod Biol* 2011;159:267–272.

Casado-Zapico S, Martin V, Garcia-Santos G, et al. Regulation of the expression of death receptors and their ligands by melatonin in haematological cancer cell lines and in leukaemia cells from patients. *J Pineal Res* 2011;50:345–355.

Casado-Zapico S, Rodriguez-Blanco J, Garcia-Santos G, et al. Synergistic antitumor effect of melatonin with several chemotherapeutic drugs on human Ewing sarcoma cancer cells: Potentiation of the extrinsic apoptotic pathway. *J Pineal Res* 2010;48:72–80.

Caspi A, Moffitt TE, Morgan J, et al. Maternal expressed emotion predicts children's antisocial behavior problems: Using monozygotic-twin differences to identify environmental effects on behavioral development. *Dev Psychol* 2004;40:149–161.

Castro N, Spengler M, Lollivier V, Wellnitz O, Meyer HHD, Bruckmaier RM. Diurnal pattern of melatonin in blood and milk of dairy cows. *Milchwiss-Milk Sci Int* 2011;66:352–353.

Castro-Silva C, de Bruin VM, Cunha GM, Nunes DM, Medeiros CA, de Bruin PF. Melatonin improves sleep and reduces nitrite in the exhaled breath condensate in cystic fibrosis-a randomized, double-blind placebo-controlled study. *J Pineal Res* 2010;48:65–71.

Catherine C, Remontet L, Noel-Baron F, et al. A dietary supplement to improve the quality of sleep: A randomized placebo controlled trial. *BMC Complementary Altern Med* 2010;10:29.

Chananda SR, Behen ME, Juhász C, et al. Significance of abnormalities in developmental trajectory and asymmetry of cortical serotonin synthesis in autism. *Int J Dev Neurosci* 2005;23:171–182.

Chen G, Huo Y, Tan DX, Liang Z, Zhang W, Zhang Y. Melatonin in Chinese medicinal herbs. *Life Sci* 2003;73:19–26.

Chugani DC, Muzik O, Behen M, et al. Developmental changes in brain serotonin synthesis capacity in autistic and non-autistic children. *Ann Neurol* 1999;45:287–295.

Chung KF, Lee CK. Over-the-counter sleeping pills: A survey of use in Hong Kong and a review of their constituents. *Gen Hosp Psychiatry* 2002;24:430–435.

Cools R, Nakamura K, Daw ND. Serotonin and dopamine: Unifying affective, activational, and decision functions. *Neuropsychopharmacology* 2011;36:98–113.

Coutts RT, Baker GB, Pasutto FM. Food stuffs as sources of psychoactive amines and their precursors: Content, significance and identification. *Adv Drug Res* 1986;15:169–232.

Cruzon G. Serotonin and appetite. *Ann NY Acad Sci* 1990;600:521–530.

Cubero J, Toribio F, Garrido M, et al. Assays of the amino acid tryptophan in cherries by HPLC-fluorescence. *Food Anal Methods* 2010;3:36–39.

Cubero J, Valero V, Sánchez J, et al. The circadian rhythm of tryptophan in breast milk affects the rhythms of 6-sulfatoxymelatonin and sleep in newborn. *Neuro Endocrinol Lett* 2005;26:657–661.

Curzon G. Brain tryptophan: Normal and disturbed control. In GA Filippini, CVL Costa, and A Bertazzo (eds), *Recent Advances in Tryptophan Research*, pp. 27–34. New York: Plenum Press, 1996.

Cutando A, Aneiros-Fernandez J, Aneiros-Cachaza J, Arias-Santiago S. Melatonin and cancer: Current knowledge and its application to oral cavity tumours. *J Oral Pathol Med* 2011;40:593–597.

Cutando A, Lopez-Valverde A, Arias-Santiago S, De Vicente J, Gomez de Diego R. Role of melatonin in cancer treatment. *Anticancer Res* 2012;32:2747–2754.

Dalton EJ, Rotondi D, Levitan RD, Kennedy SH, Brown GM. Use of slow-release melatonin in treatment-resistant depression. *J Psychiatry Neurosci* 2000;25:48–52.

Davidson RJ, Putnam KM, Larson CL. Dysfunction in the neural circuitry of emotion regulation-a possible prelude to violence. *Science* 2000;289:591–594.

Daw ND, Kakade S, Dayan P. Opponent interactions between serotonin and dopamine. *Neural Netw* 2002;15:603–616.

De Benedittus G, Massei R. 5HT precursors in migraine prophylaxis: A double-blind cross-over study with L-5-hydroxy-tryptophan versus placebo. *Clin J Pain* 1986;3:123–129.

de la Puerta C, Carrascosa-Salmoral MP, García-Luna PP, et al. Melatonin is a phytochemical in olive oil. *Food Chem* 2007;104:609–612.

De Leersnyder H, Bresson JL, de Blois MC, et al. Beta 1-adrenergic antagonists and melatonin reset the clock and restore sleep in a circadian disorder, Smith-Magenis syndrome. *J Med Genet* 2003;40:74–78.

De Simoni MG, Dal Toso G, Fodritto F, Sokola A, Algeri S. Modulation of striatal dopamine metabolism by the activity of dorsal raphe serotonergic afferences. *Brain Res* 1987;411:81–88.

Diaz BL, Llaneza PC. Endocrine regulation of the course of menopause by oral melatonin: First case report. *Menopause* 2008;5:388–392.

Dijk DJ, Cajochen C. Melatonin and the circadian regulation of sleep initiation, consolidation, structure and the sleep EEG. *J Biol Rhythms* 1997;12:627–635.

Dodick DW, Capobianco DJ. Treatment and management of cluster headache. *Curr Pain Headache Rep* 2001;5:83–91.

Dolberg OT, Hirschmann S, Grunhaus L. Melatonin for the treatment of sleep disturbances in major depressive disorder. *Am J Psychiatry* 1998;155:1119–1121.

Dominguez-Rodriguez A, Abreu-Gonzalez P, Reiter RJ. Clinical aspects of melatonin in the acute coronary syndrome. *Curr Vasc Pharmacol* 2009;7:367–373.

Doya K. Metalearning and neuromodulation. *Neural Netw* 2002;15:495–506.

Engler AC, Hadash A, Shehadeh N, Pillar G. Breastfeeding may improve nocturnal sleep and reduce infantile colic: Potential role of breast milk melatonin. *Eur J Pediatr* 2012;171(4):729–732.

Esposito E, Iacono A, Muia C, et al. Signal transduction pathways involved in protective effects of melatonin in C6 glioma cells. *J Pineal Res* 2008;44:78–87.

FDA. FDA issues dietary supplements final rule (press release). June 22, 2007. Retrieved August 4, 2009. http://www.fda.gov/NewsEvents/Newsroom/PressAnnouncements/2007/ucm108938.htm.

Fee DB, Swartz KR, Scheff N, Roberts K, Gabbita P, Scheff S. Melatonin-analog, beta-methyl-6-chloromelatonin, supplementation in spinal cord injury. *Brain Res* 2010;1340:81–85.

Fernstrom JD. Dietary amino acids and brain function. *J Am Diet Assoc* 1994;94:71–77.

Fildes JE, Yonan N, Keevil BG. Melatonin: A pleiotropic molecule involved in pathophysiological processes following organ transplantation. *Immunology* 2009;127:443–449.

Forrest CM, Mackay GM, Stoy N, Stone TW, Darlington LG. Inflammatory status and kynurenine metabolism in rheumatoid arthritis treated with melatonin. *Br J Clin Pharmacol* 2007;64:517–526.

Fulia F, Gitto E, Cuzzocrea S, et al. Increased levels of malondialdehyde and nitrite/nitrate in the blood of asphyxiated newborns: Reduction by melatonin. *J Pineal Res* 2001;31:343–349.

Gaby AR. Natural approaches to epilepsy. *Altern Med Rev* 2007;12:9–24.

Gagnier JJ. The therapeutic potential of melatonin in migraines and other headache types. *Altern Med Rev* 2001;6(4):383–389.

Garrido M, Espino J, González-Gómez D, et al. A nutraceutical product based on Jerte Valley cherries improves sleep and augments the antioxidant status in humans. *e-SPEN* 2009;4:e321–e323.

Garrido M, Paredes SD, Cubero J, et al. Jerte Valley cherry-enriched diets improve nocturnal rest and increase 6-sulfatoxy melatonin and total antioxidant capacity in the urine of middle-aged and elderly humans. *J Gerontol A Biol Sci Med Sci* 2010;65:909–914.

Garstang J, Wallis M. Randomized controlled trial of melatonin for children with autistic spectrum disorders and sleep problems. *Child Care Health Dev* 2006;32:585.

Geldmacher DS, Lerner AJ, Voci JM, Noelker EA, Somple LC, Whitehouse PJ. Treatment of functional decline in adults with Down syndrome using selective serotonin-reuptake inhibitor drugs. *J Geriatr Psychiatry Neurol* 1997;10:99–104.

Gershon MD. Serotonin and its implication for the management of irritable bowel syndrome. *Rev Gastroenterol Disord* 2003;3:25–34.

Gesch CB, Hammond SM, Hampson SE, Eves A, Crowder MJ. Influence of supplementary vitamins, minerals and essential fatty acids on the antisocial behaviour of young adult prisoners: Randomised, placebo-controlled trial. *Br J Psychiatry* 2002;181:22–28.

Giannotti F, Cortesi F, Cerquiglini A, Bernabei P. An open-label study of controlled-release melatonin in treatment of sleep disorders in children with autism. *J Autism Dev Disord* 2006;36:741–752.

González-Gómez D, Lozano M, Fernández-León MF, Ayuso MC, Bernalte MJ, Rodríguez AB. Detection and quantification of melatonin and serotonin in eight sweet cherry cultivars (*Prunus avium* L.). *Eur Food Res Technol* 2009;229:223–229.

González-Gómez D, Lozano M, Fernández-León MF, Bernalte MJ, Ayuso MC, Rodríguez AB. Sweet cherry phytochemicals: Identification and characterization by HPLC-DAD/ESI-MS in sweet-cherry cultivars grown in Valle del Jerte (Spain). *J Food Compos Anal* 2010;23:533–539.

Grant SG, Melan MA, Latimer JJ, Witt-Enderby PA. Melatonin and breast cancer: Cellular mechanisms, clinical studies and future perspectives. *Expert Rev Mol Med* 2009;11:e5.

Grant WB. Effect of interval between serum draw and follow-up period on relative risk of cancer incidence with respect to 25-hydroxyvitamin D level implications for meta-analyses and setting vitamin D guidelines. *Dermatoendocrinol* 2011;3:199–204.

Greenlee H, Hershman DL, Jacobson JS. Use of antioxidant supplements during breast cancer treatment: A comprehensive review. *Breast Cancer Res Treat* 2009;115:437–452.

Greenlee H, White E, Patterson RE, Kristal AP. Supplement use among cancer survivors in the Vitamins and Lifestyle (VITAL) study cohort. *J Altern Complement Med* 2004;10:660–666.

Grossman E, Laudon M, Zisapel N. Effect of melatonin on nocturnal blood pressure: Meta-analysis of randomized controlled trials. *Vasc Health Risk Manag* 2011;7:577–584.

Hattori A, Migitaka H, Iigo M, et al. Identification of melatonin in plants and its effects on plasma melatonin levels and binding to melatonin receptors in vertebrates. *Biochem Mol Biol Int* 1995;35:627–634.

Heine W, Radke M, Wutzke KD. The significance of tryptophan in human nutrition. *Amino Acids* 1995;9:91–205.

Heisler LK, Cowley MA, Tecott LH, et al. Activation of central melanocortin pathways by fenfluramine. *Science* 2002;297:609–611.

Hill SM, Frasch T, Xiang S, Yuan L, Duplessis T, Mao L. Molecular mechanisms of melatonin anticancer effects. *Integr Cancer Ther* 2009;8:337–346.

Hörl WH. Adjunctive therapy in anaemia management. *Nephrol Dial Transplant* 2002;17:56–59.

Hotta Y, Nagatsu A, Liu W, et al. Protective effects of antioxidative serotonin derivatives isolated from safflower against postischemic myocardial dysfunction. *Mol Cell Biochem* 2002;238:151–162.

Howatson G, Bell PG, Tallent J, Middleton B, McHugh MP, Ellis J. Effect of tart cherry juice (*Prunus cerasus*) on melatonin levels and enhanced sleep quality. *Eur J Nutr* 2012;51:909–916.

Hussain SA, Khadim HM, Khalaf BH, et al. 2006. Effects of melatonin and zinc on glycemic control in type 2 diabetic patients poorly controlled with metformin. *Saudi Med J* 2006;27:1483–1488.

Inoue T, Tsuchiya K, Koyama T. Regional changes in dopamine and 5-hydroxytryptamine activation with various intensity of physical and psychological stress in rat brain. *Pharmacol Biochem Behav* 1994;49:911–920.

Iriti M, Varoni EM, Vitalini S. Melatonin in traditional Mediterranean diets. *J Pineal Res* 2010;49:101–105.

Ivy AS, Rodriguez FG, Garcia C, Chen MJ, Russo-Neustadt AA. Noradrenergic and serotonergic blockade inhibits BDNF mRNA activation following exercise and antidepressant. *Pharmacol Biochem Behav* 2003;75:81–88.

Izykowska I, Gebarowska E, Cegielski M, Podhorska-Okolow M, Piotrowska A, Zabel M, Dziegiel P. Effect of melatonin on melanoma cells subjected to UVA and UVB radiation in in *vitro* studies. *In Vivo* 2009;23:733–738.

James SP, Mendelson WB, Sack DA, et al. The effect of melatonin on normal sleep. *Neuropsychopharmacology* 1987;1:41–44.

Jan JE, Hamilton D, Seward N, Fast DK, Freeman RD, Laudon M. Clinical trials of controlled-release melatonin in children with sleep-wake cycle disorders. *J Pineal Res* 2000;29:34–39.

Jean-Louis G, von Gizycki H, Zizi F. Melatonin effects on sleep, mood and cognition in elderly with mild cognitive impairment. *J Pineal Res* 1998;25:177–183.

Johns J, Bo S, Pratheepawanit-Johns N. Screening of Thai medicinal herbs, fruit and vegetables for dietary melatonin. FASEB Summer Research Conference: Melatonin Receptors: Actions and Therapeutics, FASEB, Denver, CO, 2011.

Jones C, Huyton M, Hindley D. Melatonin and epilepsy. *Arch Dis Child* 2005;90:1203.

Joo SS, Yoo YM. Melatonin induces apoptotic death in LNCaP cells via p38 and JNK pathways: Therapeutic implications for prostate cancer. *J Pineal Res* 2009;47:8–14.

Josefsson LG, Rask L. Cloning of a putative G-protein-coupled receptor from *Arabidopsis thaliana*. *Eur J Biochem* 1997;249:415–420.

Jou MJ, Peng TI, Yu PZ, et al. Melatonin protects against common deletion of mitochondrial DNA-augmented mitochondrial oxidative stress and apoptosis. *J Pineal Res* 2007;43:389.

Jouan PN, Pouliot Y, Gauthier SF, Laforest JP. Hormones in bovine milk and milk products: A survey. *Int Dairy J* 2006;16:1408–1414.

Jouvet M. Sleep and serotonin: An unfinished story. *Neuropsychopharmacol* 1999;21:24S–27S.

Jung-Hynes B, Schmit TL, Reagan-Shaw SR, Siddiqui IA, Mukhtar H, Ahmad N. Melatonin, a novel Sirt1 inhibitor, imparts antiproliferative effects against prostate cancer *in vitro* in culture and *in vivo* in TRAMP model. *J Pineal Res* 2011;50:140–149.

Karbownik M, Reiter RJ, Cabrera J, Garcia JJ. Comparison of the protective effect of melatonin with other antioxidants in the hamster kidney model of estradiol-induced DNA damage. *Mutat Res* 2001;474:87–92.

Kato K, Hirai K, Nishiyama K, et al. Neurochemical properties of ramelteon (TAK-375), a selective MT1/MT2 receptor agonist. *Neuropharmacol* 2005;48:301–310.

Kedziora-Kornatowska K, Szewczyk-Golec K, Czuczejko J, et al. Antioxidative effects of melatonin administration in elderly primary essential hypertension patients. *J Pineal Res* 2008;45:312–317.

Kedziora-Kornatowska K, Szewczyk-Golec K, Kozakiewicz M, et al. Melatonin improves oxidative stress parameters measured in the blood of elderly type 2 diabetic patients. *J Pineal Res* 2009;46:333–337.

Kim BC, Shon BS, Ryoo YW, Kim SP, Lee KS. Melatonin reduces x-ray irradiation-induced oxidative damages in cultured human skin fibroblasts. *J Dermatol Sci* 2001;26:194–200.

Koh W, Jeong SJ, Lee HJ, et al. Melatonin promotes puromycin-induced apoptosis with activation of caspase-3 and 5′-adenosine monophosphate activated kinase-alpha in human leukemia HL-60 cells. *J Pineal Res* 2011;50:367–373.

Koppisetti S, Jenigiri B, Terron MP, et al. Reactive oxygen species and the hypomotility of the gall bladder as targets for the treatment of gallstones with melatonin: A review. *Dig Dis Sci* 2008;53:2592–6203.

Kotlarczyk MP, Lassila HC, O'Neil CK, et al. Melatonin osteoporosis prevention study (MOPS): A randomized, double-blind, placebo-controlled study examining the effects of melatonin on bone health and quality of life in perimenopausal women. *J Pineal Res* 2012;52:414–426.

Krauchi K, Wirz-Justice A. Circadian clues to sleep onset mechanisms. *Neuropsychopharmacol* 2001;25:S92–S96.

Kucukakin B, Gfgenur I, Reiter RJ, Rosenberg J. Oxidative stress in relation to surgery: Is there a role for the antioxidant melatonin? *J Surg Res* 2009;152:338–347.

Kumarasamya Y, Middletona M, Reida RG, Naharb L, Sarkera SD. Biological activity of serotonin conjugates from the seeds of *Centaurea nigra*. *Fitoterapia* 2003;74:609–612.

Kurecki AE, Alpay F, Tanindi S, et al. Plasma trace element, plasma glutathione peroxidase, and superoxide dismutase levels in epileptic children receiving antiepileptic drug therapy. *Epilepsia* 1995;36:600–604.

Kvaskoff M, Weinstein P. Are some melanomas caused by artificial light? *Med Hypotheses* 2010;75:305–311.

Lanfumey L, Mongeau R, Cohen-Salmon C, Hamon M. Corticosteroid-serotonin interactions in the neurobiological mechanisms of stress-related disorders. *Neurosci Biobehav Rev* 2008;32:1174–1184.

Lee T, Dugoua JJ. Nutritional supplements and their effect on glucose control. *Curr Diab Rep* 2011;11:142–148.

Lemoine P, Nir T, Laudon M, Zisapel Z. Prolonged-release melatonin improves sleep quality and morning alertness in insomnia patients aged 55 years and older and has no withdrawal effects. *J Sleep Res* 2007;16:372–380.

Levy SE, Hyman SL. Complementary and alternative medicine treatments for children with autism spectrum disorders. *Child Adolesc Psychiatr Clin N Am* 2008;17:803–809.

Lidberg L, Tuck JR, Asberg M, Scalia-Tomba GP, Bertilsson L. Homicide, suicide and CSF 5-HIAA. *Acta Psychiatr Scand* 1985;71:230–236.

Linnoila VM, Virkkunen M. Aggression, suicidality, and serotonin. *J Clin Psychiatry* 1992;3:46–51.

Lissoni P, Chilelli M, Villa S, Cerizza L, Tancini G. Five years survival in metastatic non-small cell lung cancer patients treated with chemotherapy alone or chemotherapy and melatonin: A randomized trial. *J Pineal Res* 2003;35:12–15.

Lovinger DM. The role of serotonin in alcohol's effects on the brain. *Curr Separ* 1999;18:1.

Machida M, Dubousset J, Yamada T, Kimura J. Serum melatonin levels in adolescent idiopathic scoliosis prediction and prevention for curve progression: A prospective study. *J Pineal Res* 2009;46:344–348.

Maestroni GJM, Conti A. Melatonin in human breast cancer tissue: Association with nuclear grade and estrogen receptor status. *Lab Invest* 1996;75:557–561.

Maharaj DS, Glass BD, Daya S. Melatonin: New places in therapy. *Biosci Rep* 2007;27:299–320.

Maissen C, Ludin HF. Comparison of the effect of 5 hydroxytryptophan and propranolol with interval treatment of migraine. *Schweiz Med Wochenschr* 1991;121:1585–1590.

Manchester LC, Tan DX, Reiter RJ, et al. High levels of melatonin in the seeds of edible plants: Possible function in germ tissue protection. *Life Sci* 2000;67:3023–3029.

Markov M, Kolev P. Effect of melatonin on the antioxidant defence system in patients with epilepsy. *J Clin Med* 2009;2:12–15.

Mattson MP, Maudsley S, Martin B. BDNF and 5-HT: A dynamic duo in age-related neuronal plasticity and neurodegenerative disorders. *Trends Neurosci* 2004;27:589–594.

McCarty MF. Zinc and multi-mineral supplementation should mitigate the pathogenic impact of cadmium exposure. *Med Hypotheses* 2012;79:642–648.

McMullan CJ, Schernhammer ES, Rimm EB, et al. Melatonin secretion and the incidence of type 2 diabetes. *JAMA* 2013;309:1388–1396.

Mediavilla MD, Sanchez-Barcelo EJ, Tan DX, Manchester L, Reiter RJ. Basic mechanisms involved in the anti-cancer effects of melatonin. *Curr Med Chem* 2010;36:4462–4480.

Medical News Today. FDA tightens up dietary supplement manufacturing and labelling. June 26, 2007. Retrieved September 2, 2013. http://www.medicalnewstoday.com/healthnews.php?newsid=75250&nfid=crss.

Melke J, Goubran Botros H, Chaste P, et al. Abnormal melatonin synthesis in autism spectrum disorders. *Mol Psychiatry* 2008;13:90–98.

Mendelson WB. Efficacy of melatonin as a hypnotic agent. *J Biol Rhythms* 1997;12:651–656.

Mercolini L, Mandrioli R, Raggi MA. Content of melatonin and other antioxidants in grape-related foodstuffs: Measurement using a MEPS-HPLC-F method. *J Pineal Res* 2012;53:21–28.

Miczek KA, Fish EW, DeBold JF, de Almeida RMM. Social and neural determinants of aggressive behavior: Pharmacotherapeutic targets at serotonin, dopamine and γ-aminobutyric acid systems. *Psychopharmacol* 2002;163:434–458.

Mills E, Wu P, Seely D, Guyatt G. Melatonin in the treatment of cancer: A systematic review of randomized controlled trials and meta-analysis. *J Pineal Res* 2005;39:360–366.

Min KJ, Kim HS, Park EJ, Kwon TK. Melatonin enhances thapsigargin-induced apoptosis through reactive oxygen species–mediated upregulation of CCAAT-enhancer-binding protein homologous protein in human renal cancer cells. *J Pineal Res* 2012;53:99–106.

Moore T, Scarpa A, Raine A. A meta-analysis of serotonin metabolite 5-HIAA and antisocial behavior. *Aggres Behav* 2002;28:299–316.

Movsesian MA. Beta-adrenergic receptor agonists and cyclic nucleotide phosphodiesterase inhibitors: Shifting the focus from inotropy to cyclic adenosine monophosphate. *J Am Coll Cardiol* 1999;34:318–324.

Murch SJ, Simmons CB, Saxena PK. Melatonin in feverfew and other medicinal plants. *Lancet* 1997;350:1598–1599.

Nagatsu A, Zhang HL, Mizukami H. Tyrosinase inhibitory and antitumor promoting activities of compounds isolated from safflower (*Carthamus tinctorius* L.) and cotton (*Gossypium hirsutum* L.) oil cakes. *Nat Prod Res* 2000;14:153–158.

Nishimura J, Saegusa Y, Dewa Y, et al. Antioxidant enzymatically modified isoquercitrin or melatonin supplementation reduces oxidative stress-mediated hepatocellular tumor promotion of oxfendazole in rats. *Arch Toxicol* 2010;84:143–153.

Oaknin-Bendahan S, Anis Y, Nir I, Zisapel N. Effects of long-term administration of melatonin and a putative antagonist on the ageing rat. *Neuro Rep* 1995;6:785–788.

Oh EY, Ansell C, Nawaz H, Yang CH, Wood PA, Hrushesky WJ. Global breast cancer seasonality. *Breast Cancer Res Treat* 2010;123:233–243.

Olde Rikkert MG, Rigaud AS. Melatonin in elderly patients with insomnia: A systematic review. *Z Gerontol Geriatr* 2001;34:491–497.

Padumanoda J, Johns A, Sangkasat A, Tiyaworanund S. Rapid analysis of melatonin content in Thai herbs used as sleeping aids. *DARU J Pharmaceutl Sci* 2013;22:6.

Pandi-Perumal SR, Srinivasan V, Maestroni GJM, et al. Melatonin: Nature's most versatile biological signal? *FEBS J* 2006;273:2813–2838.

Paul MA, Love RJ, Hawton A, Arendt J. Sleep and the endogenous melatonin rhythm of high artic residents during summer and winter. *Physiol Behav* 2015;141:199–206.

Pearl PL, Drillings IM, Conry JA. Herbs in epilepsy: Evidence for efficacy, toxicity, and interactions. *Semin Pediatr Neurol* 2011;18:203–208.

Peled N, Shorer Z, Peled E, Pillar G. Melatonin effect on seizures in children with severe neurologic deficit disorders. *Epilepsia* 2001;42:1208–1210.

Peuhkuri K, Sihvola N, Korpela R. Dietary factors and fluctuating levels of Melatonin. *Food Nutr Res* 2012;56:17252–17260.

Piasek A, Bartoszek A, Namieśnik J. Phytochemicals that counteract the cardiotoxic side effects of cancer chemotherapy. *Postepy Hig Med Dosw* 2009;63:142–158.

Pires ML, Benedito-Silva AA, Pinto L, et al. Acute effects of low doses of melatonin on the sleep of young healthy subjects. *J Pineal Res* 2001;31:326–332.

Poeggler B, Saarela S, Reiter RJ, et al. Melatonin: A highly potent endogenous radical scavenger and electron donor; New aspects of the oxidation chemistry of this indole accessed *in vitro*. *Ann NY Acad Sci* 1994;738:419–420.

Pohanka M. Alzheimer's disease and related neurodegenerative disorders: Implication and counteracting of melatonin. *J Appl Biomed* 2011;9:185–196.

Poldinger W, Calanchini B, Schwarz W. A functional-dimensional approach to depression, comparing 5-HTP to fluvocarnine. *Psychopathology* 1991;24:53–81.

Radziuk J, Pye S. Diurnal rhythm in endogenous glucose production is a major contributor to fasting hyperglycemia in type 2 diabetes: Suprachiasmatic deficit or limit cycle behavior. *Diabetologia* 2006;49:1619–1628.

Raine A, Mellingen K, Liu J, Venables P, Mednick SA. Effects of environmental enrichment at ages 3–5 years on schizotypal personality and antisocial behavior at ages 17 and 23 years. *Am J Psychiatry* 2003;60:1627–1635.

Ramakrishna A, Giridhar P, Sankar KU, Ravishankar GA. Melatonin and serotonin profiles in beans of *Coffea* species. *J Pineal Res* 2012;52:470–476.

Reiter RJ. The pineal and its hormones in the control of reproduction in mammals. *Endocr Rev* 1980;1:109–131.

Reiter RJ. Melatonin: The chemical expression of darkness. *Mol Cell Endocrinol* 1991;79:C153–C159.

Reiter RJ. Melatonin: Medicinal chemistry and therapeutic potential. *Curr Top Med Chem* 2002;2:113–209.

Reiter RJ, Manchester LC, Tan DX. Neurotoxins: Free radical mechanisms and melatonin protection. *Curr Neuropharmacol* 2010;8:194–210.

Reiter RJ, Tan DX, Manchester LC, et al. Melatonin and reproduction revisited. *Biol Reprod* 2009;81:445–456.

Reiter RJ, Tan DX, Manchester LC, Tamura H. Melatonin defeats neurally-derived free radicals and reduces the associated neuromorphological and neurobehavioral damage. *J Physiol Pharmacol* 2007;58:5–22.

Richardson-Andrews RC. The sunspot theory of schizophrenia: Further evidence, a change of mechanism, and a strategy for the elimination of the disorder. *Med Hypotheses* 2009;72:95–98.

Robbins TW, Crockett MJ. Role of serotonin in impulsivity and compulsivity: Comparative studies in experimental animals and humans. In CP Muller, B Jacobs (eds), *Handbook of the behavioral neurobiology of serotonin*, vol. 21, pp. 415–428. London, UK: Elsevier, 2009.

Rodriguez-Naranjo MI, Gil-Izquierdo AG, Troncoso AM, Cators E, Garcia-Parilla MC. Melatonin: A new bioactive compound in wine. *J Food Compost Anal* 2011;24:603–608.

Rogers RD. The roles of dopamine and serotonin in decision making: Evidence from pharmacological experiments in humans. *Neuropsychopharmacol* 2011;36:114–132.

Rondanelli M, Opizzi A, Monteferrario F, Antoniello N, Manni R, Klersy C. The effect of melatonin, magnesium, and zinc on primary insomnia in long-term care facility residents in Italy: A double-blind, placebo-controlled clinical trial. *J Am Geriatr Soc* 2011;59:82–90.

Rossignol DA, Frye RE. Melatonin in autism spectrum disorders: A systematic review and meta-analysis. *Dev Med Child Neurol (Meta-analysis)* 2011;53:783–792.

Roth T, Seiden D, Sainati S, et al. Effects of ramelteon on patient-reported sleep latency in order adults with chronic insomnia. *Sleep Med* 2006;7:312–318.

Ryan ND, Birmaher B, Perel JM, et al. Neuroendocrine response to L-5-hydroxytryptophan challenge in prepubertal major depression: Depressed vs normal children. *Arch Gen Psychiatry* 1992;49:843–851.

Sanchez-Barcelo EJ, Mediavilla MD, Tan DX, Reiter RJ. Clinical uses of melatonin: Evaluation of human trials. *Curr Med Chem* 2010;17:2070–2095.

Scheer FA, Czeisler CA. Melatonin, sleep, and circadian rhythms. *Sleep Med Rev* 2005;9:5–9.

Schultz W, Dayan P, Montague PR. A neural substrate of prediction and reward. *Science* 1997;275:1593–1599.

Seely D, Wu P, Fritz H, et al. Melatonin as adjuvant cancer care with and without chemotherapy: A systematic review and meta-analysis of randomized trials. *Integr Cancer Ther* 2012;11:293–303.

Seidl R, Kaehler ST, Prast H, et al. Serotonin (5-HT) in brains of adult patients with Down syndrome. *J Neural Transm Suppl* 1999;57:221–232.

Shamir E, Laudon M, Barak Y, et al. Melatonin improves sleep quality of patients with chronic schizophrenia. *J Clin Psychiatry* 2000;61:373–377.

Sharma S, Sharma J. Regulation of appetite: Role of serotonin and hypothalamus. Iranian. *J Pharmacol Therap* 2012;11:73–79.

Sharma VK, Singaravel M, Subbaraj R, Chandrashekaran MK. Timely administration of melatonin accelerates reentrainment to phase-shifted light-dark cycles in the field mouse *Mus booduga*. *Chronobiol Int* 1999;16:163–710.

Shibata S, Cassone VM, Moore RY. Effects of melatonin on neuronal activity in rat suprachiasmatic nucleus *in vitro*. *Neurosci Lett* 1989;97:140–144.

Shimazaki M, Martin JL. Do herbal agents have a place in the treatment of sleep problems in long-term care? *J Am Med Dir Assoc* 2007;8:248–252.

Shiu SY, Law IC, Lau KW, Tam PC, Yip AW, Ng WT. Melatonin slowed the early biochemical progression of hormone-refractory prostate cancer in a patient whose prostate tumor tissue expressed MT1 receptor subtype. *J Pineal Res* 2003;35:177–182.

Singer C, Tractenberg RE, Kaye J, et al. A multicenter, placebo-controlled trial of melatonin for sleep disturbance in Alzheimer's disease. *Sleep* 2003;26:893–901.

Sircar R. Effect of melatonin on cocaine-induced behavioral sensitization. *Brain Res* 2000;857:295–299.

Srinivasan V, Pandi-Perumal SR, Brzezinski A, Bhatnagar KP, Cardinali DP. Melatonin, immune function and cancer. *Recent Pat Endocr Metab Immune Drug Discov* 2011;5:109–123.

Srinivasan V, Pandi-Perumal SR, Cardinali DP, et al. Melatonin in Alzheimer's disease and other neurodegenerative disorders. *Behav Brain Funct* 2006;2:15.

Srinivasan V, Spence DW, Moscovitch A, et al. Malaria: Therapeutic implications of melatonin. *J Pineal Res* 2010;48:1–8.

Srinivasan V, Spence WD, Pandi-Perumal SR, et al. Melatonin and human reproduction: Shedding light on the darkness hormone. *Gynecol Endocrinol* 2009;25:779–785.

Strasser F. Appraisal of current and experimental approaches to the treatment of cachexia. *Curr Opin Support Palliat Care* 2007;1:312–316.

Sturtz M, Cerezo AB, Cantos-Villar E, Garcia-Parrilla MC. Determination of the melatonin content of different varieties of tomatoes (*Lycopersicon esculentum*) and strawberries (*Fragaria ananassa*). *Food Chem* 2011;127:1329–1334.

Swann AC. Neuroreceptor mechanisms of aggression and its treatment. *J Clin Psychiatry* 2003;64:26–35.

Tam CW, Shiu SY. Functional interplay between melatonin receptor-mediated antiproliferative signaling and androgen receptor signaling in human prostate epithelial cells: Potential implications for therapeutic strategies against prostate cancer. *J Pineal Res* 2011;51:297–312.

Tamura H, Takasaki A, Miwa I, et al. Oxidative stress impairs oocyte quality and melatonin protects oocytes from free radical damage and improves fertilization rate. *J Pineal Res* 2008;44:280–287.

Tan DX, Manchester LC, Terron MP, Flores LJ, Reiter RJ. One molecule, many derivatives: A never-ending interaction of melatonin with reactive oxygen and nitrogen species? *J Pineal Res* 2007;42:28–42.

Tan DX, Hardeland R, Manchester LC, et al. The changing biological roles of melatonin during evolution: From an antioxidant to signals of darkness, sexual selection and fitness. *Biol Reviews* 2009;85:607–623.

Terry PD, Villinger F, Bubenik GA, Sitaraman SV. Melatonin and ulcerative colitis: Evidence, biological mechanisms, and future research. *Inflamm Bowel Dis* 2009;15:134–140.

Titus F, Davalos A, Alom J, et al. 5-hydroxytryptophan versus methysergide [*sic*] in the prophylaxis of migraine: A randomized clinical trial. *Eur Neurol* 1986;25:327–329.

Todini L, Terzano GM, Borghese A, Debenedetti A, Malfatti A. Plasma melatonin in domestic female Mediterranean sheep (Comisana breed) and goats (Maltese and Red Syrian). *Res Vet Sci* 2011;90:35–39.

Toma CD, Svoboda M, Arrich F, Ekmekcioglu C, Assadian O, Thalhammer T. Expression of the melatonin receptor (MT) 1 in benign and malignant human bone tumors. *J Pineal Res* 2007;43:206–213.

Traskman-Bendz L, Asberg M, Schalling D. Serotonergic function and suicidal behavior in personality disorders: Abdominal aortic aneurysm. *Gen Pathophysiol Mol* 1986;487:168–714.

Tütüncülar F, Eskiocak S, Basaran UN, Ekuklu G, Ayvaz S, Vatansever U. The protective role of melatonin in experimental hypoxic brain damage. *Pediatr Int* 2005;47:434–439.

Ulfberg J, Micic S, Strom J. Afternoon serum-melatonin in sleep disordered breathing. *J Intern Med* 1998;244:163–168.

Uz T, Akhisaroglu M, Ahmed R, Manev H. The pineal gland is critical for circadian period 1 expression in the striatum and for circadian cocaine sensitization in mice. *Neuropsychopharmacology* 2003;28:2117–2123.

Van Erp AM, Miczek KA. Aggressive behavior, increased accumbal dopamine, and decreased cortical serotonin in rats. *J Neurosci* 2000;20:9320–9325.

van Geijlswijk IM, Korzilius HPLM, Smits MG. The use of exogenous melatonin in delayed sleep phase disorder: A meta-analysis. *Sleep* 2010;33:1605–1614.

Veatch OJ, Pendergast JS, Allen MJ, et al. Genetic variation in melatonin pathway enzymes in children with autism spectrum disorder and comorbid sleep onset delay. *J Autism Dev Disord* 2015;45:100–110.

Virkkunen M, De Jong J, Bartko JJ, Goodwin FK, Linnoila M. Relationship of psychobiological variables to recidivism in violent offenders and impulsive fire setters: A follow-up study. *Arch Gen Psychiatry* 1989;46:600–603.

Virkkunen M, Rawlings R, Tokola R, et al. CSF biochemistries, glucose metabolism, and diurnal activity rhythms in alcoholic, violent offenders, fire setters, and healthy volunteers. *Arch Gen Psychiatry* 1994;51(1):20–27.

Wade AG, Ford I, Crawford G, et al. Efficacy of prolonged release melatonin in insomnia patients aged 55–80 years: Quality of sleep and next-day alertness outcomes. *Curr Med Res Opin* 2007;23:2597–2605.

Waldhauser F, Waldhauser M, Lieberman HR, et al. Bioavailability of oral melatonin in humans. *Neuroendocrinol* 1984;39:307–313.

Wang J, Jiang C, Li S, Zheng J. Study on analysis method of melatonin and melatonin content in corn and rice seeds. *Chin Agric Sci Bull* 2009;25:20–24.

Wang X. The antiapoptotic activity of melatonin in neurodegenerative diseases. *CNS Neurosci Ther* 2009;15:345–357.

Wang X, Wang ZH, Wu YY, et al. Melatonin attenuates scopolamine-induced memory/synaptic disorder by rescuing EPACs/miR-124/Egr1 pathway. *Mol Neurobiol* 2013;47:373–381.

Watanabe M, Kobayashi Y, Takahashi N, Kiguchi K, Ishizuka B. Expression of melatonin receptor (MT1) and interaction between melatonin and estrogen in endometrial cancer cell line. *J Obstet Gynaecol Res* 2008;34:567–573.

Whittle N, Sartori SB, Dierssen M, Lubec G, Singewald N. Fetal Down syndrome brains exhibit aberrant levels of neurotransmitters critical for normal brain development. *Pediatrics* 2007;120:e1465–e1471.

Wikner J, Svanborg E, Wetterberg L, Röjdmark S. Melatonin secretion and excretion in patients with obstructive sleep apnoea syndrome. *Sleep* 1997;20:1002–1007.

Wright B, Sims D, Smart S, et al. Melatonin versus placebo in children with autism spectrum conditions and severe sleep problems not amenable to behavior management strategies: A randomized controlled cross-over trial. *J Autism Dev Disord* 2011;41:175–184.

Wurtman RJ, Wurtman JJ. Carbohydrates and depression. *Sci Am* 1989;260:68–75.

Xi SC, Siu SW, Fong SW, Shiu SY. Inhibition of androgen-sensitive LNCaP prostate cancer growth *in vivo* by melatonin: Association of antiproliferative action of the pineal hormone with mt1 receptor protein expression. *Prostate* 2001;46:52–61.

Xu Y, Ku B, Tie L, et al. Curcumin reverses the effects of chronic stress on behavior, the HPA axis, BDNF expression and phosphorylation of CREB. *Brain Res* 2006;1122:56–64.

Yamamotova A, Pometlova M, Harmatha J, Raskova H, Rokyta R. The selective effect of N-feruloylserotonins isolated from *Leuzea carthamoides* on nociception and anxiety in rats. *J Ethno Pharm* 2007;112:368–374.

Yuksel A, Cengiz M, Seven M, Ulitin T. Erythrocyte glutathione, glutathione peroxidase, superoxide dismutase and serum lipid peroxidation in epileptic children with valproate and carbamazepine monotherapy. *J Basic Clin Physiol Pharmacol* 2000;11:73–81.

Zawilska JB, Skene DJ, Arendt J. Physiology and pharmacology of melatonin in relation to biological rhythms. *Pharmacol Rep* 2009;61:383–410.

Zemlan FP, Mulchahey J, Scharf MB, et al. The efficacy and safety of the melatonin agonist beta-methyl-6-chloromelatonin in primary insomnia: A randomised, placebo-controlled, crossover clinical trial. *J Clin Psychiatry* 2005;66:384–390.

Zesiewicz TA, Evatt ML. Potential influences of complementary therapy on motor and non-motor complications in Parkinson's disease. *CNS Drugs* 2009;23:817–835.

Zhang S, Zuo L, Gui S, Zhou Q, Wei W, Wang Y. Induction of cell differentiation and promotion of endocan gene expression in stomach cancer by melatonin. *Mol Biol Res* 2012;39:2843–2849.

Zhdanova IV, Wurtman RJ, Lynch HJ, et al. Sleep-inducing effects of low doses of melatonin ingested in the evening. *Clin Pharmacol Ther* 1995;57:552–558.

Zhdanova IV, Wurtman RJ, Morabito C, et al. Effects of low oral doses of melatonin, given 2–4 hours before habitual bedtime, on sleep in normal young humans. *Sleep* 1996;19:423–431.

Zhdanova IV, Wurtman RJ, Regan MM, et al. Melatonin treatment for age-related insomnia. *J Clin Endocrinol Metab* 2001;86:4727–4730.

Zhou JN, Liu RY, van Heerikhuize J, Hofman MA, Swaab DF. Alterations in the circadian rhythm of salivary melatonin begin during middle-age. *J Pineal Res* 2003;34:11–16.

Zhu LQ, Wang SH, Ling ZQ, Wang Q, Hu MQ, Wang JZ. Inhibition of melatonin biosynthesis activates protein kinase a and induces Alzheimer-like tau hyperphosphorylation in rats. *Chin Med Sci J* 2005;20:83–87.

Zmilacher K, Battegay R, Gastpar M. L-5-hydroxytryptophan alone and in combination with a peripheral decarboxylase inhibitor in the treatment of depression. *Neuropsychobiology* 1988;20:28–35.

31

Psycho-Oncologic Effects of Serotonin and Melatonin

Hui-Yen Chuang and Jeng-Jong Hwang
National Yang-Ming University
Taipei, Taiwan

Chun-Kai Fang
Mackay Memorial Hospital
Taipei, Taiwan

CONTENTS

ABSTRACT   The relationship between mental and physical health has been discovered and studied for thousands of years. It has been reported that patients with depression may have higher incidences of several diseases including cardiovascular diseases and cancer. On the other hand, many patients suffering from these diseases are also prescribed antidepression drugs. Depression is also often found in drug abusers.

It is known that serotonin (5-HT) functions as a principal neurotransmitter, resulting in the feeling of happiness, and has a strong correlation with depression. Moreover, 5-HT is a derivative of tryptophan, one of the essential amino acids in the diet. A lower level of serotonin is often detected in the patients with cancer and cardiovascular diseases than that found in healthy controls. Altered tryptophan metabolism in the tumor microenvironment skews the balance of neuroprotectants and neurotoxins, thus it has been suggested as one of the underlying mechanisms of depression development in cancer and other related diseases. The neurotoxins generated in the altered tryptophan metabolic pathways contribute to the depression. Furthermore, melatonin, the final product of the metabolism, decreases significantly in cancer patients as well.

Melatonin is known as a circadian regulator, controlling the sleep and reproduction, and also functions as a powerful antioxidant related to neuron protections and antitumor immunity. It is found that both patients with cancer or depression and drug abusers are often suffering with insomnia, which is highly correlated to the levels of serum melatonin. Higher cancer incidence has been reported in night workers than that in day workers, indicating that the melatonin may play a role in protecting neurons and normal cells from oxidative stresses, one of the main causes involved in the development of cardiovascular diseases, cancer, and neuronal diseases including Alzheimer's disease and depression. In this chapter, we will discuss the correlations of 5-HT and melatonin with cardiovascular diseases, cancer, and neuronal diseases, and their roles in mental and physical health from various aspects.

KEY WORDS: *amphetamine, cocaine, indoleamine 2,3-dioxygenase, ketamine, kynurenine, melatonin, methamphetamine, serotonin.*

Abbreviations

AADC:	aromatic amino acid decarboxylase
AANAT:	aralkylamine N-acetyltransferase
ADCC:	antibody-dependent cell cytotoxicity
AMPH:	amphetamine
APC:	antigen-presenting cell
ATD:	acute tryptophan deprivation
BBB:	blood–brain barrier
BEDS:	Brief Edinburgh Depression Scale
BH4:	tetrahydrobiopterin
BLA:	basolateral amygdala
C3HOM:	cyclic 3-hydroxymelatonin
CNS:	central nervous system
CSF:	cerebrospinal fluid
CuZnSOD:	copper–zinc superoxide dismutase
D2R:	dopamine D2 receptor
DAT:	dopamine transporter
DC:	dendritic cell
DOI:	(±)-2,5-dimethoxy-4-iodoamphetamine hydrochloride
EDS:	Edinburgh Depression Scale
EEG:	electroencephalogram
[^{18}F]FDG:	2-deoxy-2-[^{18}F]fluoro-D-glucose
GABA:	γ-aminobutyric acid
5-HIAA:	5-hydroxyindoleacetic acid
HADS:	Hospital Anxiety and Depression Scale
IDO:	indoleamine 2,3-dioxygenase
iNOS:	inducible nitric oxide synthase
KO:	knockout
LNAA:	large neutral amino acid
LSD:	lysergic acid diethylamide
LY341495:	2S-2-amino-2-(1S,2S-2-carboxycyclopropan-1-yl)-3-(xanth-9-yl)- propionic acid

METH:	methamphetamine
mGluR2:	metabotropic glutamate receptor 2
MDA:	malondialdehyde
MDMA:	3,4-methylenedioxymethamphetamine
MN:	micronuclei
NEI:	neuroendocrine–immune
NK:	natural killer [cells]
NMDA:	*N*-methyl-*D*-aspartate
nNOS:	neuronal nitric oxide synthase
NOR:	novel object recognition
NOS:	nitric oxide synthase
PCEs:	polychromatic erythrocytes
PET:	positron emission tomography
PBN:	α-phenyl-*N*-tert-butylnitrone
PLC:	phospholipase C
PNS:	peripheral nervous system
R-96544:	(2R,4R)-5-(2-[2-(2-[3-methoxyphenyl]ethyl)phenoxy]ethyl)-1-methyl-3-pyrrolidinol hydrochloride
REM:	rapid eye movement
RNS:	reactive nitrogen species
ROS:	reactive oxygen species
RT:	radiotherapy
SCN:	suprachiasmatic nucleus
SERT:	serotonin transporter
SPECT:	single-photon emission computed tomography
TBI:	traumatic brain injury
TDO:	tryptophan 2,3-dioxygenase
TPH:	tryptophan hydroxylase
TRAMP:	transgenic adenocarcinoma mouse prostate
Treg:	regulatory T cell
UV:	ultraviolet

Introduction

The field of *psycho-oncology* has been actively studied during the past decade. The interactions among neuron systems, tumor microenvironment, and immunity are discussed in this field. Cancer patients with compromised immunity are often diagnosed with depression, particularly those in advanced stages. Tryptophan and its metabolites including serotonin, melatonin, and kynurenines function as connectors and link these three areas together as shown in Figure 31.1.

Except for cancer patients, patients with autoimmune diseases and drug abusers also have a higher prevalence of depression than do healthy controls. Moreover, the depression is closely associated with neuron damage under various conditions including Alzheimer's disease, schizophrenia, and drug abuse. Many drugs abusers such as 3,4-methylenedioxymethamphetamine (MDMA) and ketamine abusers show extreme happiness and hallucinations soon after receiving drugs; however, their cognitive functions are impaired at later time points. It has been reported that mood, appetite, learning, and memory are related to the serotonin (5-HT) level in the brain. On the other hand, many drug abusers develop insomnia that is tightly correlated with the concentration of melatonin, a derivative converted from 5-HT. It has been reported that night workers or people with insomnia seem to have higher cancer incidence than in normal controls. Melatonin is known as a powerful antioxidant influencing antitumor immunity and protecting against neuron damage caused by ischemia/reperfusion processes. Hence, the roles of 5-HT and melatonin and their related pathways will be discussed in this chapter from different

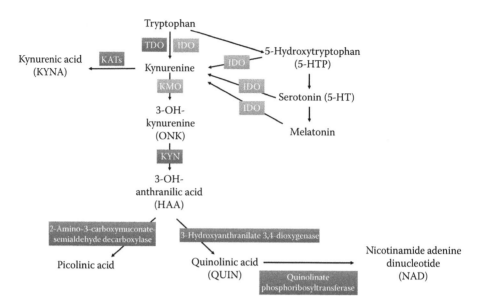

FIGURE 31.1 Tryptophan and its metabolites including serotonin, melatonin, and kynurenine function as connectors and link depression, the immune system, and cancer together. Indoleamine 2,3-dioxygenase (IDO) plays an important role in this correlation, which metabolizes tryptophan, 5-hydroxytrytophan, serotonin, and melatonin into kynurenine. TDO: tryptophan 2,3-dioxygenase; KAT: kynurenine-oxoglutarate transaminase; KMO: kynurenin 3-monooxygenase.

points of view. Moreover, how these molecules affect health in normal subjects, drug abusers, and cancer patients will also be covered.

Effects of Club Drugs on Serotonergic Neurosystem

There were around 52 million abusers of methamphetamine (METH) worldwide in 2010 according to the World Drug Report 2012 (United Nations Office on Drugs and Crime [UNODC] World Drug Report 2012). This number is about 2.5 times the number of the cocaine abusers (19 million) and opiates abusers (13 million). Drug abuse is a serious public health issue affecting teens and young adults (Ersche et al. 2012). Club drugs function as addictive substances, and result in a feeling of well-being at the beginning followed by physical tolerance, dependence, relapse, and craving due to adaptive changes in the central nervous system (CNS) to the drugs. These drugs modify the activity of various receptors including serotonin receptors, γ-aminobutyric acid type A (GABA) receptors, and *N*-methyl-*D*-aspartate (NMDA) receptors (Lüscher and Ungless 2006). Acute intoxication including cerebral hemorrhage, hyperthermia, and the serotonin syndrome has been reported in preclinic and clinic after the administration of club drugs (Freese et al. 2002; Gahlinger 2004). Excess serotonin level in the CNS or peripheral nervous system (PNS) causes the serotonin syndrome, which is rare but lethal. The symptoms including agitation, hyperreflexia, increased heart rate, and hyperthermia, which can be relieved by the withdrawal of drugs or by administrating the serotonin antagonist.

Serotonin (5-HT) plays a critical role in the acute effects and chronic addiction caused by club drugs. Thus, in this section, we will discuss how the commonly used club drugs such as METH, MDMA (Ecstasy), ketamine, and lysergic acid diethylamide (LSD) influence the serotonergic neurons.

Methamphetamine (METH)

Synthesis of METH is a one-step process from its precursor, ephedrine. Easy synthesis results in wider availability and lower cost, causing METH to become one of the most used club drugs among young adults and teens. METH induces euphoria by the rapid release of dopamine, serotonin, and norepinephrine.

METH affects the cardiovascular system (Kaye et al. 2007), disrupts monoaminogeneric systems in the brain (Kita et al. 2003), and influences oral health (Donaldson and Goodchild 2006). Most of the studies focused on the importance of dopamine, which is critical for the reward and reinforcement processes. However, decreased 5-HT synthesis and 5-HT uptake terminal disruptions are also found in the METH-treated animals. However, the exact mechanisms of disruption of monoaminergic damage remain unclear. Cadet et al. (2007) proved that the disruptions of both dopaminergic and serotonergic systems are mediated by generating enormous superoxide radicals using copper–zinc superoxide dismutase (CuZnSOD) transgenic mice and [^{125}I]RT-155 autoradiography (Hirata et al. 1995). Significant increases of [^{125}I]RT-155 were found in the striatum in both heterozygous and homozygous CuZnSOD transgenic mice, which produced less oxidative stress after receiving METH. Free radicals generated by xanthine oxidase have been shown with the selectivity on different monoamine transporters by isolating striatal synaptosomes. Except for inhibiting the dopamine transporter (DAT), METH also blocks the uptake of 5-HT transporters (SERT) rapidly and reversibly after administration. However, the reversed uptake of SERT was transient: the uptake of 5-HT decreased again 1 week after the administration (Kokoshka et al. 1998). These findings partially explain that the formation of METH-induced depression and cognitive impairment observed in METH abusers (Gibb et al. 1987; Chen et al. 2015). Interestingly, unlike dopamine-mediated neurotoxicity influences many brain regions, the serotonin-mediated neurotoxicity is limited to few regions, mainly in the striatum (Hirata et al. 1995). METH damages presynaptic 5-HTs in the forebrain region including anterior cingulate cortex, caudate nucleus, nucleus accumbens, and spetum verified by decreased (^{3}H)-paroxetine binding in these regions (Armstrong and Noguchi 2004). METH-induced cognitive impairment is due to the damage of both the dopaminergic and serotonergic systems. Damage to either dopaminergic or serotonergic neurons, alone, is inadequate to impair the cognition (Belche et al. 2005). Gross et al. (2011) treated rats with dopamine D1 antagonist, SCH23390, or dopamine D2 antagonist, sulpiride, through intrastriatal injection to investigate the interactions between dopaminergic and serotonergic neurons. The results indicate that administration of D1 or D2 antagonist attenuates the losses of both dopamine transporter and SERT of the striatum, cerebral cortex, hippocampus, and basolateral amygdala (BLA) by [^{125}I]RT-155 binding. This is the first study to demonstrate that the serotonergic terminal damage induced by METH within the cerebral cortex and amygdala are correlated to the striatal DATs (Gross et al. 2011). In addition to decreasing the SERT level, METH also inhibits the 5-HT level. Reduced 5-HT level was found within 24 h after administration of a single large dose (40 mg/kg) (Cappon et al. 2000) or multiple small doses (410 mg/kg) (Ali et al. 1994) of METH. The body temperature rose rapidly with the reduction of 5-HT (Cappon et al. 2000), and the mortality caused by hyperthermia could be reduced by decreasing the environmental temperature. The depletion of 5-HT could be recovered when the room temperature decreased to 4°C (Ali et al. 1994). Similar findings were reported by Haughey et al. (2000); decreased [^{3}H]5-HT uptake in both striatum and hippocampus were recovered when the rats were placed under a room temperature of 6°C. The authors also proposed that hyperthermia might trigger the generation of free radicals resulting in METH-induced monoaminergic neurotoxicity (Haughey et al. 2000). These studies point out that 5-HT or SERT plays some role in METH-induced neurotoxicity; however, further investigations about other roles of 5-HT and its interactions with other neurotransmitters in METH addiction are still needed.

3,4-Methylenedioxymethamphetamine (MDMA, Ecstasy)

MDMA, one of the amphetamine derivatives, can result in a rapid increase of the extracellular 5-HT level and significant reduction of 5-HT, as other amphetamines. Unlike amphetamine (AMPH) and METH, MDMA primarily affects the dopaminergic neurons, impairs more serotonergic neurons, and results in memory deficits and cognitive impairments (Cadet et al. 2007). Although MDMA is less toxic than AMPH, it has been reported to induce hyperthermia, serotonin syndrome, and even death (Mueller and Korey 1998; Parrot 2002). Besides, the dose of MDMA for generating long-term serotonergic neuron damages is much lower than that of METH (Ricaurte et al. 1985). One-time exposure of MDMA causes "midweek blues" 2 days after taking tablets, and the mood recovers to the baseline in a week as reported by some prospective surveys (Curran and Travill 1997; Parrott and Lasky 1998; Curran 2000). Because MDMA-induced happiness and relief last for only several hours, it was first thought not to induce addiction. However, accumulating evidences show that repeated MDMA exposure causes long-term serotonergic neuron damages and behavior changes. The effects were first observed in rats in the 1980s: the concentrations of 5-HT and its metabolite, 5-hydroxyindoleacetic acid (5-HIAA), were lower in the brains of MDMA-treated rats compared with those of the controls (Ricaurte et al. 1985; Schmidt et al. 1986). Moreover, Ricaurte et al. demonstrated that MDMA selectively damaged the serotonergic neurons rather than dopaminergic and norepinephrinergic neurons, and the severity of damages was in a dose-dependent manner (Ricaurte et al. 1985). Selective destruction of serotonergic neurons by MDMA is also observed in rhesus monkeys treated with MDMA repeatedly, especially in the cerebral cortex and striatum, with reduced 5-HT and 5-HIAA concentrations, as well as the lower SERT density. Surprisingly, the long-term serotonergic neuron damage in rats might recover afterward (Commins et al. 1987), but this phenomenon was not found in primates (Insel et al. 1989; Ricaurte et al. 2000).

The effects of MDMA on serotonergic neurons could also be monitored by imaging. McCann et al. (1998) used [^{11}C]McN-5652, a SERT ligand, to access the SERT levels in MDMA users and healthy controls. A negative correlation between [^{11}C]McN-5652 binding and number of MDMA exposures was found (McCann et al. 1998). Similar results in current MDMA users but not former MDMA users were also reported (Buchert et al. 2003, 2004). However, these results do not imply that the MDMA-induced serotonergic neuron damage is reversible. McCann et al. again performed [^{11}C]DASB/positron emission tomography (PET) on abstinent MDMA users, and found significant reduction of SERT binding in multiple brain regions related to memory function including dorsolateral prefrontal cortex, orbitofrontal cortex, and parietal cortex, but not DAT, which was reported to be destroyed by MDMA treatment in rats (Commins et al. 1987; McCann et al. 2008). They also found that the SERT bindings and memory function in healthy controls showed a positive correlation, which was devastated in MDMA users. These results imply that SERT reduction is critical for memory deficits in MDMA users (Buchert et al. 2004).

MDMA users show worse memory and cognition tasks than nonusers, which partially reflect the serotonergic neuron damages caused by MDMA. Recently, Bosch et al. (2013) performed 2-deoxy-2-[^{18}F]fluoro-*D*-glucose ([^{18}F]FDG)/PET on both MDMA users and nonusers, and found uptakes of [^{18}F] FDG in the dorsolateral prefrontal and parietal areas. Moreover, the glucose hypometabolism strongly overlaps with both impaired memory function and exposures of MDMA (Bosch et al. 2013). The adverse effects of MDMA are related to dosage, sex, and dosing regimen. Reneman et al. (2001) revealed that women are more susceptible to MDMA-induced toxicity assessed by [^{123}I]β-CIT/single-photon emission computed tomography (SPECT) (Reneman et al. 2001), which was further confirmed by Buchert et al. using voxel-based PET (Buchert et al. 2004).

The effects of MDMA are species-dependent with unknown mechanisms. Capela et al. (2006) first demonstrated that MDMA induced neuron toxicity and cell deaths through the 5-HT$_{2A}$ receptor and free radicals (Capela et al. 2006). In the study, the neurotoxicity and cell deaths were blocked by ketanserin, and (2R, 4R)-5-(2-[2-(2-[3-methoxyphenyl]ethyl)phenoxy]ethyl)-1-methyl-3-pyrrolidinol hydrochloride (R-96544); both are 5-HT$_{2A}$ receptor antagonists. However, another 5-HT$_{2A}$ receptor agonist, (±)-2,5-dimethoxy-4-iodoamphetamine hydrochloride (DOI), caused neurotoxicity in MDMA-treated cortical cells. Notably, the neurotoxicity induced by MDMA was also reduced by treating cells with α-phenyl-*N*-tert-butylnitrone (PBN) or NOS inhibitor, *N*ω-nitro-*L*-arginine.

The 5-HT$_{2A}$ receptor and oxidative stress are involved in MDMA-induced serotonergic neuron damages, which could be imaged by PET or SPECT. However, the morphological changes could not be concluded from the functional images, and long-term morphological imaging should be performed to understand how MDMA destroys the normal 5-HT signaling, which results in morphological changes and the impairment of memory and cognition. Whether MDMA-induced neurotoxicity is permanent or reversible is still being debated. Several articles have discussed the relationships among the use of MDMA, serotonergic neuron damage, and memory and cognitive impairments, which may provide more comprehensive information (Murphy et al. 2009; Parrot 2013; Wagner et al. 2013).

Ketamine

Ketamine is an illicit drug used as an anesthetic and acts rapidly through small binding to the plasma proteins. Ketamine also functions as an NMDA receptor antagonist, thus is used as an antidepressant also. Because of the psychotomimetic effects such as visual and auditory hallucinations induced by ketamine, its abuse has become a serious social issue worldwide since the 1970s. The extreme dissociation induced by ketamine is known as "K-hole," and the hallucinogenic effects could last for more than an hour dependent on the method of ingestion (Giannini et al. 1985; Muetzelfeldt et al. 2008).

Ketamine is thought of as a safe anesthetic in clinic, however, many deaths such as traffic accidents and drowning have been reported after the use of ketamine (Morgan et al. 2012) Most ketamine abusers are teens or young adults, and several animal studies show that ketamine could cause neural deaths followed by impaired cognition especially in neonates and children (Hayashi et al. 2002; Rudin et al. 2005; Zou et al. 2009; Brambrink et al. 2012). Unintended pregnancy is often found in ketamine abusers because hallucinations induced by ketamine heighten sexual desire. Notably, ketamine hydrochloride has good water solubility and lipid permeability; accordingly, it can cross the placenta easily and affects the brain development of fetuses. However, the role 5-HT plays in the damage of the CNS caused by ketamine has not been discussed in detail. It is known that ketamine causes nausea and vomiting through inhibiting the reuptake of 5-HT (Kohrs and Durieux 1998), which is similar to the function of mirtazapine (Chen et al. 2008; Chang et al. 2010). 5-HT has been proposed to strongly correlate to hallucinations (Lesch 1998), one of the typical symptoms of schizophrenia. Accordingly, ketamine is widely used to establish the schizophrenia animal models (Becker et al. 2003; Keilhoff et al. 2004).

Lysergic Acid Diethylamide (LSD)

Lysergic acid diethylamide (LSD) was discovered accidentally by the German chemist, Albert Hoffman, who later realized its hallucinogenic effects in 1943. Compared with other recreational drugs, LSD exhibits strong psychedelic effects with the threshold dose of 20–30 μg, compared with the dosage of most other drugs in mg. Interestingly, no death has been reported due to an overdose of LSD until now. The short-term effects of LSD include decreased alertness to danger, less attention, and impaired memory.

Unlike most serotonergic hallucinogens that show little impact on dopaminergic neurons, LSD influences not only the dopaminergic neurons (Prada et al. 1975) but also serotonergic neurons (Aghajanian et al. 1972; Bennett and Snyder 1976). LSD has a similar structure to serotonin, and the effects elicited by LSD are thought to be mediated through 5-HT signaling pathways (Bennett and Snyder 1976). Administration of LSD in rats induces behavioral syndromes including tremor, rigidity and lateral hand weaving, which are related to the central 5-HT-mediated activity (Trulson et al. 1976).

Studies show that LSD interacts with 5-HT$_{2A}$ and 5-HT$_{2C}$ receptors; nevertheless, the 5-HT$_{2A}$ receptor plays the most important role in LSD-induced effects. The activation of 5-HT$_{2A}$ receptors contributes to the hallucinogenic effects of LSD. The 5-HT$_{2A}$ receptor is a G protein–coupled receptor, and is the main excitatory receptor among all the 5-HT receptors. The activation of G protein can be determined by [^{35}S] GTPγS binding assay. Gresch et al. (2005) demonstrated that tolerance to LSD is mediated by reducing 5-HT$_{2A}$ signaling pathways in the cortex in a rat model. They also found that MDL 100907, a selective 5-HT$_{2A}$ receptor antagonist, blocked the [^{35}S]GTPγS binding stimulated by DOI, which was reduced in the medial prefrontal and anterior cingulate cortexes of chronically LSD-treated rats. Similar findings were observed by Krall et al. (2008) with a SERT knockout (KO) mouse model that has been reported to have significantly decreased density and functional activity of 5-HT$_{2A}$ receptors. Backstrom et al. (1999) showed that LSD could cause phosphoinositide hydrolysis through phospholipase C (PLC) in NIH-3T3 cells expressing rat 5-HT$_{2C}$ receptor in a dose-dependent manner. Nevertheless, LSD could not induce phosphorylation of the 5-HT$_{2C}$ receptor and the following calcium release as 5-HT (Backstrom et al. 1999).

The LSD-induced psychedelic activity could be blocked by many 5-HT$_{2A}$ receptor antagonists in part or whole. The hallucinogenic effects induced by LSD are mainly mediated selectively by 5-HT$_{2A}$ receptors. Except for affecting the 5-HT$_{2A}$ receptor, LSD also promotes the recognition of dopamine D2 receptor (D2R) and facilitates the allosteric receptor–receptor interactions of 5-HT$_{2A}$ and D2R (Borroto-Escuela et al. 2014). These findings may help the design of antipsychotic drugs for schizophrenia to lower the adverse effects caused by blockade of excess D2R (Meltzer and Stahl 1976; Miyamoto et al. 2005). 5-HT$_{2A}$ receptors activated by LSD could interact with metabotropic glutamate receptor 2 (mGluR2), and is also responsible for the neuropsychological responses as demonstrated in a mGluR2 KO mouse model (Moreno et al. 2011), and the mGlu2 receptor antagonist 2S-2-amino-2-(1S,2S-2-carboxycyclopropan-1-yl)-3-(xanth-9-yl)-propionic acid (LY341495) (Moreno et al. 2013).

Psychiatric Medication and Serotonergic Neurosystem

Serotonin in Cognition

Cognition including learning and memory is thought to be partially modulated by 5-HT. The distributions of 5-HT receptors are located in various brain areas related to learning or memory such as the hippocampus, amygdala, and cortex (Meneses 1999). Patients with lower 5-HT levels including those with depression, Alzheimer's disease, and schizophrenia are often found with impaired cognition (Rubinsztein et al. 2006; Yoon et al. 2008; Minzenberg and Carter 2012). It has been reported that 5-HT could modulate the cognitive status (Riedel et al. 2003). Tryptophan crosses the blood–brain barrier (BBB) using active transporters that are also used by large neutral amino acids (LNAAs) including leucine, isoleucine, tyrosine, phenylalanine, and valine. Accordingly, the concentration of tryptophan in the brain is associated with the transporters' competition between tryptophan and the LNAAs. LNAAs (Biskup et al. 2012; Fernstrom et al. 2013) or collagen (Evers et al. 2005) treatment is used for the deprivation of plasma tryptophan, which results in the blockade of 5-HT synthesis due to the shortage of its precursor. Acute tryptophan deprivation (ATD) has been used to study the effects of 5-HT on cognition and mood, and impaired cognition is observed in the healthy volunteers receiving ATD (Murphy et al. 2002). The impacts of 5-HT on cognition are also studied by treating subjects with extra tryptophan, and improved cognition has been reported as well (Haider et al. 2006; Haider et al. 2007).

Among all the cognitive functions, only the short-term and long-term memories have been proven to have a positive correlation with 5-HT level. It seems to be a negative correlation between 5-HT level and attention (Schmitt et al. 2000; Booji et al. 2005), though the attention is thought to be affected primarily by dopamine and norepinephrine (Bymaster et al. 2002; Swanson et al. 2006). Subjects who underwent the ATD showed impaired delayed and immediate recall and recognition in visual–verbal learning tests (Sambeth et al. 2007; Mendelsohn et al. 2009). Interestingly, these effects are more profound in women (Evers et al. 2006). The interaction between estrogen and 5-HT in cognition has been discussed in detail (Amin et al. 2005). Great cognition improvement due to tryptophan is observed in stress-vulnerable subjects rather than healthy controls (Markus et al. 2000; Markus et al. 2002). Notably, 6 weeks' administration of tryptophan (e.g., 50 or 100 mg/kg in rats) significantly enhanced both short-term and long-term memory (Haider et al. 2006; Khaliq et al. 2006; Haider et al. 2007). These findings imply that dietary supplement of tryptophan may benefit the memory function. Both 5-HT and the neurotransmitters derived from tryptophan play a role in memory.

Memory deficits are often found in patients with Alzheimer's disease and depression, as well as in the drug abusers. Moreover, the SERT expression in the brain is decreased in these particular populations compared with the healthy controls. Hence, decreased SERTs seem to be an ideal biomarker for memory deficit. SERTs handle the reuptake of 5-HT and are located on the terminals of 5-HT neurons. Selective serotonin reuptake inhibitors (SSRIs), a type of antidepressant, target the SERTs to inhibit the reuptake of 5-HT, which results in the increase of the extracellular level of 5-HT and rebalance the SERTs level. Improved cognition has been reported in patients treated with SSRIs (Schöne and Ludwig 1994; Geretsegger et al. 1994).

The correlations between SERTs and memory function have been studied using SERT knockout (KO) mice and rats since their generation in 1998 and 2006, respectively. The short-term memory is found to be impaired after ATD in the SERT KO rats evaluated by novel object recognition (NOR) task (Olivier et al. 2008). However, the absence of SERT does not affect the cognition in all aspects as reviewed by Kalueff et al. (2010). For instance, the spatial/object memory related to the hippocampus was impaired; nevertheless, the emotional memory associated with the amygdala was improved in the SERT KO mice and rats.

Among all the 14 subtypes of 5-HT receptors, 5-HT$_{1A}$, 5-HT$_4$, and 5-HT$_6$ are the most discussed 5-HT receptors in learning and memory. The 5-HT$_{1A}$ receptors are mainly expressed in the brain areas related to learning and memory such as the frontal cortex and hippocampus (Chalmers and Watson 1991), and are found decreased both in patients with Alzheimer's disease and aged people (Francis et al. 1992; Meltzer et al. 1998). The effects of 5-HT$_{1A}$ agonists on cognition ranged from improvement to impairment as the dose was increased (Kline et al. 2002; Manuel-Apolinar and Meneses 2004). Moreover, 5-HT$_{1A}$ also modulates the release of glutamine and choline, which are important for the cognition as well. Both the basal ganglia and hippocampus show the highest 5-HT$_4$ expression, and moderate the expression of 5-HT$_4$ receptors in the frontal cortex and amygdala. All of these brain areas are associated with learning and memory processes, and the expression of 5-HT$_4$ in these regions is also found decreased in Alzheimer's disease (Reynolds et al. 1995). Improved performance in various cognition tests including olfactory associative learning (Marchetti et al. 2004) and spatial memory (Fontana et al. 1997) are observed after administration of 5-HT$_4$ agonist such as RS67333. Most 5-HT$_6$ receptors are expressed in the CNS, especially in the striatum, nucleus accumbens, olfactory tubercles, hippocampus, and cortex. The importance of 5-HT$_6$ receptor in memory process has been implied by the injection of antisense oligonucleotides against 5-HT$_6$ receptor in rats, and the lengthened retention of the learned platform position was further validated using 5-HT$_6$ receptor antagonist, Ro 04-679 (Woolley et al. 2001). Several 5-HT$_6$ receptor antagonists reverse the impaired cognitive functions by improving consolidation (King et al. 2004; Hirst et al. 2006; Mitchell et al. 2006), and increase the levels of dopamine and acetylcholine in the frontal cortex of rats (Birrell and Brown 2000).

More detailed reviews can be found in the following articles (King et al. 2008; Terry et al. 2008; Pytliak et al. 2011).

Serotonin in Depression

It has been reported by the *New York Times* that one-tenth of all Americans are on an antidepressant (Rabin, 2013). According to the WHO's report, depression will become the biggest health problem by 2030 and influence millions of people (Mathers and Loncar 2006). Disability resulting from depression and related medical expenses directly and indirectly make enormous costs to society, estimated at about $90–110 billion in the EU and US (Gustavsson et al. 2011; Olesen et al. 2012; Mrazek et al. 2014).

Many antidepressants act as 5-HT reuptake inhibitors or antagonists of 5-HT receptors, indicating a strong relationship between 5-HT and depression. As a neurotransmitter, 5-HT relays messages between different brain regions to exert designated brain functions. Any imbalance in the 5-HT pathway results in bad moods, appetite disorder, anxiety, and insomnia, the symptoms of depression in clinic (Berger et al. 2009). It seems to be a bidirectional modulation of brain damage and depression. Oxidative stress–induced neuronal damage is often found in depressive subjects (Hayley et al. 2005; Eren et al. 2007a,b). It has been shown in many studies that brain injuries including traumatic brain injury (TBI), brain tumors, and stroke would increase the incidence of depression (Lipsey et al. 1983; Jorge et al. 2004; Rapoport 2012).

Moreover, depression is often discovered in patients with cancer or autoimmune diseases with poor outcomes, in which the immune system are disrupted. The balance of 5-HT is found to be disrupted in these patients and is consistent with the onset of depression (Brown and Paraskevas 1982; Maes et al. 2011). In this section, we will mainly discuss the relationship between cancer and depression, and the possible underlying mechanisms.

Clinical Observations of Cancer and Depression

The prevalence of depression in patients with cancer has been analyzed by different groups in the last decades (Raison and Miller 2003; Massie 2004; Linden et al. 2012; Li et al. 2012). However, varied results are discovered. It is agreed that depression is not induced by a single risk factor but confounding factors such as gender, ages, and medical interventions would affect the study results (Wedding et al. 2008; Linden and Girgis 2012). The patients with advanced stages of cancer show higher prevalence than those with early cancer and the healthy subjects (Breitbart et al. 1995; Hopwood and Stephens

2000; Massie 2004). Thus, depression could be used as a predictor of worse prognosis (Sotelo et al. 2014; Meyer et al. 2014), bad quality of life, and prolonged hospitalization in advanced cancer (Hopwood and Stephens 2000; Henderson et al. 2012; Reyes-Gibby et al. 2012; Que et al. 2013). The interaction between cancer prognosis and cancer might be associated with the interactions among the neuroendo-crine–immune (NEI) system, which means that the regulations of neurotransmitters, hormones, and cytokines could be affected by each other. For instance, pancreatic (Boyd and Riba 2007; Turaga et al. 2011), head and neck (Myers et al. 1999; Archer et al. 2008), and breast cancer (Reich et al. 2008; Chen et al. 2009) patients showed higher incidence of depression than other cancer types due to cytokine secreted from cancer cells or related complications.

It has been reported that 5%–26% of advanced cancer patients also suffer from major depression, and coprescribed with antidepressants (Miovic and Block 2007). Many studies also found that patients behaved better after receiving antidepressants (Rayner et al. 2011; Rodríguez et al. 2012; Laoutidis and Mathiak 2013). However, Lloyd-Williams et al. (1999) noted that the antidepressant medications given to patients are usually at inadequate doses and too late. The studies showed that most cancer patients with depression are underdiagnosed and undertreated. Depression assessment could be done by several scales as follows. The Hospital Anxiety and Depression Scale (HADS) is the most widely used, focusing on the indicators related to anhedonia and showing positive results in patients with cancer therapy (Zigmond and Snaith 1983). The sensitivity and specificity of HADS limits the usage for depression screening in cancer patients (Ibbotson et al. 1994). Adopted from postnatal depression evaluation, the Edinburg Depression Scale (EDS) demonstrates some promising results (Cox et al. 1987). The Brief Edinburg Depression Scale (BEDS) is modified from EDS and was proposed in 2007 (Lloyd-Williams et al. 2007). Though only six criteria, the BEDS shows better ability to discriminate depression in patients with advanced cancer. A newer scale with better sensitivity and specificity for depression assessment is urgently needed to improve the diagnosis rate of the cancer-related depression.

The attitude with which patients face diseases is positively correlated with the outcomes and prognosis of diseases including cancer. It is reported that positive prognosis could be expected in those patients who faced cancer as a challenge, and with a fighting spirit. By contrast, the prognosis was relatively worse when patients displayed a 'helplessness/hopelessness' attitude (Watson et al. 1999; Petticrew et al. 2002). These findings imply that physical health status could be interfered with by mental health status.

As mentioned earlier, patients with autoimmune diseases tend to develop depression due to Th1/Th2 cytokines imbalance (Martino et al. 2012). Besides, development of depression was found in patients receiving IFN-α therapy for cancer or hepatitis B/C treatments (Wichers and Maes 2002; Asnis and De La Garza 2006). Increased evidence indicates that antitumor therapies may induce or worsen depression in patients *via* increasing proinflammatory cytokine expressions (Young et al. 2014; Slavich and Irwin 2014). People and animals with elevated TNF-α, IL-1, and IL-6 levels showed sickness behavior similar to major depression (Ghia et al 2008; Hocaoglu et al. 2012; Bufalino et al. 2013). Cytokines with large molecular weight enter into the brain from the circulation through leaky regions of the blood–brain barrier or by active transport. Activated immune cells including macrophages, monocytes, and CD4+ T cells could migrate into the brain freely and subsequently release cytokines (Watkins et al. 1995; Plotkin et al. 1996; Rivest et al. 2000; D'Mello et al. 2009). Proinflammatory cytokines cause increased neuron damage, SERT and DAT expressions, and both activity and high levels of indoleamine 2,3-dioxygenase (IDO) are found in cancer patients with depression. Moreover, 5-HT availability and neurogenesis were also decreased by elevated proinflammatory cytokines (Myint and Kim 2003). These studies show that the regulations between mental and physical health are not unidirectional, but complex interactions, sug-gesting that both mental and physical medications should be introduced for cancer treatment.

Abnormal Metabolism of Tryptophan in Cancer

Though most of the tryptophan is used for protein synthesis and metabolized through the kynurenine pathway, part of the tryptophan is also metabolized through a 5-HT pathway [50] to generate the 5-HT, a neurotransmitter tightly related to depression. IDO or tryptophan 2,3-dioxygenase (TDO), expressed mainly in the liver, and tryptophan hydroxylase (TPH) are two enzymes involved in the kynurenine and the 5-HT pathways, respectively. As mentioned above, the serum level of 5-HT is found to be decreased in cancer patients, especially in those with advanced stages of the disease. Accordingly, the tryptophan metabolism in cancer patients is speculated to be changed.

Only 1% of tryptophan derived from the diet is utilized for 5-HT synthesis in the brain, and more than 90% of dietary tryptophan is metabolized through the kynurenine pathway under normal physiological condition. However, upregulation of IDO has been reported in several types of cancer, and results in the deprivation of tryptophan from circulation and the reduction of 5-HT and melatonin. IDO catalyzes all the materials from tryptophan to melatonin along the 5-HT pathway. The kynurenine formed in the kynurenine pathway shows contradictory functions as neuroprotectant and neurotoxin (Chen and Guillemin 2009), and are strongly associated with depression and other neuronal diseases.

In addition, both cell- and humoral-mediated antitumor immunity are impaired because upregulated IDO depletes the tryptophan (Prendergast 2008) and skews the cytokine expression profiles from Th1 to Th2 (Martino et al. 2012) within the tumor microenvironment.

Correlation of Serotonin and Melatonin

The serotonin (5-HT) is synthesized from tryptophan through an 5-HT pathway as previously described. Tryptophan is first converted into 5-hydroxy-L-tryptophan (5-HTP) by TPH, the rate-limiting enzyme, and its cofactor tetrahydrobiopterin (BH4) under the normal physiological condition. The aromatic amino acid decarboxylase (AADC) further converts the 5-HTP into 5-HT. There are two types of TPH: TPH-1 and TPH-2. The latter is specifically expressed in neurons. 5-HT is utilized as a precursor for melatonin (N-acetyl-5-methoxytryptamine) synthesis in vertebrate animals. 5-HT is converted to N-acetyl serotonin by aralkylamine N-acetyltransferase (AANAT), and further converted into melatonin by hydroxyindole-O-methyltransferase. Melatonin is highly related to the light/dark cycle.

BH4 also serves as the cofactor of nitric oxide synthase (NOS). The NOS could be induced by several pro-inflammatory cytokines upregulated in ischemia/reperfusion and tumor microenvironments. Overexpression of the inducible nitric oxide synthase (iNOS) would result in lower productions of 5-HT and melatonin due to the competition for TPH between iNOS and BH4. In addition, generation of nitric oxide from arginine by iNOS could be facilitated by BH4. Excess free radicals including reactive oxygen species (ROS) and reactive nitrogen species (RNS) are known to damage neurons and the immune system, and cause the depletion of tryptophan or 5-HT. Thus, the level of melatonin will indirectly regulate the synthesis of 5-HT. In other words, the 5-HT and melatonin could be modulated by each other to ensure the functions of both neurons and the immune system. The melatonin has multiple functions and serves as a powerful antioxidant that will be discussed in the later section.

Role of Serotonin and Melatonin in Cancer

Serotonin and Cancer

Upregulated IDO and the kynurenine pathway cause significant reductions of serotonin (5-HT) and melatonin in the serum of cancer patients (Bartsch et al. 1999; Capuron et al. 2002). It has been reported that serotonin exerts antitumor effects though the role of 5-HT in cancer progression remains ambiguous (Burtin et al. 1982; Manda et al. 1988). It has been discovered that with colon cancer, angiogenesis and progression can be promoted by 5-HT through an angiostatin pathway with downregulation of MMP-12 has been discovered in colon cancer (Nocito et al. 2008). Both cancer cells and endothelial cells express 5-HT receptors which can be activated by 5-HT, and promote the cancer progression. Additionally, several studies demonstrate that treatments of 5-HT receptor antagonist or TPH inhibitor block the progression of several types of cancer such as cholangiocarcinoma (Alpini et al. 2008), lung cancer (Asada et al. 2008), and pancreatic cancer (Gurbuz et al. 2014). Immune cells also play critical roles in both cancer progression and inhibition. It has been shown that 5-HT has strong impacts on chemotaxis, cytokine generation, and the proliferation of immune cells (Ahern 2011). However, most immune cells cannot synthesize 5-HT themselves. Both activated dendritic cells (DCs) and B cells uptake 5-HT from the microenvironment through the upregulation of SERT expression (Meredith et al. 2005; O'Connell et al. 2006). Since identification of SERT in T cells is not successful, it is inferred that T cells uptake 5-HT using DAT. Moreover, activated T cells display more expressions of TPH, which is one of the enzymes involved in 5-HT synthesis (O'Connell et al. 2006; León-Ponte et al. 2007).

The interactions between 5-HT and immune cells in cancer remain obscure. The antigen presentation of DCs affected by 5-HT were found to be inconsistent. Under chronic inflammation, 5-HT attenuates the antigen-presenting ability of DCs to CD8+ T cell (Branco-de-Almeida et al. 2011). By contrast, 5-HT boosts both innate and adaptive cytokine generations and enhances the stimulation of naïve T cells *via* NF-κB pathway in a gut inflammation mouse model (Li et al. 2011). Some studies indicate that 5-HT increases both the migration of immature DCs and expression of Th2 cytokine in DCs in allergic diseases (Katoh et al. 2006; Müller et al. 2009). Antigen-presenting cells (APCs) bearing Ia-antigen are specific subpopulations, and are required for the proliferation of antigen-specific T cells (Cowing et al. 1978). Ia-antigen expression and phagocytosis of macrophages induced by IFN-γ could be suppressed by 5-HT in a concentration-dependent manner of IFN-γ (Sternberg et al. 1986). 5-HT could enhance the phagocytosis of macrophages under a low level of IFN-γ, but block the phagocytosis when the IFN-γ level was high in the medium (Sternberg et al. 1987). Besides, 5-HT upregulates the expressions of $5-HT_{2B}$ and $5-HT_7$ receptors, and polarizes macrophages to differentiate into the M2 subtype (de las Casas-Engel et al. 2013). According to these results, 5-HT seems to reduce the antitumor immunity through impairing antigen presentation of DCs and macrophages, and switching the cytokine expressions to a Th2 profile. However, it has been reported that the proliferation and activation of T cells stimulated by IL-2 can be enhanced *via* the $5-HT_{1A}$ receptor, and elevates antitumor immunity substantially (Aune et al. 1994; Young and Matthews 1995). Activation of T cells could upregulate the expressions of both $5-HT_{1A}$ and $5-HT_7$ receptors (Young and Matthews 1995). A better prognosis of skin tumor and higher IFN-γ and IL-17 produced by T cells were found in TPH1$^{-/-}$ mice (Nowak et al. 2012). Secretion of IL-16 and recruitments of CD4-expressing cells (e.g., CD4+ T cell, monocyte, and DCs) by CD8+ T cells were increased by 5-HT stimulation (Laberge et al. 1996). Furthermore, the IL-16 level is tightly correlated with the cancer progression (Kovacs 2001; Compérat et al. 2010).

In summary, how 5-HT influences the immune system and cancer progression is not well elucidated due to the controversial findings and complicated immune regulation. Further studies are needed to understand the roles and functions of 5-HT in immunomodulation and cancer progression under various conditions.

Melatonin and Cancer

Melatonin is synthesized from 5-HT and shows antitumor effects in types of cancer. Melatonin induces apoptosis and necrosis in pancreatic cancer cells through changing the balance of Bcl-2/Bax (Xu et al. 2013). Melatonin functions as a novel SIRT1 inhibitor, and inhibits the proliferation of prostate cancer both *in vitro* and *in vivo* in transgenic adenocarcinoma mouse prostate (TRAMP) models (Jung-Hynes et al. 2011). Inhibition of SIRT1 by melatonin in breast cancer results in increased apoptosis through MDM2 downregualtion and p53 enhancement (Proietti et al. 2014). Melatonin blocks angiogenesis in breast cancer (Jardim-Perassi et al. 2014) and liver cancer (Carbajo-Pescador et al. 2014) *via* downregulations of VEGF-R2 and VEGF, respectively. In lung cancer, increased apoptosis, elevated lipid contents, and reduced intensities of nucleic acids and proteins with secondary structure by melatonin are also observed (Plaimee et al. 2014). Furthermore, hypoxia-induced anti-apoptosis in lung cancer can be reversed by melatonin (Lee et al. 2014). Melatonin decreases the migration and metastasis by blocking expressions of HIF-1α and rho-associated protein kinase (ROCK) in oral cancer (Goncalves et al. 2014). Lower MMP-9 and NF-κB expressions in liver cancer treated with melatonin shows less migration and invasiveness (Ordoñez et al. 2014). Inhibition of cancer-initiating cells seems to be an alternative means of tumor control, and melatonin has been proven to induce autophagy and block the proliferation of glioma-initiating cells (Martín et al. 2014).

The treatment outcomes of cancer are highly affected by the status of immune cells. It has been reported that differentiations and cytokine profiles of immune cells are influenced by free radicals,

NO, and peroxynitrite. Under normal physiological condition, both NO and ROS are used to eliminate the foreign pathogens in the host. However, in the acidic tumor microenvironment, excess NO triggers the microenvironment becoming immunosuppressive (Wink et al. 2011). Peroxynitrite in the microenvironment results in nitrotyrosinylation of antigens and cytokines, and impairs the antigen recognition abilities and infiltration of antitumor CD8+ T cells (Nagaraj et al. 2007; Molon et al. 2011). Melatonin prevents antigen or cytokine nitration by removing the NO and other free radicals presenting in the microenvironment. It has been reported that antigen presentations of macrophages could be enhanced by melatonin and followed by increased T cell proliferation (Pioli et al. 1993). Macrophages function not only as antigen-presenting cells but also as generators of superoxide and NO, which may induce DNA damage. Macrophages in cancer are often found with higher iNOS and COX-2 expressions, which could be diminished by melatonin through inhibition of NF-κB (Gilad et al. 1998; Deng et al. 2006). Both monocytes and macrophages eliminate foreign pathogens by phagocytosis (Dale et al. 2008). The numbers of monocytes in bone marrow and spleen are increased after melatonin treatment in mice (Conti et al. 2000). Both macrophage and human myeloid monocytic cells line treated with melatonin results in the higher production of antitumor cytokine, IL-12 (Cutolo et al. 1999). The number of natural killer cells (NK cells) is increased significantly after melatonin administration in cancer patients treated with or without IL-2 (Lissoni et al. 1992; Miller et al. 2006). The number, activity, and antibody-dependent cell cytotoxicity (ADCC) of NK cells are elevated after melatonin treatment (Giordano and Palermo 1991). Though the numbers or function of CD8+ T cells are not affected by melatonin, the conversion of regulatory T cells (Tregs) from CD4+ T cells is hindered by melatonin treatment (Liu et al. 2011). The differentiation of CD4+ T cells can be shifted to Th1 subtype since cytokine profiles are changed by melatonin. The correlation of serotonin, melatonin, immune system, and microenvironment to the cancer and depression is depicted in Figure 31.2.

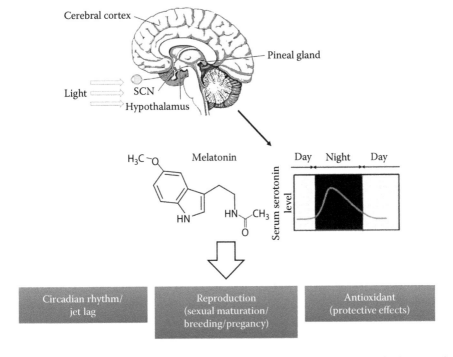

FIGURE 31.2 Health effects of melatonin on the circadian rhythm, reproduction, and anti-oxidation in human. Synthesis of melatonin is controlled by the light/dark cycle. Following the reception of light by the retina and circadian pacemaker, the suprachiasmatic nucleus (SCN) generates an inhibitory signal to depress the synthesis of melatonin. Melatonin is primarily generated by the pineal gland during the night and results in the increase of the serum melatonin level. The level of melatonin is closely correlated to the circadian rhythm, reproduction, and protection from free radicals in human.

Health Effects and Melatonin

As mentioned in the previous section, melatonin is primarily synthesized by the pineal gland, which controls the regulation of the circadian cycle and sleep. It has been reported that dysregulation of melatonin would affect reproduction in mammals. Moreover, melatonin also acts as a direct free radical scavenger and an indirect antioxidant, which are important for diminishing ischemia/reperfusion- and radiation-induced oxidative stress. The roles of melatonin in three aspects including the circadian cycle and sleep, reproduction, and scavenging of free radicals will be discussed here.

Regulation of Circadian Cycle and Sleep

The secretion of melatonin was found to be high in the dark and low under the light, in the 1960s (Axelrod and Weissbach 1960). Melatonin shows hypnotic effects through lowering the core body temperature and reducing alertness. Administration of melatonin does not alter the electroencephalogram (EEG) patterns of sleep significantly as do synthetic hypnotics such as barbiturates (Cramer et al. 1974); however, increased proportions of rapid eye movement (REM) and stage 2 sleep are found, both of which concurred with the phase shifts in the circadian pacemaker (Antón-Tay 1974).

As a lipid-soluble hormone, melatonin is directly released into the circulation and cerebrospinal fluid (CSF) after its synthesis. The synthesis of melatonin initiates from 9 to 12 weeks after birth; before that the newborns obtain melatonin from mother's milk. Interestingly, a stable sleep–wake cycle is established from three months old when the secretion of melatonin begins in humans (Attanasio et al. 1986). The secretion of melatonin increases rapidly and peaks up at five years old, then decreases afterward. The concentration of melatonin is significantly lower in those who have insomnia (Attenburrow et al. 1996; Nagtegaal et al. 1998). However, the peak concentration of melatonin is related to the quality of sleep (i.e., sleep time/time in bed) (Haimov et al. 1994). Accordingly, melatonin has been introduced for the treatment of sleep disorder or insomnia for more than 20 years (James et al. 1990; Garfinkel et al. 1995; Zhdanova et al. 2001). Melatonin is recommended to be taken in the evening or before bedtime in adults. It has been reported that melatonin-induced sleepiness is dose-dependent when the physiological concentration of serum melatonin ranges from 50–200 pg/ml, but without dose-dependency above 200 pg/ml (Dollins et al. 1993 1994).

As mentioned above, melatonin-induced hypnotic effects are mediated through thermoregulation and phase shifts of the circadian cycle. The suprachiasmatic nucleus (SCN) is known as the circadian pacemaker expressing two high-affinity G protein–coupled receptors, MT_1 and MT_2, for the melatonin (Dubocovich et al. 1996; Gillette and McArthur 1996). Modulations of sleep, core body temperature, and immune functions are controlled bidirectionally by both the SCN and melatonin. MT_1 and MT_2 affect the sleeping time by inhibiting neuron activities and shifting the phase of circadian rhythm with different ligand selectivity in the SCN (Roca et al. 1996; Dubocovich et al. 1998; Hunt et al. 2001). Melatonin receptor KO mice have been used to study the effects of melatonin since its generation in the late 1990s (Roca et al. 1996). MT_1 seems to be more prevalent for regulating the circadian cycle, as shown in several studies (Shibata et al. 1989; Stehle et al. 1989). It has been shown that MT_1 increases significantly without altering MT_2 expression in the SCN in patients with depression (Wu et al. 2013) who suffered with sleep disorders.

Except for treating sleep disorders, melatonin is also used for modulating the circadian rhythm in the blind, night shift workers, and international travelers. Most blind people have a free-running circadian cycle rather than the circadian rhythm of 24 h due to deficiency of the light zeitgeber from the retina (Miles et al. 1977; Skene and Arendt 2007). The dose of melatonin for modulating the circadian rhythm in the blind is only 0.5 mg/day for long-term purposes (Sack et al. 2000), and functions primarily on the phase shifting through MT_2 (Dubocovich et al. 2005). Melatonin not only corrects the circadian cycle but also improves the sleep quality and daytime sleepiness, differing from that of benzodiazepine or sedative hypnotics, which shows little benefit to the sleep of the blind. Jet lag is often discovered in international travelers who fly across several time zones, and the symptoms including daytime sleepiness and sleep disturbances (Comperatore and Krueger 1990). Jet lag is caused by desynchronization of the established circadian cycle and the day/night cycle of the destination. Travelers flying across more than five time zones often have jet lag, especially when traveling eastbound. Currently, the dose of melatonin applied for reducing jet lag range from 0.5 mg to 5 mg, and doses higher than 5 mg show no additional improvements. Melatonin should be taken before bedtime as darkness has fallen to achieve the best result, and a dose taken before the travel is not recommended (Herxheimer and Petrie 2001). Re-entrainment of the circadian cycle is also needed for night shift workers who are reported to have higher incidences of cancer and heart disease. Melatonin is thought to be helpful for these people; however, the optimal dose and dosing time remain unknown. Several methods have been proposed for circadian entrainment in this particular population, as follows. First, wear sunglasses that are as dark as possible when commuting to increase the release of melatonin, which would be inhibited by light stimulation. Secondly, apply artificial bright light at night during work. Crowley et al. (2003) concluded that application of the bright light results in more benefits than other strategies.

Melatonin and Reproduction

Except for modulating the circadian cycle, melatonin influences reproduction, which is related to the light/dark cycle and/or estrus cycle. Melatonin exerts antigonadotropic effects and regulates the link between reproductive activity and seasons in seasonal breeding species such as hamsters. Seasonal breeding is not observed in the Syrian hamsters that have received a pinealectomy; nevertheless, the decreased gonadotropin levels and gonadal regression reappear after melatonin injection (Tamarkin et al. 1977a). Except for testicular regression found in male hamsters, anestrus was also found in female hamsters (Tamarkin et al. 1977b). Though humans are not seasonal breeders, seasonal patterns in conception and birth rates are also found in areas near the Arctic (Rojansky et al. 1992). The hypothesis that the pineal gland could affect puberty was first proposed in 1898 by Otto Heubner. After that, many observations of precocious puberty are reported, and mostly in boys with melatonin deficiency that have impaired pineal gland functions (Kitay 1954; Waldhauser et al. 1991). The relationship between melatonin levels and sex maturation is observed in both men and women; moreover, melatonin also controls the production of sex steroids. Increased serum melatonin is detected in women with hypothalamic amenorrhea (Bergu et al. 1988; Laughlin et al. 1991; Kadva et al. 1998) and in men with hypogonadotropic hypogonadism (Puig-Domingo et al. 1992; Luboshitzky et al. 1995). In addition, melatonin modulates ovary function directly. Large amounts of melatonin are stored in the ovarian follicular fluid and this subsequently stimulates progesterone synthesis by granulosa lutein cells highly expressing melatonin receptors (Webley and Leidenberger 1986; Brzezinski et al. 1987).

It is known that melatonin levels during the night in pregnant women are higher than nonpregnant women until week 32 of gestation and diminish rapidly after delivery (Tamura et al. 2008). The circadian cycle of the fetus is controlled by maternal melatonin since melatonin is lipophilic and can across the placenta freely without alteration (Sagrillo-Fagundes et al. 2014). It has been reported that melatonin protects the placenta from oxidative stress by inducing ischemia/reperfusion through melatonin receptors expressed on the placenta in rats. Besides, reduced delivery of fetal nutrients would induce oxidative stress in the placenta and damage the placenta function. Richter et al. (2009) validated that maternal melatonin treatment improved the placental efficiency and rescued the body weight of the fetus in undernourished pregnancy by upregulating expressions of antioxidative enzymes such as Mn-SOD and catalase. Higher levels of antioxidative enzymes and melatonin are detected in women with normotensive pregnancies; nevertheless, the women with pre-eclampsia show lower serum melatonin. It is speculated that the antioxidative effects caused by melatonin reduce the oxidative damage in the placenta and endothelial cells (Tamura et al. 2008). The decreased melatonin level is correlated with the significant inhibition of AANAT expression and activity. In addition, the expressions of melatonin receptors are also reduced (Lanoix et al. 2008, 2012). Thus, the serum melatonin level might be employed as a diagnostic factor for pregnancy with preeclampsia, and the exogenous melatonin could be used as a potential treatment for the situation.

Melatonin Functions as a Free Radical Scavenger and Related Protective Effects

Melatonin is a powerful free radical scavenger and antioxidant through directly neutralizing the toxic reactants and stimulating the expressions and activities of antioxidative enzymes. Thus, melatonin exhibits the abilities to reduce the molecular damage caused by free radicals and associated oxygen- and nitrogen-based reactants. Both melatonin and its metabolites are free radical scavengers. They are able to induce the synthesis of glutathione (GSH), reduce the electron leakage from mitochondria, and limit the production of inflammatory cytokines and subsequent inflammation.

Cyclic 3-hydroxymelatonin (C3HOM), one of the metabolites of melatonin, has been shown to be more potent as a free radical scavenger and antioxidant than melatonin. C3HOM prevents the cells from apoptosis by inhibiting the release of cytochrome C from mitochondria. It is suggested that the antioxidative ability of melatonin might be partially related to C3HOM. Melatonin could work alone or with other antioxidants such as vitamins C and E synergistically, and shows high scavenging activity of peroxyl radicals generated from lipid peroxidation (Pieri et al. 1995). It has been reported that normal cells including neurons, myocardium, and hepatocytes are damaged by free radicals generated during the ischemia/reperfusion process (Hess and Manson 1984; Cao et al. 1988; Lucchesi 1990; Nauta et al. 1990). Melatonin is used as an adjuvant agent for cancer therapy through unclear underlying mechanisms. However, either chemotherapy or radiotherapy combined with melatonin exhibit better tumor control and less normal tissue damage than a single treatment. Morishima et al. (1998) reported that chemotherapeutic drugs caused cardiomyopathy;, renal injury (Parlakpinar et al. 2002) and hepatic injury (Catalá et al. 2007) could be prevented by melatonin through the inhibition of lipid peroxidation. Though the interactions between melatonin and NO are shown in few reports, the inhibition of iNOS expression and NO formation by melatonin have been presented in several models (Dong et al. 2003; Kilic et al. 2005; Ortiz et al. 2014). A systemic review and meta-analysis conducted by Seeley et al. (Seeley et al. 2012) indicated that a combination of melatonin and chemotherapy could enhance therapeutic outcomes and lowered one-year mortality. Another study also showed that melatonin prevents adverse effects caused by chemotherapy or radiotherapy and synergistically enhances the therapeutic outcomes (Lissoni et al. 1996).

Radiotherapy (RT) is utilized as a standard or an adjuvant treatment for various cancer types; however, the therapeutic efficacy of RT is limited by tolerance and related radiation-induced damage of normal tissues. Melatonin is thought to be an ideal radioprotector because of its free radical scavenging and antioxidative abilities. The radioprotection of melatonin is observed both *in vitro* and *in vivo*. 300 mg melatonin was given orally to the subjects 0 (i.e., 5–10 min), 1 and 2 h prior to the blood being drawn and 150 cGy irradiation, and the frequencies of chromosome aberration and micronuclei (MN) were determined to assess the radiation damages. Fewer chromosome aberrations and MN were detected in the blood samples from subjects that received melatonin 1 and 2 h before irradiation than in those of the 5–10 min samples (Vijayalaxmi et al. 1996). Vijayalaxmi et al. performed a total body irradiation (TBIR) using 150 cGy in another study, and the percentages of polychromatic erythrocytes (PCEs) and the incidence of MN were increased and decre ased by melatonin, respectively, in a dose-dependent manner (Vijayalaxmi et al. 1999a). Melatonin prevents the lethal effects of TBIR in a dose-dependent manner. Mortality was monitored for 30 days after receiving 815 cGy TBIR in mice (Vijayalaxmi et al. 1996b).

Melatonin has been used as the radioprotector for normal tissues and organs during RT. Liver toxicity is always a big concern for liver cancer treated with RT. Melatonin diminishes liver toxicity induced by irradiation has been reported by Taysi et al. (2003) and El-Missiry et al. (2007) using different irradiation doses. Oxidative stress markers, such as malondialdehyde (MDA) and NO, as well as the liver function markers including AST and ALT, were also reduced in the melatonin-pretreated groups (Taysi et al. 2003; El-Missiry et al. 2007). On the other hand, lung fibrosis is the main side-effect of lung irradiation. Less radiation-related injuries have been reported in mice pretreated with 200 mg melatonin before 12 Gy lung irradiation *via* melatonin-mediated downregulations of TGF-β, TNF-α and oxidative stress (Jang et al. 2013). Melatonin also diminishes radiation-induced oral mucositis through inhibition of NK-κB/NLRP3 (NLR-related protein 3 nucleotide-binding domain leucine-rich repeat containing receptors-related protein 3)-mediated inflammation and generations of ROS (Ortiz et al. 2015). Ultraviolet (UV) and ionizing radiation may induce the formation of cataracts, which can be inhibited by melatonin through increasing the levels of superoxide dismutase (SOD) and

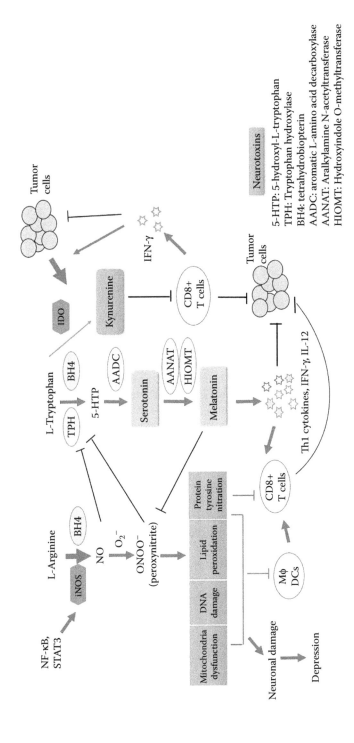

FIGURE 31.3 The serotonin, melatonin, immune system, and microenvironment to cancer and depression are closely correlated. Both serotonin and melatonin are the metabolites of tryptophan, and are closely correlated with antitumor immunity. When the levels of indoleamine 2,3-dioxygenase (IDO) and inducible nitric oxide synthase (iNOS) are elevated in the tumor microenvironment, the metabolisms of tryptophan and arginine are altered. The abnormal metabolites then further inhibit the antitumor immunity and cause neuronal damage. (From Chuang and Hwang, 2015).

glutathione peroxidase (GSH-Px) (Karslioglu et al. 2005). Dermatitis is often observed in patients treated with radiation. The results from Kim et al. (2001) imply that melatonin could protect the fibroblasts from radiation-induced apoptosis by repressing lipid peroxidation without the involvement of the p53/p21 pathway. Except radioprotection, melatonin also enhances the therapeutic outcomes of RT. Higher one-year survival rates and less radiation-induced toxicities were observed in glioblastoma treated with melatonin and RT (Lissoni et al. 1996). Improved results and protection resulting from the melatonin treatment have also been reported in chemotherapy (Lissoni et al. 1997; Sanchez-Barcelo et al. 2012). Based on these findings, melatonin has been suggested to be applied to minimize normal tissue damage and produce better therapeutic outcomes in RT. Nevertheless, the dosage and dosing timing of melatonin require further investigation (Shirazi et al. 2007; Mihandoost et al. 2014). The health effects of melatonin on the circadian rhythm, reproduction, and antioxidation are shown in Figure 31.3.

Conclusion

Serotonin (5-HT) converted from the essential amino acid tryptophan could regulate mood, appetite, learning, and memory. Melatonin is a potent antioxidant converted from 5-HT that controls the circadian cycle, sleep, and reproduction.

As mentioned above, the feeling of well-being caused by club drugs results from the abundant release of 5-HT. However, excess 5-HT may cause the serotonin syndrome that is sometimes found in drug abusers and leads to sudden death. Cognition impairment has been reported in living subjects exposed to club drugs and is not related to the period of exposure. 5-HT plays a critical role in cognitive processes, and both the 5-HT receptors and serotonin transporters (SERTs) are mainly located in the brain areas related to learning and memory including the hippocampus and amygdala. Decreased expressions of 5-HT receptors and SERTs are often found in patients with depression and Alzheimer's disease, again indicating the relationships between 5-HT and cognition.

Both the circadian cycle and sleep are regulated by melatonin, and its synthesis is controlled by the light/dark cycle. Melatonin also controls reproduction in seasonal breeders such as hamsters. Though humans are not seasonal breeders, melatonin may still play a role in conception and birth rates, especially in the Arctic regions. As an antioxidant, melatonin could remove the free radicals generated from ischemia/reperfusion processes and ionizing irradiation. Hence, melatonin is also a bioprotector preventing living subjects from oxidative stress and damages.

Both 5-HT and melatonin not only influence the neuron system but also play substantial roles in immunomodulation and anticancer effects. Abnormal levels of 5-HT and melatonin caused by altered tryptophan metabolism result in neuron damage, cognition impairment, insomnia, and compromised immunity. Furthermore, depression and cancer may develop subsequently and worsen the imbalance among 5-HT, melatonin, and other metabolites produced along the tryptophan metabolic pathways. These preclinical or clinical findings imply that the levels of 5-HT and melatonin are critical for maintaining health, and the 5-HT and melatonin do link up with the neuron system, antitumor effects, and immunity.

Acknowledgments

This review is supported by grants (MOST 103-2633-B-010-001 and NSC 102-2321-B-010-005) from the Ministry of Science and Technology, Taiwan.

REFERENCES

Aghajanian, G. K., Haigler, H. J., and Bloom, F. E. Lysergic acid diethylamide and serotonin: Direct actions on serotonin-containing neurons in rat brain. *Life Sciences*. 1972. 11: 615–622.
Ahern, G. P. 5-HT and the immune system. *Current Opinion in Pharmacology*. 2011. 11: 29–33.

Ali, S. F., Newport, G. D., Holson, R. R., Slikker Jr, W., and Bowyer, J. F. Low environmental temperatures or pharmacologic agents that produce hypothermia decrease methamphetamine neurotoxicity in mice. *Brain Research*. 1994. 658: 33–38.

Alpini, G., Invernizzi, P., Gaudio, E., Venter, J., Kopriva, S., Bernuzzi, R., Onori, P. et al. Serotonin metabolism is dysregulated in cholangiocarcinoma, which has implications for tumor growth. *Cancer Research*. 2008. 68: 9184–9193.

Amin, Z., Canli, T., and Epperson, C. N. Effect of estrogen–serotonin interactions on mood and cognition. *Behavioral and Cognitive Neuroscience Reviews*. 2005. 4: 43–58.

Anton-Tay, F. Melatonin: Effects on brain function. *Advances in Biochemical Psychopharmacology*. 1974. 11: 315–324.

Archer, J., Hutchison, I., and Korszun, A. Mood and malignancy: Head and neck cancer and depression. *Journal of Oral Pathology and Medicine*. 2008. 37: 255–270.

Armstrong, B. D., and Noguchi, K. K. The neurotoxic effects of 3,4-methylenedioxymethamphetamine (MDMA) and methamphetamine on serotonin, dopamine, and GABA-ergic terminals: An *in vitro* autoradiographic study in rats. *Neurotoxicology*. 2004. 25: 905–914.

Asada, M., Ebihara, S., Yamanda, S., Niu, K., Okazaki, T., Sora, I., and Arai, H. Depletion of serotonin and selective inhibition of 2B receptor suppressed tumor angiogenesis by inhibiting endothelial nitric oxide synthase and extracellular signal-regulated kinase 1/2 phosphorylation. *Neoplasia*. 2009. 11: 408–417.

Asnis, G. M., and De La Garza (II), R. Interferon-induced depression in chronic hepatitis C: A review of its prevalence, risk factors, biology, and treatment approaches. *Journal of Clinical Gastroenterology*. 2006. 40: 322–335.

Attenburrow, M. E., Dowling, B. A., Sharpley, A. L., and Cowen, P. J. Case-control study of evening melatonin concentration in primary insomnia. *British Medical Journal*. 1996. 312: 1263–1264.

Aune, T. M., Golden, H. W., and McGrath, K. M. Inhibitors of serotonin synthesis and antagonists of serotonin 1A receptors inhibit T lymphocyte function *in vitro* and cell-mediated immunity *in vivo*. *The Journal of Immunology*. 1994. 153: 489–498.

Axelrod, J., and Weissbach, H. Enzymatic *O*-methylation of *N*-acetylserotonin to melatonin. *Science*. 1960. 131: 1312.

Backstrom, J. R., Chang, M. S., Chu, H., Niswender, C. M., and Sanders-Bush, E. Agonist-directed signaling of serotonin 5-HT2C receptors: Differences between serotonin and lysergic acid diethylamide (LSD). *Neuropsychopharmacology*. 1999. 21: 77s–81s.

Bartsch, C., Bartsch, H., Bellmann, O., and Lippert, T. H. Depression of serum melatonin in patients with primary breast cancer is not due to an increased peripheral metabolism. *Cancer*. 1991. 67: 1681–1684.

Becker, A., Peters, B., Schroeder, H., Mann, T., Huether, G., and Grecksch, G. Ketamine-induced changes in rat behaviour: A possible animal model of schizophrenia. *Progress in Neuropsychopharmacology and Biological Psychiatry* 2003. 27: 687–700.

Belcher, A. M., O'Dell, S. J., and Marshall, J. F. Impaired object recognition memory following methamphetamine, but not p-chloroamphetamine- or d-amphetamine-induced neurotoxicity. *Neuropsychopharmacology*. 2005. 30: 2026–2034.

Bennett, J. P., and Snyder, S. H. Serotonin and lysergic acid diethylamide binding in rat brain membranes: Relationship to postsynaptic serotonin receptors. *Molecular Pharmacology*. 1976. 12: 373–389.

Berga, S. L., Mortola, J. F., and Yen, S. S. Amplification of nocturnal melatonin secretion in women with functional hypothalamic amenorrhea. *The Journal of Clinical Endocrinology and Metabolism*. 1988. 66: 242–244.

Berger, M., Gray, J. A., and Roth, B. L. The expanded biology of serotonin. *Annual Review of Medicine*. 2009. 60: 355–366.

Birrell, J. M., and Brown, V. J. Medial frontal cortex mediates perceptual attentional set shifting in the rat. *The Journal of Neuroscience*. 2000. 20: 4320–4324.

Biskup, C. S., Sánchez, C. L., Arrant, A., Van Swearingen, A. E. D., Kuhn, C., and Zepf, F. D. Effects of acute tryptophan depletion on brain serotonin function and concentrations of dopamine and norepinephrine in C57BL/6J and BALB/cJ mice. *PLoS ONE*. 2012. 7: e35916.

Booij, L., Van der Does A. J. W., Haffmans, P. M. J., Riedel, W. J., Fekkes, D., and Blom, M. J. B. The effects of high-dose and low-dose tryptophan depletion on mood and cognitive functions of remitted depressed patients. *Journal of Psychopharmacology*. 2005. 19: 267–275.

Borroto-Escuela, D. O., Romero-Fernandez, W., Narvaez, M., Oflijan, J., Agnati, L. F., and Fuxe, K. Hallucinogenic 5-HT2AR agonists LSD and DOI enhance dopamine D2R protomer recognition and signaling of D2–5-HT2A heteroreceptor complexes. *Biochemical and Biophysical Research Communications*. 2014. 443: 278–284.

Bosch, O. G., Wagner, M., Jessen, F., Khun, K. U., Joe, A., Seifritz, E., Maier, W., Biersack, H. J., and Quednow, B. B. Verbal memory deficits are correlated with prefrontal hypometabolism in ^{18}F-FDG PET of recreational MDMA users. *PLoS ONE*. 2013. 8: e61234.

Boyd, A. D., and Riba, M. Depression and pancreatic cancer. *Journal of the National Comprehensive Cancer Network*. 2007. 5: 113–116.

Brambrink, A. M., Evers, A. S., Avidan, M. S., Farber, N. B., Smith, D. J., Martin, L. D., Dissen, G. A., Creeley, C. E., and Olney, J. W.. Ketamine-induced neuroapoptosis in the fetal and neonatal rhesus macaque brain. *Anesthesiology*. 2012. 116: 372–384.

Branco-de-Almeida, L. S., Kajiya, M., Cardoso, C. R., Silve, M. J., Ohta, K., Rosalen, O. L., Franco, G. C., Han, X., Taubman, M. A., and Kawai, T. Selective serotonin reuptake inhibitors attenuate the antigen presentation from dendritic cells to effector T lymphocytes. *FEMS Immunology and Medical Microbiology*. 2011. 62: 283–294.

Breitbart, W., Bruera, E., Chochinov, H., and Lynch, M. Neuropsychiatric syndromes and psychological symptoms in patients with advanced cancer. *Journal of Pain and Symptom Management*. 1995. 10: 131–141.

Brown, J. H., and Paraskevas, F. Cancer and depression: Cancer presenting with depressive illness: An autoimmune disease? *The British Journal of Psychiatry*. 1982. 141: 227–232.

Brzezinski, A., Seibel, M. M., Lynch, H. J., Deng, M. H., and Wurtman, R. J. Melatonin in human preovulatory follicular fluid. *The Journal of Clinical Endocrinology and Metabolism*. 1987. 64: 865–867.

Buchert, R., Thomasius, R., Nebeling, B., Peterson, K., Obrocki, J., Jenicke, L., Wilke, F., Wartberg, L., Zapletalova, P., and Clausen, M. Long-term effects of "Ecstasy" use on serotonin transporters of the brain investigated by PET. *Journal of Nuclear Medicine*. 2003. 44: 375–384.

Buchert, R., Thomasius, R., Wilke, F., Nebeling, B., Peterson, K., Obrocki, J., Schulz, O., Schmidt, U., and Clausen, M. A voxel-based PET investigation of the long-term effects of "Ecstasy" consumption on brain serotonin transporters. *The American Journal of Psychiatry*. 2004. 161: 1181–1189.

Bufalino, C., Hepgul, N., Aguglia, E., and Pariante, C. M. The role of immune genes in the association between depression and inflammation: A review of recent clinical studies. *Brain, Behavior, and Immunity*. 2013. 31: 31–47.

Burtin, C., Scheinmann, P., Salomon, J. C., Lespinats, G., and Canu, P. Decrease in tumour growth by injections of histamine or serotonin in fibrosarcoma-bearing mice: Influence of H1 and H2 histamine receptors. *British Journal of Cancer*. 1982. 45: 54–60.

Bymaster, F. P., Katner, J. S., Nelson, D. L., Hemrick-Luecke, S. K., Threlkeld, P. G., Heiligenstein, J. H., Morin, S. N., Gehlert, D. R., and Perry, K. W. Atomoxetine increases extracellular levels of norepinephrine and dopamine in prefrontal cortex of rat: A potential mechanism for efficacy in attention deficit/hyperactivity disorder. *Neuropsychopharmacology*. 2002. 27: 699–711.

Cadet, J. L., Krasnova, I. N., Jayanthi, S., and Lyles, J. Neurotoxicity of substituted amphetamines: Molecular and cellular mechanisms. *Neurotoxicity Research*. 2007. 11: 183–202.

Cao, W., Carney, J. M., Duchon, A., Floyd, R. A., and Chevion, M. Oxygen free radical involvement in ischemia and reperfusion injury to brain. *Neuroscience Letters*. 1988. 88: 233–238.

Cappon, G. D., Pu, C., and Vorhees, C. V. Time-course of methamphetamine-induced neurotoxicity in rat caudate-putamen after single-dose treatment. *Brain Research*. 2000. 863: 106–111.

Capuron, L., Ravaud, A., Neveu, P. J., Miller, A. H., Maes, M., and Dantzer, R. Association between decreased serum tryptophan concentrations and depressive symptoms in cancer patients undergoing cytokine therapy. *Molecular Psychiatry*. 2002. 7: 468–473.

Carbajo-Pescador, S., Ordonez, R., Benet, M., Jover, R., Garcia-Palomo, A., Mauriz, J. L., and Gonzalez-Gallego, J. Inhibition of VEGF expression through blockade of HIF1α and STAT3 signalling mediates the anti-angiogenic effect of melatonin in HepG2 liver cancer cells. *British Journal of Cancer*. 2013. 109: 83–91.

Catala, A., Zvara, A., Puskas, L. G., and Kitajka, K. Melatonin-induced gene expression changes and its preventive effects on adriamycin-induced lipid peroxidation in rat liver. *Journal of Pineal Research*. 2007. 42: 43–49.

Chalmers, D. T., and Watson, S. J. Comparative anatomical distribution of 5-HT1A receptor mRNA and 5-HT1A binding in rat brain: A combined *in situ* hybridisation/*in vitro* receptor autoradiographic study. *Brain Research*. 1991. 561: 51–60.

Chang, F. L., Ho, S. T., and Sheen, M. J. Efficacy of mirtazapine in preventing intrathecalmorphine-induced nausea and vomiting after orthopaedic surgery*. *Anaesthesia*. 2010. 65: 1206–1211.

Chen, C. C., Lin, C. S., Ko, Y. P., Hung, Y. C., Lao, H. C., and Hsu, Y. W. Premedication with mirtazapine reduces preoperative anxiety and postoperative nausea and vomiting. *Anesthesia and Analgesia*. 2008. 106: 109–113.

Chen, C. K., Lin, S. K., Chen, Y. C., Huang, M. C., Chen, T. T., Ree, S. C., and Wang, L. J. Persistence of psychotic symptoms as an indicator of cognitive impairment in methamphetamine users. *Drug and Alcohol Dependence*. 2015. 148: 158–164.

Chen, X., Zheng, Y., Zheng, W., Gu, K., Chen, Z., Lu, W., and Shu, X. O. Prevalence of depression and its related factors among Chinese women with breast cancer. *Acta Oncologica*. 2009. 48: 1128–1136.

Chen, Y., and Guillemin, G. J. Kynurenine pathway metabolites in humans: Disease and healthy states. *International Journal of Tryptophan Research*. 2009. 2: 1–19.

Chuang H. Y. and Hwang J. J. 2015. The neuroimmune roles of indoleamine pathways in cancer. In Angel Catalá (ed.), *Indoleamines: Sources, Roles in Biological Processes and Health Effects Biochemistry Research Trends*, pp. 79–105. Hauppauge, NY: Nova.

Commins, D. L., Vosmer, G., Virus, R. M., Woolverton, W. L., Schuster, C. R., and Seiden, L. S. Biochemical and histological evidence that methylenedioxymethylamphetamine (MDMA) is toxic to neurons in the rat brain. *The Journal of Pharmacology and Experimental Therapeutics*. 1987. 241: 338–345.

Comperat, E., Roupret, M., Drouin, S. J., Camparo, P., Bitker, M. O., Houlgatte, A. et al. Tissue expression of IL16 in prostate cancer and its association with recurrence after radical prostatectomy. *The Prostate*. 2010. 70: 1622–1627.

Connell, P. J., Wang, X., Leon-Ponte, M., Griffiths, C., Pingle, S. C., and Ahern, G. P. A novel form of immune signaling revealed by transmission of the inflammatory mediator serotonin between dendritic cells and T cells. *Blood* 2006. 107: 1010–1017.

Conti, A., Conconi, S., Hertens, E., Skwarlo-Sonta, K., Markowska, M., and Maestroni, G. J. M. Evidence for melatonin synthesis in mouse and human bone marrow cells. *Journal of Pineal Research*. 2000. 28: 193–202.

Cowing, C., Pincus, S. H., Sachs, D. H., and Dickler, H. B. A Subpopulation of adherent accessory cells bearing both I-A and I-E or C subregion antigens is required for antigen-specific murine T lymphocyte proliferation. *The Journal of Immunology*. 1978. 121: 1680–1686.

Cox, J. L., Holden, J. M., and Sagovsky, R. Detection of postnatal depression. Development of the 10-item Edinburgh Postnatal Depression Scale. *The British Journal of Psychiatry*. 1987. 150: 782–786.

Cramer, H., Rudolph, J., Consbruch, U., and Kendel, K. On the effects of melatonin on sleep and behavior in man. *Advances in Biochemical Psychopharmacology*. 1974. 11: 187–191.

Crowley, S. J., Lee, C., Tseng, C. Y., Fogg, L. F., and Eastman, C. I. Combinations of bright light, scheduled dark, sunglasses, and melatonin to facilitate circadian entrainment to night shift work. *Journal of Biological Rhythms*. 2003. 18: 513–523.

Curran, H. V. Is MDMA ("Ecstasy") neurotoxic in humans? An overview of evidence and of methodological problems in research. *Neuropsychobiology*. 2000. 42: 34–41.

Curran, H. V., and Travill, R. A. Mood and cognitive effects of ±3,4-methylenedioxymethamphetamine (MDMA, "Ecstasy"): Week-end "high" followed by mid-week low. *Addiction* 1997. 92: 821–831.

Cutolo, M., Villaggio, B., Candido, F., Valenti, S., Glusti, M., Felli, L., Sulli, A., and Accardo, S. Melatonin influences interleukin-12 and nitric oxide production by primary cultures of rheumatoid synovial macrophages and THP-1 cellsa. *Annals of the New York Academy of Sciences*. 1999. 876: 246–254.

Prada, M. D., Saner, A., Burkard, W. P., Bartholini, G., and Pletscher, A. Lysergic acid diethylamide: Evidence for stimulation of cerebral dopamine receptors. *Brain Research*. 1975. 94: 67–73.

Dale, D. C., Boxer, L., and Liles, W. C. The phagocytes: Neutrophils and monocytes. *Blood*. 2008. 112: 935–945.

de las Casas-Engel, M., Domínguez-Soto, A., Sierra-Filardi, E., Bragardo, R., Nieto, C., Puig-Kroger, A., Samaniego, R. et al. Serotonin skews human macrophage polarization through HTR2B and HTR7. *The Journal of Immunology*. 2013. 190: 2301–2310.

Deng, W. G., Tang, S. T., Tseng, H. P., and Wu, K. K. Melatonin suppresses macrophage cyclooxygenase-2 and inducible nitric oxide synthase expression by inhibiting p52 acetylation and binding. *Blood*. 2006. 108: 518–524.

D'Mello, C., Le, T., and Swain, M. G. Cerebral microglia recruit monocytes into the brain in response to tumor necrosis factoralpha signaling during peripheral organ inflammation. *The Journal of Neuroscience*. 2009. 29: 2089–2102.

Dollins, A., Lynch, H., Wurtman, R., Deng, M. H., Kischka, K. U., Gleason, R. E., and Lieberman, H. R. Effect of pharmacological daytime doses of melatonin on human mood and performance. *Psychopharmacology*. 1993. 112: 490–496.

Dollins, A. B., Zhdanova, I. V., Wurtman, R. J., Lynch, H. J., and Deng, M. H. Effect of inducing nocturnal serum melatonin concentrations in daytime on sleep, mood, body temperature, and performance. *Proceedings of the National Academy of Sciences*. 1994. 91: 1824–1828.

Donaldson, M., and Goodchild, J. H. Oral health of the methamphetamine abuser. *American Journal of Health System Pharmacy*. 2006. 63: 2078–2082.

Dong, W. G., Mei, Q., Yu, J. P., Xu, J. M., Xiang, L., and Xu, Y. Effects of melatonin on the expression of iNOS and COX-2 in rat models of colitis. *World Journal of Gastroenterology*. 2003. 9: 1307–1311.

Dubocovich, M. L., Benloucif, S., and Masana, M. I. Melatonin receptors in the mammalian suprachiasmatic nucleus. *Behavioural Brain Research*. 1996. 73: 141–147.

Dubocovich, M. L., Hudson, R. L., Sumaya, I. C., Masana, M. I., and Manna, E. Effect of MT1 melatonin receptor deletion on melatonin-mediated phase shift of circadian rhythms in the C57BL/6 mouse. *Journal of Pineal Research*. 2005. 39: 113–120.

Dubocovich, M. L., Yun, K., Al-Ghoul, W. M., Benloucif, S., and Masana, M. I. Selective MT2 melatonin receptor antagonists block melatonin-mediated phase advances of circadian rhythms. *Federation of American Societies for Experimental Biology Journal*. 1998. 12: 1211–1220.

El-Missiry, M. A., Fayed, T. A., El-Sawy, M. R., and El-Sayed, A. A. Ameliorative effect of melatonin against gamma irradiation–induced oxidative stress and tissue injury. *Ecotoxicology and Environmental Safety*. 2007. 66: 278–286.

Eren, İ., Nazıroğlu, M., and Demirdaş, A. Protective effects of lamotrigine, aripiprazole and escitalopram on depression-induced oxidative stress in rat brain. *Neurochemical Research*. 2007a. 32: 1188–1195.

Eren, I., Nazıroğlu, M., Demirdaş, A., Celik, O., Uğuz, A. C., Altunbaşak, A., Ozmen, I., and Uz, E. Venlafaxine modulates depression-induced oxidative stress in brain and medulla of rat. *Neurochemical Research*. 2007b. 32: 497–505.

Ersche, K. D., Jones, P. S., Williams, G. B., Turton, A. J., Robbins, T. W., and Bullmore, E. T. Abnormal brain structure implicated in stimulant drug addiction. *Science*. 2012. 335: 601–604.

Evers, E. A., Tillie, D. E., van der Veen, F. M., Lieben, C. K., Jolles, J., Deutz, N. E., Schmidt, J. A. Effects of a novel method of acute tryptophan depletion on plasma tryptophan and cognitive performance in healthy volunteers. *Psychopharmacology*. 2005. 178: 92–99.

Evers, E. A., van der Veen, F. M., Jolles, J., Deutz, N. E., and Schmitt, J. A. Acute tryptophan depletion improves performance and modulates the BOLD response during a Stroop task in healthy females. *NeuroImage*. 2006. 32: 248–255.

Fernstrom, J. D., Langham, K. A., Marcelino, L. M., Irvine, Z. L., Fernstrom, M. H., and Kaye, W. The ingestion of different dietary proteins by humans induces large changes in the plasma tryptophan ratio, a predictor of brain tryptophan uptake and serotonin synthesis. *Clinical Nutrition (Edinburgh, Scotland)*. 2013. 32: 1073–1076.

Fontana, D. J., Daniels, S. E., Wong, E. H., Clark, R. D., and Eglen, R. M. The effects of novel, selective 5-hydroxytryptamine (5-HT)4 receptor ligands in rat spatial navigation. *Neuropharmacology*. 1997. 36: 689–696.

Francis, P. T., Pangalos, M. N., and Bowen, D. M. Animal and drug modelling for Alzheimer synaptic pathology. *Progress in Neurobiology*. 1992. 39: 517–545.

Freese, T. E., Miotto, K., and Reback, C. J. The effects and consequences of selected club drugs. *Journal of Substance Abuse Treatment*. 2002. 23: 151–156.

Gahlinger, P. M. Club drugs: MDMA, gamma-hydroxybutyrate (GHB), Rohypnol, and ketamine. *American Family Physician*. 2004. 69: 2619–2626.

Garfinkel, D., Laudon, M., Nof, D., and Zisapel, N. Improvement of sleep quality in elderly people by controlled-release melatonin. *Lancet*. 1995. 346: 541–544.

Geretsegger, C., Bohmer, F., and Ludwig, M. Paroxetine in the elderly depressed patient: Randomized comparison with fluoxetine of efficacy, cognitive and behavioural effects. *International Clinical Psychopharmacology*. 1994. 9: 25–29.

Ghia, J.-E., Blennerhassett, P., and Collins, S. M. Impaired parasympathetic function increases susceptibility to inflammatory bowel disease in a mouse model of depression. *The Journal of Clinical Investigation*. 2008. 118: 2209–2218.

Giannini, A. J., Loiselle, R. H., Giannini, M. C., and Price, W. A. Phencyclidine and the dissociatives. *Psychiatric Medicine*. 1985. 3: 197–217.

Gibb, J. W., Stone, D. M., Stahl, D. C., and Hanson, G. R. The effects of amphetamine-like designer drugs on monoaminergic systems in rat brain. *NIDA Research Monograph*. 1987. 76: 316–321.

Gilad, E., Wong, H. R., Zingarelli, B., Virag, L., O,Connor, M., Salzman, A. L., and Szabo, C. Melatonin inhibits expression of the inducible isoform of nitric oxide synthase in murine macrophages: Role of inhibition of NFκB activation. *The FASEB Journal*. 1998. 12: 685–693.

Gillette, M. U., and McArthur, A. J. Circadian actions of melatonin at the suprachiasmatic nucleus. *Behavioural Brain Research*. 1996. 73: 135–139.

Giordano, M., and Palermo, M. S. Melatonin-induced enhancement of antibody-dependent cellular cytotoxicity. *Journal of Pineal Research*. 1991. 10: 117–121.

Goncalves Ndo, N., Rodrigues, R. V., Jardim-Perassi, B. V., Moschetta, M. G., Lopes, J. R., Colombo, J., and Zuccari, D. A. Molecular markers of angiogenesis and metastasis in lines of oral carcinoma after treatment with melatonin. *Anti-cancer Agents in Medicinal Chemistry*. 2014. 14: 1302–1311.

Gresch, P. J., Smith, R. L., Barrett, R. J., and Sanders-Bush, E. Behavioral tolerance to lysergic acid diethylamide is associated with reduced serotonin-2A receptor signaling in rat cortex. *Neuropsychopharmacology*. 2005. 30: 1693–1702.

Gross, N. B., Duncker, P. C., and Marshall, J. F. Striatal dopamine D1 and D2 receptors: Widespread influences on methamphetamine-induced dopamine and serotonin neurotoxicity. *Synapse*. 2011. 65: 1144–1155.

Gurbuz, N., Ashour, A. A., Alpay, S. N., and Ozpolat, B. Down-regulation of 5-HT 1B and 5-HT 1D receptors inhibits proliferation, clonogenicity and invasion of human pancreatic cancer cells. *PLoS ONE*. 2014. 9: e105245.

Gustavsson, A., Svensson, M., Jacobi, F., Allgulander, C., Alonso, J., Beghi, E., Dodel, R., et al. Cost of disorders of the brain in Europe 2010. *European Neuropsychopharmacology*. 2011. 21: 718–779.

Haider, S., Khaliq, S., Ahmed, S. P., and Haleem, D. J. Long-term tryptophan administration enhances cognitive performance and increases 5HT metabolism in the hippocampus of female rats. *Amino Acids*. 2006. 31: 421–425.

Haider, S., Khaliq, S., and Haleem, D. J. Enhanced serotonergic neurotransmission in the hippocampus following tryptophan administration improves learning acquisition and memory consolidation in rats. *Pharmacological Reports*. 2007. 59: 53–57.

Haimov, I., Laudon, M., Zisapel, N., Souroujon, M., Nof, D., Shlitner, A., Herer, P., Tzischinsky, O., and Lavie, P. Sleep disorders and melatonin rhythms in elderly people. *British Medical Journal*. 1994. 309: 167.

Haughey, H. M., Fleckenstein, A. E., Metzger, R. R., and Hanson, G. R. The effects of methamphetamine on serotonin transporter activity: Role of dopamine and hyperthermia. *Journal of Neurochemistry*. 2000. 75: 1608–1617.

Hayashi, H., Dikkes, P., and Soriano, S. G. Repeated administration of ketamine may lead to neuronal degeneration in the developing rat brain. *Pediatric Anesthesia*. 2002. 12: 770–774.

Hayley, S., Poulter, M. O., Merali, Z., and Anisman, H. The pathogenesis of clinical depression: Stressor- and cytokine-induced alterations of neuroplasticity. *Neuroscience*. 2005. 135: 659–678.

Henderson, V. P., Clemow, L., Massion, A. O., Hurley, T. G., Druker, S., and Hébert, J. R. The effects of mindfulness–based stress reduction on psychosocial outcomes and quality of life in early-stage breast cancer patients: A randomized trial. *Breast Cancer Research and Treatment*. 2012. 131: 99–109.

Herxheimer, A., and Petrie, K. J. Melatonin for preventing and treating jet lag. *The Cochrane Database of Systematic Reviews*. 2001. 1: CD001520.

Hess, M. L., and Manson, N. H. Molecular oxygen: Friend and foe. The role of the oxygen free radical system in the calcium paradox, the oxygen paradox and ischemia/reperfusion injury. *Journal of Molecular and Cellular Cardiology*. 1984. 16: 969–985.

Hirata, H., Ladenheim, B., Rothman, R. B., Epstein, C., and Cadet, J. L. Methamphetamine-induced serotonin neurotoxicity is mediated by superoxide radicals. *Brain Research*. 1995. 677: 345–347.

Hirst, W. D., Stean, T. O., Rogers, D. C., Sunter, D., Pugh, P., Moss, S. F., Bromidge, S. M., et al. SB-399885 is a potent, selective 5-HT6 receptor antagonist with cognitive enhancing properties in aged rat water maze and novel object recognition models. *European Journal of Pharmacology.* 2006. 553: 109–119.

Hocaoglu, C., Kural, B., Aliyazicioglu, R., Deger, O., and Cengiz, S. IL-1beta, IL-6, IL-8, IL-10, IFN-gamma, TNF-alpha and its relationship with lipid parameters in patients with major depression. *Metabolic Brain Disease.* 2012. 27: 425–430.

Hopwood, P., and Stephens, R. J. Depression in patients with lung cancer: Prevalence and risk factors derived from quality-of-life data. *Journal of Clinical Oncology.* 2000. 18: 893–903.

Hunt, A. E., Al-Ghoul, W. M., Gillette, M. U., and Dubocovich, M. L. Activation of MT(2) melatonin receptors in rat suprachiasmatic nucleus phase advances the circadian clock. *American Journal of Physiology Cell Physiology.* 2001. 280: C110–C118.

Ibbotson, T., Maguire, P., Selby, P., Priestman, T., and Wallace, L. Screening for anxiety and depression in cancer patients: The effects of disease and treatment. *European Journal of Cancer.* 1994. 30a: 37–40.

James, S. P., Sack, D. A., Rosenthal, N. E., and Mendelson, W. B. Melatonin administration in insomnia. *Neuropsychopharmacology.* 1990. 3: 19–23.

Jang, S. S., Kim, H. G., Lee, J. S., Han, J. M., Park, H. J., Huh, G. J., and Son, C. G. Melatonin reduces X-ray radiation–induced lung injury in mice by modulating oxidative stress and cytokine expression. *International Journal of Radiation Biology.* 2013. 89: 97–105.

Jardim-Perassi, B. V., Arbab, A. S., Ferreira, L. C., Borin, T. F., Varma, N. R., Iskander, A. S., Shankar, A., Ali, M. M., and de Campos Zuccari, D. A. Effect of melatonin on tumor growth and angiogenesis in xenograft model of breast cancer. *PLoS ONE.* 2014. 9: e85311.

Jorge, R. E., Robinson, R. G., Moser, D., Tateno, A., Crespo-Facorro, B., and Arndt, S. Major depression following traumatic brain injury. *Archives of General Psychiatry.* 2004. 61: 42–50.

Jung-Hynes, B., Schmit, T. L., Reagan-Shaw, S. R., Siddiqui, I. A., Mukhtar, H., and Ahmad, N. Melatonin, a novel Sirt1 inhibitor, imparts antiproliferative effects against prostate cancer *in vitro* in culture and *in vivo* in TRAMP model. *Journal of Pineal Research.* 2011. 50: 140–149.

Kadva, A., Djahanbakhch, O., Monson, J., Di, W. L., and Silman, R. Elevated nocturnal melatonin is a consequence of gonadotropin-releasing hormone deficiency in women with hypothalamic amenorrhea. *The Journal of Clinical Endocrinology and Metabolism.* 1998. 83: 3653–3662.

Kalueff, A. V., Olivier, J. D., Nonkes, L. J., and Homberg, J. R. Conserved role for the serotonin transporter gene in rat and mouse neurobehavioral endophenotypes. *Neuroscience and Biobehavioral Reviews.* 2010. 34: 373–386.

Karslioglu, I., Ertekin, M. V., Taysi, S., Koçar, I., Sezen, O., Gepdiremen, A., Koç, M., and Bakan, N. Radioprotective effects of melatonin on radiation-induced cataract. *Journal of Radiation Research* 2005. 46: 277–282.

Katoh, N., Soga, F., Nara, T., Tamagawa-Mineoka, R., Nin, M., Kotani, H., Masuda, K., and Kishimoto, S. Effect of serotonin on the differentiation of human monocytes into dendritic cells. *Clinical and Experimental Immunology.* 2006. 146: 354–361.

Kaye, S., McKetin, R., Duflou, J., and Darke, S. Methamphetamine and cardiovascular pathology: A review of the evidence. *Addiction.* 2007. 102: 1204–1211.

Keilhoff, G., Bernstein, H. G., Becker, A., Grecksch, G., and Wolf, G. Increased neurogenesis in a rat ketamine model of schizophrenia. *Biological Psychiatry.* 2004. 56: 317–322.

Khaliq, S., Haider, S., Ahmed, S. P., Perveen, T., and Haleem, D. J. Relationship of brain tryptophan and serotonin in improving cognitive performance in rats. *Pakistan Journal of Pharmaceutical Sciences.* 2006. 19: 11–15.

Kilic, E., Kilic, U., Reiter, R. J., Bassetti, C. L., and Hermann, D. M. Tissue-plasminogen activator–induced ischemic brain injury is reversed by melatonin: Role of iNOS and Akt. *Journal of Pineal Research.* 2005. 39: 151–155.

Kim, B. C., Shon, B. S., Ryoo, Y. W., Kim, S. P., and Lee, K. S. Melatonin reduces X-ray irradiation–induced oxidative damages in cultured human skin fibroblasts. *Journal of Dermatological Science.* 2001. 26: 194–200.

King, M. V., Marsden, C. A., and Fone, K. C. A role for the 5-HT(1A), 5-HT4 and 5-HT6 receptors in learning and memory. *Trends in Pharmacological Sciences.* 2008. 29: 482–492.

King, M. V., Sleight, A. J., Woolley, M. L., Topham, I. A., Marsden, C. A., and Fone, K. C. 5-HT6 receptor antagonists reverse delay-dependent deficits in novel object discrimination by enhancing consolidation: An effect sensitive to NMDA receptor antagonism. *Neuropharmacology*. 2004. 47: 195–204.

Kita, T., Wagner, G. C., and Nakashima, T. Current research on methamphetamine-induced neurotoxicity: Animal models of monoamine disruption. *Journal of Pharmacological Sciences*. 2003. 92: 178–195.

Kitay, J. I. Pineal lesions and precocious puberty: A review. *The Journal of Clinical Endocrinology and Metabolism*. 1954. 14: 622–625.

Kline, A. E., Yu, J., Massucci, J. L., Zafonte, R. D., and Dixon, C. E. Protective effects of the 5-HT1A receptor agonist 8-hydroxy-2-(di-n-propylamino)tetralin against traumatic brain injury–induced cognitive deficits and neuropathology in adult male rats. *Neuroscience Letters*. 2002. 333: 179–182.

Kohrs, R., and Durieux, M. E. Ketamine: Teaching an old drug new tricks. *Anesthesia and Analgesia*. 1998. 87: 1186–1193.

Kokoshka, J. M., Metzger, R. R., Wilkins, D. G., Gibb, J. W., Hanson, G. R., and Fleckenstein, A. E. Methamphetamine treatment rapidly inhibits serotonin, but not glutamate, transporters in rat brain. *Brain Research*. 1998. 799: 78–83.

Kovacs, E. The serum levels of IL-12 and IL-16 in cancer patients. Relation to the tumour stage and previous therapy. *Biomedicine and Pharmacotherapy*. 2001. 55: 111–116.

Krall, C. M., Richards, J. B., Rabin, R. A, and Winter, J. C. Marked decrease of LSD-induced stimulus control in serotonin transporter knockout mice. *Pharmacology Biochemistry and Behavior*. 2008. 88: 349–357.

Laberge, S., Cruikshank, W. W., Beer, D. J., and Center, D. M. Secretion of IL-16 (lymphocyte chemoattractant factor) from serotonin-stimulated CD8+ T cells *in vitro*. *The Journal of Immunology*. 1996. 156: 310–315.

Lanoix, D., Beghdadi, H., Lafond, J., and Vaillancourt, C. Human placental trophoblasts synthesize melatonin and express its receptors. *Journal of Pineal Research*. 2008. 45: 50–60.

Lanoix, D., Guerin, P., and Vaillancourt, C. Placental melatonin production and melatonin receptor expression are altered in preeclampsia: New insights into the role of this hormone in pregnancy. *Journal of Pineal Research*. 2012. 53: 417–425.

Laoutidis, Z. G., and Mathiak, K. Antidepressants in the treatment of depression/depressive symptoms in cancer patients: A systematic review and meta-analysis. *BMC Psychiatry*. 2013. 13: 140.

Laughlin, G. A., Loucks, A. B., and Yen, S. S. Marked augmentation of nocturnal melatonin secretion in amenorrheic athletes, but not in cycling athletes: Unaltered by opioidergic or dopaminergic blockade. *The Journal of Clinical Endocrinology and Metabolism*. 1991. 73: 1321–1326.

Lee, Y. J., Lee, J. H., Moon, J. H., and Park, S. Y. Overcoming hypoxic-resistance of tumor cells to TRAIL-induced apoptosis through melatonin. *International Journal of Molecular Sciences*. 2014. 15: 11941–11956.

León-Ponte, M., Ahern, G. P., and Connell, P. J. Serotonin provides an accessory signal to enhance T-cell activation by signaling through the 5-HT7 receptor. *Blood*. 2006. 109: 3139–3146.

Lesch, K. P. Hallucinations: Psychopathology meets functional genomics. *Molecular Psychiatry*. 1998. 3: 278–281.

Li, M., Fitzgerald, P., and Rodin, G. Evidence-based treatment of depression in patients with cancer. *Journal of Clinical Oncology*. 2012. 30: 1187–1196.

Li, N., Ghia, J.-E., Wang, H., McClemmens, J., Cote, F., Suehiro, Y., Mallet, J., and Khan, W. I. Serotonin activates dendritic cell function in the context of gut inflammation. *The American Journal of Pathology*. 2011. 178: 662–671.

Linden, W., and Girgis, A. Psychological treatment outcomes for cancer patients: What do meta-analyses tell us about distress reduction? *Psychooncology*. 2012. 21: 343–350.

Linden, W., Vodermaier, A., MacKenzie, R., and Greig, D. Anxiety and depression after cancer diagnosis: Prevalence rates by cancer type, gender, and age. *Journal of Affective Disorders*. 2012. 141: 343–351.

Lipsey, J. R., Robinson, R. G., Pearlson, G. D., Rao, K., and Price, T. R. Mood change following bilateral hemisphere brain injury. *The British Journal of Psychiatry*. 1983. 143: 266–273.

Lissoni, P., Meregalli, S., Nosetto, L., Barni, S., Tancini, G., Fossati, V., and Maestroni, G. Increased survival time in brain glioblastomas by a radioneuroendocrine strategy with radiotherapy plus melatonin compared to radiotherapy alone. *Oncology*. 1996. 53: 43–46.

Lissoni, P., Paolorossi, F., Ardizzoia, A., Barni, S., Chilelli, M., Mancuso, M., Tancini, G., Conti, A., and Maestroni, G. A randomized study of chemotherapy with cisplatin plus etoposide *versus* chemoendocrine therapy with cisplatin, etoposide and the pineal hormone melatonin as a first-line treatment of advanced non-small cell lung cancer patients in a poor clinical state. *Journal of Pineal Research.* 1997. 23: 15–19.

Lissoni, P., Tisi, E., Barni, S., Ardizzola, A., Rovelli, F., Rescaldani, R., and Ballabio, D., Benenti, C., Angeli, M., and Tancini, G. Biological and clinical results of a neuroimmunotherapy with interleukin-2 and the pineal hormone melatonin as a first line treatment in advanced non-small cell lung cancer. *British Journal of Cancer.* 1992. 66: 155–158.

Liu, H., Xu, L., Wei, J. E., Xie, M. R., Wang, S. E., and Zhou, R. X. Role of CD4+ CD25+ regulatory T cells in melatonin-mediated inhibition of murine gastric cancer cell growth *in vivo* and *in vitro. Anatomical Record.* 2011. 294: 781–788.

Lloyd-Williams, M., Friedman, T., and Rudd, N. A survey of antidepressant prescribing in the terminally ill. *Palliative Medicine.* 1999. 13: 243–248.

Lloyd-Williams, M., Shiels, C., and Dowrick, C. The development of the Brief Edinburgh Depression Scale (BEDS) to screen for depression in patients with advanced cancer. *Journal of Affective Disorders.* 2007. 99: 259–264.

Lucchesi, B. R. Myocardial ischemia, reperfusion and free radical injury. *The American Journal of Cardiology.* 1990. 65: 14i–23i.

Lüscher, C., and Ungless, M. A. The mechanistic classification of addictive drugs. *PLoS Med.* 2006. 3: e437.

Maes, M., Leonard, B. E., Myint, A. M., Kubera, M., and Verkerk, R. The new '5-HT' hypothesis of depression: Cell-mediated immune activation induces indoleamine 2,3-dioxygenase, which leads to lower plasma tryptophan and an increased synthesis of detrimental tryptophan catabolites (TRYCATs), both of which contribute to the onset of depression. *Progress in Neuro-psychopharmacology & Biological Psychiatry.* 2011. 35: 702–721.

Manda, T., Nishigaki, F., Mori, J., and Shimomura, K. Important role of serotonin in the antitumor effects of recombinant human tumor necrosis factor-α in mice. *Cancer Research.* 1988. 48: 4250–4255.

Manuel-Apolinar, L., and Meneses, A. 8-OH-DPAT facilitated memory consolidation and increased hippocampal and cortical cAMP production. *Behavioural Brain Research.* 2004. 148: 179–184.

Marchetti, E., Chaillan, F. A., Dumuis, A., Bockaert, J., Soumireu-Mourat, B., and Roman, F. S. Modulation of memory processes and cellular excitability in the dentate gyrus of freely moving rats by a 5-HT4 receptor's partial agonist, and an antagonist. *Neuropharmacology.* 2004. 47: 1021–1035.

Markus, C. R., Olivier, B., and de Haan, E. H. Whey protein rich in alpha-lactalbumin increases the ratio of plasma tryptophan to the sum of the other large neutral amino acids and improves cognitive performance in stress-vulnerable subjects. *The American Journal of Clinical Nutrition.* 2002. 75: 1051–1056.

Markus, C. R., Olivier, B., Panhuysen, G. E., Van Der Gugten, J., Alles, M. S., Tuiten, A. et al. The bovine protein alpha-lactalbumin increases the plasma ratio of tryptophan to the other large neutral amino acids, and in vulnerable subjects raises brain serotonin activity, reduces cortisol concentration, and improves mood under stress. *The American Journal of Clinical Nutrition.* 2000. 71: 1536–1544.

Martín, V., Sanchez-Sanchez, A. M., Puente-Moncada, N., Gomez-Lobo, M., Alvarex-Vega, M. A., Antolin, I., and Rodriguez, C. Involvement of autophagy in melatonin-induced cytotoxicity in glioma-initiating cells. *Journal of Pineal Research.* 2014. 57: 308–316.

Martino, M., Rocchi, G., Escelsior, A., and Fornaro, M. Immunomodulation mechanism of antidepressants: Interactions between serotonin/norepinephrine balance and Th1/Th2 balance. *Current Neuropharmacology.* 2012. 10: 97–123.

Massie, M. J. Prevalence of depression in patients with cancer. *Journal of the National Cancer Institute Monographs.* 2004. 32: 57–71.

Mathers, C. D., and Loncar, D. Projections of global mortality and burden of disease from 2002 to 2030. *PLoS Med.* 2006. 3: e442.

McCann, U., Szabo, Z., and Vranesic, M., Palermo, M., Mathews, W. B., Ravert, H. T., Dannals, R. F., and Ricaurte, G. A. Positron emission tomographic studies of brain dopamine and serotonin transporters in abstinent (±)3,4-methylenedioxymethamphetamine ("Ecstasy") users: Relationship to cognitive performance. *Psychopharmacology.* 2008. 200: 439–450.

McCann, U. D., Szabo, Z., Scheffel, U., Dannals, R. F., and Ricaurte, G. A. Positron emission tomographic evidence of toxic effect of MDMA ("Ecstasy") on brain serotonin neurons in human beings. *Lancet.* 1998. 352: 1433–1437.

Meltzer, C. C., Smith, G., DeKosky, S. T., Pollock, B. G., Mathis, C. A., Moore, R. Y., Kupfer, D. J., Reynolds (III), C. F. Serotonin in aging, late-life depression, and Alzheimer's disease: The emerging role of functional imaging. *Neuropsychopharmacology.* 1998. 18: 407–430.

Meltzer, H. Y., and Stahl, S. M. The dopamine hypothesis of schizophrenia: A review. *Schizophrenia Bulletin.* 1976. 2: 19–76.

Mendelsohn, D., Riedel, W. J., and Sambeth, A. Effects of acute tryptophan depletion on memory, attention and executive functions: A systematic review. *Neuroscience and Biobehavioral Reviews.* 2009. 33: 926–952.

Meneses, A. 5-HT system and cognition. *Neuroscience and Biobehavioral Reviews.* 1999. 23: 1111–1125.

Meredith, E. J., Holder, M. J., Chamba, A., Challa, A., Drake-Lee, A., Bunce, C. M., Drayson, M. T., et al. The serotonin transporter (SLC6A4) is present in B-cell clones of diverse malignant origin: Probing a potential anti-tumor target for psychotropics. *The FASEB Journal.* 2005. 19: 1187–1189.

Meyer, F., Fletcher, K., Prigerson, H. G., Braun, I. M., and Maciejewski, P. K. Advanced cancer as a risk for major depressive episodes. *Psychooncology.* 2014. 24 (9): 1080–1087.

Mihandoost, E., Shirazi, A., Mahdavi, S. R., and Aliasgharzadeh, A. Can melatonin help us in radiation oncology treatments? *Biomed Research International.* 2014. 2014: 578137.

Miles, L. E., Raynal, D. M., and Wilson, M. A. Blind man living in normal society has circadian rhythms of 24.9 hours. *Science.* 1977. 198: 421–423.

Miller, S. C., Pandi, P. S. R, Esquifino, A. I., Cardinali, D. P., and Maestroni, G. J. M. The role of melatonin in immuno-enhancement: Potential application in cancer. *International Journal of Experimental Pathology.* 2006. 87: 81–87.

Minzenberg, M. J., and Carter, C. S. Developing treatments for impaired cognition in schizophrenia. *Trends in Cognitive Sciences.* 2012. 16: 35–42.

Miovic, M., and Block, S. Psychiatric disorders in advanced cancer. *Cancer.* 2007. 110: 1665–1676.

Mitchell, E. S., Hoplight, B. J., Lear, S. P., and Neumaier, J. F. BGC20-761, a novel tryptamine analog, enhances memory consolidation and reverses scopolamine-induced memory deficit in social and visuospatial memory tasks through a 5-HT6 receptor–mediated mechanism. *Neuropharmacology.* 2006. 50: 412–420.

Miyamoto, S., Duncan, G. E., Marx, C. E., and Lieberman, J. A. Treatments for schizophrenia: A critical review of pharmacology and mechanisms of action of antipsychotic drugs. *Molecular Psychiatry.* 2005. 10: 79–104.

Molon, B., Ugel, S., Del Pozzo, F., Soldani, C., Zilio, S., Avelia, D., De Palma, A., et al. Chemokine nitration prevents intratumoral infiltration of antigen-specific T cells. *The Journal of Experimental Medicine* 2011. 208: 1949–1962.

Moreno, J. L., Holloway, T., Albizu, L., Sealfon, S. C., and Gonzalez-Maeso, J. Metabotropic glutamate mGlu2 receptor is necessary for the pharmacological and behavioral effects induced by hallucinogenic 5-HT2A receptor agonists. *Neuroscience Letters.* 2011. 493: 76–79.

Moreno, J. L., Holloway, T., Rayannavar, V., Sealfon, S. C., and Gonzalez-Maeso, J. Chronic treatment with LY341495 decreases 5-HT(2A) receptor binding and hallucinogenic effects of LSD in mice. *Neuroscience Letters.* 2013. 536: 69–73.

Morgan, C. J. A., Curran, H. V., and The Independent Scientific Committee on Drugs. Ketamine use: A review. *Addiction.* 2012. 107: 27–38.

Morishima, I., Matsui, H., Mukawa, H., Hyashi, K., Toki, Y., Okumura, K., Ito, T., and Hayakawa, T. Melatonin, a pineal hormone with antioxidant property, protects against adriamycin cardiomyopathy in rats. *Life Sciences.* 1998. 63: 511–521.

Mrazek, D. A., Hornberger, J. C., Altar, C. A., and Degtiar, I. A review of the clinical, economic, and societal burden of treatment-resistant depression: 1996–2013. *Psychiatric Services.* 2014. 65: 977–987.

Mueller, P. D., and Korey, W. S. Death by "Ecstasy": The serotonin syndrome? *Annals of Emergency Medicine.* 1998. 32: 377–380.

Muetzelfeldt, L., Kamboj, S. K., Rees, H., Taylor, J., Morgan, C. J., and Curran, H. V. Journey through the K-hole: Phenomenological aspects of ketamine use. *Drug and Alcohol Dependence.* 2008. 95: 219–229.

Müller, T., Dürk, T., Blumenthal, B., Grimm, M., Cicko, S., Panther, E., Sorichter, S., et al. 5-hydroxytryptamine modulates migration, cytokine and chemokine release and T-cell priming capacity of dendritic cells *in vitro* and *in vivo*. *PLoS ONE*. 2009. 4: e6453.

Murphy, F. C., Smith, K. A., Cowen, P. J., Robbins, T. W., and Sahakian, B. J. The effects of tryptophan depletion on cognitive and affective processing in healthy volunteers. *Psychopharmacology*. 2002. 163: 42–53.

Murphy, P. N., Wareing, M., Fisk, J. E., and Montgomery, C. Executive working memory deficits in abstinent ecstasy/MDMA users: A critical review. *Neuropsychobiology*. 2009. 60: 159–175.

Myers, E. N., De Boer, M. F., McCormick, L. K., Pruyn, J. F. A., Ryckman, R. M., and van den Borne, B. W. Physical and psychosocial correlates of head and neck cancer: A review of the literature. *Otolaryngology: Head and Neck Surgery*. 1999. 120: 427–436.

Myint, A. M., and Kim, Y. K. Cytokine-serotonin interaction through IDO: A neurodegeneration hypothesis of depression. *Medical Hypotheses*. 2003. 61: 519–525.

Nagaraj, S., Gupta, K., Pisarev, V., Kinarsky, L., Sherman, S., Kang, L., Herber, D., Schneck, J., and Gabrilovich, D. I. Altered recognition of antigen is a mechanism of CD8+ T cell tolerance in cancer. *Nature Medicine*. 2007. 13: 828–835.

Nagtegaal, E., Peeters, T., Swart, W., Smits, M., Kerkhof, G., and van der Meer, G. Correlation between concentrations of melatonin in saliva and serum in patients with delayed sleep phase syndrome. *Therapeutic Drug Monitoring*. 1998. 20:181–183.

Nauta, R. J., Tsimoyiannis, E., Uribe, M., Walsh, D. B., Miller, D., and Butterfield, A. Oxygen-derived free radicals in hepatic ischemia and reperfusion injury in the rat. *Surgery, Gynecology and Obstetrics*. 1990. 171: 120–125.

Nocito, A., Dahm, F., Jochum, W., Jang, J. H., Georgieve, P., Bader, M., Graf, R., and Clavien, P. A. Serotonin regulates macrophage-mediated angiogenesis in a mouse model of colon cancer allografts. *Cancer Research*. 2008. 68: 5152–5158.

Nowak, E. C., de Vries, V. C., Wasiuk, A., Ahonen, C., Bennett, K., A. Le Mercier, I., Ha, D. G., and Noelle, R. J. Tryptophan hydroxylase-1 regulates immune tolerance and inflammation. *The Journal of Experimental Medicine*. 2012. 209: 2127–2135.

O'Connell, P. J., Wang, X., Leon-Ponte, M., Griffiths, C., Pingle, S. C., and Ahern, G. P. A novel form of immune signaling revealed by transmission of the inflammatory mediator serotonin between dendritic cells and T cells. *Blood*. 2006. 107: 1010–1017.

Olesen, J., Gustavsson, A., Svensson, M., Wittchen, H. U., and Jonsson, B. The economic cost of brain disorders in Europe. *European Journal of Neurology*. 2012. 19: 155–162.

Olivier, J. D., Jans, L. A., Korte-Bouws, G. A., Korte, S. M., Deen, P. M., Cools, A. R., Ellenbroek, B. A., and Blokland, A. Acute tryptophan depletion dose dependently impairs object memory in serotonin transporter knockout rats. *Psychopharmacology*. 2008. 200: 243–254.

Ordoñez, R., Carbajo-Pescador, S., Prieto-Dominguez, N., García-Palomo, A., González-Gallego, J., and Mauriz, J. L. Inhibition of matrix metalloproteinase-9 and nuclear factor kappa B contribute to melatonin prevention of motility and invasiveness in HepG2 liver cancer cells. *Journal of Pineal Research*. 2014. 56: 20–30.

Ortiz, F., Acuna-Castroviejo, D., Doerrier, C., Dayoub, J. C., Lopez, L. C., Venegas, C., Garcia, J. A., et al. Melatonin blunts the mitochondrial/NLRP3 connection and protects against radiation-induced oral mucositis. *Journal of Pineal Research*. 2015. 58: 34–49.

Ortiz, F., Garcia, J. A., Acuna-Castroviejo, D., Doerrier, C., Lopez, A., Venegas, C., Volt, H., Luna-Sanchez, M., Lopez, L. C., and Escames, G. The beneficial effects of melatonin against heart mitochondrial impairment during sepsis: Inhibition of iNOS and preservation of nNOS. *Journal of Pineal Research*. 2014. 56: 71–81.

Parlakpinar, H., Sahna, E., Ozer, M. K., Ozugurlu, F., Vardi, N., and Acet, A. Physiological and pharmacological concentrations of melatonin protect against cisplatin-induced acute renal injury. *Journal of Pineal Research*. 2002. 33: 161–166.

Parrott, A. C. Recreational Ecstasy/MDMA, the serotonin syndrome, and serotonergic neurotoxicity. *Pharmacology, Biochemistry, and Behavior*. 2002. 71: 837–844.

Parrott, A. C. MDMA, serotonergic neurotoxicity, and the diverse functional deficits of recreational "Ecstasy" users. *Neuroscience and Biobehavioral Reviews*. 2013. 37: 1466–1484.

Parrott, A. C., and Lasky, J. Ecstasy (MDMA) effects upon mood and cognition: Before, during and after a Saturday night dance. *Psychopharmacology*. 1998. 139: 261–268.

Petticrew, M., Bell, R., and Hunter, D. Influence of psychological coping on survival and recurrence in people with cancer: Systematic review. *British Medical Journal*. 2002. 325 (7372): 1066.

Pieri, C., Moroni, F., Marra, M., Marcheselli, F., and Recchioni, R. Melatonin is an efficient antioxidant. *Archives of Gerontology and Geriatrics*. 1995. 20: 159–165.

Pioli, C., Caroleo, M. C., Nistico, G., and Doriac, G. Melatonin increases antigen presentation and amplifies specific and non specific signals for T-cell proliferation. *International Journal of Immunopharmacology*. 1993. 15: 463–468.

Plaimee, P., Weerapreeyakul, N., Thumanu, K., Tanthanuch, W., and Barusrux, S. Melatonin induces apoptosis through biomolecular changes, in SK-LU-1 human lung adenocarcinoma cells. *Cell Proliferation*. 2014. 47: 564–577.

Plotkin, S. R., Banks, W. A., and Kastin, A. J. Comparison of saturable transport and extracellular pathways in the passage of interleukin-1 alpha across the blood–brain barrier. *Journal of Neuroimmunology*. 1996. 67: 41–47.

Prendergast, G. C. Immune escape as a fundamental trait of cancer: Focus on IDO. *Oncogene*. 2008. 27: 3889–3900.

Proietti, S., Cucina, A., Dobrowolny, G., D'Anselmi, F., Dinicola, S., Masiello, M. G., Pasqualato, A., et al. Melatonin down-regulates MDM2 gene expression and enhances p53 acetylation in MCF-7 cells. *Journal of Pineal Research*. 2014. 57: 120–129.

Puig-Domingo, M., Webb, S. M., Serrano, J., Peinado, M. A., Corcoy, R., Ruscalieda, J., Reiter, R. J., and de Leiva, A. Brief report: Melatonin-related hypogonadotropic hypogonadism. *The New England Journal of Medicine*. 1992. 327: 1356–1359.

Pytliak, M., Vargova, V., Mechirova, V., and Felsoci, M. Serotonin receptors: From molecular biology to clinical applications. *Physiological Research/Academia Scientiarum Bohemoslovaca*. 2011. 60: 15–25.

Que, J. C., Sy Ortin, T. T., Anderson, K. O., Gonzalez-Suarez, C. B., Feeley, T. W., and Reyes-Gibby, C. C. Depressive symptoms among cancer patients in a Philippine tertiary hospital: Prevalence, factors, and influence on health-related quality of life. *Journal of Palliative Medicine*. 2013. 16: 1280–1284.

Rabin R. C. A glut of antidepressants. *The New York Times*. 2013, August 12th.

Raison, C. L., and Miller, A. H. Depression in cancer: New developments regarding diagnosis and treatment. *Biological Psychiatry*. 2003. 54: 283–294.

Rapoport, M. J. Depression following traumatic brain injury: Epidemiology, risk factors and management. *CNS Drugs*. 2012. 26: 111–121.

Rayner, L., Price, A., Evans, A., Valsraj, K., Hotopf, M., and Higginson, I. J. Antidepressants for the treatment of depression in palliative care: Systematic review and meta-analysis. *Palliative Medicine*. 2011. 25: 36–51.

Reich, M., Lesur, A., and Perdrizet-Chevallier, C. Depression, quality of life and breast cancer: A review of the literature. *Breast Cancer Research and Treatment* 2008. 110: 9–17.

Reneman, L., Booij, J., de Bruin, K., Reitsma, J. B., de Wolff, F. A., Gunning, W. B., den Heeton, G. J., and van den Brink, W. Effects of dose, sex, and long-term abstention from use on toxic effects of MDMA (ecstasy) on brain serotonin neurons. *The Lancet*. 2001. 358: 1864–1869.

Reyes-Gibby, C. C., Anderson, K. O., Morrow, P. K., Shete, S., and Hassan, S. Depressive symptoms and health-related quality of life in breast cancer survivors. *Journal of Women's Health*. 2012. 21: 311–318.

Reynolds, G. P., Mason, S. L., Meldrum, A., de Keczer, S., Parnes, H., Eglen, R. M., and Wong, E. H. 5-hydroxytryptamine (5-HT)4 receptors in post mortem human brain tissue: Distribution, pharmacology and effects of neurodegenerative diseases. *British Journal of Pharmacology*. 1995. 114: 993–998.

Ricaurte, G., Bryan, G., Strauss, L., Seiden, L., and Schuster, C. Hallucinogenic amphetamine selectively destroys brain serotonin nerve terminals. *Science*. 1985. 229: 986–988.

Ricaurte, G. A., Yuan, J., and McCann, U. D. (+/−)3,4-Methylenedioxymethamphetamine ("Ecstasy")-induced serotonin neurotoxicity: Studies in animals. *Neuropsychobiology*. 2000. 42: 5–10.

Richter, H. G., Hansell, J. A., Raut, S., and Giussani, D. A. Melatonin improves placental efficiency and birth weight and increases the placental expression of antioxidant enzymes in undernourished pregnancy. *Journal of Pineal Research*. 2009. 46: 357–364.

Riedel, W. J., Sobczak, S., and Schmitt, J. J. Tryptophan modulation and cognition. In Allegri, G., Costa, C. L., Ragazzi, E., Steinhart, H., and Varesio, L., (editors), *Developments in Tryptophan and Serotonin Metabolism*. Boston, Massachusetts: Springer US. 2003: 207–213.

Rivest, S., Lacroix, S., Vallieres, L., Nadeau, S., Zhang, J., and Laflamme, N. How the blood talks to the brain parenchyma and the paraventricular nucleus of the hypothalamus during systemic inflammatory and infectious stimuli. *Proceedings of the Society for Experimental Biology and Medicine Society for Experimental Biology and Medicine.* 2000. 223: 22–38.

Roca, A. L., Godson, C., Weaver, D. R., and Reppert, S. M. Structure, characterization, and expression of the gene encoding the mouse Mel1a melatonin receptor. *Endocrinology.* 1996. 137: 3469–3477.

Rodriguez, V. B., Orgaz, B. P., Bayon, C., Palau, A., Torres, G., Hospital, A., Benito, G., Dieguez, M., and Fernandez Liria, A. Differences in depressed oncologic patients' narratives after receiving two different therapeutic interventions for depression: A qualitative study. *Psychooncology.* 2012. 21: 1292–1298.

Rojansky, N., Brzezinski, A., and Schenker, J. G. Seasonality in human reproduction: An update. *Human Reproduction.* 1992. 7: 735–745.

Rubinsztein, J. S., Michael, A., Underwood, B. R., Tempest, M., and Sahakian, B. J. Impaired cognition and decision-making in bipolar depression but no "affective bias" evident. *Psychological Medicine.* 2006. 36: 629–639.

Rudin, M., Ben-Abraham, R., Gazit, V., Tendler, Y., Tashlykov, V., and Katz, Y. Single-dose ketamine administration induces apoptosis in neonatal mouse brain. *Journal of Basic and Clinical Physiology and Pharmacology.* 2005. 16: 231–243.

Sack, R. L., Brandes, R. W., Kendall, A. R., and Lewy, A. J. Entrainment of free-running circadian rhythms by melatonin in blind people. *The New England Journal of Medicine.* 2000. 343: 1070–1077.

Sagrillo-Fagundes, L., Soliman, A., and Vaillancourt, C. Maternal and placental melatonin: Actions and implication for successful pregnancies. *Minerva Ginecologica.* 2014. 66: 251–266.

Sambeth, A., Blokland, A., Harmer, C. J., Killkens, T. O., Nathan, P. J., Porter, R. J., Schmidt, J. A., et al. Sex differences in the effect of acute tryptophan depletion on declarative episodic memory: A pooled analysis of nine studies. *Neuroscience and Biobehavioral Reviews.* 2007. 31: 516–529.

Sanchez-Barcelo, E. J., Mediavilla, M. D., Alonso-Gonzalez, C., and Reiter, R. J. Melatonin uses in oncology: Breast cancer prevention and reduction of the side effects of chemotherapy and radiation. *Expert Opinion on Investigational Drugs.* 2012. 21: 819–831.

Schmidt, C. J., Wu, L., and Lovenberg, W. Methylenedioxymethamphetamine: A potentially neurotoxic amphetamine analogue. *European Journal of Pharmacology.* 1986. 124: 175–178.

Schmitt, J. A., Jorissen, B. L., Sobczak, S., van Boxtel, M. P., E. Hogervorst, N. E. Deutz, and Reidel, W. J. Tryptophan depletion impairs memory consolidation but improves focussed attention in healthy young volunteers. *Journal of Psychopharmacology.* 2000. 14: 21–29.

Schöne, W., and Ludwig, M. Paroxetine in the treatment of depression in geriatric patients: A double-blind comparative study with fluoxetine. *Fortschritte der Neurologie-Psychiatrie.* 1994. 62 (Suppl. 1): 16–18.

Shibata, S., Cassone, V. M., and Moore, R. Y. Effects of melatonin on neuronal activity in the rat suprachiasmatic nucleus *in vitro. Neuroscience Letters.* 1989. 97: 140–144.

Shirazi, A., Ghobadi, G., and Ghazi-Khansari, M. A radiobiological review on melatonin: A novel radioprotector. *Journal of Radiation Research.* 2007. 48: 263–272.

Skene, D. J., and Arendt, J. Circadian rhythm sleep disorders in the blind and their treatment with melatonin. *Sleep Medicine.* 2007. 8: 651–655.

Slavich, G. M., and Irwin, M. R. From stress to inflammation and major depressive disorder: A social signal transduction theory of depression. *Psychological Bulletin.* 2014. 140: 774–815.

Sotelo, J. L., Musselman, D., and Nemeroff, C. The biology of depression in cancer and the relationship between depression and cancer progression. *International Review of Psychiatry.* 2014. 26: 16–30.

Stehle, J., Vanecek, J., and Vollrath, L. Effects of melatonin on spontaneous electrical activity of neurons in rat suprachiasmatic nuclei: An *in vitro* iontophoretic study. *Journal of Neural Transmission.* 1989. 78: 173–177.

Sternberg, E. M., Trial, J., and Parker, C. W. Effect of serotonin on murine macrophages: Suppression of Ia expression by serotonin and its reversal by 5-HT2 serotonergic receptor antagonists. *The Journal of Immunology.* 1986. 137: 276–282.

Sternberg, E. M., Wedner, H. J., Leung, M. K., and Parker, C. W. Effect of serotonin (5-HT) and other monoamines on murine macrophages: Modulation of interferon-gamma induced phagocytosis. *The Journal of Immunology.* 1987. 138: 4360–4365.

Swanson, C. J., Perry, K. W., Koch-Krueger, S., Katner, J., Svensson, K. A., and Bymaster, F. P. Effect of the attention deficit/hyperactivity disorder drug atomoxetine on extracellular concentrations of norepinephrine and dopamine in several brain regions of the rat. *Neuropharmacology.* 2006. 50: 755–760.

Tamarkin, L., Hollister, C. W., Lefebvre, N. G., and Goldman, B. D. Melatonin induction of gonadal quiescence in pinealectomized Syrian hamsters. *Science.* 1977a. 198: 953–955.

Tamarkin, L., Lefebvre, N. G., Hollister, C. W., and Goldman, B. D. Effect of melatonin administered during the night on reproductive function in the Syrian hamster. *Endocrinology.* 1977b. 101: 631–634.

Tamura, H., Nakamura, Y., Terron, M. P., Flores, L. J., Manchester, L. C., Tan, D. X., Sugino, N., and Reiter, R. J. Melatonin and pregnancy in the human. *Reproductive Toxicology.* 2008. 25: 291–303.

Tan, D. X., Hardeland, R., Manchester, L. C., Galano, A., and Reiter, R. J. Cyclic-3-hydroxymelatonin (C3HOM), a potent antioxidant, scavenges free radicals and suppresses oxidative reactions. *Current Medicinal Chemistry.* 2014. 21: 1557–1565.

Taysi, S., Koc, M., Buyukokuroglu, M. E., Altinkaynak, K., and Sahin, Y. N. Melatonin reduces lipid peroxidation and nitric oxide during irradiation-induced oxidative injury in the rat liver. *Journal of Pineal Research.* 2003. 34: 173–177.

Terry Jr, A. V., Buccafusco, J. J., and Wilson, C. Cognitive dysfunction in neuropsychiatric disorders: Selected serotonin receptor subtypes as therapeutic targets. *Behavioural Brain Research.* 2008. 195: 30–38.

Trulson, M. E., Ross, C. A., and Jacobs, B. L. Behavioral evidence for the stimulation of CNS serotonin receptors by high doses of LSD. *Psychopharmacology Communications.* 1976. 2: 149–164.

Turaga, K. K., Malafa, M. P., Jacobsen, P. B., Schell, M. J., and Sarr, M. G. Suicide in patients with pancreatic cancer. *Cancer.* 2011. 117: 642–647.

Vijayalaxmi, Meltz, M. L., Reiter, R. J., and Herman, T. S. Melatonin and protection from genetic damage in blood and bone marrow: Whole-body irradiation studies in mice. *Journal of Pineal Research.* 1999a. 27: 221–225.

Vijayalaxmi, Meltz, M. L., Reiter, R. J., Herman, T. S., and Kumar, K. S. Melatonin and protection from whole-body irradiation: Survival studies in mice. *Mutation Research.* 1999b. 425: 21–27.

Vijayalaxmi, Reiter, R. J., Herman, T. S., and Meltz, M. L. Melatonin and radioprotection from genetic damage: *In vivo/in vitro* studies with human volunteers. *Mutation Research.* 1996. 371: 221–228.

Wagner, D., Becker, B., Koester, P., Gouzoulis-Mayfrank, E., and Daumann, J. A prospective study of learning, memory, and executive function in new MDMA users. *Addiction.* 2013. 108: 136–145.

Waldhauser, F., Boepple, P. A., Schemper, M., Mansfield, M. J., and Crowley Jr, W. F. Serum melatonin in central precocious puberty is lower than in age-matched prepubertal children. *The Journal of Clinical Endocrinology and Metabolism.* 1991. 73: 793–796.

Watkins, L. R., Goehler, L. E., Relton, J. K., Tartaglia, N., Silbert, L., Martin, D., and Maier, S. F. Blockade of interleukin-1 induced hyperthermia by subdiaphragmatic vagotomy: Evidence for vagal mediation of immune-brain communication. *Neuroscience Letters.* 1995. 183: 27–31.

Watson, M., Haviland, J. S., Greer, S., Davidson, J., and Bliss, J. M. Influence of psychological response on survival in breast cancer: A population-based cohort study. *Lancet.* 1999. 354: 1331–1336.

Webley, G. E., and Leidenberger, F. The circadian pattern of melatonin and its positive relationship with progesterone in women. *The Journal of Clinical Endocrinology and Metabolism.* 1986. 63: 323–328.

Wedding, U., Koch, A., Rohrig, B., Pientka, L., Sauer, H., Hoffken, K. et al. Depression and functional impairment independently contribute to decreased quality of life in cancer patients prior to chemotherapy. *Acta Oncologica (Stockholm, Sweden).* 2008. 47: 56–62.

Wichers, M., and Maes, M. The psychoneuroimmuno-pathophysiology of cytokine-induced depression in humans. *The International Journal of Neuropsychopharmacology.* 2002. 5: 375–388.

Wink, D. A., Hines, H. B., Cheng, R. Y. S., Switzer, C. H., Flores-Santana, W., Vitek, M. P., Ridnour, L. A., and Colton, C. A. Nitric oxide and redox mechanisms in the immune response. *Journal of Leukocyte Biology.* 2011. 89: 873–891.

Woolley, M. L., Bentley, J. C., Sleight, A. J., Marsden, C. A., and Fone, K. C. A role for 5-HT6 receptors in retention of spatial learning in the Morris water maze. *Neuropharmacology.* 2001. 41: 210–219.

Wu, Y. H., Ursinus, J., Zhou, J. N., Scheer, F. A., Ai-Min, B., Jockers, R., van Heerikhuize, J., and Swaab, D. F. Alterations of melatonin receptors MT1 and MT2 in the hypothalamic suprachiasmatic nucleus during depression. *Journal of Affective Disorders.* 2013. 148: 357–367.

Xu, C., Wu, A., Zhu, H., Fang, H., Xu, L., Ye, J., and Shen, J. Melatonin is involved in the apoptosis and necrosis of pancreatic cancer cell line SW-1990 via modulating of Bcl-2/Bax balance. *Biomedicine and Pharmacotherapy.* 2013. 67: 133–139.

Yoon, J. H., Minzenberg, M. J., Ursu, S., Ryan Walter, B. S., Wendelken, C., Ragland, J. D., and Carter, C. S. Association of dorsolateral prefrontal cortex dysfunction with disrupted coordinated brain activity in schizophrenia: Relationship with impaired cognition, behavioral disorganization, and global function. *The American Journal of Psychiatry*. 2008. 165: 1006–1014.

Young, J. J., Bruno, D., and Pomara, N. A review of the relationship between proinflammatory cytokines and major depressive disorder. *Journal of Affective Disorders*. 2014. 169: 15–20.

Young, M. R., and Matthews, J. P. Serotonin regulation of T-cell subpopulations and of macrophage accessory function. *Immunology*. 1995. 84: 148–152.

Zhdanova, I. V., Wurtman, R. J., Regan, M. M., Taylor, J. A., Shi, J. P., and Leclair, O. U. Melatonin treatment for age-related insomnia. *The Journal of Clinical Endocrinology and Metabolism*. 2001. 86: 4727–4730.

Zigmond, A. S., and Snaith, R. P. The hospital anxiety and depression scale. *Acta Psychiatrica Scandinavica*. 1983. 67: 361–370.

Zou, X., Patterson, T. A., Divine, R. L., Sadovova, N., Zhang, X., Hanig, J. P., Paule, M. G., Slikker Jr, W., and Wang, C. Prolonged exposure to ketamine increases neurodegeneration in the developing monkey brain. *International Journal of Developmental Neuroscience*. 2009. 27: 727–731.

32

Melatonin Disruption in Autism and Its Regulation by Natural Supplements

Ann Mary Alex and Moinak Banerjee
Rajiv Gandhi Center for Biotechnology
Kerala, India

CONTENTS

ABSTRACT Autism is a complex disorder, influenced by genetic and environmental factors. The major endophenotypes of autism are sensory and circadian dysfunction and immune dysregulation. The two major observable phenotypes of circadian dysfunctions are difficulties in sleeping and cognitive deficits. Melatonin seems to be a central molecule, as it functions vary from antioxidant properties and immune regulation to maintenance of sleep and circadian architecture, and many others. This chapter looks into the various biochemical and genetic evidences for the association of melatonin with autism. A lower melatonin level in autism has been reported by several investigators. Maternal levels of melatonin can also influence the development of the disease. Therefore, it is imperative to maintain a normal level of all the molecules that can influence the synthesis of melatonin. In this chapter we demonstrate ways and means to maintain melatonin levels by dietary modification, which can intervene at various steps in the melatonin synthesis and maintain the resultant melatonin levels. We also suggest that understanding the molecular mechanism involved in the biosynthetic process of melatonin in autism might help in addressing phenotypic variants in the disease.

KEY WORDS: *autism, melatonin, serotonin, tryptophan, sleep, nutrition, circadian rhythm.*

Abbreviations

AANAT: *N*-acetyl transferase
ASD: autism spectrum disorders
ASMT: acetylserotonin *O*-methyltransferase

DDC:	dopa decarboxylase
EC:	enterochromaffin cells
GI:	gastrointestinal
HIOMT:	hydroxyindole-O-methyltrasferase
IFNγ:	interferon gamma
IL1:	interleukin
NK:	natural killer cells
MAOA:	monoamine oxidase A
REM:	rapid eye movement
SAM:	S-adenosyl methionine
TNFα:	tumor necrosis factor alpha
TPH:	tryptophan hydroxylase

Introduction

Autism is a neurodevelopmental disorder that is characterized by deficits in social communication and interaction and shows restricted and repetitive behaviour. It manifests within the first three years of life. It is a highly heterogenous disorder with the major biological endophenotypes being circadian and sensory dysfunction, neurodevelopmental delay, immune dysfunction, and stereotypic movement (Sacco et al. 2010). It is a complex disorder with both genetic and environmental factors contributing toward the etiology. A recent twin study in a UK cohort suggests a high heritability of 56%–95% for the disorder suggesting a strong genetic component in the pathophysiology of the disease (Colvert et al. 2015). The global prevalence estimates a median of 62/10,000 from the data from different populations (Elsabbagh et al. 2012). The risk is three to four times higher in males than females. A number of environmental factors such as vitamin and mineral deficiencies, immunological background, and exposure to various toxic agents also influence the development and severity of autistic symptoms (Dietert et al. 2011).

One of the major endophenotypes associated with autism spectrum disorders (ASD) is circadian and sensory dysfunction. The two major observable phenotypes of circadian dysfunctions are difficulties in sleeping and cognitive deficits. Sleep disturbances are observed in 56%–83% of the children with ASD. Alterations in melatonin levels can influence the sleep–wake cycle. Melatonin production is known to be influenced by circadian rhythms and the monoamine metabolic pathway. Melatonin and its metabolites take part in a vast array of biological functions making this a very important molecule in the proper functioning and maintenance of the physiology. Melatonin (5-hydroxy-N-acetyltryptamine) is a hormone that is endogenously produced in mammals during the dark period. This indolamine is present in both the plant and animal kingdoms with identical structure in both the systems (Tan et al. 2012). In mammals, the hormone is primarily produced in the pineal gland. It is also synthesized in a small amount by peripheral systems such as retinal cells, enterochromaffin (EC) cells in the gut, immune system cells, Harderian gland, cerebellum, airway epithelium, liver, kidney, adrenals, thymus, thyroid, pancreas, ovary, carotid body, placenta, and endometrium (Kvetnoy 1999; Huether 1993).

Melatonin Synthesis in Humans

L-tryptophan is the starting amino acid in the biosynthesis of melatonin. This amino acid cannot be synthesized in humans and hence needs to supplemented through diet. It is converted to 5-hydroxy tryptophan by the enzyme tryptophan hydroxylase (TPH), which is the rate-limiting enzyme in the synthesis of serotonin. TPH has two isoforms: TPH1, which is active in the periphery, the enterochromaffin cells of the GI tract and pineal gland; and TPH2, which is specifically expressed in the brain regions. Tetrahydrobiopterin and iron act as cofactors of TPH. Vitamin D has been shown to activate the transcription of TPH2 in the brain and repress TPH1 in other tissues (Patrick and Ames 2014). Dopa decarboxylase (DDC) further converts the product into serotonin (Slominski et al. 2008). Vitamin B6 is a cofactor in this process (Calderón-Guzmán et al. 2004). Serotonin is a neurotransmitter involved in many behavioral cognitive and physiological functions. It exhibits a diurnal rhythm with maximum production during the daytime, the level dropping to less than 80% after dark. The level of serotonin is regulated by the genes involved in the synthesis, transport, and reuptake of receptor proteins. It is also modulated by the epigenetic modifications such as DNA methylation, which in turn is influenced by factors such as diet. A tryptophan-rich diet can lead to a higher synthesis of serotonin in the brain as tryptophan can cross the blood–brain barrier. Serotonin is synthesized in the neurons in brain and EC cells in the GI tract. 95% of the serotonin is produced in the gut. Platelets have a high affinity for serotonin and accumulate the neurotransmitter produced from the GI cells. The majority of the serotonin is catabolized by monoamine oxidase (Kim and Camilleri 2000).

Serotonin is converted to *N*-acetyl serotonin by *N*-acetyl transferase (AANAT), which is the rate-limiting enzyme in this pathway, and further to melatonin by acetylserotonin *O*-methyltransferase (ASMT), previously known as hydroxyindole-*O*-methyltrasferase (HIOMT) (Slominski et al. 2008). Calcium has been reported to be required for the activity of AANAT (Zawilska 1992), and magnesium increases the activity of ASMT (Billyard et al. 2006). Calcium can regulate AANAT activity by affecting the intracellular cyclic AMP content, which is also critical in the regulation of melatonin biosynthesis (Zawilska 1992). These evidences suggest that external supplementation of calcium and magnesium sources can enhance the melatonin content. In addition to these environmental sources, a methionine-rich diet can also influence the epigenetic modulation of melatonin synthesis. The reaction catalyzed by ASMT requires a methyl group, which is provided by metabolic processing of homocysteine and methionine into *S*-adenosyl methionine (Fournier et al. 2002). The melatonin level has its highest peak at nighttime and lowers during daytime showing a reverse diurnal pattern as that of serotonin. Melatonin is metabolized primarily in the liver by CYP1A2 and secondarily in the kidney (Ma 2005; Lane and Moss 1985). Homocysteine has also been reported to have circadian rhythmicity that is under the control of circadian genes (Paul et al. 2014). Alteration at any step in the metabolic processing either endogenously or exogenously can decide a healthy or diseased state in autism.

Functions of Melatonin

Melatonin production is controlled by photoperiod and it has a vast array of biological functions. The secretion peaks in the middle of the night (2–4 a.m.) and gradually falls in the later half of the night (Brzezinski 1997). The physiological roles of melatonin in humans are in the regulation of diurnal rhythm, sleep, mood, immunity, reproduction, intestinal motility, and metabolism (Arendt et al. 1999; Skwarlo-Sonta 2002; Delagrange et al. 2003). Laboratory studies and clinical trials suggest that melatonin may also possess neuroprotective (Thomas and Mohanakumar 2004), cardioprotective (Chen et al. 2003), and anticancer (Panzer and Viljoen 1997) properties. Melatonin has been evaluated as a therapeutic agent in a number of conditions, including sleep disturbance, jet lag, and metastatic cancer (Lissoni et al. 1989; Herxheimer and Petrie 2002; Singer et al. 2003).

The effect of melatonin is produced partly by binding to receptor proteins, and partly because of the antioxidant property it exerts by interacting with other proteins. Membrane receptors of melatonin, MT1, MT2, and putative MT3 binding site have been shown to activate multiple signaling pathways (Barrenetxe et al. 2004; Serle et al. 2004). Melatonin can pass through biological membranes and bind to nuclear receptors also. Melatonin secreted by the placental tissue can regulate intracellular processes (e.g., G-proteins), the activity of second messengers (e.g., cAMP, IP3, Ca2+) and have an important role in gene expression regulation during neurodevelopment (de Faria Poloni et al. 2011). These suggest the plethora of activities that are regulated by the molecule.

Role in Circadian Rhythm and Sleep

The various physiological and behavioral functions in humans exhibit a rhythmic pattern maintained by the pacemaker circadian clock located in the suprachiasmatic nucleus in the hypothalamus. It controls the sleep–wake cycle, regulation of body temperature, blood pressure, synthesis and secretion of several hormones, and so on. Sleep is an active process that is finely and reliably regulated by two parallel mechanisms—circadian regulation and sleep homeostasis. Sleep homeostasis is a regulatory mechanism based on prior duration of continuous sleep and wakefulness. Melatonin release and timing of sleep is closely related. Melatonin affects the sleep propensity and also the duration and quality of sleep. It has been shown to trigger the onset of sleep (Brzezinski 1997; Brzezinski et al. 2005). Exogenous melatonin has been found to improve the percentage of rapid eye movement (REM) sleep (Kunz et al. 2004).

Role in Immune System

Melatonin is secreted by some key immunocompetent cells and has numerous roles in the immune system. It acts as an immune buffer, acting as a stimulant under basal or immunosuppressive conditions or as an anti-inflammatory compound in acute inflammatory conditions (Guerrero and Reiter 2002). Studies have reported close relation between the night time peaks of melatonin and proliferation of progenitor cells for macrophages and granulocytes (Kuci et al. 1988). NK activity enhancement (Lissoni et al. 1985) and modulation of immune mediators such as cytokines was observed on supplementation of melatonin (Carrillo-Vico et al. 2006). The various functions of the hormone are mediated through the two different receptors (MT1 and MT2) on the membrane and also intracellular receptors.

Role as an Antioxidant

Melatonin functions as an antioxidant through direct and indirect routes. It has the capacity to directly scavenge the oxygen- and nitrogen-based free radicals (Hardeland et al. 1993; Allegra et al. 2003; Ng et al. 2000). It acts as an indirect antioxidant by enhancing the activity of other antioxidant enzymes (Reiter et al. 2000; Rodriguez et al. 2004). It can also increase the production of an antioxidant compound, glutathione, by stimulating the enzyme gamma glutamylcysteine synthase (Urata et al. 1999) and can also help in protecting antioxidant enzymes from oxidation.

Disrupted Melatonin in Autism

Several studies have suggested a difference in the sleep architecture in ASD. There have been many reports on the decreased nighttime melatonin level in autistic children. This can be correlated to the physiological abnormalities such as sleep onset latency, lower total sleep time (Limoges 2005; Malow and McGrew 2008), and lower REM sleep (Buckley et al. 2010), which is observed in patients with autism. How melatonin is likely to influence the observable phenotypes of circadian dysfunctions resulting in difficulties in sleeping and cognitive deficits is demonstrated in Figure 32.1.

Recent studies have focused on the immune parameters in the development of autism. Decreased levels of IgG, IgA and higher IgE/IgG are reported in some studies pertaining to the humoral immunity (Grether et al. 2010; Wasilewska et al. 2012). Multiple studies have reported an elevated level of tumor necrosis factor alpha (TNFα), interferon gamma (IFNγ), cytokines such as interleukin (IL1β, IL-5, IL-8, IL-12p70, IL-13, IL-17) I, and the PBMC of autistic patients (Ashwood et al. 2011a,b; Manzardo et al. 2012). The imbalance in the levels may be attributed to the lessened modulation of cytokines by melatonin.

A large number of studies have been done on the oxidative stress status in autism. Oxidative stress occurs as a result of an imbalance between the endogenous antioxidants and production of reactive oxygen species (Valko et al. 2007). Oxidative stress is high in patients with autism when compared to controls, with the level of prooxidative markers shown to be increased while there is a reduction in the antioxidant molecule glutathione, and other antioxidation enzymes (Rossignol and Frye 2014). The reduced melatonin-mediated antioxidant effect may be a contributing factor for this.

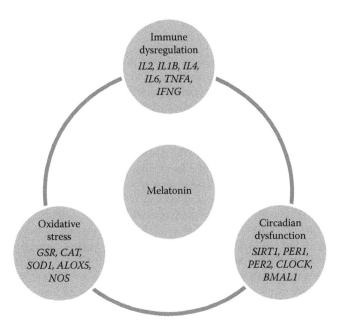

FIGURE 32.1 Melatonin is known to influence the molecular mechanism of observable phenotypes of circadian dysfunctions resulting in difficulties in sleeping and cognitive deficits.

Evidence from Biochemical Studies

Abnormalities in levels of tryptophan, serotonin, melatonin, and its metabolites have been implicated in autism. Two studies reported a reduced urinary level of tryptophan in autistic children (Evans et al. 2009; Kałuzna-Czaplinska et al. 2010). There have been consistent reports on hyperserotonemia in platelets in autistic individuals (Anderson 2002; Cook et al. 1989; Leboyer et al. 1999; Hranilovic et al. 2007; Mulder et al. 2004). Interestingly, a significantly lower serotonin level was observed in the autistic brain when compared to normal individuals (McBride 1989; Chugani et al. 1999). Genetic studies using genes involved in serotonin synthesis, receptors, and transporters in autism helped further in understanding the background of the aberrations in serotonin synthesis in these individuals (Huang and Santangelo 2008; Orabona et al. 2009; Tordjman et al. 2001; Coon et al. 2005; Yang et al. 2012). A study reports a polymorphism in the serotonin transporter to be associated with a modification of the severity of autistic behavior rather than with risk of autism (Tordjman et al. 2001).

There have been several reviews on the reports of sleep–wake rhythm disturbances, in autistic individuals (Glickman 2010; Tordjman et al. 2005), as well as improvement in sleep in autistic patients when melatonin was administered (Doyen et al. 2011; Rossignol and Frye 2011; Guénolé et al. 2011). In addition, alterations in serotonin in autism in both the central and peripheral systems have been widely reported (Richdale 1999; Anderson 2002; Nakamura et al. 2010). Melatonin is produced from serotonin in two steps. The altered level may impact the level of melatonin in the systems.

Prior studies of melatonin production in autistic disorder were often limited by small sample sizes and were not entirely consistent, but all reports had a consensus on abnormalities in the melatonin production. Disrupted melatonin level mainly decreased nocturnal melatonin levels when compared to normal controls have been reported in many studies (Melke et al. 2008; Ritvo et al. 1993; Nir et al. 1995; Kulman et al. 1999; Tordjman et al. 2005; Mulder et al. 2010; Tordjman et al. 2012). A significant decrease in the plasma melatonin level of autistic probands and their relatives was reported in a study, with 51% of probands and 31% of mothers of autistic patients showing a lower melatonin level than normal controls (Pagan et al. 2014).

Evidence from Genetic Data

The disrupted melatonin synthesis in autism has prompted many studies on the genetic variants in the genes involved in the pathway. All the genes involved in the synthesis of melatonin from tryptophan have been studied as candidate genes in both family–based and case control association studies. An ethnic specific association was observed which indicated that none of these markers could be replicated in the global population with equal confidence. This might imply genetic or phenotypic variants in different ethnic populations. TPH1 was reported to be regulating the 5-HT neurons in the developing brain. Mutations in this gene affected the 5-HT level in the developing brain but were unaffected in the adult brain in mouse studies. The abnormalities in the developing brain can contribute to psychiatric conditions (Nakamura 2006). A study in an Indian population reported a strong association of TPH2 variant with ASD (Singh et al. 2013). A nominal association was observed for TPH2 (Coon et al. 2005) and haplotypic association in TPH1 with whole blood serotonin (Cross et al. 2008). Studies in the DDC gene show a strong association with ASD risk in a case control study, suggesting it might be a risk allele in ASD etiology (Toma et al. 2013). An earlier report with a smaller sample size reported no association of two deletion variants in the DDC gene with functional significance in autism (Lauritsen et al. 2002). MAOA variants have been reported to have a sexual dimorphic effect in ASD with the risk alleles showing differential effect in both males and females. In MAOA SNP rs6323 the low activity T allele posed a higher risk in males, but not in females (Verma et al. 2014).

A significant decrease in AANAT gene expression was observed in a subgroup of ASD patients with severe language impairment (Hu et al. 2009; Hu and Steinberg 2009). Most of the studies involving the melatonin pathway have focused on the ASMT gene. A study by Melke et al. (2008) reported a positive association of two promoter polymorphisms (rs4446909 and rs598681) with ASD. The mutant allele of the variants resulted in lower transcript level expression of the enzyme. Decreased expression of ASMT was also correlated to the lower melatonin level in autistic probands, and also in parents with the rs4446909 and rs598681 variant. Apart from the common variations, a number of rare variants including nonsynonymous and splice site mutations have been reported in the gene in autistic probands. *In vitro* studies have shown functional significance of a splice cite variant with low transcript level (Melke et al. 2008). Further studies had failed to duplicate the association with ASD risk in larger populations. A microduplication in the exon 8 of ASMT gene was observed in another study. The duplications were present in 9 children with ASD who also presented sleep disturbances (Cai et al. 2008).

Mutations in the melatonin receptors altering the functional properties of melatonin receptors (MT1 and MT2) have been reported in individuals with ASD (Chaste et al. 2010). But a positive association could not be found, suggesting they may not be major risk factors for ASD but may contribute to the overall risk in the presence of other pathogenic variants. These genetic variants in the melatonin biosynthetic pathway do indicate that alterations in melatonin or its upstream precursors do have a genetic basis. Understanding this genetic basis of alteration of melatonin or its upstream precursors in autism will help in resolving the appropriate concerns, rather than global concerns.

Melatonin Supplementation in Diet

Melatonin is involved in a plethora of functions ranging from maintenance of circadian rhythm, the sleep–wake cycle, boosting immunity, and protection against oxidation. Exogenously supplied melatonin also performs similar functions and hence dietary supplement can be used to boost deficiencies. Although the concentrations are lower in natural products, having no side effects is a benefit. Melatonin augmentation can be done both directly or indirectly by providing appropriate nutritional supplements required to boost the melatonin synthesis. A number of food items have been found that boost the melatonin or melatonin synthesis pathway (Figure 32.2).

Depending on the mutational background of the genes involved in the melatonin synthesis of an individual, appropriate supplements can be provided to boost the melatonin system. One of the reasons for the lower melatonin synthesis can be the reduced availability of L-tryptophan. It is an essential amino

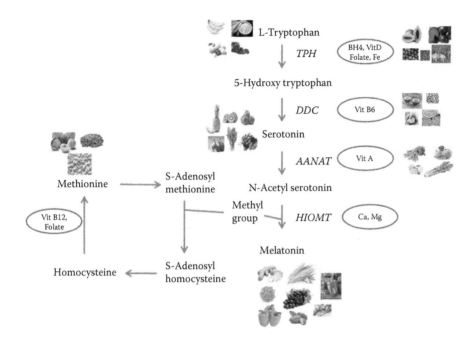

FIGURE 32.2 The melatonin biosynthetic pathway. Nutritional supplements required at specific steps in the biosynthetic pathway to boost the melatonin synthesis.

acid that has to be supplied through diet. It can cross the blood–brain barrier through the large neutral amino acid transporter (Dingerkus et al. 2012) and help induce the synthesis of melatonin. Tryptophan-rich sources that are commonly available include oats, bananas, dried prunes, peanuts, chocolate, and fruits such as cherries (Garrido et al. 2012; Cubero et al. 2009; Richard et al. 2009). Tryptophan can be utilized more efficiently from a carbohydrate-rich diet than a protein-rich diet, as in a protein-rich diet other large amino acids compete with tryptophan in crossing the blood–brain barrier and hence reduce the availability of tryptophan in the brain (Fernstrom 1985). If the enzyme TPH is found to be defective or mutated, it will be ideal to supplement the diet with the downstream molecules such as serotonin and melatonin. If the activity of the enzyme is reduced, it is ideal to enrich the next component in the pathway exogenously. Banana, plantain, pineapple, kiwi, plums, and tomato are food sources with high serotonin content. Avacados, dates, grapefruits, honeydew melon, olives, broccoli, eggplant, figs, spinach, and cauliflower have moderate serotonin content in them (Feldman and Lee 1985). The final product of the pathway, melatonin, can also be obtained from various food sources. The dietary supplement can complement the action of the endogenously produced melatonin. Foods rich in melatonin include corn, rice, barley, grains, and ginger (Badria 2002).

A number of vitamin and mineral deficiencies also can affect this pathway. Folate deficiency can impair the synthesis of tetrahydrobiopterin (BH4), which is a cofactor of the enzyme TPH (Coppen et al. 1989). Folate, which is the bioavailable form of vitamin B9, is found in many plants sources such as lentils, beans, fruits such as papaya, grapefruit, strawberries, and avocado. They are also present in vegetables such as spinach, asparagus, beets, carrot, and cauliflower, and in seeds and nuts such as almonds and peanuts. Vitamin D can activate the transcription of TPH2 in brain. A common source of vitamin D in humans is exposure of skin to sunlight (Holick et al. 1980). Edible mushrooms have been reported to be sources of vitamin D2 and the dietary supplement improves the serum 25-hydroxyvitamin D concentrations (Outila et al. 1999). Vitamin B6 is a cofactor for the enzyme DDC, which is the final enzyme in the synthesis of serotonin (Footitt 2013). It is abundantly found in rice bran, pistachio nuts, sunflower seeds, sweet potato, spinach, banana, and avocado. Vitamin A deficiency has been reported to affect the activity of AANAT enzyme and hence melatonin level (Ashton et al. 2015). Carrot, mango, watermelon, apricot, plum, and celery are rich in vitamin A. Calcium and magnesium ions have been reported to affect the activity of ASMT enzyme (Peuhkuri et al. 2012). A methyl donor, *S*-adenosyl methionine (SAM), supplies the methyl group to *N*-acetyl serotonin in the final step of melatonin synthesis. SAM is synthesized from amino acid methionine and upon loss of methyl the group is converted to homocysteine, which is reconverted to methionine. Methionine is absorbed from the diet as it is an essential amino acid. It is available in plant sources such as white beans, soybeans, sunflower seeds, oats, walnuts, hazelnuts, and so on. This cycle requires vitamin B12 and folate for the maintenance. Supplementation of these vitamins and minerals through diet can support the healthy regulation of melatonin biosynthesis.

Maternal low melatonin levels reported in autistic cases can contribute to the etiology of the disease. Hence supplementation in pregnant women and autistic children is important. As melatonin is a small lipophilic molecule that can cross biological membranes, maternal melatonin can pass through to the fetal brain where it can perform antioxidant functions. Although many natural sources of melatonin are available caution should be taken in administration to autistic patients, as they are more prone to gastrointestinal (GI) problems. Therefore, a diet involving high protein and adequate slow-digesting carbohydrates, with limited fat content, is suggested. A gluten-free and casein-free diet is also followed by many and has been reported to alleviate the symptoms and severity. The results have been inconsistent and the evidence for the effect of the intervention is poor (Millward et al. 2008). Another study based on parental reports from 387 parents or caregivers suggested that the difference in the GI and immunological background may affect the efficacy of the intervention in different patients (Pennesi and Klein 2012).

Conclusion

The major endophenotypes of autism are sensory and circadian dysfunction, and immune dysregulation. Melatonin seems to be a central molecule in controlling these endophenotypic variants, as it has functions of antioxidant properties, immune regulation, maintenance of sleep and circadian architecture, and many others. Therefore, it is imperative to maintain the normal levels of all the molecules that can influence the synthesis of melatonin. In this chapter we demonstrate ways and means to maintain melatonin levels by dietary modification that can intervene at various steps in the melatonin synthesis and maintain the resultant melatonin levels. We also suggest that understanding the molecular mechanism involved in the biosynthetic process of melatonin in autism spectrum disorders might help in addressing phenotypic variants in the disease.

REFERENCES

Allegra, M., Reiter, R. J., Tan, D. X., Gentile, C., Tesoriere, L., and Livrea, M. A. The chemistry of melatonin's interaction with reactive species. *J Pineal Res.* 2003. 34: 1–10.

Anderson, G. M. Genetics of childhood disorders: XLV. Autism, part 4: Serotonin in autism. *J Am Acad Child Adolesc Psychiatry.* 2002. 41: 1513–1516.

Arendt, J., Middleton, B., Stone, B., and Skene, D. Complex effects of melatonin: Evidence for photoperiodic responses in humans? *Sleep.* 1999. 22: 625–635.

Ashton, A. J., Stoney, P. N., and McCaffery, P. J. Investigating the role of vitamin A in melatonin production in the pineal gland. *Proc Nutr Soc.* 2015. 74: E195.

Ashwood, P., Krakowiak, P., Hertz-Picciotto, I., Hansen, R., Pessah, I., and Van de Water, J. Elevated plasma cytokines in autism spectrum disorders provide evidence of immune dysfunction and are associated with impaired behavioral outcome. *Brain Behav Immun.* 2011a. 25: 40–45.

Ashwood, P., Krakowiak, P., Hertz-Picciotto, I., Hansen, R., Pessah, I. N., and Van de Water, J. Altered T cell responses in children with autism. *Brain Behav Immun.* 2011b. 25: 840–849.

Badria, F. A. Melatonin, serotonin, and tryptamine in some egyptian food and medicinal plants. *J Med Food.* 2002. 5: 153–157.

Barrenetxe, J., Delagrange, P., and Martínez, J. A. Physiological and metabolic functions of melatonin. *J Physiol Biochem.* 2004. 60: 61–72.

Billyard, A. J., Eggett, D. L., and Franz, K. B. Dietary magnesium deficiency decreases plasma melatonin in rats. *Magnes Res.* 2006. 19: 157–161.

Brzezinski, A. Melatonin in humans. *N Engl J Med*. 1997. 336: 186–195.

Brzezinski, A., Vangel, M. G., Wurtman, R. J., Norrie, G., Zhdanova, I., Ben-Shushan, A., and Ford, I. Effects of exogenous melatonin on sleep: A meta-analysis. *Sleep Med Rev*. 2005. 9: 41–50.

Buckley, A. W., Rodriguez, A. J., Jennison, K., Buckley, J., Thurm, A., Sato, S., and Swedo, S. Rapid eye movement sleep percentage in children with autism compared with children with developmental delay and typical development. *Arch Pediatr Adolesc Med*. 2010. 164: 1032–1037.

Cai, G., Edelmann, L., Goldsmith, J. E., Cohen, N., Nakamine, A., Reichert, J. G., Hoffman, E. J., et al. Multiplex ligation-dependent probe amplification for genetic screening in autism spectrum disorders: Efficient identification of known microduplications and identification of a novel microduplication in ASMT. *BMC Med Genomics*. 2008. 1: 50.

Calderón-Guzmán, D., Hernández-Islas, J. L., Espitia-Vázquez, I., Barragán-Mejía, G., Hernández-García, E., Santamaría-del Ángel, D., and Juárez-Olguín, H. Pyridoxine, regardless of serotonin levels, increases production of 5-hydroxytryptophan in rat brain. *Arch Med Res*. 2004. 35: 271–274.

Carrillo-Vico, A., Reiter, R. J., Lardone, P. J., Herrera, J. L., Fernandez-Montesinos, R., Guerrero, J. M., and Pozo, D. The modulatory role of melatonin on immune responsiveness. *Curr Opin Investig Drugs*. 2006. 7: 423–431.

Chaste, P., Clement, N., Mercati, O., Guillaume, J. L., Delorme, R., Botros, H. G., Pagan, C., et al. Identification of pathway-biased and deleterious melatonin receptor mutants in autism spectrum disorders and in the general population. *PLoS ONE*. 2010. 5: e11495.

Chen, Z., Chua, C. C., Gao, J., Hamdy, R. C., and Chua, B. H. L. Protective effect of melatonin on myocardial infarction. *Am J Physiol Heart Circ Physiol*. 2003. 284: H1618–H1624.

Chugani, D. C., Muzik, O., Behen, M., Rothermel, R., Janisse, J. J., Lee, J., and Chugani, H. T. Developmental changes in brain serotonin synthesis capacity in autistic and nonautistic children. *Ann Neurol*. 1999. 45: 287–295.

Colvert, E., Tick, B., McEwen, F., Stewart, C., Curran, S. R., Woodhouse, E., Gillan, N., et al. Heritability of autism spectrum disorder in a UK population–based twin sample. *JAMA Psychiatry*. 2015. 72: 415–423.

Cook Jr, E., Leventhal, B., Heller, W., Metz, J., Wainwright, M., and Freedman, D. Autistic children and their first-degree relatives: Relationships between serotonin and norepinephrine levels and intelligence. *J Neuropsy Clin Neurosci*. 1989. 2: 268–274.

Coon, H., Dunn, D., Lainhart, J., Miller, J., Hamil, C., Battaglia, A., Tancredi, R., Leppert, M. F., Weiss, R., and McMahon, W. Possible association between autism and variants in the brain-expressed tryptophan hydroxylase gene (TPH2). *Am J Med Genet*. 2005. 135B: 42–46.

Coppen, A., Swade, C., Jones, S. A., Armstrong, R. A., Blair, J. A., and Leeming, R. J. Depression and tetrahydrobiopterin: The folate connection. *J Affect Disord*. 1989. 16: 103–107.

Cross, S., Kim, S. J., Weiss, L. A., Delahanty, R. J., Sutcliffe, J. S., Leventhal, B. L., Cook Jr, E. H., and Veenstra-Vanderweele, J. Molecular genetics of the platelet serotonin system in first-degree relatives of patients with autism. *Neuropsychopharmacology*. 2008. 33: 353–360.

Cubero, J., Toribio, F., Garrido, M., Hernández, M. T., Maynar, J., Barriga, C., and Rodríguez, A. B. Assays of the amino acid tryptophan in cherries by HPLC-fluorescence. *Food Anal Methods*. 2009. 3: 36–39.

de Faria Poloni, J., Feltes, B. C., and Bonatto, D. Melatonin as a central molecule connecting neural development and calcium signaling. *Funct Integr Genomics*. 2011. 11: 383–388.

Delagrange, P., Atkinson, J., Boutin, J. A., Casteilla, L., Lesieur, D., Misslin, R., Pellissier, S., Pénicaud, L., and Renard, P. Therapeutic perspectives for melatonin agonists and antagonists. *J Neuroendocrin*. 2003. 15: 442–448.

Dietert, R. R., Dietert, J. M., and Dewitt, J. C. Environmental risk factors for autism. *Emerg Health Threats J*. 2011. 20: 7111.

Dingerkus, V. L. S., Gaber, T. J., Helmbold, K., Bubenzer, S., Eisert, A., Sánchez, C. L., and Zepf, F. D. Acute tryptophan depletion in accordance with body weight: Influx of amino acids across the blood–brain barrier. *J Neural Transm*. 2012. 119: 1037–1045.

Doyen, C., Mighiu, D., Kaye, K., Colineaux, C., Beaumanoir, C., Mouraeff, Y., Rieu, C., Paubel, P., and Contejean, Y. Melatonin in children with autistic spectrum disorders: Recent and practical data. *Eur Child Adoles Psy*. 2011. 20: 231–239.

Elsabbagh, M., Divan, G., Koh, Y. J., Kim, Y. S., Kauchali, S., Marcín, C., Montiel-Nava, C., et al. Global prevalence of autism and other pervasive developmental disorders. *Autism Res*. 2012. 5: 160–179.

Evans, T., Perry, G., Smith, M., Salomon, R., McGinnis, W., Sajdel-Sulkowska, E., and Zhu, X. Evidence for oxidative damage in the autistic brain. In oxidative stress, inflammation, and immune abnormalities. In Chauhan, A., Chauhan, V., and Brown, W. T., (editors). *Autism: Oxidative Stress, Inflammation and Immune Abnormalities.* Boca Raton, Florida: Taylor and Francis. 2009. 35–46.

Feldman, J. M., and Lee, E. M. Serotonin content of foods: Effect on urinary excretion of 5-hydroxyindoleacetic acid. *Am J Clin Nutr.* 1985. 42: 639–643.

Fernstrom, J. D. Dietary effects on brain serotonin synthesis: Relationship to appetite regulation. *Am J Clin Nutr.* 1985. 42: 1072–1082.

Footitt, E. J. Vitamin B6 and serotonin metabolism in neurological disorders of childhood. PhD dissertation (London: University College London). 2013.

Fournier, I., Ploye, F., Cottet-Emard, J. M., Brun, J., and Claustrat, B. Folate deficiency alters melatonin secretion in rats. *J Nutr.* 2002. 132: 2781–2784.

Garrido, M., Espino, J., Toribio-Delgado, A. F., Cubero, J., Maynar-Mariño, J. I., Barriga, C., Paredes, S. D., and Rodríguez, A. B. A Jerte Valley Cherry–based product as a supply of tryptophan. *Int J Trytophan Res.* 2012. 5: 9.

Glickman, G. Circadian rhythms and sleep in children with autism. *Neurosci Biobehav Rev.* 2010. 34: 755–768.

Grether, J. K., Croen, L. A., Anderson, M. C., Nelson, K. B., and Yolken, R. H. Neonatally measured immunoglobulins and risk of autism. *Autism Res.* 2010. 3: 323–332.

Guénolé, F., Godbout, R., Nicolas, A., Franco, P., Claustrat, B., and Baleyte, J. M. Melatonin for disordered sleep in individuals with autism spectrum disorders: Systematic review and discussion. *Sleep Med Rev.* 2011. 15: 379–387.

Guerrero, J. M., and Reiter, R. J. Melatonin-immune system relationships. *Curr Top Med Chem.* 2002. 2: 167–179.

Hardeland, R., Reiter, R. J., Poeggeler, B., and Tan, D. X. The significance of the metabolism of the neurohormone melatonin: Antioxidative protection and formation of bioactive substances. *Neurosci Biobehav Rev.* 1993. 17: 347–357.

Herxheimer, A., and Petrie, K. J. Melatonin for the prevention and treatment of jet lag. *Cochrane Database Syst Rev.* 2002. 2.

Holick, M., MacLaughlin, J., Clark, M., Holick, S., Potts, J., Anderson, R., Blank, I., Parrish, J., and Elias, P. Photosynthesis of previtamin D3 in human skin and the physiologic consequences. *Science.* 1980. 210: 203–205.

Hranilovic, D., Bujas-Petkovic, Z., Vragovic, R., Vuk, T., Hock, K., and Jernej, B. Hyperserotonemia in adults with autistic disorder. *J Autism Dev Disord.* 2007. 37: 1934–1940.

Hu, V. W., Sarachana, T., Kim, K. S., Nguyen, A., Kulkarni, S., Steinberg, M. E., Luu, T., Lai, Y., and Lee, N. H. Gene expression profiling differentiates autism case-controls and phenotypic variants of autism spectrum disorders: Evidence for circadian rhythm dysfunction in severe autism. *Autism Res.* 2009. 2: 78–97.

Hu, V. W., and Steinberg, M. E. Novel clustering of items from the autism diagnostic interview––revised to define phenotypes within autism spectrum disorders. *Autism Res.* 2009. 2: 67–77.

Huang, C. H., and Santangelo, S. L. Autism and serotonin transporter gene polymorphisms: A systematic review and meta-analysis. *Am J Med Genet.* 2008. 147B: 903–913.

Huether, G. The contribution of extrapineal sites of melatonin synthesis to circulating melatonin levels in higher vertebrates. *Experientia.* 1993. 49: 665–670.

Kałuzna-Czaplinska, J., Michalska, M., and Rynkowski, J. Determination of tryptophan in urine of autistic and healthy children by gas chromatography/mass spectrometry. *Med Sci Monit.* 2010. 16: CR488–CR492.

Kim, D. Y., and Camilleri, M. Serotonin: A mediator of the brain-gut connection. *Am J Gastroenterol.* 2000. 95: 2698–2709.

Kuci, S., Becker, J., Veit, G., Handgretinger, R., Attanasio, A., Bruchelt, G., Treuner, J., Niethammer, D., and Gupta, D. Circadian variations in the immunomodulatory role of the pineal gland. *Neuro Endocrinol Lett.* 1988. 10: 65–79.

Kulman, G., Lissoni, P., Rovelli, F., Roselli, M. G., Brivio, F., and Sequeri, P. Evidence of pineal endocrine hypofunction in autistic children. *Neuro Endocrinol Lett.* 1999. 21: 31–34.

Kunz, D., Mahlberg, R., Müller, C., Tilmann, A., and Bes, F. Melatonin in patients with reduced REM sleep duration: Two randomized controlled trials. *J Clin Endocrinol Metab.* 2004. 89: 128–134.

Kvetnoy, I. M. Extrapineal melatonin: Location and role within diffuse neuroendocrine system. *Histochem J.* 1999. 31: 1–12.

Lane, E. A., and Moss, H. B. Pharmacokinetics of melatonin in man: First pass hepatic metabolism. *J Clin Endocrinol Metab.* 1985. 61: 1214–1216.

Lauritsen, M. B., Børglum, A. D., Betancur, C., Philippe, A., Kruse, T. A., Leboyer, M., and Ewald, H. Investigation of two variants in the DOPA decarboxylase gene in patients with autism. *Amer J Med Genet.* 2002. 114: 466–470.

Leboyer, M., Philippe, A., Bouvard, M., Guilloud-Bataille, M., Bondoux, D., Tabuteau, F., Feingold, J., Mouren-Simeoni, M. C., and Launay, J. M. Whole blood serotonin and plasma beta-endorphin in autistic probands and their first-degree relatives. *Biol Psychiatry.* 1999. 45: 158–163.

Limoges, E. Atypical sleep architecture and the autism phenotype. *Brain.* 2005. 128: 1049–1061.

Lissoni, P., Barni, S., Crispino, S., Tancini, G., and Fraschini, F. Endocrine and immune effects of melatonin therapy in metastatic cancer patients. *Euro J Can Clin Oncol.* 1989. 25: 789–795.

Lissoni, P., Marelli, O., Mauri, R., Resentini, M., Franco, P., Esposti, D., Esposti, G., et al. Ultradian chrono-modulation by melatonin of a placebo effect upon human killer cell activity. *Chronobiologia.* 1985. 13: 339–343.

Ma, X. Metabolism of melatonin by human cytochromes p450. *Drug Metab Dispos.* 2005. 33: 489–494.

Malow, B. A., and McGrew, S. G. Sleep disturbances and autism. *Sleep Med Clin.* 2008. 3: 479–488.

Manzardo, A. M., Henkhaus, R., Dhillon, S., and Butler, M. G. Plasma cytokine levels in children with autistic disorder and unrelated siblings. *Int J Dev Neurosci.* 2012. 30: 121–127.

McBride, P. A. Serotonergic responsivity in male young adults with autistic disorder. *Arch Gen Psychiatry.* 1989. 46: 213.

Melke, J., Goubran Botros, H., Chaste, P., Betancur, C., Nygren, G., Anckarsäter, H., Rastam, M., et al. Abnormal melatonin synthesis in autism spectrum disorders. *Mol Psychiatry.* 2008. 13: 90–98.

Millward, C., Ferriter, M., Calver, S., and Connell-Jones, G. Gluten- and casein-free diets for autistic spectrum disorder. *Cochrane Database Syst Rev.* 2008. 2: CD003498.

Mulder, E. J., Anderson, G. M., Kema, I. P., de Bildt, A., van Lang, N. D. J., den Boer, J. A., and Minderaa, R. B. Platelet serotonin levels in pervasive developmental disorders and mental retardation: Diagnostic group differences, within-group distribution, and behavioral correlates. *J Am Acad Child Adolesc Psychiatry.* 2004. 43: 491–499.

Mulder, E. J., Anderson, G. M., Kemperman, R. F. J., Oosterloo-Duinkerken, A., Minderaa, R. B., and Kema, I. P. Urinary excretion of 5-hydroxyindoleacetic acid, serotonin and 6-sulphatoxymelatonin in normoserotonemic and hyperserotonemic autistic individuals. *Neuropsychobiology.* 2010. 61: 27–32.

Nakamura, K. Late developmental stage-specific role of tryptophan hydroxylase 1 in brain serotonin levels. *J Neurosci.* 2006. 26: 530–534.

Nakamura, K., Sekine, Y., Ouchi, Y., Tsujii, M., Yoshikawa, E., Futatsubashi, M., Tsuchiya, K. J., et al. Brain serotonin and dopamine transporter bindings in adults with high-functioning autism. *Arch Gen Psychiatry.* 2010. 67: 59.

Ng, T. B., Liu, F., and Zhao, L. Antioxidative and free radical scavenging activities of pineal indoles. *J Neural Transm.* 2000. 107: 1243–1251.

Nir, I., Meir, D., Zilber, N., Knobler, H., Hadjez, J., and Lerner, Y. Brief report: Circadian melatonin, thyroid-stimulating hormone, prolactin, and cortisol levels in serum of young adults with autism. *J Autism Dev Disord.* 1995. 25: 641–654.

Orabona, G., Griesi-Oliveira, K., Vadasz, E., Bulcão, V., Takahashi, V., Moreira, E., Furia-Silva, M., Ros-Melo, A., Dourado, F., and Matioli, R. HTR1B and HTR2C in autism spectrum disorders in Brazilian families. *Brain Res.* 2009. 1250: 14–19.

Outila, T. A., Mattila, P. H., Piironen, V. I., and Lamberg-Allardt, C. J. Bioavailability of vitamin D from wild edible mushroom (*Cantharellus tubaeformis*) as measured with a human bioassay. *Am J Clin Nutr.* 1999. 69: 95–98.

Pagan, C., Delorme, R., Callebert, J., Goubran-Botros, H., Amsellem, F., Drouot, X., Boudebesse, C., et al. The serotonin-*N*-acetylserotonin–melatonin pathway as a biomarker for autism spectrum disorders. *Transl Psychiatry.* 2014. 4: e479.

Panzer, A., and Viljoen, M. The validity of melatonin as an oncostatic agent. *J Pineal Res.* 1997. 22: 184–202.

Patrick, R. P., and Ames, B. N. Vitamin D hormone regulates serotonin synthesis. Part 1: Relevance for autism. *FASEB J.* 2014. 28: 2398–2413.

Paul, B., Saradalekshmi, K. R., Alex, A. M., and Banerjee, M. Circadian rhythm of homocysteine is hCLOCK genotype dependent. *Mol Biol Rep.* 2014. 41: 3597–3602.

Pennesi, C. M., and Klein, L. C. Effectiveness of the gluten-free, casein-free diet for children diagnosed with autism spectrum disorder: Based on parental report. *Nutr Neurosci.* 2012. 15l: 85–91.

Peuhkuri, K., Sihvola, N., and Korpela, R. Dietary factors and fluctuating levels of melatonin. *Food Nutr Res.* 2012. 56.

Reiter, R. J., Tan, D. X., Osuna, C., and Gitto, E. Actions of melatonin in the reduction of oxidative stress. *J Biomed Sci.* 2000. 7: 444–458.

Richard, D. M., Dawes, M. A., Mathias, C. W., Acheson, A., Hill-Kapturczak, N., and Dougherty, D. M. L-tryptophan: Basic metabolic functions, behavioral research and therapeutic indications. *Int J Tryptophan Res.* 2009. 2: 45.

Richdale, A. L. Sleep problems in autism: Prevalence, cause, and intervention. *Dev Med Child Neurol.* 1999. 41: 60–66.

Ritvo, E. R., Ritvo, R., Yuwiler, A., Brothers, A., Freeman, B. J., and Plotkin, S. Elevated daytime melatonin concentrations in autism: A pilot study. *Eur Child Adolesc Psychiatry.* 1993. 2: 75–78.

Rodriguez, C., Mayo, J. C., Sainz, R. M., Antolin, I., Herrera, F., Martin, V., and Reiter, R. J. Regulation of antioxidant enzymes: A significant role for melatonin. *J Pineal Res.* 2004. 36: 1–9.

Rossignol, D. A., and Frye, R. E. Melatonin in autism spectrum disorders: A systematic review and meta-analysis. *Dev Med Child Neurol.* 2011. 53: 783–792.

Rossignol, D. A., and Frye, R. E. Evidence linking oxidative stress, mitochondrial dysfunction, and inflammation in the brain of individuals with autism. *Front Physiol.* 2014. 5: 150.

Sacco, R., Curatolo, P., Manzi, B., Militerni, R., Bravaccio, C., Frolli, A., Lenti, C., et al. Principal pathogenetic components and biological endophenotypes in autism spectrum disorders. *Autism Res.* 2010. 3: 237–252.

Serle, J. B., Wang, R. F., Peterson, W. M., Plourde, R., and Yerxa, B. R. Effect of 5-MCA-NAT, a putative melatonin MT3 receptor agonist, on intraocular pressure in glaucomatous monkey eyes. *J Glaucoma.* 2004. 13: 385–388.

Singer, C., Tractenberg, R. E., Kaye, J., Schafer, K., Gamst, A., Grundman, M., Thomas, R., and Thal, L. J. A multicenter, placebo-controlled trial of melatonin for sleep disturbance in Alzheimer's disease. *Sleep.* 2003. 26: 893.

Singh, A. S., Chandra, R., Guhathakurta, S., Sinha, S., Chatterjee, A., Ahmed, S., Ghosh, S., and Rajamma, U. Genetic association and gene–gene interaction analyses suggest likely involvement of ITGB3 and TPH2 with autism spectrum disorder (ASD) in the Indian population. *Prog Neuropsychopharmacol Biol Psychiatry.* 2013. 45: 131–143.

Skwarlo-Sonta, K. Melatonin in immunity: Comparative aspects. *Neuro Endocrinol Lett.* 2002. 23: 61–66.

Slominski, A., Tobin, D. J., Zmijewski, M. A., Wortsman, J., and Paus, R. Melatonin in the skin: Synthesis, metabolism and functions. *Trends Endocrinol Metab.* 2008. 19: 17–24.

Tan, D. X., Hardeland, R., Manchester, L. C., Korkmaz, A., Ma, S., Rosales-Corral, S., and Reiter, R. J. Functional roles of melatonin in plants, and perspectives in nutritional and agricultural science. *J Exp Botany.* 2012. 63: 577–597.

Thomas, B., and Mohanakumar, K. P. Melatonin protects against oxidative stress caused by 1-methyl-4-phenyl-1,2,3,6-tetrahydropyridine in the mouse nigrostriatum. *J Pineal Res.* 2004. 36: 25–32.

Toma, C., Hervás, A., Balmaña, N., Salgado, M., Maristany, M., Vilella, E., Aguilera, F., et al. Neurotransmitter systems and neurotrophic factors in autism: Association study of 37 genes suggests involvement of DDC. *World J Biol Psychiatry.* 2013. 14: 516–527.

Tordjman, S., Anderson, G. M., Bellissant, E., Botbol, M., Charbuy, H., Camus, F., Graignic, R., et al. Day and nighttime excretion of 6-sulphatoxymelatonin in adolescents and young adults with autistic disorder. *Psychoneuroendocrinology.* 2012. 37: 1990–1997.

Tordjman, S., Anderson, G. M., Pichard, N., Charbuy, H., and Touitou, Y. Nocturnal excretion of 6-sulphatoxymelatonin in children and adolescents with autistic disorder. *Biol Psychiatry.* 2005. 57: 134–138.

Tordjman, S., Gutknecht, L., Carlier, M., Spitz, E., Antoine, C., Slama, F., Carsalade, V., et al. Role of the serotonin transporter gene in the behavioral expression of autism. *Mol Psychiatry.* 2001. 6: 434–439.

Urata, Y., Honma, S., Goto, S., Todoroki, S., Iida, T., Cho, S., Honma, K., and Kondo, T. Melatonin induces γ-glutamylcysteine synthetase mediated by activator protein-1 in human vascular endothelial cells. *Free Radic Biol Med.* 1999. 27: 838–847.

Valko, M., Leibfritz, D., Moncol, J., Cronin, M. T. D., Mazur, M., and Telser, J. Free radicals and antioxidants in normal physiological functions and human disease. *Int J Biochem Cell Biol.* 2007. 39: 44–84.

Verma, D., Chakraborti, B., Karmakar, A., Bandyopadhyay, T., Singh, A. S., Sinha, S., Chatterjee, A., et al. Sexual dimorphic effect in the genetic association of monoamine oxidase A (MAOA) markers with autism spectrum disorder. *Prog Neuropsychopharmacol Biol Psychiatry.* 2014. 50: 11–20.

Wasilewska, J., Kaczmarski, M., Stasiak-Barmuta, A., Tobolczyk, J., and Kowalewska, E. Low serum IgA and increased expression of CD23 on B lymphocytes in peripheral blood in children with regressive autism aged 3–6 years old. *Arch Med Sci.* 2012. 2: 324–331.

Yang, S. Y., Yoo, H. J., Cho, I. H., Park, M., and Kim, S. A. Association with tryptophan hydroxylase 2 gene polymorphisms and autism spectrum disorders in Korean families. *Neurosci Res.* 2012. 73: 333–336.

Zawilska, J. B. The role of calcium in the regulation of melatonin biosynthesis in the retina. *Acta Neurobiol Exp.* 1992. 52: 265–265.

33

Melatonin in Oral Health

Vijay Kumar Chava
Narayana Dental College and Hospital
Andhra Pradesh, India

CONTENTS

ABSTRACT   The purpose of this chapter is to summarize the findings accumulated concerning the "hormone of darkness", melatonin, in the last few years. Based on the various origins and functional potentials of melatonin the recent medical and dental research is geared up toward the prevention of free radical–mediated diseases and maintenance of oral health by using specific nutrient supplementation. Recently, studies have detected the secretion of melatonin from salivary glands and gingival crevicular fluid (GCF), further emphasizing its local activity. Thus, within our confines, the effects of melatonin in oral health were reviewed, adding a note on its properties, functions, and future.

KEY WORDS: *antioxidant, anti-inflammatory, immunosuppression, cell protector, neurohormone, saliva; osseo-integration; regeneration.*

Abbreviations

AANAT: aryl-alkyl-amine *N*-acetyltransferase
BIC: bone-to-implant contact
DNA: deoxyribonucleic acid
GCF: gingival crevicular fluid
HIOMT: hydroxy-indole-*O*-methyltransferase
HOB-M: human osteoblast mandibular cells
IL-2: interleukin 2
mRNA: messenger ribonucleic acid
Me 1a R: melatonin 1a receptor

OPG: osteoprotegerin
RANK: receptor activator of nuclear factor kappa B
RANKL: receptor activator of nuclear factor kappa B ligand
ROS: reactive oxygen species
SCN: suprachiasmatic nucleus
SV-HFO: human fetal osteoblastic cellular line.

Introduction

Melatonin is an indoleamine neurotrasmitter substance synthesized from the essential amino acid tryptophan that regulates various human functions (Kalsbeek and Buijs 2002). The effects of melatonin was described in 1917, but was first isolated and structurally identified in 1958 by an American dermatologist, Aaron Lerner, and his coworkers as an amphibian skin-lightening factor present in extracts of bovine pineal glands (Lerner et al. 1958). Melatonin derives its name from serotonin, based on its ability to blanch the skin of amphibians (Mc Cord and Allen 2001). The unique, highly lipophilic nature of melatonin makes it accessible to every cell in living systems exhibiting important physiological roles. It is widely distributed in animals and plants. In humans it is found in high concentrations, in bone marrow, intestine, and at the subcellular level in the mitochondria and nucleus and other body fluids (Reiter et al. 2009). The multiple diverse functions of melatonin are identified to be mediated by specific receptors and play a significant role in the protection of oral activity (Cutando et al. 2007a). Consideration of its profound systemic effects and the discovery of this mammalian neurohormone in plants and animal sources (Illnerova et al. 1993) stimulated the quest for further investigation.

Biological Functions

Literature has shown that melatonin is involved in numerous physiological and pathophysiological processes including control of sleep, circadian rhythm, retinal physiology, seasonal reproductive cycle, oncostatic potential, regulation of immune response, antioxidation and free radical scavenging, mitochondrial respiration, cardiovascular function and blood pressure control, bone metabolism, and gastrointestinal physiology. Other actions of this hormone include the inhibition of dopamine release in the hypothalamus and retina. It is also involved in pubertal development and the aging process (Rodella et al. 2012).

Circadian Rhythm

The rhythmic production of melatonin is a consequence of neural impulses from the biological clock; that is, from the suprachiasmatic nucleus (SCN) and hypothalamus (Buijis and Kalsbeek 2001). It is known as the "chemical expression of darkness" as most of its synthesis occurs during nighttime (Klein and Moore 1979). In healthy individuals, peak serum melatonin levels are seen between 12:00 a.m–2:00 a.m. and 2:00–4:00 a.m., with minimum secretion occurring during the day between 12.00 p.m and 2.00 p.m. (Simonneaux and Ribelayga 2003). Following its secretion unbound melatonin diffuses passively into the saliva and oral mucosa to enter the oral cavity (Reiter 1986). So, salivary melatonin represents the percentage of free melatonin (Czesnikiewicz-Guzik et al. 2007). It is discharged from the individual acinar cells of the salivary glands due to the contraction of the myoepithelial elements of the acini (Reiter et al. 2015). Salivary melatonin levels (2–4 pg/ml) (Laasko et al. 1990) forms 24%–33% of the plasma melatonin levels (Cutando et al. 2003). Factors such as smoking, exposure to light, alcohol consumption, and aging lower the levels of salivary melatonin with no variability in terms of gender (Burgess and Fogg 2008). A substantial increase in concentration of melatonin in the oral cavity was observed when oral liquid products or the sublingual tablets were used for various purposes. Other factors that affect the amount of melatonin in saliva include the type of food eaten before saliva collection; fruits, grains, vegetables, nuts, and commonly used liquids in day-to-day life may significantly improve the oral health

(Reiter et al. 2015). Recent evidence has observed its presence in GCF (Srinath et al. 2010). GCF melatonin levels are 60% lower than serum levels and 30% lower than salivary levels. Measurement of salivary melatonin is a reliable technique to monitor the circadian rhythms of melatonin.

Properties and Functions of Melatonin

Since its discovery, melatonin is shown to have a variety of important functions in all species. These systemic and oral functions are indeed related to its hormonal properties (Chava and Sirisha 2012) as shown in Table 33.1.

Effects of Melatonin in Oral Cavity

Current research portrays melatonin as a cell protector rather than a hormone, due to its presence in extra pineal organs, including the oral cavity (Vakkuri et al. 1985). It passively diffuses into the oral cavity, and is released into the saliva (Vakkuri 1985). A significant correlation between concentrations of melatonin in saliva and serum was reported with a general conclusion that melatonin concentration is a reliable index of serum melatonin levels (Nowak et al. 1987).

TABLE 33.1

Properties and Functions of Melatonin

Functions	Mechanism of Action
Antioxidant	Direct effects: Neutralizes a variety of ROS such as OH, ROO, H_2O_2, and O2 (non receptor–mediated).
	Interacts with the lipid bilayers and stabilizes mitochondrial membranes thereby improving the electron transport chain.
	Indirect effects: Regulates nitric oxide production by interaction with nitric oxide synthesis.
	Increases gene expression and enzyme activities of glutathione, promoting removal of H_2O_2 and super oxide dismutase. (Receptor-mediated.)
Anti-inflammatory	Inhibits inflammatory enzyme COX-2 by binding to the active sites of COX-1 and COX-2.
Immunomodulatory function	Promotes endogenic production of IL-2 concomitant with serum melatonin concentration.
	Activates CD+4 lymphocytes by increasing the production of IL-2 and IFN-γ.
	Modulates immune functions by activating on CD+4 cells and monocytes.
Anticancer effect	Stimulates production of IL-2, thereby stimulates the activity of natural killer cells which intervene in the cytolytic mechanisms.
	Amplifies the antitumor activity of IL-2.
	Scavenges ROS, which are secondary messengers in the signaling pathway leading to cell division.
	Inhibits the growth of oral squamous cell carcinoma cells.
	Involved in the pathogenesis of oral precancerous lesions such as lichen planus and leukoplakia, by free radical scavenging action.
Antiviral effect	Induces production of IL-1β, which is useful in treating viral infections.
Antimycotic effect	Enhances phagocytic function and reduces oxidative stress originating during candidiasis.
Bone remodeling	Promotes osteoblast differentiation and bone formation.
	Stimulates synthesis of type-1 collagen fibers in human osteoblasts *in vitro.*
	Inhibits bone resorption by interfering with the activity of osteoclasts through its indirect antioxidant and direct free radical scavenging action.
	Increases genic expression of bone sialoprotein and other protein markers thereby reducing osteoblast differentiation time.
	Downregulates receptor activator of nuclear factor kappa B ligand (RANKL)–mediated osteoclast formation and activation.

Source: Data from Chava, V. K., and Sirisha, K. *Int J Dent.* 2012. 720185: 1–9.

Role of Melatonin in Oral Cancerous Lesions

Melatonin plays a pivotal role in many inflammatory processes of the oral cavity and thus is effective in treating pathologies (Gómez-Moreno et al. 2010) such as squamous cell carcinoma and epidermoid carcinoma. In relation to oral cancer, it is speculated that exogenous restoration of melatonin receptor 1 A inhibited the growth of oral squamous cell carcinoma cells, lacking its expression (Chaiyarit et al. 2005). By its actions against ROS, melatonin may protect against precancerous oral diseases such as leukoplakia and lichen planus (Chaiyarit et al. 2005; Taubman et al. 2005). Melatonin also counteracts the negative effects of immunosuppressive drug therapy by acting on T-helper lymphocytes, lymphokines such as gamma interferon, and interleukin 2 (IL-2) (Cutando and Silvestre 1995).

Role of Melatonin in Tooth Development

Melatonin may play a role in physiological tooth development and growth by regulation of odontogenic cells in tooth germs (Kumasaka et al. 2010). Immunohistochemical analysis revealed that melatonin 1a receptor (Me 1a R) was expressed in secretory ameloblasts, the cells of stratum intermedium and stellate reticulum, external dental epithelial cells, odontoblasts, and dental sac cells. Seasonal variation in the release of melatonin was associated with the changes in development of caries in hamsters (Mechin and Toury 1973).

Role of Melatonin in Inflammatory Conditions of the Oral Cavity

GCF melatonin levels play a significant role in periodontal disease. The pathogenesis of periodontitis is related to increased oxidative stress, which in turn leads to tissue damage and bone loss. The antioxidative and free radical scavenging action of melatonin may reduce this tissue damage (Cutando et al. 2006). In addition, the positive effects of melatonin and its derivatives on inflammatory mediators and bone cells may be beneficial in improving periodontal health. They also proved that salivary melatonin levels may vary according to the degree of periodontal disease. A negative association was found between salivary melatonin levels and periodontal disease severity. Consequently, decreased saliva and melatonin production with age predisposes older individuals to increased risk of oral and periodontal diseases. Dental procedures including tooth extraction may result in local inflammation and oxidative stress in the oral cavity. Upon local administration, the antioxidant action of melatonin may be useful in counteracting this oxidative stress (Cutando et al. 2007a,b).

Role of Melatonin in Osseointegration and Regeneration

Melatonin was reported to stimulate the growth of new bone around implants in the tibia of rat (Calvo-Guirado et al. 2010). It has biological significance in terms of osseointegration around implants. Two weeks after implant insertion melatonin significantly increased all parameters of osseointegration such as bone-to-implant contact (BIC), total periimplant bone, and interthread bone, as well as new bone formation. Osseointegration was reported to be more effective when melatonin was mixed with collagenized porcine bone (Cutando et al. 2008). Melatonin at micromolecular concentration promotes the proliferation of human osteoblast mandibular cells (HOB-M) and cells of a human fetal osteoblastic cellular line (SV-HFO). This effect is dose dependent and is maximum at concentrations of 50 micrometers (Nakade et al. 1999). At concentrations ranging from 5–500 micrometers melatonin lowers the expression of messenger ribonucleic acid (mRNA) from the receptor activator of nuclear factor kappa B (RANK) and increases levels of both osteoprotegerin (OPG) as well as mRNA from the OPG in preostoblastic cell lines (Koyama et al. 2002). Due to these actions on bone its use as a biomimetic agent during endosseous dental implant surgery is proposed (Simon and Watson 2002).

Role in Reducing Toxicity Due to Dental Materials

Several cytotoxic and genotoxic effects of dental methacrylate monomers promote oxidative processes. Melatonin's antioxidative action may protect against these effects by reducing oxidative deoxyribonucleic acid (DNA) damage induced by methacrylates. Melatonin when used as a component of dental materials exhibited biocompatibility without altering the properties of these dental materials. Alternatively, the regular use of melatonin oral rinse may reduce the side effects of methacrylate monomers (Blasiak et al. 2011).

Role in Modification of Salivary Components

Novel evidences proposed that melatonin induced protein synthesis in the rat parotid gland and thereby affects glandular activity. This effect is MT1- and MT2 receptor–mediated and is primarily dependent on nitric oxide generation via the activity of neuronal type nitric oxide synthase. This enzyme probably originated from parenchymal cells of the parotid gland. This novel action of melatonin may increase its clinical implications in the treatment of xerostomia, caries, periodontitis, oral mucosal infections, salivary gland inflammation, and wound healing (Cevik-Aras et al. 2011). Melatonin stimulates both innate and adaptive immune responses and modulates enzymes with proinflammatory actions while limiting the production of inflammatory mediators including cytokines and leukotrienes (Reiter et al. 2015).

The above finding put forth the quest for new sources of melatonin in the oral cavity. Studies have evaluated the expression of melatonin-synthesizing enzyme aryl-alkyl-amine *N*-acetyltransferase (AANAT) and hydroxy-indole-*O*-methyltransferase (HIOMT) mRNA in rat submandibular gland by quantitative reverse transcription–polymerase chain reaction. According to the observations, AANAT was expressed in the epithelial cells of striated ducts in rat salivary glands and AANAT, HIONT, and melatonin were expressed in the epithelial cells of striated ducts in human submandibular glands. In addition, the expression of the most potent melatonin receptor (melatonin1a) in rat buccal mucosa was confirmed (Shimozuma et al. 2011). Melatonin at physiological concentrations increased nerve growth factor synthesis in mouse submandibular gland (Pongsa-Asawapaiboon et al. 1998.). Thus, these oral sources of melatonin may alter salivary melatonin levels and thereby promote better oral health. So, without doubt all these actions have important consequences at the time of treatment of our high-risk dental patients who have in one way or the other an altered immunological system.

Conclusion

Traditionally, melatonin was considered as the principal secretory hormone of the pineal gland, with profound systemic effects. Novel evidences brought to light that this "chemical of darkness" has oral sources, and implications. Based on the well-clarified role of ROS, lipid peroxidation products, and antioxidant potentials of melatonin, the recent medical and dental research is geared up toward the prevention of free radical–mediated diseases and maintenance of oral health by using specific nutrient supplementation. Further studies are needed to focus the attention on therapeutic uses of melatonin as a coadjuvant in oral hygiene aids and as an antimicrobial in local therapy to promote it as a natural inhibitor of inflammation. Oral sources of melatonin may alter salivary melatonin levels and thereby promote better oral health. Use of melatonin as a nutraceutical through the intake of melatonin-rich plants seems to be promising in the future.

REFERENCES

Blasiak, J., Kasznicki, J., Drzewoski, J., Pawlowska, E., Szczepanska, J., and Reiter, R. J. 2011. Perspectives on the use of melatonin to reduce cytotoxic and genotoxic effects of methacrylate based dental materials. *J Pineal Res.* 51: 157–162.

Buijis, R. M., and Kalsbeek, A. 2001. Hypothalamic integration of central and peripheral clocks. *Nat Rev Neurosci.* 2: 521–526.

Burgess, H. J., and Fogg, L. F. 2008. Individual differences in the amount and timing of salivary melatonin secretion. *PLOS One.* 3: 3055–3064.

Calvo-Guirado, J. L., Ramírez-Fernández, M. P., Gómez-Moreno, G., Maté-Sánchez, J. E., Delgado-Ruiz, R., Guardia, J., López-Marí, L., et al. 2010. Melatonin stimulates the growth of new bone around implants in the tibia of rabbits. *J Pineal Res.* 49: 356–363.

Cevik-Aras, H., Godoy, T., and Ekstrom, J. 2011. Melatonin-induced protein synthesis in the rat parotid gland. *J Physiol Pharmacol.* 62 (1): 95–99.

Chaiyarit, P., Ma, N., Hiraku, Y., Pinlaor, S., Yongvanit, P., Jintakanon, D., Murata, M., Oikawa, S., and Kawanishi, S. 2005. Nitrative and oxidative DNA damage in oral lichen planus in relation to human oral carcinogenesis. *Cancer Sci.* 96: 553–559.

Chava, V. K., and Sirisha, K. 2012. Melatonin: A novel indolamine in oral health and disease. *Int J Dent.* 720185: 1–9.

Cutando, A., and Silvestre, F. J. 1995. Melatonin: Implications at the oral level. *Bull Group Int Rech Sci Stomatol Odontol* 38: 3–4: 81–86.

Cutando, A., Galindo, P., Gómez-Moreno, G., Arana, C., Bolaños, J., Acuña-Castroviejo, D., and Wang, H. L. 2006. Relationship between salivary melatonin and severity of periodontal disease. *J Periodontol.* 77: 1533–1538.

Cutando, A., Gomez-Morena, G., Arana, C., Castroviejo-Acuna, D., and Reiter, J. R. 2007a. Melatonin: Potential functions in the oral cavity. *J Periodontol.* 78: 1094–1102.

Cutando, A., Gomez-Moreno, G., Arana, C., Escames, G., and Acuña-Castroviejo, D. 2007b. Melatonin reduces oxidative stress because of tooth removal. *J Pineal Res.* 42: 419–420.

Cutando, A., Gómez-Moreno, G., Arana, C., Muñoz, F., Lopez-Peña, M., Stephenson, J., and Reiter, R. J. 2008. Melatonin stimulates osteointegration of dental implants. *J Pineal Res.* 45: 174–179.

Cutando, A., Gomez-Moreno, G., Villalba, J., Ferrera, M. J., Escames, G., and Acuna-Castroveizo, D. 2003. Relationship between salivary melatonin levels and periodontal status in diabetic patients. *J Pineal Res.* 35: 239–244.

Czesnikiewicz-Guzik, M., Konturek, S. J., Loster, B., Wisniewska, G., and Majewski, S. 2007. Melatonin and its role in oxidative stress related diseases of the oral cavity. *J Physiol Pharmacol.* 58: 5–19.

Gómez-Moreno, G., Guardia, J., Ferrera, M. J., Cutando, A., and Reiter, R. J. 2010. Melatonin in diseases of the oral cavity. *Oral Dis.* 16: 242–247.

Illnerova, H., Buresova, M., and Presl, J. 1993. Melatonin rhythm in human milk. *J Clin Endocrinol Metab.* 77: 838–841.

Kalsbeek, A., and Buijs, R. M. 2002. Output pathways of the mammalian suprachaismatic nucleus: Coding circadian time by transmitter selection and specific targeting. *Cell Tissue Res.* 309: 109–118.

Klein, D. C., and Moore, R. Y. 1979. Pineal *N*-acetyltransferase and hydroxyindole-*O*-methyltransferase: Control by the retinohypothalamic tract and the suprachaismatic nucleus. *Brain Res.* 174: 245–262.

Koyama, H., Nakade, O., Takada, A., Kakur, T., and Lau, K. M. 2002. Melatonin at pharmacological doses increases bone mass by suppressing resorption through down-regulation of the RANKL-mediated osteoclast formation and activation. *J Bone Miner Res.* 17: 1219–1229.

Kumasaka, S., Shimozuma, M., Kawamoto, T., Mishima, K., Tokuyama, R., Kamiya, Y., Davaadorj, P., Saito, I., and Satomura, K. 2010. Possible involvement of melatonin in tooth development: Expression of melatonin 1a receptor in human and mouse tooth germs. *Histochem Cell Biol.* 133: 577–584.

Laasko, M. L., Porkka-Heiskanen, P. T., Alila, A., Stenberg, D., and Johansson, G. 1990. Correlation between salivary and serum melatonin: Dependence on serum melatonin levels. *J Pineal Res.* 9: 39–50.

Lerner, A. B., Case, J. D., Takahashi, Y., Lee, T. H., and Mori, W. 1958. Isolation of melatonin, the pineal factor that lightens melanocytes. *J Am Chem Soc.* 80: 10:2587.

McCord, C. P., and Allen, F. P. 2001. Evidence associating pineal gland function with alterations in pigmentation. *J Exp Zool.* 23: 207–224.

Mechin, J. C., and Toury, C. 1973. Action of cariogenic diet on fixation and retention of skeleton and teeth strontium in rats. *Rev Odontostomatol.* 20: 55–59.

Nakade, O., Koyama, H., Ariji, H., Yajima, A., and Kakur, T. 1999. Melatonin stimulates proliferation of type-I collagen synthesis in human bone cells *in vitro. J Pineal Res.* 27: 106–110.

Nowak, R., McMillen, I. C., Redman, J., and Short, R. V. 1987. The correlation between serum and salivary melatonin concentrations and urinary 6-hydrxymelatonin sulphate excretion rates: Two non-invasive techniques for monitoring human circadian rhythmicity. *Clin Endocrinol.* 27: 445–452.

Pongsa-Asawapaiboon, A., Asavaritikrai, P., Withyachumnarnkul, B., and Sumridthong, A. 1998. Melatonin increases nerve growth factor in mouse submandibular gland. *J Pineal Res.* 24: 73–77.

Reiter, R. J. 1986. Normal patterns of melatonin levels in the pineal gland and body fluids of humans and experimental animals. *Neural Transm Suppl.* 21: 35–54.

Reiter, R. J., Paredes, S. D., Manchester, L. C., and Tan, D. X. 2009. Reducing oxidative/nitrosative stress: A newly discovered genre for melatonin. *Crit Rev Biochem Mol Biol.* 44: 175–200.

Reiter, R. J., Rosales-Corral, S. A., Liu, X. Y., Acuna-Castroviejo, D., Escames, G., and Tan, D. X. 2015. Melatonin in the oral cavity: Physiological and pathological implications. *J Periodontal Res.* 50 (1): 9–17.

Rodella, L. F., Lambanca, M., and Foglio, E. 2012. Melatonin in dentistry. In Watson, Ronald Ross (editor), *Melatonin in the Promotion of Health.* 219–231. Boca Raton, Florida: CRC Press.

Shimozuma, M., Tokuyama, R., Tatehara, S., Umeki, H., Ide, S., Mishima, K., Saito, I., and Satomura, K. 2011. Expression and cellular localization of melatonin-synthesizing enzymes in rat and human salivary glands. *Histochem Cell Bio.* 135 (4): 389–396.

Simon, Z., and Watson, P. A. 2002. Biomimetic dental implants: New ways to enhance osteointegration. *J Can Dent Assoc.* 68: 286–288.

Simonneaux, V., and Ribelayga, C. 2003. Generation of the melatonin endocrine message in mammals: A review of the complex regulation of melatonin synthesis by norepinephrine, peptides and other pineal transmitters. *Pharmacol Rev.* 55: 325–395.

Srinath, R., Acharya, A. B., and Thakur, S. L.2010. Salivary and gingival crevicular fluid melatonin in periodontal health and disease. *J Periodontol.* 81 (2): 277–83.

Taubman, M. A., Valverde, P., Han, X., Kawai, T. 2005. Immune response: The key to bone resorption in periodontal disease. *J Periodontol.* 76: 2033–2041.

Vakkuri, O. 1985. Diurnal rhythm of melatonin in human saliva. *Acta Physiol Scand.* 124: 409–412.

Vakkuri, O., Leppaluoto, J., and Kauppila, A. 1985. Oral administration and distribution of melatonin in human serum, saliva and urine. *Life Sci.* 37: 489–495.

.

Section VI

Receptors, Transporters, and Signaling

34

Multi-Layered Architecture of Serotonin 2A Receptor Signaling

Shishu Pal Singh, Mitradas M. Panicker, and Shuchita Soman
National Centre for Biological Sciences (TIFR), GKVK Campus
Bengaluru, India

CONTENTS

ABSTRACT   Serotonin is an important monoamine found in a number of organisms including plants, fungi, invertebrates, and vertebrates. In mammals, it is considered an important neurotransmitter and found in the central nervous system (CNS) as well as the periphery. It exerts its effects primarily via a family of G protein–coupled receptors. In this chapter, signaling via one of its most studied receptors, the 5-HT2A, is discussed. Its canonical signaling pathways, cell-dependent effects, and functional selectivity are addressed. Interaction with other endogenous neurotransmitters, particularly dopamine (DA), is also of importance in determining its function. More recent results of homodimerization of the receptors as well as heterodimerization with other receptors add to the complexity of the behavior of this receptor.

KEY WORDS: *serotonin, 5-HT$_{2A}$, signaling, ligand promiscuity, functional selectivity, biased agonism, receptor dimerization.*

Abbreviations

5-HT: serotonin
AA: arachidonic acid
ADHD: attention deficit hyperactivity disorder
Akt: serine/threonine-specific protein kinase, protein kinase B
BRET: bioluminescence resonance energy transfer
CaM: calmodulin
cAMP: cyclic adenosine monophosphate
CHO: Chinese hamster ovary
CNS: central nervous system
DA: dopamine
DAG: di-acyl glycerol

DOB:	2,5-dimethoxy-4-bromoamphetamine
DOI:	2,5-dimethoxy-4-iodoamphetamine
ERK:	extracellular signal-regulated kinase
FRET:	fluorescence resonance energy transfer
GPCR:	G protein–coupled receptor
IP$_3$:	inositol 3-phosphate
JAK:	janus kinase
KO:	knock out
LSD:	lysergic acid diethylamide
MAPK:	mitogen-associated protein kinase
MEK:	mitogen-activated protein kinase
mTOR:	mammalian target of rapamycin (serine/threonine protein kinase)
NAD(P)H:	nicotinamide adenine dinucleotide phosphate
NMDA:	*N*-methyl-*D*-aspartate
NO:	nitrous oxide
OCD:	obsessive compulsive disorder
P^{38}:	mitogen-activated protein kinase
PG:	prostaglandin
PI3K:	phosphoinositide 3-kinase
PKC:	protein kinase C
PLA2:	phospholipase A2
PLC:	phospholipase C
PLD:	phospholipase D
PP2A:	protein phosphatase 2A
Rho, Ras, Raf:	small GTPases
RO:	reactive oxygen
S6K1:	p70 ribosomal protein S6 kinase 1
SN:	*substantia nigra*
Src:	tyrosine-protein kinase Src (Src from sarcoma)
STAT:	signal transducer and activator of transcription
TCB:	(4-bromo-3, 6-dimethoxybenzocyclobuten-1-yl) methylamine hydrobromide
VTA:	ventral tegmental area
βArr2:	β-arrestin 2

Introduction

Serotonin (5-HT) plays a very important role in mammalian behavior and physiology. Evolutionarily an ancient molecule, it has been found even in plants, where its function is mostly proposed to be as a precursor to auxin, an important antioxidant and hormone (Akula et al. 2011). In fungi it provides protection from UV toxicity and stimulates cell proliferation (Belenikina et al. 1991). It is found in many invertebrates such as *C. elegans*, *Drosophila*, *Aplysia*, and crustaceans as well (Paupardin-Tritsch and Gerschenfeld 1973; Valles and White 1988; Johansson 1991; Chase and Koelle 2007). In vertebrates, serotonin acts both as a neurotransmitter and neuromodulator in the CNS and as a signaling molecule in the periphery. In humans it has been implicated in a multitude of disorders such as schizophrenia, bipolar disorder, OCD, depression, attention deficit hyperactivity disorder (ADHD), and so on. Serotonin effects its actions via a family of G protein–coupled receptors (GPCRs) and a ligand-gated ion channel. These are classified into 14 receptor subtypes in mammals: 1A, 1B, 1D, 1E, 1F, 2A, 2B, 2C, 3, 4, 5A, 5B, 6, and 7 (Peroutka 1997), and homologous receptors in other animals (Kroeze and Roth 2006). GPCRs are transmembrane receptors that set up a canonical downstream signaling pathway via second messengers such as IP$_3$, cyclic adenosine monophosphate (cAMP), di-acyl glycerol (DAG), and arachidonic acid (AA). GPCRs also comprise over 50% of the targets of drug discovery research (Imming et al. 2006).

Historically, using second messenger and signal transduction studies, the serotonin G protein–coupled receptor (GPCR) subtypes have been divided into four main categories: 5-HT$_1$, 5-HT$_2$, 5-HT$_5$, and 5-HT$_6$ (which includes 5-HT$_4$ and 5-HT$_7$ receptors) (Glennon et al. 1995; Peroutka 1997). Of these, the 5-HT$_1$ subfamily is coupled primarily to the G$_{i/o}$ subunit, which inhibits adenylyl cyclase resulting in lower cAMP concentrations; the 5-HT$_2$ subfamily is coupled to the G$\alpha_{q/11}$ subunit and activates the phospholipase C–protein kinase C (PLC–PKC) pathway, leading to accumulation of IP$_3$ and Ca^{2+}; and the 5-HT$_6$ subfamily increases adenylyl cyclase activity and thereby increases cAMP levels. The 5-HT$_5$ subfamily currently is not known to couple to any signaling pathway and the second messenger it uses remains unknown (Glennon et al. 1995; Peroutka 1997). The 5-HT$_3$ is the lone receptor that is not a GPCR, but a ligand-gated ion channel. However, with the emergence of the idea of functional selectivity or biased agonism, the above characterization appears over-simplified. Functional selectivity, specifically at the 5-HT$_{2A}$, will be discussed further in this chapter.

Among the receptors, the 5-HT$_{2A}$, which belongs to the 5-HT$_2$ subfamily, is an important receptor expressed in the brain as well as the periphery. The 5-HT$_{2A}$ is a group A GPCR. Group A GPCRs are rhodopsin-like membrane proteins with seven transmembrane helices that bind to G proteins intracellularly. A variety of techniques have determined it to be expressed in the neocortex, the olfactory bulb, the hippocampus, subiculum, the ventral tegmental area (VTA) and SN, and the dorsal horns of the spinal cord, and so on, in the CNS in humans (Andrade et al. 2015). In the periphery, it is expressed in the coronary arteries, epicardium, platelets, renal artery, peripheral lymphocytes, spleen, and uterus (Cordeaux et al. 2009; Andrade et al. 2015). It has been implicated in a wide range of functions including learning and memory, neurogenesis, and has also proven to be one of the primary targets for antipsychotics (Iqbal and van Praag 1995; Soudijn 1997). It has been reported to modulate the symptoms of nicotine withdrawal as well as cocaine addiction (Lesurtel et al. 2006; Zaniewska et al. 2015), and is also responsible for the hallucinogenic activity of compounds such as lysergic acid diethylamide (LSD), 2,5-dimethoxy-4-iodoamphetamine (DOI), DOB, and so on (Ronken and Olivier 1997; Roth 1998). Peripherally, it is involved in liver regeneration, inflammation, hypertension, and the contraction of smooth muscle walls (Ronken and Olivier 1997; Walther et al. 2003; Lesurtel et al. 2006; Yu et al. 2008). Studies have also shown a correlation between 5-HT$_{2A}$ polymorphism (T102C) and increased platelet aggregation (Shimizu et al. 2003). The serotonin 2A receptor also plays an important role in the neuromodulatory functions of serotonin, especially the release of DA in the brain. Activation of the 5-HT$_{2A}$ leads to increased dopaminergic function and release, which is opposite to the effect exerted by 5-HT$_{2C}$ on dopaminergic function (Esposito et al. 2008).

5-HT$_{2A}$ Signal Transduction, Ligands, and Ligand Promiscuity

Activation of 5-HT$_{2A}$ can induce diverse signaling pathways which can be G-protein dependent or independent and cell- or tissue-specific (see Figure 34.1). (Raymond et al. 2001; Bockaert et al. 2006; Turner et al. 2006; Cowen 2007; Millan et al. 2008; Masson et al. 2012; Zhang and Stackman 2015). In most tissues and cells, 5-HT$_{2A}$ activates PLC-β via G$\alpha_{q/11}$ which leads to the generation of IP$_3$ and DAG, in turn, causing release of Ca^{2+} from intracellular stores and activation of PKC (Conn and Sanders-Bush 1984, 1985; Roth et al. 1984, 1986; Takuwa et al. 1989; Tamir et al. 1992; Berg et al. 1998; Briddon et al. 1998; Hagberg et al. 1998; Grotewiel and Sanders-Bush 1999). Besides PLC-β, 5-HT$_{2A}$ also activates phospholipase A2 (PLA$_2$) and phospholipase D (PLD). 5-HT$_{2A}$ coupling to PLA$_2$ and subsequent AA mobilization has been shown in CHO cells (Berg et al. 1996, 1998), 1C11 cells (Tournois et al. 1998), C6 glioma cells (Garcia and Kim 1997), and NIH3T3 cells (Kurrasch-Orbaugh et al. 2003a, 2003b). The 5-HT$_{2A}$ has been shown to couple to PLD in cultured rat renal mesangial cells (Kurscheid-Reich et al. 1995), COS-7 cells (Mitchell et al. 1998; Robertson 2003) and bovine pulmonary artery smooth muscle cells (Liu and Fanburg 2008).

Though 5-HT$_{2A}$ activation leads to increase in inositol triphosphate (IP$_3$), it is also known to regulate cAMP formation in specific cell types. It activates cAMP formation in A1A1 cells by two distinct

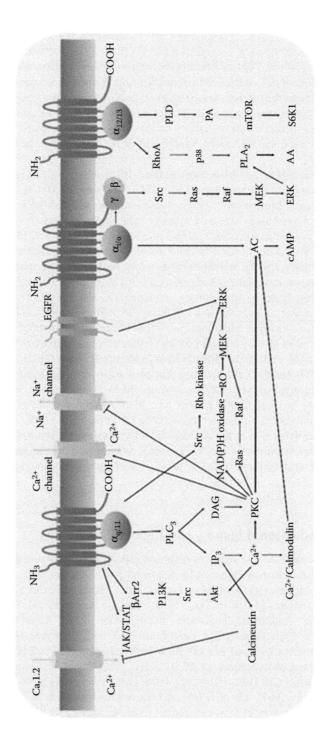

FIGURE 34.1   5-HT$_{2A}$ signaling pathway.

pathways: protein kinase C (PKC) and calcium/calmodulin (Berg et al. 1994). In FRTL-5 thyroid cells, the $5\text{-}HT_{2A}$ increased cAMP, an effect which was pertussis toxin–sensitive (Tamir et al. 1992). In rat renal mesangial cells, $5\text{-}HT_{2A}$ inhibited forskolin-stimulated cAMP accumulation in intact cells and inhibited AC activity in washed membranes. The $5\text{-}HT_{2A}$-mediated inhibition did not involve PLC, Ca^{2+}, or PKC but was pertussis toxin–sensitive indicating involvement of $G_{i/o}$-proteins (Garnovskaya et al. 1995).

Activation of $5\text{-}HT_{2A}$ can lead to extracellular signal-regulated kinase (ERK) activation in many neuronal and non-neuronal cell types via diverse intracellular signaling mechanisms. In tracheal smooth muscle cells, $5\text{-}HT_{2A}$-mediated ERK–mitogen-associated protein kinase (MAPK) activation involves the protein kinase C/Raf-1 pathway (Hershenson et al. 1995). In rat aortic tissue, activation of ERK-MAPK involves Src and is independent of PKC, PI3K, or the EGF receptor tyrosine kinase (Florian and Watts 1998; Watts 1998; Banes et al. 1999). In renal mesangial cells, ERK activation by $5\text{-}HT_{2A}$ has been shown to require reactive oxygen species production via an NAD(P)H oxidase-like enzyme along with PKC activation (Grewal et al. 1999; Greene et al. 2000). In PC-12 cells, $5\text{-}HT_{2A}$ could activate ERK, independent of PKC, and required Ca^{2+}/calmodulin and tyrosine kinases (Quinn 2002). In rat skeletal muscle myoblasts (Guillet-Deniau et al. 1997), rat vascular smooth muscle cells (Banes et al. 2005), BeWo and JEG-3 cells (Oufkir et al. 2010), and rat frontal cortex (Singh et al. 2010) the $5\text{-}HT_{2A}$ has been shown to activate the janus kinase/signal transducer and activator of transcription (JAK/STAT) pathway.

$5\text{-}HT_{2A}$ can also regulate ion channels and activation of L-type Ca^{2+} channels in specific cell types (Eberle-Wang et al. 1994; Jalonen et al. 1997; Watts 1998). However, $5\text{-}HT_{2A}$-mediated activation of L-type Ca^{2+} channels should be inferred cautiously because of the overlap of the pharmacology of L-type Ca^{2+}-channel blockers and $5\text{-}HT_2$ receptor antagonists (Okoro 1999). $5\text{-}HT_{2A}$ has been shown to regulate both voltage-dependent and voltage-independent Ca^{2+} channels (Nakaki et al. 1985; Eberle-Wang et al. 1994; Hagberg et al. 1998). IP_3-mediated elevated cytosolic Ca^{2+} levels upon $5\text{-}HT_{2A}$ activation have been shown to cause opening of Ca^{2+}-activated K^+ channels in C6 glioma cells (Bartrup and Newberry 1994) and Ca^{2+}-activated Cl^- channels in *Xenopus* oocytes (Montiel et al. 1997). Closing of K^+ channels upon activation of $5\text{-}HT_2$ receptors expressed in *Xenopus* oocytes has also been reported (Parker et al. 1990). In rat cortical astrocytes, the $5\text{-}HT_{2A}$ activates two types of Ca^{2+} channels: L-type Ca^{2+} channel and small conductance Ca^{2+}-activated K^+ channel (Jalonen et al. 1997). In prefrontal cortex (PFC) pyramidal neurons, $5\text{-}HT_{2A}$ activation has been shown to suppress membrane $Ca_v1.2$ L-type Ca^{2+} currents via a $G_{\alpha q}$-mediated $PLC\beta/IP_3$/calcineurin signaling pathway (Day et al. 2002), and also reduce amplitude of rapidly inactivating Na^+ currents and persistent Na^+ currents by a PKC-dependent mechanism (Carr et al. 2002).

$5\text{-}HT_{2A}$ can either stimulate or inhibit nitric oxide (NO) synthesis in certain tissues and cells. Acceleration of gastrointestinal transit (GIT) by escin Ib in mice is mediated by the release of endogenous prostaglandins (PGs) and NO, which has been shown to be $5\text{-}HT_{2A}$ dependent (Matsuda et al. 2000), whereas in C6 glioma cells, $5\text{-}HT_{2A}$ activation inhibited the production of NO in response to cytokines (Miller and Gonzalez 1998). In renal mesangial cells, $5\text{-}HT_{2A}$ has been shown to induce the production of H_2O_2 and superoxide through the action of an NAD(P)H oxidase-like enzyme (Grewal et al. 1999; Greene et al. 2000).

Being a major signaling target of Ca^{2+} mobilization in most cells types, calmodulin (CaM) has also been shown to be involved in $5\text{-}HT_{2A}$ signaling. In A1A1 cells, $5\text{-}HT_{2A}$-mediated cAMP formation has been shown to be both PKC-dependent and calcium/calmodulin-dependent (Berg et al. 1994). Upregulation of $5\text{-}HT_{2A}$ in cerebellar granule cells by agonists was shown to involve calcium influx and a calmodulin-dependent pathway (Chen et al. 1995). In renal mesangial cells, the regulation of 5-HT-mediated Cox-2 mRNA expression involves CaM kinase along with PKC and p42/44 MAPK (Stroebel and Goppelt-Struebe 1994; Goppelt-Struebe et al. 1999). In prefrontal pyramidal cells, activation of $5\text{-}HT_{2A}$ by LSD and DOB which inhibited *N*-methyl-*D*-aspartate (NMDA) receptor-mediated transmission involved Ca2+/CaM-KII-dependent signal transduction pathway (Arvanov et al. 1999). Turner and Raymond (2005) have shown that CaM coimmunoprecipitates with the $5\text{-}HT_{2A}$ in NIH-3T3 fibroblasts in an agonist-dependent manner and that the receptor contains two putative CaM binding regions (Turner and Raymond 2005). In A1A1v cells, $5\text{-}HT_{2A}$-mediated transamidation of Rac1 by transglutaminase requires phospholipase C, Ca2+, and calmodulin signaling (Dai et al. 2011).

5-HT$_{2A}$ has also been linked to a variety of transport processes. 5-HT$_{2A}$ activates the Na$^+$–H$^+$ exchanger in renal mesangial cells (Saxena et al. 1993; Garnovskaya et al. 1995), and in airway smooth muscle stimulates the Na$^+$–K$^+$ pump possibly through the Na$^+$–H$^+$ exchanger (Rhoden et al. 2000). 5-HT$_{2A}$ transfected into NIH3T3 fibroblasts also activates Na$^+$/K$^+$/2Cl$^-$ cotransport (Mayer and Sanders 1994).

Besides G proteins, 5-HT$_{2A}$ also couples to β-arrestin2 (βArr2) and stimulates Akt phosphorylation via the βArr2/phosphoinositide 3-kinase (PI3K)/Src/Akt cascade (Schmid and Bohn 2010). In PC12 cells, 5-HT$_{2A}$ agonists activate Akt through endogenously expressed 5-HT$_{2A}$ via receptor-mediated increases in intracellular Ca^{2+} (Johnson-Farley et al. 2005).

Identified originally as the "classical" 5-HT$_2$ receptor due to its lower affinity for 5-HT, 5-HT$_{2A}$ binds to a number of ligands which include antipsychotics as well as psychedelics (Peroutka and Snyder 1979; Julius et al. 1990; van Wijngaarden and Soudijn 1997). However, most of these ligands are nonselective for 5-HT$_{2A}$ and bind other serotonin receptor subtypes and/or other GPCRs. Ligands that act as 5-HT$_{2A}$ agonists include DOI, DOB, LSD, and αMe-5-HT, and those that act as 5-HT$_{2A}$ antagonists/ inverse agonists include MDL 100907, ketanserin, risperidone, ritanserin, olanzapine, and clozapine (van Wijngaarden and Soudijn 1997).

Interestingly, 5-HT$_{2A}$ can also be activated by other endogenous biogenic amines like dopamine, tryptamine, tyramine, norepinephrine, β-PEA, and octopamine, a phenomenon referred as ligand promiscuity (Woodward et al. 1992; Bhattacharyya 2004; Bhattacharyya et al. 2006; Raote et al. 2007, 2013; Bhattacharya 2010; Raote 2013). Activation of 5-HT$_{2A}$ by DA has been shown in *Xenopus* oocytes as well as HEK293 cells where DA has been found to act as a partial agonist on rat as well as human 5-HT$_{2A}$ (Woodward et al. 1992; Bhattacharyya 2004; Bhattacharyya et al. 2006; Raote et al. 2007, 2013; Bhattacharya 2010; Raote 2013). 5-HT$_{2A}$ when primed with very low levels of 5-HT requires lower concentrations of DA to activate and cause internalization of the receptor (Bhattacharyya 2004; Bhattacharyya et al. 2006; Bhattacharya 2010; Raote 2013). Besides DA, tryptamine, tyramine, norepinephrine, β-PEA, and octopamine, but not epinephrine and histamine, have also been found to activate rat 5-HT$_{2A}$ with varied efficacy and potency (Raote 2013).

Functional Selectivity

An important facet of signaling via the serotonin 2A receptor is the ability of the receptor to activate different signaling pathway(s) depending on the ligand it binds. This property is called "ligand directed trafficking of receptor stimulus," "biased agonism," or "functional selectivity" (Berg and Clarke 2006). Classical receptor-ligand theory proposes that the effect of a ligand, when it interacts with a particular receptor, is dependent on a few factors such as agonist concentration, affinity for the ligand, and the intrinsic efficacy of the agonist which is affected by the receptor density. According to this model, a particular receptor is associated with a particular signal transduction pathway, which would then be activated or inhibited by the action of the ligand on the receptor. The ability of ligands to bind to a receptor, also called their *affinity*, and their ability to bring about a response from the receptor, also called *efficacy*, were thought to be intrinsic to the ligand and hence incontrovertible (Kenakin 2004). Based on these principles, ligands have been classified as agonists (partial or full), inverse agonists (partial or full), or antagonists. However, experiments have shown that the effects of ligands on a receptor can vary; either by the activation of noncanonical signaling pathways, or because of different cellular milieus. The intrinsic efficacy of a particular ligand was defined as "the capacity of a drug to promote a change in receptor conformation that leads to increased affinity/efficacy of the activated receptor for the next molecule in the signal transduction cascade (e.g., G protein)" (Berg and Clarke 2006). It is therefore possible to imagine an array of conformations available to the functional receptor. It can therefore be hypothesized that certain conformations from this pool of conformations would be preferred by certain ligands and hence also stabilized on binding to the receptor. These new conformations of the receptor now make available to the receptor different intracellular docking molecules and mechanism, and confer the ability to participate/activate signaling cascades different from the canonical signaling cascade normally set up by the cognate ligand. Hence, this property of receptor signaling is called ligand-directed trafficking of receptor stimulus, or biased agonism, or functional selectivity (Kenakin 1995). Functional selectivity has been shown for multiple receptors, even in the serotonin receptor family, such as, 5-HT_{1A}, 5-HT_{2B}, and 5-HT_{2C} (Berg et al. 1998; Pauwels and Colpaert 2003; Moya et al. 2007; Miller et al. 2009; Chilmonczyk et al. 2015).

The canonical signaling pathway set up by 5-HT_{2A} involved the activation of PLC (phospholipase C) through G_q, which leads to IP_3 and DAG accumulation and activation of PKC, as well as accumulation of AA through activation of PLA2. Work by Berg and Clarke has shown that hallucinogenic phenethylamine and phenylisopropylamine derivatives show functional selectivity at the human 5-HT_{2A}. They measured IP (IP_3) and AA accumulation in CHO cells expressing 5-HT_{2A} to measure the bias of the ligands to preferentially activate one pathway over another. The compound 2,5-dimethoxyphenylisopropylamine, only activated the PLC pathway, whereas the phenethylamine derivative, 2,5- dimethoxy-4-nitrophenethylamine, activates only the PLA2 pathway and does not set up signaling via the PLC-PKC pathway (Moya et al. 2007). Drug-elicited head-twitch/head-shake response has been used as a preliminary test for determining the hallucinogenic property of a particular drug. It is well established that this behavior requires 5-HT_{2A} (Corne and Pickering 1967). When Berg et al. injected rats with both the phenethylamine derivatives as well as phenylisopropylamine derivatives they observed that the phenethylamine derivatives failed to induce head-shake in rats while the phenylisopropylamine derivatives were effective head-shake inducers (Moya et al. 2007).

Molecular modeling has suggested some interactions that may be relevant in functional selectivity. Homology modeling and ligand docking study of a fully activated 5-HT_{2A} with 5-HT as a balanced ligand (causes a balanced response for both pathways) and 2C-N (2,5-dimethoxy-4-nitrophenethylamine), a compound with partial efficacy for AA pathway but no efficacy for the IP_3 pathway, has been suggested. Molecular dynamic studies suggest that serotonin forms a hydrogen bond between the nitrogen of its indole ring and residue S5.46 and another one between its hydroxyl substituent and residue N6.55. This numbering is based on the Ballesteros–Weinstein nomenclature where the letter denotes the amino acid, the first number denotes the helix number and the second number denotes the residue position relative to the most conserved residue in the helix (given the number 50) as determined by reference sequences (Isberg et al. 2015). In comparison, 2C-N forms a contact between its nitro group and residue N6.55 in helix 6, which is common with the contacts formed by serotonin, and hence it was proposed that this interaction would be necessary for setting up the AA pathway, which is also common between 5-HT and 2C-N. In this study, the researchers also went on to characterize potential ligands for biased agonism which were modified forms of 5-HT. Of the selected ligands, MetI has a diminished capacity for hydrogen bonding at residue S5.46 due to a methyl substitution at the amine of the indole group, but interaction with the N6.55 residue is intact, thus making the ligand biased for AA signaling over IP_3 signaling. The other ligand used in the study was MetT, which has a methyl substitution at the hydroxyl group which would disrupt the ability to hydrogen bond with residue N6.55 thus preferring IP accumulation over AA accumulation. On testing these ligands on CHO cells expressing 5-HT_{2A}, it was observed that, as predicted, the MetI compound showed preferred activation of the AA pathway over the IP_3 pathway, whereas, MetT showed no AA accumulation at the concentrations used, and only showed IP_3 accumulation (Martí-Solano et al. 2015).

Another study had also predicted that conformations of the intracellular loop 2 of the 5-HT_{2A} are involved in the hallucinogenic or nonhallucinogenic outcomes of the action of particular ligands on the receptor (Perez-Aguilar et al. 2014).

Work by Raote et al. has extended the idea of functional selectivity of the 5-HT_{2A} further to the endocytosis and recycling of the receptor. Internalization of the receptor by 5-HT and DOI required phosphorylation of the receptor by PKC and dephosphorylation of the receptor by PP2A, whereas internalization by DA (an endogenous ligand) and clozapine (an inverse agonist/antagonist) was PKC-independent. The time required for recycling of the receptor back to the cell surface was also observed to be variable and ligand-dependent. Receptors internalized by 5-HT and DA recycled back to the cell surface in ~3 h, whereas receptors internalized by DOI and clozapine required much longer, that is, ~8 h (Raote et al. 2013). This indicates that signals for recycling and the cellular machineries associated with receptors internalized by different ligands can differ (Figure 34.2).

Functional selectivity via the 5-HT_{2A} has also been observed when expressed in different cellular contexts. Tryptamine, an agonist at the 5-HT_{2A}, shows bias toward the PLC-IP pathway as compared to the

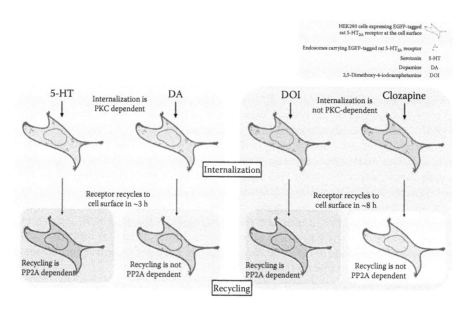

FIGURE 34.2 Functional selectivity in the internalization and recycling of the 5-HT$_{2A}$.

PLA2-AA pathway when activating 5-HT$_{2A}$ in NIH-3T3 cells (Kurrasch-Orbaugh et al. 2003a), whereas this bias is reversed in favor of the PLA2-AA pathway when 5-HT$_{2A}$ is expressed in CHO cells (Berg et al. 1998). This strengthens the idea that a receptor, sampling all available receptor conformations, can associate with a number of downstream signaling pathways, and that the ligands acting on the receptor preferentially stabilize some of these conformations to set up distinct signaling pathways.

Dimerization

While it has been sufficient so far to view the receptor as a monomer while trying to understanding the process of activation and signaling, an aspect of the serotonin 2A receptor which has come to light only in the last decade is its ability to oligomerize with itself and dimerize with other transmembrane receptors. This results in profound effects in its signaling, and this aspect remains to be explored further. The oligomerization and its subsequent effects have been studied using a variety of methods. Most of these studies have employed homo- or heterofluorescence resonance energy transfer (FRET) as well as bioluminescence resonant energy transfer (BRET) techniques to establish the presence of the receptor(s) multimers, and probe the effects of these interactions. Competitive ligand binding experiments in cell lines expressing recombinant human 5-HT$_{2A}$ lacking constitutive activity gave biphasic binding curves. Previous studies have shown that such biphasic curves are the hallmark of receptor dimerization (Albizu et al. 2006). When binding curves were plotted for clozapine *versus* haloperidol displacing radioactive 2,5-dimethoxy-4-bromoamphetamine (DOB, a 5-HT$_{2A}$ agonist) at the 5-HT$_{2A}$, clozapine showed biphasic curves. This biphasic nature of the curves of clozapine's action was further demonstrated when clozapine's ability to inhibit IP$_3$ or AA accumulation was studied. Clozapine showed a monophasic inhibitory curve for the inhibition of IP$_3$ accumulation, whereas it showed a biphasic inhibitory curve for inhibition of AA release (Lo et al. 2009). Furthermore, the 5-HT$_{2C}$ was known to dimerize and it shows high sequence homology with the 5-HT$_{2A}$. Using FRET and immunoprecipitation, the serotonin 2A receptor was shown to also homodimerize in HEK293 cells (Lo et al. 2009). This adds another layer of complexity to signaling via the 5-HT$_{2A}$. Furthermore, *in situ* hybridization showed overlapping presence of the 5-HT$_{2A}$ and mGLuR2 mRNA in mouse somatosensory cortex layer V cells. The expression of 5-HT$_{2A}$ was necessary for normal expression of mGluR2 in the cortex, as knocking out 5-HT$_{2A}$ led to lowered expression of mGluR2 in the cortex (González-Maeso et al. 2008). Hence, a direct interaction mechanism between 5-HT$_{2A}$ and mGluR2 was examined and it was shown that the 5-HT$_{2A}$ dimerizes with the mGLuR2. This was also shown to affect the downstream signaling through a signaling complex. The 5-HT$_{2A}$ activates both G$_{q/11}$ and G$_{i/o}$. Coexpression of mGluR2 led to increased affinity for G$_{i/o}$, whereas activation of the mGluR2 by an agonist led to increased affinity for G$_{q/11}$. The transmembrane helices 4 and 5 of mGluR2 were also shown to be responsible for dimerization with the 5-HT$_{2A}$. The presence of this complex was shown not only in HEK293 cells, but also in membrane preparations from mouse primary cultures (González-Maeso et al. 2008). This complex is necessary for the cellular and behavioral effects of hallucinogens, such as DOI, which are agonists at 5-HT$_{2A}$. DOI is a 5-HT$_{2A}$ agonist and a known hallucinogen and its effects are considered psychomimetic (Aghajanian and Marek 2000). Knocking out mGluR2 results in loss of hallucinogenic activity of DOI via 5-HT$_{2A}$ (Moreno et al. 2011) and the absence of the drug-induced head-twitch/shake response, which is used as a discriminatory test for hallucinogenic activity in rodent models (Corne and Pickering 1967). In mice which were mGluR2 knockouts, DOI-injected mice also show absence of the head-twitch response (Moreno et al. 2011). Expression of early growth response protein 2 (EGR-2), which has been shown to be a hallucinogen-specific transcript expressed in the brain (González-Maeso et al. 2007), was also abolished in mGluR2 knock out (KO) mice when treated with DOI (Moreno et al. 2011). Thus, this complex and its signaling are now considered a novel target for therapeutics against psychosis.

The serotonergic and dopaminergic systems interact at a variety of levels. DA and serotonin systems are known to modulate each other's functions in the CNS. These two systems have also been implicated in various psychiatric disorders such as anxiety, depression, schizophrenia, and so on. Typical antipsychotics target DA D$_2$ receptor, whereas atypical antipsychotics target 5-HT$_{2A}$. These two receptors have now been shown to physically interact, that is, dimerize. The 5-HT$_{2A}$ has been shown to dimerize with the DA D$_2$ receptor. Using immunofluorescence labeling, colocalization of these two receptors has been shown in the medial prefrontal cortex and *pars reticulate* of the *substantia nigra* (SN) in the rat brain (Lukasiewicz et al. 2010). Using FRET-based assays; the action of agonists such as DOI on the 5-HT$_{2A}$-D$_2$ dimer has also been investigated. In HEK293 cells expressing 5-HT$_{2A}$ and D$_2$ and in mouse brain striatal membranes, application of D$_2$ agonist quinpirole increased the affinity of DOI for the 5-HT$_{2A}$. This increase in affinity also translated into increased efficacy of DOI to cause IP$_3$ production, in the presence of the D$_2$ agonist. This increased efficacy was, however, not seen for serotonin, but was observed only for DOI. MK-801 (an NMDA antagonist) is used in pharmacological models for schizophrenia in mice (Reimherr et al. 1986). MK801-induced locomotor effects are countered by the action of haloperidol, a typical antipsychotic acting via the D$_2$. However, using 5-HT$_{2A}$ KO mice, it was shown that expression of 5-HT$_{2A}$ is necessary for the effect of haloperidol in ameliorating the effects of MK801 (Albizua et al. 2012). Another study has shown that the rise in intracellular calcium on activation of G$_{q/11}$ via 5-HT$_{2A}$ by 5-HT and (4-bromo-3, 6-dimethoxybenzocyclobuten-1-yl) methylamine hydrobromide (TCB) was enhanced when the accompanying D$_2$ was also activated using quinpirole. Conversely, activating the 5-HT$_{2A}$ by agonists such as 5-HT or TCB led to reduced potency of quinpirole to activate the G$_{i/0}$ signaling via D$_2$R (Borroto-Escuela et al. 2010). Using bioinformatics approaches, the amino acid triplets LLT (Leu-Leu-Thr) and AIS (Ala-Ile-Ser) have been proposed as being responsible for the receptor–receptor interactions. The presence of both these amino acid triplets has been shown in the 5-HT$_{2A}$-D$_2$R heterodimer (Borroto-Escuela et al. 2010).

Conclusion

In conclusion, this review aims to highlight emerging paradigms in GPCR biology through the lens of the 5-HT$_{2A}$ and to bring into greater focus the various aspects of 5-HT$_{2A}$ signaling, which deviate from the classical receptor–ligand interaction theory. The 5-HT$_{2A}$ is an especially interesting model because it is implicated in various disorders such as schizophrenia, bipolar disorder, obsessive compulsive disorder (OCD), sleep apnea, and cocaine addiction (Peroutka and Snyder 1980; Iqbal and van Praag 1995; Mintun et al. 2004; Adams et al. 2005). Antagonists of the 5-HT$_{2A}$, often used as drugs, affect both physiology of the CNS and behavior (Soudijn 1997). In the periphery, this receptor has also been implicated in liver regeneration, platelet aggregation, pulmonary hypertension, and so on (Walther et al. 2003; Lesurtel et al. 2006; Dempsie 2008).

It is important to keep in mind that most of the studies with the 5-HT_{2A} that have investigated receptor signaling or receptor dimerization have used non-neural cell lines, and often the receptor is over-expressed. Moreover, studies demonstrating functional selectivity have focused on the ligands under consideration and not on the cellular milieu. Such studies, by virtue of design, cannot take into account that the receptor is not functioning in its natural cellular milieu, is most often overexpressed, and cannot reflect differential localization of receptors within the cell, all of which can affect downstream signaling. Recent studies have also modulated the receptor intracellularly using molecules such as pepducins and intrabodies. Pepducins are lipid-modified peptides, which are cell permeable and can be designed to correspond to specific intracellular regions of GPCRs. Studies with chemokine receptor CXCR4 have used pepducins to stimulate biased signaling through this receptor, but the exact mechanism remains unknown (Shukla 2014). Similarly, intrabodies, which are single-chain antibodies, have been expressed intracellularly to bind to the cytoplasmic portions of the receptors. Work with the β-adrenergic receptor 2 has shown that intrabodies targeting the cytoplasmic segments of this receptor caused biased receptor response and inhibited cAMP production or βArr2-dependent responses (Shukla 2014). The discoveries of properties such as biased agonism, heterodimerization, and intracellular biasing of receptors have thrown up more questions, and opened up new and interesting areas of inquiry.

Acknowledgments

The authors are grateful for financial support (SS, MMP, SS) and funding for the laboratory (MMP) from the National Centre of Biological Sciences (Tata Institute of Fundamental Research), Bangalore, and discussions with members of the MMP laboratory.

REFERENCES

Adams, K. H., Hansen, E. S., Pinborg, L. H., Hasselbalch, S. G., Svarer, C., Holm, S., Bolwig, T. G., and Knudsen, G. M. Patients with obsessive-compulsive disorder have increased 5-HT2A receptor binding in the caudate nuclei. *Int J Neuropsychopharmacol.* 2005. 8: 391–401.

Aghajanian, G. K., and Marek, G. J. Serotonin model of schizophrenia: Emerging role of glutamate mechanisms. *Brain Res Rev.* 2000. 31: 302–312.

Akula, R., Giridhar, P., and Ravishankar, G. A. Phytoserotonin. *Plant Signal Behav.* 2011. 6 (6): 800–809.

Albizu, L., Balestre, M.-N., Breton, C., Pin, J.-P., Manning, M., Mouillac, B., Barberis, C., and Durroux, T. Probing the existence of G protein–coupled receptor dimers by positive and negative ligand-dependent cooperative binding. *Mol Pharmacol.* 2006. 70 (5): 1783–1791.

Albizua, L., Holloway, T., González-Maeso, J., and Sealfon, S. C. Functional crosstalk and heteromerization of serotonin 5-HT2a and dopamine D2 receptors. *Neuropharmacology.* 2012. 29 (6): 997–1003.

Andrade, R., Barnes, N. M., Baxter, G., Bockaert, J., Branchek, T., Cohen, M. L., Dumuis, A., et al. 5-Hydroxytryptamine receptors: 5-HT2A receptor. 2015. Last modified on 04/08/2015. Accessed on 23/05/2016. IUPHAR/BPS Guide to PHARMACOLOGY, http://www.guidetopharmacology.org/GRAC/ObjectDisplayForward?objectId=6.

Arvanov, V. L., Liang, X., Russo, A., and Wang, R. Y. LSD and DOB: Interaction with 5-HT(2A) receptors to inhibit NMDA receptor-mediated transmission in the rat prefrontal cortex. *Eur J Neurosci.* 1999. 11 (9): 3064–3072.

Banes, A., Florian, J. A., and Watts, S. W. Mechanisms of 5-hydroxytryptamine2A receptor activation of the mitogen activated protein kinase pathway in vascular smooth muscle. *J Pharmacol Exp Ther.* 1999. 291 (3): 1179–1187.

Banes, A. K. L., Shaw, S. M., Tawfik, A., Patel, B. P., Ogbi, S., Fulton, D., and Marrero, M. B. Activation of the JAK/STAT pathway in vascular smooth muscle by serotonin. *Am J Physiol Cell Physiol.* 2005. 288 (4): C805–C812.

Bartrup, J. T., and Newberry, N. R. 5-HT2A Receptor-mediated outward current in c6 glioma cells is mimicked by intracellular IP3 release. *Neuroreport.* 1994. 5 (10): 1245–1248.

Belenikina, N. S., Strakhovskaya, M. G., and Fraikin, G. Y. Near-UV activation of yeast growth. *J Photochem Photobiol B Biol.* 1991. 10 (1–2): 51–55.

Berg, K. A., and Clarke, W. P. Agonist-directed trafficking of 5-HT receptor-mediated signal transduction. In Roth, B. L. (editor), *The Serotonin Receptors: From Molecular Pharmacology to Human Therapeutics (The Receptors).* 2006. 207–235. Totowa, New Jersey: Humana Press.

Berg, K. A., Clarke, W. P., Chen, Y., Ebersole, B. J., McKay, R. D., and Maayani, S. 5-hydroxytryptamine type 2A receptors regulate cyclic AMP accumulation in a neuronal cell line by protein kinase C-dependent and calcium/calmodulin-dependent mechanisms. *Mol Pharmacol.* 1994. 45 (5): 826–836.

Berg, K. A., Maayani, S., and Clarke, W. P. 5-Hydroxytryptamine2C receptor activation inhibits 5-hydroxy-tryptamine1B-like receptor function *via* arachidonic acid metabolism. *Mol Pharmacol.* 1996. 50 (4): 1017–1023.

Berg, K. A., Maayani, S., Goldfarb, J., Scaramellini, C., Leff, P., and Clarke, W. P. Effector pathway–dependent relative efficacy at serotonin type 2A and 2C receptors: Evidence for agonist-directed trafficking of receptor stimulus. *Mol Pharmacol.* 1998. 54 (1): 94–104.

Bhattacharyya, S. Internalization and recycling of the serotonin 2A receptor in non-neuronal and neuronal cells. PhD Thesis. 2004. Bangalore, India: Tata Institute of Fundamental Research.

Bhattacharya, A. Dissection of the internalization and recycling of the human serotonin 2A (5-HT2A) receptor. PhD Thesis. 2010. Bangalore, India: Tata Institute of Fundamental Research.

Bhattacharyya, S., Raote, I., Bhattacharya, A., Miledi, R., and Panicker, M. M. Activation, internalization, and recycling of the serotonin 2A receptor by dopamine. *Proc Natl Acad Sci.* 2006. 103 (41): 15248–15253.

Bockaert, J., Claeysen, S., Bécamel, C., Dumuis, A., and Marin, P. Neuronal 5-HT metabotropic receptors: Fine-tuning of their structure, signaling, and roles in synaptic modulation. *Cell Tissue Res.* 2006. 326 (2): 553–572.

Borroto-Escuela, D. O., Romero-Fernandez, W., Tarakanov, A. O., Marcellino, D., Ciruela, F., Agnati, L. F., and Fuxe, K. Dopamine D2 and 5-hydroxytryptamine 5-HT2A receptors assemble into functionally interacting heteromers. *Biochem Biophys Res Commun.* 2010. 401 (4): 605–610.

Briddon, S. J., Leslie, R. A., and Elliott, J. M. Comparative desensitization of the human 5-HT 2A and 5-HT 2C receptors expressed in the human neuroblastoma cell line SH-SY5Y. *Br J Pharmacol.* 1998. 125 (4): 727–734.

Carr, D. B., Cooper, D. C., Ulrich, S. L., Spruston, N., and Surmeier, D. J. Serotonin receptor activation inhibits sodium current and dendritic excitability in prefrontal cortex via a protein kinase C-dependent mechanism. *J Neurosci.* 2002. 22 (16): 6846–6855.

Chase, D. L., and Koelle, M. Biogenic amine neurotransmitters in *C. elegans. WormBook.* 2007. 1–15.

Chen, H., Li, H., and Chuang, D. M. Role of second messengers in agonist up-regulation of 5-HT2A (5-HT2) receptor binding sites in cerebellar granule neurons: Involvement of calcium influx and a calmodulin-dependent pathway. *J Pharmacol Exp Ther* 1995. 275(2): 674–680.

Chilmonczyk, Z., Bojarski, A. J., Pilc, A., and Sylte, I. Functional selectivity and antidepressant activity of serotonin 1A receptor ligands. *Int J Mol Sci.* 2015. 16 (8): 18474–18506.

Conn, P. J., and Sanders-Bush, E. Selective 5HT-2 Antagonists inhibit serotonin stimulated phosphatidylino-sitol metabolism in cerebral cortex. *Neuropharmacology.* 1984. 23 (8): 993–996.

Conn, P. J., and Sanders-Bush, E. Serotonin-stimulated phosphoinisitide turnover: Mediation by the S2 binding site in rat cerebral cortex but not in subcortical regions. *J Pharmacol Exp Ther.* 1985. 234 (1): 195–203.

Cordeaux, Y., Pasupathy, D., Bacon, J., Charnock-Jones, D. S., and Smith, G. C. S. Characterization of serotonin receptors in pregnant human myometrium. *J Pharmacol Exp Ther.* 2009. 328 (3): 682–691.

Corne, S. J., and Pickering, R. W. A possible correlation between drug-induced hallucinations in man and a behavioural response in mice. *Psychopharmacologia.* 1967. 11: 65–78.

Cowen, D. S. Serotonin and neuronal growth factors: A convergence of signaling pathways. *J Neurochem.* 2007. 101 (5): 1161–1171.

Dai, Y., Dudek, N. L., Li, Q., and Muma, N. A. Phospholipase C, Ca2+, and calmodulin signaling are required for 5-HT2A receptor-mediated transamidation of RAC1 by transglutaminase. *Psychopharmacology.* 2011. 213 (2–3): 403–412.

Day, M., Olson, P. A., Platzer, J., Striessnig, J., and Surmeier, D. J. Stimulation of 5-HT(2) receptors in prefrontal pyramidal neurons inhibits Ca(v)1.2 L type Ca(2+) currents via a PLCbeta/IP3/calcineurin signaling cascade. *J Neurophysiol.* 2002. 87 (5): 2490–2504.

Dempsie, Y. Pulmonary hypertension: Therapeutic targets within the serotonin system. 2008. 455–462.

Eberle-Wang, K., Braun, B. T., and Simansky, K. J. Serotonin contracts the isolated rat pylorus via a 5-HT2-like receptor. *Am J Physiol.* 1994. 266: R284–R291.

Esposito, E., and Matteo, V. Di Giovanni, G. Serotonin–dopamine interaction: An overview. *Prog Brain Res.* 2008. 172: 3–6.

Florian, J. A., and Watts, S. W. Integration of mitogen-activated protein kinase kinase activation in vascular 5-hydroxytryptamine2A receptor signal transduction. *J Pharmacol Exp Ther.* 1998. 284 (1): 346–355.

Garcia, M. C., and Kim, H. Mobilization of arachidonate and docosahexaenoate by stimulation of the 5-HT 2A receptor in rat C6 glioma cells. *Brain Res.* 1997. 768 (1–2): 43–48.

Garnovskaya, M. N., Nebigil, C. G., Arthur, J. M., Spurney, R. F., Raymond, J. R., Spurney, F., and Carolina, N. 5-hydroxytryptamine2A receptors expressed in rat renal mesangial cells inhibit cyclic amp accumulation. *Mol Pharmacol.* 1995. 48 (2): 230–237.

Glennon, R. A., Westkaemper, R. B., and Dukat, M. Serotonin receptor subtypes and ligands. *Psychopharmacol Fourth Gener Prog.* 1995. 1–23.

González-Maeso, J., Ang, R., Yuen, T., Chan, P., Noelia, V., López-Giménez, J. F., Zhou, M., et al. Identification of a serotonin/glutamate receptor complex implicated in psychosis. *Nature.* 2008. 452 (7183): 93–97.

González-Maeso, J., Weisstaub, N. V., Zhou, M., Chan, P., Ivic, L., Ang, R., Lira, A., et al. Hallucinogens recruit specific cortical 5-HT2A receptor-mediated signaling pathways to affect behavior. *Neuron.* 2007. 53 (3): 439–452.

Goppelt-Struebe, M., Hahn, A., Stroebel, M., and Reiser, C. O. A. Independent regulation of cyclo-oxygenase 2 expression by P42/44 mitogen-activated protein kinases and Ca2+/calmodulin dependent kinase. *Biochem J.* 1999. 339: 329–334.

Greene, E. L., Houghton, O., Collinsworth, G., Garnovskaya, M. N., Nagai, T., Sajjad, T., Bheemanathini, V., Grewal, J. S., Paul, R. V., and Raymond, J. R. 5-HT(2A) receptors stimulate mitogen-activated protein kinase via H(2)O(2) generation in rat renal mesangial cells. *Am J Physiol Renal Physiol.* 2000. 278 (4): F650–F658.

Grewal, J. S., Mukhin, Y. V., Garnovskaya, M. N., Raymond, J. R., and Greene, E. L. Serotonin 5-HT2A receptor induces TGF-BETA1 expression in mesangial cells via ERK: Proliferative and fibrotic signals. *Am J Physiol.* 1999. 276. (6 Pt 2): F922–F930.

Grotewiel, M. S., and Sanders-Bush, E. Differences in agonist-independent activity of 5-HT2A and 5-HT2C receptors revealed by heterologous expression. *Naunyn Schmiedebergs Arch Pharmacol.* 1999. 359 (1): 21–27.

Guillet-Deniau, I., Burnol, A. F., and Girard, J. Identification and localization of a skeletal muscle secrotonin 5-HT2A receptor coupled to the JAK/STAT pathway. *J Biol Chem.* 1997. 272 (23): 14825–14829.

Hagberg, G. B., Blomstrand, F., Nilsson, M., Tamir, H., and Hansson, E. Stimulation of 5-HT2A receptors on astrocytes in primary culture opens voltage-independent Ca2+ channels. *Neurochem Int.* 1998. 32 (2): 153–162.

Hershenson, M. B., Chao, T. S., Abe, M. K., Gomes, I., Kelleher, M. D., Solway, J., and Rosner, M. R. Histamine antagonizes serotonin and growth factor–induced mitogen-activated protein kinase activation in bovine tracheal smooth muscle cells. *J Biol Chem.* 1995. 270 (34): 19908–19913.

Imming, P., Sinning, C., and Meyer, A. Drugs, their targets and the nature and number of drug targets. *Nat Rev Drug Discov.* 2006. 5 (10): 821–834.

Iqbal, N., and van Praag, H. M. The role of serotonin in schizophrenia. *Eur Neuropsychopharmacol.* 1995. 5 Suppl. (118): 11–23.

Isberg, V., de Graaf, C., Bortolato, A., Cherezov, V., Katritch, V., Marshall, F. H., Mordalski, S., et al. Generic GPCR residue numbers: Aligning topology maps while minding the gaps. *Trends Pharmacol Sci.* 2015. 36 (1): 22–31.

Jalonen, T. O., Margraf, R. R., Wielt, D. B., Charniga, C. J., Linne, M. L., and Kimelberg, H. K. Serotonin induces inward potassium and calcium currents in rat cortical astrocytes. *Brain Res.* 1997. 758 (1–2): 69–82.

Johansson, K. U. I. Identification of different types of serotonin-like immunoreactive olfactory interneurons in 4 infraorders of decapod crustaceans. *Cell Tissue Res.* 1991. 264 (2): 357–362.

Johnson-Farley, N. N., Kertesy, S. B., Dubyak, G. R., and Cowen, D. S. Enhanced activation of AKT and extracellular-regulated kinase pathways by simultaneous occupancy of Gq-coupled 5-HT2A receptors and Gs-coupled 5-HT7A receptors in PC12 cells. *J Neurochem.* 2005. 92 (1): 72–82.

Julius, D., Huang, K. N., Livelli, T. J., Axel, R., and Jessell, T. M. The 5HT2 receptor defines a family of structurally distinct but functionally conserved serotonin receptors. *Proc Natl Acad Sci.* 1990. 87: 928–932.

Kenakin, T. Agonist-receptor efficacy. II. Agonist trafficking of receptor signals. *Trends Pharmacol Sci.* 1995. 16 (7): 232–238.

Kenakin, T. Principles: Receptor theory in pharmacology. *Trends Pharmacol Sci.* 2004. 25 (4): 186–192.

Kroeze, W. K., and Roth, B. L. Molecular biology and genomic organization of G protein– coupled serotonin receptors. In Roth, B. L. (editor), *The Serotonin Receptors: From Molecular Pharmacology to Human Therapeutics (The Receptors).* 2006. 1–38. Totowa, New Jersey: Humana Press,

Kurrasch-Orbaugh, D. M., Parrish, J. C., Watts, V. J., and Nichols, D. E. A complex signaling cascade links the serotonin2A receptor to phospholipase A2 activation: The involvement of map kinases. *J Neurochem.* 2003a. 86 (4): 980–991.

Kurrasch-Orbaugh, D. M., Watts, V. J., Barker, E. L., and Nichols, D. E. Phospholipase C and phospholipase A2 signaling pathways have different receptor reserves. *J Pharmacol Exp Ther.* 2003b. 304 (1): 229–237.

Kurscheid-Reich, D., Throckmorton, D. C., and Rasmussen, H. Serotonin activates phospholipase D in rat mesangial cells. *Am J Physiol.* 1995. 268 (6 Pt 2): F997–F1003.

Lesurtel, M., Graf, R., Aleil, B., Walther, D. J., Tian, Y., Jochum, W., Gachet, C., Bader, M., and Clavien, P.-A. Platelet-derived serotonin mediates liver regeneration. *Science.* 2006. 312 (5770): 104–107.

Liu, Y., and Fanburg, B. L. Phospholipase D signaling in serotonin-induced mitogenesis of pulmonary artery smooth muscle cells. *Am J Physiol Lung Cell Mol Physiol.* 2008. 295 (3): L471–L478.

Lo, J. F., Pad, J. F., Cadavid, M. I., Vilaro, M. T., Mengod, G.; et al. Evidence for distinct antagonist-revealed functional states of 5HT2A homodimers. *Mol Pharmacol.* 2009. 75 (6): 1380–1391.

Lukasiewicz, S., Polit, A., Kędracka-Krok, S., Wędzony, K., Maćkowiak, M., and Dziedzicka-Wasylewska, M. Heterodimerization of serotonin 5-HT(2A) and dopamine D(2) receptors. *Biochim Biophys Acta.* 2010. 1803 (12): 1347–1358.

Martí-Solano, M., Iglesias, A., De Fabritiis, G., Sanz, F., Brea, J., Loza, M. I., Pastor, M., and Selent, J. Detection of new biased agonists for the serotonin 5-HT 2A receptor: Modeling and experimental validation. *Mol Pharmacol.* 2015. 87 (4): 740–746.

Masson, J., Emerit, M. B., Hamon, M., and Darmon, M. Serotonergic signaling: Multiple effectors and pleiotropic effects. *Wiley Interdiscip Rev Membr Transp Signal.* 2012. 1 (6): 685–713.

Matsuda, H., Li, Y., and Yoshikawa, M. Possible involvement of 5-HT and 5-HT2 receptors in acceleration of gastrointestinal transit by escin Ib in mice. *Life Sci.* 2000. 66 (23): 2233–2238.

Mayer, S. E., and Sanders, B. E. 5-hydroxytryptamine type 2A and 2C receptors linked to Na+/K+/Cl- cotransport. *Mol Pharmacol.* 1994. 45 (5): 991–996.

Millan, M. J., Marin, P., Bockaert, J., and Mannoury la Cour, C. Signaling at G-protein–coupled serotonin receptors: Recent advances and future research directions. *Trends Pharmacol Sci* 2008. 29: 454–464.

Miller, K. J., and Gonzalez, H. A. Serotonin 5-HT2A receptor activation inhibits cytokine-stimulated inducible nitric oxide synthase in C6 glioma cells. *Ann NY Acad Sci.* 1998. 861. 169–173.

Miller, K. J., Wu, G. Y., Varnes, J. G., Levesque, P., Li, J., Li, D., Robl, J. A., Rossi, K. A., and Wacker, D. A. Position 5.46 of the serotonin 5-HT2A receptor contributes to a species-dependent variation for the 5-HT2C agonist (R)-9-ethyl-1,3,4,10b-tetrahydro-7-trifluoromethylpyrazino[2,1-A]isoindol-6(2H)-one: Impact on selectivity and toxicological evaluation. *Mol Pharmacol.* 2009. 76 (6): 1211–1219.

Mintun, M. A., Sheline, Y. I., Moerlein, S. M., Vlassenko, A. G., Huang, Y., and Snyder, A. Z. Decreased hippocampal 5-HT2A receptor binding in major depressive disorder: *In vivo* measurement with [18F] altanserin positron emission tomography. *Biol Psychiatry.* 2004. 55. 217–224.

Mitchell, R., McCulloch, D., Lutz, E., Johnson, M., MacKenzie, C., Fennell, M., Fink, G., Zhou, W., and Sealfon, S. C. Rhodopsin-family receptors associate with small G proteins to activate phospholipase D. *Nature.* 1998. 392 (6674): 411–414.

Montiel, C., Herrero, C. J., García-Palomero, E., Renart, J., García, A. G., and Lomax, R. B. Serotonergic effects of dotarizine in coronary artery and in oocytes expressing 5-HT2 receptors. *Eur J Pharmacol.* 1997. 332 (2): 183–193.

Moreno, J. L., Holloway, T., Albizu, L., Sealfon, S. C., and González-Maeso, J. Metabotropic glutamate Mglu2 receptor is necessary for the pharmacological and behavioral effects induced by hallucinogenic 5-HT2A receptor agonists. *Neurosci Lett.* 2011. 493 (3): 76–79.

Moya, P. R., Berg, K. A., Gutiérrez-Hernandez, M. A., Sáez-Briones, P., Reyes-Parada, M., Cassels, B. K., and Clarke, W. P. Functional selectivity of hallucinogenic phenethylamine and phenylisopropylamine derivatives at human 5-hydroxytryptamine (5-HT)2A and 5-HT2C receptors. *J Pharmacol Exp Ther.* 2007. 321 (3): 1054–1061.

Nakaki, T., Roth, B. L., Chuang, D. M., and Costa, E. phasic and tonic components in 5-HT2 receptor-mediated rat aorta contraction: Participation of Ca2+ channels and phospholipase C *J Pharmacol Exp Ther.* 1985. 234 (2): 442–446.

Okoro, E. O. Overlap in the pharmacology of l-type Ca2+-channel blockers and 5-HT2 receptor antagonists in rat aorta. *J Pharm Pharmacol.* 1999. 51 (8): 953–957.

Oufkir, T., Arseneault, M., Sanderson, J. T., and Vaillancourt, C. The 5-HT2A serotonin receptor enhances cell viability, affects cell cycle progression and activates MEK-ERK1/2 and JAK2-STAT3 signalling pathways in human choriocarcinoma cell lines. *Placenta.* 2010. 31 (5): 439–447.

Parker, I., Panicker, M. M., and Miledi, R. Serotonin receptors expressed in xenopus oocytes by mRNA from brain mediate a closing of K+ membrane channels. *Mol Brain Res.* 1990. 7 (1): 31–38.

Paupardin-Tritsch, D., and Gerschenfeld, H. M. Transmitter role of serotonin in identified synapses in aplysia nervous system. *Brain Res.* 1973. 58 (2): 529–534.

Pauwels, P. J., and Colpaert, F. C. Ca2+ responses in chinese hamster ovary-k1 cells demonstrate an atypical pattern of ligand-induced 5-HT1A receptor activation. *J Pharmacol Exp Ther.* 2003. 307 (2): 608–614.

Perez-Aguilar, J. M., Shan, J., LeVine, M. V., Khelashvili, G., and Weinstein, H. A functional selectivity mechanism at the serotonin-2A GPCR involves ligand-dependent conformations of intracellular loop 2. *J Am Chem Soc.* 2014. 136 (45): 16044–16054.

Peroutka, S. J. 5-Hydroxytryptamine receptor subtypes. In Olivier, B., van Wijngaarden, I., and Soudijn, W. (editors), *Serotonin Receptors and Their Ligands. Pharmacochemistry Library.* 1997. 27: 3–13. Amsterdam, Elsevier.

Peroutka, S. J., and Snyder, S. H. Multiple serotonin receptors: Differential binding of [3H]5-hydroxytryptamine, [3H]lysergic acid diethylamide and [3H]spiroperidol. *Mol Pharmacol.* 1979. 16 (3): 687–699.

Peroutka, S. J., and Snyder, S. H. Long-term antidepressant treatment decreases spiroperidol-labeled serotonin receptor binding. *Science.* 1980. 210 (4465): 88–90.

Quinn, J. C. Activation of extracellular-regulated kinase by 5-hydroxytryptamine2A receptors in PC12 cells is protein kinase C-independent and requires calmodulin and tyrosine kinases. *J Pharmacol Exp Ther.* 2002. 303 (2): 746–752.

Raote, I. Functional selectivity in serotonin receptor 2A (5-HT2A) signaling. PhD Thesis. 2013. Bangalore, India: Tata Institute of Fundamental Research.

Raote, I., Bhattacharya, A., and Panicker, M. M. Serotonin 2A (5-HT2A) receptor function: Ligand-dependent mechanisms and pathways. In Chattopadhyay A. (editor), *Serotonin Receptors in Neurobiology.* 2007. 1–17. Boca Raton, FL: CRC Press.

Raote, I., Bhattacharyya, S., and Panicker, M. M. Functional selectivity in serotonin receptor 2A (5-HT2A) endocytosis, recycling, and phosphorylation. *Mol Pharmacol.* 2013. 83 (1): 42–50.

Raote I., Bhattacharya A., Panicker M. M. Serotonin 2A (5-HT2A) receptor function: Ligand-dependent mechanisms and pathways. In: Chattopadhyay A. (editor), *Serotonin Receptors in Neurobiology.* 2007. 105–132. Boca Raton, FL: CRC Press.

Ronken, E., and Olivier, B. The 5-HT2-type receptor family. In Olivier, B., van Wijngaarden, I., and Soudijn, W. (editors), *Serotonin Receptors and Their Ligands. Pharmacochemistry Library.* 1997. 199–213. Amsterdam, Elsevier.

Raymond, J. R., Mukhin, Y. V., Gelasco, A., Turner, J., Collinsworth, G., Gettys, T. W., Grewal, J. S., and Garnovskaya, M. N. Multiplicity of mechanisms of serotonin receptor signal transduction. *Pharmacol Ther.* 2001. 92 (2–3): 179–212.

Reimherr, F. W., Wood, D. R., and Wender, P. H. The use of MK-801, a novel sympathomimetic, in adults with attention deficit disorder, residual type. *Psychopharmacol Bull.* 1986. 22 (1): 237–242.

Rhoden, K. J., Dodson, A. M., and Ky, B. Stimulation of the Na(+)–K(+) pump in cultured guinea pig airway smooth muscle cells by serotonin. *J Pharmacol Exp Ther.* 2000. 293 (1): 107–112.

Robertson, D. N. Selective interaction of ARF1 with the carboxy-terminal tail domain of the 5-HT2A receptor. *Mol Pharmacol.* 2003. 64 (5): 1239–1250.

Ronken, E., and Olivier, B. The 5-HT2-type receptor family. In Olivier, B., van Wijngaarden, I., and Soudijn, W. (editors), *Serotonin Receptors and Their Ligands. Pharmacochemistry Library.* 1997. 199–214. Elsevier Science E-books.

Roth, B. 5-Hydroxytryptamine2-family receptors (5-hydroxytryptamine2A, 5-hydroxytryptamine2B, 5-hydroxytryptamine2C) where structure meets function. *Pharmacol Ther.* 1998. 79 (3): 231–257.

Roth, B., Nakaki, T., Chuang, D.-M., and Costa, E. Aortic recognition sites for serotonin (5HT) are coupled to phospholipase C and modulate phosphatidylinositol turnover. *Neuropharmacology.* 1984. 23 (10): 1223–1225.

Roth, B. L., Nakaki, T., Chuang, D. M., and Costa, E. 5-hydroxytryptamine2 receptors coupled to phospholipase C in rat aorta: Modulation of phosphoinositide turnover by phorbol ester. *J Pharmacol Exp Ther.* 1986. 238 (2): 480–485.

Saxena, R., Saksa, B. A., Fields, A. P., and Ganz, M. B. Activation of Na/H exchanger in mesangial cells is associated with translocation of PKC isoforms. *Am J Physiol.* 1993. 265 (1 Pt 2): F53–F60.

Schmid, C. L., and Bohn, L. M. Serotonin, but not *N*-methyltryptamines, activates the serotonin 2A receptor via a ß-arrestin2/Src/Akt signaling complex *in vivo. J Neurosci.* 2010. 30 (40): 13513–13524.

Shimizu, M., Kanazawa, K., Matsuda, Y., Takai, E., Iwai, C., Miyamoto, Y., Hashimoto, M., Akita, H., and Yokoyama, M. Serotonin-2A receptor gene polymorphisms are associated with serotonin-induced platelet aggregation. *Thromb Res.* 2003. 112 (3): 137–142.

Shukla, A. K. Biasing GPCR signaling from inside. *Sci Signal.* 2014. 7 (310): pe3.

Singh, R. K., Jia, C., Garcia, F., Carrasco, G. A., Battaglia, G., and Muma, N. A. Activation of the JAK-STAT pathway by olanzapine is necessary for desensitization of serotonin2A receptor–stimulated phospholipase C signaling in rat frontal cortex but not serotonin2A receptor–stimulated hormone release. *J Psychopharmacol.* 2010. 24 (7): 1079–1088.

Soudijn, W. 5-HT2 receptor antagonists: (Potential) therapeutics. In Olivier, B., van Wijngaarden, I., and Soudijn, W. (editors), *Serotonin Receptors and Their Ligands. Pharmacochemistry Library.* 1997. 27. 215–217. Amsterdam, Elsevier.

Stroebel, M., and Goppelt-Struebe, M. Signal transduction pathways responsible for serotonin-mediated prostaglandin G/H synthase expression in rat mesangial cells. *J Biol Chem.* 1994. 269 (37): 22952–22957.

Takuwa, N., Ganz, M., Takuwa, Y., Sterzel, R. B., and Rasmussen, H. Studies of the mitogenic effect of serotonin in rat renal mesangial cells. *Am J Physiol.* 1989. 257 (3 Pt 2): F431–F439.

Tamir, H., Hsiung, S. C., Yu, P. Y., Liu, K. P., Adlersberg, M., Nunez, E. A., and Gershon, M. D. Serotonergic signalling between thyroid cells: Protein kinase C and 5-HT2 receptors in the secretion and action of serotonin. *Synapse.* 1992. 12 (2): 155–168.

Tournois, C., Mutel, V., Manivet, P., Launay, J.-M., and Kellermann, O. Cross-talk between 5-hydroxytryptamine receptors in a serotonergic cell line. Involvement of arachidonic acid metabolism. *J Biol Chem.* 1998. 273 (28): 17498–17503.

Turner, J., Gelasco, A., Ayiku, H., Coaxum, S., Arthur, J., and Garnovskaya, M. 5-HT Receptor signal transduction pathways. In Roth, B. L. (editor), *The Serotonin Receptors: From Molecular Pharmacology to Human Therapeutics (The Receptors).* 2006. 143–206. Totowa, New Jersey: Humana Press.

Turner, J. H., and Raymond, J. R. Interaction of calmodulin with the serotonin 5-hydroxytryptamine2A receptor. A putative regulator of G protein coupling and receptor phosphorylation by protein kinase C. *J Biol Chem.* 2005. 280 (35): 30741–30750.

Valles, A. M., and White, K. Serotonin-containing neurons in *Drosophila melanogaster*: Development and distribution. *J Comp Neurol.* 1988. 268 (3): 414–428.

van Wijngaarden, I., and Soudijn, W. 5-HT2A, 5-HT2B and 5-HT2C receptor ligands. In Olivier, B., van Wijngaarden, I., and Soudijn, W. (editors), *Serotonin Receptors and Their Ligands. Pharmacochemistry Library.* 1997. 215–217. Amsterdam, Elsevier.

Walther, D. J., Peter, J., Winter, S., Höltje, M., Paulmann, N., Grohmann, M., Vowinckel, J., et al. Serotonylation of small GTpases is a signal transduction pathway that triggers platelet alpha-granule release. *Cell.* 2003. 115 (7): 851–862.

Watts, S. W. Activation of the mitogen-activated protein kinase pathway via the 5-HT2A receptor. *Ann NY Acad Sci.* 1998. 861: 162–168.

Woodward, R. M., Panicker, M. M., and Miledi, R. Actions of dopamine and dopaminergic drugs on cloned serotonin receptors expressed in xenopus oocytes. *Proc Natl Acad Sci.* 1992. 89 (10): 4708–4712.

Yu, B., Becnel, J., Zerfaoui, M., Rohatgi, R., Boulares, a H., and Nichols, C. Serotonin 5-hydroxytryptamine 2A receptor activation suppresses tumor necrosis factor-α–induced inflammation with extraordinary potency. *J Pharmacol Exp Ther.* 2008. 327 (2): 316–323.

Zaniewska, M., Alenina, N., Wydra, K., Fröhler, S., Kuśmider, M., McCreary, A. C., Chen, W., Bader, M., and Filip, M. Discovering the mechanisms underlying serotonin (5-HT) 2A and 5-HT 2C receptor regulation following nicotine withdrawal in rats. *J Neurochem.* 2015. 134 (4): 704–716.

Zhang, G., and Stackman, R. W. The role of serotonin 5-HT2A receptors in memory and cognition. *Front Pharmacol.* 2015. 6: 225.

35

Serotonin Transporters in Theranostics of Neuropsychiatric Disorders

B. Swathy and Moinak Banerjee
Rajiv Gandhi Center for Biotechnology
Kerala, India

CONTENTS

ABSTRACT   The serotonin transporter (SERT or 5HTT) is a type of monoamine transporter protein that plays a vital role in serotonergic neurotransmission. It transports serotonin from the synaptic cleft to the presynaptic neuron, thereby terminating the action of serotonin and aiding in its recycling. Thus, SERT's function is to maintain the homeostatic levels of serotonin in the extracellular space. Disruption of the normal functioning of SERT is implicated in various neurological conditions, including schizophrenia, autism, depression, attention deficit hyperactivity disorder, and anxiety. Being a major player in the serotonin pathway, SERT is a major target site for many therapeutic drugs and a diagnostic marker for schizophrenia, depression, and other psychiatric conditions. The potential role of SERT in therapeutics and diagnostics in the area of neuropsychiatry is explored in this chapter. The genetic variabilities in SERT can also predispose people to disease susceptibility and treatment response. Dietary supplements can also modulate the neurotransmitter reuptake processes at the synapse.

KEY WORDS: *serotonin transporter, schizophrenia, neuropsychiatry, depression.*

Abbreviations

[¹¹C] DASB:	[^{11}C]-3-amino-4-(2-dimethylaminomethyl-phenylsulfanyl)-benzonitrile)
[11C]-MADAM:	(11)C-N,N-dimethyl-2-(2-amino-4-methylphenylthio)benzylamine
[123I]beta-CIT:	[(123)I]beta-CIT (2beta-carbomethoxy-3beta-(4-iodophenyl)tropane)
123I-ADAM:	(123)I-labeled 2-((2-((dimethylamino)methyl) phenyl)thio)-5-iodophenylamine
5HT:	serotonin
5HTTLPR:	serotonin transporter–linked polymorphic region
ADHD:	attention deficit hyperactivity disorder

DNMT:	DNA methyltransferase
GAD:	generalized anxiety disorder
MDD:	major depressive disorder
NaSSA:	noradrenergic and specific serotonergic antidepressant
NAT:	noradrenaline transporters
OCD:	obsessive compulsive disorder
PET:	positron emission tomography
PMAT:	plasma membrane monoamine transporter
PMDD:	premenstrual dysphoric disorder
PTSD:	post-traumatic stress disorder
SAMe:	S-adenosyl methionine
SARI:	serotonin antagonist and reuptake inhibitor
SLC6A4:	solute carrier family 6 member 4
SMS:	serotonin modulator and stimulator
SNP:	single-nucleotide polymorphism
SNRI:	serotonin–norepinephrine reuptake inhibitor
SPECT:	single-photon emission computed tomography
SSRI:	selective serotonin reuptake inhibitor
VNTR:	variable number tandem repeat

Introduction

Serotonin is a monoamine neurotransmitter involved in regulating a wide range of psychological and biological tasks such as mood, appetite, sleep, memory, and sexual function. It is synthesized from the amino acid L-tryptophan by enzymes L-tryptophan hydroxylase and amino acid decarboxylase. Serotonin is mainly stored in the enterochromaffin cells in the mucosa of the gastrointestinal tract, in serotonergic neurons of the central nervous system, and in platelets. Serotonin secreted in the brain activates various serotonin (5HT) receptors and subsequently it is reuptaken by presynaptic neurons via serotonin transporters (Figure 35.1).

The termination of serotonin function is accomplished by the reuptake of serotonin from the synapse by various monoamine transporters, mainly the serotonin transporter (SERT or 5HTT). In the brain, about 95% of serotonin is uptaken by SERT, although other transporters such as plasma membrane monoamine transporter (PMAT) also contribute significantly. SERT acts by a Na^+/Cl^--dependent process. In the first step, sequential binding of sodium ion, serotonin, and chloride ion to SERT occurs, followed by the release of serotonin inside the cells. Intracellular potassium ion then binds to SERT, reverting its position, allowing it to bind to another serotonin molecule. Imbalanced Na^+/K^+ homeostasis can result in altered SERT function.

Disruption of the normal functioning of SERT leads to altered serotonin levels in the synapse, affecting various psychological and biological processes. Various medical conditions associated with SERT dysfunction include disturbances in mood and behavior, memory and learning, and the sleep–wake cycle, which are primarily manifested in neuropsychiatric disorders. Genetic and epigenetic variability in the SERT gene are reported to be associated with many neurological conditions, including schizophrenia, depression, and autism. Members of the serotonin pathway, specifically SERT, are a primary target of various drugs, including antidepressants and mood stabilizers. Here, we focus on the clinical implications of genetic research in line with variations of the SERT gene. The potential role of SERT in diagnostics and therapeutic strategies are also discussed.

SERT as a Target for Treatment

Owing to the central role played by SERT in the serotonin pathway, it can serve as a key target for antidepressant therapies. Selective serotonin reuptake inhibitors (SSRIs) and serotonin–norepinephrine

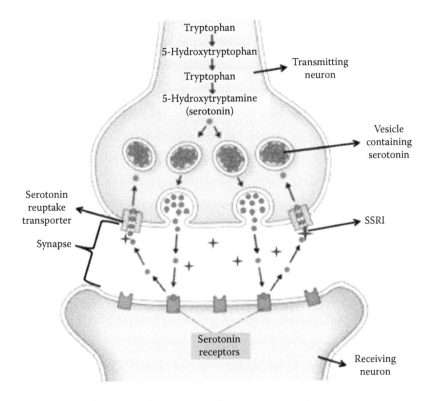

FIGURE 35.1 Serotonin synthesis, transmission and reuptake and the role of SSRI.

reuptake inhibitors (SNRIs) are two major classes of second-generation antidepressants that work by targeting serotonin transporters. Some of the commonly used SSRI and SNRIs that have been approved by the FDA are listed in Table 35.1. Other classes of antidepressant drugs such as serotonin antagonist and reuptake inhibitors (SARIs), serotonin modulators and stimulators (SMSs), and noradrenergic and specific serotonergic antidepressants (NaSSAs) act as antagonists for various serotonin receptors and inhibit the reuptake of serotonin.

Selective serotonin reuptake inhibitors or serotonin-specific reuptake inhibitors (SSRIs) are a widely used class of antidepressant drugs. They block the reabsorption of serotonin into the presynaptic cell, thus increasing the extracellular level of serotonin in the synapse, making it available for postsynaptic receptors. They offer a highly specific mechanism of action compared with other classes of antidepressants. SSRIs are considered the first-line treatment for major depressive disorder (MDD), obsessive compulsive disorder (OCD), generalized anxiety disorder (GAD), nerve pain, panic disorders, post-traumatic stress disorder (PTSD), and premenstrual dysphoric disorder (PMDD). Compared with other classes of antidepressant drugs, SSRIs offer the lowest risk of side effects. With some drugs in this class, side effects such as nausea, headache, akathisia, sexual dysfunction, weight gain, and anticholinergic effects are reported.

Serotonin–noradrenaline reuptake inhibitors (SNRIs) are a class of antidepressant drugs that block the reuptake of both serotonin and noradrenaline by binding to SERT and the noradrenaline transporter (NAT). Most of these drugs affect serotonin to a more significant extent than norepinephrine. SNRIs are used in the treatment of MDD, OCD, anxiety disorders, attention deficit hyperactivity disorder (ADHD), and chronic neuropathic pain (including neuropathies and fibromyalgia). SNRIs are prescribed only when a person fails to respond to an SSRI. They are associated with many adverse side effects, including nausea, dizziness, dry mouth, constipation, insomnia, loss of appetite, and sexual problems.

TABLE 35.1

Commonly used SSRIs and SNRIs approved by the US Food and Drug Administration (FDA) and Their Therapeutic Uses

Class of Drug	Drug (Brand Name)	MDD	OCD	GAD	PMDD	PTSD	Panic Disorder	Social Anxiety	Chronic Pain Syndrome
SSRI	Citalopram (Celexa)	✓	✓						
	Escitalopram (Lexapro)	✓		✓					
	Fluoxetine (Prozac)	✓	✓		✓		✓		
	Fluvoxamine (Luvox)	✓	✓						
	Paroxetine (Paxil, Pexeva)	✓	✓	✓	✓	✓	✓	✓	
	Sertraline (Zoloft)	✓	✓		✓	✓	✓	✓	
	Trazodone (Oleptro)	✓							
SNRI	Desvenlafaxine (Pristiq)	✓							
	Duloxetine (Cymbalta)	✓	✓	✓					✓
	Levomilnacipran (Fetzima)	✓							
	Milnacipran (Savella, Ixel, Dalcipran, Toledmin)	✓							✓
	Venlafaxine (Efexor, Effexor)	✓		✓					✓

Impact of Genetic Variabilities in SERT in Disease and Treatment Response

In humans, SERT is encoded by the solute carrier family 6 member 4 (*SLC6A4*) gene, which is found on chromosome 17 in location 17q11.1–q12. Mutations in this gene are known to alter serotonin transporter function and are associated with complex neurological phenotypes and treatment response. Some of the functionally important mutations reported in *SLC6A4* linked to various serotonin-related disorders, including depression, OCD, schizophrenia, and autism spectrum disorders, are discussed below (Figure 35.2).

5HTTLPR (serotonin transporter–linked polymorphic region) is a 44 base pair insertion–deletion polymorphism located in the promoter region of *SLC6A4*. Several studies have reported a strong association of 5HTTLPR with common mental disorders such as PTSD (Lee et al., 2005; Xie et al., 2009), suicide (Lin and Tsai, 2004), OCD (Hu et al., 2006), depression (Willeit et al., 2003; Antypa et al., 2010), autism (Kluck et al., 1997; Yirmiya et al., 2001), childhood aggression (Beitchman et al., 2006), and anxiety-related personality traits (Sen et al., 2004). Near to the 5HTTLPR region, two single-nucleotide polymorphisms (SNPs) are reported: rs25531 (A/G) and rs25532(C/T). Allele-specific associations of 5HTTLPR/rs25531 with ADHD and autism spectrum disorder (Gadow et al., 2013) and haplotypic associations of 5HTTLPR/rs25531/rs25532 with Tourette's syndrome are reported (Moya et al., 2013). Allelic and genotypic associations of 5HTTLPR, rs2066713, and STin2 polymorphisms have been studied and

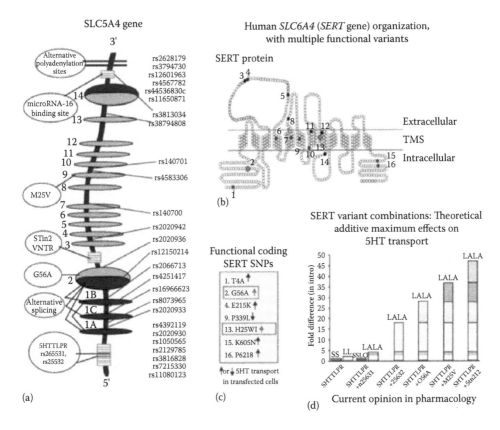

FIGURE 35.2 Human *SLC6A4* (SERT gene) organization with multiple functional variants. (Adapted from Murphy DL and Moya PR, *Current Opinion in Pharmacology* 11, 3–10, 2011. With permission.)

a haplotype linking these three risk alleles is reported to be associated with schizophrenia in a South Indian population (Vijayan et al., 2009).

I425V is a rare functional variant located on exon 9 that produces a gain-of-function of SERT (Ozaki et al., 2003). This SNP is reported to be associated with susceptibility to OCD (Delorme et al., 2005; Grados et al., 2007). *In vitro* studies have shown that this variant causes constitutive activation of transport activity (Kilic et al., 2003). Gly56Ala in exon 2 is yet another gain-of-function mutation reported in OCD and autism that is associated with increased transporter activity (Sutcliffe et al., 2005; Prasad et al., 2009). A VNTR located in intron 2 known as STin2 shows association with schizophrenia (Fan and Sklar, 2005), bipolar affective puerperal psychosis (Coyle et al., 2000), and depression (Ogilvie et al., 1996). A recent study has shown that novel mutations in the *SLC29A4* plasma membrane monoamine transporter (PMAT) gene is linked to autism spectrum disorder (Adamsen et al., 2014).

Several studies have investigated the relationship between genetic polymorphisms in SERT and responses to drug treatment. An association between the 5HTTLPR genotype and the therapeutic effects of various SSRIs including fluvoxamine (Smeraldi et al., 1998; Yoshida et al., 2002), citalopram, (Arias et al., 2003), and paroxetine (Zanardi et al., 2000) has been reported. Differential responses to the anti-psychotic drug clozapine (Arranz et al., 2000) and the mood stabilizer lithium (Serretti, 2002) have been reported with the 5HTTLPR allelic state. Variations in 5HTTLPR have been reported to be associated with differential remission rates following SSRI medication (Mrazek et al., 2009). It is also shown to influence the antidepressant response to serotonergic drug treatments and resultant sleep deprivation (Benedetti et al., 2003) and the pathogenesis of antidepressant-induced mania in bipolar disorder (Mundo et al., 2001).

Impact of Epigenetic Variabilities in SERT in Disease and Treatment Response

Apart from genetic variations, environmental influences mediated by epigenetic mechanisms can also contribute to altered *SLC6A4* expression and liability to various disease conditions. The increased DNA promoter methylation status of *SLC6A4* is linked to reduced gene expression. The state of *SLC6A4* promoter methylation is altered in depression (Zhao et al., 2013; Kim et al., 2013; Wankerl et al., 2014), bipolar disorder (Sugawara et al., 2011), and childhood aggression (Wang et al., 2012). A similar observation was reported in another study that showed hypermethylation of *SLC6A4* to be associated with childhood adversities and poor clinical presentation, but not with treatment response (Kang et al., 2013). A recent study indicated the role of *SLC6A4* methylation in mediating antidepressant treatment response. This study in Caucasians showed hypomethylation of the SERT transcriptional control region that impairs response to SSRI treatment in patients with MDD, possibly via increased SERT expression and consecutively decreased serotonin availability (Domschke et al., 2014). These emerging studies on the epigenetic control of *SLC6A4* do indicate that both endogenous and exogenous control of the epigenome should also be considered while understanding a disease or evaluating treatment response.

Pharmacoepigenomics Target for SSRIs

Recently, with the advent of pharmacoepigenomics, it has been noted that drugs are known to mediate therapeutic effects by various epigenetic mechanisms. Drugs that target the methylation and histone modification status of the *SLC6A4* promoter are reported to have therapeutic potential. Various currently used SSRIs are also known to induce epigenetic effects. Escitalopram administration is associated with decreases in DNA methylation in the promoter of P11(S100A10) in a genetic model of depression (Melas et al., 2012). This study has also shown that escitalopram-induced hypomethylation is associated with a reduction in the mRNA levels of two DNA methyltransferases, *DNMT1* and *DNMT3a*. Paroxetine inhibits *DNMT1* activity through the modulation of G9a protein levels (Nicole et al., 2012). Fluoxetine is shown to increase microRNA miR-16 levels, which in turn reduces SERT expression, thus establishing the role of microRNAs in the therapeutic actions of SSRIs (Baudry et al., 2010). The emerging role of pharmacoepigenomics is providing newer dimensions in our understanding of drug response and its side effects.

Dietary Modulation of SSRIs

Apart from drugs, dietary nutrients can also modulate the neurotransmitter reuptake processes at the synapse. For instance, methionine and S-adenosyl methionine (SAM) are serotonin reuptake inhibitors. Studies have shown that SAM can serve as an adjunctive treatment strategy for SSRI nonresponders with MDD (Papakostas et al., 2010). SAM is synthesized from methionine, which is absorbed from the diet as an essential amino acid. It is available in plant sources such as white beans, soybeans, sunflower seeds, oats, walnuts, hazelnuts, and so on. This cycle requires vitamin B12 and folate for the maintenance. Supplementation of these vitamins and minerals through diet can support the healthy regulation of serotonin.

Folate is shown to enhance the antidepressant action of fluoxetine. Co-administration of folic acid with fluoxetine increased response rates in patients with major depression, especially women (Coppen and Bailey, 2000). Folic acid alone or combined with fluoxetine, 17-beta estradiol, or neuropeptide Y produced antidepressant-like actions in ovariectomized rats subjected to the forced-swim test (Molina-Hernández and Téllez-Alcántara, 2011). The efficacy of folinic acid (leucovorin) as an adjunct among SSRI-refractory patients with MDD is noted (Alpert et al., 2002). Randomized trials have shown that nutritional supplementations such as vitamin B12, zinc, L-methylfolate, magnesium, omega-3 fatty acid, tryptophan, and 5-hydroxytryptophan have a significant effect in treating major depression (Syed et al., 2013; Ranjbar et al., 2013; Papakostas et al., 2012; Eby and Eby, 2010; Gertsik et al., 2012; Shaw et al., 2002). The side effects of various antidepressants can be counteracted by these nutrient therapies. Therefore, the possible interactions of antidepressants with nutrient supplementations must be taken into account.

In Vivo Imaging of SERT

In vivo assessment of SERT can serve as a diagnostic marker for major depression and other psychiatric disorders and also for antidepressant treatment. Radiotracer-based functional imaging techniques such as positron emission tomography (PET) and single-photon emission computed tomography (SPECT) help to diagnose and monitor disease mechanisms in neuropsychiatric disorders and to elucidate the effects of drug action (Brooks, 2005).

For PET studies, the widely used SERT radioligands are [^{11}C]-3-amino-4-(2-dimethylaminomethyl-phenylsulfanyl)-benzonitrile) ([^{11}C] DASB) and (11)C-N,N-dimethyl-2-(2-amino-4-methylphenylthio) benzylamine ([11C]-MADAM) (Houle et al., 2000; Lundberg et al., 2005). For SPECT, radioligands such as [(123)I]beta-CIT (2beta-carbomethoxy-3beta-(4-iodophenyl)tropane) ([123I]beta-CIT) and (123)I-labeled 2-((2-((dimethylamino)methyl) phenyl)thio)-5-iodophenylamine (123I-ADAM) are used (Kuikka et al., 1993; Oya et al., 2000).

SERT imaging using [^{11}C] DASB PET can be successfully used to investigate MDD and anti-depressant occupancy (Meyer et al., 2004a,b, 2007; Gryglewski et al., 2014). SPECT studies have shown the reduction of SERT availability in OCD and Tourette's syndrome (Stengler-Wenzke et al., 2004; Müller-Vahl et al., 2005). *In vivo* PET imaging of serotonin transporters with 4-[18F]-ADAM can be used to detect serotonergic neuron loss, monitoring the progression of Parkinson's disease and overseeing the effectiveness of therapy (Weng et al., 2013). [^{11}C] DASB PET studies can measure the SERT binding potential in various brain regions in patients with MDD, the magnitude of which can provide information about serotonin levels (Meyer et al., 2004a,b). Assessment of the SERT occupancy of various SSRIs using these imaging techniques provides information about the therapeutic effects of the drugs (Pirker et al., 1995; Meyer et al., 2001; Kent et al., 2002; Klein et al., 2006, 2007; Baldinger et al., 2014). SERT imaging in rat brain with 123I-ADAM and small-animal SPECT has proved a useful tool for preclinical studies during antidepressant drug development (Hwang et al., 2007).

Conclusion

SERT plays a critical role in the maintenance of serotonergic neurotransmission. Any genetic and epigenetic alterations occurring in this transporter have various clinical implications. SERT can serve as a diagnostic marker in the pathogenesis of neuropsychiatric disorders and response to medications. The drugs that target SERT, such as SSRIs and SNRIs, have proved to be a remarkably important advance in the field of psychopharmacology. The newly developing areas of pharmacoepigenomics and nutri-epigenomics can provide new insights into antidepressant drug action and drug response. It could also provide the means to individualize antidepressant drug therapy in future.

REFERENCES

Adamsen D, Ramaekers V, Ho HT, Britschgi C, RüfenachtV, Meili, D, Bobrowski, E, Philippe P, Nava C, Van Maldergem, L, Bruggmann R, Walitza S, Wang J, Grünblatt E, Thöny B. Autism spectrum disorder associated with low serotonin in CSF and mutations in the SLC29A4 plasma membrane monoamine transporter (PMAT) gene. *Molecular Autism* 2014;5:43.

Alpert JE, Mischoulon D, Rubenstein GE, Bottonari K, Nierenberg AA, Fava M. Folinic acid (leucovorin) as an adjunctive treatment for SSRI-refractory depression. *Annals of Clinical Psychiatry* 2002;14:33–38.

Antypa N, Van der Does A. Serotonin transporter gene, childhood emotional abuse and cognitive vulnerability to depression. *Genes, Brain, and Behavior* 2010;9:615–620.

Arias B, Catalán R, Gastó C, Gutiérrez B, Fañanás, L. 5-HTTLPR polymorphism of the serotonin transporter gene predicts non-remission in major depression patients treated with citalopram in a 12-weeks follow up study. *Journal of Clinical Psychopharmacology* 2003;23:563–567.

Arranz MJ, Munro J, Birkett J, Bolonna A, Mancama D, Sodhi M, Lesch K, Meyer J, Sham P, Collier D. Pharmacogenetic prediction of clozapine response. *The Lancet* 2000;355:1615–1616.

Baldinger P, Kranz GS, Haeusler D, Savli M, Spies M, Philippe C, Hahn A, Höflich A, Wadsak W, Mitterhauser, M. Regional differences in SERT occupancy after acute and prolonged SSRI intake investigated by brain PET. *Neuroimage* 2014;88:252–262.

Baudry A, Mouillet-Richard S, Schneider B, Launay JM, Kellermann O. miR-16 targets the serotonin transporter: A new facet for adaptive responses to antidepressants. *Science* 2010;329:1537–1541.

Beitchman JH, Baldassarra L, Mik H, Hons B, Vincenzo De Luca M, King N, Bender D, Ehtesham S, Kennedy JL. Serotonin transporter polymorphisms and persistent, pervasive childhood aggression. *American Journal of Psychiatry* 2006;163:1103–1105.

Benedetti F, Colombo C, Serretti A, Lorenzi C, Pontiggia A, Barbini B, Smeraldi E. Antidepressant effects of light therapy combined with sleep deprivation are influenced by a functional polymorphism within the promoter of the serotonin transporter gene. *Biological Psychiatry* 2003;54:687–692.

Brooks DJ. Positron emission tomography and single-photon emission computed tomography in central nervous system drug development. *NeuroRx* 2005;2:226–236.

Coppen A, Bailey J. Enhancement of the antidepressant action of fluoxetine by folic acid: A randomised, placebo controlled trial. *Journal of Affective Disorders* 2000;60:121–130.

Coyle N, Jones I, Robertson E, Lendon C, Craddock N. Variation at the serotonin transporter gene influences susceptibility to bipolar affective puerperal psychosis. *The Lancet* 2000;356:1490–1491.

Delorme R, Betancur C, Wagner M, Krebs MO, Gorwood P, Pearl P, Nygren G, Durand CM, Buhtz F, Pickering P. Support for the association between the rare functional variant I425V of the serotonin transporter gene and susceptibility to obsessive compulsive disorder. *Molecular Psychiatry* 2005;10:1059–1061.

Domschke K, Tidow N, Schwarte K, Deckert J, Lesch KP, Arolt V, Zwanzger P, Baune BT. Serotonin transporter gene hypomethylation predicts impaired antidepressant treatment response. *International Journal of Neuropsychopharmacology* 2014;17:1167–1176.

Eby GA, Eby KL. Magnesium for treatment-resistant depression: A review and hypothesis. *Medical Hypotheses* 2010;74:649–660.

Fan JB, Sklar P. Meta-analysis reveals association between serotonin transporter gene STin2 VNTR polymorphism and schizophrenia. *Molecular Psychiatry* 2005; 10:928–938.

Gadow KD, DeVincent CJ, Siegal VI, Olvet DM, Kibria S, Kirsch SF, Hatchwell E. Allele-specific associations of 5-HTTLPR/rs25531 with ADHD and autism spectrum disorder. *Progress in Neuro-Psychopharmacology and Biological Psychiatry* 2013;40:292–297.

Gertsik L, Poland RE, Bresee C, Rapaport MH. Omega-3 fatty acid augmentation of citalopram treatment for patients with major depressive disorder. *Journal of Clinical Psychopharmacology* 2012;32:61–64.

Grados M, Samuels J, Shugart Y, Willour V, Wang Y, Cullen B, Bienvenu O, Hoehn-Saric R, Valle D, Liang K. Rare plus common SERT variants in obsessive-compulsive disorder. *Molecular Psychiatry* 2007;12:422–423.

Gryglewski G, Lanzenberger R, Kranz GS, Cumming P. Meta-analysis of molecular imaging of serotonin transporters in major depression. *Journal of Cerebral Blood Flow and Metabolism* 2014;34:1096–1103.

Houle S, Ginovart N, Hussey D, Meyer J, Wilson A. Imaging the serotonin transporter with positron emission tomography: Initial human studies with [11C] DAPP and [11C] DASB. *European Journal of Nuclear Medicine* 2000;27:1719–1722.

Hu, XZ, Lipsky RH, Zhu G, Akhtar LA, Taubman J, Greenberg BD, Xu K, Arnold PD, Richter MA, Kennedy JL. Serotonin transporter promoter gain-of-function genotypes are linked to obsessive-compulsive disorder. *American Journal of Human Genetics* 2006;78:815–826.

Hwang LC, Chang CJ, Liu HH, Kao HC, Lee SY, Jan ML, Chen CC. Imaging the availability of serotonin transporter in rat brain with 123I-ADAM and small-animal SPECT. *Nuclear Medicine Communications* 2007;28:615–621.

Kang HJ, Kim JM, Stewart R, Kim SY, Bae KY, Kim SW, Shin IS, Shin MG, Yoon JS. Association of *SLC6A4* methylation with early adversity, characteristics and outcomes in depression. *Progress in Neuro-Psychopharmacology and Biological Psychiatry* 2013;44:23–28.

Kent JM, Coplan JD, Lombardo I, Hwang DR, Huang Y, Mawlawi O, Van Heertum RL, Slifstein M, Abi-Dargham A, Gorman JM. Occupancy of brain serotonin transporters during treatment with paroxetine in patients with social phobia: A positron emission tomography study with [11C] McN 5652. *Psychopharmacology* 2002;164:341–348.

Kilic F, Murphy DL, Rudnick G. A human serotonin transporter mutation causes constitutive activation of transport activity. *Molecular Pharmacology* 2003;64:440–446.

Kim JM, Stewart R, Kang HJ, Kim SW, Shin IS, Kim HR, Shin MG, Kim JT, Park MS, Cho KH. A longitudinal study of *SLC6A4* DNA promoter methylation and poststroke depression. *Journal of Psychiatric Research* 2013;47:1222–1227.

Klein N, Sacher J, Geiss-Granadia T, Attarbaschi T, Mossaheb N, Lanzenberger R, Poetzi C, Holik A, Spindelegger C, Asenbaum S. *In vivo* imaging of serotonin transporter occupancy by means of SPECT and [123I] ADAM in healthy subjects administered different doses of escitalopram or citalopram. *Psychopharmacology* 2006;188:263–272.

Klein N, Sacher J, Geiss-Granadia T, Mossaheb N, Attarbaschi T, Lanzenberger R, Spindelegger C, Holik A, Asenbaum S, Dudczak R. Higher serotonin transporter occupancy after multiple dose administration of escitalopram compared to citalopram: An [123I] ADAM SPECT study. *Psychopharmacology* 2007;191:333–339.

Kluck SM, Poustka F, Benner A, Lesch KP, Poustka A. Serotonin transporter (5-HTT) gene variants associated with autism? *Human Molecular Genetics* 1997;6:2233–2238.

Kuikka JT, Bergström KA, Vanninen E, Laulumaa V, Hartikainen P, Länsimies E. Initial experience with single-photon emission tomography using iodine-123-labelled 2 β-carbomethoxy-3β-(4-iodophenyl) tropane in human brain. *European Journal of Nuclear Medicine* 1993;20:783–786.

Lee HJ, Lee MS, Kang RH, Kim H, Kim SD, Kee BS, Kim YH, Kim YK, Kim JB, Yeon BK. Influence of the serotonin transporter promoter gene polymorphism on susceptibility to posttraumatic stress disorder. *Depression and Anxiety* 2005;21:135–139.

Lin PY, Tsai G. Association between serotonin transporter gene promoter polymorphism and suicide: Results of a meta-analysis. *Biological Psychiatry* 2004;55:1023–1030.

Lundberg J, Odano I, Olsson H, Halldin C, Farde L. Quantification of 11C-MADAM binding to the serotonin transporter in the human brain. *Journal of Nuclear Medicine* 2005;46:1505–1515.

Melas PA, Rogdaki M, Lennartsson A, Björk K, Qi H, Witasp A, Werme M, Wegener G, Mathé AA, Svenningsson P. Antidepressant treatment is associated with epigenetic alterations in the promoter of P11 in a genetic model of depression. *International Journal of Neuropsychopharmacology* 2012;15:669–679.

Meyer JH. Imaging the serotonin transporter during major depressive disorder and antidepressant treatment. *Journal of Psychiatry & Neuroscience* 2007;32:86–102.

Meyer JH, Houle S, Sagrati S, Carella A, Hussey DF, Ginovart N, Goulding V, Kennedy J, Wilson AA. Brain serotonin transporter binding potential measured with carbon11–labeled DASB positron emission tomography: Effects of major depressive episodes and severity of dysfunctional attitudes. *Archives of General Psychiatry* 2004a;61:1271–1279.

Meyer JH, Wilson AA, Ginovart N, Goulding V, Hussey D, Hood K, Houle, S. Occupancy of serotonin transporters by paroxetine and citalopram during treatment of depression: A [11C] DASB PET imaging study. *American Journal of Psychiatry* 2001;158:1843–1849.

Meyer JH, Wilson AA, Sagrati S, Hussey D, Carella A, Potter WZ, Ginovart N, Spencer EP, Cheok A, Houle S. Serotonin transporter occupancy of five selective serotonin reuptake inhibitors at different doses: An [11C] DASB positron emission tomography study. *American Journal of Psychiatry* 2004b;161:826–835.

Molina-Hernández M, Téllez-Alcántara NP. Fluoxetine, 17-β estradiol or folic acid combined with intra-lateral septal infusions of neuropeptide Y produced antidepressant-like actions in ovariectomized rats forced to swim. *Peptides* 2011;32:2400–2406.

Moya PR, Wendland JR, Rubenstein LM, Timpano KR, Heiman GA, Tischfield JA, King RA, Andrews AM, Ramamoorthy S, McMahon, FJ. Common and rare alleles of the serotonin transporter gene, *SLC6A4*, associated with Tourette's disorder. *Movement Disorders* 2013;28:1263–1270.

Mrazek D, Rush A, Biernacka J, O'Kane D, Cunningham J, Wieben E, Schaid D, Drews M, Courson V, Snyder, K. *SLC6A4* variation and citalopram response. *American Journal of Medical Genetics Part B: Neuropsychiatric Genetics* 2009;150:341–351.

Müller-Vahl KR, Meyer GJ, Knapp WH, Emrich HM, Gielow P, Brücke T, Berding G. Serotonin transporter binding in Tourette syndrome. *Neuroscience Letters* 2005;385:120–125.

Mundo E, Walker M, Cate T, Macciardi F, Kennedy JL. The role of serotonin transporter protein gene in antidepressant-induced mania in bipolar disorder: Preliminary findings. *Archives of General Psychiatry* 2001;58:539–544.

Nicole Z, Jurgen Z, Tatjana P, Shuang Y, Florian H, Theo R. Antidepressants inhibit DNA methyltransferase 1 through reducing G9a levels. *Biochemical Journal* 2012;448:93–102.

Ogilvie A, Battersby S, Fink G, Harmar A, Goodwin G, Bubb V, Smith CD. Polymorphism in serotonin transporter gene associated with susceptibility to major depression. *The Lancet* 1996;347:731–733.

Oya S, Choi SR, Hou C, Mu M, Kung MP, Acton PD, Siciliano M, Kung HF. 2-((2-((dimethylamino) methyl) phenyl) thio)-5-iodophenylamine (ADAM): An improved serotonin transporter ligand. *Nuclear Medicine and Biology* 2000;27:249–254.

Ozaki N, Goldman D, Kaye W, Plotnicov K, Greenberg B, Lappalainen J, Rudnick G, Murphy D. Serotonin transporter missense mutation associated with a complex neuropsychiatric phenotype. *Molecular Psychiatry* 2003;8:933–936.

Papakostas GI, Mischoulon D, Shyu I, Alpert JE, Fava M. S-adenosyl methionine (SAMe) augmentation of serotonin reuptake inhibitors for antidepressant nonresponders with major depressive disorder: A double-blind, randomized clinical trial. *American Journal of Psychiatry* 2010;167:942–948.

Papakostas GI, Shelton RC, Zajecka JM, Etemad B, Rickels K, Clain A, Baer L, Dalton ED, Sacco GR, Schoenfeld D. L-methylfolate as adjunctive therapy for SSRI-resistant major depression: Results of two randomized, double-blind, parallel-sequential trials. *American Journal of Psychiatry* 2012;169:1267–1274.

Pirker W, Asenbaum S, Kasper S, Walter H, Angelberger P, Koch G, Pozzera A, Deecke L, Podreka I, Brücke T. β-CIT SPECT demonstrates blockade of 5HT-uptake sites by citalopram in the human brain *in vivo. Journal of Neural Transmission: General Section* 1995;100:247–256.

Prasad HC, Steiner JA, Sutcliffe JS, Blakely RD. Enhanced activity of human serotonin transporter variants associated with autism. *Philosophical Transactions of the Royal Society B: Biological Sciences* 2009;364:163–173.

Ranjbar E, Kasaei MS, Mohammad-Shirazi M, Nasrollahzadeh J, Rashidkhani B, Shams J, Mostafavi SA, Mohammadi MR. Effects of zinc supplementation in patients with major depression: A randomized clinical trial. *Iranian Journal of Psychiatry* 2013;8:73.

Sen S, Burmeister M, Ghosh D. Meta-analysis of the association between a serotonin transporter promoter polymorphism (5-HTTLPR) and anxiety-related personality traits. *American Journal of Medical Genetics Part B: Neuropsychiatric Genetics* 2004;127:85–89.

Serretti A. Lithium long-term treatment in mood disorders: Clinical and genetic predictors. *Pharmacogenomics* 2002;3:117–129.

Shaw K, Turner J, Del Mar C. Are tryptophan and 5-hydroxytryptophan effective treatments for depression? A meta-analysis. *Australian and New Zealand Journal of Psychiatry* 2002;36:488–491.

Smeraldi E, Zanardi R, Benedetti F, Bella DD, Perez J, Catalano M. Polymorphism within the promoter of the serotonin transporter gene and antidepressant efficacy of fluvoxamine. *Molecular Psychiatry* 1998;3:508–511.

Stengler-Wenzke K, Müller U, Angermeyer MC, Sabri O, Hesse S. Reduced serotonin transporter–availability in obsessive–compulsive disorder (OCD). *European Archives of Psychiatry and Clinical Neuroscience* 2004;254:252–255.

Sugawara H, Iwamoto K, Bundo M, Ueda J, Miyauchi T, Komori A, Kazuno A, Adati N, Kusumi I, Okazaki, Y. Hypermethylation of serotonin transporter gene in bipolar disorder detected by epigenome analysis of discordant monozygotic twins. *Translational Psychiatry* 2011;1:e24.

Sutcliffe JS, Delahanty RJ, Prasad HC, McCauley JL, Han Q, Jiang L, Li C, Folstein SE, Blakely RD. Allelic heterogeneity at the serotonin transporter locus (*SLC6A4*) confers susceptibility to autism and rigid-compulsive behaviors. *American Journal of Human Genetics* 2005;77:265–279.

Syed EU, Wasay M, Awan S. Vitamin B12 supplementation in treating major depressive disorder: A randomized controlled trial. *Open Neurology Journal* 2013;7:44–48.

Vijayan NN, Iwayama Y, Koshy LV, Natarajan C, Nair C, Allencherry PM, Yoshikawa T, Banerjee, M. Evidence of association of serotonin transporter gene polymorphisms with schizophrenia in a South Indian population. *Journal of Human Genetics* 2009;54:538–542.

Wang D, Szyf M, Benkelfat C, Provençal N, Turecki G, Caramaschi D, Côté SM, Vitaro F, Tremblay RE, Booij L. Peripheral *SLC6A4* DNA methylation is associated with *in vivo* measures of human brain serotonin synthesis and childhood physical aggression. *PloS One* 2012;7:e39501.

Wankerl M, Miller R, Kirschbaum C, Hennig J, Stalder T, Alexander N. Effects of genetic and early environmental risk factors for depression on serotonin transporter expression and methylation profiles. *Translational Psychiatry* 2014;4:e402.

Weng SJ, Shiue CY, Huang WS, Cheng CY, Huang SY, Li I, Tao CC, Chou TK, Liao MH, Chang YP. PET imaging of serotonin transporters with 4-[18F]-ADAM in a Parkinsonian rat model. *Cell Transplantation* 2013;22:1295–1305.

Willeit M, Praschak-Rieder N, Neumeister A, Zill P, Leisch F, Stastny J, Hilger E, Thierry N, Konstantinidis A, Winkler D. A polymorphism (5-HTTLPR) in the serotonin transporter promoter gene is associated with DSM-IV depression subtypes in seasonal affective disorder. *Molecular Psychiatry* 2003; 8:942–946.

Xie P, Kranzler HR, Poling J, Stein MB, Anton RF, Brady K, Weiss RD, Farrer L, Gelernter J. Interactive effect of stressful life events and the serotonin transporter 5-HTTLPR genotype on posttraumatic stress disorder diagnosis in 2 independent populations. *Archives of General Psychiatry* 2009;66:1201–1209.

Yirmiya N, Pilowsky T, Nemanov L, Arbelle S, Feinsilver T, Fried I, Ebstein RP. Evidence for an association with the serotonin transporter promoter region polymorphism and autism. *American Journal of Medical Genetics* 2001;105:381–386.

Yoshida K, Ito K, Sato K, Takahashi H, Kamata M, Higuchi H, Shimizu T, Itoh K, Inoue K, Tezuka T. Influence of the serotonin transporter gene-linked polymorphic region on the antidepressant response to fluvoxamine in Japanese depressed patients. *Progress in Neuro-Psychopharmacology and Biological Psychiatry* 2002;26:383–386.

Zanardi R, Benedetti F, Di Bella D, Catalano M, Smeraldi, E. Efficacy of paroxetine in depression is influenced by a functional polymorphism within the promoter of the serotonin transporter gene. *Journal of Clinical Psychopharmacology* 2000;20:105–107.

Zhao J, Goldberg J, Bremner JD, Vaccarino V. Association between promoter methylation of serotonin transporter gene and depressive symptoms: A monozygotic twin study. *Psychosomatic Medicine* 2013;75:523–529.

36

Physiology of Melatonin, Melatonin Receptors, and Their Role in the Regulation of Reproductive Behavior in Animals

Vijay Kumar Saxena
ICAR-Central Sheep and Wool Research Institute
Rajasthan, India

CONTENTS

ABSTRACT Melatonin is an important hormone of the neuroendocrine system having a wide range of functions, from being an antioxidant to the regulation of the circadian and rhythmic annual seasonal cycles. Melatonin levels in terms of dynamic secretions from the pineal gland vary between day and night, providing information about the environmental day length for the hypothalamic–pituitary–gonadal (HPG) axis. The endogenous circadian rhythm of melatonin is generated in the suprachiasmatic nucleus (SCN) of the hypothalamus. The mammalian pineal gland has a rich network of noradrenergic innervations. Noradrenergic stimulations as outputs from the supracervical ganglion increase severalfold during the nighttime. The nocturnal regulation of melatonin secretion is based prominently on the activation of protein kinase A, which thereby brings about the phosphorylation of the cAMP response–binding protein (CREB). MT1 and MT2 are the two prominent G-coupled melatonin receptors, having seven transmembrane domains, although both receptors differ in their binding affinities to melatonin. The seasonal control of reproduction in animals is an adaptive strategy to prevent the birth of offspring at times of inclement weather and reduced availability of nutrients. Seasonality in domesticated animals is a serious impediment to maximizing production. Hormones such as progestins, eCG, hCG, and PGF2-alpha analog have been used to try to evade seasonality in addition to utilizing the *ram effect*. Recently, dopamine antagonists have been successfully used to bring about cyclicity in anestrus sheep in non-breeding season. The development of a marker for out-of-season breeding in sheep is a current area of research, especially in relation to SNP markers of the MTNR1A gene.

KEY WORDS: *melatonin, pineal gland, melatonin receptors, MTNR1A, photoperiodism, seasonality.*

Abbreviations

5HT: 5-hydroxytryptamine
5HTP: 5-hydroxytryptophan
AANAT: arylalkylamine N-acetyltransferase
cAMP: cyclic adenosine monophosphate
CNS: central nervous system
CRE: cAMP response element
CREB: cAMP response element–binding protein
eCG: equine chorionic gonadotropin
GIT: gastrointestinal tract
GPCR: G protein–coupled receptor
hCG: human chorionic gonadotropin
HIOMT: hydroxyindole-O-methyltransferase
HPG: hypothalamic–pituitary–gonadal axis
MT1:   melatonin receptor 1
MT2:   melatonin receptor 2
PGF2α: prostaglandin F2 alpha
PKA: protein kinase A
ROR: retinoid orphan receptor
RZR: retinoid Z receptor
SCG: supracervical ganglion
SCN: suprachiasmatic nucleus

Introduction

The pineal gland, which is the synthesizing glandular machinery of melatonin, had long been considered a third eye, owing to earlier prevailing thought that it was the seat of the imaginary soul. By the end of the nineteenth century, owing to the efforts of modern scientific organized research conducted in the field of pineal biology, the anatomy, histology, nerval innervations, and embryology of the pineal gland was intensively studied, and it was found to be a photoreceptor organ with a close resemblance to the epiphyses of the lower vertebrates. The physiological functions of the pineal gland began to be investigated in the twentieth century. Demonstrations of the positive correlation in the incidences of precocious puberty and the presence of pineal tumors in patients indicated the role of the pineal gland in reproduction. In order to discover its putative therapeutic value, a group of American dermatologists (Lerner et al., 1958) were trying to find out the unknown compound present in the pineal gland that was able to blanch the skin chromatophores of amphibians. They were able to isolate and characterize the methoxy derivative of serotonin from bovine pineal tissue and named it *melatonin*. It was a wonderful discovery, as the compound has brought about a dynamic transformation in the role of the pineal gland from a vestige to a legitimate component of the endocrine system.

Chemistry and Synthesis of Melatonin

Melatonin (N-acetyl-5-methoxytryptamin) is predominantly synthesized by the pineal gland, although extrapineal sources such as the gastrointestinal tract (GIT), certain areas of the central nervous system (CNS), and several leukocytes have also been shown to produce it in less significant amounts under specific conditions (Pandi Perumal et al., 2006; Hardeland 2008; Hardeland et al., 2011). Tryptophan, which is an aromatic amino acid with an indole ring, is the synthetic precursor of all the pineal 5-methoxyindoles. The amino acid is metabolized by the enzyme tryptophan hydroxylase into 5-hydroxytryptophan (5HTP). This reaction takes place in the mitochondria of pinealocytes. 5HTP is converted into 5-hyrodroxytryptamine

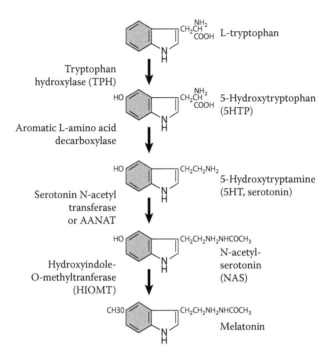

FIGURE 36.1 Chemistry of melatonin synthesis.

(5HT) by the aromatic amino acid decarboxylase in the cytosol. 5HT is acetylated into N-acetylserotonin by the enzyme arylalkylamine N-acetyltransferase (AANAT) and N-acetylserotonin is subsequently O-methylated by the enzyme hydroxyindole-O-methyltransferase (HIOMT) into melatonin (N-acetyl-5-methoxytryptamine) (Figure 36.1). Cytological studies of the pineal gland showed greater metabolic activity of the gland during the nighttime (Quay, 1956; Mogler, 1958). Melatonin production was thereafter found to increase profoundly during the nighttime, which was evident in the pineal glands of rats having 100–200 pg of melatonin during the daytime, increasing to 1000–2000 ng during the nighttime. Melatonin production in the pineal glands of humans ranges from less than 5 pg/mL during the daytime to a peak of 30–150 pg/mL in the middle of the night (Lynch et al., 1975).

Thus, melatonin production by the pineal gland follows a specific circadian rhythm, with maximum production triggered by the darkness of night (Stehle et al., 2011). The endogenous circadian rhythm of melatonin is generated in the suprachiasmatic nucleus (SCN) of the hypothalamus and entrained principally by the photoperiodic variability of light–dark cycle, through the retino–hypothalamic tract embedded in the optic nerve. The mammalian retina is responsive to the blue light fragment (460–480 nm) of visible light, owing to the presence of five specialized subtypes of intrinsically photosensitive retinal ganglionic cells (ipRGC) (Lucas et al., 2014; Zhoa et al., 2014). Neural messages to the pineal gland arrive via the supracervical postganglionic fibers. This is a specialized feature of the pineal gland, whereby sectioning of the sympathetic nerve fibers abolishes melatonin-induced circadian rhythmicity and mimics the effect of pinealectomy (Reiter and Hester, 1966). Another notable difference of the pineal gland from other glands of the endocrine system is that it does not significantly alter its hormonal secretions in response to the feedback effect of peripherally derived hormonal signals (Cano et al., 2008).

Nocturnal Regulation of Melatonin Synthesis

The mammalian pineal gland has a rich network of noradrenergic innervations. Noradrenergic stimulations as outputs from the supracervical ganglion increase severalfold during the nighttime. The noradrenergic signaling transmitted through both alpha-1 and beta-1 adrenergic receptors results in a large-scale increase of cyclic adenosine monophosphate (cAMP) levels and thereby activates protein kinase A (PKA). The activation of PKA brings about the phosphorylation of the cAMP response element–binding protein (CREB). CREB is a cellular transcription factor that binds to certain DNA sequences called *cAMP response elements* (CREs), thereby modulating the transcription of underlying downstream genes. P-CREB enhances the expression of genes coding for the enzymes of melatonin synthesis such as AANAT and HIOMT, as these enzymes have putative CRE binding sites in their promoter region (Baler et al., 1997). PKA also directly brings about the phosphorylation of the AANAT protein at Thr-31 residue, which provides it with enhanced stability against degradation by the proteosomal activity (Ganguly et al., 2001). The marked nocturnal increase in AANAT activity following activation by $\beta 1/\alpha 1$-AR/cAMP/PKA induces multifold increases in the conversion of 5HT into melatonin. Cessation of norepinephrine release blocks activation of the $\beta 1/\alpha 1$-AR/cAMP/PKA pathway owing to exposure to light following the end of night (Simonneaux and Ribelayga, 2003). During the day, owing to a lack of posttranslational modification, the AANAT protein is destroyed by the proteosomal activity, leading to reduced melatonin production (Figure 36.2).

Melatonin Receptors and Their Physiology

Melatonin exerts most of its physiological functions on target cells by interacting with G protein–coupled receptors (GPCRs). In mammals, MT1 and MT2 are the predominant GPCRs to have been characterized and studied for their localization. Expressions of the MT1 receptor (also described as MTNR1A) have been predominantly observed in the pars tuberalis of the pituitary (Morgan et al., 1994), in the SCN of the hypothalamus, and in other organs of the brain, as well as in certain peripheral tissues. The MT1 subtype has been implicated in the transmission of the circadian effects of melatonin and reproduction. MT2 (also described as MTNR1B) is highly expressed in the human retina (Reppert et al., 1995), where it has been proposed to play a key role in the physiology of this tissue. The MT1 receptor is 350 amino acids in length, whereas the MT2 receptor consists of 363, and they share nearly 60% homology with each other (Dubocovich and Markowska, 2005). Both the melatonin receptors have the characteristic structure of GPCRs, with seven transmembrane (TM) alpha-helices and four intracellular and four extracellular domains. GPR50 is another G-coupled orphan melatonin-like receptor, which shows 45% homology to MT1 and MT2, although it does not bind melatonin effectively. GPR50 cells were found to be richly present in the dorsomedial hypothalamus, periventricular nucleus, and median eminence. MT3 is another putative melatonin receptor that was identified on pharmacological grounds, with a lower melatonin affinity in comparison with MT1 and MT2 and showing rapid ligand association/dissociation kinetics. It was later identified as the quinone reductase 2 enzyme, showing functional homology with other quinone reductases in human tissues (Nosjean et al., 2000).

There is another class of nuclear melatonin receptors belonging to the retinoid orphan receptor (ROR)/ retinoid Z receptor (RZR) family. The superfamily members that bind melatonin contain in their protein sequence an NH_2 terminal domain, a DNA-binding domain, a ligand binding domain at the carboxyl terminal end, a zinc double finger, and a hinge region. Members of these families have been suggested to play a role in the immunostimulatory effects of melatonin in B and T lymphocytes (Fildes et al., 2009), as well as in the indirect antioxidant effects of melatonin.

Role of Melatonin in Seasonal Control of Reproduction

The pineal gland is a major component of the endocrine system, which allows animals to sense photoperiodic variability through dynamic modulation in the secretion of melatonin. The duration of melatonin secretion is an important parameter signaling the environmental day length to the neuroendocrine system, thereby providing input for the prospective regulation of neuroendocrine, and thus the fine-tuning of reproduction. Seasonal breeding animals such as sheep, deer, horses, yaks, and hamsters time their

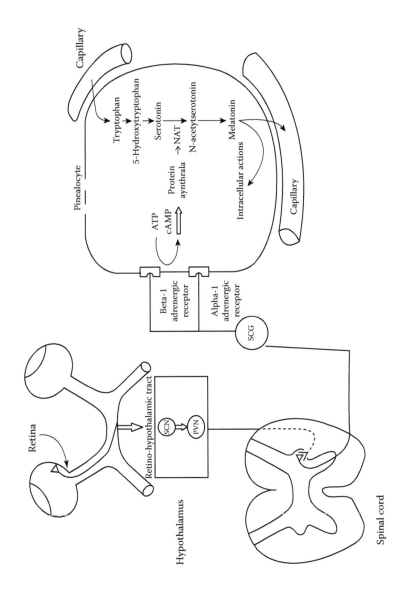

FIGURE 36.2 Mechanism of melatonin synthesis and its control.

reproductive cycles using photoperiodic cues. Long-day breeding animals such as horses and several rodents are sexually less active during the winter months (when nocturnal melatonin levels are at their peak), while short-day animals like sheep and deer are sexually most active during the same time. Thus, melatonin per se is neither pro-gonadotrophic nor anti-gonadotrophic, but serves as a medium of transferring the message of environmental timing to the reproductive hypothalamic–pituitary–gonadal (HPG) axis. Syrian hamsters, which are long-day breeders, undergo significant regression in their gonadal weight during the winter months and the phenomenon can be averted if prior pinealectomy is performed on them. Both the duration of melatonin secretion during the night and its amplitude play a role in the transmission of photoperiodic response, although the duration has a predominant role. Annual variations in the amplitude of melatonin secretion are evident in animals maintained in natural habitat–like conditions. Other environmental factors that also vary periodically throughout the year, such as temperature, humidity, nutritional adequacy, and availability, can be utilized and integrated to modulate the secretion of melatonin by animals. Among other nonphotic factors, environmental temperature stands apart in terms of its relative importance in modulating the perception of photoperiodism by animals—perhaps by altering the metabolism of the pineal gland, for example. A decrease in temperature increases the activity of the pineal enzymatic machinery in rats (Nir et al., 1975) and increases the amplitude of melatonin secretion in Syrian hamsters (Brainard et al., 1982). The pars tuberalis of the anterior lobe of the pituitary gland has the maximum density of melatonin receptors and is supposed to be the seat of melatonin-controlled seasonal rhythmicity, by altering the expression profile of clock genes. The seasonal control of reproduction is imperative in certain animals to prevent the birth of offspring at times of inclement weather and reduced availability of nutrients, which are dangerous for their survival. Seasonal reproduction serves as a natural contraception method (Lincoln and Short, 1980) and is a viable adaptation, especially in feral conditions.

Seasonality in Sheep

Sheep as such are short-day breeders, being reproductively most active when nocturnal melatonin levels begin increasing, owing to reduced day length. Sheep breeds originating from temperate climatic conditions prevailing generally at higher latitudes demonstrate considerable seasonality in their reproductive behavior, recorded by monitoring the estrus behavior of sheep throughout the year by keeping them bare and not allowing them to breed. The reproductive season in sheep of temperate climatic conditions generally extends from the early summer/autumn to the end of winter, with the anestrus period coinciding with the spring season. Breeds of tropical and subtropical regions are either aseasonal or intermittently polyestrous (Rosa and Bryant, 2003); they have a tendency to breed throughout the year owing to lesser photodependence (Poulton et al., 1987). The breeding period in most ewes starts in early summer/autumn and ends in winter, when day lengths begin to increase. The duration of nocturnal melatonin secretion regulates the pulsatile secretion of GnRH from the hypothalamus, which in turn effects the release of gonadotrophin from the pituitary (Karsch et al., 1988; Malpaux et al., 2001). In addition, it has been found that the specific effects of short/long days are not permanent due to the phenomenon of photorefractoriness, which occurs when animals are continuously exposed to constant photoperiods for a long time. This is why a sheep when exposed to prolonged short-day conditions after winter solstice will not continue its cyclicity for an extended period and enters into anestrus (Robinson and Karsch, 1984).

Seasonality in sheep is a serious impediment to maximizing production. Over the years, several strategies and methods have been used to try to expand the breeding season in ewes, with limited success. The *ram effect* has been used to induce cyclicity in anestrus ewes by exposing them to males after a period of forced isolation (Chemineau et al., 2006; Ungerfedl et al., 2004). Other commonly used hormonal strategies to induce cyclicity include the use of synthetic slow-release progestins in the form of microimplants, PGF2-alpha analogues such as cloprostenol, equine chorionic gonadotropins (eCGs), and human chorionic gonadotropins (hCGs). Melatonin implants have also been used to induce estrus in anestrus ewes (Kaya et al., 1998), to hasten the breeding season (Croker et al., 1982), and to increase the conception rate in ewes (Gómez Brunet et al., 1995). Recently, sulpiride, a dopamine antagonist, has been found to be useful in inducing cyclicity in anestrus ewes during the nonbreeding season (Saxena et al., 2015a). Sulpiride is a substituted benzamide with predominantly D2-blocking activity.

Another important direction in which research is heading is to find out the genetic basis of out-of-season breeding in ewes. Several studies have confirmed that there is a relation between reproductive performance, especially with respect to out-of-season breeding in ewes, and polymorphisms in the MTNR1A gene. Two single-nucleotide polymorphisms at positions 606 and 612 of the coding sequence have been found to be associated with seasonal reproductive activity in ewes (Pelletier et al., 2000; Chu et al., 2006; Mateescu et al., 2009; Carcangiu et al., 2009). The MTNR1A gene consists of two exons spanned by an intron, and most of the structural part of the receptor is coded by exon 2. The melatonin receptor genes of several Indian sheep breeds were characterized and studied for their variability in terms of gene and genotype frequencies at the 606 and 612 loci of the coding sequence (Saxena et al., 2014, 2015b). We have found that the genotype frequencies of the Indian tropical sheep breeds are strikingly different from the temperate sheep breeds, as well as the studied subtemperate sheep breeds of the country (Saxena et al., 2015c,d). However, a basic and functional characterization in terms of the causal mechanism behind the role of these SNPs is still elusive and needs to be probed.

Conclusion

Melatonin is an important neuroendocrine hormone regulating circadian and rhythmic annual seasonal cycles in animals. Melatonin is predominantly secreted by the pineal gland in proportion to the period of darkness. The endogenous circadian rhythm of melatonin is generated in the suprachiasmatic nucleus (SCN) of the hypothalamus and entrained principally by the photoperiodic variability of the light–dark cycle, through the retino–hypothalamic tract embedded in the optic nerve. The mammalian pineal gland has a rich network of noradrenergic innervations. Noradrenergic stimulations as outputs from the supracervical ganglion increase severalfold during the nighttime. Marked nocturnal increases in AANAT activity following activation by β1/α1-AR/cAMP/PKA induces multifold increases in the conversion of 5HT into melatonin. Melatonin exerts most of its physiological functions on target cells by interacting with GPCRs. In mammals, MT1 and MT2 are the predominant GPCRs. There is another class of nuclear melatonin receptor belonging to the ROR/RZR family. Melatonin per se is neither pro-gonadotrophic nor anti-gonadotrophic, but serves as a medium of transferring the message of environmental timing to the reproductive HPG axis. Seasonal breeding animals such as sheep, deer, horses, yaks, and hamsters time their reproductive cycles using photoperiodic cues. Sheep as such are short-day breeders, being reproductively most active when nocturnal melatonin levels begin to increase, owing to reduced day length. Evading seasonality in livestock animals, especially sheep, is imperative for maximizing production. The development of a marker for out-of-season breeding in sheep is a current area of research, especially in relation to SNP markers of the MTNR1A gene.

REFERENCES

Balcr R, Covington S, Klein DC (1997). The rat arylalkylamine N-acetyltransferase gene promoter: cAMP activation via a cAMP-responsive element-CCAAT complex. *J Biol Chem* 272: 6979–6985.

Brainard GC, Petterborg LJ, Richardson BA, Reiter RJ (1982). Pineal melatonin in Syrian hamster: Circadian and seasonal rhythms in animals maintained under laboratory and natural conditions. *Neuroendocrinology* 35: 342–348.

Cano P, Jimenez-Ortega V, Larrad A, Reyes Toso CF, Cardinali DP, Esquifino AI (2008). Effect of high-fat diet on 24-h pattern of circulating levels of prolactin, luteinizing hormone, testosterone, corticosterone, thyroid-stimulating hormone and glucose and pineal melatonin content, in rats. *Endocrine* 33: 118–125.

Carcangiu V, Mura MC, Vacca GM, Pazzola M, Dettori ML, Luridiana S, Bini PP (2009). Polymorphism of the melatonin receptor *MTNR1A* gene and its relationship with seasonal reproductive activity in the Sarda sheep breed. *Anim Reprod Sci.* 116: 65–72.

Chemineau P, Pellicer-Rubio MT, Lassoued N, Khaldi G, Monniaux D (2006). Male induced short oestrus and ovarian cycles in sheep and goats: A working hypothesis. *Reprod Nutr Dev* 46: 417–429.

Chu MX, Cheng DX, Liu WZ, Fang L, Ye SC (2006). Association between melatonin receptor 1A gene and expression of reproductive seasonality in sheep. *Asian–Aust J Anim Sci* 19: 1079–1084.

Croker KP, Butler LG, John MA, McColm SC (1982). Induction of ovulation and cyclic activity in anoestrous ewes with testosterone treated wethers and ewes. *Theriogenology* 17: 349–354.

Dubocovich ML, Markowska M (2005). Functional MT1 and MT2 melatonin receptors in mammals. *Endocrine* 27, 101–110.

Fildes JE, Yonan N, Keevil BG (2009). Melatonin: A pleiotropic molecule involved in pathophysiological processes following organ transplantation. *Immunology* 127: 443–449.

Ganguly S, Gastel JA, Weller JL, Schwartz C, Jaffe H, Namboodiri MA, Coon SL, Hickman AB, Rollag M, Obsil T, Beauverger P, et al. (2001). Role of a pineal cAMP-operated arylalkylamine N-acetyltransferase/14–3-3-binding switch in melatonin synthesis. *Proc Natl Acad Sci USA* 98: 8083–8088.

Gómez Brunet A, López Sebastian A, Picazo RA, Cabellos B, Goddard S (1995). Reproductive response and LH secretion in ewes treated with melatonin implants and induced to ovulate with the ram effect. *Anim Reprod Sci* 39: 23–34.

Hardeland R (2008). Melatonin, hormone of darkness and more: Occurrence, control mechanisms, actions and bioactive metabolites. *Cell Mol Life Sci* 65(13): 2001–2018.

Hardeland R, Cardinali DP, Srinivasan V, Spence DW, Brown GM, Pandi-Perumal SR (2011). Melatonin: A pleiotropic, orchestrating regulator molecule. *Prog Neurobiol* 93(3): 350–384.

Karsch FJ, Malpaux B, Wayne NL, and Robinson JE (1988). Characteristics of the melatonin signal that provide the photoperiodic code for timing seasonal reproduction in the ewe. *Reprod Nutr Dev* 28: 459–472.

Kaya A, Ataman MB, Coya K, Karaca F, Aksoy M, Yildiz C, Ergin A (1988). Konya Merinosu koyunlarinda melatonin, progesteron-PMSG ve koc etkisiuygulamalarinin erken anostrus doneminde bazi ureme parametrelerine etkileri. [The effect of combinations of melatonin and the ram effect, progesterone and PMSG and the ram effect on the onset of ovarian activity and some reproductive traits in Central Anatolian Merino ewes early in the anestrous season.] *Hayvancilik Arastirma Dergisi* 8: 5–10.

Lerner AB, Case JD, Takahashi Y, Lee TH, Mori W (1958). Isolation of melatonin, the pineal factor that lightens melanocytes. *J Am Chem Soc* 80: 2587–2592.

Lincoln GA, Short RV (1980). Seasonal breeding: Nature's contraceptive. *Rec Prog Norm Res* 36: 1–52.

Lucas RJ, Peirson SN, Berson DM, Brown TM, Cooper HM, Czeisler CA, Figuero MG, Gamlin PD, Lockley SW, O'Hagan JB, Price LL, et al. (2014). Measuring and using light in the melanopsin age. *Trends Neurosci* 37: 1–9.

Lynch HJ, Wurtman RJ, Moskowitz MA, Archer MC, Ho MH (1975). Daily rhythm in human urinary melatonin. *Science* 187: 169–171.

Malpaux B, Migaud M, Tricoire H, Chemineau P (2001). Biology of mammalian photoperiodism and the critical role of the pineal gland and melatonin. *J Biol Rhythms* 16: 336–347.

Mateescu RG, Lunsford AK, Thonney ML (2009). Association between melatonin receptor 1A gene polymorphism and reproductive performance in Dorset ewes. *J Anim Sci* 87(8): 2485–2488.

Mogler KH (1958). Das Endokrine System des Syrischen Goldhamster unter Berücksichtigung des Nartürlichen und Experimentellen Winterschläf. *Z Morphol Oekol Tiere* 47: 267–308.

Morgan PJ, Barrett P, Howell HE, Helliwell R (1994). Melatonin receptors: Localization, molecular pharmacology and physiological significance. *Neurochem Int* 24: 101–46.

Nir I, Hirschmann N, Sulman FG (1975). The effect of heat on rat pineal hydroxyindole-O-methyltransferase activity. *Experientia* 31: 867–868.

Nosjean O, Ferro M, Coge F, Beauverger P, Henlin JM, Lefoulon F, Fauchere JL, Delagrange P, Canet E, Boutin JA (2000). Identification of the melatonin-binding site MT3 as the quinone reductase 2. *J Biol Chem* 275: 31311–31317.

Pandi-Perumal SR, Srinivasan V, Maestroni GJM, Cardinali DP, Poeggeler B, Hardeland R (2006). Melatonin: Nature's most versatile biological signal? *FEBS Journal* 273 (13): 2813–2838.

Pelletier J, Bodin L, Hanocq E, Malpaux B, Teyssier J, Thimonier J, Chemineau P (2000). Association between expression of reproductive seasonality and alleles of the gene for Mel(1a) receptor in the ewe. *Biol. Reprod.* 62: 1096–1101.

Poulton AL, Symons AM, Kelly MI, Arendt J (1987). Intraruminal soluble glass boluses containing melatonin can induce early onset of ovarian activity in ewes. *J Reprod Fert* 80: 235–239.

Quay WB (1956). Volumetric and cytologic variation in the pineal body of *Peromycus leucopus* (Rodentia) with respect to sex, captivity and daylength. *J Morphol* 98: 471–495.

Reiter RJ, Hester RJ (1966). Interrelationships of the pineal gland, the superior cervical ganglia and the photoperiod in the regulation of the endocrine systems of hamsters. *Endocrinology* 79: 1168–1170.

Reppert SM, Godson C, Mahle CD, Weaver DR, Slaugenhaupt SA, Gusella JF (1995). Molecular characterization of a second melatonin receptor expressed in human retina and brain: The Mel1b melatonin receptor. *Proc Natl Acad Sci USA* 92: 8734–8738.

Robinson JE, Karsch FJ (1984). Refractoriness to inductive day lengths terminates the breeding season of the Suffolk ewe. *Biol Reprod* 31: 656–663.

Rosa HJD, Bryant MJ (2003). Seasonality of reproduction in sheep. *Small Ruminant Res* 48: 155–171.

Saxena VK, Jha BK, Meena AS, Naqvi SMK (2014). Sequence analysis and identification of new variations in the coding sequence of melatonin receptor gene (MTNR1A) of Indian Chokla sheep breed. *Meta Gene* 2: 450–458.

Saxena VK, Jha BK, Meena AS, Naqvi SMK (2015a). Characterization of MTNR1A gene in terms of genetic variability in a panel of sub-temperate and sub-tropical Indian sheep breeds. *J Genetics* 94: 715–721.

Saxena VK, Jha BK, Meena AS, Narula HK, Kumar D, Naqvi SMK (2015b). Assessment of genetic variability in the coding sequence of melatonin receptor gene (MTNR1A) in tropical arid sheep breeds of India. *Reprod Domestic Anims* 50: 517–521.

Saxena VK, Jha BK, Meena AS, Narula HK, Kumar D, Naqvi SMK (2015c). Melatonin receptor 1A (MTNR1A) gene sequence characterization and SNP identification in tropical sheep breeds of India. *Recep Clinical Investigation* 2(3).

Saxena VK, De Kalyan, Kumar D, Naqvi SMK, Krishnaswamy N, Tiwari AK (2015d). Induction of ovulation in anestrus ewes by blocking the negative feedback of estrogen using a dopamine receptor antagonist. *Theriogenology* 84: 1362–1366.

Simonneaux V, Ribelayga C (2003). Generation of the melatonin endocrine message in mammals: A review of the complex regulation of melatonin synthesis by norepinephrine, peptides, and other pineal transmitters. *Pharmacol Rev* 55: 325–395.

Stehle JH, Saade A, Rowashdeh O, Ackermann K, Jilg A, Sebesteny T, Maronde E (2011). A survey of molecular details in the human pineal gland in the light of phylogeny, structure, function and chronobiological diseases. *J Pineal Res* 51: 17–43.

Ungerfedl R, Fosberg M, Rubianes E (2004). Overview of the response of anoestrus ewes to the ram effect. *Reprod Fertility Dev* 16: 479–490.

Zhoa X, Stafford BK, Godin AL, King WM, Wong KY (2014). Photoresponse diversity among the five types of intrinsically photosensitive retinal ganglion cells. *J Physiol* 592: 1619–1636.

Index

Milton Keynes UK
Ingram Content Group UK Ltd.
UKHW052025141024
449569UK00016B/707